Emerging Trends in Cell and Gene Therapy

Michael K. Danquah • Ram I. Mahato
Editors

Emerging Trends in Cell and Gene Therapy

Springer

Editors
Michael K. Danquah
Pharmaceutical Sciences
Chicago State University
Illinois
USA

Ram I. Mahato
Department of Pharmaceutical Sciences
College of Pharmacy
University of Nebraska Medical Center
Omaha
USA

ISBN 978-1-4939-6293-8 ISBN 978-1-62703-417-3 (eBook)
DOI 10.1007/978-1-62703-417-3
Springer New York Heidelberg Dordrecht London

Printed on acid-free paper

Springer is part of Springer Science+Business Media (www.springer.com)

To my parents George and Elizabeth: my lifelong teachers, who have devoted their lives to making the world a better place by caring for others and whose constant encouragement and sublime standards serve to inspire me. To the loving memory of Gabriel: my brother and hero. For a brief moment in time you lit up our world. To Jennifer: for layers of love and extraordinary patience.

Michael K. Danquah

I dedicate this book to my wife Subhashini; my children Kalika and Vivek for their love and support; my late mother Sarswati for believing in me; and to my students and mentors who have always helped me in my quest for learning and in achieving higher goals.

Ram I. Mahato

Preface

Emerging Trends in Cell and Gene Therapy is meant for those who seek the golden thread that runs through the fields of cell therapy, gene therapy, and tissue engineering, yet have found other books too specialized to do so. This book aims to arm basic scientists and clinicians with this golden thread so they are better positioned to address the debilitating diseases presently plaguing mankind.

Cell and gene therapies are promising approaches for treating genetic and acquired diseases. To date, numerous biological barriers and ethical issues have limited their clinical translation. Nonetheless, active research in cell and gene therapy in both academia and industry is continually providing fresh insight that promises to bring these potentially potent therapies to our doorstep. While there are several books already available covering cell and gene therapy, most of these deal with both subject areas separately. Furthermore, many of these books only address various aspects such as fundamental principles and delivery or application of cell or gene therapy. This current situation has the tendency of leaving the interested readers with a fragmented understanding regarding these two areas and the flexible and powerful therapeutic platforms which can be developed when various aspects of cell and gene therapy are combined. Hence, there is a great demand from the scientific community for a book providing a holistic perspective on novel and important areas at the interface of cell and gene therapy, as well as potential synergistic therapeutic benefit obtained when both therapeutic approaches are combined with delivery strategies. Here is what this book offers you.

First, it is broadly organized to provide critical and in-depth review in the following three key areas: (1) basic biological aspects of stem cell sources, differentiation, and engineering, (2) application of stem cells and gene therapy to specific human disease, and (3) utilization of biomaterials and stem cells in regenerative medicine. This arrangement allows the readers to observe the common theme involved in the integration of cell, gene therapy, and tissue engineering and how it can be used to guide future research.

Second, this book covers a range of topics including recent advances in embryonic stem cell engineering towards tailored lineage differentiation, the human amniotic membrane as a potential tissue and cell source for cell therapy and regenerative medicine, emerging strategies for the selection of vectors, delivery techniques and therapeutic targets for gene transfer to the heart, application of microfluidics to study stem cell dynamics, biomimetic multiscale topography for cell alignment, and spinal cord repair by means of tissue engineered scaffolds. The contents of *Emerging Trends in Cell and Gene Therapy* are contributed by leading international research and clinical experts and therefore represent current understanding, practice, and state of the fields of cell therapy, gene therapy, and tissue engineering. Hence, this book offers, in a single volume, the required comprehensive understanding regarding the connecting thread running through cell therapy, gene therapy, and tissue engineering for veterans and newcomers to the field.

About the Authors

Michael K. Danquah graduated with BS in Chemical Engineering from Kwame Nkrumah University of Science and Technology, Ghana. He received his MS in Chemical Engineering from The University of Kentucky, Lexington, KY, and PhD in Pharmaceutics and Drug Delivery from The University of Tennessee Health Science Center, Memphis, TN, USA. He is currently an Assistant Professor in the Department of Pharmaceutical Sciences, Chicago State University, Chicago. His research interests include small molecule and gene-based combination therapy for treating cancer, design and synthesis of novel biomaterials, and application of nano-biotechnology for enhanced drug delivery.

Ram I. Mahato is the Professor and Chair of the Department of Pharmaceutical Sciences at the University of Nebraska Medical Center, Omaha, USA. Previously, he was serving as a tenured Professor of Pharmaceutics and Drug Delivery in the Department of Pharmaceutical Sciences at the University of Tennessee Health Science Center, Memphis. Dr. Mahato has served as a Research Assistant Professor at the University of Utah; Senior Scientist at GeneMedicine, Inc.; and as a postdoctoral fellow at the University of Southern California, Washington University, and Kyoto University. He received PhD in Pharmaceutics and Drug Delivery from the University of Strathclyde (Glasgow, Great Britain). He has two issued United States patents and is author or coauthor of more than 106 papers and book chapters. He has edited nine journal theme issues and six books and written one textbook. He is a Feature Editor for *Pharmaceutical Research* and a member of the editorial board of eight journals. He is an AAPS and CRS Fellow and regular member of the BTSS Study section of the National Institutes of Health. His research includes delivery and targeting of small molecules, oligonucleotides, siRNA, and genes for treating diabetes, fibrosis, and cancer.

Contents

Contributors

Stelios T. Andreadis, PhD Department of Chemical and Biological Engineering, University at Buffalo, The State University of New York, Amherst, NY, USA

Eishi Ashihara Department of Clinical and Translational Physiology, Kyoto Pharmaceutical University, Kyoto, Japan

Francisco J. Blanco Garcia Servicio de Reumatología, INIBIC-Hospital Universitario A Coruña, A Coruña, Spain

Osteoarticular and Aging Research Laboratory, Hospital Universitário A Coruña, A Coruña, Spain

Charles R. Bridges Department of Thoracic and Cardiovascular Surgery, Sanger Heart and Vascular Institute, Carolinas Healthcare System, Charlotte, NC, USA

Shousong Cao Department of Medicine, Roswell Park Cancer Institute, Buffalo, NY, USA

Central South University, Changsha, Hunan, China

Huaying Chen Faculty of Engineering, Graduate School of Biomedical Engineering, The University of New South Wales, Sydney, NSW, Australia

School of Materials Science and Engineering, Shandong University, Ji'nan, China

Qizhi Chen Department of Materials Engineering, Monash University, Clayton, VIC, Australia

Xiongbiao Chen, PhD Division of Biomedical Engineering, University of Saskatchewan, Saskatoon, SK, Canada

Department of Mechanical Engineering, University of Saskatchewan, Saskatoon, SK, Canada

Mukai Chimutengwende-Gordon Division of Surgery, University College of London, London, UK

Rosa Chung Sansom Institute for Health Research, School of Pharmacy and Medical Sciences, University of South Australia, Adelaide, SA, Australia

Long-Jun Dai, PhD Department of Surgery, University of British Columbia, Vancouver, BC, Canada

Francisco J. de Toro Departamento de Medicina, INIBIC-Universidad de A Coruña, A Coruña, Spain

Servicio de Reumatología, INIBIC-Hospital Universitario A Coruña, A Coruña, Spain

Margot Den Hondt, MD Department of Development and Regeneration, Cell and Gene Therapy Applications Research Unit, Katholieke Universiteit Leuven, University Hospitals, Leuven, Belgium

Silvia Diaz-Prado Departamento de Medicina, INIBIC-Universidad de A Coruña, A Coruña, Spain

Yanan Du Department of Biomedical Engineering, School of Medicine, Tsinghua University, Beijing, China

Abir O. El Sadik Basic Sciences Department, King Saud Bin AbdulAziz University for Health Sciences, Riyadh, Saudi Arabia

Anthony S. Fargnoli Department of Thoracic and Cardiovascular Surgery, Sanger Heart and Vascular Institute, Carolinas Healthcare System, Charlotte, NC, USA

Isaac Fuentes-Boquete Departamento de Medicina, INIBIC-Universidad de A Coruña, A Coruña, Spain

Keiichi Fukuda, MD, PhD Cardiology Division, Department of Internal Medicine, Keio University school of Medicine, Tokyo, Japan

Yuji Haraguchi Institute of Advanced Biomedical Engineering and Science, TWIns, Tokyo Women's Medical University, Shinjuku-ku, Tokyo, Japan

Jean-Philippe Hauzeur, MD, PhD Department of Rheumatology, University of Liège, CHU Sart Tilman, Brussels, Belgium

Benoit Hendrickx, MD, PhD Department of Cardiovascular Sciences, Molecular and Vascular Biology Research Unit, KU Leuven, Leuven, Belgium

Department of Development and Regeneration, Cell and Gene Therapy Applications Research Unit, Katholieke Universiteit Leuven, University Hospitals, Leuven, Belgium

Aji Huang Department of Breast Surgery, Affiliated Cancer Hospital, Medical Center of Fudan University, Shanghai, China

Manrong Jiang Department of Anatomy and Cell Biology, Indiana University School of Medicine-Fort Wayne, Fort Wayne, IN, USA

Emmanuel Katsanis Department of Pediatrics, University of Arizona, Tucson, AZ, USA

Michael G. Katz Department of Thoracic and Cardiovascular Surgery, Sanger Heart and Vascular Institute, Carolinas Healthcare System, Charlotte, NC, USA

John Kessler Feinberg School of Medicine, Northwestern University, Chicago, IL, USA

Departments of Neurology and Molecular Pharmacology and Biological Chemistry, Feinberg School of Medicine, Northwestern University, Chicago, IL, USA

Aleem Ahmed Khan Habeebullah's Memorial Centre for Stem Cell Sciences, Centre for Liver Research and Diagnostics, Hyderabad, AP, India

Wasim S. Khan, MBChB, MSc, MRCS, PhD Division of Surgery, University College of London, London, UK

Division of Surgery, Institute of Orthopaedic and Musculoskeletal Sciences, Royal National Orthopaedic Hospital, University College London, Stanmore, Middlesex, UK

Nicolas Larmonier Department of Pediatrics, University of Arizona, Tucson, AZ, USA

Feng Li, PhD Department of Pharmaceutical Sciences, School of Pharmacy, Hampton University, Hampton, VA, USA

Hong Liu Department of Otolaryngology, Head & Neck Surgery, Xiangya Hospital, Central South University, Changsha, Hunan, China

Zhiyong Liu Departments of Developmental Neurobiology, St. Jude Children's Research Hospital, Memphis, TN, USA

Jie Luo Department of Neurosurgery, Taihe Hospital, Hubei University of Medicine, Shiyan, China

Aernout Luttun, PhD Department of Cardiovascular Sciences, Molecular and Vascular Biology Research Unit, Endothelial Cell Biology Unit, Katholieke Universiteit Leuven, Leuven, Belgium

Taira Maekawa Department of Transfusion Medicine and Cell Therapy, Kyoto University Hospital, Kyoto, Japan

Kara E. McCloskey Graduate Program in Biological Engineering and Small-scale Technologies, University of California, Merced, CA, USA

School of Engineering, University of California, Merced, CA, USA

Panagiotis Mistriotis Bioengineering Laboratory, Department of Chemical and Biological Engineering, University at Buffalo, The State University of New York, Amherst, NY, USA

Emma Muiños-Lopez Servicio de Reumatología, INIBIC-Hospital Universitario A Coruña, A Coruña, Spain

Mitsushige Murata Cardiology Division, Department of Internal Medicine, Keio University school of Medicine, Tokyo, Japan

T.V. Ramakrishna Murthy Habeebullah's Memorial Centre for Stem Cell Sciences, Centre for Liver Research and Diagnostics, Hyderabad, AP, India

Robert E. Nordon Graduate School of Biomedical Engineering, The University of New South Wales, Sydney, NSW, Australia

Hilary North Department of Neurology, Feinberg School of Medicine, Northwestern University, Chicago, IL, USA

Teruo Okano, PhD Institute of Advanced Biomedical Engineering and Science, TWIns, Tokyo Women's Medical University, Shinjuku-ku, Tokyo, Japan

Syed Ameer Basha Paspala PAN Research Foundations, Hyderabad, AP, India

Department of Neurology, CARE, The Institute of Medical Sciences, Hyderabad, AP, India

Yvonne Peck Division of BioEngineering, School of Chemical and Biomedical Engineering, Nanyang Technological University, Singapore, Singapore

M. Ian Phillips, PhD, DSc Norris Professor of Life Sciences, Center for Rare Disease Therapies, Keck Graduate Institute, Claremont, CA, USA

Louella A. Pritchette Department of Thoracic and Cardiovascular Surgery, Sanger Heart and Vascular Institute, Carolinas Healthcare System, Charlotte, NC, USA

David J. Schreyer, PhD Division of Biomedical Engineering, University of Saskatchewan, Saskatoon, SK, Canada

Department of Anatomy and Cell Biology, University of Saskatchewan, Saskatoon, SK, Canada

Tatsuya Shimizu Institute of Advanced Biomedical Engineering and Science, TWIns, Tokyo Women's Medical University, Shinjuku-ku, Tokyo, Japan

Russell Simpson Cardiovascular Division, King's College London BHF Centre, London, UK

Hong Sun Department of Otolaryngology, Head & Neck Surgery, Xiangya Hospital, Central South University, Changsha, Hunan, China

Xu-Yong Sun Department of Organ Transplantation, Institute of Transplant Medicine, Nanning, China

Sandeep Vishwakarma PAN Research Foundation, Hyderabad, AP, India

Centre for Liver Research and Diagnostics, Deccan College of Medical Sciences, Hyderabad, AP, India

Dong-An Wang, PhD Division of BioEngineering, School of Chemical and Biomedical Engineering, Nanyang Technological University, Singapore, Singapore

Jingyu Wang Department of Biomedical Engineering, School of Medicine, Tsinghua University, Beijing, China

Mindan Wang, BSc Division of Biomedical Engineering, College of Engineering, University of Saskatchewan, Saskatoon, SK, Canada

Garth L. Warnock Department of Surgery, University of British Columbia, Vancouver, BC, Canada

Kristoff Verdonck, MD Department of Cardiovascular Sciences, Molecular and Vascular Biology Research Unit, KU Leuven, Leuven, Belgium

Jan J. Vranckx, MD, PhD Department of Development and Regeneration, Cell and Gene Therapy Applications Research Unit, Katholieke Universiteit Leuven, University Hospitals, Leuven, Belgium

Cory J. Xian, PhD Sansom Institute for Health Research, School of Pharmacy and Medical Sciences, University of South Australia, Adelaide, SA, Australia

Qingbo Xu Cardiovascular Division, King's College London BHF Centre, London, UK

Zhen Xu Department of Biomedical Engineering, School of Medicine, Tsinghua University, Beijing, China

Masayuki Yamato Institute of Advanced Biomedical Engineering and Science, TWIns, Tokyo Women's Medical University, Shinjuku-ku, Tokyo, Japan

Ningning Yang, PhD Department of Pharmaceutical Sciences, Manchester College School of Pharmacy, Fort Wayne, IN, USA

Shinsuke Yuasa, MD, PhD Cardiology Division, Department of Internal Medicine, Keio University school of Medicine, Tokyo, Japan

Department of Cardiology, Center for Integrated Medical Research, Keio University School of Medicine, Tokyo, Japan

Chapter 1
The Mechanism of Stem Cell Differentiation into Smooth Muscle Cells

Russell Simpson and Qingbo Xu

Abstract Stem cells represent one of the most promising areas in biological and medical research for the treatment of vascular disease; by taking advantage of their unique ability to undergo unlimited self-renewal and to differentiate into specific cell lineages, they potentially provide an unlimited cell source for vascular tissue repair and for the construction of engineered vessels. Emerging evidence indicates that the mobilisation and recruitment of circulating or tissue-resident stem/progenitor cells give rise to smooth muscle cells (SMCs) which participate in numerous cardiovascular diseases such as atherosclerosis. Understanding the regulatory mechanisms that control smooth muscle differentiation and their recruitment from vascular progenitors is essential for stem cell therapy for vascular diseases and regenerative medicine. In this chapter, we examine the differentiation process of SMCs from pluripotent stem cells, highlighting the environmental cues and signalling pathways that control phenotypic modulation within the vasculature. We highlight the potential targets for promoting/inhibiting SMC differentiation and discuss their application for vessel-tissue engineering and treatment of cardiovascular pathologies.

Keywords Stem cell • Stem cell differentiation • Atherosclerosis • Epigenetic modification • MicroRNA

R. Simpson • Q. Xu (✉)
Cardiovascular Division, King's College London BHF Centre,
125 Cold Harbour Lane, London SE5 9NU, UK
e-mail: qingbo.xu@kcl.ac.uk

M.K. Danquah, R.I. Mahato (eds.), *Emerging Trends in Cell and Gene Therapy*,
DOI 10.1007/978-1-62703-417-3_1, © Springer Science+Business Media New York 2013

1.1 Introduction

Blood vessels are composed mainly of two types of cells, endothelial cells that line the lumen and vascular smooth muscle cells (SMCs) that form the structure of the media [1]. Vascular SMCs refers to the particular type of smooth muscle found within and composing the majority of the wall of blood vessels. In addition to providing structural integrity within the vasculature, the main role of SMC is to regulate blood flow and pressure in blood vessels, a mechanism that is responsible for the redistribution of the blood within the body to areas where it is needed. Vascular smooth muscle can contract or relax through highly regulated contractile machinery which in the differentiated cell is composed of specific contractile proteins. A host of human diseases including cancer, atherosclerosis, hypertension and restenosis [2, 3] can be directly attributed in part to dysfunctionality of SMCs. Deciphering the cellular and molecular mechanisms which control the differentiation and phenotypic plasticity of SMCs is vital to develop new strategies to prevent and ameliorate these diseases particularly those effecting vasculogenesis. The limited lifespan of adult vascular SMCs and the difficulty in obtaining adult and mature arteries from patients present limitations for constructing autologous human vessels in vitro to regenerate a diseased adult cardiovascular system. Finding alternative cell sources to obtain large amounts of functional SMCs for development of vascular tissue engineering has generated much interest and research in the clinical use of stem cells.

Stem cells are characterised by the unique capacity for unlimited growth and self-renewal whilst maintaining the potential to differentiate into specialised cells. Generally stem cells can be divided into embryonic stem cells and tissue-resident or adult stem cells [4, 5]. Aside from their origin, the major distinction between different forms of stem cells is their "pluripotency", that is to say their ability to develop into any cell type from the three germ layers endoderm (interior stomach lining, gastrointestinal tract, lungs), mesoderm (muscle, bone, blood, urogenital) or ectoderm (epidermal tissues and nervous system) [6–9]. Embryonic stem cells (ESCs) are the pluripotent derivatives of the inner cell mass of blastocytes, hollow sphere-shaped embryos of 200–250 cells [5, 10]. They are the most promising pluripotent stem cell sources and give rise to all types of mature tissue cells in the human body [8, 9]. The isolation of the first ESCs from mouse embryos [11] led to the revolutionary knockout mouse technology which is still widely used today [12]. Alternatively, adult stem cells are derived from blood, bone marrow, vessel wall and other tissues, but unlike ESCs, they display variable capacities for differentiation and are not pluripotent in the true sense of the word [13]. Other stem cells of non-human sources are embryonic germ cells derived from the gonad ridge of primordial germ cells and recently discovered post-implantation epiblast-derived stem cells in mouse [14, 15]. Mesoangioblasts have also been characterised recently as stem cells that can differentiate into SMC [16, 17].

Elucidating the underlying mechanisms for stem cell differentiation has been a considerable challenge for researchers. Yamamoto et al. demonstrated that mechanical force produced by fluid flow can induce ESC differentiation into endothelial cells [18], whilst Wang et al. [19] revealed that shear stress induced and suppressed angiogenic growth factors and SMC-associated growth factors, respectively. In addition to shear stress, growth factors and cytokines have been shown to directly regulate ESC differentiation [13], and the expression levels of cytokines and growth factors are likewise altered during differentiation of mesenchymal stem cells, for example [20]. Coculture of mouse neural stem cells with human endothelial-like cells gives rise to neural stem cells that have the potential to form capillary networks [21], highlighting the role of cytokines in stem cell differentiation.

In the last several years, a major achievement has been the ability to differentiate ESCs into vascular endothelial cells, SMCs and cardiomyocytes in vitro, providing not only an understanding of the development process but also a potential source for cardiovascular tissue repair [22]. The limited lifespan of adult vascular smooth muscle cells and difficulty in sourcing them present challenges for constructing human vessels in vitro to replace diseased or injured vasculature. The progress of SMC differentiation from stem cells has led to increased interest in their clinical potential to create tissue-engineered vascular grafts to treat terminal cardiovascular diseases. Furthermore, accumulating evidence indicates that the mobilisation and recruitment of circulating or tissue-resident progenitor cells that give rise to SMCs can participate in many vascular diseases including atherosclerosis, angioplasty restenosis and neointima hyperplasia after arterial injury and transplant arteriosclerosis [5, 23, 24]. Hence, in recent years, much effort has been made to understand the regulatory mechanisms which promote stem cell and progenitor cell differentiation towards SMC lineage for improving current therapeutic avenues for cardiovascular disease and vascular tissue engineering.

1.2 Smooth Muscle Cell Phenotypic Switching in Atherosclerosis

Arterial SMCs normally reside in the arterial wall in a differentiated contractile state where they provide structural support to the vasculature and control blood pressure and blood flow through highly regulated contractile mechanisms. Differentiated SMCs in adult blood vessels proliferate at an extremely low rate, exhibit low synthetic activity and express a unique repertoire of ion channels, signalling molecules and contractile proteins required for the cell's contractile function [25, 26]. Differentiated SMCs express a variety of SMC-specific contractile and contractile-associated proteins that contribute to these functions including SM-myosin heavy chain [27, 28], SM22α [29], calponin [29, 30] and SM α-actin [3, 31, 32]. Although this repertoire is specifically expressed in the fully differentiated SMC, most of these markers are expressed at least transiently in other cells during repair or pathological conditions [33], making identification of mature SMCs problematic.

Differentiation of SMCs is necessary for maturation and remodelling of the vasculature [34–36], and in addition, they secrete important components of the extracellular matrix (ECM) such as elastin and collagen, which assist in regulating mechanical properties of blood vessels [37, 38]. Unlike the cardiac and skeletal muscle cells, adult SMCs demonstrate remarkable plasticity, and in response to vascular injury, during remodelling to changes in blood flow or in different disease states, SMCs in the arterial wall can undergo profound and reversible phenotypic alterations, a process called "phenotypic switching" [39] (reviewed by Owens [25]). These dedifferentiated or "synthetic" SMCs are characterised by decreased SMC differentiation marker gene expression and increased SMC proliferation, migration, ECM synthesis [40, 41], contractile SMCs and can synthesise up to 25–46 times more collagen [42, 43] probably as a result of increased responsiveness to growth factors. Differentiation and phenotypic modulation of SMCs are controlled by a dynamic array of extrinsic cues. The fact that vascular SMCs are not terminally differentiated and retain the ability to modulate their phenotype to changing environmental cues likely evolved in higher organisms as it conferred a survival mechanism for vascular repair. Paradoxically, an unfortunate consequence of this plasticity is that it allows rapid adaptation to fluctuating environmental cues during development and progression of vascular diseases; asthma, hypertension, cancer and development of irreversible atherosclerotic lesions have all been shown to be attributed in part to phenotypic switching [39, 40, 44]. Hence, because it is believed that transition to the "synthetic" state facilitates many of the pathogenic roles of SMCs, an understanding of the factors regulating SMC differentiation is paramount for treatment strategies [45]. Whilst much is known regarding factors and mechanisms that control SMC differentiation in cultured cells, we still have an incomplete knowledge of the transcription regulatory mechanisms that ultimately regulate SMC phenotypic switching in vivo, and this is by no means made easier by the plasticity of this cell type or the fact that SMCs derive from multiple precursors throughout the embryo [46]. Unlike cardiac and skeletal muscle cells, during embryonic development, SMCs are derived from numerous distinct populations of precursor cells. Coronary artery SMCs in the vasculature, for example, are derived from proepicardial cells, whereas the aortic arch and thoracic aorta contain SMCs which have originated from the neural crest [46]. It is this origin-associated diversity which may account for the distinct structural and functional properties analogous with SMCs [46] such as the variant expression of contractile proteins with SMCs from various tissues [47, 48].

A major challenge has been to elucidate not only the environmental cues that regulate phenotypic switching in SMCs but how these processes become disrupted in disease states. A further complexity is that the precise nature of phenotypic switching is highly variable in these different diseases, with changes in atherosclerosis involving profound changes in SMC morphology, function and gene expression patterns, compared with the much more subtle changes in contractility associated with asthma and hypertension [40], for example. Moreover, the precise role of the SMC varies greatly depending on the stage of these diseases, and this is best illustrated in atherosclerosis which is probably the best-known example of a disease in which SMC phenotype switching plays a critical role.

Arteriosclerosis is an overlying term covering all pathologies in which arteries become harder and less elastic. Arteriosclerosis is characterised by SMC hyperplasia or hypertrophy and matrix protein accumulation in the intima or media or both, with or without lipid deposition, resulting in thickening and stiffness of the arterial wall [49]. Arteriosclerosis includes spontaneous atherosclerosis, accelerated (transplant) arteriosclerosis, vein graft atherosclerosis and restenosis after percutaneous transluminal coronary angioplasty [50]. Atherosclerosis, the most common form of arteriosclerosis, is a disease responsible for over 55 % of all deaths in Western civilisation [51]. In atherosclerosis lesions, the three major cell components are the SMCs, which are the most abundant cell type around the necrotic core, and the lymphocytes (intracellular and extracellular lipid) [52]. It has been estimated that up to 70 % of lesion development mass is made up of SMCs or SMC products such as ECM [25, 53]. Atherosclerosis is a progressive disease characterised by the formation of atheromatous plaques within the walls of large- and medium-sized arteries. Early lesions, otherwise known as fatty streaks, may occur in the intima as early as childhood and develop into plaques with a lipid-rich core within the central portion of the thickened intima in adults. The characteristic feature of the advanced atherosclerotic plaque is irregular thickening of the arterial intima by inflammatory cells, extracellular lipid (atheroma) and fibrous tissue (sclerosis) [54]. A large part of the lesions comprise seemingly inert and acellular fibrous tissue, but there is often a distinct and highly cellular fibrous cap which arises from the migration and proliferation of vascular smooth muscle cells and from matrix deposition [53]. The fibrous cap undoubtedly contributes something to luminal encroachment, but its importance has recently been emphasised as a strong determinant of the likelihood of plaque rupture at later stages. Rupture leads to the release of lipids which results in a signal cascade that leads to thrombus formation [53, 55–57], thereby contributing to arterial occlusions, coronary disease, myocardial infarction and stroke. It is now known that within the fibrous cap of advanced atherosclerotic plaques, SMCs may play either a beneficial role or detrimental role in determining plaque stability, depending on the cells' phenotypic state [58, 59]. In their synthetic state, SMCs are the primary cells responsible for stabilising fibrous caps by virtue of their proliferation and production of extracellular proteins. However, in response to environmental signals that are poorly characterised, these cells can become apoptotic and activate expression of matrix metalloproteinases and inflammatory mediators that can act together in promoting end-stage disease events such as plaque rupture and thrombosis [58, 59]. It had been argued that the accumulation of smooth muscle cells in the tunica intima was a negative feature of plaque progression [51, 60]. Recently, however, pathologists and cardiologists have come to see the formation and survival of a fibrous cap consisting of smooth muscle cells and connective tissue as a good thing, as part of an attempt by the vessel wall to encapsulate the toxic products accumulating in the necrotic core [61]. It is known that medial SMCs and those within arteriosclerotic lesions differ dramatically and there has been extensive work made in an attempt to study this phenotypic switching between normal and diseased states [62, 63]. During formation of arteriosclerosis, it is believed that before SMCs can migrate from the media into intima, a transition in their phenotype is required [64]. Medial non-proliferating SMCs have a contractile phenotype which they need to maintain vascular tone. When SMCs proliferate, they take on a synthetic phenotype which is associated with modulated gene expression and generation of proteins. The prevailing theory for the pathogenesis of arteriosclerosis suggests that during atherosclerotic plaque or neointima formation or both, SMCs from the media migrate to the intima and assume the synthetic phenotype, proliferate, produce extracellular matrix and participate in fibrous cap formation [51, 53]. According to this view, intimal SMCs in transplant arteriosclerotic lesions should originate from the donor vessels; however, there is now growing evidence to support the recipient origin of SMCs in neointimal lesions in animal models [65–68], whilst it has been argued that SMCs in human transplant arteriosclerosis are derived from both donors and recipients [5].

There is now growing evidence that stem cells and smooth muscle progenitor cells also contribute to arteriosclerosis by differentiating into SMCs in the intima [65, 67–72]. Derivation from these different sources may be the main reason as to why SMCs in arteriosclerotic lesions display a diversity of phenotypes, characteristics and behaviours. Since this is an important issue for understanding the pathogenesis of arteriosclerosis, the sections that follow concentrate on smooth muscle origins and the mechanism of SMC differentiation from stem cells.

1.3 Smooth Muscle Progenitors

It is now appreciated that adult stem cells are present in a host of tissues and organs (Fig. 1.1) [73, 74]. SMC accumulation in the intima is a key event in the development of arteriosclerosis [75], and as described above, the most accepted theory had been that the majority of intimal SMC are derived from the media of the vessel [76]. This long-standing dogma is being revisited following the discovery that different sources of cells may be responsible for smooth muscle accumulation in atherosclerosis. Emerging evidence has demonstrated the existence of a population of vascular stem/progenitor cells in a variety of tissues including circulating bone marrow-derived stem cells [67, 77] and/or resident $Sca1^{+}$ adventitial cells [74, 78]. There is also evidence demonstrating that SMC or SMC-like cells may be derived from a variety of sources, including transdifferentiation of endothelial cells [79] and adventitial fibroblasts [80–82] as well as medial SMC [83]. Specifically, bone marrow- and vessel wall-derived progenitors have been shown to have the ability to differentiate into SMCs which can participate in angiogenesis and vascular remodelling [84–88]. Furthermore, these cells may be directly or indirectly involved in cardiovascular disease development [89, 90] and participate in atherosclerotic plaque development and neointima formation [74, 91–95]. The lack of definitive SMC lineage-tracing studies in the context of atherosclerosis and problems in pinpointing phenotypically modulated SMC within lesions that have attenuated SMC marker genes and/or induced expression of markers of alternative cell types, that is,

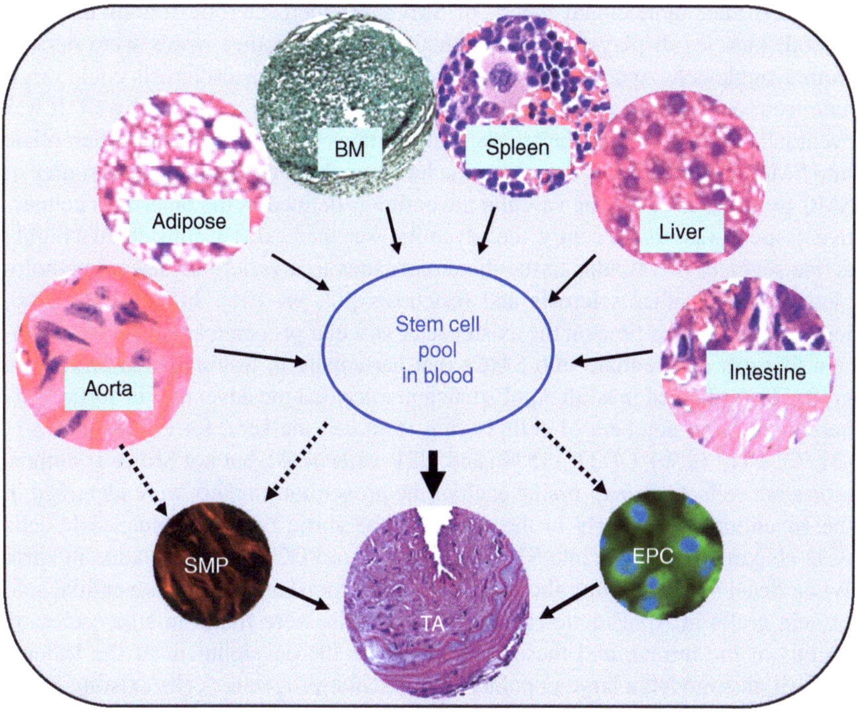

Fig. 1.1 Stem/progenitor cell origins. Stem/progenitor cells could be released from arterial wall, adipose tissue, bone marrow (*BM*), spleen, liver and intestine into blood, where they form circulating stem cell pool in blood. Smooth muscle progenitors (*SMPs*) and endothelial progenitor cells (*EPC*) accumulate within the intima, where they differentiate into SMCs contributing to the lesion formation of arteriosclerosis

macrophages, raise major questions regarding the contributions of SMC at all stages of atherogenesis. The precise frequency and roles of progenitor cell-derived SMCs in arteriosclerosis remain uncertain, but it is however widely agreed that progenitors can contribute to SMC accumulation in lesions, depending on the differential degrees of vessel damage [1]. Yet, there is still uncertainty about the origin and residency sites of smooth muscle progenitors in vivo, and given the innate heterogeneity of SMCs, it is not surprising that there is conflicting data. It was demonstrated that hematopoietic stem cells could give rise to arterial SMCs after injection into the border zone of experimental myocardial infarcts in mice [69]. In native atherosclerosis, Sata et al. demonstrated that SMCs in atherosclerotic plaques were shown to originate from bone marrow progenitors, implying that SMCs were derived from hematopoietic stem cells [67]. One group showed the majority of neointimal SMCs within plaques of experimental atherosclerosis in sex-matched chimeric scenarios and transgenic bone marrow transplant settings are derived from the bone marrow [66]. Other investigators failed to identify bone marrow-derived SMCs in atherosclerosis [68, 83, 96]. Early on, Benditt and Benditt [97] described their monoclonal theory of SMCs in atherosclerotic lesions in which smooth muscles displayed a monoclonal origin or in other words were derived from a single cell. According to this theory, SMCs in arteriosclerosis could originate from one stem/progenitor cell that may be present in the arterial wall. It was eventually discovered that the arterial wall contains stem cells that can differentiate into SMCs [13]. Recently, the adventitia has been the focus as a potential source of SMC progenitors [74]. The vascular adventitia is defined as the outermost connective tissue of vessels. Recently, the adventitia was increasingly considered a highly active segment of vascular tissue that contributes to a variety of disease pathologies, including atherosclerosis and restenosis [82, 98–101]. In 2004, Hu et al. reported for the first time on the existence of vascular progenitor cells in the adventitia that can differentiate into SMCs that participate in lesion formation in vein grafts. They showed in adult ApoE-deficient mice that the adventitia in aortic roots harboured large numbers of cells having stem cell markers, for example, Sca-1^+ (21 %), c-kit^+ (9 %), $CD34^+$ (15 %) and $Flk1^+$ cells (4 %), but not SSEA-1^+ embryonic stem cells. Cells expressing each of the progenitor markers were identified in the adventitia, particularly in the region of the aortic root. Isolated Sca-1^+ cells were able to differentiate into SMCs in response to PDGF-BB stimulation in vitro. When Sca-1^+ cells carrying the LacZ gene were transferred to the adventitial side of vein grafts in ApoE-deficient mice, β-gal^+ cells were found in atherosclerotic lesions of the intima, and these cells enhanced the development of the lesions. Thus, in this model, a large population of vascular progenitor cells existing in the adventitia could differentiate into SMCs that contribute to atherosclerosis [74]. These findings indicated that ex vivo expansion of these progenitor cells may have implications for cellular, genetic and tissue engineering approaches to vascular disease.

Progenitor cells can participate in the pathogenesis of arteriosclerosis by SMC accumulation and inducing narrowing of the lumen, but this is not the whole picture. Reports have demonstrated that injection of smooth muscle progenitor cells was shown to reduce the progression of atherosclerotic plaques in the early stages, providing evidence that the recruitment of these smooth muscle progenitor cells can promote plaque stabilisation [102]. Healthy patients were shown to demonstrate increased numbers of peripheral blood-derived progenitor cells that express smooth muscle markers compared with those patients with acute coronary syndrome, illustrating the potential benefit of SMC progenitors. Interestingly Simper et al. [103] reports that smooth muscle progenitor cells in circulating blood are characterised by ECM and matricellular proteins that were unique to the profile of vascular smooth muscle progenitor cells and aortic SMCs; however, they exhibited reduced proteases and inflammatory cytokines [103]. This suggests that circulating smooth muscle progenitor cells may also prove instrumental in alleviating atherosclerosis and/or plaque stabilisation. Resident vascular stem/progenitor cells may play an important role in the pathogenesis of atherosclerosis; however, regardless of the SMC source, the principle of local environmental cues impacting the pattern of gene expression and behaviour of these cells applies.

1.4 Smooth Muscle Cell Differentiation Mechanism

SMC differentiation from stem cells is a multifaceted process and still remains a poorly defined process. The differentiation of ESCs and vascular progenitors into vascular-specific cell lineages, that is, SMCs, is dependent on several factors, including the microenvironment, mechanic forces, cytokines or growth factors, ECM and communication with adjacent cells. The search for the transcription mechanisms that regulate SMC gene expression and differentiation has been hindered by properties intrinsic to these cells, namely, their plasticity and different embryological origins. During embryonic development, vascular SMCs can originate from at least five different sources of progenitors, including serosal mesothelium, neural crest, proepicardium, secondary heart field and somites. Not only do these distinct populations differ in vessel locality, they exhibit additional distinctions in SMC function [46, 104]. How vascular SMCs respond to environmental cues including growth factors is also lineage specific and is shown to vary depending on their developmental origin [46]. As mentioned above, this is further complicated by the fact that SMCs display phenotypic modulation in vitro and in vivo, and even in adult organisms, SMCs are not terminally differentiated [105] and are capable of switching between a secretory and contractile phenotype [106]. Cultured SMCs could rarely be stably maintained and are limited in the capacity for regulatory mechanism and pathway studies [104]; hence, extensive work has been focused on exploring the molecular mechanisms of SMC differentiation through inducible in vitro SMC differentiation systems. Despite the aforementioned challenges, in recent years, several well-established in vitro models which study SMC differentiation from stem cells have become available (for review, see [10]). Major advances have been made in the last decade to differentiate SMCs from mouse embryonic stem cells and other types of adult stem cells [107–110]. Accumulating evidence from these different systems [10] has revealed that stem cell-SMC differentiation is orchestrated by a precise coordinated molecular network that can be regulated by changes in environmental cues, activation of signal transduction pathways and altered gene expression regulated by transcriptional (co)factors, microRNSa and chromosome structural modifiers [111, 112]. The next section does not cover all recognised aspects of the mechanisms regulating SMC differentiation but rather highlights the novel mechanisms recently identified as underlying stem cell differentiation in SMCs.

1.5 Microenvironment and Integrins in SMC Differentiation

Progenitor cells that reside in the vascular tissue, in particular, are likely to play a direct or indirect role in the pathology of atherosclerosis. Intriguingly, recent evidence demonstrates an important link between smooth muscle, endothelial and hematopoietic cells through their origins from common progenitors in embryonic and adult tissue [113]. Furthermore, these vascular progenitor cells have the potential to differentiate either into endothelial cells to repair damaged endothelium or into SMCs to participate in neointimal lesions. Smooth muscle progenitors have a more heterogeneous and indefinite embryonic origin, which provides different sources for distinct SMC populations in the vessel wall [46]. It is therefore suggested that the microenvironment in which the progenitor cells reside is a vital component of their differentiation into SMCs.

Materials that lie between cells, the matrix components, have major instructive roles for cellular activities. One emerging theme is that stem cell fate and differentiation decisions are largely dependent on the dynamic interplay between stem cells and the stem cell niche, the microenvironment where the cell is localised [114]. The stem cell niche is a specific anatomic location that regulates how stem cells participate in tissue generation, maintenance and repair. The niche saves stem cells from depletion, while protecting the host from overexuberant stem cell proliferation [115]. An important component of the niche is the ECM. This ECM is the defining component of connective tissue that surrounds and supports cells, but its functions extend from its role as a scaffold to mediating responses of physiological and pathophysiological signals [116]. The components of the ECM, although appearing amorphous by light microscopy, form a highly organised interlocking mesh of glycosaminoglycans, proteoglycans, glycoproteins, peptide growth factors and structural proteins such as collagen and to a lesser extent elastin [117]. ECM has been shown to play an important role in homeostasis, embryonic development, tissue morphogenesis and various signalling pathways among almost all vertebrates [116]. It is implicated in cell migration, growth, differentiation and cell adherence [118] via external signals [114, 119]. Furthermore, studies have demonstrated that ECM can modify bioactivities of cytokines and growth factors, namely, TGF-β and PDGF [119]. Likewise the ECM itself can respond to a variety of differentiation signals provided by their local environments. Proteins and growth factors that reside in the ECM can secrete various matrix-altering agents, including proteases, that alter ECM-encoded differentiation signals [120]. Furthermore, stem cells are able to alter the very matrix signals acting upon them in a feedback system [121]. Such cell-ECM interactions have been reported to functionally affect the differentiation of mesenchymal stem cells into vascular cells [122]. When seeded on endothelial cell matrix, mesenchymal stem cells were found to induce matrix alterations which depleted the factors responsible for endothelial cell differentiation, yet activated factors that predispose differentiation towards SMCs [112].

Collagens are the most abundant proteins found in the animal kingdom and represent one of the primary components of the ECM. There are at least 12 types of collagen; types I, II and III are the most abundant and form fibrils of similar structure. Type IV collagen forms a two-dimensional reticulum and is a major component of the basal lamina. As one of the most important components of the ECM in the vascular wall, there is substantial evidence to indicate a role for collagen in stem cell differentiation. Yamashita et al. [123] and Sone et al. [124] demonstrated that $VEGFR2^+$ progenitor cells isolated from mouse stem cells could differentiate into SMCs using collagen IV as coated medium. They also demonstrated that human embryonic stem cells could be differentiated into functional SMCs using collagen IV [123, 124], and other studies have proposed that collagen type IV plays a role in the early stage of F9 stem cell differentiation and embryogenesis [125]. The functional role of collagen type IV in SMC differentiation in ESCs is still unclear. Recently Xiao et al. found collagen IV can promote ESCs to differentiate into stem cell antigen-1-positive (Sca-1^{+}) progenitor cells which could then give rise to SMCs [7]. It is recognised that a highly purified cell population is a key issue for successful tissue engineering. In this study the authors demonstrated that continued culture of differentiated ESC-derived SMCs for >30 days could achieve large numbers of functional SMCs with high purity (<95 %). Importantly these SMCs only expressed high levels of SMC markers and not others such as endothelial cell-specific marker (CD144), leukocyte common antigen (CD45) and Mac-1. Furthermore, it was reported that collagen IV was a crucial component of stem cell-SMC differentiation in non-collagen-IV-coated plates due to autocrine production of collagen IV. Pretreatment of ESCs with antibodies against collagen IV significantly inhibited SMC marker expression [7]. SMC differentiation is associated with changes in basement membrane composition from fibronectin, which supports SMC proliferation, to collagen IV and laminin, which promote SMC differentiation [126–128]. Vessel injury can lead to the degradation and induction of matrix components such as fibronectin and collagen I, events which are likely to contribute to SMC differentiation [129].

ECM is mediated largely by the integrin family of cell surface adhesion receptors. Integrins belong to a family of non-covalently associated heterodimeric cell surface receptors composed of α- and β-subunits [130]. To date 18 α- and 8 β-integrin subunits have been described, and they can combine to form up to 24 different heterodimers [131], and the mechanisms by which integrins regulate cell growth are well documented [132]. The interplay between ECM components and integrins offers an important function in various biological processes, including progenitor cell homing [133, 134], cell attachment, spreading, proliferation, survival, morphogenesis and gene expression [135–138] by influencing the balance between stem cell renewal and differentiation [139]. Integrins have been shown to perform important roles in differentiation of mesoderm-derived lineages including myofibroblasts [140] and myocytes [141]. High surface expression of β1 integrin and moderate levels of α1 and low levels of αv and β3 were reported in circulating smooth muscle progenitor cells in human peripheral blood [134]. Kogata et al. demonstrated that integrin-linked kinase, a very weak serine/threonine kinase, was found to bind to the cytoplasmic tail of integrin β receptor and negatively regulate RhoA activity in SMCs. Deletion of integrin-linked kinase in PDGF receptor-β expressing cells in vivo led to attenuated SMC investment and hypercontractility [142]. Conversely, Wu et al. showed integrin-linked kinase negatively regulates SMC differentiation markers in airway tissue. They demonstrated that overexpression and suppression of integrin-linked kinase decreased and increased SMC differentiation, respectively, but interestingly following artery injury, integrin-linked kinase expression was attenuated, and conversely this resulted in phenotypic switching from differentiation to proliferation and neointimal hyperplasia [143]. These data collectively demonstrate that integrin-linked kinase and integrins interact with ECM to control SMC differentiation and proliferation, but additional studies are required. Studies have demonstrated that ECM-integrins play an important role in differentiation of SMCs from stem cells since ECM-integrin interaction negatively regulates ESC self-renewal [144]. The activation of integrin receptors by tyrosine phosphorylation of β-subunits is essential for their function, whereby signal transmission through these complexes can effect various aspects of cell physiology, including SMC differentiation [131]. Many studies on SMC differentiation have highlighted the dependence of this process on the interactions of α1β1 and αv integrins with collagen IV.

Collagen-related integrins include $\alpha_1\beta_1$, $\alpha_2\beta_1$, $\alpha_3\beta_1$, $\alpha_{10}\beta_1$ and $\alpha_{11}\beta_1$ [145] in which $\alpha_1\beta_1$ is abundant on SMCs. Xiao et al. demonstrated that high levels of α_1, αv, β_1 and β_3 were associated with both Sca-1^+ progenitor cells and ES-derived SMCs [7]. The authors demonstrated that inhibition of the integrin pathway by blockade with R&G peptides and blocking antibodies against the specific integrins significantly inhibited SMC differentiation. These findings clearly indicate the importance of the collagen IV-integrin signalling pathway for SMC differentiation from stem cells [145].

After tethering to collagen IV, stem/progenitor cells may need additional environmental stimuli to initiate differentiation of SMCs. Many reports in the literature demonstrate that soluble mitogenic growth factors TGF-β [146, 147] and PDGF-BB [77, 148] are necessary for SMC differentiation in vivo and in vitro. The initial signal is sensed by progenitor cells via their cell surface receptors, namely, TGF-β and PDGF receptors [40, 112]. Signal transmission through tyrosine phosphorylation events is then responsible for cellular activities such as SMC differentiation [131].

1.6 Regulation of SMC Differentiation by TGF-β

Transforming growth factor (TGF)-β_1 is a potent multifunctional cytokine that has been shown to play a role in modulating vascular development and maintenance by controlling the growth, differentiation and matrix deposition of SMCs. The classic signalling pathway for TGF-β involves the Smad family of transcriptional activators [149, 150]. The receptor-associated R-Smad, Smad2 and Smad3 are phosphorylated directly by the TGF-β type 1 receptor kinase, after which they hetero-oligomerise with Smad4, translocate to the nucleus and bind to specific DNA sequences where they initiate target gene transactivation, either alone or in association with other transcriptional partners [151]. TGF-β null mice report loss of early SMC coating around the nascent dorsal aorta [152, 153], suggesting that TGF-β signalling pathway may be required for SMC differentiation. Consistent with these results, studies in mice with gene knockout of the type 1 TGF-β receptor, TGF-β receptor II, TGF-β_1 or SMADs resulted in early embryonic lethality due to the defects in vasculogenesis, angiogenesis, hematopoiesis and mesenchymal apoptosis [153–156]. These data demonstrated that homeostatically regulated TGF-β signalling is crucial and required for normal vasculature development and formation. TGF-β_1 was shown to induce a variety of SMC differentiation marker genes including SM α-actin, SM-myosin heavy chain and calponin and led to immature bloods vessel-derived

SMCs [157, 158]. Sinha et al. [147] further demonstrated that SMC-specific gene expression was downregulated by a soluble truncated TGF-β type II receptor, an anti-TGF-β_1 antibody or small interfering (si)RNAs directed against Smad2 or Smad3, providing direct evidence that that TGF-β_1 signalling through Smad2 and Smad3 plays an important role in the development of SMCs from totipotential ESCs.

Additional studies have reported that TGF-β strongly stimulates SMC differentiation marker gene expression in a number of cell types, including mesenchymal and ESCs, aortic SMCs lung fibroblasts, 10T1/2 and Monc-1 [147, 158–161]. Qui et al. [162] specifically proved that Smad3 can physically interact with serum response factor (SRF) to facilitate differentiation to SMC lineage.

1.7 PDGFs and SMC Differentiation

PDGF-BB can induce phenotypic switching in vitro in cultured SMCs [163, 164]. The fact that PDGF-BB is released following vessel injury [165] coupled with the observation that PDGF-BB signalling inhibition suppresses neointimal growth [166] suggested that it is involved in SMC phenotype in vivo. It is highly expressed by endothelial cells and was demonstrated in one study to be paramount for recruitment and proliferation of SMCs within the maturing vasculature [167]. The mechanism by which PDGF-BB regulates SMC differentiation involves multiple, overlapping signalling pathways. Activation of tyrosine kinase receptors such as PDGF receptor-β triggers the ras/Raf/MEK/ERK kinase cascade, resulting in phosphorylation of Elk-1 and induction of multiple early response genes which are SRF dependent. However, there is opposing data as to the effects of PDGF in SMC differentiation. PDGF-BB has been described as a factor that can induce rather profound suppression of SMC markers genes [168, 169]. The work from the Owens laboratory proposes that PDGF-BB acts as a suppressor of SMC differentiation. They propose a mechanism whereby PDGF-BB induces levels of Krüppel-like factor 4 which in turn suppresses myocardin expression, interfering with SRF/myocardin factor binding to the SMC-specific promoters with subsequent silencing of SMC marker genes [170, 171]. In addition to downregulation of SMC differentiation, PDGF-BB was also shown to stimulate SMC phenotypic modulation by enhancing SMC migration and proliferation in arterial injury models [166, 172]. In contrast, there is amassing evidence to suggest that PDGF-BB can *promote* stem cells or progenitor cells to differentiate into SMC phenotype [7, 74, 173]. Hu et al. [74] confirmed that in vascular SMC progenitor cells isolated from the adventitia, exogenously applied PDGF-BB drove SMC differentiation. Xiao et al. [7] further examined the role of PDGF receptor-mediated SMC differentiation in mouse stem cell-derived Sca-1^+ progenitors. Following exogenously applied PDGF-BB in the presence of 10 % FBS, the authors failed to observe a significant upregulation of SMC marker genes; they did however demonstrate marked inhibition of SMC differentiation when a siRNA specific for PDGF receptor-β was utilised. In another study, PDGF-BB 2 was shown to directly promote mouse ESC differentiation into SMCs in the absence of FBS [174]. Collectively these data support a role for PDGF receptor pathway-mediated stem cell-SMC phenotypic transformation.

1.8 Epigenetic Modifications and HDAC Signalling

There has been extensive progress in recent years in clarifying the complex mechanisms that control SMC differentiation and phenotypic plasticity, and now it is realised that numerous layers of epigenetic modification play a crucial role. Chromatin is a dynamic polymer mass of genetic material composed of DNA and proteins that condense to form chromosomes during eukaryotic cell division. Its structure is regulated by both epigenetic (e.g. DNA methylation, histone modifications or histone-binding proteins) and trans-acting DNA-binding proteins (e.g. transcription factors/repressors or polymerase machinery) modifications [175]. The importance of these epigenetic modifications has been demonstrated in the development of cardiovascular disease [176, 177]. Here we discuss how epigenetic mechanisms play a key role in SMC differentiation, as well as in phenotypic switching in response to vascular injury or atherosclerotic disease.

Alterations in chromatin conformation are critical for controlling how accessible genomic DNA is to sequence-specific transcriptional activators/repressors. The mechanism underlying these mechanisms is not clear, but stem cells have a unique chromatin structure, often reflecting a globally more active chromatin state than "normal" cells. As differentiation advances, chromatin changes to a repressed and inactive state [178]. The current opinion held by most researchers in the chromatin field is that histone modifications are crucial. Studies have described a "histone bivalent" model which regulates ESC status by controlling gene expression for lineage-specific genes, which are silent in pluripotent ESCs, but expressed on differentiation [179–182]. The epigenetic status must be abolished in pluripotent stem cells to trigger development and subsequent cell differentiation. Specific residues in the *N*-terminal tails of histones are prone to reversible modifications including methylation, acetylation, phosphorylation [183] and proteolysis [184]. The homeostasis of histone acetylation and deacetylation is known to control the expression of genes through alterations in chromosome assembly or disassembly and through interactions with transcription factors [185, 186].

Histone methylation is regulated by two families of proteins called histone methyltransferases and demethylases [187] and has been shown to play an essential role in SMC differentiation (for review see [44, 188]). Acetylation of histones are carried out by histone acetylases (HATs), whilst deacetylation is carried out by histone deacetylases (HDACs) [189]. HATs and HDACs modify the acetylation state of histones in opposing ways. During SMC differentiation, the tails of histone proteins associated with the promoters of SMC-selective genes such as those encoding SM α-actin and SM-myosin heavy chain are posttranslationally modified through alterations such as acetylation of histones 3 and 4 and dimethylation of lysine 4 and 79 on histone 3 [108, 190]. These modifications are believed to open up the chromatin within these promoters to allow binding of SRF-myocardin complexes to CArG box elements and to drive expression of SMC-selective genes [188]. Regulation of SMC gene expression is reliant on the binding affinity of SRF to CArG box DNA sequences and myocardin within intact chromatin [106]. Deacetylation carried out by HDACs removes the acetyl groups from lysine residues in histones, and this results in a particular region of chromatin to be condensed, leading to suppression of gene expression. Recently, the role of HDACs as key mediators in differentiation of stem cells towards specific lineages has been highlighted [191–193].

To date, 18 mammalian HDACs have been discovered and characterised into four different classes according to sequence homology. The most relevant HDACs in SMC differentiation are classes 1 and 11. Class II HDACs have been shown to upregulate and downregulate SMC-specific genes via interaction with myocardin [193]. The discovery of myocardin in the Olson laboratory has been hailed as one of the most exciting advances in the field of SMC differentiation in the past decade [194]. Myocardin is a specific coactivator of SRF shown to bind to the CArG element located within promoters or the intron sequence of SMC differentiation genes [112] that is exclusively expressed in cardiac and differentiated SMCs [195]. It had been shown to be vital for early SMC formation during embryogenesis. A mutation leading to myocardin loss of function is lethal in mouse embryos and is characterised by deficient SMC components in perivascular cells of the dorsal aorta [196]. Studies in knockout mice showing attenuated SMC-positive cells and staining in vessel wall and dorsal aorta of embryos suggested there was a direct link between SMC differentiation and HDAC7 [197]. Further studies by Margariti et al. [198] revealed that upregulation of HDAC7 splicing mediates PDGF-BB-induced SMC differentiation from ESCs by modulating the SRF-myocardin complex. Normally, HDAC7 is presented as a partially spliced isoform lacking the first 22 amino acids because it contains a 57-bp intron [199]. This short isoform of HDAC7 when bound to MEF2C leads to downregulation of SMC gene markers and furthermore inhibits activation of cytoplasmic spliced HDAC7, resulting in differentiation of non-SMC lineages. Stimulus with PDGF-BB acts to remove this intron, allowing the full-length HDAC7 to be activated in the cytoplasm. This sliced HDAC7 translocates to the nucleus where it interacts with SRF, driving its binding affinity to gene promoter and coactivation of myocardin. The resulting SRF-myocardin complex is recruited to the SM22-α promoter and promotes SMC gene expression [198]. Additionally activated HDAC7 prevents recruitment of HDAC2 and HDAC5 to the promoter where they can inhibit SMC differentiation [200]. The net result is ESC differentiation towards a SMC lineage. The involvement of HDAC in the signal pathways of stem cell differentiation into SMCs has been illustrated in a schematic figure (Fig. 1.2). Further studies by Zhang et al. [201] have demonstrated that Sp1 plays an important role in the regulation of HDAC7. They report that mutation of the Sp1 site within the PDGF-BB responsive element or direct knockdown of Sp1 abrogated PDGF-BB-induced HDAC7 upregulation and SMC differentiation gene expression in differentiating ES cells.

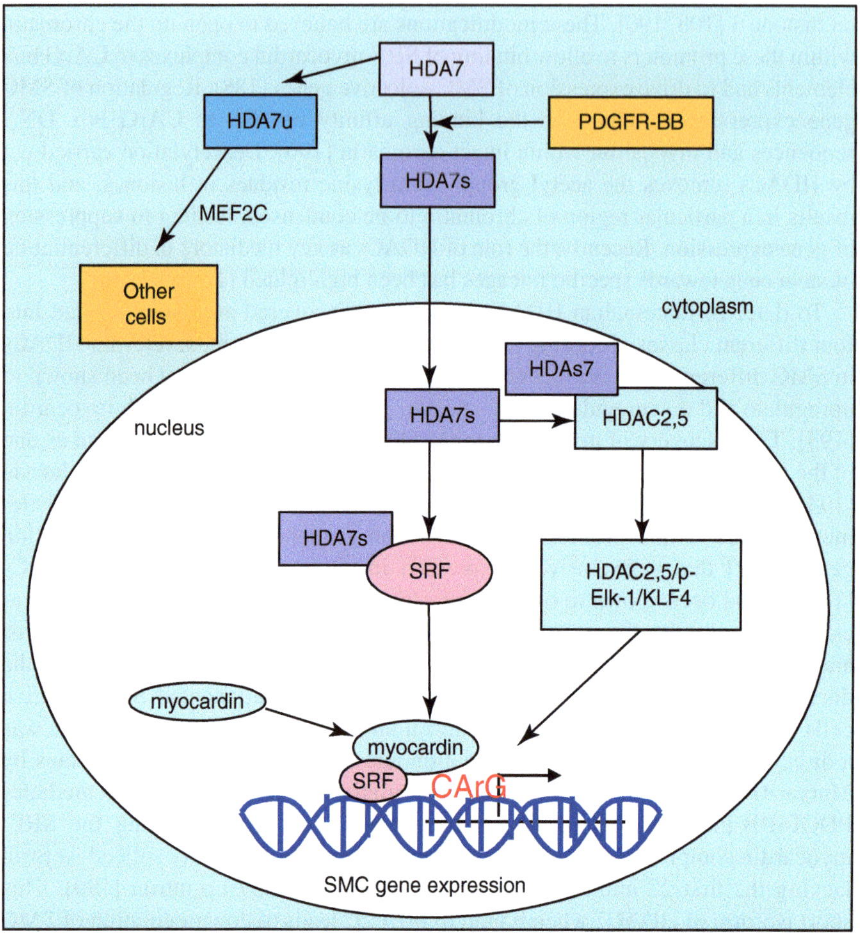

Fig. 1.2 Regulatory role of HDACs in SMC differentiation. At the early stage of SMC differentiation, HDAC7 is expressed as a full form (HDACu), which binds to MEF2C and inhibits MEF2C activity, resulting in stem cell differentiation to other cell types. In response to PDGF-BB stimulation, HDAC7 is spliced (*HDAC7s*) and translocated to the nucleus. Then, HDAC7s enhances the binding of SRF to CArG and the association with myocardin. HDAC7s interacts with HDAC2 and 5, preventing them from binding to SMC promoter. This results in the SMC differentiation from stem cells

1.9 Nox4 and Nrf3 in SMC Differentiation

Reactive oxygen species (ROS) are highly reactive molecules that are generated following interaction of integrins, extracellular matrix and cytokines. They act as second messengers and mediate a host of cellular processes including vascular physiology and pathogenesis including hypertension, restenosis and atherosclerosis [202]. Previous reports have shown that ROS is involved in proliferation, migration and differentiation of vascular SMCs [203, 204], and even more recently, the role of ROS has emerged as an important mediator of SMC differentiation from ESCs [205, 206]. The main source of ROS in the cardiovascular system is nicotinamide adenine dinucleotide phosphate oxidases (NADPH oxidases, Noxs). The classical Nox complex is made up a membrane-bound cytochrome b558 (composed of one gp91phox and one p22phox subunit) which makes up the catalytic core of the enzyme and four cytosolic regulatory subunits (p47phox, p67phox, p40phox and Rac). When these translate to the cytochrome b558, the enzyme is activated [112]. They are classified by their differing isoforms of the catalytic Nox subunit [207]. These isoforms include Nox1–5 and Duoxn1 and 2 [208, 209]. Two major Nox isoforms (Nox1 and Nox4) are located in human and rodent aortic SMCs [203]. Nox1 has been shown to mediate signal transduction and is important in SMC hypertrophy and cell proliferation, whilst Nox4 expression is unregulated at the end of neointima formation during differentiation phase in carotid injury-induced restenosis and atherosclerosis [210, 211]. Nox4 is found in all vascular cells but mainly resides in the media of vessel walls [212]. Unlike other isoforms, Nox4 is not found in the plasma membrane. Following its activation through interaction with $p22^{phox}$ 4 on internal membranes, the Nox4 complex generates ROS [213–215]. This is comprised of H_2O_2 and production of O_2–. Recently Xiao et al. [205] demonstrated that Nox4-derived H_2O_2 is integral to the differentiation of ESCs into SMCs. Silencing of Nox4 suppressed differentiation, whilst sustained Nox4 signalling enhanced differentiation of SMC gene markers. The authors demonstrated that autocrine TGF-1β indirectly generated ROS via Nox4 activation. Nox4 translocation from the cytoplasm to the nucleus resulted in upregulation of H_2O_2 which in turn led to induction and phosphorylation of SRF and its translocation into the nucleus. Phosphorylated SRF binds to the CArG element on the promoter-enhancer regions of SMC-specific genes, recruiting myocardin to the promoter to form a SRF-myocardin complex. This complex was shown to be essential for regulating early-stage Nox4-mediated stem cell differentiation. Furthermore, in late-stage differentiation, nuclear Nox4 associates with SMC filaments, which facilitate maintenance of SMC phenotype [189]. Meanwhile, Nox4-derived O_2 has been shown to increase SRF-mediated gene transcription activation through a p38 MAPK-dependent pathway [204]. Ultimately these events promote SMC differentiation.

Nuclear factor erythroid 2-related factor (Nrf)3 is a member of the cap'n'collar family of transcription factors [216]. Nrf3 is now considered to be a key transcription factor in regulating SMC differentiation by modulating the balance of ROS generation. Pepe et al. [217] recently demonstrated that Nrf3 is crucial for stem cell differentiation towards SMCs. Usually Nrf3 resides in the endoplasmic reticulum (ER); then during the early stages of SMC differentiation having translocated to the nucleus following ER stress, Nrf3 can directly bind to the promoter region of SMC-specific genes (i.e. SM α-actin and SM22α) that promote the formation of the SRF-myocardin complex. Cytoplasmic Nrf3 on the other hand is able to promote Nox4-mediated ROS production which drives SMC differentiation. Nrf3 is able to repress antioxidant responsive element-mediated gene expression of antioxidant enzymes, such as the NAD(P)H:quinone oxidoreductase 1 (NQO1) [218] and peroxiredoxin 6 [219]. During embryonic development, Nrf3 also plays a role in mesodermal layer determination [220]. The involvement of Nox4/Nrf3 in the signal pathways of stem cell differentiation into SMCs has been illustrated in a schematic figure (Fig. 1.3).

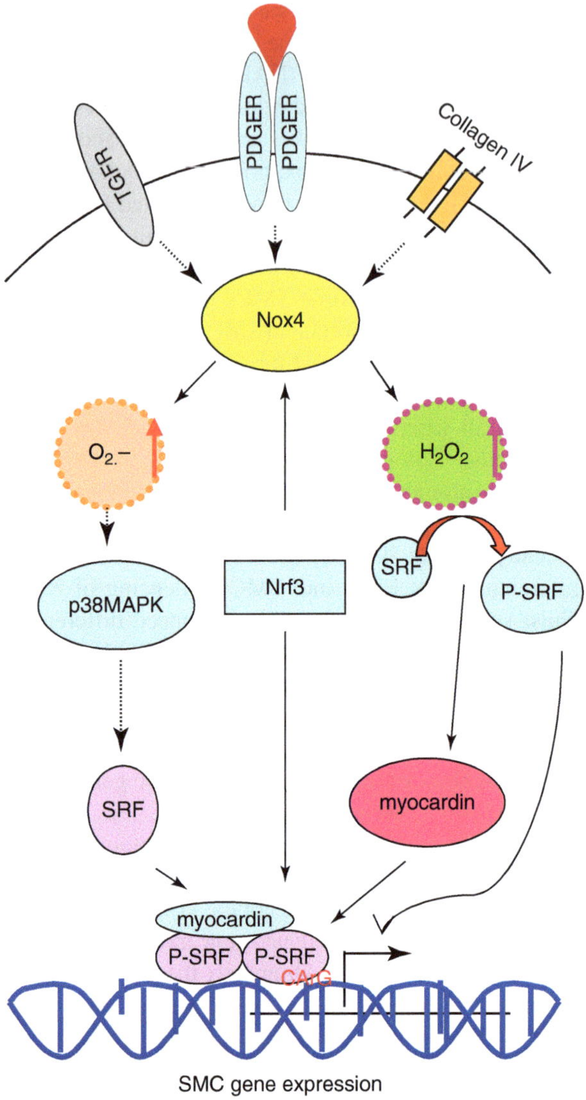

Fig. 1.3 Regulatory role of Nox4 in SMC differentiation. PDGF-BB and TGF-β1 can bind to their receptors in the presence of interactions between collagen IV and integrins, which indirectly induce Nox4 expression. Activated Nox4 generates ROS (H_2O_2 and $O_2.^-$). Nox4-derived H_2O_2 upregulates SRF gene transcription and protein translation, phosphorylates SRF in the cytoplasm and drives activated SRF to translocate into the nucleus from cytoplasm. Phosphorylated SRF binds to CArG elements within the promoter-enhancer region of SMC-specific genes, recruits coactivator myocardin and other transcription factors and then regulates SMC differentiation. Meanwhile, Nox4-derived $O_2.^-$ activates indirectly HDAC7, increases SRF-mediated gene transcription activation and further drives SMC differentiation. Furthermore, Nrf3 is involved in both Nox4 expression and direct interaction with transcription factors for SMC gene expression

1.10 MicroRNA and SMC Differentiation

MicroRNAs (miRNAs) are a class of endogenous, highly conserved, single-stranded non-coding small (~22 nucleotide) RNAs which play important roles in widespread cellular function such as development, differentiation, proliferation, migration and apoptosis [221, 222]. Several studies have reported that miRNAs can regulate cardiogenesis and angiogenesis during embryonic development, which makes them potential therapeutic targets in cardiovascular disease [223, 224]. miRNAs are negative or positive post-transcriptional regulators that bind to complementary sequences on target messenger RNA transcripts (mRNAs), usually resulting in translational repression or target degradation and gene silencing [225, 226]. The human genome may encode over 1,000 miRNAs, which may target about 60 % of mammalian genes and are abundant in many human cell types [227–229]. miRNAs are first transcribed into primary miRNAs (pri-miRNAs), after which nuclear cleavage of the pri-miRNA by drosha Rnase III and cofactor DGCR8/Pasha generates an intermediate state (pre-miRNA) which is transported from the nucleus to the cytoplasm via an exportin-5 and RanGTP-dependent mechanism [225]. Here pre-miRNA is further processed by another RNase III enzyme Dicer to form mature miRNAs. One strand of mature miRNA base pair imperfectly binds to target mRNA specifically at the 3′ untranslated region and forms a nuclease complex known as the RNA-induced silencing complex which silences mRNA primarily via mRNA translational repression [221, 230] and through their degradation by argonaute-catalysed mRNA cleavage [231, 232]. It is now known that during the differentiation process from ESCs to SMCs, miRNAs are highly regulated [233]. MicroRNAs play a central role in regulating the self-renewal and differentiation programme of stem cells [234, 235]. Many microRNAs display the ability to initiate the switch from pluripotent to a lineage-specific state by selectively suppressing pluripotent factors. Most recently, work has demonstrated that as well as facilitating SMC differentiation that miR-145 also represses pluripotency in human ESCs [236, 237]. Xu et al. showed that it could repress OCT4, SOX2 and Kruppel-like factor 4, the core pluripotency factors [237]. Recently it was demonstrated that miR-145 facilitated SMC differentiation from neural crest stem cells [236] and was downregulated in atherosclerotic vessels. It has also been shown to play a role in phenotypic switching of SMC differentiation and regulation of blood pressure in mouse models [238]. miR-145 is reported to highly integrate into a transcriptional network and acts as a critical switch in SMC differentiation [236], and its overexpression in human ES-pre-SMCs is a promising method to obtain functional mature SMCs from human ESCs, which can be utilised for reliable experimental research in the fields of atherosclerosis, hypertension and other vascular diseases [239]. miR-143/145 cluster is believed to regulate SMC differentiation from stem/progenitor cells and in addition show involvement in SMC phenotypic switch [240]. miR-143 and miR-145 interact with SRF which not only regulates cytoskeletal remodelling but can regulate phenotypic switching of SMCs during cardiovascular disease [238]. In miR-143/145 double knockout mice, the aorta and femoral artery were characterised by reduced contractile vascular SMCs and increased synthetic vascular SMCs and inhibition in SMC-specific differentiation markers [241, 242].

Other miRNAs that have been shown to regulate SMC differentiation include miR-1 and miR-10a via KLF4 and HDAC4, respectively [233, 243]. miR-221 has also been reported to be involved in SMC phenotypic switching by mediating the effect of PDGF. Specifically following PDGF treatment, miR-221 attenuates expression of c-Kit and p27Kip1 which leads to decreased levels of myocardin which promotes switching of SMCs from a contractile to a synthetic and less contractile phenotype [244].

MicroRNAs are now considered to be the missing linkers in SMC differentiation mechanism. They are central to the differentiation and dedifferentiation of SMCs, and therefore, miRNAs represent as master regulators for controlling phenotypic switching. Since failure of SMCs to acquire and maintain the contractile phenotype is thought to contribute to many cardiovascular diseases such as arteriosclerosis, the ability to direct the activities of miRNAs offers an alternative strategy for regulating SMC differentiation and phenotypic modulation from stem/progenitor cells for treatment of cardiovascular pathologies.

1.11 Perspective in Therapeutic Potential

Cell differentiation from stem cells is intricate and still a poorly defined activity. Stem cell research provides a unique opportunity for understanding the molecular mechanisms of cell differentiation towards SMCs in vitro and in vivo. The SMC is a fascinating cell type that can exhibit a wide range of different phenotypes in development and disease. Unlike other cell lineage differentiation, SMCs are not terminally differentiated, and this plasticity makes elucidating the underlying mechanism especially complex. Although extensive research has revealed that many signal pathways and molecules, such as SRF-myocardin complex, collagen IV-integrins, TGF-β1, PDGFs, HDAC7, Nox4-H_2O_2 and micro-145/143, orchestrate SMC differentiation, the mechanistic networks that govern stem/progenitor transition to SMC lineage remain unclear. Stem cells have a role in vascular repair. Following vascular injury/disease, stem cells derived from different sources may participate in SMC accumulation, and hence the fate of stem cell differentiation into SMCs is a key issue for the progression of arteriosclerosis (Fig. 1.4). The need for deciphering these regulatory mechanisms is now ever more prevalent since stem cell research could be vital not only to further our understanding of the pathogenesis of disease but also for the development of cell-based therapies and tissue engineering. When translated to the field of vascular disease, the potential therapeutic uses of donor-derived or patient-derived stem cells offer broad potential in treatment of cardiovascular disease. Stem cell-based therapeutic vascular (re)generation show great promise for treatment of cardiovascular ischemic diseases which are currently clinically challenging. Inducing angiogenesis in ischemic tissues safely and locally via pharmacological interventions is difficult, so new gene- or cell-based therapies are being explored. Stem cells have been shown to initiate vasculogenesis and angiogenesis processes in clinical ischemic conditions. Several authors have demonstrated that stem cells can give rise to vascularised "biotissues" which can be utilised for transplantation and in vitro tissue-based toxicology studies [111]. Phase 1 and phase 2 clinical trials in patients with myocardial or limb ischemia have recently investigated the benefits of several putative vascular progenitor cells [245]. Myocardial infarction has been clinically tested with bone marrow cell transfer therapy, but the trial results are inconsistent [246]. miR-based

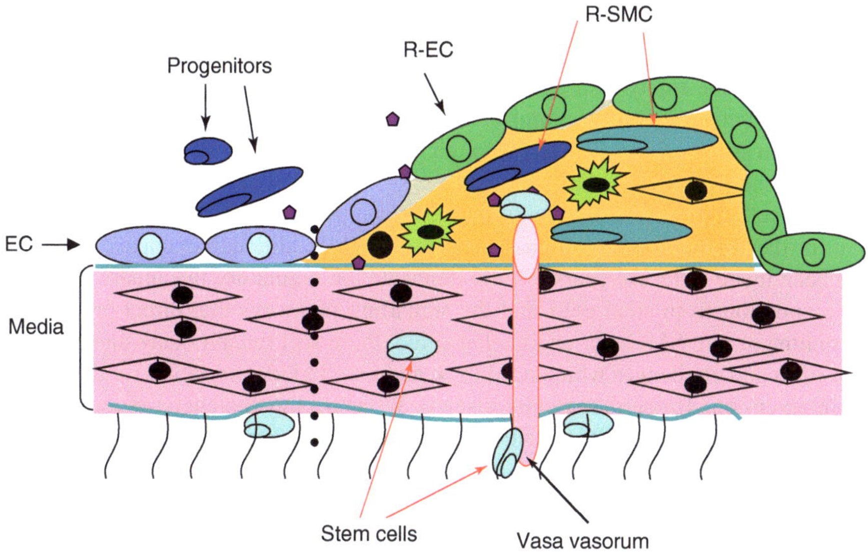

Fig. 1.4 Schematic representation for stem cells contributing to arteriosclerosis. Endothelial cells (*EC*) covering the early neointimal lesions are derived from stem/progenitor cells of the recipient (*R*). Functions and differentiating abilities of progenitor cells may also be influenced by risk factors and local environment, resulting in endothelial dysfunction. Smooth muscle progenitor cells in blood may migrate into the lesions. Meanwhile, stem cells presented in the media and adventitia can migrate into the lesions via vasa vasorum. These cells differentiate into neo-SMCs (SMC) within arteriosclerotic lesions, which are different from medial SMCs. This process repeats several times, leading to the formation of arteriosclerosis

therapy has been proved effectively in animal models of several cardiovascular diseases, including cardiac hypertrophy, myocardial infarction, heart failure and artery injury [240].

One groundbreaking new technology which has emerged in recent years has been the generation of induced pluripotent stem cells (iPSCs). First produced from mouse fibroblasts in 2006 by Yamanaka's group [247], they hold great potential as an alternative source for vascular cells for vascular regeneration, because they provide unlimited source of pluripotent stem cells that can be used in replacement therapy without the therapeutic limitations of ESCs and adult progenitor cells such as ethical concerns (for ESCs) and immunogenicity/allograft rejection (for progenitor cells). Gene delivery systems carrying reprogramming transcription factors (e.g. Oct4, Sox2, Klf4, cMYC, Lin28 and Nanog) are introduced into an adult somatic cell. Upon overexpression of the reprogramming factors, somatic cells undergo reprogramming to induce pluripotency and can be expanded exponentially and maintained in a pluripotent phenotype indefinitely or differentiated to all cell types including SMCs [248, 249]. In combination with other technologies such as tissue engineering, it is feasible that whole tissues such as arteries could be grown from iPSCs. As of yet, iPSC-derived vascular cells have not been tested in vivo. Before this technology can be applied in a clinical setting, more basic studies and translation research will need to be performed. Better characterisation of identity and thorough evaluation of the safety of iPSCs are needed before they can show potential therapeutic use, but there is great promise, and it is predicted that vascular derivatives of pluripotent stem cells particularly through iPSC exploitation will be used for tissue replacement strategies [111].

Stem cells represent a promising therapeutic approach for regenerative medicine and for the treatment of cardiovascular diseases [24]. Although extensive progress has been made in recent years to fully delineate the regulatory machinery of SMC differentiation from stem cells and the signalling pathways that direct progenitor commitment into SMCs (or endothelial cells), there still remain many unanswered questions and challenges which need to be addressed to identify effective therapeutic interventions for cardiovascular disease. What are the epigenetic programming mechanisms that direct ESC transition to SMC lineage and which changes are stable during SMC phenotypic switching during vascular injury, disease and repair? What is the molecular switch that directs stem/progenitor differentiation to SMCs? What other potential molecular targets mediate SMC differentiation, such as the newly identified microRNAs? What are the epigenetic mechanisms that allow transcription factor access to the SMC-specific genes? In answering these and other important questions, investigators ultimately aim to design drugs which target vascular progenitor cells and manipulate trafficking to the intima to impede atherosclerotic plaque formation or better yet direct progenitors to the cell type that is beneficial for the vessel wall. For now, further research into the biology of stem cells and their differentiation into SMCs is needed to take advantage of their regenerative properties to provide therapeutic strategies in cardiovascular disease.

Acknowledgments This work was supported by grants from the British Heart Foundation and Oak Foundation.

References

1. Campagnolo P, Wong MM, Xu Q (2011) Progenitor cells in arteriosclerosis: good or bad guys? Antioxid Redox Signal 15(4):1013–1027
2. Huang H et al (2005) Differentiation from embryonic stem cells to vascular wall cells under in vitro pulsatile flow loading. J Artif Organs 8(2):110–118
3. Mack CP (2011) Signaling mechanisms that regulate smooth muscle cell differentiation. Arterioscler Thromb Vasc Biol 31(7):1495–1505
4. Mathur A, Martin JF (2004) Stem cells and repair of the heart. Lancet 364(9429):183–192
5. Xu Q (2008) Stem cells and transplant arteriosclerosis. Circ Res 102(9):1011–1024
6. Daley GQ (2010) Stem cells: roadmap to the clinic. J Clin Invest 120(1):8–10
7. Xiao Q et al (2007) Stem cell-derived Sca-1+ progenitors differentiate into smooth muscle cells, which is mediated by collagen IV-integrin alpha1/beta1/alphav and PDGF receptor pathways. Am J Physiol Cell Physiol 292(1):C342–C352
8. Kiskinis E, Eggan K (2010) Progress toward the clinical application of patient-specific pluripotent stem cells. J Clin Invest 120(1):51–59
9. Thomson JA et al (1998) Embryonic stem cell lines derived from human blastocysts. Science 282(5391):1145–1147

10. Xie C et al (2011) Smooth muscle cell differentiation in vitro: models and underlying molecular mechanisms. Arterioscler Thromb Vasc Biol 31(7):1485–1494
11. Martin GR (1981) Isolation of a pluripotent cell line from early mouse embryos cultured in medium conditioned by teratocarcinoma stem cells. Proc Natl Acad Sci USA 78(12): 7634–7638
12. Hooper M et al (1987) HPRT-deficient (Lesch-Nyhan) mouse embryos derived from germline colonization by cultured cells. Nature 326(6110):292–295
13. Margariti A, Zeng L, Xu Q (2006) Stem cells, vascular smooth muscle cells and atherosclerosis. Histol Histopathol 21(9):979–985
14. Brons IG et al (2007) Derivation of pluripotent epiblast stem cells from mammalian embryos. Nature 448(7150):191–195
15. Tesar PJ et al (2007) New cell lines from mouse epiblast share defining features with human embryonic stem cells. Nature 448(7150):196–199
16. Brunelli S, Cossu G (2005) A role for MSX2 and necdin in smooth muscle differentiation of mesoangioblasts and other mesoderm progenitor cells. Trends Cardiovasc Med 15(3):96–100
17. Brunelli S et al (2004) Msx2 and necdin combined activities are required for smooth muscle differentiation in mesoangioblast stem cells. Circ Res 94(12):1571–1578
18. Yamamoto K et al (2005) Fluid shear stress induces differentiation of Flk-1-positive embryonic stem cells into vascular endothelial cells in vitro. Am J Physiol Heart Circ Physiol 288(4): H1915–H1924
19. Wang H et al (2005) Shear stress induces endothelial differentiation from a murine embryonic mesenchymal progenitor cell line. Arterioscler Thromb Vasc Biol 25(9):1817–1823
20. Kim DH et al (2005) Gene expression profile of cytokine and growth factor during differentiation of bone marrow-derived mesenchymal stem cell. Cytokine 31(2):119–126
21. Wurmser AE et al (2004) Cell fusion-independent differentiation of neural stem cells to the endothelial lineage. Nature 430(6997):350–356
22. Gepstein L (2002) Derivation and potential applications of human embryonic stem cells. Circ Res 91(10):866–876
23. Xiao Q, Roberts N, Jahangiri M, Xu Q (2007) Stem cells, progenitor cells and vascular diseases. In: Fong CA (ed) Stem cell research development. NOVA Science Publishers, New York. pp 5–54
24. Adams B, Xiao Q, Xu Q (2007) Stem cell therapy for vascular disease. Trends Cardiovasc Med 17(7):246–251
25. Owens GK (1995) Regulation of differentiation of vascular smooth muscle cells. Physiol Rev 75(3):487–517
26. Somlyo AP, Somlyo AV (2003) Ca^{2+} sensitivity of smooth muscle and nonmuscle myosin II: modulated by G proteins, kinases, and myosin phosphatase. Physiol Rev 83(4):1325–1358
27. Babij P, Kelly C, Periasamy M (1991) Characterization of a mammalian smooth muscle myosin heavy-chain gene: complete nucleotide and protein coding sequence and analysis of the 5′ end of the gene. Proc Natl Acad Sci USA 88(23):10676–10680
28. Madsen CS et al (1998) Smooth muscle-specific expression of the smooth muscle myosin heavy chain gene in transgenic mice requires 5′-flanking and first intronic DNA sequence. Circ Res 82(8):908–917
29. Duband JL et al (1993) Calponin and SM 22 as differentiation markers of smooth muscle: spatiotemporal distribution during avian embryonic development. Differentiation 55(1):1–11
30. Miano JM et al (2000) Serum response factor-dependent regulation of the smooth muscle calponin gene. J Biol Chem 275(13):9814–9822
31. Mack CP, Owens GK (1999) Regulation of smooth muscle alpha-actin expression in vivo is dependent on CArG elements within the 5′ and first intron promoter regions. Circ Res 84(7):852–861
32. Gabbiani G et al (1981) Vascular smooth muscle cells differ from other smooth muscle cells: predominance of vimentin filaments and a specific alpha-type actin. Proc Natl Acad Sci USA 78(1):298–302
33. Li L et al (1996) SM22 alpha, a marker of adult smooth muscle, is expressed in multiple myogenic lineages during embryogenesis. Circ Res 78(2):188–195

34. Carmeliet P (2000) Mechanisms of angiogenesis and arteriogenesis. Nat Med 6(4):389–395
35. Hanahan D (1997) Signaling vascular morphogenesis and maintenance. Science 277(5322): 48–50
36. Hungerford JE, Little CD (1999) Developmental biology of the vascular smooth muscle cell: building a multilayered vessel wall. J Vasc Res 36(1):2–27
37. Wagenseil JE, Mecham RP (2009) Vascular extracellular matrix and arterial mechanics. Physiol Rev 89(3):957–989
38. Li DY et al (1998) Elastin is an essential determinant of arterial morphogenesis. Nature 393(6682):276–280
39. Alexander MR, Owens GK (2012) Epigenetic control of smooth muscle cell differentiation and phenotypic switching in vascular development and disease. Annu Rev Physiol 74:13–40
40. Owens GK, Kumar MS, Wamhoff BR (2004) Molecular regulation of vascular smooth muscle cell differentiation in development and disease. Physiol Rev 84(3):767–801
41. Campbell GR, Campbell JH (1990) The phenotypes of smooth muscle expressed in human atheroma. Ann N Y Acad Sci 598:143–158
42. Ang AH et al (1990) Collagen synthesis by cultured rabbit aortic smooth-muscle cells. Alteration with phenotype. Biochem J 265(2):461–469
43. Geng YJ, Libby P (2002) Progression of atheroma: a struggle between death and procreation. Arterioscler Thromb Vasc Biol 22(9):1370–1380
44. Owens GK (2007) Molecular control of vascular smooth muscle cell differentiation and phenotypic plasticity. Novartis Found Symp 283:174–191; discussion 191–193
45. Pidkovka NA et al (2007) Oxidized phospholipids induce phenotypic switching of vascular smooth muscle cells in vivo and in vitro. Circ Res 101(8):792–801
46. Majesky MW (2007) Developmental basis of vascular smooth muscle diversity. Arterioscler Thromb Vasc Biol 27(6):1248–1258
47. Herring BP et al (2006) Regulation of myosin light chain kinase and telokin expression in smooth muscle tissues. Am J Physiol Cell Physiol 291(5):C817–C827
48. Hoggatt AM, Simon GM, Herring BP (2002) Cell-specific regulatory modules control expression of genes in vascular and visceral smooth muscle tissues. Circ Res 91(12):1151–1159
49. Stary HC (1989) Evolution and progression of atherosclerotic lesions in coronary arteries of children and young adults. Arteriosclerosis 9(1 Suppl):I19–I32
50. Xu Q (2004) Mouse models of arteriosclerosis: from arterial injuries to vascular grafts. Am J Pathol 165:1–10
51. Ross R (1999) Atherosclerosis – an inflammatory disease. N Engl J Med 340(2):115–126
52. Xu QB et al (1990) Immunology of atherosclerosis: cellular composition and major histocompatibility complex class II antigen expression in aortic intima, fatty streaks, and atherosclerotic plaques in young and aged human specimens. Clin Immunol Immunopathol 56(3):344–359
53. Ross R (1993) The pathogenesis of atherosclerosis: a perspective for the 1990s. Nature 362(6423):801–809
54. Newby AC, Zaltsman AB (1999) Fibrous cap formation or destruction – the critical importance of vascular smooth muscle cell proliferation, migration and matrix formation. Cardiovasc Res 41(2):345–360
55. Ross R (1995) Growth regulatory mechanisms and formation of the lesions of atherosclerosis. Ann N Y Acad Sci 748:1–4; discussion 4–6
56. Ross R (1993) Atherosclerosis – an inflammatory disease. N Engl J Med 340(2):115–126
57. Ross R et al (1984) Human atherosclerosis. I. Cell constitution and characteristics of advanced lesions of the superficial femoral artery. Am J Pathol 114(1):79–93
58. Galis ZS, Khatri JJ (2002) Matrix metalloproteinases in vascular remodeling and atherogenesis: the good, the bad, and the ugly. Circ Res 90(3):251–262
59. Galis ZS et al (1994) Increased expression of matrix metalloproteinases and matrix degrading activity in vulnerable regions of human atherosclerotic plaques. J Clin Invest 94(6):2493–2503
60. Wissler RW (1991) Update on the pathogenesis of atherosclerosis. Am J Med 91(1B):3S–9S
61. Schwartz SM, Virmani R, Rosenfeld ME (2000) The good smooth muscle cells in atherosclerosis. Curr Atheroscler Rep 2(5):422–429

62. Ross R, Glomset JA (1976) The pathogenesis of atherosclerosis (first of two parts). N Engl J Med 295(7):369–377
63. Ross R, Glomset JA (1976) The pathogenesis of atherosclerosis (second of two parts). N Engl J Med 295(8):420–425
64. Campbell JH, Campbell GR (1994) The role of smooth muscle cells in atherosclerosis. Curr Opin Lipidol 5(5):323–330
65. Hillebrands JL et al (2001) Origin of neointimal endothelium and alpha-actin-positive smooth muscle cells in transplant arteriosclerosis. J Clin Invest 107(11):1411–1422
66. Shimizu K et al (2001) Host bone-marrow cells are a source of donor intimal smooth- muscle-like cells in murine aortic transplant arteriopathy. Nat Med 7(6):738–741
67. Sata M et al (2002) Hematopoietic stem cells differentiate into vascular cells that participate in the pathogenesis of atherosclerosis. Nat Med 8(4):403–409
68. Hu Y et al (2002) Smooth muscle cells in transplant atherosclerotic lesions are originated from recipients, but not bone marrow progenitor cells. Circulation 106(14):1834–1839
69. Han CI, Campbell GR, Campbell JH (2001) Circulating bone marrow cells can contribute to neointimal formation. J Vasc Res 38(2):113–119
70. Li J et al (2001) Vascular smooth muscle cells of recipient origin mediate intimal expansion after aortic allotransplantation in mice. Am J Pathol 158(6):1943–1947
71. Saiura A et al (2001) Circulating smooth muscle progenitor cells contribute to atherosclerosis. Nat Med 7(4):382–383
72. Hu Y et al (2002) Both donor and recipient origins of smooth muscle cells in vein graft atherosclerotic lesions. Circ Res 91(7):e13–e20
73. Asahara T et al (1999) Bone marrow origin of endothelial progenitor cells responsible for postnatal vasculogenesis in physiological and pathological neovascularization. Circ Res 85(3):221–228
74. Hu Y et al (2004) Abundant progenitor cells in the adventitia contribute to atherosclerosis of vein grafts in ApoE-deficient mice. J Clin Invest 113(9):1258–1265
75. Ross R (1986) The pathogenesis of atherosclerosis – an update. N Engl J Med 314(8): 488–500
76. Ross R, Glomset JA (1973) Atherosclerosis and the arterial smooth muscle cell: proliferation of smooth muscle is a key event in the genesis of the lesions of atherosclerosis. Science 180(4093):1332–1339
77. Simper D et al (2002) Smooth muscle progenitor cells in human blood. Circulation 106(10): 1199–1204
78. Hoofnagle MH, Wamhoff BR, Owens GK (2004) Lost in transdifferentiation. J Clin Invest 113(9):1249–1251
79. DeRuiter MC et al (1997) Embryonic endothelial cells transdifferentiate into mesenchymal cells expressing smooth muscle actins in vivo and in vitro. Circ Res 80(4):444–451
80. Li G et al (2000) Direct in vivo evidence demonstrating neointimal migration of adventitial fibroblasts after balloon injury of rat carotid arteries. Circulation 101(12):1362–1365
81. Scott NA et al (1996) Identification of a potential role for the adventitia in vascular lesion formation after balloon overstretch injury of porcine coronary arteries. Circulation 93(12): 2178–2187
82. Sartore S et al (2001) Contribution of adventitial fibroblasts to neointima formation and vascular remodeling: from innocent bystander to active participant. Circ Res 89(12):1111–1121
83. Tanaka K et al (2003) Diverse contribution of bone marrow cells to neointimal hyperplasia after mechanical vascular injuries. Circ Res 93(8):783–790
84. Abedin M, Tintut Y, Demer LL (2004) Mesenchymal stem cells and the artery wall. Circ Res 95(7):671–676
85. Aicher A, Zeiher AM, Dimmeler S (2005) Mobilizing endothelial progenitor cells. Hypertension 45(3):321–325
86. Hirschi KK, Majesky MW (2004) Smooth muscle stem cells. Anat Rec A Discov Mol Cell Evol Biol 276(1):22–33
87. Urbich C, Dimmeler S (2004) Endothelial progenitor cells functional characterization. Trends Cardiovasc Med 14(8):318–322

88. Xu Q (2007) Progenitor cells in vascular repair. Curr Opin Lipidol 18(5):534–539
89. Anversa P et al (2007) Concise review: stem cells, myocardial regeneration, and methodological artifacts. Stem Cells 25(3):589–601
90. Xu Q (2006) The impact of progenitor cells in atherosclerosis. Nat Clin Pract Cardiovasc Med 3(2):94–101
91. Dimmeler S, Zeiher AM (2004) Vascular repair by circulating endothelial progenitor cells: the missing link in atherosclerosis? J Mol Med 82(10):671–677
92. Foteinos G et al (2008) Rapid endothelial turnover in atherosclerosis-prone areas coincides with stem cell repair in apolipoprotein E-deficient mice. Circulation 117(14):1856–1863
93. Hibbert B, Chen YX, O'Brien ER (2004) c-kit-immunopositive vascular progenitor cells populate human coronary in-stent restenosis but not primary atherosclerotic lesions. Am J Physiol Heart Circ Physiol 287(2):H518–H524
94. Sata M (2003) Circulating vascular progenitor cells contribute to vascular repair, remodeling, and lesion formation. Trends Cardiovasc Med 13(6):249–253
95. Wassmann S et al (2006) Improvement of endothelial function by systemic transfusion of vascular progenitor cells. Circ Res 99(8):e74–e83
96. Bentzon JF et al (2006) Smooth muscle cells in atherosclerosis originate from the local vessel wall and not circulating progenitor cells in ApoE knockout mice. Arterioscler Thromb Vasc Biol 26(12):2696–2702
97. Benditt EP, Benditt JM (1973) Evidence for a monoclonal origin of human atherosclerotic plaques. Proc Natl Acad Sci USA 70(6):1753–1756
98. Shi Y et al (1996) Adventitial myofibroblasts contribute to neointimal formation in injured porcine coronary arteries. Circulation 94(7):1655–1664
99. Zalewski A, Shi Y (1997) Vascular myofibroblasts. Lessons from coronary repair and remodeling. Arterioscler Thromb Vasc Biol 17(3):417–422
100. Rey FE, Pagano PJ (2002) The reactive adventitia: fibroblast oxidase in vascular function. Arterioscler Thromb Vasc Biol 22(12):1962–1971
101. Wilcox JN et al (1996) The role of the adventitia in the arterial response to angioplasty: the effect of intravascular radiation. Int J Radiat Oncol Biol Phys 36(4):789–796
102. Zoll J et al (2008) Role of human smooth muscle cell progenitors in atherosclerotic plaque development and composition. Cardiovasc Res 77(3):471–480
103. Simper D et al (2010) Comparative proteomics profiling reveals role of smooth muscle progenitors in extracellular matrix production. Arterioscler Thromb Vasc Biol 30(7):1325–1332
104. Yoshida T, Owens GK (2005) Molecular determinants of vascular smooth muscle cell diversity. Circ Res 96(3):280–291
105. Gittenberger-de Groot AC et al (1999) Smooth muscle cell origin and its relation to heterogeneity in development and disease. Arterioscler Thromb Vasc Biol 19(7):1589–1594
106. Kawai-Kowase K, Owens GK (2007) Multiple repressor pathways contribute to phenotypic switching of vascular smooth muscle cells. Am J Physiol Cell Physiol 292(1):C59–C69
107. Blank RS et al (1995) A retinoic acid-induced clonal cell line derived from multipotential P19 embryonal carcinoma cells expresses smooth muscle characteristics. Circ Res 76(5):742–749
108. Manabe I, Owens GK (2001) Recruitment of serum response factor and hyperacetylation of histones at smooth muscle-specific regulatory regions during differentiation of a novel P19-derived in vitro smooth muscle differentiation system. Circ Res 88(11):1127–1134
109. Drab M et al (1997) From totipotent embryonic stem cells to spontaneously contracting smooth muscle cells: a retinoic acid and db-cAMP in vitro differentiation model. FASEB J 11(11):905–915
110. Jain MK et al (1998) In vitro system for differentiating pluripotent neural crest cells into smooth muscle cells. J Biol Chem 273(11):5993–5996
111. Kane NM et al (2011) Pluripotent stem cell differentiation into vascular cells: a novel technology with promises for vascular re(generation). Pharmacol Ther 129(1):29–49
112. Xiao Q et al (2010) The mechanism of stem cell differentiation into smooth muscle cells. Thromb Haemost 104(3):440–448

113. Bollerot K, Pouget C, Jaffredo T (2005) The embryonic origins of hematopoietic stem cells: a tale of hemangioblast and hemogenic endothelium. Acta Pathol Microbiol Immunol Scand Suppl 113(11–12):790–803
114. Fuchs E, Tumbar T, Guasch G (2004) Socializing with the neighbors: stem cells and their niche. Cell 116(6):769–778
115. Scadden DT (2006) The stem-cell niche as an entity of action. Nature 441(7097):1075–1079
116. Poschl E et al (2004) Collagen IV is essential for basement membrane stability but dispensable for initiation of its assembly during early development. Development 131(7):1619–1628
117. Narang AS, Varia S (2011) Role of tumor vascular architecture in drug delivery. Adv Drug Deliv Rev 63(8):640–658
118. Hutchings H, Ortega N, Plouet J (2003) Extracellular matrix-bound vascular endothelial growth factor promotes endothelial cell adhesion, migration, and survival through integrin ligation. FASEB J 17(11):1520–1522
119. Walker MR, Patel KK, Stappenbeck TS (2009) The stem cell niche. J Pathol 217(2):169–180
120. Behonick DJ, Werb Z (2003) A bit of give and take: the relationship between the extracellular matrix and the developing chondrocyte. Mech Dev 120(11):1327–1336
121. Lozito TP et al (2009) Human mesenchymal stem cells express vascular cell phenotypes upon interaction with endothelial cell matrix. J Cell Biochem 107(4):714–722
122. Flaim CJ, Chien S, Bhatia SN (2005) An extracellular matrix microarray for probing cellular differentiation. Nat Methods 2(2):119–125
123. Yamahara K et al (2008) Augmentation of neovascularization [corrected] in hindlimb ischemia by combined transplantation of human embryonic stem cells-derived endothelial and mural cells. PLoS One 3(2):e1666
124. Sone M et al (2007) Pathway for differentiation of human embryonic stem cells to vascular cell components and their potential for vascular regeneration. Arterioscler Thromb Vasc Biol 27(10):2127–2134
125. Kawasaki H et al (2002) Generation of dopaminergic neurons and pigmented epithelia from primate ES cells by stromal cell-derived inducing activity. Proc Natl Acad Sci USA 99(3): 1580–1585
126. Hedin U et al (1988) Diverse effects of fibronectin and laminin on phenotypic properties of cultured arterial smooth muscle cells. J Cell Biol 107(1):307–319
127. Thyberg J, Hultgardh-Nilsson A (1994) Fibronectin and the basement membrane components laminin and collagen type IV influence the phenotypic properties of subcultured rat aortic smooth muscle cells differently. Cell Tissue Res 276(2):263–271
128. Hirst SJ, Twort CH, Lee TH (2000) Differential effects of extracellular matrix proteins on human airway smooth muscle cell proliferation and phenotype. Am J Respir Cell Mol Biol 23(3):335–344
129. Raines EW (2000) The extracellular matrix can regulate vascular cell migration, proliferation, and survival: relationships to vascular disease. Int J Exp Pathol 81(3):173–182
130. Hynes RO (2002) Integrins: bidirectional, allosteric signaling machines. Cell 110(6):673–687
131. Bouvard D et al (2001) Functional consequences of integrin gene mutations in mice. Circ Res 89(3):211–223
132. Luo M, Guan JL (2010) Focal adhesion kinase: a prominent determinant in breast cancer initiation, progression and metastasis. Cancer Lett 289(2):127–139
133. Chavakis E et al (2005) Role of beta2-integrins for homing and neovascularization capacity of endothelial progenitor cells. J Exp Med 201(1):63–72
134. Deb A et al (2004) Integrin profile and in vivo homing of human smooth muscle progenitor cells. Circulation 110(17):2673–2677
135. Boudreau N, Bissell MJ (1998) Extracellular matrix signaling: integration of form and function in normal and malignant cells. Curr Opin Cell Biol 10(5):640–646
136. Danen EH, Yamada KM (2001) Fibronectin, integrins, and growth control. J Cell Physiol 189(1):1–13
137. Hata RI (1996) Where am I? How a cell recognizes its positional information during morphogenesis. Cell Biol Int 20(1):59–65

138. Ramirez F, Rifkin DB (2003) Cell signaling events: a view from the matrix. Matrix Biol 22(2):101–107
139. Watt FM (2002) Role of integrins in regulating epidermal adhesion, growth and differentiation. EMBO J 21(15):3919–3926
140. Lygoe KA et al (2004) AlphaV integrins play an important role in myofibroblast differentiation. Wound Repair Regen 12(4):461–470
141. Rohwedel J et al (1998) Loss of beta1 integrin function results in a retardation of myogenic, but an acceleration of neuronal, differentiation of embryonic stem cells in vitro. Dev Biol 201(2):167–184
142. Kogata N et al (2009) Integrin-linked kinase controls vascular wall formation by negatively regulating Rho/ROCK-mediated vascular smooth muscle cell contraction. Genes Dev 23(19):2278–2283
143. Wu Y et al (2008) Integrin-linked kinase regulates smooth muscle differentiation marker gene expression in airway tissue. Am J Physiol Lung Cell Mol Physiol 295(6):L988–L997
144. Hayashi Y et al (2007) Integrins regulate mouse embryonic stem cell self-renewal. Stem Cells 25(12):3005–3015
145. Nykvist P (2000) Distinct recognition of collagen subtypes by alpha(1)beta(1) and alpha(2) beta(1) integrins. Alpha(1)beta(1) mediates cell adhesion to type XIII collagen. J Biol Chem 275(11):8255–8261
146. Chen S, Lechleider RJ (2004) Transforming growth factor-beta-induced differentiation of smooth muscle from a neural crest stem cell line. Circ Res 94(9):1195–1202
147. Sinha S et al (2004) Transforming growth factor-beta1 signaling contributes to development of smooth muscle cells from embryonic stem cells. Am J Physiol Cell Physiol 287(6): C1560–C1568
148. Sone M et al (2003) Different differentiation kinetics of vascular progenitor cells in primate and mouse embryonic stem cells. Circulation 107(16):2085–2088
149. Takimoto T et al (2010) Smad2 and Smad3 are redundantly essential for the TGF-beta-mediated regulation of regulatory T plasticity and Th1 development. J Immunol 185(2):842–855
150. Yoshimura A, Wakabayashi Y, Mori T (2010) Cellular and molecular basis for the regulation of inflammation by TGF-beta. J Biochem 147(6):781–792
151. Moustakas A, Souchelnytskyi S, Heldin CH (2001) Smad regulation in TGF-beta signal transduction. J Cell Sci 114(Pt 24):4359–4369
152. Li DY et al (1999) Defective angiogenesis in mice lacking endoglin. Science 284(5419): 1534–1537
153. Oh SP et al (2000) Activin receptor-like kinase 1 modulates transforming growth factor-beta 1 signaling in the regulation of angiogenesis. Proc Natl Acad Sci USA 97(6):2626–2631
154. Dickson MC et al (1995) Defective haematopoiesis and vasculogenesis in transforming growth factor-beta 1 knock out mice. Development 121(6):1845–1854
155. Oshima M, Oshima H, Taketo MM (1996) TGF-beta receptor type II deficiency results in defects of yolk sac hematopoiesis and vasculogenesis. Dev Biol 179(1):297–302
156. Yang X et al (1999) Angiogenesis defects and mesenchymal apoptosis in mice lacking SMAD5. Development 126(8):1571–1580
157. Bjorkerud S (1991) Effects of transforming growth factor-beta 1 on human arterial smooth muscle cells in vitro. Arterioscler Thromb 11(4):892–902
158. Hautmann MB, Madsen CS, Owens GK (1997) A transforming growth factor beta (TGFbeta) control element drives TGFbeta-induced stimulation of smooth muscle alpha-actin gene expression in concert with two CArG elements. J Biol Chem 272(16):10948–10956
159. Kurpinski K et al (2010) Transforming growth factor-beta and notch signaling mediate stem cell differentiation into smooth muscle cells. Stem Cells 28(4):734–742
160. Jeon ES et al (2006) Sphingosylphosphorylcholine induces proliferation of human adipose tissue-derived mesenchymal stem cells via activation of JNK. J Lipid Res 47(3):653–664
161. Hirschi KK, Rohovsky SA, D'Amore PA (1998) PDGF, TGF-beta, and heterotypic cell-cell interactions mediate endothelial cell-induced recruitment of 10T1/2 cells and their differentiation to a smooth muscle fate. J Cell Biol 141(3):805–814

162. Qiu P, Feng XH, Li L (2003) Interaction of Smad3 and SRF-associated complex mediates TGF-beta1 signals to regulate SM22 transcription during myofibroblast differentiation. J Mol Cell Cardiol 35(12):1407–1420
163. Wang Z et al (2004) Myocardin and ternary complex factors compete for SRF to control smooth muscle gene expression. Nature 428(6979):185–189
164. Holycross BJ et al (1992) Platelet-derived growth factor-BB-induced suppression of smooth muscle cell differentiation. Circ Res 71(6):1525–1532
165. Uchida K et al (1996) Expression of platelet-derived growth factor B-chain in neointimal smooth muscle cells of balloon injured rabbit femoral arteries. Atherosclerosis 124(1):9–23
166. Ferns GA et al (1991) Inhibition of neointimal smooth muscle accumulation after angioplasty by an antibody to PDGF. Science 253(5024):1129–1132
167. Betsholtz C et al (2004) Role of platelet-derived growth factor in mesangium development and vasculopathies: lessons from platelet-derived growth factor and platelet-derived growth factor receptor mutations in mice. Curr Opin Nephrol Hypertens 13(1):45–52
168. Blank RS, Owens GK (1990) Platelet-derived growth factor regulates actin isoform expression and growth state in cultured rat aortic smooth muscle cells. J Cell Physiol 142:635–642
169. Corjay MH, Blank RS, Owens GK (1990) Platelet-derived growth factor-induced destabilization of smooth muscle alpha-actin mRNA. J Cell Physiol 145(3):391–397
170. Liu Y et al (2005) Kruppel-like factor 4 abrogates myocardin-induced activation of smooth muscle gene expression. J Biol Chem 280(10):9719–9727
171. Regan CP et al (2000) Molecular mechanisms of decreased smooth muscle differentiation marker expression after vascular injury. J Clin Invest 106(9):1139–1147
172. Jawien A et al (1992) Platelet-derived growth factor promotes smooth muscle migration and intimal thickening in a rat model of balloon angioplasty. J Clin Invest 89(2):507–511
173. Gerecht-Nir S et al (2003) Human embryonic stem cells as an in vitro model for human vascular development and the induction of vascular differentiation. Lab Invest 83(12):1811–1820
174. Xiao Q et al (2006) Sca-1+ progenitors derived from embryonic stem cells differentiate into endothelial cells capable of vascular repair after arterial injury. Arterioscler Thromb Vasc Biol 26(10):2244–2251
175. Cheung P, Allis CD, Sassone-Corsi P (2000) Signaling to chromatin through histone modifications. Cell 103(2):263–271
176. Hiltunen MO et al (2002) DNA hypomethylation and methyltransferase expression in atherosclerotic lesions. Vasc Med 7(1):5–11
177. Turunen MP, Aavik E, Yla-Herttuala S (2009) Epigenetics and atherosclerosis. Biochim Biophys Acta 1790(9):886–891
178. Wiblin AE et al (2005) Distinctive nuclear organisation of centromeres and regions involved in pluripotency in human embryonic stem cells. J Cell Sci 118(Pt 17):3861–3868
179. Gillespie RF, Gudas LJ (2007) Retinoid regulated association of transcriptional co-regulators and the polycomb group protein SUZ12 with the retinoic acid response elements of Hoxa1, RARbeta(2), and Cyp26A1 in F9 embryonal carcinoma cells. J Mol Biol 372(2):298–316
180. Kashyap V, Gudas LJ (2010) Epigenetic regulatory mechanisms distinguish retinoic acid-mediated transcriptional responses in stem cells and fibroblasts. J Biol Chem 285(19):14534–14548
181. Azuara V et al (2006) Chromatin signatures of pluripotent cell lines. Nat Cell Biol 8(5):532–538
182. Bernstein BE et al (2006) A bivalent chromatin structure marks key developmental genes in embryonic stem cells. Cell 125(2):315–326
183. Schones DE, Zhao K (2008) Genome-wide approaches to studying chromatin modifications. Nat Rev Genet 9(3):179–191
184. Duncan EM et al (2008) Cathepsin L proteolytically processes histone H3 during mouse embryonic stem cell differentiation. Cell 135(2):284–294
185. Wu J et al (2001) TUP1 utilizes histone H3/H2B-specific HDA1 deacetylase to repress gene activity in yeast. Mol Cell 7(1):117–126
186. Yang XJ, Seto E (2003) Collaborative spirit of histone deacetylases in regulating chromatin structure and gene expression. Curr Opin Genet Dev 13(2):143–153

187. Zhou B et al (2011) Role of histone deacetylases in vascular cell homeostasis and arteriosclerosis. Cardiovasc Res 90(3):413–420
188. McDonald OG, Owens GK (2007) Programming smooth muscle plasticity with chromatin dynamics. Circ Res 100(10):1428–1441
189. Zhang L (2012) An updated view on stem cell differentiation into smooth muscle cells. Vascul Pharmacol 56(5–6):280–287
190. McDonald OG et al (2006) Control of SRF binding to CArG box chromatin regulates smooth muscle gene expression in vivo. J Clin Invest 116(1):36–48
191. Dressel U et al (2001) A dynamic role for HDAC7 in MEF2-mediated muscle differentiation. J Biol Chem 276(20):17007–17013
192. Kato H, Tamamizu-Kato S, Shibasaki F (2004) Histone deacetylase 7 associates with hypoxia-inducible factor 1alpha and increases transcriptional activity. J Biol Chem 279(40):41966–41974
193. Sterner DE, Berger SL (2000) Acetylation of histones and transcription-related factors. Microbiol Mol Biol Rev 64(2):435–459
194. Wang D et al (2001) Activation of cardiac gene expression by myocardin, a transcriptional cofactor for serum response factor. Cell 105(7):851–862
195. Du KL et al (2003) Myocardin is a critical serum response factor cofactor in the transcriptional program regulating smooth muscle cell differentiation. Mol Cell Biol 23(7):2425–2437
196. Li S et al (2003) The serum response factor coactivator myocardin is required for vascular smooth muscle development. Proc Natl Acad Sci USA 100(16):9366–9370
197. Chang S et al (2006) Histone deacetylase 7 maintains vascular integrity by repressing matrix metalloproteinase 10. Cell 126(2):321–334
198. Margariti A et al (2009) Splicing of HDAC7 modulates the SRF-myocardin complex during stem-cell differentiation towards smooth muscle cells. J Cell Sci 122(Pt 4):460–470
199. Zhou B et al (2011) Splicing of histone deacetylase 7 modulates smooth muscle cell proliferation and neointima formation through nuclear beta-catenin translocation. Arterioscler Thromb Vasc Biol 31(11):2676–2684
200. Yoshida T, Gan Q, Owens GK (2008) Kruppel-like factor 4, Elk-1, and histone deacetylases cooperatively suppress smooth muscle cell differentiation markers in response to oxidized phospholipids. Am J Physiol Cell Physiol 295(5):C1175–C1182
201. Zhang L et al (2010) Sp1-dependent activation of HDAC7 is required for platelet-derived growth factor-BB-induced smooth muscle cell differentiation from stem cells. J Biol Chem 285(49):38463–38472
202. Taniyama Y, Griendling KK (2003) Reactive oxygen species in the vasculature: molecular and cellular mechanisms. Hypertension 42(6):1075–1081
203. Clempus RE, Griendling KK (2006) Reactive oxygen species signaling in vascular smooth muscle cells. Cardiovasc Res 71(2):216–225
204. Su B et al (2001) Redox regulation of vascular smooth muscle cell differentiation. Circ Res 89(1):39–46
205. Xiao Q et al (2009) Embryonic stem cell differentiation into smooth muscle cells is mediated by Nox4-produced H_2O_2. Am J Physiol Cell Physiol 296(4):C711–C723
206. Yin X et al (2006) Proteomic analysis reveals higher demand for antioxidant protection in embryonic stem cell-derived smooth muscle cells. Proteomics 6(24):6437–6446
207. Cave AC et al (2006) NADPH oxidases in cardiovascular health and disease. Antioxid Redox Signal 8(5–6):691–728
208. Touyz RM et al (2002) Expression of a functionally active gp91phox-containing neutrophil-type NAD(P)H oxidase in smooth muscle cells from human resistance arteries: regulation by angiotensin II. Circ Res 90(11):1205–1213
209. Lassegue B, Clempus RE (2003) Vascular NAD(P)H oxidases: specific features, expression, and regulation. Am J Physiol Regul Integr Comp Physiol 285(2):R277–R297
210. Sorescu D et al (2002) Superoxide production and expression of nox family proteins in human atherosclerosis. Circulation 105(12):1429–1435
211. Szocs K et al (2002) Upregulation of Nox-based NAD(P)H oxidases in restenosis after carotid injury. Arterioscler Thromb Vasc Biol 22(1):21–27

212. Clempus RE et al (2007) Nox4 is required for maintenance of the differentiated vascular smooth muscle cell phenotype. Arterioscler Thromb Vasc Biol 27(1):42–48
213. Sumimoto H, Miyano K, Takeya R (2005) Molecular composition and regulation of the Nox family NAD(P)H oxidases. Biochem Biophys Res Commun 338(1):677–686
214. Martyn KD et al (2006) Functional analysis of Nox4 reveals unique characteristics compared to other NADPH oxidases. Cell Signal 18(1):69–82
215. Deliri H, McNamara CA (2007) Nox 4 regulation of vascular smooth muscle cell differentiation marker gene expression. Arterioscler Thromb Vasc Biol 27(1):12–14
216. Kobayashi A et al (1999) Molecular cloning and functional characterization of a new Cap'n' collar family transcription factor Nrf3. J Biol Chem 274(10):6443–6452
217. Pepe AE et al (2010) Crucial role of nrf3 in smooth muscle cell differentiation from stem cells. Circ Res 106(5):870–879
218. Sankaranarayanan K, Jaiswal AK (2004) Nrf3 negatively regulates antioxidant-response element-mediated expression and antioxidant induction of NAD(P)H:quinone oxidoreductase1 gene. J Biol Chem 279(49):50810–50817
219. Chowdhury I et al (2009) Oxidant stress stimulates expression of the human peroxiredoxin 6 gene by a transcriptional mechanism involving an antioxidant response element. Free Radic Biol Med 46(2):146–153
220. Etchevers HC (2005) The cap 'n' collar family member NF-E2-related factor 3 (Nrf3) is expressed in mesodermal derivatives of the avian embryo. Int J Dev Biol 49(2–3):363–367
221. Rana TM (2007) Illuminating the silence: understanding the structure and function of small RNAs. Nat Rev Mol Cell Biol 8(1):23–36
222. Schickel R et al (2008) MicroRNAs: key players in the immune system, differentiation, tumorigenesis and cell death. Oncogene 27(45):5959–5974
223. Bonauer A et al (2009) MicroRNA-92a controls angiogenesis and functional recovery of ischemic tissues in mice. Science 324(5935):1710–1713
224. Cordes KR, Srivastava D (2009) MicroRNA regulation of cardiovascular development. Circ Res 104(6):724–732
225. Bartel DP (2009) MicroRNAs: target recognition and regulatory functions. Cell 136(2):215–233
226. Kusenda B et al (2006) MicroRNA biogenesis, functionality and cancer relevance. Biomed Pap Med Fac Univ Palacky Olomouc Czech Repub 150(2):205–215
227. Bentwich I et al (2005) Identification of hundreds of conserved and nonconserved human microRNAs. Nat Genet 37(7):766–770
228. Friedman RC et al (2009) Most mammalian mRNAs are conserved targets of microRNAs. Genome Res 19(1):92–105
229. Lim LP et al (2003) The microRNAs of *Caenorhabditis elegans*. Genes Dev 17(8):991–1008
230. Hutvagner G, Zamore PD (2002) A microRNA in a multiple-turnover RNAi enzyme complex. Science 297(5589):2056–2060
231. Guo H et al (2010) Mammalian microRNAs predominantly act to decrease target mRNA levels. Nature 466(7308):835–840
232. Lim LP et al (2005) Microarray analysis shows that some microRNAs downregulate large numbers of target mRNAs. Nature 433(7027):769–773
233. Huang H (2010) miR-10a contributes to retinoid acid-induced smooth muscle cell differentiation. J Biol Chem 285(13):9383–9389
234. Ivey KN, Srivastava D (2010) MicroRNAs as regulators of differentiation and cell fate decisions. Cell Stem Cell 7(1):36–41
235. Martinez NJ, Gregory RI (2010) MicroRNA gene regulatory pathways in the establishment and maintenance of ESC identity. Cell Stem Cell 7(1):31–35
236. Cordes KR (2009) miR-145 and miR-143 regulate smooth muscle cell fate and plasticity. Nature 460(7256):705–710
237. Xu N et al (2009) MicroRNA-145 regulates OCT4, SOX2, and KLF4 and represses pluripotency in human embryonic stem cells. Cell 137(4):647–658
238. Xin M et al (2009) MicroRNAs miR-143 and miR-145 modulate cytoskeletal dynamics and responsiveness of smooth muscle cells to injury. Genes Dev 23(18):2166–2178

239. Yamaguchi S et al (2011) The role of microRNA-145 in human embryonic stem cell differentiation into vascular cells. Atherosclerosis 219(2):468–474
240. Cheng Y et al (2009) MicroRNA-145, a novel smooth muscle cell phenotypic marker and modulator, controls vascular neointimal lesion formation. Circ Res 105(2):158–166
241. Boettger T et al (2009) Acquisition of the contractile phenotype by murine arterial smooth muscle cells depends on the Mir143/145 gene cluster. J Clin Invest 119(9):2634–2647
242. Elia L et al (2009) The knockout of miR-143 and -145 alters smooth muscle cell maintenance and vascular homeostasis in mice: correlates with human disease. Cell Death Differ 16(12): 1590–1598
243. Xie C et al (2011) MicroRNA-1 regulates smooth muscle cell differentiation by repressing Kruppel-like factor 4. Stem Cells Dev 20(2):205–210
244. Davis BN et al (2009) Induction of microRNA-221 by platelet-derived growth factor signaling is critical for modulation of vascular smooth muscle phenotype. J Biol Chem 284(6): 3728–3738
245. Kumar AH, Caplice NM (2010) Clinical potential of adult vascular progenitor cells. Arterioscler Thromb Vasc Biol 30(6):1080–1087
246. Wollert KC, Drexler H (2010) Cell therapy for the treatment of coronary heart disease: a critical appraisal. Nat Rev Cardiol 7(4):204–215
247. Takahashi K, Yamanaka S (2006) Induction of pluripotent stem cells from mouse embryonic and adult fibroblast cultures by defined factors. Cell 126(4):663–676
248. Shimizu RT et al (1995) The smooth muscle alpha-actin gene promoter is differentially regulated in smooth muscle versus non-smooth muscle cells. J Biol Chem 270(13):7631–7643
249. Landerholm TE et al (1999) A role for serum response factor in coronary smooth muscle differentiation from proepicardial cells. Development 126(10):2053–2062

Chapter 2
Recent Advances in Embryonic Stem Cell Engineering Toward Tailored Lineage Differentiation

Zhen Xu, Jingyu Wang, and Yanan Du

Abstract Embryonic stem cell-based drug screening and therapeutics provide unique opportunities for drug discovery, tissue engineering, and regenerative medicine. Despite the great promise, a major limitation in translation of embryonic stem cells (ESCs) technology to clinical applications is how to direct their differentiation into tailored lineage commitment. This lineage commitment is precisely controlled by the ESC microenvironment in vivo. Engineering strategies to reconstruct a biomimetic microenvironment offer useful tools for guiding ESC differentiation in vitro. The purpose of this chapter is to summarize and examine the latest literatures describing application of engineering approaches to control ESC differentiation. We review recent studies and techniques that focus on physical strategies (e.g., geometrical constraint, mechanical force, extracellular matrix stiffness, and topography) and biochemical approaches (e.g., genetic engineering, immobilized growth factors, coculture) and highlight the significance of creating three-dimensional (3D) microenvironment for directed ESC differentiation. The perspectives in engineering ESC microenvironments are also discussed for future advancement of this emerging field.

Keywords Embryonic stem cells • Differentiation • Microenvironment • Engineering

The authors Zhen Xu and Jingyu Wang are contributed equally to this chapter.

Z. Xu • J. Wang • Y. Du (✉)
Department of Biomedical Engineering, School of Medicine, Tsinghua University, B-310, Medical Science Building, Beijing 100084, China
e-mail: duyanan@tsinghua.edu.cn

M.K. Danquah, R.I. Mahato (eds.), *Emerging Trends in Cell and Gene Therapy*,
DOI 10.1007/978-1-62703-417-3_2,

2.1 Introduction

Since the successful derivation of embryonic stem cells (ESCs) from the inner cell mass of mouse embryos in 1981 [1, 2], ESCs have been anticipated to play an increasingly important role in stem cell biology, drug discovery, and regenerative medicine [3]. Unlike adult stem cells, which have limited self-renewal capability and are committed to differentiate into particular downstream lineages of their originated tissues, ESCs possess the ability of nearly unlimited self-renewal and differentiation into all downstream cellular lineages [4, 5].

To translate ESC research into successful applications in biomedicine, two major challenges must be overcome: (1) reproducible and large-scale expansion of stem cells in vitro without the loss of their stemness and (2) direct differentiation into desired cellular lineages with high efficiency and efficacy [6]. It is desirable to expand ESCs in their pluripotent state to acquire adequate quantity and direct ESC differentiation into specific cell lineages tailored for cell-based applications. Incomplete or uncontrolled ESC differentiation will lead to heterogeneous cell populations which will be tumorigenic rather than therapeutic in cell therapy and undermine the reliability of stem cell-based in vitro models [7].

During embryonic development, ESCs reside within a complex three-dimensional (3D) microenvironment providing both biophysical and biochemical support. Increasing evidences demonstrate that ESC microenvironment or ESC niche is responsible for regulating ESC behaviors. Soluble factors, extracellular matrix (ECM), and surrounding cells are major components to constitute ESC niche which define the biomechanical characteristics, geometric configuration, and activation of signaling pathways of ESCs. These cues work synergistically on regulating the ESCs' fate both spatially and temporally in a highly precise manner [8]. Control over ESC differentiation is under extensive investigation via optimizing of the parameters of the ESC's microenvironment. The increasing integration of material science, chemistry, mechanics, electronic as well as nano-/microscale technologies into stem cell biology offers comprehensive and powerful tools to engineer biomimetic microenvironment to overcome the challenges in stem cell engineering.

Due to the unique features of ESCs and the vast amount of work done in the stem cell engineering, we only reviewed the advancement in ESC engineering for tailored lineage differentiation in this chapter. Specifically, we focus on studies published within the recent 5 years (2006–2011), which may provide clues for the state-of-the-art development in this emerging field. Both physical and biochemical cues as well as the 3D features within the engineered stem cell niche are emphasized to establish amenable microenvironment for inducing ESC differentiation into desired cellular lineages (Fig. 2.1). Among physical cues, the influence of mechanical force, geometrical cues, and substrate stiffness/topology are illustrated, while four biochemical cues, genetic engineering, immobilized growth factors, synthetic small molecules, and coculture with supporting cells are introduced as representative tools to regulate ESC differentiation. In addition, the use of 3D culture systems is acquiring increasing attention, which presents closer mimicry to the natural stem cell niche than the conventional two-dimensional (2D) culture configurations [9].

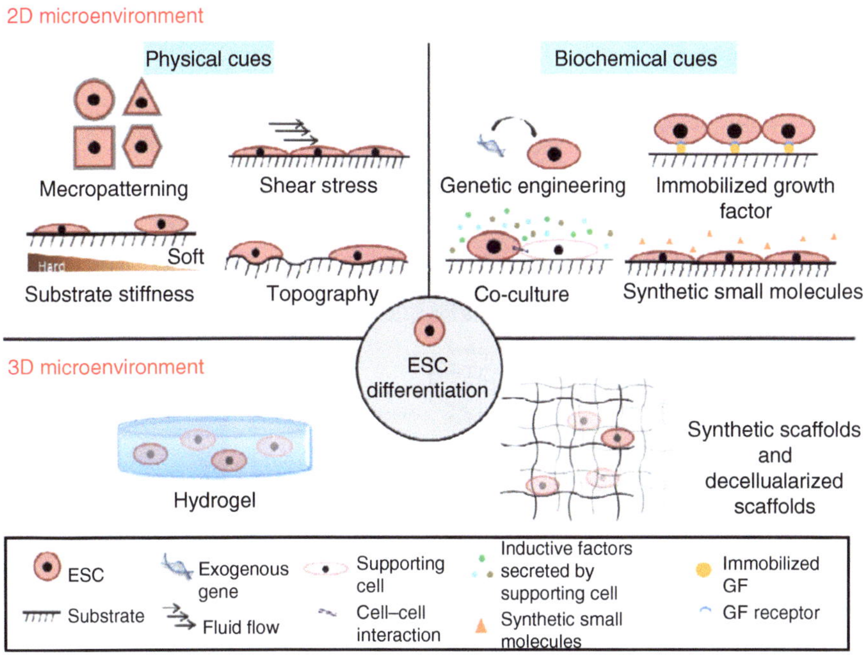

Fig. 2.1 Schematic illustration of engineering stem cell niches for tailored cellular differentiation in 2D (*top*) and 3D (*bottom*) microenvironment. ESC differentiation can be modulated on 2D substrates by various physical cues, such as micropatterning, shear stress, substrate stiffness and topography, as well as biochemical cues, such as genetic engineering, immobilized growth factors, coculture, and synthetic small molecules. Reconstruction of the complexity and multicomponent of ESC niche can be obtained in 3D hydrogel, synthetic scaffolds, and decellularized scaffolds

Hydrogel, synthetic and decellularized scaffolds, are highlighted as 3D culture strategies for inducing optimized ESC differentiation. The perspectives in stem cell engineering are also discussed to highlight its huge potential in stem cell biology and regenerative medicine.

2.2 Engineering ESC Niche for Tailored Cellular Differentiation

2.2.1 Physical Strategies to Optimize ESC Niche

Accumulating evidences in the literature reveal the vital role played by physical stimulus in controlling ESC differentiation [10, 11]. Here we focus on three aspects of physical stimuli for controlling ESC differentiation, namely, (1) geometrical constraint (i.e., the shape and size of ESCs colony), (2) external mechanical stimulation, and (3) the physical cues present in ECM (i.e., matrix stiffness and topography).

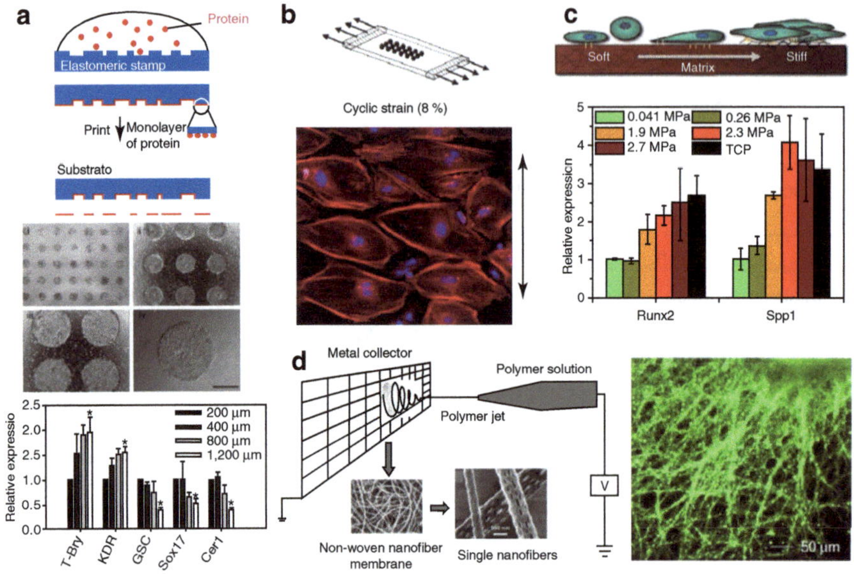

Fig. 2.2 Regulation of ESC differentiation via physical strategies on 2D substrate. (**a**) Geometrical constraint to regulate ESC differentiation: schematic overview of microcontact printing (*top*). Micropatterned hESC colonies and increased endoderm marker expression as colony size decreased (*bottom*) [13, 14]. (**b**) Cyclic strain as mechanical stimulation: schema of cyclic strain generation (*top*) and mESC differentiation into vascular lineage (*bottom*) [15, 16]. (**c**) Matrix stiffness of the polymeric substrate: schematic drawing of cell morphology in response to varied matrix stiffness (*top*) and osteogenic preference of mESCs on stiffer substrate [17, 18]. (**d**) Nano-topographical stimulus exerted by electrospun nano-fibers: schematic of apparatus used to generate nano-fibers (*left*) and hESC neuronal differentiation in response to nano-fibers (*right*) [19, 20] (Images are reproduced with the permission)

2.2.1.1 Geometrical Constraint

Cellular niche in vivo involves well-regulated geometrical features related to the specialized functions played by tissues such as the fibrous muscular tissue and hexagon liver lobule. Micropatterning techniques offer powerful tools to enable spatial control of ESC microenvironment with finely tuned geometrical features [12].

Due to its simplicity, 2D micropatterned substrates fabricated by microcontact printing have been extensively explored for ESC differentiation. Microcontact printing is used to generate predetermined patterns of biomolecules on the substrate surface with high accuracy. In microcontact printing, a template (usually made from PDMS) with desired patterns is prepared by soft lithography which is used to "stamp" molecules of interest as "ink" to transfer the patterning to a substrate [13] (Fig. 2.2a, top). Size-controlled ESC colonies prepared by this technique enable independent investigation of geometrical effects on ESC differentiation.

Zandstra's group investigated the size-dependent behaviors of micropatterned hESCs with diameter range from 200 to 800 μm. In serum- and growth factor-free medium, hESC colonies on larger micropatterns were demonstrated with increased self-renewal capacity, while hESCs with smaller colony sizes were inclined to differentiate toward extraembryonic endoderm. The difference between small and large hESC colonies was related to the suppression of bone morphogenetic protein (BMP) signaling. In large colonies, the expression of two BMP inhibitors, growth and differentiation factor, and Lefty-B were both upregulated, resulting in suppression of Smad1-mediated BMP signaling, which has been known to maintain hESCs in undifferentiated state. Furthermore, inhibition of growth and differentiation factors by siRNA resulted in phosphorylated Smad1 activation and enhanced hESC differentiation toward extraembryonic endoderm [21]. The same research team further revealed that this size-dependent control of cell fate occurs even in the presence of inductive factors (activin A and BMP2). They prepared hESC colonies with size ranging from 200 to 1,200 μm by microcontact printing and found that expression of endoderm-associated genes Sox 17, GSC, and Cer1 increased with decreasing colony size, while the expression of mesoderm markers Brachyury and KDR was greatest in larger colony. Their results demonstrated that large hESC colonies are inclined to differentiate into mesoderm, while small colonies prefer to result in definitive endoderm destiny. In this study, only the TGF-β pathway activators were used to induce endoderm and mesoderm differentiation. It was discussed that besides the TGF-β pathway, there may exist other signaling pathways such as PI3K signaling and Wnt signaling responsible for this colony-size-mediated effects [14] (Fig. 2.2a, bottom). As for mouse ESCs (mESCs), Sasaki et al. developed a novel approach to form size-controlled embryonic bodies (EBs, 3D aggregates) derived from micropatterned mESC colonies. They investigated the optimal diameter of ESC colonies (arranging from 100 to 400 μm) for cardiac differentiation. Cardiogenesis of mESCs was found to be maximal for embryonic bodies derived from ESC colonies with diameter of 200 μm [22].

2.2.1.2 External Mechanical Stimulation

Mechanical force applied externally (e.g., cyclic strain or shear stress) is proved as one of the powerful engineering approaches in regulating the proliferation, differentiation, regeneration, and homeostasis of stem cells. The effect of mechanical stimulation is especially phenomenal for adult stem cells (e.g., adipose-derived stem cells or mesenchyme stem cells (MSC)) [23]. In contrast, fewer studies so far have focused on the role of mechanical force in regulating ESC differentiation. It is known that ESCs derived from inner cell mass of the embryo are residing in an aqueous environment and likely to be exposed to various forms of fluid shears and strains during embryonic development [24]. Both cyclic strain and shear stress have been well recognized as important regulators in controlling cardiovascular functions, great efforts have been made to elucidate these mechanical stimulations in regulating ESCs toward differentiation in cardiovascular lineages.

Saha et al. applied cyclic strain generated by a uniaxial mechanical strain-loading device on ESC-cultured elastic polymer substrate (PDMS). A programmable microcomputer was applied to tune the amplitude and frequency of stretching. As evidenced by upregulation of Oct4 and SSEA-4 expression, applying cyclic biaxial stretch to the elastic substrate at 10 % strain (as defined by 10 % deformation) and at rates of 6–30 cycles/min could inhibit hESC differentiation [25]. However, cyclic strain was reported to promote mESC differentiation into vascular smooth muscle cell (VSMC). Four to 12 % of cyclic strain was shown to promote mESC differentiation toward VSMC lineage as indicated by the expression of α-actin and SM-MHC. The underlying mechanism of the mechanical stimulation was related to PDGF receptor-β-mediated signaling pathways since blocking the growth factor receptor-β on ESCs could damage the effect of the cyclic strain-induced differentiation toward VSMC lineage [16] (Fig. 2.2b, bottom). Using a similar stretching method, Heo et al. demonstrated that exposure of mouse ES cells to cyclic strain direct mESCs into cardiomyogenesis. Cyclic strain could activate PI3K pathway and then upregulate the expression of Cx43 and Nkx2.5 which are key factors required for cardiomyocyte differentiation [26]. Magnetic twisting cytometry, a well-established method for applying controlled and precise local mechanical stresses of physiologic magnitudes to single live cell, has been applied to generate cyclic strain in an attempt to drive mESC differentiation in a single cell resolution. By attaching a 4 μm-RGD-coated magnetic bead on the apical surface of the mESCs and applying a weak, oscillatory stress (17.5 Pa at 0.3 Hz) for 60 min, OCT3/4 expression was downregulated by about 35 % within 24 h and by about 50 % within 72 h. The reduced expression of Oct3/4, a hallmark of mESC differentiation, indicated a local mechanical stimulation through a focal adhesion might be sufficient to drive an individual mESC to differentiate. The high spatial resolution of this local stimulation approach enabled local differentiation of an individual cell while keeping nearby cells undifferentiated [27].

While cyclic strains are usually applied to live cells mediated by solid materials, shear stress is mainly applied through fluidic shear which mimic the blood flow in the vascular system. There is no surprise that shear stress is of particular interest in directing ESCs toward vascular cell lineage. mESCs were subjected to shear stress in a fluidic device formed by two parallel plates upregulated the expression of vascular endothelial cell(EC)-specific markers such as PECAM-1 and VE-cadherin, while the expression of mural cell marker (SMA), blood cell marker (CD3), or epithelial cell marker (keratin) remained at constant level. The results revealed that shear stress promotes tailored differentiation of mESCs into the vascular endothelial lineage [28]. While specific mechanoreceptors for shear sensing of ESCs are still unknown, some mechanistic insights were unveiled by Voldman et al. [29]. They cultured mESCs in a multiplex microfluidic device, where shear stresses could be varied by >1,000 times (0.016–16 dyn/cm^2). mESCs responded to the fluidic shear of the entire range with significant upregulation of the epiblast marker Fgf5 compared to the static culture. HSPGs were identified to be one of the molecular components involved in stem cell mechanosensing. As a major proteoglycan on the cell surface, HSPGs are known as shear-sensing element in ESs. Investigation of downstream mediator of HSPGs to transduce shear stress will lead to further elucidation on shear mechanotransduction mechanisms in stem cells.

2.2.1.3 Physical Properties of Matrix

Stem cells are not only sensitive to externally applied mechanical stimulation but also susceptible to the inherited physical properties of the ECM such as stiffness, topology, and hydrophobicity. Studies of 2D cultures revealed that substrate stiffness acts as an important factor in the process of tissue formation. In addition, it has been demonstrated that the stiffness of 2D substrate has an effect on stem cell spreading and cytoskeleton assembly [30]. The effect of substrate stiffness on stem cell behaviors has received extensive attention, especially for adult stem cells. Several research groups have reported that MSC can sense the stiffness of substrate and differentiate into various downstream lineages, and the underlying mechanisms have been extensively explored [31, 32]. Only recently, the effects of substrate stiffness as physical cues to control ESCs' fate especially hESC differentiation have started gaining more attention, and the underlying mechanism still remained elusive.

When mESCs are cultured on flexible PDMS substrate with varying stiffness from 0.041 to 2.7 MPa, early mesendoderm differentiation markers such as Brachyury, Mixl1, and Eomes were upregulated on stiffer in comparison to softer substrates. Moreover, osteogenic differentiation of mESCs was also enhanced on stiffer substrate. These findings highlight the important role played by substrate stiffness in regulating both early and terminal differentiation of mESCs. The author's proposed explanation was that stiffer substrates more closely mimic the natural ECM around migrating mesendoderm cells residing in the early embryo and therefore are more effective in directing mESCs toward specific differentiation [18] (Fig. 2.2c, bottom).

Similarly, hESCs could also be able to sense and respond to the stiffness of underlying substrate. Park et al. revealed that PET with optimal stiffness could inhibit hESC differentiation and promote maintenance of hESC self-renewal. No expression of ectoderm, mesoderm, or endoderm markers was observed in the hESCs cultured on the PET substrate with stiffness of 0.345 GPa. In addition, Rho/ROCK signaling pathway, one of the ECM stiffness-based signaling pathways, was downregulated when cells were cultured on PET membranes with stiffness of 0.345 GPa. However, it remains unclear whether downregulation of ROCK expression is the cause or consequence of inhibition of hESC differentiation [33].

In addition to substrate stiffness, ESCs could also respond to the topographical features of the substrate, which provide noninvasive, non-biochemical means for controlling ESC differentiation. The tissue surface is usually not smooth or flattened but covered with grooves, ridges, pits, pores, and the fibrillar meshwork of the ECM, composed predominantly of intertwined collagen and elastin fibers with diameters ranging from 10 to 300 nm [34]. ESC differentiation is strongly affected by the physical interactions between cells and local topographical features and has been documented in terms of both controlled differentiation and retention of self-renewal/proliferative capabilities.

Electrospinning is a facile and effective technique to generate nanoscale topographical structures of biomaterials. Electrospun nano-fibrous materials have been applied as an attractive substrate to regulate ESC differentiation due to the nanoscaled morphology of the nano-fibers which are analogous to the structure of protein fibrils and fibers in natural ECM [19] (Fig. 2.2d, left). Massumi et al. prepared electrospun PLGA nano-fibrous scaffolds with different roughness, height distribution, and alignments which were proved as effective topological cues to promote mESC differentiation into mesodermal-derived cells and germ cells. In contrast, it inhibited the derivation of endodermal cell lineages. The findings demonstrated that topological cues such as roughness and alignments can promote mESC differentiation toward a specific cellular lineage [35]. hESCs were also able to sense the topographical cues within biomimetic nanostructures and then exhibit tailored lineage differentiation. hESCs displayed favorable interactions with the electrospun nano-fibers, establishing spreading outgrowths and connections to adjacent cells and attaching to individual nano-fibers. A great amount of cells cultured on nano-fibers were stained positive for the early neuronal marker β-tubulin III, the mature neuronal marker MAP2ab, and the dopaminergic marker tyrosine hydroxylase, but with little or no staining of the astrocyte marker GFAP. In contrast, when cultured on flat substrate without electrospun nano-fibers, the number of hESCs stained with GFAP greatly increased under the same differentiation conditions. The results revealed that the physical cues induced by the nano-topographical substrates could direct hESCs toward neuronal lineage, while conventional flat substrate led to astrocyte differentiation with the aid of neuronal differentiation medium [20] (Fig. 2.2d, right).

Besides electrospinning, nano-topological features can be also generated by other techniques such as phase separation and nano-patterning, which are proven as powerful tools to regulate ESC differentiation. Smith et al. prepared the nano-fibrous PLLA matrix by phase separation and investigated mESC differentiation on nano-fibrous matrix vs. flat matrix. Even without osteogenic supplements, NF matrix was able to induce the expression of osteogenic markers, such as osteocalcin and bone sialoprotein. However, when mESCs were cultured on flat PLLA surface, both osteogenic supplements and BMP2 were required as addition to the culture media for reaching the same level of osteogenic differentiation as cultured on the NF matrix [36]. The same research group further showed enhanced osteogenic differentiation of hESC-derived osteogenic progenitor cells on nano-fibrous matrix which expressed higher levels of osteogenic markers (Rux2 and osteocalcin) and reduced level of neuronal marker (TUJ1) under osteogenic differentiation conditions [37].

The emergence of surface micro- and nano-patterning techniques has enabled researchers to investigate ESC behaviors on micro- and nanostructures. Lee et al. fabricated nanoscale patterned ridge/groove arrays with spacing of 350 nm and height of 500 nm by UV-assisted capillary force lithography. In the absence of any differentiation-inducing agent, the nano-patterned ridge/groove arrays could promote differentiation of hESCs to neuronal lineage (neurons) after 5 days with highly upregulated expression of neuronal differentiation marker NeuroD1 compared to the hESCs cultured on flat matrix. Furthermore, expression levels of the endoderm marker GATA6 and the mesoderm marker DCN were lower in the hESCs on the nano-pattern surface than in the hESCs on the flat surface. This study provides evidence that the topological cues such as nano-patterned ridge/groove arrays alone can effectively induce hESC differentiation into neural lineage with high specificity [38].

2.2.2 *Engineering Biochemical Cues to Induce ESC Differentiation*

In parallel with the physical approaches as externally applied manipulator for stem cell fates, latest advances in biochemical engineering have diversified the portfolios of methods in inducing ESC differentiation via biochemical cues in addition to the traditional inductive approach which mainly resorts to soluble factors supplemented in the culture medium. Herein we summarized the representative progresses of engineered biochemical cues in four aspects to provide new insight into directed ESC differentiation through (1) genetic engineering, (2) immobilized growth factors, (3) coculture, and (4) synthetic small molecules.

2.2.2.1 Genetic Engineering

Due to its readiness for gene manipulation, genetic engineering is increasingly utilized for directly editing and modifying the genome or epigenetic inheritance of ESCs, in order to enhance expression of specific proteins for promoting stem cell differentiation. The success of this powerful tool has been demonstrated in coaxing ESC differentiation to neuron, vascular, and hepatic lineages.

Lmx1a, an important transcription factor in neuron development, was transfected into ESCs in an attempt to promote its expression level and ESC differentiation into mesencephalic dopamine (mesDA) neurons [39]. Parkinson's disease is characterized by progressive degeneration of mesDA neurons. Therefore, ESC-derived mesDA neurons are promised as a potential cell source for therapeutic treatment. mesDA neurons are usually obtained by induction with signaling growth factors during central nervous system development (e.g., FGF8, Wnt) but with only limited differentiation efficiency. The forced expression of Lmx1a in both mESCs and hESCs transfected using lentiviral vectors could effectively promote the generation of mesDA neurons. Upon Lmx1a transfection, 75–95 % of mESC-derived neurons expressed molecular and physiological properties of primary mesDA neurons in vitro. When transplanted into 6-hydroxy dopamine lesioned neonatal rats, these ESC-derived cells (ESDCs) integrated and innervated the striatum similarly to primary mesDA neurons. Thus, the enriched generation of functional mesDA neurons by forced expression of Lmx1a may be of future importance in cell replacement therapy of Parkinson's disease.

In another example, VEGF, a critical growth factor for ECs, was transiently transfected to stem cells for promoting angiogenesis [40]. Using PBAE, a family of hydrolytically biodegradable polymers that can condense DNA to form nanoparticles, nonviral nanoparticles were developed to deliver human VEGF gene to hESDCs. Genetically-engineered stem cells demonstrated markedly enhanced human VEGF production, cell viability, and engraftment into target tissues. Results of 2 weeks implantation showed that the scaffolds seeded with VEGF-expressing cells led to two- to fourfold higher vessel densities compared with non-engineered control cells. These results indicate that transfection of VEGF to hESCs through biodegradable polymer nanoparticles may be therapeutic tools for vascularizing tissue constructs and treating ischemic disease.

The efficacy to directly engineer ESC at the gene level has been also demonstrated for hepatic differentiation. Takayama et al. adopted sequential transduction technology to transfect HNF4α gene into hESDCs through adenovirus vector to promote hepatic maturation for application in drug toxicity prediction [41]. They first obtained hepatoblasts derived from ESC by SOX17 and HEX transduction. Afterward, they overexpressed stage-specific HNF4α, a master regulator of liver-specific gene expression in hepatoblasts, and found that the introduction of this single gene can efficiently promote stem cell hepatic differentiation and maturation (Fig. 2.3a). The differentiation efficacy was ~80 % characterized by CYP, ASGPR1, or ALB expression. Since ectopic expression of HNF4α, a leading regulator of the epithelial phenotype, in fibroblast is known to induce mesenchymal-to-epithelial transition (MET), the mechanism behind this highly efficient hepatic maturation of ESCs may be related to MET activation in culture. Recently, this technology has been commercialized by ReproCELL Inc. in Japan to generate hepatocyte-like cells as substitute of primary hepatocytes for drug testing.

2.2.2.2 Immobilized Growth Factors

Growth factors are cell-secreted molecules recruited in stimulating cellular growth, expansion, and differentiation. Soluble growth factors are commonly used as supplements in basal culture medium for controlling stem cell fate. Considering the high cost and short half-life of growth factors, surface engineering approaches have been adopted to immobilize growth factors on the solid substrate for providing continuous stimuli for stem cell differentiation without repeated supplement in the soluble form.

Minato et al. reported a strategy for cardiac differentiation of ESCs using substrate immobilization of fusion proteins comprised of IGFBP4 and elastin-like polypeptides [42]. Damaged cardiac tissues do not normally regenerate because cardiomyocytes cannot proliferate in adults. Hence, the ESC-derived cardiomyocytes have high potential as substitute cell source for transplantation. Soluble IGFBP4 was reported to be a promoter of ESC differentiation toward cardiac lineage through Wnt/β-catenin signaling inhibition. In the fusion protein, IGBP4 acted as functional domain for cardiac differentiation, while elastin-like polypeptides were used for substrate binding. These elastin-like polypeptides could stably adsorb to substrates such as polystyrene dishes through hydrophobic interactions at 37 °C. While at lower temperatures, they would detach from the substrates due to increased

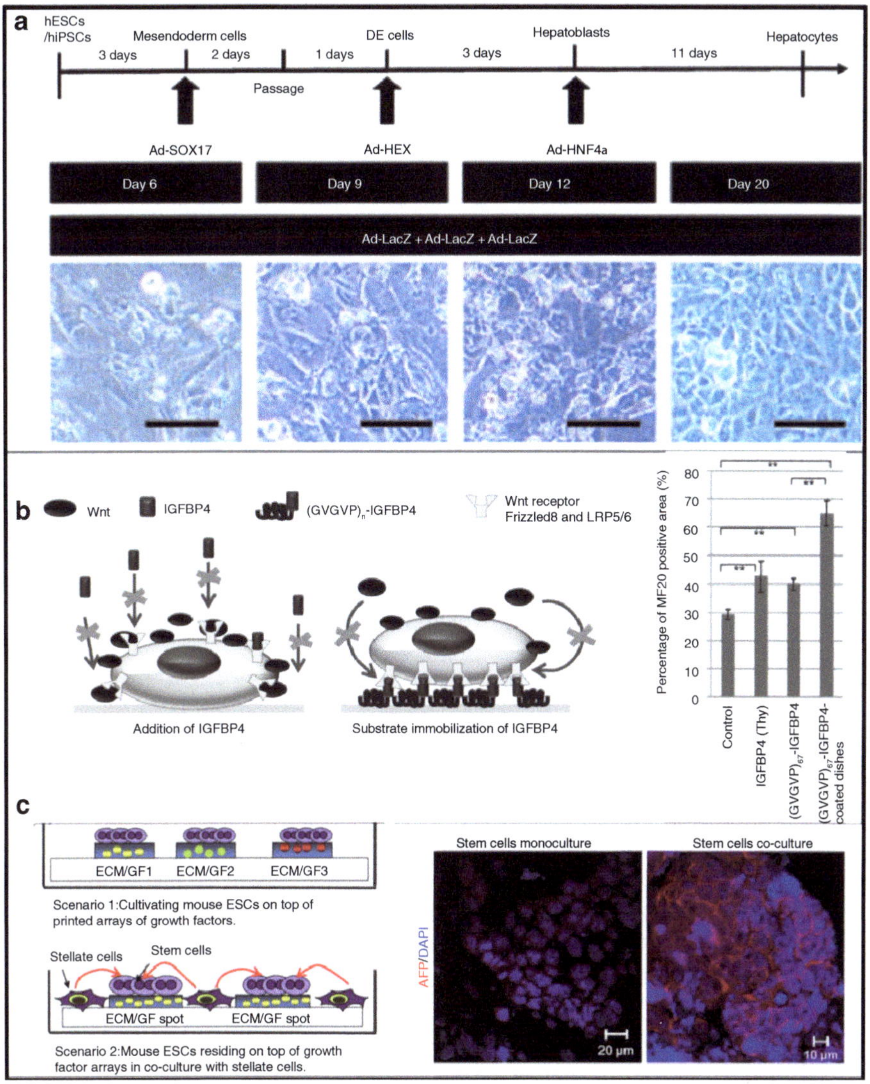

Fig. 2.3 Biochemical cues induced ESC differentiation on 2D substrate. (**a**) The procedures (*top*) and sequential morphological changes (*bottom*) of ESC during hepatic differentiation [41]. The ESCs were transfected with three factors of SOX17, HEX, and HNF4α. (**b**) *Left*: Schema of the inhibitory effect on Wnt/β-catenin signaling after addition of IGFBP4 and substrate immobilization of IGFBP4 [42]. *Right*: Improved cardiac differentiation of ESC indicated by quantification of MF20 (anti-α myosin heavy chain) immunostaining positive area in the DAPI staining area. (**c**) *Left*: Schema of patterned coculture system of mESC and stellate cells for hepatic differentiation [43]. *Right*: Immunostaining for intracellular AFP (hepatic progenitor marker) of mESC cocultures was stronger than monoculture, indicating higher differentiation efficiency (Images are reproduced with the permission)

hydrophilicity which can be potentially utilized for cell harvesting without addition of proteolytic enzymes. IGFBP4-immobilized polystyrene dishes have been shown to effectively induce cardiomyocyte differentiation of ESCs by strong and continuous inhibition of Wnt/β-catenin signaling (Fig. 2.3b). These findings suggest that soluble factor-immobilized substrate can be a useful platform to effectively and economically induce ESC differentiation in regenerative medicine and tissue engineering toward various tissue lineages (e.g., heart, liver, and neuron).

Since directly binding of growth factors to target substrate may lead to growth factor deactivation, intermediate binders are often used to protect the active sites and improve the binding density. Employing heparin as adapter molecule, Lam et al. [44] demonstrated that immobilized bFGF and EGF had different effects on hESC differentiation into neural cells compared to both growth factors in their soluble forms. To combine the biochemical and biophysical cues, bFGF and EGF were either physically adsorbed or bound via heparin on PLLA nano-fibrous scaffold-coated plates. The adsorbed EGF and bFGF did not effectively enhance axon growth. In contrast, immobilization of bFGF or EGF onto nano-fibers via heparin as the adapter molecule significantly promoted axon growth. This study elucidated the effect of immobilized bFGF and EGF in neural differentiation and axon growth and demonstrated an effective surface engineering approach to immobilize active bFGF and EGF onto aligned nano-fibers, which provide potential cell source for neural tissue regeneration.

2.2.2.3 Coculture

Due to the short half-life and rapid consumption by cellular uptake, inductive factors are unable to stimulate stem cell behaviors consistently. Coculturing ESCs with supporting cells recapitulates key features of the natural stem cell niche which usually contains multiple cell types. Coculture systems can promote the differentiation of ESCs via a paracrine signaling pathway, where the inductive factors are continuously secreted by supporting cells [45].

Based on the findings that hepatic stellate cells (HSCs) could secrete factors to stimulate hepatocyte proliferation [46], Revzin et al. investigated the effect of coculture of ESC with HSCs on hepatic differentiation of ESCs [43]. For in vitro culture, mESCs were encapsulated in patterned hydrogel spots (diameter = 500 μm) combined with collagen I and fibronectin, while HSCs were seeded around and functioned with the ESCs through paracrine signaling (Fig. 2.3c). The comparison showed that coculturing method could support a better performance of hepatic differentiation. Heterotypic cocultures will be broadly applicable for identifying the composition of the microenvironment niche for ESC differentiation into various tissue types.

Lee et al. demonstrated that the coculture of hepatic cells could enhance chondrogenesis of ESCs [47]. mESC-derived EBs were cocultured with hepatic cells in 3D bilayered hydrogels. EBs were aggregates of ESCs and were formed in liquid suspension culture. After 3-week coculture with hepatic cells, ESDCs revealed a fourfold increase of GAG level compared with ESDCs cultured alone. This result was supported by real-time PCR analysis, which demonstrated an 80-fold increase in aggrecan expression in cocultured ESDCs. Additionally, type IIB collagen expression was observed only with cocultured ESDCs, and immunohistochemical analysis resulted in significantly more positive type II collagen staining with cocultured ESDCs. Moreover, at day 21, gene expression of other lineages in HEPA-cocultured ESDCs was either comparable to or lower than those of ESDCs cultured alone. These results indicated that coculture of ESDCs with hepatic cells significantly enhanced specific chondrogenic differentiation of ESDCs.

ECs represent one major component of the embryonic pancreatic niche and play a key role in facilitating the differentiation of ESCs toward insulin-producing β-cells in vivo through the activation of Wnt signaling pathway. Talavera-Adame et al. developed a coculture system of mouse EBs and human microvascular ECs (HMECs) to investigate whether interaction of ECs with EBs in coculture promotes differentiation of pancreatic progenitors and insulin-producing cells [48]. EBs were obtained from hanging drop culture method. An increase in the expression of the pancreatic markers PDX-1, Ngn3, Nkx6.1, proinsulin, GLUT-2, and Ptf1a was observed at the interface between EBs and ECs. No expression of these markers was found at the periphery of EBs cultured without ECs or those cocultured with mouse embryonic fibroblasts (MEFs). These results indicate that the differentiation of EBs to pancreatic progenitors and insulin-producing cells can be enhanced by ECs in vitro.

2.2.2.4 Synthetic Small Molecules

Although most intrinsic signaling molecules in natural stem cell microenvironment have been proven to be effective in regulating pluripotent cell fate, these natural inductive factors still lack specificity to induce ESC differentiation to a defined lineage. Synthetic small molecules can potentially meet this challenge to precisely regulate ESC behaviors. The synthetic small molecules of interests can be screened out from compound libraries with tens of thousands of candidates which can be derived from modifications of the chemical backbones of defined natural molecules with various functional residues. The small molecule libraries exhibited diverse and tunable biochemical properties, which enabled high-throughput screening of functional candidates for more specific induction of ESC differentiation. The identification of small molecules would be useful in understanding the underlying molecular mechanisms of these processes and providing a more controllable method for induced stem cell differentiation [49]. Moreover, batch production of synthetic small molecules presents advantage in cost reduction and facilitates the large-scale manufacture of ESCs and their derivatives.

Li et al. identified that synergistic inhibition of GSK3, TGF-β, and Notch signaling pathways by synthetic small molecules could efficiently and quickly induce neuronal differentiation from monolayer-cultured hESCs [50]. In a chemically defined medium supplemented with two small synthetic molecules named CHIR99021 and SB431542, hESCs were converted into homogenous primitive neuroepithelium referred as primitive neuron stem cells (NSC) within 1 week. CHIR99021 was identified by screening of combinatorial libraries of substituted dihydropyrimidines, a compound family that inhibited human GSK-3 at low micromole concentrations [51]. And SB431542 was identified by screening of compound collection for inhibitors of ALK5 [52]. NSCs derived from ESCs in vitro were proved potent in further maturation toward midbrain and hindbrain neuronal subtypes when transplanted in mice. This work provides a working protocol for induced differentiation of ESCs toward NSC with broad differentiation potential, as well as a valuable tool to study the early molecular events initiating human neuron induction.

Gonzalez et al. developed a stepwise strategy to identify a chemically defined method to generate cardiomyocytes from monolayer-cultured hESC with high efficiency and homogeneity [53]. Through activation of Wnt signaling pathway, over 90 % hESCs treated with CHIP99021 (in a concentration of 10 uM) were induced to primitive steak cells. They further identified that a combination of three known small synthetic molecules, namely, IWR-1-endo (a Wnt antagonist), purmorphamine (an Shh signaling agonist), and SB431542 (an inhibitor of activin-like kinase) would effectively induce primitive steak cells differentiation into cardiac cell lineage. The receptors to mediate the interactions with the small synthetic molecules and downstream signal pathways are under exploration to further understand the underlying mechanism for optimizing the cardiac differentiation

2.2.3 Controlling ESC Fate in 3D Microenvironment

During embryonic development, the differentiation of embryo into three germ layers and determined lineages is tightly regulated by cell-matrix and cell-cell interactions with highly spatial and temporal precision in a 3D manner. To mimic the 3D architecture and biological role of the ECM, there is an increasing interest in developing engineering approaches that enable modulation of the behaviors of stem cells [54]. This approach is based on a premise that cellular responses to environmental factors are predictable [10], and the 3D culture models could permit recapitulation of embryonic development in vitro to a degree of complexity which is not achievable in a 2D culture system [54]. There are emerging trends to utilize 3D microenvironment as stem cell niche to support long-term self-renewal and directed differentiation of ESCs in a feeder-free condition. Herein, we highlight some of the latest advances on ESC differentiation in three widely used 3D microenvironments, namely, (1) hydrogel, (2) engineered tissue scaffold, and (3) decellularized scaffold.

2.2.3.1 Hydrogel

Hydrogels are water-swollen, cross-linked polymeric structures with high water content, whose gelation can be induced through pH or temperature changes, ion-ion interactions, covalent bonding, non-covalent interactions, or polymerization [55].

Hydrogels can encompass biological functional entities such as cells, tissues, organs, or entire organisms in 3D culture and closely mimic natural tissues for its soft and rubbery consistence [56].

Hyaluronic acid (HA) is one of the major components of natural ECM and proved to play critical role in maintaining ESCs in undifferentiated state during embryogenesis in vivo [57]. In an attempt to examine the function of HA as engineered stem cell microenvironment, Sharon Gerecht et al. synthesized photopolymerized methacrylated HA hydrogel for cultivation of hESC in a 3D configuration. When encapsulated in HA hydrogel disks and cultured in MEF-conditioned medium, hESCs preserved undifferentiated state, normal karyotype, as well as pluripotency. ESC differentiation could be induced in situ within the same hydrogel simply by altering to endothelia growth medium supplemented with VEGF. Cell sprouting and elongation were observed after 48 h incubation. And hyaluronidase, together with collagenase, could be used for releasing and harvesting of the encapsulated cells. It is concluded that HA hydrogels, with their developmentally relevant composition and tunable physical properties, provide a unique microenvironment for the self-renewal and differentiation of hESCs.

As supporting matrices for ESCs development, hydrogels are usually immobilized on substrates such as tissue culture plates or glass slides. Alternatively, cell-laden hydrogels in the form of microbeads or microcapsules can suspend in aqueous environment with improved mobility and mass transfer of nutrients, oxygen, and stimuli. Chayosumrit et al. established a 3D model to culture and induce hESC differentiation by encapsulating cells in calcium alginate microcapsules [58]. The encapsulated hESCs exhibited improved survival and proliferation after treatment of Y27632 (a ROCK inhibitor). Then hESC clusters were directly induced to definitive endoderm cells that held higher marker expression of mesendoderm (Brachyury >70-fold), definitive endoderm (SOX17 >300-fold, FOXA2 >800-fold, and CXCR4 >100-fold), and primitive gut tube (HNF1b >120-fold) as compared with the undifferentiated hESCs (Fig. 2.4a). These data showed that microcapsules could support the differentiation of hESCs into definitive endoderm in 3D and could have potential application for immune isolation and prevention of teratoma formation of hESCs during transplantation.

2.2.3.2 Engineered Tissue Scaffold

3D scaffolds have been widely applied in tissue engineering as cell carriers for transplantation which improve initial cell retention, survival, differentiation, and host integration [60]. Numerous scaffolds of synthetic or natural origin are under development for designated properties such as minimized cytotoxicity, good biocompatibility, defined porosity pore sizes, and interconnectivity [61], which are expected to actively participate in promoting the efficacy of the ESC differentiation.

Liu et al. built a biodegradable polymer scaffold composed of PLLA and PGA using a nonwoven textile process to study hepatic differentiation of mESCs [62]. Cells derived from 5-day-cultured EB were resuspended in culture medium-matrigel

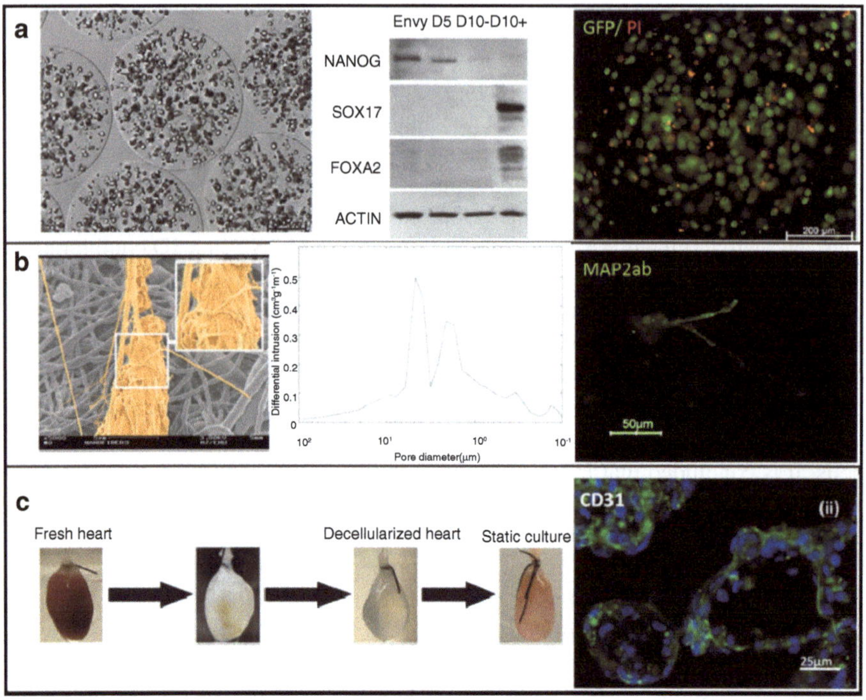

Fig. 2.4 Induced ESC differentiation in 3D microenvironment. (**a**) Morphology (*left*) and viability (*right*) of encapsulated hESCs in alginate microcapsules [58]. Western blot analysis (*middle*) showed downregulation of NANOG (pluripotent marker) and expression of SOX17 and FOXA2 (definitive endoderm markers) on day 10 in Y27632 (Rock inhibitor)-treated samples. (**b**) *Left*: SEM micrographs (pseudo-colored) of the hESCs in 3D electrospun PU scaffolds (5,000× magnification) [20]. *Middle*: Bimodal distribution of pore diameter of PU scaffolds displayed peaks at 5–6 and 1 μm. *Right*: Fluorescence images showed hESC-derived neuron cells after 47 days of culture in scaffold indicated by MAP2ab (mature neural marker). (**c**) *Left*: Characterization of heart during the decellularization process and reseeding [59]. *Right*: Immunostaining with CD31 (endothelial marker, *green*) antibody on the reseeded decellularized heart sections for hESC-differentiated cells. DAPI (*blue*) was used to stain the nuclei of the cells (Images are reproduced with the permission)

mixed solution and transferred to the scaffold. The 3D differentiated hepatocyte-like cells were able to express several liver-specific markers and proteins, secrete ALB, store glycogen, and allow the uptake of low-density lipoproteins. The results provide an alternate method for promoting functional hepatic differentiation of ESCs as potential cell source for clinical use. However, cell death occurred during longer term culture, especially at the center of scaffold, due to poor nutrient exchange. The mass transfer issues illustrated here for bulky 3D scaffold highlight a trend and demand to use micro-fabrication techniques for engineering microscale 3D scaffolds with well-defined architectural and diffusion properties.

Using the salt-leaching technique, Zoldan et al. built porous scaffold from PLLA/PLGA/PCA/PEGDA with varied stiffness to investigate the influence of scaffold

Table 2.1 Scaffold stiffness modulates ESC fate [63]

Scaffold stiffness (MPa)	ESC fate within scaffolds
>6	Keep undifferentiated state
1.5–6	Mesodermal differentiation
0.1–1	Endoderm differentiation
<0.1	Ectodermal differentiation

stiffness on germ layer specification during embryogenesis [63]. Through a broad range of gene analysis and protein expression test, they found that the scaffold stiffness plays a critical role in stimulating ESC spontaneous differentiation toward different germ layers, as summarized in the diagram below (Table 2.1). The underlying mechanism may be related so that scaffold stiffness mimic natural mechanical forces experienced during gastrulation-related cell movement, ultimately directing cell differentiation.

Fibrous scaffold, prepared by electrospinning, have been identified as a promising candidate for cardiac, endothelial, and neural tissue transplantation by providing a special guidance for cell elongation. Carlberg et al. prepared electrospun scaffolds composed of biocompatible aromatic polyether-based polyurethane resin to induce neuron differentiation of hESCs [20]. These scaffolds were fabricated in ~150 μm thick and with ~84 % porosity (Fig. 2.4b). Then hESCs were seeded into scaffolds and allowed for neuron differentiation up to 47 days. hESCs in scaffolds displayed favorable interaction with the fibers of substrate. Cells cultured in 3D fibrous scaffolds exhibited improved outgrowths, more established connections to neighboring cells, and better attachment to individual fibers when compared with cells cultured on 2D substrate. Consequently, electrospun polyurethane scaffolds showed great potential as a substrate for hESC propagation and neuronal differentiation.

2.2.3.3 Decellularized Scaffold

In clinical transplantation, xenogeneic and allogeneic cellular antigens are recognized as foreign tissues or organs by the host and induce an immunological rejection, while components of the naturally derived ECMs are generally well conserved and tolerated [64]. More and more studies suggest that the properties of ECM, such as the composition, 3D ultrastructure, and surface topology, all contribute to cell expansion, differentiation, and constructive remodeling responses of host tissue [65]. ECMs in tissue and whole organs can be directly derived through decellularization processes while maintaining the biochemical and physical properties of the original ECMs. The standard principle of obtaining a decellularized scaffold, recruiting physical and chemical treatments, is to efficiently remove all cellular and nuclear material while minimizing any adverse effect on the composition, biological activity, and mechanical integrity of the remaining ECM [64, 66]. With biocompatible physical and biochemical cues, decellularized scaffolds emerge as a promising substrate to induce ESC differentiation and a potential candidate of clinical tissue or organ transplantation.

Cortiella et al. made the first attempt to produce and use whole lung decellularized scaffold as a matrix to support development of engineered lung tissue from mESCs [67]. The heterogeneous cell population of the lung is supported by a unique organ-specific ECM network consisting of collagen and elastin as well as other matrix components critical to lung function. They found that a combination of mechanical, enzymatic, and physical processes provided the most efficient and gentle decellularization method to remove cells without significant loss of natural lung ECM and structural features. The comparison of ESC behavior was performed between decellularized lung scaffold with Gelfoam, Matrigel, and collagen hydrogel matrix on mESC attachment, differentiation, and subsequent formation of complex tissue. Decellularized lung scaffold allowed better retention of cells with improved differentiation into epithelial and endothelial cell lineages compared to all the other purified matrices. These findings support the utility of decellularized lung scaffold as a matrix for engineering lung tissue and highlight the critical role played by matrix or scaffold-associated cues in guiding ESC differentiation toward lung-specific lineages.

Ng et al. explored the differentiation potential of hESCs in decellularized hearts under static culture [59]. Upregulation of various cardiac-specific markers such as cTnT, Nkx-2.5, Myl2, Myl7, Myh6, and CD31 was showed after 2 weeks of culture. Implantation of decellularized constructs in SCID mice revealed the persistence of cardiac marker-expressing cells and visually vascular network, but no beating function was observed. These results indicate that the intact ECM components and preserved mechanical properties of the decellularized heart had directed differentiation of the stem/progenitor cells into the cardiac lineage. However, further investigation on obtaining mature and functional myocardial cells is required for pushing it as cell therapy for treating cardiovascular diseases.

2.3 Conclusion and Perspectives

Despite the enormous advances in the ESC biology, several challenges still prevent their promised application in regenerative medicine. These challenges include how to precisely control the ESC self-renewal and lineage commitment, how to massively harvest functional ESDCs, as well as how to apply the differentiated cells in vivo for safe and effective therapy. Engineering approaches enable recreation of the complexity of ESC natural microenvironment through controlling cell-ECM, cell-cell, and cell-signaling factor interactions in a highly spatial and temporal manner. Here we reviewed the latest advancement in engineering stem cell microenvironment to induce tailored cellular differentiation of ESC with high efficiency and specification. Both physical strategies (e.g., geometrical constraint, shear stress, substrate stiffness, and topography) and biochemical approaches (e.g., genetic engineering, immobilized growth factors, coculture, and synthetic small molecules) have been demonstrated as powerful tools to manipulate the stem cell niche; meanwhile, 3D engineered microenvironment mimicking the natural dimensionality of the ESC niche shows great potency to improve the efficacy of the ESC differentiation.

To ultimately confront the above-mentioned challenges, future endeavors in ESC engineering toward high-efficient and tailored lineage differentiation are envisioned to relate to the following aspects:

1. Integration of multiple engineering approaches, rather than counting on individual one, in order to recreate the complexity and multicomponents in natural stem cell niche and elicit synergistic effects on genetic and epigenetic properties of the ESCs for improved lineage differentiation
2. Extensive application of nano-/micro-fabrication technologies to engineer fine-tuned stem cell niche with highly spatial and temporal resolution
3. Identification and application of synthetic, chemically defined microenvironment as substitute for naturally derived components to achieve reproducible and animal component-free induction of ESC differentiation (such as using combination of synthetic small molecules to replace natural growth factors or taking polymeric substrate modified with bioactive chemical ligands to replace natural ECM)
4. Comparison of the influence exerted by engineering approaches on generic ESC line derived from embryo and personalized pluoripotent cells generated by induced pluripotent stem cells (iPSCs) technology on their differentiation, potency, genetic/epigenetic difference, and immune response/oncogenicity as implants in vivo.

References

1. Evans M, Kaufman M (1981) Establishment in culture of pluripotent cells from mouse embryos. Nature 292:154–156
2. Martin GR (1981) Isolation of a pluripotent cell line from early mouse embryos cultured in medium conditioned by teratocarcinoma stem cells. Proc Natl Acad Sci USA 78(12): 7634–7638
3. Murry CE, Keller G (2008) Differentiation of embryonic stem cells to clinically relevant populations: lessons from embryonic development. Cell 132:661–680
4. Stojkovic M, Lako M, Strachan T, Murdoch A (2004) Derivation, growth and applications of human embryonic stem cells. Reproduction 128:259–267
5. Edalat F, Bae H, Manoucheri S, Cha HM, Khademhosseini A (2012) Engineering approaches toward deconstructing and controlling the stem cell environment. Ann Biomed Eng 40(6): 1301–1315
6. Keller G (2005) Embryonic stem cell differentiation: emergence of a new era in biology and medicine. Genes Dev 19(10):1129–1155
7. Toh YC, Blagovic K, Voldman J (2010) Advancing stem cell research with microtechnologies: opportunities and challenges. Integr Biol (Camb) 2(7–8):305–325
8. Zhang H, Dai S, Bi J, Liu KK (2011) Biomimetic three-dimensional microenvironment for controlling stem cell fate. Interface Focus 1(5):792–803
9. Discher DE, Mooney DJ, Zandstra PW (2009) Growth factors, matrices, and forces combine and control stem cells. Science 324(5935):1673–1677
10. Burdick JA, Vunjak-Novakovic G (2009) Engineered microenvironments for controlled stem cell differentiation. Tissue Eng Part A 15(2):205–219
11. Guilak F, Cohen DM, Estes BT, Gimble JM, Liedtke W, Chen CS (2009) Control of stem cell fate by physical interactions with the extracellular matrix. Cell Stem Cell 5(1):17–26
12. Dickinson LE, Kusuma S, Gerecht S (2011) Reconstructing the differentiation niche of embryonic stem cells using biomaterials. Macromol Biosci 11(1):36–49

13. Bernard A, Renault JP, Delamarche E (2000) Microcontact printing of proteins. Adv Mater 12(14):1067–1070
14. Lee LH, Peerani R, Ungrin M, Joshi C, Kumacheva E, Zandstra P (2009) Micropatterning of human embryonic stem cells dissects the mesoderm and endoderm lineages. Stem Cell Res 2(2):155–162
15. Brown TD (2000) Techniques for mechanical stimulation of cells in vitro: a review. J Biomech 33:3–14
16. Shimizu N, Yamamoto K, Ando J (2008) Cyclic strain induces mouse embryonic stem cell differentiation into vascular smooth muscle cells by activating PDGF receptor β. J Appl Physiol 104(3):766–772
17. Wells RG (2008) The role of matrix stiffness in regulating cell behavior. Hepatology 47(4): 1394–1400
18. Evans ND, Minelli C, Gentleman E, LaPointe V, Patankar SN, Kallivretaki M, Chen X, Roberts CJ, Stevens MM (2009) Substrate stiffness affects early differentiation events in embryonic stem cells. Eur Cell Mater 18:1–13; discussion 13–14
19. Huang ZM, Zhang YZ, Kotaki M, Ramakrishna S (2003) A review on polymer nanofibers by electrospinning and their applications in nanocomposites. Compos Sci Technol 63(15):2223–2253
20. Carlberg B, Axell M, Kuhn HG (2009) Electrospun polyurethane scaffolds for proliferation and neuronal differentiation of human embryonic stem cells. Biomed Mater 4(4):045004–045010
21. Peerani R, Rao BM, Bauwens C, Yin T, Wood GA, Nagy A, Kumacheva E, Zandstra PW (2007) Niche-mediated control of human embryonic stem cell self-renewal and differentiation. EMBO J 26(22):4744–4755
22. Sasaki D, Shimizu T, Masuda S, Kobayashi J, Itoga K, Tsuda Y, Yamashita JK, Yamato M, Okano T (2009) Mass preparation of size-controlled mouse embryonic stem cell aggregates and induction of cardiac differentiation by cell patterning method. Biomaterials 30(26):4384–4389
23. Bayati V, Sadeghi Y, Shokrgozar MA, Haghighipour N, Azadmanesh K, Amanzadeh A, Azari S (2011) The evaluation of cyclic uniaxial strain on myogenic differentiation of adipose-derived stem cells. Tissue Cell 43(6):359–366
24. Dawson E, Mapili G, Erickson K, Taqvi S, Roy K (2008) Biomaterials for stem cell differentiation. Adv Drug Deliv Rev 60(2):215–228
25. Saha S, Ji L, de Pablo JJ, Palecek SP (2006) Inhibition of human embryonic stem cell differentiation by mechanical strain. J Cell Physiol 206(1):126–137
26. Heo JS, Lee JC (2011) β-catenin mediates cyclic strain-stimulated cardiomyogenesis in mouse embryonic stem cells through ROS-dependent and integrin-mediated PI3K/Akt pathways. J Cell Biochem 112(7):1880–1889
27. Chowdhury F, Na S, Li D, Poh YC, Tanaka TS, Wang F, Wang N (2010) Material properties of the cell dictate stress-induced spreading and differentiation in embryonic stem cells. Nat Mater 9(1):82–88
28. Yamamoto K, Sokabe T, Ando J (2005) Fluid shear stress induces differentiation of Flk-1-positive embryonic stem cells into vascular endothelial cells in vitro. Am J Physiol Heart Circ Physiol 288(4):1915–1924
29. Voldman YC, Toh J (2011) Fluid shear stress primes mouse embryonic stem cells for differentiation in a self-renewing environment via heparan sulfate proteoglycans transduction. FASEB J 25:1208–1217
30. Georges PC, Janmey PA (2005) Cell type-specific response to growth on soft materials. J Appl Physiol 98(4):1547–1553
31. Engler AJ, Sen S, Sweeney HL, Discher DE (2006) Matrix elasticity directs stem cell lineage specification. Cell 126(4):677–689
32. Park JS, Chu JS, Tsou AD, Diop R, Tang Z, Wang A, Li S (2011) The effect of matrix stiffness on the differentiation of mesenchymal stem cells in response to TGF-beta. Biomaterials 32(16):3921–3930
33. Lee S, Kim J, Park TJ, Shin Y, Lee SY, Han YM, Kang S, Park HS (2011) The effects of the physical properties of culture substrates on the growth and differentiation of human embryonic stem cells. Biomaterials 32(34):8816–8829

34. Chai C, Leong KW (2007) Biomaterials approach to expand and direct differentiation of stem cells. Mol Ther 15(3):467–480
35. Massumi M, Abasi A, Soleiman M (2012) The effect of topography on differentiation fates of matrigel-coated mouse embryonic stem (mES) cells cultured on PLGA nanofibrous scaffolds. Tissue Eng Part A 18(5–6):1–12
36. Smith LA, Liu X, Hu J, Ma PX (2009) The influence of three-dimensional nanofibrous scaffolds on the osteogenic differentiation of embryonic stem cells. Biomaterials 30(13):2516–2522
37. Smith LA, Liu X, Hu J, Ma PX (2010) The enhancement of human embryonic stem cell osteogenic differentiation with nano-fibrous scaffolding. Biomaterials 31(21):5526–5535
38. Lee MR, Kwon KW, Jung H, Kim HN, Suh KY, Kim K, Kim KS (2010) Direct differentiation of human embryonic stem cells into selective neurons on nanoscale ridge/groove pattern arrays. Biomaterials 31(15):4360–4366
39. Friling S, Andersson E, Thompson LH, Jonssonc ME, Hebsgaard JB, Nanou E, Alekseenko Z, Marklund U, Kjellander S, Volakakis N, Hovatta O, El Manira A, Bjorklund A, Perlmann T, Ericson J (2009) Efficient production of mesencephalic dopamine neurons by Lmx1a expression in embryonic stem cells. Proc Natl Acad Sci USA 106(18):7613–7618
40. Yang F, Cho SW, Son SM, Bogatyrev SR, Singh D, Green JJ, Mei Y, Park S, Bhang SH, Kim BS, Langer R, Anderson DG (2010) Genetic engineering of human stem cells for enhanced angiogenesis using biodegradable polymeric nanoparticles. Proc Natl Acad Sci USA 107(8):3317–3322
41. Takayama K, Inamura M, Kawabata K, Katayama K, Higuchi M, Tashiro K, Nonaka A, Sakurai F, Hayakawa T, Kusuda Furue M, Mizuguchi H (2012) Efficient generation of functional hepatocytes from human embryonic stem cells and induced pluripotent stem cells by HNF4alpha transduction. Mol Ther 20(1):127–137
42. Minato A, Ise H, Goto M, Akaike T (2012) Cardiac differentiation of embryonic stem cells by substrate immobilization of insulin-like growth factor binding protein 4 with elastin-like polypeptides. Biomaterials 33(2):515–523
43. Tuleuova N, Lee JY, Lee J, Ramanculov E, Zern MA, Revzin A (2010) Using growth factor arrays and micropatterned cocultures to induce hepatic differentiation of embryonic stem cells. Biomaterials 31(35):9221–9231
44. Lam HJ, Patel S, Wang AJ, Chu J, Li S (2010) In vitro regulation of neural differentiation and axon growth by growth factors and bioactive nanofibers. Tissue Eng Part A 16(8):2641–2648
45. Duester G (2008) Retinoic acid synthesis and signaling during early organogenesis. Cell 134(6):921–931
46. Deng X, Chen YX, Zhang X, Zhang JP, Yin C, Yue HY, Lin Y, Han ZG, Xie WF (2008) Hepatic stellate cells modulate the differentiation of bone marrow mesenchymal stem cells into hepatocyte-like cells. J Cell Physiol 217(1):138–144
47. Lee HJ, Yu C, Chansakul T, Varghese S, Hwang NS, Elisseeff JH (2008) Enhanced chondrogenic differentiation of embryonic stem cells by coculture with hepatic cells. Stem Cells Dev 17(3):555–563
48. Talavera-Adame D, Wu G, He Y, Ng TT, Gupta A, Kurtovic S, Hwang JY, Farkas DL, Dafoe DC (2011) Endothelial cells in coculture enhance embryonic stem cell differentiation to pancreatic progenitors and insulin-producing cells through BMP signaling. Stem Cell Rev 7(3):532–543
49. Ding S, Wu TYH, Brinker A, Peters EC, Hur W, Gray NS, Schultz PG (2003) Synthetic small molecules that control stem cell fate. Proc Natl Acad Sci USA 100(13):7632–7637
50. Li WL, Sun W, Zhang Y, Wei WG, Ambasudhan R, Xia P, Talantova M, Lin TX, Kim J, Wang XL, Kim WR, Lipton SA, Zhang K, Ding S (2011) Rapid induction and long-term self-renewal of primitive neural precursors from human embryonic stem cells by small molecule inhibitors. Proc Natl Acad Sci USA 108(20):8299–8304
51. Ring DB, Johnson KW, Henriksen EJ, Nuss JM, Goff D, Kinnick TR, Ma ST, Reeder JW, Samuels I, Slabiak T, Wagman AS, Hammond MEW, Harrison SD (2003) Selective glycogen synthase kinase 3 inhibitors potentiate insulin activation of glucose transport and utilization in vitro and in vivo. Diabetes 52(3):588–595

52. Callahan JF, Burgess JL, Fornwald JA, Gaster LM, Harling JD, Harrington FP, Heer J, Kwon C, Lehr R, Mathur A, Olson BA, Weinstock J, Laping NJ (2002) Identification of novel inhibitors of the transforming growth factor beta1 (TGF-beta1) type 1 receptor (ALK5). J Med Chem 45(5):999–1001
53. Gonzalez R, Lee JW, Schultz PG (2011) Stepwise chemically induced cardiomyocyte specification of human embryonic stem cells. Angew Chem Int Ed 50(47):11181–11185
54. Kraehenbuehl TP, Langer R, Ferreira LS (2011) Three-dimensional biomaterials for the study of human pluripotent stem cells. Nat Methods 8(9):731–736
55. Kasko AM, Wong DY (2010) Two-photon lithography in the future of cell-based therapeutics and regenerative medicine: a review of techniques for hydrogel patterning and controlled release. Future Med Chem 2(11):1669–1680
56. Lieleg O, Ribbeck K (2011) Biological hydrogels as selective diffusion barriers. Trends Cell Biol 21(9):543–551
57. Gerecht S, Burdick JA, Ferreira LS, Townsend SA, Langer R, Vunjak-Novakovic G (2007) Hyaluronic acid hydrogel for controlled self-renewal and differentiation of human embryonic stem cells. Proc Natl Acad Sci USA 104(27):11298–11303
58. Chayosumrit M, Tuch B, Sidhu K (2010) Alginate microcapsule for propagation and directed differentiation of hESCs to definitive endoderm. Biomaterials 31(3):505–514
59. Ng SLJ, Narayanan K, Gao SJ, Wan ACA (2011) Lineage restricted progenitors for the repopulation of decellularized heart. Biomaterials 32(30):7571–7580
60. Schussler O, Chachques JC, Mesana TG, Suuronen EJ, Lecarpentier Y, Ruel M (2010) 3-Dimensional structures to enhance cell therapy and engineer contractile tissue. Asian Cardiovasc Thorac Ann 18(2):188–198
61. Borzacchiello A, Gloria A, Mayol L, Dickinson S, Miot S, Martin I, Ambrosio L (2011) Natural/synthetic porous scaffold designs and properties for fibro-cartilaginous tissue engineering. J Bioact Compat Polym 26(5):437–451
62. Liu T, Zhang SC, Chen X, Li GQ, Wang YJ (2010) Hepatic differentiation of mouse embryonic stem cells in three-dimensional polymer scaffolds. Tissue Eng Part A 16(4):1115–1122
63. Zoldan J, Karagiannis ED, Lee CY, Anderson DG, Langer R, Levenberg S (2011) The influence of scaffold elasticity on germ layer specification of human embryonic stem cells. Biomaterials 32(36):9612–9621
64. Gilbert TW, Sellaro TL, Badylak SF (2006) Decellularization of tissues and organs. Biomaterials 27(19):3675–3683
65. Crapo PM, Gilbert TW, Badylak SF (2011) An overview of tissue and whole organ decellularization processes. Biomaterials 32(12):3233–3243
66. Elder BD, Eleswarapu SV, Athanasiou KA (2009) Extraction techniques for the decellularization of tissue engineered articular cartilage constructs. Biomaterials 30(22):3749–3756
67. Cortiella J, Niles J, Cantu A, Brettler A, Pham A, Vargas G, Winston S, Wang J, Walls S, Nichols JE (2010) Influence of acellular natural lung matrix on murine embryonic stem cell differentiation and tissue formation. Tissue Eng Part A 16(8):2565–2580

Chapter 3
Human Amniotic Membrane: A Potential Tissue and Cell Source for Cell Therapy and Regenerative Medicine

Silvia Diaz-Prado, Emma Muiños-Lopez, Isaac Fuentes-Boquete, Francisco J. de Toro, and Francisco J. Blanco Garcia

Abstract The human amniotic membrane (HAM) is the innermost membrane surrounding the fetus. HAM is a highly abundant and readily available tissue that is becoming appreciated as an alternative to adult bone marrow mesenchymal stem cells (BM-MSCs) useful for cell therapy and regenerative medicine. This tissue provides high efficiency in noninvasive and safe MSC recovery with no intrusive procedures. HAM contains two cell types from different embryological origins: human amnion epithelial cells (hAECs), derived from the embryonic ectoderm, and human amnion mesenchymal stromal cells (hAMSCs), derived from the embryonic mesoderm. hAMSCs and hAECs are immune-privileged cells that can be isolated without the sacrifice of human embryos, avoiding immunological rejection problems and the ethical conflict of using human embryonic stem cells (hESCs). Regarding their immunophenotype, both cell types demonstrate the expression of the common well-defined human mesenchymal and embryonic stem cell markers and the absence of hematopoietic markers. Moreover, both cell populations have similar multipotential for in vitro differentiation into all three germ layers: ectoderm,

S. Diaz-Prado • I. Fuentes-Boquete
Departamento de Medicina, INIBIC-Universidad de A Coruña, A Coruña, Spain

E. Muiños-Lopez
Servicio de Reumatología, INIBIC-Hospital Universitario A Coruña, A Coruña, Spain

F.J. de Toro
Departamento de Medicina, INIBIC-Universidad de A Coruña, A Coruña, Spain

Servicio de Reumatología, INIBIC-Hospital Universitario A Coruña, A Coruña, Spain

F.J. Blanco Garcia (✉)
Servicio de Reumatología, INIBIC-Hospital Universitario A Coruña, A Coruña, Spain

Osteoarticular and Aging Research Laboratory, Hospital Universitário A Coruña, C/As Xubias S/N, 15006, A Coruña, Spain
e-mail: fblagar@sergas.es

M.K. Danquah, R.I. Mahato (eds.), *Emerging Trends in Cell and Gene Therapy*, DOI 10.1007/978-1-62703-417-3_3, © Springer Science+Business Media New York 2013

mesoderm, and endoderm lineages. Indeed, the potential application of amnion-derived cells in a variety of diseases, in particular those associated with degenerative processes, is under clinical or preclinical investigation. The HAM has other biological properties important for tissue engineering, including anti-fibrosis, anti-inflammatory, anti-scarring, antimicrobial, as well as adequate mechanical properties and low immunogenicity. Therefore, amnion allografts are widely applied in ophthalmology, plastic surgery, dermatology, and gynecology. In this chapter, the localization, isolation, characterization, and differentiation potential of amnion-derived cells are discussed. Moreover, the potential clinical applications of either amnion-derived cells or the whole HAM are also reviewed.

Keywords Human amniotic membrane • Adult bone marrow mesenchymal stem cells • Human amnion mesenchymal stromal cells • Cell therapy • Regenerative medicine

3.1 Mesenchymal Stem Cell Concept

Mesenchymal stem cells (MSCs) are multipotent non-hematopoietic progenitors located within the stroma of the bone marrow and other organs that are phenotypically characterized by the expression of several markers (e.g., CD73, CD90, and CD105) and the lack of expression of CD14 or CD11b, CD19 or CD79α, CD34, CD45, and HLA-DR surface molecules [1, 2]. According to a proposal of the International Society for Cellular Therapy [3], three criteria define all types of stem cells: self-renewal, multipotency, and the ability to reconstitute a tissue in vivo. Since there are no specific markers for MSCs, the main criteria for their identification are adherence to the plastic of the tissue culture flask, fibroblast-like morphology, prolonged capacity for proliferation, and the capacity to differentiate into cells of mesodermal lineage in vitro. MSCs are classified, according to the developmental stage from which they are obtained, into embryonic, fetal, or adult stem cells. hESCs are pluripotent and could give rise to all specialized cell types of the organism. However, the tumorigenicity of these cells and technical and ethical considerations limit their availability. In contrast, adult stem cells are rare cells thought to be present in all tissues and responsible for maintaining the homeostasis of the specific tissue [4]. These cells, previously thought to be limited in potential, have been shown to differentiate into multiple mesoderm-type lineages, including chondrocytes, osteoblasts, adipocytes, tenocytes, myotubes, astrocytes, and hematopoietic-supporting stroma [5–7], and also into cell types of ectodermal (e.g., neurons) and endodermal (e.g., hepatocytes) origin [8].

These cells have been isolated from several tissues such as bone marrow [2, 9], articular cartilage [10], synovial membrane [11, 12], perichondrium [13], periosteum [14], connective tissue of dermis and skeletal muscle [15], adipose tissue [16, 17], peripheral blood [18–20], liver [21], lung [22], placenta [5, 23–25], umbilical cord [26–28], umbilical cord blood [29], amniotic fluid [23, 25, 30], and amniotic membrane [31–33]. Moreover, the list of tissues with the potential for tissue engineering is increasing because of recent progress in stem cell biology [34].

Cell therapy using MSCs is a new clinical approach for the treatment of a large number of genetic and degenerative human diseases, including hematopoietic and immune system disorders, diabetes, heart failures, chronic liver injuries, and neurodegenerative disorders. The recent use of autologous or allogenic stem cells has been suggested as an alternative therapeutic approach for cartilage treatment [35, 36]. Human MSCs are probably responsible for normal tissue renewal as well as for response to injury [37–39]. Stem cell transplantation uses cells isolated from small tissue samples, proliferated in culture, to obtain the appropriate number for clinical applications. The use of autologous MSCs avoids immunological rejection problems and the ethical conflict of using hESCs. For these reasons, MSCs are a promising cell resource for tissue engineering and cell-based therapies [38]. The interest in MSCs and their possible application in cell therapy have resulted in a better understanding of the basic biology of these cells. Due to the low number of MSCs that can be isolated from a tissue sample, culture expansion is necessary to obtain adequate cell numbers for clinical purposes and for the analysis of molecular mechanisms.

The bone marrow is the traditional tissue source used for obtaining adult MSCs, but it has a number of disadvantages. The most important limitations are the accessibility and that the procedure required for obtaining this type of tissue is invasive, painful, and associated with morbidity. In addition, the number of cells obtained is low and the potential for proliferation and differentiation declines with donor age [40, 41]. Therefore, the identification of alternative sources of MSCs for both therapeutic and research purposes would be beneficial.

The HAM or amnion has recently emerged as another novel and alternative source of stem cell populations. The HAM is the innermost membrane surrounding the fetus. Because it arises from embryonic epiblast cells prior to gastrulation, it has been suggested that it may retain a reservoir of stem cells throughout pregnancy [42].

3.2 Human Amniotic Membrane or Amnion

The placenta is a structure of fetal-maternal origin with a round shape, 15–20 cm in diameter, and 2–3 cm in thickness [43]. The thickness of the full-term amnion varies between humans and depends on the location of the sample. HAM functions as a filter and preventive shock absorber that protects against infections, traumas, and toxins. This organ is involved in the maintaining fetal tolerance and allows nutrient uptake and gas exchange with the mother but also contains a high number of progenitor cells or stem cells. Moreover, the volume of term placenta makes it an attractive source of stem cells, since as an average human term placenta weighs more than 590 g [44]. HAM develops from extraembryonic tissue and consists of both a fetal component (the chorionic plate) and a maternal component (the deciduas) that are comprised of an epithelial monolayer, a thick basement membrane,

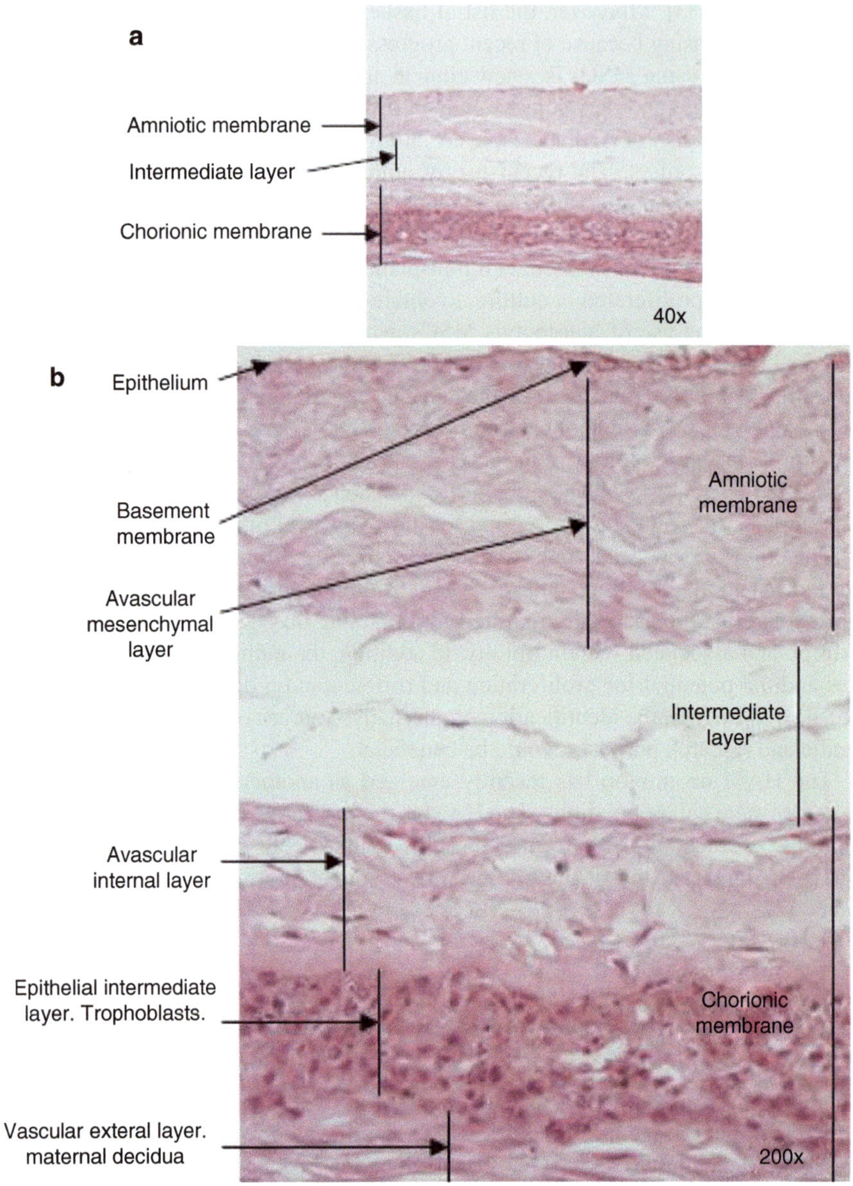

Fig. 3.1 Structure of the fetal membrane at term stained with hematoxylin and eosin (HE). Original magnification: 40× (**a**) and 200× (**b**)

and an avascular stroma [45, 46] (Fig. 3.1). The amnion is a thin (up to 2 mm), avascular, strong, elastic, translucent, and semipermeable fetal membrane attached to the chorionic membrane. Both the amnion and chorion form the amniotic sac filled with amniotic fluid, providing and protecting the fetal environment. The outer layer, the chorion, consists of trophoblastic chorionic and mesenchymal tissues. The inner layer, the amnion, consists of a single layer of ectodermally derived epithelium uniformly arranged on the basement membrane, which is one of the thickest membranes found in any human tissue, and a collagen-rich mesenchymal layer [47]. This mesenchymal layer can be subdivided into the compact layer, forming the main fibrous skeleton of the HAM, the fibroblast layer, and an intermediate layer, which is also called the spongy layer or *zona spongiosa* [45]. Resistance to rupture of HAM is provided by the collagen present in the basement membrane of the amnion. Spontaneous premature rupture of the fetal membranes complicates 1–4 % of the pregnancies. This is due to multiple factors such as infection and genetic predisposition. These premature ruptures are associated with elevated expression levels of relaxins, low expression levels of extracellular matrix (ECM) proteins synthesized by the fetal membranes, or to degradation of these proteins by induced matrix metalloproteinases (MMPs) and subsequent ECM remodeling [48].

The two layers of the amniotic membrane originate at day 8–9 after fertilization, when implantation of the blastocyst has occurred. The inner cell mass of the blastocyst differentiates into two layers, the epiblast and the hypoblast; both layers form the bilaminar embryonic disc. The epiblast gives rise to the three germ layers (ectoderm, mesoderm, and endoderm) and the amniotic epithelium [49].

3.3 Localization of Human Amniotic Membrane-Derived Cells

The localization of HAM-derived cells was examined by our group [31, 32]. We assessed the co-localization of different stem cell markers in histological sections of amniotic membrane by means of immunofluorescence assays. In particular, we studied the co-localization of the CD44, CD90, CD105, and CD271 markers.

Our group did not observe any cells in which co-localization of three and/or four stem cell markers occurred. However, we frequently observed co-localization of double markers, for example, we found CD105 co-located with CD90, CD44 co-located with CD90, and CD271 co-localized with CD44 (Fig. 3.2). Most cells labeled with the different stem cell markers were hAMSCs from the thick basement membrane, although in some membranes we observed hAECs, derived from the embryonic ectoderm, that were labeled only for the CD105 marker. hAMSCs are derived from embryonic mesoderm [50] and are sparsely distributed in the stroma underlying the amnion epithelium [51]. On the other hand, hAECs form a continuous monolayer of embryonic ectodermally derived epithelium uniformly arranged on the basement membrane in contact with the amniotic fluid.

The immunofluorescence results of our group [31, 32] indicated that the HAM contains at least two different cell types having stem cell characteristics and that

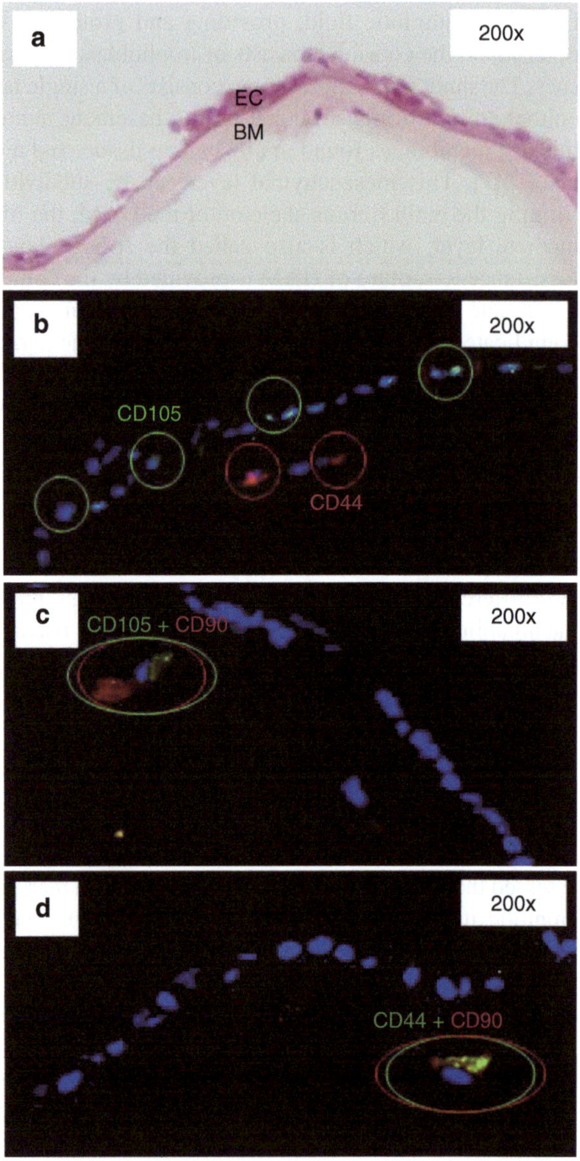

Fig. 3.2 Localization of HAM-derived cells, hAMSCs and hAECs, in healthy HAMs. Representative section of the HAM stained with HE (**a**), indicating the epithelial cells from the extraembryonic ectoderm (*EC*) and the thick basement membrane (*BM*). Immunofluorescence analysis of stem cell marker expression of human amnion cells, nuclei were counterstained with 4′,6-diamidino-2-phenylindole (DAPI) (**b**–**d**). Representative images of hAECs, positive for CD105, and hAMSCs, positive for CD44 (**b**). Representative photos of hAMSCs positive for CD105-CD90 (**c**) and CD44-90 (**d**). Original magnifications: 200× (Images taken from Díaz-Prado et al. [32])

these cells are located in the basement membrane and in the single layer of ectodermally derived epithelium. These common and well-defined human MSCs markers were previously described for bone marrow MSCs. Moreover, we showed that hAECs are positive for the epithelial marker cytokeratin 7, which confirms its epithelial nature (Fig. 3.3).

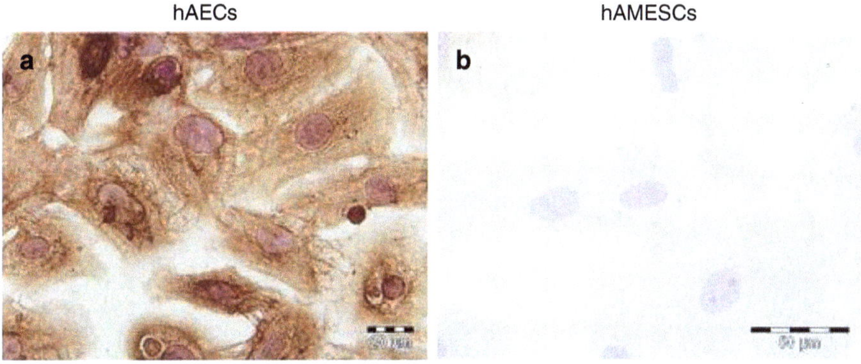

Fig. 3.3 hAECs (**a**) and hAMSCs (**b**) were stained with CK7 antibody by means of immunohistochemistry

3.4 Human Amniotic Membrane as a Source of Stem Cells

HAM expresses only moderate levels of major histocompatibility complex (MHC) class I antigens and MHC class II antigens on its surface. HAM-isolated cells have anti-inflammatory properties. Moreover, there was no evidence of tumorigenicity when isolated human amniotic cells were transplanted into human volunteers or into patients in an attempt to correct lysosomal storage diseases [52–54]. Therefore, hAECs and hAMSCs seem to be immune-privileged cells and suitable for allotransplantation and regenerative medicine [40, 55].

Because fetal tissues are routinely discarded postpartum, HAMs have proved to be abundant, inexpensive, and easily obtained with a virtually limitless availability [45, 47, 56–58]. Therefore, the HAM represents a very useful source of progenitor cells for a variety of applications. Because human embryos are not sacrificed for the isolation of progenitor cells from HAMs, the current controversies associated with the use of hESCs can be avoided [43, 55, 56, 58]. Given the minimal ethical and legal issues associated with HAM cell usage, further investigation into their functional potentials in vivo is warranted.

HAM is becoming appreciated as an alternative to bone marrow for adult MSCs for regenerative medicine. This tissue provides high efficiency in MSC recovery with no intrusive procedures [33]. Moreover, harvesting cells from the HAM is noninvasive and safe. A major advantage of cells isolated from the HAM is that they are harvested after birth and can be cryogenically stored to be available in a timely manner for patient therapy after being thawed and expanded for use in tissue engineering, cell transplantation, and gene therapy.

MSCs from first-, second- and third-trimester placental compartments, including the amnion, chorion, decidua parietalis, and decidua basalis, were isolated and represent less than 1 % of the cells present in the human placenta [22, 33, 59, 60].

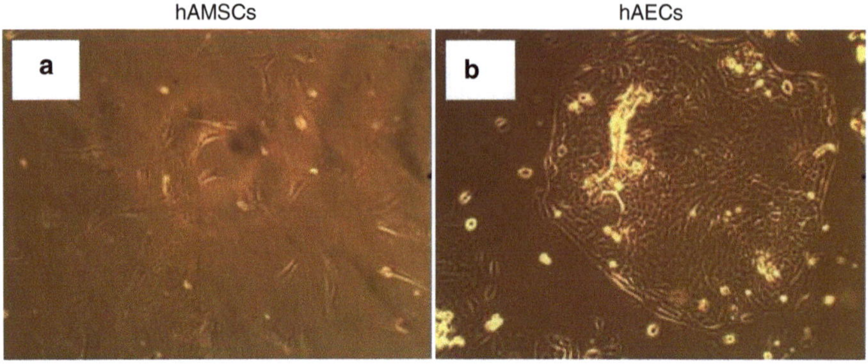

Fig. 3.4 Morphology of cultured hAMSCs (**a**) and hAECs (**b**) isolated from healthy HAM. Original magnifications: 100×

HAM contains two different cell types from different embryological origin [33, 61]: hAECs, derived from embryonic ectoderm [50] which form a continuous monolayer that contacts the amniotic fluid, and hAMSCs, derived from embryonic mesoderm [50] which are sparsely distributed in the stroma that underlies the amnion epithelium [51]. Both hAECs and hAMSCs secrete various antiangiogenic and anti-inflammatory proteins such as interleukin (IL)-1 receptor antagonist; activin A; tissue inhibitors of metalloproteinases (TIMP-1, TIMP-2, TIMP-3, and TIMP-4); and IL-10 which are deposited within the amniotic membrane stroma [62]. Hyaluronic acid may act as a ligand for CD44 and may entrap inflammatory cells in the stroma.

Some papers reported the isolation of HAM-derived cells, from the mesenchymal and epithelial regions of the amnion, from the full-term amnion after its manual separation from the chorion. Bailo et al. [63] isolated and characterized amnion and chorion cells from human term placenta suggesting that both kinds of cells may represent an advantageous source of progenitor cells with potential applications in a variety of cell therapy and transplantation procedures. For this purpose, different methods to isolate HAM-derived cells have been published [33, 59, 63–67].

All these protocols start with a mechanical separation of the amniotic membrane from the underlying chorion through the spongy layer [43]. This step is followed by a digestion with trypsin, dispase, or other digestive enzymes, in different concentrations and for different periods of time, to release the hAECs from the basal membrane. hAMSCs can be subsequently released through subsequent digestion with collagenase [49], alone or combined with DNAase [23].

Regardless of the morphological features of human amnion-derived cells, hAMSCs show plastic adherence and fibroblast-like growth usually observed with MSCs from bone marrow (Fig. 3.4). After 3–4 weeks of hAMSCs culture, it is possible to obtain a population of adherent mesenchymal cells morphologically identical to MSCs isolated from bone marrow. These stromal cells are easy to expand in vitro for at least 9 passages without morphological changes. Furthermore, their immunophenotypic characterization demonstrates the presence of common well-defined human MSC markers previously described for bone marrow (CD90, CD44, CD73, CD166, CD105, CD29) with the absence of the hematopoietic markers CD34 and CD45 and the concomitant lack of fibroblast marker [68, 69]. The absence of hematopoietic or monocytic marker gene expression excludes the possibility that the observed plasticity of these cells is due to contamination with stem cells from fetal or cord blood or with embryonic fibroblasts. This antigen expression pattern is consistent with the data previously published in cells isolated from the amnion and other regions of the term placenta [5, 51, 63]. hAMSCs are also positive for pluripotency markers such as Oct4 (octamer-binding protein 4), NANOG, SOX2 (SRY-related HMG-box gene 2), and REX-1 [49], but positivity for embryonic stem cell markers, SSEA-3 or SSEA-4, remains debated [49]. hAMSCs may be considered as superior to adult MSCs in their differentiation and proliferation capacity due to their higher OCT4 mRNA levels [33]. Moreover, hAMSCs also express low levels of HLA-A, HLA-B, and HLA-C, but do not express HLA-DR, indicating that these stromal cells may be useful in clinical transplantation procedures [49].

On the other hand, isolated hAECs are small-size cells that are easy to expand in vitro for at least three passages without morphological changes; they display epithelial morphologies and grow into a tightly packed, cobblestone monolayer in culture [70] (Fig. 3.4). These cells generally have a central or eccentric nucleus, one or two nucleoli, and abundant cytoplasm, usually vacuolated [66]. hAECs are positive for desmin and vimentin [58]. These epithelial cells also reveal an antigen expression profile characteristic of culture-expanded MSCs [51], since they are positive for the same markers as for hAMSCs. Primary hAECs seem to contain class IA and class II HLAs, consistent with a low risk of tissue rejection [42]. They do not express HLA-A, HLA-B, and HLA-C belonging to class I of the MHC and HLA-DR and HLA-DQ belonging to the class II MHC [45, 65]. When these cells follow pancreatic or hepatic differentiation, but not cardiogenic differentiation, express a significant percentage of class IA but not class II HLAs [71]. In addition, hAECs secrete a number of immunosuppressive factors that target the innate and adaptive immune systems, which may support survival following transplantation [70]. Evidence for long-term self-renewal is not still available for hAECs, probably may be due to the absence of telomerase that limits their ability to divide in culture.

Phenotypes of the two cell populations (Fig. 3.5), hAMSCs and hAECs, are maintained from passage 0 to passage 9 [32]. It is important to notice that although both populations show similar signature regarding cell surface receptor expression pattern, they show many differences with regard to cell shape and cell arrangement [32, 51]. These same findings were previously described by Bilic et al. [51]. These investigators isolated these two populations and concluded that hAECs and hAMSCs in culture exhibited and maintained a similar marker profile of mesenchymal progenitors. hAECs also express surface markers of undifferentiation normally present on embryonic stem and germ cells such as SSEA-4 and STRO-1. Both embryonic stem cell markers are present in more quantity in hAECs than in hAMSCs [32, 42, 51], possibly indicating that hAECs could be at a more early state of undifferentiation. In this regard, Ilancheran et al. [42] also showed that hAECs

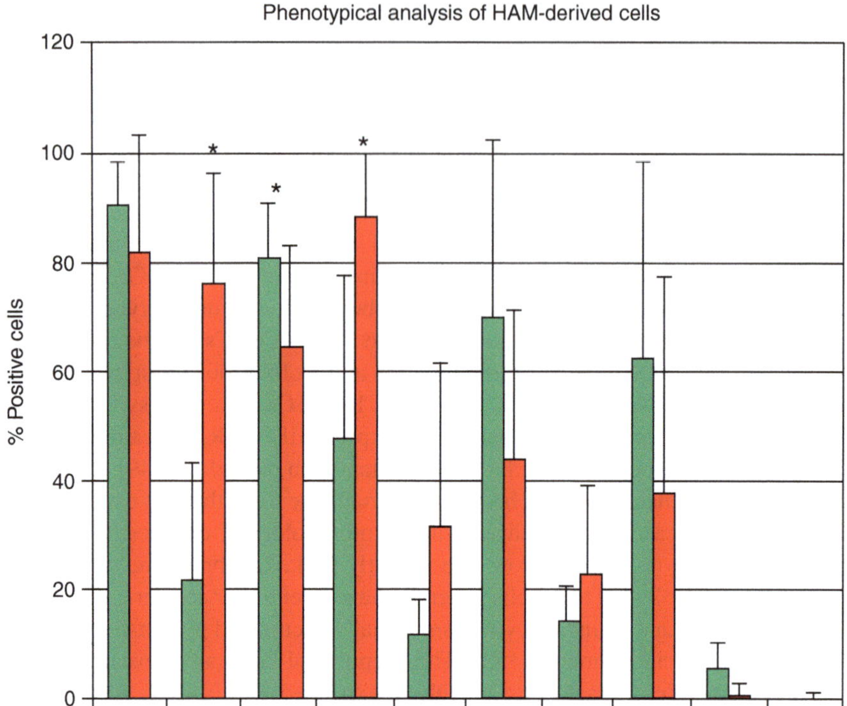

	CD29	CD44	CD73	CD90	CD105	CD168	CD117	SSEA-4	STR0-1	CD34
hAECs	90.73	21.74	80.88	47.54	11.52	70.58	14.31	61.91	5.14	0.02
hAMSCs	82.16	76.11	64.34	88.44	31.83	44.36	22.78	37.82	1.29	0.175

hAECs hAMSCs

Fig. 3.5 Analysis of hematopoietic and standard adult stem cell markers on hAECs and hAMSCs. *Means *P* value <0.05 (Mann–Whitney *U*-test)

expressed SSEA-3 (stage-specific embryonic antigen 3), SSEA-4, TRA-1–60 (tumor rejection antigen) and TRA-1–81, and other antigens such as the ABCG 2/ BCRP (a member of the ATP-binding cassette superfamily), CD9, CD24, CD90, CD117, E-cadherin, integrin α6 and β1, and c-met (receptor growth factor of the hepatocyte) [43, 45]. It has to be noted that initially isolated, hAECs are not homogenously positive for all these antibodies. Some surface markers such as CCR4- and CD117-positive cells are very rare, while others such as CD9 and integrin α6 and β1 are expressed on virtually 100 % of the cells, indicating that hAECs are a heterogeneous cell population with respect to cell surface profiling [67]. These epithelial cells also express Oct4, NANOG, SOX2, REX-1, FGF4, Lefty-A, and TDGF-1gene products associated with pluripotent embryonic stem cells [49, 67, 72, 73]. When hAECs are cultured as an adherent monolayer for several weeks, small spheroids are evidenced over the cobblestone pavement of epithelial cells. These cell clusters express SSEA-3, SSEA-4, TRA 1–60, and TRA 1–80 stem cell-specific cell surface antigens. Moreover, the stem cell molecular marker genes Oct4 and NANOG are also expressed in the small cell clusters, suggesting that hAECs form embryonic body-like structures that maintain their stem cell nature in culture [73].

hAECs and hAMSCs can be grown in Dulbecco's modified Eagle's medium (DMEM) supplemented with 20 % fetal bovine serum (FBS) and 1 % penicillin-streptomycin (P/E) and seeded into culture flasks. Moreover, hAECs could be cultured with or without the addition of growth factors such as epidermal growth factor (EGF) or basic fibroblast growth factor (bFGF) and mostly in the absence of leukemia inhibitor factor (LIF) [48]. Both populations should be expanded in a humidified 5 % CO_2 atm at 37 °C. After the isolation of both cell types, it is advisable to perform immunohistochemical stainings (e.g., for cytokeratin 7, CK7) to demonstrate the purity of both populations. In this regard, only hAECs may be positive for this or other epithelial markers. In monolayer cultures, these hAECs are positive for low molecular weight cytokeratins, confirming their epithelial nature. Moreover, although initially they are vimentin-negative, hAECs become vimentin-positive during cell culture. Vimentin-positive hAECs remain positive for cytokeratins, indicating that in vitro culture may induce dedifferentiation of these epithelial cells [67].

Recently, Parolini et al. [49] published a comparison of key features of HAM-derived cells and human BM-MSCs. These authors postulated that BM-MSCs have a higher cell doubling time than hAECs, while for the hAMSC this time was not reported yet. Regarding the maximum number of passages, it ranges from 5 to 10 for hAMSCs, 10–20 for BM-MSCs, and 30 for hAECs. But there is a contradiction with the passage number at which HAM-derived cells stop proliferation. Based on the literature, proliferation slows down with every passage and cells settle into senescence until proliferation ceases. For example, Miki et al. [66] and Parolini et al. [65] state that hAECs grow normally for 2–6 passages before proliferation ceases. On the contrary, Bilic et al. [51] confirmed that hAECs and hAMSCs proliferation almost stops beyond passage 5, whereas Toda et al. [58] postulated that hAECs senescence is reached at lower passages, P3 or P4. However, Alviano et al. [33] and Soncini et al. [59] indicated that hAMSCs are easily expanded in vitro for at least 15 passages without any visible morphological alterations, but they used cells not exceeding P4 for cell characterization and multilineage differentiation potential studies.

Another comparison between placental cells and BM-MSCs was the aim of the paper published by Barlow et al. [5]. These authors compared human placenta-derived MSCs (the placental tissue included amnion, chorion, and decidua) and human bone marrow-derived MSC in terms of cell characteristics, optimal growth conditions, mesodermal lineage differentiation, and in vivo safety specifically to determine if human placenta-derived MSCs could represent a source of human MSC for clinical trials. They demonstrated that both populations were similar in terms of growth condition requirements and in terms of subsequent biological characterization. However, both populations differed with respect to their proliferation capabilities at different seeding densities. In this regard, human bone marrow-derived MSCs proliferated more slowly than human placenta-derived MSCs in every experiment.

Also the latter had greater long-term growth ability than the former. Moreover, MSCs from both sources exhibited similar morphology, size, and cell surface phenotype, and mesodermal differentiation ability with the exception that human placenta-derived MSC consistently appeared less able to differentiate to the adipogenic lineage. In line with the results obtained, these authors suggested that human placenta is an acceptable alternative source for human MSC.

All published protocols to obtain HAM-derived cells yield hAMSCs, but no studies have compared their efficacy in the isolation. Our group [31] compared two protocols, described in the literature by Alviano et al. [33] and Soncini et al. [59], for the isolation of hAMSCs from the HAM. Alviano's protocol involved three digestions (one mechanical and two enzymatic), whereas Soncini's protocol used only two enzymatic digestions. This study included the comparison of hAMSCs, isolated using both methodologies, in terms of their phenotypic characterization and their in vitro potential for differentiation toward osteogenic, adipogenic, and chondrogenic mesodermal lineages. Both protocols allowed the successful isolation and culture of cells attached to the culture flask with fibroblast-like cell morphology from full-term placenta. These cells showed similar immunophenotype but with differences in cell yield and in the in vitro differentiation potential into the main mesodermal lineages. In particular, quantitative studies showed that Soncini's protocol typically showed an increase in the hAMSCs isolation yield of almost tenfold with regard to Alviano's protocol. Also, the former protocol allowed the isolation and expansion of a larger number of cells in a very short time period. This ready and rapid availability of cells is one criterion required of a source of MSCs for it to be considered for cell transplantation. Therefore, the differences found using both protocols should be taken into account when using these cells for cell therapy.

3.5 Differentiation Potential of Human Amniotic Membrane-Derived Cells

Placental MSCs have been shown to differentiate into chondrogenic, osteogenic, endothelial, hepatocytic, myogenic, and neurogenic lineages, with some differences among cell types depending on the placental tissue sources [8, 33, 40, 43, 50, 56, 68, 69, 74, 75] (Figs. 3.6 and 3.7).

hAMSCs differentiation to neuronal lineage has been demonstrated by the fact that these cells express neuronal markers (nestin, Musashi 1, neuron-specific enolase, neurofilament medium, microtubule-associated protein [MAP]-2 and Neu-N) and glial (GFAP) markers, after their culture in specific neural-induction media [50, 60, 75, 76].

Tamagawa et al. [74] showed that hAMSCs were able to differentiate into cells with characteristics of hepatocytes. In this regard, native cells expressed typical hepatocytic mRNA such as albumin, CK (cytokeratin) 18, α-fetoprotein, α1-antitrypsin, and HNF-4α, but only glucose-6-phosphatase and ornithine transcarbamylase expression and glycogen storage were observed after in vitro hepatic induction.

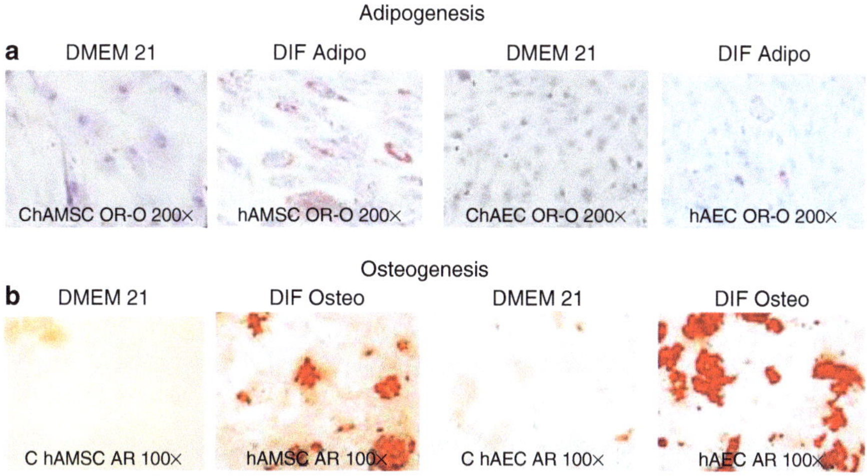

Fig. 3.6 Adipogenic (*DIF Adipo*) and osteogenic (*DIF Osteo*) differentiation of human amnion mesenchymal stromal cells (*hAMSCs*) and human amnion epithelial cells (*hAECs*) with their respective controls (*C hAMSC* and *C hAEC*) grown for 21 days in Dulbecco's Modified Eagle Medium (*DMEM 21*). The presence of adipocytes was assessed by detection of lipid drops using Oil Red O (*OR-O*) stain (**a**). The presence of the calcium deposits characteristic of osteoblasts was detected using Alizarin Red (*AR*) stain (**b**). Original magnifications: 200× (**a**) and 100× (**b**) (Images taken from Díaz-Prado et al. [32])

Regarding hAMSC differentiation toward mesodermal lineage, In't Anker et al. [22] demonstrated the potential of hAMSCs to differentiate into osteogenic and adipogenic cells. After osteogenic differentiation, hAMSCs suffered morphologic changes and showed calcium deposits when they were stained with von Kossa's dye. On the other hand, and after adipogenic differentiation, hAMSCs become multi-vacuolated cells that were stained with Oil Red O stain. Later, Portmann-Lanz et al. [60] showed the capacity of these stromal cells for differentiation to chondrogenic and myogenic lineages. Chondrogenic differentiation of these cells was demonstrated by the presence of abundant collagen in the ECM by means of Alcino's blue dye. Myogenic differentiation of hAMSCs has been determined by RT-PCR since Portmann-Lanz et al. [60] demonstrated the mRNA expression of myogenic transcription factors such as MyoD and myogenin and the protein expression of desmin in hAMSCs cultured in differentiation media. Alviano et al. [33] confirmed these results and also were the first to demonstrate the angiogenic differentiation potential of these cells. This latter study revealed that hAMSCs, after culture in induction media with VEGF, expressed endothelial-specific markers such as the receptors of the vascular endothelial growth factor 1 and 2 (FLT-1, KDR), ICAM-1, as well as the appearance of CD34 and von Willebrand Factor (vWF)-positive cells.

Regarding cardiomyogenic potential, it has been demonstrated that hAMSCs expressed cardiac-specific genes such as GATA4, MLC-2a (myosin light chain), MLC-2v, cTnI, and cTnT [77, 78] after cardiomyogenic induction. Zhao et al. [77]

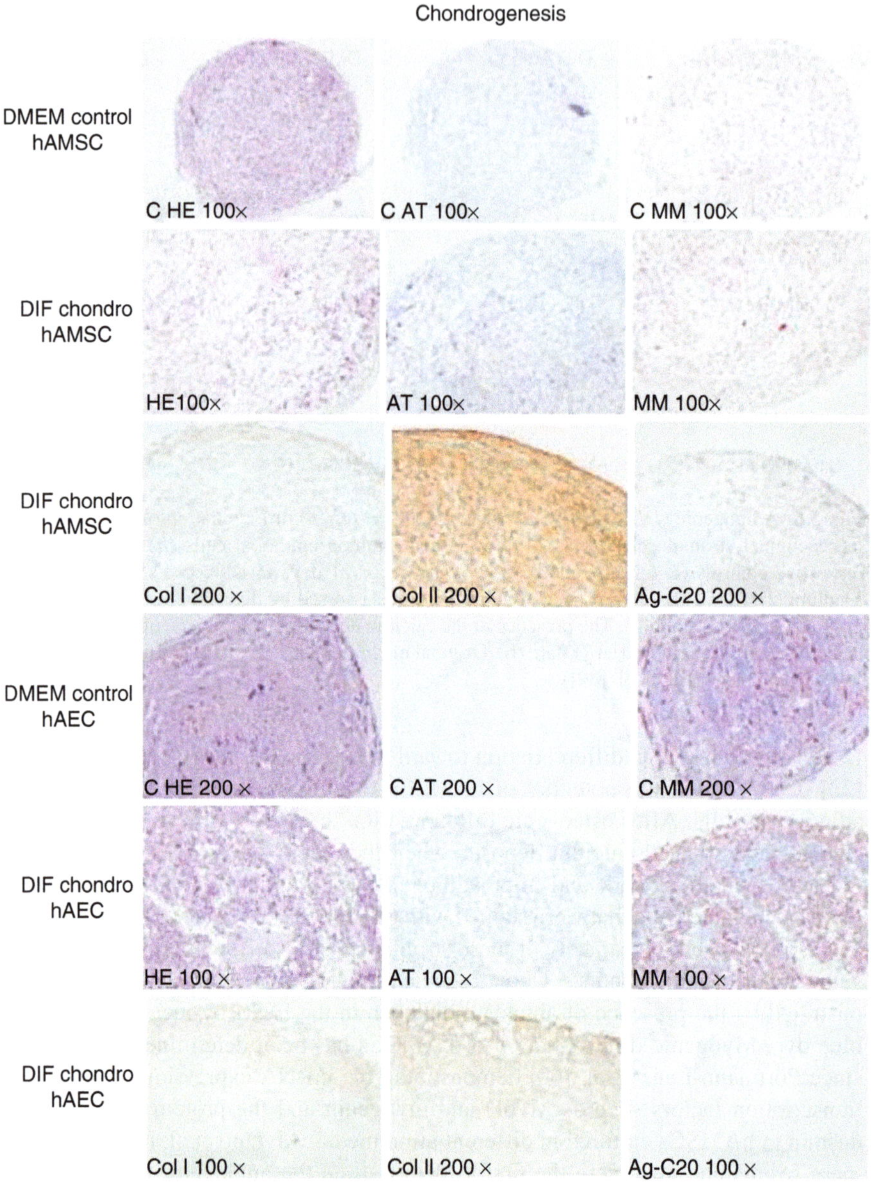

Fig. 3.7 Chondrogenic differentiation (*DIF Chondro*) of human amnion mesenchymal stromal cells (*hAMSCs*) and human amnion epithelial cells (*hAECs*) and their respective controls (*C hAMSC* and *hAEC*) grown for 21 days in Dulbecco's Modified Eagle Medium (*DMEM*). Micropellets were stained with HE, Masson's trichrome (*MM*), and toluidine blue (*AT*) for proteoglycans. Immunodetection of Agg (*Ag-C20*) and collagen type II (*Col II*) was performed to detect molecules characteristic of hyaline cartilage. Immunodetection for collagen type I (*Col I*) was also assessed. Original magnifications: 100× and 200× (Images taken from Díaz-Prado et al. [32])

showed that after hAMSCs transplantation into the myocardial infarcts in rat hearts, these cells survived in the scar tissue for at least 2 months and differentiated into cardiomyocyte-like cells. On the other hand, spontaneous differentiation of hAMSCs toward myofibroblasts has also been observed after their culture in standard medium (DMEM/FBS) within 2 passages [79].

The ability of hAECs to differentiate into cardiomyocytic, myocytic, osteocytic, adipocytic (mesodermal), pancreatic, hepatic (endodermal), neural, and astrocytic (neuroectodermal) cells in vitro has been established [42, 43, 73, 80]. However, in contrast with embryonic stem cells, hAECs did not form tumors up to 7 months posttransplantation in SCID/Beige mice [42, 73]. The capacity of hAECs to differentiate into cell types from all three germ layers may be associated with the fact that the hAECs are directly derived from the epiblast and thus may retain the plasticity of pregastrulation embryonic stem cells.

The pluripotency of hAECs was supported by the study of Tamagawa et al. [81]. The ultimate approach to determine the pluripotency of amniotic epithelium-derived stem cells is to generate chimeric animals by injecting the single stem cell into a blastocyst. If the stem cell contributes all germ layer cells in the chimeric animal, pluripotency will be confirmed [67]. Tamagawa et al. [81] created a xenogeneic chimera with hAECs and mouse embryonic stem cells in vitro. This chimera gives rise to cells of all germ layers, confirming the in vitro pluripotency of hAECs. Later studies have corroborated the ability of hAECs to in vitro differentiate into cells from the three germ layers [42, 43, 73, 80].

hAECs have characteristics of neural progenitor cells since freshly epithelial cells constitutively express a number of neural genes, including neuron-specific enolase (NSE), NF-M, and myelin basic protein (MBP), perhaps suggesting a predilection for neural differentiation [70]. Exposure of hAECs to all-trans-retinoic acid and FGF4 resulted in adoption of an elongated, neural morphology and enhanced expression of some differentiation markers for neural stem such as nestin and GAD (glutamate decarboxylase). Differentiation to astrocyte-like and oligodendrocyte-like cells was also evidenced by expression of glial fibrillary acidic protein (GFAP) and cyclic nucleotide phosphodiesterase (CNP), respectively [73]. Kakishita et al. [82] and Elwan and Sakuragawa [83] demonstrated the differentiation of the epithelial cells to neural cells (ectodermal lineage) with capacity to synthesize and release acetylcholine, catecholamines, neurotrophic factors, activin, noggin, and dopamine, suggesting their possible utility in the treatment of neural degenerative diseases. In this regard, several studies have already been published showing promising results in animal models with Parkinson's disease and mucopolysaccaridosis type VII. Studies of intracerebral grafting of hAECs for the treatment of a mouse model of Parkinson's disease showed that these epithelial cells can synthesize and release catecholamine and neurotrophic factors such as nerve growth factor, neurotrophin-3, and brain-derived neurotrophic factor [82, 84, 85]. Kosuga et al. [86] suggested that transplantation of hAECs transduced with adenoviral vectors can be employed for the treatment of congenital lysosomal storage disorders.

Hepatic differentiation (endodermal lineage) of hAECs was reported by Sakuragawa et al. [87]. They demonstrated albumin and α-fetoprotein production from cultured hAECs, and when LacZ-labeled human hAECs were transplanted into the liver of an immunodeficient mouse, the transplanted cells were found to integrate into the hepatic plate. Some reports demonstrated that these epithelial cells also displayed other functional properties associated with hepatocytes, such as glycogen storage and expression liver-enriched transcription factors, such as hepatocyte nuclear factor (HNF)-3γ and HNF4α, CCAAT/enhancer-binding protein (C/EBP-α and C/EBP-β), and several of the drug-metabolizing genes (cytochrome P450) [73, 88, 89]. Some papers showed albumin production and induction of early markers of hepatic differentiation of hAECs after the addition of specific growth factors to the culture media such as FGF-2, hepatocyte growth factor, oncostatin M, and heparin sodium salt [88]. These findings suggest the potential utility of hAECs to restore hepatic tissues that have been diseased or injured.

Differentiation of hAECs to another endodermal lineage, pancreatic, was reported. Wei et al. [90] cultured these epithelial cells in the presence of nicotinamide to induce pancreatic differentiation, and they observed that the treated cells initiated the expression of multiple pancreatic genes, including the transcription factor Pax-6 and the hormones glucagon and insulin. Subsequent transplantation of these insulin-expressing cells in the spleen of diabetic SCID mice normalized the levels of serum glucose for several months after the transplant, indicating the therapeutic potential of hAECs to treat diabetes mellitus type I. Later, Miki et al. [73] showed by RT-PCR analysis that, after pancreatic differentiation, hAECs express pancreatic α- and β-cell markers such as the transcription factors PDX-1 (pancreatic duodenum homeobox 1), PAX-6 (paired box homeotic gene 6), and NKX2.2 (NK2 transcription factor-related locus 2) and the mature hormones insulin and glucagon.

The differentiation of hAECs to cardiac cells (mesodermal lineage) was first evaluated by Miki et al. [73]. They demonstrated by RT-PCR that cardiac-specific genes atrial and ventricular myosin light chain 2 (MLC-2A and MLC-2V, respectively) and the transcription factors GATA-4 and Nkx 2.5 are expressed or induced in hAECs cultured in media supplemented with ascorbic acid 2-phosphate for 14 days. The immunohistochemical analysis of alpha-actinin expression showed a staining pattern very similar to the one reported for hESC-derived cardiomyocytes.

Differentiation of hAECs to another mesodermal lineages was reported by Ilancheran et al. [42], who showed that native hAECs can differentiate into cells with a phenotype and marker characteristic of mesodermal-derived myocytes, osteocytes, and adipocytes.

3.6 Preclinical Studies of Amnion-Derived Cells Applications

There are a limited number of studies showing results of preclinical investigations using amnion-derived cells [91]. New research focusing on alternative therapeutic applications is currently in progress.

Some reports suggested the beneficial effects of primary hECs in lung fibrosis when they were transplanted into a mouse model. After the transplantation, these cells expressed surfactant proteins and displayed lamellar bodies indicating their differentiation into type II pneumocytes in vivo. hECS transplantation reduced collagen deposition, induced its degradation, and overall reduced fibrosis in the injured lungs [92]. Regarding liver fibrosis, hECs transplantation also showed a significant reduction in the number of hepatic cells producing collagen [93].

Cell therapy using hAECs was assessed for the treatment of pancreatic diseases. In rat models of insulin-dependent diabetes mellitus, transplanted hAECs were able to normalize blood glucose level, since they were able to differentiate into pancreatic β-cells in vivo [90]. On the other hand, and for the treatment of muscle diseases, when hAMSCs were transplanted into a mouse model of Duchenne muscular dystrophy, they underwent myogenic differentiation or fusion with host muscle cells [94].

Moreover, the differentiation potential of hAECs into neurons and glial cells was investigated by several groups for the treatment of neurological disorders which affect both the spinal cord and the brain. For example, for the treatment of Parkinson's disease, hAECs transplanted into an immunosuppressed rat model of Parkinson's disease produced dopaminergic and other diffusible molecules with trophic and beneficial activities on dopaminergic neurons [82, 84]. In case of ischemic stroke, hAECs transplanted into ischemic rats resulted in an improvement of behavioral dysfunction and reduction of infarct volume. These beneficial effects probably could be due to the hECs differentiation toward neurogenic lineage in vivo and to the paracrine actions of the neurotrophic factors secreted by these amnion epithelial cells [95]. Moreover, hAECs have been investigated to treat spinal cord injury. When these cells were transplanted into a monkey or rat models, hAECs prevented degeneration of axotomized neurons and exerted neurotrophic effects, in part due to the release of neurotrophic factors by hAECs [96].

3.7 Clinical Application of Human Amniotic Membrane as Scaffold

HAM has been reported for the first time as a biological dressing to heal skin wounds a century ago [97]. Davis was the first to report the use of fetal membranes as surgical materials in skin transplantations. Later, other surgery applications for HAM have been reported, such as its use as a biological dressing for skin wound treatment, chronic leg ulcers, and burn injuries. Since the 1940s, the use of de-epithelialized HAM has been well documented in ophthalmology for the treatment of Stevens-Johnson syndrome, cicatricial pemphigoid, acute thermal and alkali burns, pterygium surgery, and limbal stem cell transplantation among others [98–102]. HAMs have also been used as biologic dressings for plastic surgery, dermatology, and gynecology procedures [103–107]. In management of open wounds, HAM provides a clean and closed wound in the shortest time possible; it avoids fluid, nutrient, and heat loss; prevents wound infection and pain; and reduces mobility. The amnion adheres firmly to an exposed surface. Moreover, HAM can provide a healthy new substrate suitable for reepithelization and epithelial healing [47]. These properties enable surgeons to apply the graft on various tissue surfaces without need for suturing or application of secondary dressings. Immediately after grafting, the process of biodegradation begins and the membrane self-dissolves over a period of time from days to 3–4 weeks depending on the characteristics of the wound, the presence or absence of coexisting pathogens, the polarization of the applied graft, and the type of graft applied.

Importantly, full-term placentas are evaluated after the birth of the baby and are discarded at the hospital as medical waste. Therefore, HAMs are inexpensive and easily obtained with an availability that is virtually limitless, negating the need for mass tissue banking [45, 47, 57, 58]. The HAM possesses clinical considerable advantages to make it potentially attractive as a biomaterial. It is antimicrobial, antifibrosis, antiangiogenic, and antitumorigenic and has acceptable mechanical properties. It also reduces pain and inflammation, inhibits scarring, enhances wound healing and epithelialization, has analgesic properties, acts as an anatomical and vapor barrier, and modulates angiogenesis, all important requirements for tissue engineering [45]. Several growth factors, such as TGF-β, βFGF, EGF, TGF-α, keratinocyte growth factor, and hepatocyte growth factor, produced from amniotic membrane, are involved in some of these processes [4]. All these characteristics are not shared by other natural or synthetic polymers, highlighting the clinical advantages of HAM as a scaffold compared to other biocompatible products. Also, amnion shows little or no immunogenicity, and the immune response against the graft, if there is, is slight and ineffective, so it does not represent transplantation risks. On the contrary, chorion shows high immunogenicity, and for this reason, it is not used as biomaterial for transplantation purposes. It is important to note that HAM has been approved as a medical material by the Food and Drug Administration [67].

Nowadays, HAMs are used as allograft in general surgery for reconstructions, as an autograft in neonatal reconstruction surgery and as a scaffold in tissue engineering research [48]. The low cost of amnion graft preparation and the very good clinical results in multipurpose applications have made it a viable alternative to other natural (i.e., preserved human skin) and synthetic wound dressings [108]. Moreover, for all the clinical applications, HAM is usually preserved and stored using different methods such as cryopreservation, irradiation, air drying, lyophilization, or glycerol preservation.

3.8 Summary

The HAM, an abundant, inexpensive, and readily obtained tissue that is discarded postpartum, represents a valuable cell and tissue source of great interest in the field of cell therapy and regenerative medicine. Both cell populations isolated from HAM, hAMSCs and hAECs, show an antigen expression profile characteristic of culture-expanded MSCs and differentiation potential into ectodermal, mesodermal, and endodermal lineages. hAMSCs, hAECs, and HAM fragments were used in preclinical studies to treat pancreatic, muscle, vascular, lung, and liver diseases. However, more studies are needed to demonstrate the potential effects of either amnion-derived cells or amnion allografts in animal models of different diseases in the hope of increasing their future clinical applications.

Acknowledgments Our research was supported by Servizo Galego de Saúde, Xunta de Galicia (PS07/84), Cátedra Bioiberica de la Universidade da Coruña, and Instituto de Salud Carlos III CIBER BBN CB06-01-0040; Ministerio Ciencia en Innovacion PLE2009-0144; Fondo Investigacion Sanitaria-PI 08/2028; and funds from FEDER (European Community). Rede Galega de Terapia Celular (REDICENT, CN2012/142).

References

1. Mrugala D, Dossat N, Ringe J, Delorme B, Coffy A, Bony C, Charbord P, Häupl T, Daures J-P, Noël D, Jorgensen C (2009) Gene expression profile of multipotent mesenchymal stromal cells: identification of pathways common to TGFβ3/BMP2-induced chondrogenesis. Cloning Stem Cells 11:61–76
2. Kastrinaki M-C, Andreakou I, Charbord P, Papadaki HA (2008) Isolation of human bone marrow mesenchymal stem cells using different membrane markers: comparison of colony/cloning efficiency, differentiation potential, and molecular profile. Tissue Eng Part C Methods 14:333–339
3. Dominici M, Le Blanc K, Mueller I, Slaper-Cortenbach I, Marini F, Krause D, Deans R, Keating A, Prockop DJ, Horwitz E (2006) Minimal criteria for defining multipotent mesenchymal stromal cells. The International Society for Cellular Therapy position statement. Cytotherapy 8:315–317
4. Yu SJ, Soncini M, Kaneko Y, Hess DC, Parolini O, Borlongan CV (2009) Amnion: a potent graft source for cell therapy in stroke. Cell Transplant 18:111–118
5. Barlow S, Brooke G, Chatterjee K, Price G, Pelekanos R, Rossetti T, Doody M, Venter D, Pain S, Gilshenan K, Atkinson K (2008) Comparison of human placenta- and bone marrow-derived multipotent mesenchymal stem cells. Stem Cells Dev 17:1095–1108
6. Minguell JJ, Conget P, Erices A (2000) Biology and clinical utilization of mesenchymal progenitor cells. Braz J Med Biol Res 33:881–887
7. Caplan AI (1991) Mesenchymal stem cells. J Orthop Res 9:641–650
8. Pasquinelli G, Tazzari P, Ricci F, Vaselli C, Buzzi M, Conte R (2007) Ultrastructural characteristics of human mesenchymal stromal (stem) cells derived from bone marrow and term placenta. Ultrastruct Pathol 31:23–31
9. Yoo JU, Barthel TS, Nishimura K, Solchaga L, Caplan AI, Goldberg VM, Johnstone B (1998) The chondrogenic potential of human bone-marrow-derived mesenchymal progenitor cells. J Bone Joint Surg Am 80:1745–1757
10. Alsalameh S, Amin R, Gemba T, Lotz M (2004) Identification of mesenchymal progenitor cells in normal and osteoarthritic human articular cartilage. Arthritis Rheum 50:1522–1532
11. Fickert S, Fiedler J, Brenner RE (2003) Identification, quantification and isolation of mesenchymal progenitor cells from osteoarthritic synovium by fluorescence automated cell sorting. Osteoarthritis Cartilage 11:790–800
12. De Bari C, Dell'Acio F, Tylzanowski P, Luyten FP (2001) Multipotent mesenchymal stem cells from adult human synovial membrane. Arthritis Rheum 44:1928–1942
13. Dounchis JS, Goomer RS, Harwood FL, Khatod M, Coutts RD, Amiel D (1997) Chondrogenic phenotype of perichondrium-derived chondroprogenitor cells is influenced by transforming growth factor-beta 1. J Orthop Res 15:803–807

14. Nakahara H, Bruder SP, Haynesworth SE, Holecek JJ, Baber MA, Goldberg VM, Caplan AI (1990) Bone and cartilage formation in diffusion chambers by subcultured cells derived from the periosteum. Bone 11:181–188
15. Young HE, Steele TA, Bray RA, Hudson J, Floyd JA, Hawkins K, Thomas K, Austin T, Edwards C, Cuzzourt J, Duenzl M, Lucas PA, Black AC Jr (2001) Human reserve pluripotent mesenchymal stem cells are present in the connective tissues of skeletal muscle and dermis derived from fetal, adult, and geriatric donors. Anat Rec 264:51–62
16. Zuk PA, Zhu M, Ashjian P, De Ugarte DA, Huang JI, Mizuno H, Alfonso ZC, Fraser JK, Benhaim P, Hedrick MH (2002) Human adipose tissue is a source of multipotent stem cells. Mol Biol Cell 13:4279–4295
17. Zuk PA, Zhu M, Mizuno H, Huang J, Futrell JW, Katz AJ, Benhaim P, Lorenz HP, Hedrik MH (2001) Multilineage cells from human adipose tissue: implications for cell-based therapies. Tissue Eng 7:211–228
18. Villaron EM, Almeida J, Lopez-Holgado N, Alcoceba M, Sánchez-Abarca LI, Sanchez-Guijo FM, Alberca M, Pérez-Simon JA, San Miguel JF, Del Cañizo MC (2004) Mesenchymal stem cells are present in peripheral blood and can engraft after allogenic haematopoietic stem cell transplantation. Haematologica 89:1421–1427
19. Kuznetsov SA, Mankani MH, Gronthos S, Satomura K, Bianco P, Robey PG (2001) Circulating skeletal stem cells. J Cell Biol 153:1133–1140
20. Zvaifler NJ, Marinova-Mutafchieva L, Adams G, Edwards CJ, Moss J, Burger JA, Maini RN (2000) Mesenchymal precursor cells in the blood of normal individuals. Arthritis Res 2:477–488
21. Le Blanc K, Götherström C, Ringdén O, Hassan M, McMahon R, Horwitz E, Anneren G, Axelsson O, Nunn J, Ewald U, Nordén Lindeberg S, Jansson M, Dalton A, Aström E, Westgren M (2005) Fetal mesenchymal stem-cell engraftment in bone after in utero transplantation in a patient with severe osteogenesis imperfecta. Transplantation 79:1607–1614
22. In't Anker PS, Scherjon SA, Kleijburg-van der Keur C, de Groot-Swings GM, Claas FH, Fibbe WE, Kanhai HH (2004) Isolation of mesenchymal stem cells of fetal or maternal origin from human placenta. Stem Cells 22:1338–1345
23. Steigman SA, Fauza DO (2007) Isolation of mesenchymal stem cells from amniotic fluid and placenta. Curr Protoc Stem Cell Biol Chapter 1:Unit 1E.2
24. Matikainen T, Laine J (2005) Placenta-an alternative source of stem cells. Toxicol Appl Pharmacol 207(2 Suppl):544–549
25. Fauza D (2004) Amniotic fluid and placental stem cells. Best Pract Res Clin Obstet Gynaecol 18:877–891
26. Samuel GN, Kerridge IH, O'Brien TA (2008) Umbilical cord blood banking: public good or private benefit? Med J Aust 188:533–535
27. Baksh D, Yao R, Tuan RS (2007) Comparison of proliferative and multilineage differentiation potential of human mesenchymal stem cells derived from umbilical cord and bone marrow. Stem Cells 25:1384–1392
28. McGuckin CP, Forraz N, Baradez MO, Navran S, Zhao J, Urban R, Tilton R, Denner L (2005) Production of stem cells with embryonic characteristics from human umbilical cord blood. Cell Prolif 38:245–255
29. Mareschi K, Biasin E, Piacibello W, Aglietta M, Madon E, Fagioli F (2001) Isolation of human mesenchymal stem cells: bone marrow versus umbilical cord blood. Haematologica 86:1099–1100
30. You Q, Cai L, Zheng J, Tong X, Zhang D, Zhang Y (2008) Isolation of human mesenchymal stem cells from third-trimester amniotic fluid. Int J Gynaecol Obstet 103:149–152
31. Díaz-Prado S, Muíños-López E, Hermida-Gómez T, Rendal-Vázquez ME, Fuentes-Boquete I, de Toro FJ, Blanco FJ (2011) Isolation and characterization of mesenchymal stem cells from human amniotic membrane. Tissue Eng Part C Methods 17:49–59
32. Díaz-Prado S, Muíños-López E, Hermida-Gómez T, Rendal-Vázquez ME, Fuentes-Boquete I, de Toro FJ, Blanco FJ (2010) Multilineage differentiation potential of cells isolated from the human amniotic membrane. J Cell Biochem 111:846–857

33. Alviano F, Fossati V, Marchionni C, Arpinati M, Bonsi L, Franchina M, Lanzoni G, Cantoni S, Cavallini C, Bianchi F, Tazzari PL, Pasquinelli G, Foroni L, Ventura C, Grossi A, Bagnara GP (2007) Term amniotic membrane is a high throughput source for multipotent mesenchymal stem cells with ability to differentiate into endothelial cells in vitro. BMC Dev Biol 7:11
34. Bianco P, Robey PG (2001) Stem cells in tissue engineering. Nature 414:118–121
35. Jung DI, Ha J, Kang BT, Kim JW, Quan FS, Lee JH, Woo EJ, Park HM (2009) A comparison of autologous and allogenic bone marrow-derived mesenchymal stem cell transplantation in canine spinal cord injury. J Neurol Sci 285:67–77
36. Koga H, Shimaya M, Muneta T, Nimura A, Morito T, Hayashi M, Suzuki S, Ju YJ, Mochizuki T, Sekiya I (2008) Local adherent technique for transplanting mesenchymal stem cells as a potential treatment of cartilage defect. Arthritis Res Ther 10:R84
37. Hombach-Klonisch S, Panigrahi S, Rashedi I, Seifert A, Alberti E, Pocar P, Kurpisz M, Schulze-Osthoff K, Mackiewicz A, Los M (2008) Adult stem cells and their trans-differentiation potential–perspectives and therapeutic applications. J Mol Med 86:1301–1314
38. Pittenger MF (2008) Mesenchymal stem cells from adult bone marrow. Methods Mol Biol 449:27–44
39. Tsai MS, Hwang SM, Chen KD, Lee YS, Hsu LW, Chang YJ, Wang CN, Peng HH, Chang YL, Chao AS, Chang SD, Lee KD, Wang TH, Wang HS, Soong YK (2007) Functional network analysis on the transcriptomes of mesenchymal stem cells derived from amniotic fluid, amniotic membrane, cord blood, and bone marrow. Stem Cells 25:2511–2523
40. Wei JP, Nawata M, Wakitani S, Kametani K, Ota M, Toda A, Konishi I, Ebara S, Nikaido T (2009) Human amniotic mesenchymal cells differentiate into chondrocytes. Cloning Stem Cells 11:19–25
41. Ilancheran S, Moodley Y, Manuelpillai U (2009) Human fetal membranes: a source of stem cells for tissue regeneration and repair? Placenta 30:2–10
42. Ilancheran S, Michalska A, Peh G, Wallace EM, Pera M, Manuelpillai U (2007) Stem cells derived from human fetal membranes display multilineage differentiation potential. Biol Reprod 77:577–588
43. Insausti CL, Blanquer M, Bleda P, Iniesta P, Majado MJ, Castellanos G, Moraleda JM (2010) The amniotic membrane as a source of stem cells. Histol Histopathol 25:91–98
44. Abdulrazzak H, Moschidou D, Jones G, Guillot PV (2010) Biological characteristics of stem cells from foetal, cord blood and extraembryonic tissues. J R Soc Interface 7:S689–S706
45. Niknejad H, Peirovi H, Jorjani M, Ahmadiani A, Ghanavi J, Seifalian AM (2008) Properties of the amniotic membrane for potential use in tissue engineering. Eur Cell Mater 15:88–99
46. Jin CZ, Park SR, Choi BH, Lee KY, Kang CK, Min BH (2007) Human amniotic membrane as a delivery matrix for articular cartilage repair. Tissue Eng 13:693–702
47. Wilshaw SP, Kearney JN, Fisher J, Ingham E (2006) Production of an acellular amniotic membrane matrix for use in tissue engineering. Tissue Eng 12:2117–2129
48. Dovebra MP, Pereira PNG, Deprest J, Zwijsen A (2010) On the origin of amniotic stem cells: of mice and men. Int J Dev Biol 54:761–777
49. Parolini O, Soncini M, Evangelista M, Schmidt D (2009) Amniotic membrane and amniotic fluid-derived cells: potential tools for regenerative medicine? Regen Med 4:275–291
50. Tamagawa T, Ishiwata I, Ishikawa H, Nakamura Y (2008) Induced in vitro differentiation of neural-like cells from human amnion-derived fibroblast-like cells. Hum Cell 21:38–45
51. Bilic G, Zeisberger SM, Mallik AS, Zimmermann R, Zisch AH (2008) Comparative characterization of cultured human term amnion epithelial and mesenchymal stromal cells for application in cell therapy. Cell Transplant 17:955–968
52. Sakuragawa N, Yoshikawa H, Sasaki M (1992) Amniotic tissue transplantation: clinical and biochemical evaluations for some lysosomal storage diseases. Brain Dev 14:7–11
53. Scaggiante B, Pineschi A, Sustersich M, Andolina M, Agosti E, Romeo D (1987) Successful therapy of Niemann–Pick disease by implantation of human amniotic membrane. Transplantation 44:59–61
54. Akle CA, Adinolfi M, Welsh KI, Leibowitz S, McColl I (1981) Immunogenicity of human amniotic epithelial cells after transplantation into volunteers. Lancet 2:1003–1005

55. Kim SS, Song CK, Shon SK, Lee KY, Kim CH, Lee MJ, Wang L (2009) Effects of human amniotic membrane grafts combined with marrow mesenchymal stem cells on healing of full-thickness skin defects in rabbits. Cell Tissue Res 336:59–66
56. Chang Y-J, Hwang S-M, Tseng C-P, Cheng F-C, Huang S-H, Hsu L-F, Hsu L-W, Tsai M-S (2010) Isolation of mesenchymal stem cells with neurogenic potential from the mesoderm of the amniotic membrane. Cells Tissues Organs 192:93–105
57. Hennerbichler S, Reichl B, Pleiner D, Gabriel C, Eibl J, Redl H (2007) The influence of various storage conditions on cell viability in amniotic membrane. Cell Tissue Bank 8:1–8
58. Toda A, Okabe M, Yoshida T, Nikaido T (2007) The potential of amniotic membrane/amnion-derived cells for regeneration of various tissues. J Pharmacol Sci 105:215–228
59. Soncini M, Vertua E, Gibelli L, Zorzi F, Denegri M, Albertini A, Wengler GS, Parolini O (2007) Isolation and characterization of mesenchymal cells from human fetal membranes. J Tissue Eng Regen Med 1:296–305
60. Portmann-Lanz CB, Schoeberlein A, Huber A, Sager R, Malek A, Holzgreve W, Surbek DV (2006) Placental mesenchymal stem cells as potential autologous graft for pre- and perinatal neuroregeneration. Am J Obstet Gynecol 194:664–673
61. Wolbank S, Peterbauer A, Fahrner M, Hennerbichler S, van Griensven M, Stadler G, Redl H, Gabriel C (2007) Dose-dependent immunomodulatory effect of human stem cells from amniotic membrane: a comparison with human mesenchymal stem cells from adipose tissue. Tissue Eng 13:1173–1183
62. Hao Y, Ma DH, Hwang DG, Kim WS, Zhang F (2000) Identification of antiangiogenic and antiinflammatory proteins in human amniotic membrane. Cornea 19:348–352
63. Bailo M, Soncini M, Vertua E, Signoroni PB, Sanzone S, Lombardi G, Arienti D, Calamani F, Zatti D, Paul P, Albertini A, Zorzi F, Cavagnini A, Candotti F, Wengler GS, Parolini O (2004) Engraftment potential of human amnion and chorion cells derived from term placenta. Transplantation 78:1439–1448
64. Dazzi F, Marelli-Berg F (2008) Mesenchymal stem cells for graft-versus-host disease: close encounters with T cells. Eur J Immunol 38:1479–1482
65. Parolini O, Alviano F, Bagnara GP, Bilic G, Bühring HJ, Evangelista M, Hennerbichler S, Liu B, Magatti M, Mao N, Miki T, Marongiu F, Nakajima H, Nikaido T, Portmann-Lanz CB, Sankar V, Soncini M, Stadler G, Surbek D, Takahashi TA, Redl H, Sakuragawa N, Wolbank S, Zeisberger S, Zisch A, Strom SC (2008) Concise review: isolation and characterization of cells from human term placenta: outcome of the first international workshop on placenta derived stem cells. Stem Cells 26:300–311
66. Miki T, Marongiu F, Ellis E, Strom S (2007) Isolation of amniotic epithelial stem cells. Curr Protoc Stem Cell Biol 3:1E.3.1–1E.3.9
67. Miki T, Strom SC (2006) Amnion-derived pluripotent/multipotent stem cells. Stem Cell Rev 2:133–142
68. Mihu CM, Rus Ciuc D, Sorit u O, Su man S, Mihu D (2009) Isolation and characterization of mesenchymal stem cells from the amniotic membrane. Rom J Morphol Embryol 50: 73–77
69. Kobayashi M, Yakuwa T, Sasaki K, Sato K, Kikuchi A, Kamo I, Yokoyama Y, Sakuragawa N (2008) Multilineage potential of side population cells from human amnion mesenchymal layer. Cell Transplant 17:291–301
70. Marcus AJ, Woodbury D (2008) Fetal stem cells from extra-embryonic tissues: do not discard. J Cell Mol Med 12:730–742
71. Pappa KI, Anagnou NP (2009) Novel sources of fetal stem cells: where do they fit on the developmental continuum? Regen Med 4:423–433
72. Miki T, Mitamura K, Ross MA, Stolz DB, Strom SC (2007) Identification of stem cell marker-positive cells by immunofluorescence in term human amnion. J Reprod Immunol 75:91–96
73. Miki T, Lehmann T, Cai H, Stolz DB, Strom SC (2005) Stem cell characteristics of amniotic epithelial cells. Stem Cells 23:1549–1559

74. Tamagawa T, Oi S, Ishiwata I, Ishikawa H, Nakamura Y (2007) Differentiation of mesenchymal cells derived from human amniotic membranes into hepatocyte-like cells in vitro. Hum Cell 20:77–84
75. Sakuragawa N, Kakinuma K, Kikuchi A, Okano H, Uchida S, Kamo I, Kobayashi M, Yokoyama Y (2004) Human amnion mesenchyme cells express phenotypes of neuroglial progenitor cells. J Neurosci Res 78:208–214. Erratum in: J Neurosci Res 2005; 79:725
76. Kim J, Kang HM, Kim H, Kim MR, Kwon HC, Gye MC, Kang SG, Yang HS, You J (2007) Ex vivo characteristics of human amniotic membrane-derived stem cells. Cloning Stem Cells 9:581–594
77. Zhao P, Ise H, Hongo M, Ota M, Konishi I, Nikaido T (2005) Human amniotic mesenchymal cells have some characteristics of cardiomyocytes. Transplantation 79:528–535
78. Tanaka M, Chen Z, Bartunkova S, Yamasaki N, Izumo S (1999) The cardiac homeobox gene Csx/Nkx2.5 lies genetically upstream of multiple genes essential for heart development. Development 126:1269–1280
79. Li W, He H, Chen YT, Hayashida Y, Tseng SC (2008) Reversal of myofibroblasts by amniotic membrane stromal extract. J Cell Physiol 215:657–664
80. Miki T, Marongiu F, Ellis EC, Dorko K, Mitamura K, Ranade A, Gramignoli R, Davila J, Strom SC (2009) Production of hepatocyte-like cells from human amnion. Methods Mol Biol 481:155–168
81. Tamagawa T, Ishiwata I, Saito S (2004) Establishment and characterization of a pluripotent stem cell line derived from human amniotic membranes and initiation of germ layers in vitro. Hum Cell 17:125–130
82. Kakishita K, Elwan MA, Nakao N, Itakura T, Sakuragawa N (2000) Human amniotic epithelial cells produce dopamine and survive after implantation into the striatum of a rat model of Parkinson's disease: a potential source of donor for transplantation therapy. Exp Neurol 165:27–34
83. Elwan MA, Sakuragawa N (1997) Evidence for synthesis and release of catecholamines by human amniotic epithelial cells. Neuroreport 8:3435–3438
84. Kakishita K, Nakao N, Sakuragawa N, Itakura T (2003) Implantation of human amniotic epithelial cells prevents the degeneration of nigral dopamine neurons in rats with 6-hydroxydopamine lesions. Brain Res 980:48–56
85. Uchida S, Inanaga Y, Kobayashi M, Hurukawa S, Araie M, Sakuragawa N (2000) Neurotrophic function of conditioned medium from human amniotic epithelial cells. J Neurosci Res 62:585–590
86. Kosuga M, Takahashi S, Sasaki K, Enosawa S, Li XK, Okuyama S, Fujino M, Suzuki S, Yamada M, Matsuo N, Sakuragawa N, Okuyama T (2000) Phenotype correction in murine mucopolysaccharidosis type VII by transplantation of human amniotic epithelial cells after adenovirus-mediated gene transfer. Cell Transplant 9:687–692
87. Sakuragawa N, Enosawa S, Ishii T, Thangavel R, Tashiro T, Okuyama T, Suzuki S (2000) Human amniotic epithelial cells are promising transgene carriers for allogenic cell transplantation into liver. J Hum Genet 45:171–176
88. Takashima S, Ise H, Zhao P, Akaike T, Nikaido T (2004) Human amniotic epithelial cells possess hepatocyte-like characteristics and functions. Cell Struct Funct 29:73–84
89. Davila JC, Cezar GG, Thiede M, Strom S, Miki T, Trosko J (2004) Use and application of stem cells in toxicology. Toxicol Sci 79:214–223
90. Wei JP, Zhang TS, Kawa S, Aizawa T, Ota M, Akaike T, Kato K, Konishi I, Nikaido T (2003) Human amnion-isolated cells normalize blood glucose in streptozotocin-induced diabetic mice. Cell Transplant 12:545–552
91. Parolini O, Caruso M (2011) Review: preclinical studies on placenta-derived cells and amniotic membrane: an update. Placenta 32(Suppl 2):S186–S195
92. Moodley Y, Ilancheran S, Samuel C, Vaghjiani V, Atienza D, Williams ED, Jenkin G, Wallace E, Trounson A, Manuelpillai U (2010) Human amnion epithelial cell transplantation abrogates lung fibrosis and augments repair. Am J Respir Crit Care Med 182:643–651

93. Manuelpillai U, Tchongue J, Lourensz D, Vaghjiani V, Samuel CS, Liu A, Williams ED, Sievert W (2010) Transplantation of human amnion epithelial cells reduces hepatic fibrosis in immunocompetent CCl4-treated mice. Cell Transplant 19:1157–1168
94. Kawamichi Y, Cui CH, Toyoda M, Makino H, Horie A, Takahashi Y, Matsumoto K, Saito H, Ohta H, Saito K, Umezawa A (2010) Cells of extraembryonic mesodermal origin confer human dystrophin in the mdx model of Duchenne muscular dystrophy. J Cell Physiol 223:695–702
95. Liu T, Wu J, Huang Q, Hou Y, Jiang Z, Zang S, Guo L (2008) Human amniotic epithelial cells ameliorate behavioural dysfunction and reduce infarct size in the rat middle cerebral artery occlusion model. Shock 29:603–611
96. Wu ZY, Hui GZ, Lu Y, Wu X, Guo LH (2006) Transplantation of human amniotic epithelial cells improves hindlimb function in rats with spinal cord injury. Chin Med J (Engl) 119:2101–2107
97. Davis JW (1910) Skin transplantation with a review of 550 cases at the John Hopkins Hospital. Johns Hopkins Med J 15:307
98. Tsai RJ, Tsai RY (2010) Ex vivo expansion of corneal stem cells on amniotic membrane and their outcome. Eye Contact Lens 36:305–309
99. Sangwan VS, Burman S, Tejwani S, Mahesh SP, Murthy R (2007) Amniotic membrane transplantation: a review of current indications in the management of ophthalmic disorders. Indian J Ophthalmol 55:251–260
100. Gomes JA, Romano A, Santos MS, Dua HS (2005) Amniotic membrane use in ophthalmology. Curr Opin Ophthalmol 16:233–240
101. Dua HS, Gomes JA, King AJ, Maharajan VS (2004) The amniotic membrane in ophthalmology. Surv Ophthalmol 49:51–77
102. Grueterich M, Espana EM, Tseng SC (2003) Ex vivo expansion of limbal epithelial stem cells: amniotic membrane serving as a stem cell niche. Surv Ophthalmol 48:631–646
103. Tejwani S, Kolari RS, Sangwan VS, Rao GN (2007) Role of amniotic membrane graft for ocular chemical and thermal injuries. Cornea 26:21–26
104. Rinastiti M, Harijadi, Santoso AL, Sosroseno W (2006) Histological evaluation of rabbit gingival wound healing transplanted with human amniotic membrane. Int J Oral Maxillofac Surg 35:247–251
105. Santos MS, Gomes JAP, Hofling-Lima AL, Rizzo LV, Romano AC, Belfort R Jr (2005) Survival analysis of conjunctival limbal grafts and amniotic membrane transplantation in eyes with total limbal stem cell deficiency. Am J Ophthalmol 140:223–230
106. Meller D, Pires RT, Mack RJ, Figueiredo F, Heiligenhaus A, Park WC, Prabhasawat P, John T, McLeod SD, Steuhl KP, Tseng SC (2000) Amniotic membrane transplantation for acute chemical or thermal burns. Ophthalmology 107:980–989
107. Morton KE, Dewhurst CJ (1986) Human amnion in the treatment of vaginal malformations. Br J Obstet Gynaecol 93:50–54
108. Díaz-Prado S, Rendal-Vázquez ME, Muíños López E, Hermida-Gómez T, Rodríguez-Cabarcos M, Fuentes-Boquete I, de Toro FJ, Blanco FJ (2010) Potential use of the human amniotic membrane as a scaffold in human articular cartilage repair. Cell Tissue Bank 11:183–195

Chapter 4
Novel Strategies Applied to Provide Multiple Sources of Stem Cells as a Regenerative Therapy for Parkinson's Disease

Abir O. El Sadik

Abstract Advanced innovative approaches were implemented in the field of stem cell researches providing great hope for Parkinson's disease patients. Recent methods were used in the production and application of tissue regeneration and cellular therapy using the integration of nanotechnology and tissue engineering to improve the therapeutic benefits of cell replacement. Several studies were done to determine the optimal stem cell type that can offer an efficient future treatment for Parkinson's disease. Multiple signaling factors were investigated for the induction of dopamine neurons from several sources of stem cells, such as the early developing mouse embryo that provides mouse embryonic stem cells, the early developing human embryo that provides the human embryonic stem cells, and adult brain that provides adult neural stem cells. Reprogrammed somatic cells were also used to provide induced pluripotent stem cells and mesenchymal stem cells that differentiate into neural cells. Nevertheless, further studies are recommended to identify the ideal conditions and specific factors that allow stem cells to be applied efficiently for the treatment of Parkinson's disease.

Keywords Parkinson's disease • Induced pluripotent stem cells • Mesenchymal stem cells • Stem cell therapy

A.O. El Sadik
Basic Sciences Department, King Saud Bin AbdulAziz University for Health Sciences, 3660, Riyadh 11481, Saudi Arabia
e-mail: abgh50@yahoo.com

M.K. Danquah, R.I. Mahato (eds.), *Emerging Trends in Cell and Gene Therapy*,

DOI 10.1007/978-1-62703-417-3_4,

4.1 Introduction

The human brain is a complex mystery of enormous amount of information. Raveling its complicated structures and functions is essential to understand the causes of its various disorders and to reach the proper therapy of its diseases. Many efficient scientific advances have been reported for the treatment of neural diseases. One of the most common serious neurodegenerative disorders is Parkinson's disease (PD). It is attributed to selective loss of dopamine (DA) producing neurons in the substantia nigra [1]. Current medications, such as levodopa (LDOPA), the metabolic precursor of dopamine, are not sufficient to retard DA neuron degeneration. In addition, they produce several side effects, such as motor complications, sleep disturbances, and mood disorders. In contrast, it was recorded that LDOPA hastens the loss of dopamine nerve terminals in the nigrostriatum [2]. Over the last decades, surgical approaches, from the ablative surgeries of subthalamotomy and pallidotomy to the more refined deep brain stimulation, attracted the interest as a conventional therapy for PD in many countries. Subthalamic nucleus and globus pallidus deep brain stimulation were considered an alternative surgical treatment for dyskinesias and motor fluctuations which are major complications in long-term LDOPA therapy. However, it could not be proved that surgical treatment affects the disease progression, and it is suitable only for a limited number of patients [3]. The discovery of neurotrophins and their protective effects on DA neurons was documented in animal models of PD [4]. Neurotrophic factors, which are protein growth factors that regulate neuronal maturation, proliferation, branching, and synaptic plasticity, activate the pathway of cell signaling leading to enhancing the survival, growth, and regeneration of DA neurons. When provided continuously, the neurotrophic factors produce neuroprotective and neuroregenerative effects with reduction of DA cell death. For example, glial cell line-derived neurotrophic factor (GDNF), neurturin, mesencephalic astrocyte-derived neurotrophic factor (MANF), and cerebral dopamine neurotrophic factor (CDNF) were shown to act more effectively than other neurotrophins in protecting striatal neurons exerting more neuroprotective effects and improving the clinical outcome [5]. Nevertheless, they could not pass through the blood–brain barrier. In addition, they are difficult to be administrated clinically as they should be infused directly into the target area. Furthermore, the cannula produces significant side effects with limited penetration capacity [6, 7].

An alternative treatment, which has been tested for the treatment of PD, was the cellular therapy. The autologous grafts have been examined to treat neural defects. Growth-promoting effects were exerted on the mesencephalic DA neurons by some released molecules from the co-grafted Schwann cells with increase in the DA levels [8]. Moreover, autologous Schwann cells bridge graft-enhanced DA axonal elongation and exerted tyrosine hydroxylase (TH)-positive phenotype in the surrounding astroglial cells [9]. However, several side effects were recorded, such as neuroma formation, mismatch of donor-site neuronal size with the recipient site, and lack of clinical recovery, and the most important is the short age of the cells as they were taken from the patient himself [10]. A second type of cellular therapy is the intrastriatal transplantation of human fetal nigral cells. Beneficial results were accomplished when DA neurons extracted from primary ventral mesencephalon were grafted in the lesioned striatum [11]. Integration of the transplanted cells with the host neurons was demonstrated by increase in the length of their axons and dendrites and the development of afferent and efferent projections between them. In addition, DA was found to be released regularly with remarkable functional recovery as proved by positron emission tomography scanning of the activated areas by the grafted cells and prolonged clinical assessments [12]. Besides, it was concluded that the grafted tissues can overcome the immune rejection even without immunosuppression [13]. The quality of dissected tissue, the age of donor fetus, limited tissue accessibility, and the storage conditions after dissection affected greatly the behavior of grafted cells. Moreover, the resulting dense hyperdopaminergic areas secreting excess DA from the grafted cells [14, 15] and serotonergic hyperinnervation [16] were recorded to induce severe dyskinesias. These results were also supported by the finding of Lewy bodies in the transplanted DA neurons in the substantia nigra of a postmortem brain of a 68 years old man, indicating progressive neurodegeneration of the donor cells [14]. It could be concluded that the disease process not only affects the endogenous brain cells but also the new grafted cells. Furthermore, the fetal grafted cells probably contained fibroblasts, astrocytes, or other types of neurons or glial cells, in addition to DA neurons, that affected the grafted cells behavior, survival, and clinical outcome. It could be suggested that reported improvement was due to replacement of the aged brain cells by the new grafted cells rather than stimulation of the brain's own neurorestorative mechanism. Therefore, searching for an alternative source of cells was very essential to provide safe and efficient materials as a cellular therapy and to achieve a better prognosis for PD patients. However, several questions must be resolved while reaching the best graft strategy, such as choosing the optimal cell source matching with the right patients, avoiding immunorejection, regulating the side effects, and achieving the objective of the best functional benefit for the patient to live a better life. This necessity guided the researchers to develop advanced innovative approaches in the field of stem cell population as the great promise for patients suffering from neurodegenerative diseases especially PD.

4.2 Stem Cell Therapy

The common knowledge of the absence of rebuilding phenomenon in the nervous system has been established on the idea that the nerve cells have lost the property of proliferation after the embryonic period of life. This task was dramatically changed after the evidence of regeneration in the central nervous system and the presence of endogenous neural stem cells capable of proliferation, surviving, and integration into the existing functional neuronal circuits. The mammalian brain retains a lifelong capacity of neurogenesis. New neurons are generated from the subventricular zone (SVZ) of lateral ventricles and the subgranular zone of the dentate gyrus of the hippocampus. The persistence of progenitor cells in these areas, possessing the functional characteristics of neuroregeneration, is sustained throughout adult life. The most active brain tissue providing neural stem cells (NSCs) is the SVZ. These NSCs have the potency for neurogenesis of neural and glial cells. The neuronal precursors migrate from the SVZ through the rostral migratory stream to their final destination in the olfactory bulb, the granular cell layer, or to the striatum [17, 18]. It was documented that the neural progenitor cells are reduced in the subependymal zone, subgranular zone, and olfactory bulb in PD postmortem patients [19]. Another source of neurogenesis is the administration of exogenous stem cells to restore the lost neural cells. New intercellular connections could be established due to the multiplication of axonal collaterals and dendritic branches which is referred as neuroplasticity. Neuroplasticity allows the neural cells to adapt to environmental enhancement. This adaptation is the key point of continuous learning, modification of cognitive abilities, and lifelong memory formation that are progressively impaired in late stages of PD [20]. Preclinical researches in animal models proved that transplanted neural stem cells not only replace lost neurons but also successfully enhance the endogenous neurogenesis and they can produce neurotrophins allowing neuroprotection to the degenerating neurons. However, many critical concerns should be investigated before the applications of stem cell therapy in humans such as (1) selection of the optimal duration for the transplantation of stem cells to the appropriate patients, (2) control the migration of the grafted cells to the required precise location, (3) tracking the migration cells by efficient advanced techniques, (4) finding the ideal appropriate sources of stem cells that have pluripotency to differentiate into specific targeted cell types, and (5) the long-term survival and behavior of transplanted stem cells within the recipient microenvironmental tissues. On the other hand, many issues still remain to be revealed concerning the challenges facing this promising therapy, such as immune rejection, tumor formation, and the underlying intrinsic and extrinsic stresses affecting the endogenous neurons. One of these major stresses is the aging effects of the old diseased brain in PD patients as the grafted cells can undergo the pathological phenotypes of the disease-related neurons [21, 22]. NSC renewal in the aging brain is a critical area that needs further investigations. Understanding the intrinsic and extrinsic aging effects that influence the endogenous and transplanted stem cells should be implemented. All these considerations still challenge the conventional methods used in tissue regeneration. So, a main concern was to find a solution for these problems.

One of the extremely promising advancement used in the field of tissue regeneration is nanotechnology. High expectations are now predicted from the applications of nanotechnology in biomedical sciences. The approach of engineering techniques to the nanometer scale can interact with the biological systems at the molecular level. Nanostructures could be applied with high specificity in controlling stem cell behavior and guiding the interactions between cultured and grafted stem cells and the surrounding extracellular matrix. Cytocompatible biomimetic nanomaterials provided efficient substitutes improving the cell functions. Nanomaterials designated in a complex network of three-dimensional extracellular matrix of nanoscale fibers simulate the natural human tissue dimensions and architectures which successfully improve the cultured stem cell performance [23]. Concerning NSCs, nanostructures were designed with cytocompatibility properties necessary for neuronal growth. They possess some mechanical properties that support neural tissue regeneration and electrical properties that regulate neuronal behavior and stimulate neural tissue repair. Novel nanofibers and nanotubes were fabricated possessing biodegradable and biocompatible characteristics enhancing neural tissue engineering. Moreover, nanosurfaces were designed with different nanotopographies that guide cellular adhesion, spreading, morphology, and differentiation. Nanosurfaces arranged in different geometrical configurations influence the NSCs to produce various chemical and physical signals through the interactions with the specific structured extracellular matrix. In addition, these nanotopographies create a complex functional cytoskeletal organization and attachment promoting the behavior of the cultured stem cells. However, several issues should be considered before the wide applications of nanotechnology for human patients. This consideration is of great importance, particularly in NSCs, and should be fulfilled accurately, such as the reported toxic degradative products. Other issues are the maintenance of nanomaterials to their physical properties for the lifelong period with controlled biodegradability and the interactions with biological molecules that interfere with their activities.

4.2.1 Mouse Embryonic Stem Cells (ESCs)

ESCs are a promising tool for understanding the molecular and cellular control of embryonic development and the mechanisms involved in the progression of the diseases. This versatile field promoted the development of various types of cells used for replacement therapies. Mouse ESCs were shown to differentiate into neurons that could be transplanted into the brain and form functional connections with the endogenous neural tissue. They were demonstrated to reinnervate the striatal neurons with partial recovery of the motor deficit resulting from DA deficiency [24]. Cultured mouse ESCs were exposed to a series of growth factors that induced their development into neurons such as the basic fibroblast growth factor (FGF2). The addition of some extrinsic signaling factors was proved to enhance the induction of midbrain DA neurons. Moreover, these signaling factors raised the expression of the rate-limiting enzyme TH in DA synthesis with an efficiency of 30 %. Sonic hedgehog (SHH), a glycoprotein secreted from the floor plate cells, and fibroblast growth factor 8 (FGF8), secreted from the mid-hindbrain boundary, are signaling factors used to promote the expression of specific markers of DA neurons. The resulting cells were able to secrete DA in response to depolarization with the appearance of the electrophysiologic properties of the neurons [25]. Although, other studies demonstrated that these 2 factors, SHH and FGF8, could not provide DA neurons with the sufficient midbrain phenotype [26]. Another extrinsic signaling factor, Wnt1, was recorded to be essential for midbrain DA neuron specification with SHH and FGF8 [27]. Additionally, En-1, a transcription factor expressed in midbrain identity, caused approximately all the DA neurons to show a midbrain phenotype. Ptx3 is another factor that was successfully used for synthesis and metabolism of DA [28, 29]. A homeodomain transcription factor, Lmx1a, was reported to be an essential determinant of DA neurons during embryonic selectively expressed in proliferating DA progenitors and induced in response to early signaling in the ventral midbrain [30, 31]. Generation of stably differentiated mouse ESCs was achieved using a nestin enhancer (NesE). NesE is a vector that was proved to control the expression of Lmx1a. Consequently, Lmx1a can promote mouse ESCs to differentiate into DA neurons. Coupling of NesE – Lmx1a in transplanted mouse ESCs – exhibited more surviving TH-positive neurons. These neurons had the ability to innervate the striatum in a significant manner and resembled the endogenous DA neurons [31]. However, additional signaling factors are still needed for effective generation of efficient transplantable DA neurons. While the chemically inducing methods have established significant strategies for the production of DA neuronal traits from mouse ESCs, the coculturing method also was recorded to give efficient results. Cocultured ESCs with stromal cells (PA6 cells) revealed enhancement of neural differentiation and midbrain patterning [32].

4.2.2 Human ESCs

Novel techniques applied for the differentiation of human ESCs toward specific cell types have raised the hope for cell replacement therapies in PD patients. Despite these extensive studies to differentiate human ESCs to DA neuron, several issues, such as immune rejection and tumor overgrowth, needed to be resolved before transplantation of these cells into the human brain. The provision of safe and pure differentiated human ESCs with the reduction of risk factors should be established in the application of tissue regeneration to the human patients. Recent progress in the identification of external signals directed the differentiation of human ESCs into DA neurons. Thirty percent TH-expressing DA neurons were derived after the use of FGF8 and SHH. They were unsatisfied results that needed further investigations [33, 34]. Immune rejection was noted to be the marked influence producing low survival rate of TH-positive neurons. An additional challenge was the appearance of apoptotic behavior of the transplanted cells with the ectopic nonneural protein expression [35, 36]. Another problem revealed, after transplantation of human ESCs, was teratoma formation arising from non-differentiated stem cells. These cells still maintain their proliferation capacity resulting in tumor appearance [37]. Several studies were recorded to overcome the tumor activity such as genetic modification [38], selective apoptosis of tumor-inducing cells [39], and using sorting technology to purify the cells prior to transplantation [40]. One of the sorting techniques was the magnetic sorting of early postmitotic neurons using specific antibodies. This mechanism allowed for isolation of neurons and elimination of contaminating cells [41]. Another trial was done by coculturing human ESCs with immortalized mesencephalic astrocytes [42]. The results proved that the combination of FGF8 and SHH alone was not enough to produce sufficient midbrain DA neurons with less expression of midbrain markers such as En-1 and Ptx3 in the generated DA neurons. The mostly amazing reason for the less midbrain patterning was the differentiation of neuroepithelial cells derived from human ESCs into mainly a forebrain phenotype compared with that derived from mouse ESCs [43]. Recently, an advanced step was implemented to overcome the propagation of stem cells on mouse feeder layers. Several protocols used the feeder-free support systems [44, 45]. These approaches employed the complex media containing serum, multiple reagents, cell-conditioned media, or coculturing with PA6 mouse stromal cells [46, 47]. Other studies used media additives such as B27 and Matrigel® that involved undefined components with hormones and growth products of animal origin [45]. However, these studies faced again the problem of immune rejection, as these animal components contain immunogenic antigens that could be incorporated into the human ESCs [48]. An additional challenge confronted by the researchers was the prolonged time needed for the human ESCs to sufficiently express DA traits. This long duration provided the chance to the cultured cells to produce extensively branched plexuses of processes inducing irreversible mechanical damage to the cultured cells during manipulations [45].

A rapid protocol was investigated using only chemically defined human reagents in a serum-free media cultured with 1 mM dibutyryl-cAMP (DBcAMP). The study was done to examine the capacity of several well-characterized (H9, BG01) and several new uncharacterized (HUES7, HUES8) human ESC lines to differentiate into DA neurons. Cultured cells demonstrated within 3 weeks β-tubulin III-positive cells with the expression of DA traits such as TH, L-aminodecarboxylase acid (AACD), Ptx3, Lmx1b, Nurr1, and dopamine transporter (DAT). The 4 cell lines revealed variable degree of TH differentiation: H9, 60 %; HUES7, 78 %; HUES8, 81 %; and BG01, 56 %. The cells acquired a permanent pattern of DA phenotype proved by the maintenance of TH expression in the cells 5 days after removal of DBcAMP from the media. Furthermore, within 2–3 weeks, differentiated DA traits appeared in 6 hydroxy-dopamine (6-OHDA)-treated rats after transplantation of these cell lines in their striata which provided great promises after using only human-derived reagents [49]. Although, several investigations should be implemented to ensure the improvement and efficiency of DA neuron differentiation in brain animals, such as the survival rate of transplanted TH cells. It was shown that the percentage of functional TH-positive cells, which had survived in vivo, was lower than that generated in culture prior to transplantation, as well as the need of feeder layers for the cultured cells to maintain their survival. In addition, the significant data of functional efficiency of the transplanted cells should be recorded precisely.

A novel method was applied deriving more functional TH-positive neurons from differentiated human ESCs up to 86 % of the total cultured neurons. These results improved the rate of functional therapy and reduced the potential side effects mainly the teratoma formation resulting from the residual undifferentiated ESCs. Pure spherical neural masses (SNMs) were generated and expanded for longer durations with the maintenance of their differentiation capacity. At the time of use, the SNMs could be progressed into DA neurons within a short time, approximately 2 weeks [50]. The advantage of feeder-free support system for culturing and derivation of DA neurons from SNMs preserved the duration and the effort needed for handling feeder cells and removed the risk of contamination of pathogens and unwanted cells. Although transplantation of neurons into the striatum of animals produced significant functional recovery, full restoration of motor deficit still need further studies. Functional improvement resulting from cell replacement depended greatly on the endogenous DA neurons. Partial repair of the original connections with the neurons in the striatum should exist to produce efficient regression of motor deficit. Moreover, it was proved that transplanted cells can survive and provide more clinical improvement in younger patients [14]. Transplanted cells into old brains of PD patients experienced the same extrinsic stresses affecting endogenous neurons. Furthermore, the transplanted cells progressed into the pathological phenotypes of the diseased endogenous DA neurons. Additionally, aging astrocytes demonstrated detrimental effects on endogenous and transplanted neurons due to the secretion of many toxic factors [51]. Therefore, several issues, concerning the intrinsic and extrinsic aging factors that influence the diseased as well as the transplanted cells, still require further investigations. ESCs remain the most reliable stem cell source of DA neurons. The mouse ESC-derived DA neurons were proved to be favorable than the human ESCs, according to their results in vitro and in vivo. In addition, the ethical controversies of using human ESCs are considered to be a great limitation for their use in a large scale. Overcoming the challenges, such as the purity, teratoma formation, immune rejection, reduced survival rate, and limited functional recovery, and supplying sufficient number of TH-positive cells are recommended for efficient use of human ESCs to treat PD patients.

4.2.3 Adult NSCs

One of the potential advantages of the use of adult neural stem cells is the ability of autologous cell transplantation keeping away of the ethical issues. Another advantage is the capacity of adult NSCs to generate a large number of cells that could be used in the replacement of the diseased DA neurons. Moreover, NSCs proved to have a reparable effect that promotes the long-term survival of transplanted cells. They provide multiple support mechanisms such as anti-inflammatory, angiogenic and neurogenic capacity, and sufficient neuroprotective influence that enhance the survival of the endogenous neurons as well as the exogenous transplanted cells [52, 53]. NSCs could be obtained from the mouse embryo around the embryonic day (E)14-E15 and from human embryo at the 13th week [54]. These neural stem cells are so beneficial as they retain the multipotency pattern that gives them the ability to produce the three major cell lineages of the central nervous system: the neurons, oligodendrocytes, and astrocytes. Undifferentiated human fetal NSCs transplanted into the substantia nigra compacta of a PD primate model elicited successful survival rate. They showed marked integration capacity and consequently promoted the recovery of motor and behavioral deficits. However, the percentage of the cells that differentiated into DA neurons was very low (1–5 %), compared to the ESCs that could produce DA neurons at higher levels [55]. Moreover, expansion of NSCs, isolated from the developing ventral mesencephalon in vitro prior to transplantation, limited their ability to maintain the DA phenotype [56]. Rat fetal NSCs have also been shown to differentiate into DA neurons in the PD brain models with significant recovery of the motor deficits [57, 58]. Multiple growth factors were used significantly to induce rat fetal NSC differentiation and proliferation such as FGF2 epidermal growth factor (EGF), interleukin-1β, interleukin-11, and GDNF [59]. In addition, it was proved that rat fetal NSCs transplanted in diseased brains can precisely reach the target regions revealing high migratory ability and affinity. However, the restrictive proliferative capacity of the fetal NSCs to generate DA neurons and the lack of evidence of the level of improvement of the functional integration and phenotypic stability shifted the interest of many researchers from the fetal to the adult NSCs.

Adult stem cells solved many problems due to their multiple advantages. They are candidates for autologous transplantation therapies that could be harvested and reused with the same patient with no risk of tissue rejection as well as the large variety of tissues that they could be isolated from. The potential sources of adult NSc are the central nervous system [60], human olfactory mucosa that gives a favorable outcome [61], dental pulp cells cocultured with hippocampal and mesencephalic rat neurons [62], umbilical cord [63], and bone marrow [64]. Furthermore, mouse transplanted into PD rat striatum expressed DA enzymes TH and aromatic AADC with different neuronal markers [65]. Although, the rate of the cells expressed TH in these grafts was significantly low, highlighting the need of further approaches to optimize the factors affecting the quantities of generated DA neurons. To overcome this challenge, C17.2 NSCs obtained from the external germinal layer of mouse neonatal cerebellum were cultured at different levels of confluence (30 to >100 %). The level of confluence was shown to markedly affect the fate of transplanted cells. Flattened polygonal cells appeared in the low-confluence (<50 %) cultures. They migrated efficiently in the brain, although they failed to express TH. On the other hand, polygonal cells and fusiform cells were demonstrated in the high-confluence (>100 %) cultures with the expression of TH after spontaneous differentiation. It was suggested that growth factors were secreted in highly confluent cells. These growth factors promoted the cell survival and maturation and changed their morphology and potentiality. Moreover, not only the high confluence enhanced the properties of transplanted cells but also high passage as well. Maintaining the NSCs for 12–20 passages resulted in expression of TH in all the transplanted cells in 65 % of the grafts [66]. These findings suggested the existence of multiple factors that make the grafted cells to respond to TH-binding cues and direct them to develop the appropriate receptors and signaling molecules. Another suggestion was that the loss of a particular cell type, such as DA neurons in PD patients, guided the transplanted NSCs in the proper differentiation [57, 67]. These studies focused the light to more understand the endogenous stresses that affected the diseased cells as well as the growth factors secreted in the lesioned area that could influence the survival, migration, and differentiation of transplanted stem cells. These growth factors could be secreted from injured neurons and reactive neuroglial cells. Other local injury-induced agents, such as cytokines, are inflammatory agents secreted from monocytes and macrophages [58]. The recognition of these factors was a great advance to the study of stem cell therapy, promoting the fate and behavior of transplanted cells.

FGF8, SHH, and Wnt1 are necessary factors for normal DA neurons development from adult NSCs. Pitx3, Nurr1, En-1, En-2, Lmx1a, Lmx1b, Msx1, and Ngn2 are transcription factors essential for DA neuron differentiation [29]. One of these factors, which is expressed exclusively within the central nervous system in DA neurons of the substantia nigra, zona compacta and ventral tegmental area, is the paired-like homeobox protein Pitx3 [68]. This factor was proved to be essential for the specification and survival of DA neurons especially in the substantia nigra [69, 70]. Moreover, coculture neurospheres (NSs) expressing Pitx3 resulted in significant increase in the TH-positive neurons. Coculturing of NSs with only Nurr1 was not sufficient to induce TH-positive in NSCs. Interestingly, when cultured in contact with astrocytes from older embryos, Nurr1 overexpressing group of cells could elicit the correct signals and express the TH [71]. These investigations reported that the regional specification is very essential to be identified as these signaling molecules are not diffusible but highly labile and contact mediated [72]. Furthermore, a marked number of the transplanted human NSCs progeny was differentiated into astrocytes in the recovered PD primates. They expressed neuroreparative factors, such as GDNF, and revealed a homeostatic regulation to the microenvironment [73]. It was proved that the transplanted cells affect greatly the endogenous microenvironment of the lesioned brain. On the other hand, the microenvironment of the diseased brain influences the phenotype of the transplanted cells. The disease process that affects the host DA neurons can adversely affect, by the same stresses, the transplanted new cells [74, 75]. Therefore, it was recommended that the neurodegenerative process of the microenvironment should be controlled and stabilized to promote the ability of the transplanted cells to progress toward the target results. For this reason, several researches were done to compare the fate of transplanted cells in the early and late stages of the disease. It was recorded that, in the early stage of PD disease, the neurons possessed neuroreparative and neuroprotective characteristics supporting both the endogenous and transplanted DA nigrostriatal cells. Subsequently, no tumors or overgrowth formation was revealed in the grafted cells with the absence of dyskinesias. Moreover, stimulation of the release of endogenous precursors, within the brain tissue, was successfully demonstrated. It was found that the transplanted adult NSCs in the early stage of diseased patients had the capacity of secreting neuroregenerative and neuroprotective growth factors [55, 76]. These growth factors enhance the migration of progenitor cells from the areas that generated persistently NSCs in the fetal and adult brain like the SVZ. In addition, these growth factors were capable of activation of neurogenesis, not only in the transplanted cells but also in the endogenous DA neurons [77, 78]. Interactions between transplanted NSCs and the endogenous neurons were shown to be critical for the behavior and the fate of the grafted cells, proving that the internal environment influences neurogenesis, neuroprotection, and neural repair [79, 80]. Transplanted cells expressing GDNF, SHH, and stromal cell-derived factor 1-alpha (SDF-1α), in rat model of PD, had the capacity of efficiently stimulate endogenous NSC survival, proliferation, neural differentiation, and migration as well as nigrostriatal protection. These results were accompanied by significant survival and maintenance of TH-expressed transplanted neural cells in the striatum and substantia nigra. Furthermore, it was observed that the transplanted cells stimulated the release of growth factors and chemokines that are capable of plasticity induction in the host cells promoting endogenous neurogenesis [80]. Transplantation of human NSCs cloned by v-myc gene transfer (HB1.F3 cells) induced significant endogenous neurogenesis developing the therapeutic options for PD. Although the results recorded were about 140 % increase in endogenous neurogenesis, no migration, expression of TH-positive neurons, or neuroprotective effects of the endogenous NSCs were shown. Nevertheless, these NSCs were demonstrated to have the ability to produce several growth factors and multiple plasticity promoting factors, such as brain-derived neurotrophic factor (BDNF), GDNF, neurotrophin-3, and nerve growth factor (NGF) [81, 82]. These factors, especially GDNF and SHH, were reported to spread significantly into sufficient distances in the striatum to produce a stimulatory effect on the neural cells in the SVZ. Moreover, SDF-1α, an important chemokine, was demonstrated to mediate NSC migration to long distances toward injury sites [83, 84] in association with its neuroprotective activity [85]. Other studies indicated that GDNF possesses trophic actions and neuroprotective mechanisms on DA neurons of the striatum and the substantia nigra [86]. In addition, GDNF was capable of promoting SVZ neurogenesis and migration of newly born neuroblasts into the striatum [83]. Another factor, the SHH, has been shown to be essential for the survival, proliferation, and differentiation of developing DA neurons. Moreover, SHH has a neuroprotective action against toxic stresses in the adult brain. It has a chemoattractive influence on SVZ-derived neuronal progenitors in vitro, and it regulates the behavior of stem cells in the postnatal as well as the adult brain [87]. Furthermore, activation of toll-like receptors was proved to enhance neurogenesis and NSCs interactions with the induction of neuroprotective mediators' production [88]. An important addition to the previous study demonstrated the presence of synergistic interactions between the endogenous and exogenous transplanted NSCs. The endogenous NSCs proximal to the transplanted cells expressed SHH which had a neuroprotective and neural regulating capacities on the grafted NSCs. The combination of SHH expressed in the endogenous NSCs and that in the transplanted cells had a great synergistic effect on the neuroprotective action. Moreover, it was proved that expressed SHH in the endogenous NSCs could promote the survival rate of the exogenous transplanted cells [80]. It could be concluded that DA neuron transcriptional factors should be expressed in the NSCs either simultaneously or consequently with the investigation of the influence of these factors in the control of DA neuron survival, proliferation, differentiation, and migration. Several elements should be addressed such as the neuroprotective effects of endogenous NSCs precursors, their enhancement on the survival and fate of transplanted NSCs, and the synergistic interactions between the exogenous and endogenous neurons and their efficiency to produce a therapeutic development to the grafted cells.

4.2.4 Induced Pluripotent Stem Cells (iPSCs)

iPSCs derived from adult somatic cells provide new promises for cell replacement therapy for neurodegenerative diseases especially PD. They represent new prospects for stem cell sources of DA neurons. Initially iPSCs were derived from mouse embryonic and adult fibroblasts by overexpression of transcription factors such as OCT4, SOX2, NANO6, KLF4, c-MYC, and LIN28 [89, 90]. Currently, only reprogrammed fibroblasts have been used to generate iPSCs from PD patients. Lentiviral [91] and retroviral [89] vectors have been widely used to generate human iPSCs that differentiated to DA neurons expressing TH. ESC-like properties were demonstrated in the transplanted cells by the expression of surface markers, gene expression profiles, and formation of embryoid bodies [92, 93]. Mouse iPSCs were differentiated to DA neurons as well as glial cells by retroviral transduction of the transcription factors. Injection of these neurons into the cerebral ventricles was followed by migration of the grafted cells into various brain regions. The cells communicated efficiently with the endogenous neurons by synaptic integration with generation of active action potentials. Moreover, high numbers of TH-positive neurons were expressed in PD rat models with complex morphologies. In addition, the neurons were positive for En-1, VMAT2, and DAT. Clinically; the tested rats showed marked behavior recovery 4 weeks after iPSCs transplantation [94]. Reprogrammed mouse somatic cells were studied without stable integration through the use of transient transfection or adenoviral infection. These methods were tried to deliver reprogramming factors. Nevertheless, lower efficiency of these methods was the cause of the nonsatisfactory results [95, 96]. However, new studies have been recommended to generate more safe iPSCs with less rate of tumor formation by using nonviral methods. Recent studies were accomplished to generate human iPSCs that are free of the reprogramming factors. These factor-free human iPSCs were capable to show pluripotent ESC-like properties and provide a more reliable source of cells [97]. These promising strategies have been shown to overcome the challenge of the residual transgene expression in virus-carrying human iPSCs that was proved to affect their molecular characteristics. In the mean time, several researches demonstrated the integration of the viruses into the genome that represents a major limitation that alters iPSCs differentiation and produce tumor overgrowth [98]. Moreover, other elements could induce degeneration of the reprogrammed cells such as unknown genetic factors generated in the PD patients. It could be concluded that advanced strategies are needed for the implementation of the use of iPSCs in cell replacement therapy to determine the safety issues necessary for long-term PD treatment.

4.2.5 Mesenchymal Stem Cells (MSCs)

MSCs represent an alternative source of autologous adult stem cells. They have the advantage of being highly accessible. They could be harvested from the patient's bone marrow, cord blood, or peripheral blood. In addition, MSCs were shown to have the ability of multipotency. They are capable of differentiation to osteogenic, adipogenic, and endothelial lineages. Furthermore, they could differentiate to hepatocyte-like cells, erythroid cells, and neurons [99]. Moreover, multiple studies reported evidences that MSCs could exhibit neuronal features, express several neural markers, and differentiate into DA neurons [100]. Electrophysiological functioning DA neurons were generated from human MSCs using SHH, basic FGF, and FGF8 [86]. MSCs expressed multiple specific neuronal markers and transcription factors when expanded on a large scale and induced to differentiate with a specific induction medium. NTFs are essential transcription factors, such as BDNF, NGF, and GDNF, that provoked significant results in neural survival, endogenous cell proliferation, and nerve fiber regeneration [101]. In addition, they demonstrated neuroprotective ability that could slow the rate of neural degeneration and stimulation of endogenous neural regeneration with immunoregulatory characteristics [102]. Interestingly, it was shown that MSCs could release soluble factors that promoted their immunosuppressive activity. Moreover, MSCs represented an anti-inflammatory action which enhanced their neuroprotective capacity inducing clinical improvement after transplantation. Furthermore, MSCs were demonstrated to produce neurotrophic factors that were responsible for the functional recovery more than the neuron replacement effects. Consequently, long-term functional improvement with the absence of side effects in animal models must be established prior to the clinical application of MSCs cell replacement therapy.

References

1. Dauer W, Przedborski S (2003) Effect Parkinson's disease: mechanisms and models. Neuron 39(6):889–909
2. Fahn S, Oakes D, Shoulson I et al (2004) Levodopa and the progression of Parkinson's disease. N Engl J Med 351(24):2498–2508
3. Limousin P, Martinez-Torres I (2008) Deep brain stimulation for Parkinson's disease. Neurotherapeutics 5(2):309–319
4. Eslamboli A, Georgievska B, Ridley RM et al (2005) Continuous low-level glial cell line-derived neurotrophic factor delivery using recombinant adeno-associated viral vectors provides neuroprotection and induces behavioral recovery in a primate model of Parkinson's disease. J Neurosci 25(4):769–777. doi:10.1523/JNEUROSCI.4421-04.2005
5. Enciu AM, Nicolescu MI, Manole CG et al (2011) Neuroregeneration in neurodegenerative disorders. BMC Neurol 11:75. doi:10.1186/1471-2377-11-75
6. Nutt JG, Burchiel KJ, Comella CL et al (2003) Randomized, double-blind trial of glial cell line-derived neurotrophic factor (GDNF) in PD. Neurology 60(1):69–73
7. Patel NK, Gill SS (2007) GDNF delivery for Parkinson's disease. Acta Neurochir Suppl 97(Pt 2):135–154. doi:10.1007/978-3-211-33081-4_16
8. Collier TJ, Elsworth JD, Taylor JR (1994) Peripheral nerve-dopamine neuron co-grafts in MPTP-treated monkeys: augmentation of tyrosine hydroxylase-positive fiber staining and dopamine content in host systems. Neuroscience 61(4):875–889
9. Xu B, Jiang CC, Zhang L et al (2004) Therapeutic study of autologous Schwann cells' bridge graft into the brain of hemiparkinsonian monkey. Zhonghua Yi Xue Za Zhi 84(4):318–322
10. Yu X, Bellamkonda RV (2003) Tissue-engineered scaffolds are effective alternatives to autografts for bridging peripheral nerve gaps. Tissue Eng 9(3):421–430

11. Lane EL, Handley OJ, Rosser AE et al (2008) Potential cellular and regenerative approaches for the treatment of Parkinson's disease. Neuropsychiatr Dis Treat 4(5):835–845
12. Bjorklund A, Dunnett SB, Brundin P et al (2003) Neural transplantation for the treatment of Parkinson's disease. Lancet Neurol 2(7):437–445
13. Olanow CW, Goetz CG, Kordower JH et al (2003) A double-blind controlled trial of bilateral fetal nigral transplantation in Parkinson's disease. Ann Neurol 54(3):403–414
14. Freed CR, Greene PE, Breeze RE et al (2001) Transplantation of embryonic dopamine neurons for severe Parkinson's disease. N Engl J Med 344(10):710–719
15. Winkler C, Kirik D, Bjorklund A (2005) Cell transplantation in Parkinson's disease: how can we make it work? Trends Neurosci 28(2):86–92
16. Politis M, Wu K, Loane C et al (2010) Serotonergic neurons mediate dyskinesia side effects in Parkinson's patients with neural transplants. Sci Transl Med 2(38):38–46
17. Galli R, Gritti A, Bonfanti L et al (2003) Neural stem cells: an overview. Circ Res 92:598–608
18. Lee E, Son H (2009) Adult hippocampal neurogenesis and related neurotrophic factors. BMB Rep 42(5):239–244
19. Höglinger GU, Rizk P, Muriel MP et al (2004) Dopamine depletion impairs precursor cell proliferation in Parkinson disease. Nat Neurosci 7:726–735
20. Thickbroom GW, Mastaglia FL (2009) Plasticity in neurological disorders and challenges for noninvasive brain stimulation (NBS). J Neuroeng Rehabil 6:4
21. Olanow CW, Kordower JH, Lang AE et al (2009) Dopaminergic transplantation for Parkinson's disease: current status and future prospects. Ann Neurol 66(5):591–596
22. Gross RE, Watts RL, Hauser RA et al (2011) Intrastriatal transplantation of microcarrier-bound human retinal pigment epithelial cells versus sham surgery in patients with advanced Parkinson's disease: a double-blind, randomised, controlled trial. Lancet Neurol 10(6):509–519
23. Gelain F, Bottai D, Vescovi A et al (2006) Designer self-assembling peptide nanofiber scaffolds for adult mouse neural stem cell 3-dimensional cultures. PLoS One 1(1):e119
24. Kim JH, Auerbach JM, Rodríguez-Gómez JA et al (2002) Dopamine neurons derived from embryonic stem cells function in an animal model of Parkinson's disease. Nature 418(6893): 50–56
25. Lee SH, Lumelsky N, Studer L et al (2000) Efficient generation of midbrain and hindbrain neurons from mouse embryonic stem cells. Nat Biotechnol 18(6):675–679
26. Zhao S, Nichols J, Smith AG (2004) SoxB transcription factors specify neuroectodermal lineage choice in ES cells. Mol Cell Neurosci 27(3):332–342
27. Prakash N, Brodski C, Naserke T et al (2006) A Wnt1-regulated genetic network controls the identity and fate of midbrain-dopaminergic progenitors in vivo. Development 133(1):89–98
28. Rodriguez-Gomez JA, Lu JQ, Velasco I et al (2007) Persistent dopamine functions of neurons derived from embryonic stem cells in a rodent model of Parkinson disease. Stem Cells 25(4):918–928
29. Smidt MP, Burbach JP (2007) How to make a mesodiencephalic dopaminergic neuron. Nat Rev Neurosci 8(1):21–32
30. Andersson E, Tryggvason U, Deng Q et al (2006) Identification of intrinsic determinants of midbrain dopamine neurons. Cell 124(2):393–405
31. Friling S, Andersson E, Thompson LH et al (2009) Efficient production of mesencephalic dopamine neurons by Lmx1a expression in embryonic stem cells. Proc Natl Acad Sci U S A 106(18):7613–7618
32. Kawasaki H, Mizuseki K, Nishikawa S et al (2000) Induction of midbrain dopaminergic neurons from ES cells by stromal cell-derived inducing activity. Neuron 28(1):31–40
33. Perrier AL, Tabar V, Barberi T et al (2004) From the cover: derivation of midbrain dopamine neurons from human embryonic stem cells. Proc Natl Acad Sci U S A 101(34):12543–12548
34. Park CH, Lee SH (2007) Efficient generation of dopamine neurons from human embryonic stem cells. Methods Mol Biol 407:311–322
35. Park CH, Minn YK, Lee JY et al (2005) In vitro and in vivo analyses of human embryonic stem cell-derived dopamine neurons. J Neurochem 92(5):1265–1276
36. Zeng X, Cai J, Chen J et al (2004) Dopaminergic differentiation of human embryonic stem cells. Stem Cells 22(6):925–940

37. Brederlau A, Correia AS, Anisimov SV et al (2006) Transplantation of human embryonic stem cell-derived cells to a rat model of Parkinson's disease: effect of in vitro differentiation on graft survival and teratoma formation. Stem Cells 24(6):1433–1440
38. Parish CL, Parisi S, Persico MG et al (2005) Cripto as a target for improving embryonic stem cell-based therapy in Parkinson's disease. Stem Cells 23(4):471–476
39. Bieberich E, Silva J, Wang G et al (2004) Selective apoptosis of pluripotent mouse and human stem cells by novel ceramide analogues prevents teratoma formation and enriches for neural precursors in ES cell-derived neural transplants. J Cell Biol 167(4):723–734
40. Pruszak J, Sonntag KC, Aung MH et al (2007) Markers and methods for cell sorting of human embryonic stem cell-derived neural cell populations. Stem Cells 25(9):2257–2268
41. Schmandt T, Meents E, Gossrau G et al (2005) High-purity lineage selection of embryonic stem cell-derived neurons. Stem Cells Dev 14(1):55–64
42. Roy NS, Cleren C, Singh SK et al (2006) Functional engraftment of human ES cell-derived dopaminergic neurons enriched by coculture with telomerase-immortalized midbrain astrocytes. Nat Med 12(11):1259–1268
43. Pankratz MT, Li XJ, Lavaute TM et al (2007) Directed neural differentiation of human embryonic stem cells via an obligated primitive anterior stage. Stem Cells 25(6):1511–1520
44. Beattie GM, Lopez AD, Bucay N et al (2005) Activin A maintains pluripotency of human embryonic stem cells in the absence of feeder layers. Stem Cells 23(4):489–495
45. Gerrard L, Rodgers L, Cui W (2005) Differentiation of human embryonic stem cells to neural lineages in adherent culture by blocking bone morphogenetic protein signaling. Stem Cells 23(9):1234–1241
46. Itsykson P, Ilouz N, Turetsky T et al (2005) Derivation of neural precursors from human embryonic stem cells in the presence of noggin. Mol Cell Neurosci 30(1):24–36
47. Lu J, Hou R, Booth CJ et al (2006) Defined culture conditions of human embryonic stem cells. Proc Natl Acad Sci U S A 103(15):5688–5693
48. Martin MJ, Muotri A, Gage F et al (2005) Human embryonic stem cells express an immunogenic nonhuman sialic acid. Nat Med 11(2):228–232
49. Iacovitti L, Donaldson AE, Marshall CE et al (2007) A protocol for the differentiation of human embryonic stem cells into dopaminergic neurons using only chemically defined human additives: studies in vitro and in vivo. Brain Res 1127(1):19–25
50. Cho MS, Lee YE, Kim JY et al (2008) Highly efficient and large-scale generation of functional dopamine neurons from human embryonic stem cells. Proc Natl Acad Sci U S A 105(9):3392–3397
51. Lee S, Jung J, Park S et al (2011) Histone deacetylase regulates high mobility group A2-targeting microRNAs in human cord blood-derived multipotent stem cell aging. Cell Mol Life Sci 68(2):325–336
52. Einstein O, Ben-Menachem-Tzidon O, Mizrachi-Kol R et al (2006) Survival of neural precursor cells in growth factor-poor environment: implications for transplantation in chronic disease. Glia 53(4):449–455
53. Lee JP, Jeyakumar M, Gonzalez R et al (2007) Stem cells act through multiple mechanisms to benefit mice with neurodegenerative metabolic disease. Nat Med 13(4):439–447
54. Su P, Loane C, Politis M (2011) The use of stem cells in the treatment of Parkinson's disease. Insci J 1(3):136–156
55. Redmond DE, Bjugstad KB, Teng YD et al (2007) Behavioral improvement in a primate Parkinson's model is associated with multiple homeostatic effects of human neural stem cells. Proc Natl Acad Sci U S A 104(29):12175–12180
56. Chung S, Shin BS, Hwang M et al (2006) Neural precursors derived from embryonic stem cells, but not those from fetal ventral mesencephalon, maintain the potential to differentiate into dopaminergic neurons after expansion in vitro. Stem Cells 24(6):1583–1593
57. Nishino H, Hida H, Takei N et al (2000) Mesencephalic neural stem (progenitor) cells develop to dopaminergic neurons more strongly in dopamine-depleted striatum than in intact striatum. Exp Neurol 164(1):209–214
58. Jonakait GM, Wen Y, Wan Y et al (2000) Macrophage cell-conditioned medium promotes cholinergic differentiation of undifferentiated progenitors and synergizes with nerve growth factor action in the developing basal forebrain. Exp Neurol 161(1):285–296

59. Liu W, Wang X, Lu G et al (2007) Dopaminergic regeneration by neurturin-overexpressing c17.2 neural stem cells in a rat model of Parkinson's disease. Mol Neurodegener 2:19. doi:10.1186/1750-1326-2-19
60. Ma DK, Bonaguidi MA, Ming G et al (2009) Adult neural stem cells in the mammalian central nervous system. Cell Res 19(6):672–682
61. Murrell W, Wetzig A, Donnellan M et al (2008) Olfactory mucosa is a potential source for autologous stem cell therapy for Parkinson's disease. Stem Cells 26(8):2183–2192
62. Apel C, Forlenza OV, de Paula VJ et al (2008) The neuroprotective effect of dental pulp cells in models of Alzheimer's and Parkinson's disease. J Neural Transm 116(1):71–78
63. Buzańska L, Jurga M, Stachowiak EK et al (2006) Neural stem-like cell line derived from a nonhematopoietic population of human umbilical cord blood. Stem Cells Dev 15(3):391–406
64. D'Ippolito G, Diabira S, Howard GA et al (2004) Marrow-isolated adult multilineage inducible (MIAMI) cells, a unique population of postnatal young and old human cells with extensive expansion and differentiation potential. J Cell Sci 117:2971–2981
65. Yang M, Stull ND, Berk MA et al (2002) Neuronal stem cells spontaneously express dopaminergic traits after transplantation into the intact or 6-hydroxydopamine-lesioned rat. Exp Neurol 177(1):50–60
66. Yang M, Donaldson AE, Jiang Y et al (2003) Factors influencing the differentiation of dopaminergic traits in transplanted neural stem cells. Cell Mol Neurobiol 23(4–5):851–864
67. Doering LC, Snyder EY (2000) Cholinergic expression by a neural stem cell line grafted to the adult medial septum/diagonal band complex. J Neurosci Res 61(6):597–604
68. Nunes I, Tovmasian LT, Silva RM et al (2003) Pitx3 is required for development of substantia nigra dopaminergic neurons. Proc Natl Acad Sci U S A 100(7):4245–4250
69. Simeone A (2005) Genetic control of dopaminergic neuron differentiation. Trends Neurosci 28(2):62–65
70. Maxwell SL, Ho HY, Kuehner E et al (2005) Pitx3 regulates tyrosine hydroxylase expression in the substantia nigra and identifies a subgroup of mesencephalic dopaminergic progenitor neurons during mouse development. Dev Biol 282(2):467–479
71. Kessaris N, Pringle N, Richardson WD (2008) Specification of CNS glia from neural stem cells in the embryonic neuroepithelium. Philos Trans R Soc Lond B Biol Sci 363(1489): 71–85
72. O'Keeffe FE, Scott SA, Tyers P et al (2008) Induction of A9 dopaminergic neurons from neural stem cells improves motor function in an animal model of Parkinson's disease. Brain 131 (Pt 3):630–641
73. Freeman MR (2006) Sculpting the nervous system: glial control of neuronal development. Curr Opin Neurobiol 16(1):119–125
74. Lindvall O, Bjrklund A (2004) Cell therapy in Parkinson's disease. NeuroRx 1(4):382–393
75. Sanberg PR (2007) Neural stem cells for Parkinson's disease: to protect and repair. Proc Natl Acad Sci U S A 104(29):11869–11870
76. Yasuhara T, Shingo T, Date I (2007) Glial cell line-derived neurotrophic factor (GDNF) therapy for Parkinson's disease. Acta Med Okayama 61(2):51–56
77. Cooper O, Isacson O (2004) Intrastriatal transforming growth factor alpha delivery to a model of Parkinson's disease induces proliferation and migration of endogenous adult neural progenitor cells without differentiation into dopaminergic neurons. J Neurosci 24(41):8924–8931
78. Mohapel P, Frielingsdorf H, Haggblad J et al (2005) Platelet-derived growth factor (PDGF-BB) and brain-derived neurotrophic factor (BDNF) induce striatal neurogenesis in adult rats with 6-hydroxydopamine lesions. Neuroscience 132(3):767–776
79. Jiao J, Chen DF (2008) Induction of neurogenesis in nonconventional neurogenic regions of the adult central nervous system by niche astrocyte produced signals. Stem Cells 26(5): 1221–1230
80. Madhavan L, Daley BF, Paumier KL et al (2009) Transplantation of subventricular zone neural precursors induces an endogenous precursor cell response in a rat model of Parkinson's disease. J Comp Neurol 515(1):102–115

81. Yasuhara T, Matsukawa N, Hara K et al (2006) Transplantation of human neural stem cells exerts neuroprotection in a rat model of Parkinson's disease. J Neurosci 26(48): 12497–12511
82. Ryu JK, Kim J, Cho SJ et al (2004) Proactive transplantation of human neural stem cells prevents degeneration of striatal neurons in a rat model of Huntington disease. Neurobiol Dis 16(1):68–77
83. Kobayashi T, Ahlenius H, Thored P et al (2006) Intracerebral infusion of glial cell line-derived neurotrophic factor promotes striatal neurogenesis after stroke in adult rats. Stroke 37(9):2361–2367
84. Robin AM, Zhang ZG, Wang L et al (2006) Stromal cell-derived factor 1alpha mediates neural progenitor cell motility after focal cerebral ischemia. J Cereb Blood Flow Metab 26(1):125–134
85. Shyu WC, Lin SZ, Yen PS et al (2008) Stromal cell-derived factor-1 alpha promotes neuroprotection, angiogenesis, and mobilization/homing of bone marrow-derived cells in stroke rats. J Pharmacol Exp Ther 324(2):834–849
86. Ericson C, Georgievska B, Lundberg C (2005) Ex vivo gene delivery of GDNF using primary astrocytes transduced with a lentiviral vector provides neuroprotection in a rat model of Parkinson's disease. Eur J Neurosci 22(11):2755–2764
87. Palma V, Lim DA, Dahmane N et al (2005) Sonic hedgehog controls stem cell behavior in the postnatal and adult brain. Development 132(2):335–344
88. Rolls A, Shechter R, London A et al (2007) Toll-like receptors modulate adult hippocampal neurogenesis. Nat Cell Biol 9(9):1081–1088
89. Takahashi K, Tanabe K, Ohnuki M et al (2007) Induction of pluripotent stem cells from adult human fibroblasts by defined factors. Cell 131(5):861–872
90. Park IH, Arora N, Huo H et al (2008) Disease-specific induced pluripotent stem cells. Cell 134(5):877–886
91. Yu J, Vodyanik MA, Smuga-Otto K et al (2007) Induced pluripotent stem cell lines derived from human somatic cells. Science 318(5858):1917–1920
92. Dimos JT, Rodolfa KT, Niakan KK et al (2008) Induced pluripotent stem cells generated from patients with ALS can be differentiated into motor neurons. Science 321(5893): 1218–1221
93. Ebert AD, Yu J, Rose FF Jr et al (2009) Induced pluripotent stem cells from a spinal muscular atrophy patient. Nature 457(7227):277–280
94. Wernig M, Zhao JP, Pruszak J et al (2008) Neurons derived from reprogrammed fibroblasts functionally integrate into the fetal brain and improve symptoms of rats with Parkinson's disease. Proc Natl Acad Sci U S A 105(15):5856–5861
95. Okita K, Nakagawa M, Hyenjong H et al (2008) Generation of mouse induced pluripotent stem cells without viral vectors. Science 322(5903):949–953
96. Stadtfeld M, Nagaya M, Utikal J et al (2008) Induced pluripotent stem cells generated without viral integration. Science 322(5903):945–949
97. Soldner F, Hockemeyer D, Beard C et al (2009) Parkinson's disease patient derived induced pluripotent stem cells free of viral reprogramming factors. Cell 136(5):964–977
98. Markoulaki S, Hanna J, Beard C et al (2009) Transgenic mice with defined combinations of drug-inducible reprogramming factors. Nat Biotechnol 27(2):169–171
99. Jiang C (2009) Stem cell research: from molecular physiology to therapeutic applications. Sci China C Life Sci 52(7):597–598
100. Woodbury D, Schwarz EJ, Prockop DJ (2000) Adult rat and human bone marrow stromal cells differentiate into neurons. J Neurosci Res 61(4):364–370
101. Mahmood A, Lu D, Chopp M (2004) Marrow stromal cell transplantation after traumatic brain injury promotes cellular proliferation within the brain. Neurosurgery 55(5): 1185–1193
102. Barry FP, Murphy JM (2004) Mesenchymal stem cells: clinical applications and biological characterization. Int J Biochem Cell Biol 36(4):568–584

Chapter 5
Hair Follicle: A Novel Source of Stem Cells for Cell and Gene Therapy

Panagiotis Mistriotis and Stelios T. Andreadis

Abstract The adult body harbors powerful reservoirs of stem cells that enable tissue regeneration under homeostatic conditions or in response to disease or injury. The hair follicle is a readily accessible mini organ within the skin and contains stem cells from diverse developmental origins that are shown to have surprisingly broad differentiation potential. In this chapter, we discuss the biology of the hair follicle with particular emphasis on the various stem cell populations residing within the tissue. We summarize the existing knowledge on putative hair follicle stem cell markers, the differentiation potential, and technologies to isolate and expand distinct stem cell populations. We also discuss the potential of hair follicle stem cells for drug and gene delivery, tissue engineering, and regenerative medicine. We propose that the abundance of stem cells with broad differentiation potential and the ease of accessibility make the hair follicle an ideal source of stem cells for gene and cell therapies.

Keywords Hair follicle • Stem cells • Tissue engineering • Regenerative medicine Gene therapy • Drug delivery • Reprogramming

---

P. Mistriotis
Bioengineering Laboratory, Department of Chemical and Biological Engineering,
University at Buffalo, The State University of New York,
908 Furnas Hall, Amherst, NY 14260-4200, USA

S.T. Andreadis, PhD (✉)
Department of Chemical and Biological Engineering,
Department of Biomedical Engineering, and Center of Excellence in Bioinformatics
and Life Sciences, University at Buffalo, The State University of New York,
908 Furnas Hall, Amherst, NY 14260-4200, USA
e-mail: sandread@buffalo.edu

M.K. Danquah, R.I. Mahato (eds.), *Emerging Trends in Cell and Gene Therapy*,
DOI 10.1007/978-1-62703-417-3_5,

5.1 Introduction

The hair follicle (HF) is a dynamic "mini" organ supporting important biological functions of the body. HFs protect against cold and potential injuries; they also have an important sensory and immunologic functions in addition to affecting the social behavior of a person [1, 2].

HFs are easily accessible and contain stem cells from diverse developmental origins that continuously self-renew, differentiate, regulate hair growth, and contribute to skin homeostasis. Hair follicle stem cells have been shown to be highly proliferative in vitro and multipotent [3–5] that allows engineering a variety of different tissues for organ replacement and regenerative medicine. In addition, genetic engineering of the hair follicle stem cells in vivo has shown promising results, suggesting that treatment of genetic diseases of skin and hair via the hair follicle may be feasible. This chapter summarizes the existing literature regarding the differentiation potential of hair follicle stem cells, their putative markers, the common isolation methods, and their application in cell and gene therapies.

5.2 Hair Follicle Biology

HF is part of the pilosebaceous unit that contains the sebaceous gland, the apocrine gland, and the arrector pili muscle. The HF is composed of two main compartments: the upper part includes the infundibulum and the isthmus, whereas the bulb, matrix, and dermal papilla comprise the lower part. The exact position of the lower part varies during hair cycling. Adjacent to the lower portion of the infundibulum lies the sebaceous gland that waterproofs the skin by secreting sebum. The bulge is a part of the isthmus that is believed to be the stem cell reservoir, which regenerates the HF during hair growth. Cells migrate from the bulge toward the bulb, where they proliferate and differentiate in order to produce the hair shaft and all the epithelial cells that constitute the HF [2]. Finally, the inner and outer root sheaths are composed mainly of keratinocytes surrounding the hair shaft (Fig. 5.1).

HF undergo numerous cycles of growth and retraction throughout life. This dynamic process in adult life has three distinct phases, that is, anagen, catagen, and telogen, each regulated by different signals. Anagen is the growing phase. Stem cells that are located in the bulge region differentiate to all hair lineages, resulting to hair elongation. The duration of anagen in human body varies depending on the anatomic location of the follicle. On the scalp, anagen may last as long as 8 years resulting in long hair, but in other places such as the eyebrow, anagen maybe as short as 3 months. Catagen is the regression phase. At this stage the majority of the HF cells undergo apoptosis, causing reduction of the lower compartment that brings the dermal papilla cells close to the bulge cells. Exchange of signals between the papilla and the bulge regulates the duration of catagen. The cells that escape apoptosis during this phase comprise the reservoir that leads to the next anagen. Telogen is the last phase of the HF cycle, also known as resting phase. In telogen, cells enter

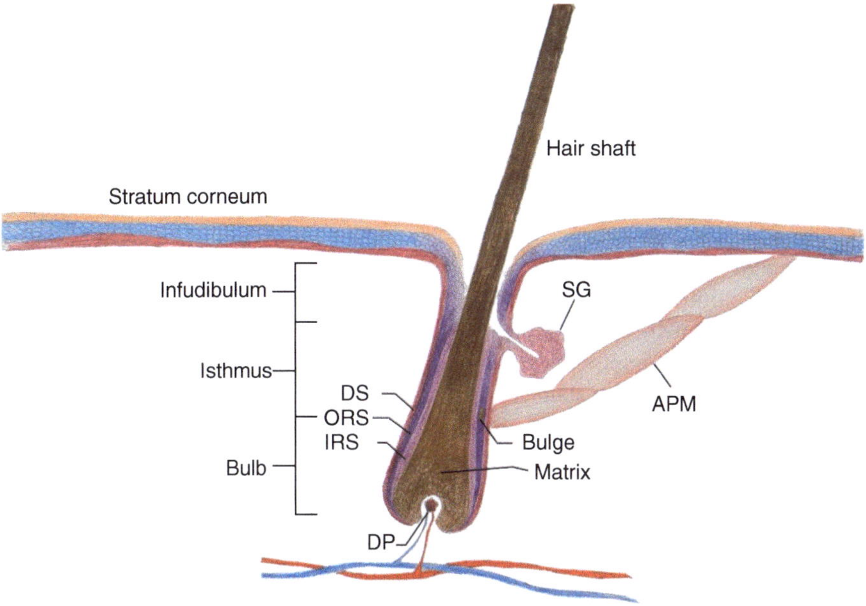

Fig. 5.1 Schematic drawing of hair follicle. *DS* dermal sheath, *ORS* outer root sheath, *IRS* inner root sheath, *DP* dermal papilla, *SG* sebaceous gland, *APM* arrector pili muscle. The illustration is not drawn in scale

a quiescent state waiting for the necessary signals to restart the cycle via the anagen. It is argued that 5–15 % of HF in the scalp remain in telogen [1, 2].

5.3 Location and Differentiation Potential of Hair Follicle Stem Cells

Stem cells can be characterized by three unique properties: self-renewal, capacity to differentiate into one or multiple cell types, and the ability to form tissues in vivo. Based on their differentiation potential, they can be categorized as totipotent, pluripotent, multipotent, and unipotent. Totipotent cells can reproduce all the cells of a living organism including the extraembryonic tissues. Pluripotent cells differ from the totipotent cells in their inability to form the extraembryonic tissues necessary for proper growth of the embryo. Multipotent stem cells have even more restricted differentiation potential, but they can still differentiate into more than one cell type. The lowest in the hierarchy are the unipotent stem cells or progenitor cells, which can generate only one cell type. Stem cells can also be classified into embryonic or adult according to their origin [6]. Although embryonic stem cells have broader differentiation potential, adult stem cells can be isolated from the patient, directly overcoming possible immune rejection after transplantation.

Adult stem cells in vivo reside in multiple tissues usually in a well-protected microenvironment called niche. Examples of stem cell populations that are surrounded by niche include the intestinal stem cells, the neural stem cells, and the HF stem cells. This way, the body holds a powerful reservoir of cells that can readily respond in case of emergency, such as an injury. Some adult stem cells like mesenchymal stem cells (MSC) have multilineage differentiation potential so that a single cell can contribute to the regeneration of multiple tissues such as fat, bone, cartilage, and muscle [7].

Although in vitro stem cells proliferate markedly in response to appropriate signals in the culture media, in vivo they remain quiescent until they are coaxed to proliferate and/or differentiate only when needed, for example, in case of injury. Under homeostatic conditions, the stem cell pool is maintained through asymmetric division, where the parent stem cell divides into two cells with varying differentiation potential: one retaining the stem cell characteristics (self-renewal) and the other assuming a more differentiated phenotype (differentiation). However, expansion of the stem cell pool would require symmetric division, where two stem cells are generated from a parental stem cell [8].

Similar to other organs, the HF contains a rich stem cell pool that resides in different anatomic locations within the HF. As a result, some scientists call the HF as a stem cell "zoo" [9]. In the next chapter, we will present the different stem cell populations, and we will elaborate on their broad differentiation potential.

5.3.1 *Bulge and Hair Germ*

Due to the complex architecture of the HF, the location of the stem cell reservoir remained elusive for many years. Initial studies reasoned that stem cells resided in the bulb [10], but this hypothesis was abandoned, as removal of the bulb did not inhibit the generation of new hair follicles [11]. In the early 1990s, Cotsarelis and colleagues were the first to propose that stem cells reside in the bulge area of HF. They took advantage of the fact that in vivo stem cells cycle very slowly so that long time after administration of tritiated thymidine, only the cells that retain the label (label-retaining cells) are the slow-cycling stem cells [12, 13]. Several years later, this finding was verified using transgenic mice that were engineered to express the fusion protein histone H2B-GFP under the keratin-5 promoter in a tetracycline-regulatable manner. As a result skin cells expressed GFP except when the mice were fed doxycycline, which suppressed GFP expression. The fast-cycling cells lost the GFP, whereas the slow-cycling stem cells retained it. These label-retaining cells were localized in the bulge region of hair follicles [14]. Furthermore, tracing studies with transgenic mice expressing the LacZ transgene under the control of either keratin-15 or Lgr5 promoter further supported the bulge activation hypothesis, which states that during anagen, stem cells from the bulge migrate in the bulb region where they are induced to proliferate and differentiate to all epithelial cell types of the HF [15, 16].

Notably, transplantation of keratin-15+ or Lgr5+ cells along with dermal fibroblasts in the dermis of nude mice generated new HF with high efficiency [15, 16]. Interestingly, damage of the bulge from autoimmune disease lichen planopilaris resulted in permanent hair loss [17], further highlighting the importance of bulge-derived stem cells for hair regeneration. In addition to hair regeneration, bulge stem cells were found to contribute to wound healing following skin injury by migrating and differentiating to epidermal keratinocytes [18, 19]. However, they are not necessary for the maintenance of the epidermis (ablation of the cells does not affect the homeostasis of the epidermis), and in the long run, they fail to stay at the sites of injury [19]. Additional studies have shown robust multipotency of bulge stem cells in vivo, where they were found to participate in angiogenesis, and in vitro where they were coaxed to differentiate into neurons, glial cells, melanocytes, keratinocytes, and mesenchymal cells [20–24].

Although it is widely accepted that the bulge harbors stem cells, the exact stem cell population is still under debate. Jaks and colleagues challenged the notion of label-retaining cells as the true stem cell population in HF, as $Lgr5^+$ cells can regenerate the whole follicle but do not coincide with the label-retaining cells of the bulge. The same study reported that $Lgr5^+$ cells were found in the hair germ, a region between the dermal papilla and the bulge, which remains discrete during telogen but overlaps with the matrix during anagen [16]. Others believe that the hair germ originates from the bulge and contributes to the generation of the new HF in the beginning of anagen [25]. In agreement, Greco and colleagues showed that the transcriptional profile of hair germ cells resembles that of bulge cells. They also found that hair germ cells proliferate faster than bulge cells and respond first to the dermal papilla signals at the late telogen. However, they also lose their proliferative capacity faster than bulge cells during long-term expansion in vitro [26].

5.3.2 *Isthmus/Infundibulum*

Cells located above the bulge are believed to retain multipotent properties. Studies have reported that they can differentiate not only into the epithelial lineages of the HF but also into the sebaceous gland and the epidermis. However, it is yet not known whether these cells represent a unique stem cell population, or a subset of bulge stem cells, or even progenitors with limited differentiation capacity.

Isolated cells from the area between the bulge and the sebaceous gland were found to be distinct from the bulge-derived stem cells since they did not express bulge-specific markers such as keratin-15 and CD34. Although they maintained their high clonogenic potential in vitro, they were also actively proliferating in vivo – in contrast to the notion that in vivo stem cells are the slow-cycling, label-retaining cells [27]. Similarly Jensen and colleagues reported that cells isolated from the upper bulge region and were not quiescent in vivo could generate new follicles after implantation, suggesting that stem cells need not be slow-cycling cells in vivo in order to be multipotent [28].

Although, during homeostasis, bulge-derived cells do not contribute to the generation of epidermis [15, 16, 19], several studies showed that cells derived from a region above the bulge can give rise to epidermis and persist there for a long time following injury [28–30].

5.3.3 *Sebaceous Gland*

There are two theories with regard to the origins of the sebaceous gland. The first asserts that stem cells residing in the bulge region migrate and give rise to resident gland cells. This theory is supported by transplantation studies showing that bulge cells generated functional sebaceous gland in vivo [15, 16]. The second theory suggests that stem cells located above the bulge differentiate into sebocytes [29–31]. Horsley and colleagues identified a unique cell population in the region of sebaceous gland that expresses the transcription factor Blimp1 and has unipotent differentiation potential into sebocytes. Loss of Blimp1 in HF resulted in activation of bulge cells, which may suggest a possible connection between bulge and sebaceous gland. The same study also showed that implanted bulge stem cells could give rise into Blimp1^{+} cells [31].

5.3.4 *Dermal Papilla and Dermal Sheath*

Dermal papilla (DP) and dermal sheath (DS) are cell populations within the HF that are believed to contain stem cells. Whereas bulge cells originate from ectoderm, DP and DS cells are derived from mesoderm, and they are known to regulate hair cycling by exchanging signals with the bulge [2]. Multiple studies showed that DP and DS cells have broad differentiation potential. In a pioneering study, Lako and colleagues demonstrated that DP and DS cells could reconstitute multiple lineages of the hematopoietic system in lethally irradiated mice [32]. Rat and human HF-derived DP and DS cells could also be induced to differentiate toward the myogenic, osteogenic, chondrogenic, and adipogenic lineage resembling bone marrow mesenchymal stem cells [3–5, 33, 34]. A recent study showed that DP/DS stem cells are the precursors of dermal stem cells and contribute to dermal maintenance and wound healing [35].

5.4 Putative Hair Follicle Stem Cell Markers

The majority of the studies in HF have been conducted in murine models. However, there are several differences that have to be taken into account between human and murine models, and conclusions derived from experiments with mice models do not

Table 5.1 Common stem cell markers and their location within hair follicle

Species	Marker	Location	References
Mice	CD34	Bulge	[36]
	TCF3	Bulge	[37, 38]
	NFATC1	Bulge	[39]
	Nestin	Bulge	[20]
	Label-retaining cells	Bulge	[12]
	K15	Bulge, hair germ	[15]
	Lgr5	Bulge, hair germ	[16]
	Lhx2	Bulge, hair germ, early hair progenitors	[40]
	Sox-9	Bulge, early hair progenitors	[41]
	MTS24	Upper bulge	[27]
	a6LowCD34–Sca-1–	Upper bulge	[28]
	Lrig1	Upper bulge	[42]
	Lgr6	Upper bulge	[30]
	Blimp 1	Upper bulge	[31]
	Sox-2	Dermal papilla, dermal sheath	[35, 43]
	Versican	Dermal papilla	[44]
	Alkalinephosphatase	Dermal papilla, hair germ	[45]
	Nexin	Dermal papilla	[46]
	CD133	Dermal papilla	[47]
Human	Mesenchymal stem cell markers	Dermal sheath	[4]
	CK15	Bulge/isthmus	[48–50]
	CD 200	Bulge/isthmus	[48–50]
	CK19	Bulge/isthmus	[48–50]

necessarily apply in human HF cells. Whereas humans have only two types of hair (vellus and heavily pigmented hairs), mice are endowed with several distinct hair types (pelage, vibrissae, cilia, hairs on the tail, ear, genital, perianal area, nipples, and around the feet). In addition, the biological cycles of human and mouse HF are different; while human HF cycle independently after birth, mouse HF cycle in synchrony [2]. Finally, the biological markers characterizing the stem cell populations in human and mouse are strikingly different. Table 5.1 summarizes the most common markers of HF stem cells based on the species they are derived from and the location where they are expressed.

5.4.1 Murine Hair Follicles

5.4.1.1 Bulge

Several markers have been proposed to characterize murine bulge stem cells. In addition to keratin-15 and Lgr5, CD34 is co-expressed with keratin-15 and has also been proposed as a potential stem cell marker of the bulge. CD34+ cells are relatively quiescent and have higher clonogenic potential in vitro as compared to CD34⁻ cells [36]. Several transcription factors have been identified in the bulge region including Tcf3, Sox-9, Lhx2, and NFATc1. Tcf3 was shown to maintain the undifferentiated cell state by repressing numerous genes that induce sebaceous gland and HF differentiation [38]. Another key transcription factor that is expressed in the bulge area is Sox-9. Sox-9$^+$ cells are first detected during the formation of hair placode, the precursor of HF during prenatal life. The cells co-localize with early label-retaining cells, which subsequently give rise to bulge stem cells. Notably, deletion of Sox-9 decreased the proliferation of bulge stem cells, impaired the generation of proliferative matrix cells, and resulted in inhibition of HF morphogenesis [41, 51].

Similar to Sox-9, Lim-homeodomain transcription factor, Lhx2 is also expressed during hair placode formation as was seen by microarray analysis in the P-cadherin$^+$ cells that mark early hair progenitors. In postnatal life, Lhx2 is expressed in the bulge and suppresses differentiation, prompting some investigators to hypothesize that it may be required for stem cell maintenance [40]. However, a recent study challenged this notion and reported that Lhx2 is required for the induction of anagen and not for the maintenance of stem cells [52]. The fourth bulge-specific transcription factor is NFATc1, which is regulated by the intracellular levels of calcium. Under high calcium conditions, NFATc1 is dephosphorylated and translocates to the nucleus, where it downregulates cyclin-dependent kinase 4 and suppresses proliferation in bulge region. As a result, downregulation of NFATc1 leads to activation of bulge-derived stem cells [39]. Interestingly, NFATc1-expressing cells coincide only partially with CD34$^+$, Tcf3$^+$, Lhx2$^+$, and Sox-9$^+$ cells in the bulge region, suggesting there is no unique marker of bulge stem cells but rather a group of transcription factors that regulate stem cell maintenance and activation through a series of complex and dynamic interactions.

Finally, other studies provided evidence that nestin is expressed in the mouse bulge stem cells. Transgenic mice expressing GFP under the nestin promoter showed that nestin-positive cells are located in the bulge region during telogen but in the upper two thirds of the outer root sheath during anagen. In vivo these cells participated in the formation of new blood vessels, and in vitro they could be coaxed to differentiate into neurons, glial cells, smooth muscle cells, melanocytes, and keratinocytes, demonstrating the multipotency of hair follicle stem cells [20, 21, 53].

5.4.1.2 Upper Bulge

Several markers have been identified over the years that target putative murine stem cells in the upper bulge region. Lgr6, an orphan G protein-coupled receptor, is expressed in the region immediately above the bulge. Lgr6$^+$ cells were shown to play a critical role in the formation of HF, sebaceous gland, and epidermis during development [30]. MTS24, a cell surface glycoprotein, also marked potential stem cells in a region above the bulge. MTS24$^+$ cells exhibited increased colony-forming capacity as compared to MTS24$^-$ cells and showed similar gene expression profile with CD34$^+$ bulge cells [27]. However, the differentiation potential of these cells was not examined. In addition, cells residing in the upper isthmus were shown to be multipotent as they could form HF, sebaceous gland, and epidermis after implantation. These cells expressed low levels of integrin α6, were negative for the hematopoietic markers CD34 and Sca-1, and exhibited distinct gene expression profile as compared to bulge cells [28]. Finally, another putative stem cell marker characterizing the region right above the bulge is transmembrane protein leucine-rich repeats and immunoglobulin-like domain protein 1 or Lrig1. Lrig1 was shown to regulate epidermal growth factor signaling by promoting the degradation of epidermal growth factor receptor [54] and to keep cells in this region in a quiescent state [42]. Indeed, in vivo Lrig1$^+$ cells appeared to be quiescent and multipotent, two of the main attributes of stem cells [29].

5.4.1.3 Dermal Papilla and Dermal Sheath

The DP and DS are known to induce HF generation by interacting with epidermal stem cells [55]. In 1999, Kishimoto et al. reported that cells in DP express the proteoglycan versican, which is usually present in the condensed mesenchyme. The same group employed the versican promoter to express either LacZ or GFP and found that when implanted on the back of nude mice along with keratinocytes, the versican$^+$ cells could reconstitute the HF but versican$^-$ cells could not [44]. Others observed that nexin-1, a protease inhibitor, was highly expressed in DP during anagen and that the nexin-1 expression level correlated with the rate of hair growth [46]. Similarly, the expression of alkaline phosphatase – an enzyme expressed in bone cells and embryonic stem cells – correlated with hair growth and was also highly expressed in DP during anagen, suggesting a positive correlation between hair induction and alkaline phosphatase activity [45]. Finally, CD133 was expressed in DP cells during HF development, but its expression was greatly diminished after birth. Nevertheless, when co-implanted with embryonic epithelial cells, CD133+ cells enabled generation of HF in vivo [47]. Interestingly, a subpopulation of CD133+ Sox2+ cells within the DP was shown to be essential for the formation of particular types of hair such as awl/auchene follicles [43]. Rendl and colleagues compared the transcriptional profile of five distinct cell populations within the HF, namely, melanocytes, dermal papilla, matrix, outer root sheath, and dermal fibroblasts. This approach successfully identified several genes and signaling pathways that were unique to each population and need to be further explored in the future [56].

5.4.2 Human Hair Follicles

Murine HFs have been largely explored with respect to stem cell markers; however, human HF have remained unexplored. In contrast to murine bulge, the human bulge cannot be identified as a distinct anatomic projection, rendering isolation of bulge cells very challenging. Screening a number of markers in vivo, Kloepper and colleagues identified CD200 and keratin-15 and keratin-19 as putative bulge stem cell markers, although their location is not restricted to the bulge but extends to a wider area of isthmus as well. In contrast to the mouse, human bulge does not express CD34, nestin, or Lhx2 [48]. In a more recent study keratin-15^{high}/$CD200^{+}$/$CD34^{-}$/$CD271^{-}$ bulge-derived cells showed increased clonogenic potential as compared to keratin-15^{low}/$CD200^{+}$/$CD34^{-}$/$CD271^{-}$ cells [50]. In agreement, CD200-expressing cells that were isolated from a population of label-retaining cells using laser capture microdissection showed increased clonogenic potential in vitro [49]. However, multipotency of CD200+ cells has not been examined. More recently our laboratory reported that DP/DS cells display a cell surface profile characteristic of mesenchymal stem cells being positive for CD90, CD44, CD49b, CD105, and CD73 [4, 5]. In addition, these cells are clonally multipotent as they can differentiate in fat, bone, cartilage, and smooth muscle with high efficiency [5].

5.5 Methods for Isolating Hair Follicle Stem Cells

Three techniques have been routinely used for the isolation of putative stem cells from the HF: microdissection, enzymatic digestion, and fluorescence-activated cell sorting (FACS). In the following, we describe each technique and elaborate on their advantages and disadvantages.

5.5.1 Microdissection

Microdissection is a technique that has been commonly applied for the isolation of cells from DP [11, 32, 33, 57, 58] as well as the bulge [49, 59]. This technique requires the use of fine forceps and blades for the isolation of the area of interest. Subsequently the isolated areas are transferred into tissue culture plates, where the cells migrate out of the tissue and proliferate in the presence of appropriate culture medium.

For DP cell isolation, application of pressure on the suprabulbar region by forceps was shown to compress the bulb and facilitate removal of the connective tissue sheath surrounding the DP, which is subsequently detached from the epithelium using a scalpel blade [60]. Finally, a highly reliable technique that has been used for isolating human bulge cells is laser capture microdissection [49]. A thermolabile membrane is placed on top of the sample, and the area of interest is targeted by laser, which melts the membrane locally marking the cells that are subsequently separated [61]. The major advantage of microdissection is that this approach preserves the whole tissue, thereby increasing the efficiency of cell isolation. However, this technique is quite laborious and requires experienced technicians.

5.5.2 *Enzymatic Digestion*

Another approach that has been employed for isolation of HF stem cells involves enzymatic digestion of the follicle from the surrounding dermis, usually with dispase or collagenase. The incubation time and concentration of enzymes used vary depending on the amount of extracellular matrix present around the follicle. Generally collagenase treatment requires few hours of incubation at 37 °C whereas dispase needs overnight treatment [3–5, 23, 49]. Others use a combination of enzymes to isolate DP cells. Specifically, dispase was employed initially to remove the follicle from the cutaneous fat, followed by collagenase D to digest the dermal sheath and isolate the DP. The remaining dermal sheath fibroblasts could be removed by low-speed centrifugation of the DP [62]. Enzyme digestion is a simple method of HF stem cells isolation but with little control over the type of cells that are obtained leading to possible variations between different isolations.

5.5.3 *Fluorescence-Activated Cell Sorting*

FACS is a common method for isolating stem cells, especially from murine HF [15, 16, 20, 28, 30, 43, 44, 47]. Fluorescently labeled antibodies are used to tag the cell surface, and cells are sorted based on fluorescence intensity, which is proportional to the expression level of the particular target receptor. FACS can also be applied for isolating cells based on markers that are not expressed on the cell surface. Our group made use of the smooth muscle alpha-actin (αSMA) promoter-driven GFP to isolate a homogeneous population of smooth muscle cells (SMC) from ovine and human HF-MSC [33, 34]. FACS yields highly purified cell populations that can be further expanded or directly analyzed for mRNA or protein expression. Regrettably, lack of reliable stem cell markers hampers use of this method in sorting human HF stem cells.

5.6 Hair Follicle Stem Cells for Tissue Engineering and Cell Therapy

5.6.1 *Tissue-Engineered Vascular Grafts*

Cardiovascular disease is the leading cause of death in USA as being reported by American Heart Association. In 2006 heart diseases accounted for more than 600,000 deaths. Almost half of the deaths were caused by coronary heart diseases, and 400,000 surgical bypass operations were performed highlighting the importance of an artificial arterial substitute (www.americanheart.org). A functional

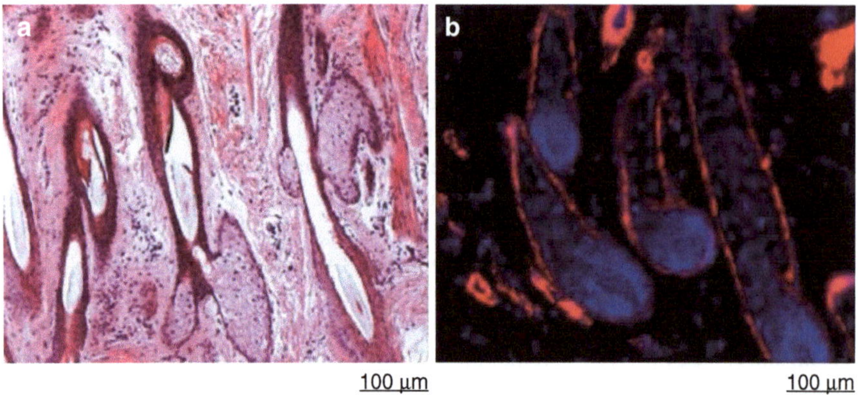

Fig. 5.2 Cells comprising the dermal sheath of hair follicle are positive for αSMA. (**a**) H&E staining from neonatal ovine dermis. (**b**) Immunohistochemistry showing αSMA^+ cells in the dermal sheath of hair follicles (Image taken from Peng et al. [63])

arterial graft should contain both endothelial cells (ECs) and SMCs. ECs line the lumen of a vessel, endow it with thromboresistant properties, and are selectively permeable to substances circulating in the blood. SMCs form the medial layer of an artery and are mainly responsible for the dilatation and constriction of the vascular wall in response to vasoactive agonists.

Our laboratory showed that DS cells of ovine and human HF stained positive for αSMA, a marker of SMC (Fig. 5.2). This finding prompted us to hypothesize that functional SMC can be derived from HF. To this end, HF were transduced with a lentivirus encoding for GFP under the control of the αSMA promoter, and GFP+ cells were sorted out using flow cytometry. We found that both ovine and human HF-derived SMC exhibited significantly higher proliferation and clonogenic potential compared to vascular SMC. In addition, tissue-engineered vascular grafts prepared from HF-derived SMCs displayed high reactivity in response to vasoactive agonists and generated significant mechanical force as shown by compaction of fibrin hydrogels [3–5, 63]. More recent studies in our laboratory showed that these vascular grafts could be implanted into the arterial circulation of an ovine animal model where they remained patent for at least 3 months [64] (Row S. et al., 2013, manuscript in preparation), suggesting that the HF may be a readily accessible source of stem cells for cardiovascular tissue regeneration and cell therapies.

5.6.2 *Tissue Engineering of Cartilage, Bone, and Fat*

In addition to myogenic differentiation, rodent DP/DS cells have the capacity to differentiate into the osteogenic, chondrogenic, and adipogenic lineage, similar to bone marrow-derived MSCs [33, 34]. Extending these studies, we demonstrated that human HF cells also possess multilineage differentiation potential [4]. We also showed that single clones give rise to all four lineages, strongly indicating that human HF-MSC represent a true stem cell population and not a mixed population of progenitors with uni-lineage differentiation potential [5]. These results suggest that human HF can be an easily accessible source of true MSC that could be employed for regeneration of bone and cartilage for the replacement of joints or for meniscus repair.

5.6.3 Skin Regeneration

Several studies suggested that HF cells migrate to the epidermis during homeostasis and to a larger extent following skin injury [18, 19, 30, 35, 65], suggesting that HF cells could be used to generate the epidermis and enhance wound healing. Indeed, Hoeller and colleagues reported generation of bioengineered skin by introducing fibroblasts and HF tissue into the dermis. Interestingly, epidermal keratinocytes migrated out of the hair follicle and developed multiple layers of epidermis and stratum corneum [66]. In addition, HF-derived melanocytes have been used to develop a pigmented skin equivalent [67]. Most importantly, transplantation of tissue-engineered skin from HF-derived stem cells was shown to enhance healing of ulcers and burns significantly [68–70]. Notably, when hair buds were introduced into bioengineered skin before implantation, they sped up and guided nerve regeneration, suggesting that HF may recover the lost sense of touch [71].

5.6.4 Nerve Regeneration

Mouse HF-derived nestin^{+}/K15^{-} stem cells have the capacity to differentiate into neurons in vitro, suggesting a possible application to nerve regeneration in a variety of central and peripheral nervous system diseases [22]. Indeed, Amoh and colleagues transplanted mouse HF nestin + stem cells into a severed sciatic nerve or spinal cord, where they differentiated into Schwann cells and promoted nerve regeneration [53, 72]. The same group also reported that human HF stem cells have the capacity to restore the function of injured nerves [73, 74]. HF-derived neuronal and Schwann cells have also been introduced into acellular sciatic nerve conduit, where they exhibited long-term survival and significant electrophysiological properties in vitro but failed to induce repeated potentials [75].

5.6.5 Engineering Functional Hair Follicle

An important application of HF stem cells is bioengineering of HF to restore abnormal hair loss (alopecia). Common forms of alopecias include (a) the androgenetic alopecia which results from the miniaturization of the hair; (b) the alopecia areata, which results from an autoimmune response that damages the hair follicle; and (c) permanent alopecia which can be caused, for example, from severe trauma [1].

Bioengineering a HF has been a topic of intense scientific research over many years. To date two strategies have been developed to achieve this goal. The first approach includes the transplantation of intact HF from a HF-rich area into the bald area. This technique requires initially the surgical excision of a thin strip of scalp that contains dense HF and subsequently the isolation of the individual follicles and implantation back to the bald scalp [76]. Although transplantation of whole follicles is considered as the gold standard for hair restoration, studies demonstrated that segments of the HF can also induce hair growth after transplantation [57, 77–82]. Transplantation of a truncated human HF after amputating the bulb has shown hair renewal suggesting bulb reformation possibly from the DS compartment [77, 78, 82]. Interestingly, transplantation of intact DP and/or DS into murine models demonstrated mesenchymal interaction with the host epithelium and subsequent hair induction as shown with the transplantation of both murine [57, 79] and human dermal compartments [81]. However, in contrast to human DS when human DP was transplanted into human skin, it failed to induce hair regeneration [80].

In severe cases of alopecias, the number of available HF is not sufficient to restore the bald site. On the other hand, HF stem cells can be expanded in culture into large numbers that may be sufficient to cover the whole area and result in hair restoration. Jahoda and colleagues were the first to report that implantation of DP cells resulted in the hair growth in mice [55]. Although the hair-inductive properties of DP cells were lost after long-term expansion in vitro, coculture with keratinocytes or in keratinocyte-conditioned medium could maintain the inductive properties of DP cells for almost 70 passages [83]. Similar to DP, DS cells were also found to induce HF growth [84]. Finally, HF restoration was enhanced by the mixture of bulge/hair germ stem cells from adult HF with neonatal dermal cells [15, 16, 29, 30, 85]. Notably, when mixed with embryonic mouse dermal and epidermal cells, mouse bone marrow-derived cells differentiated into HF cells, suggesting hair-inductive properties of bone marrow cells [86]. Although the results with mouse models are very encouraging, the significance of these findings in large animal models or humans has yet to be demonstrated.

5.6.6 Drug Delivery Through the Hair Follicle

Skin is an easily accessible organ that has been widely considered as a unique target for drug delivery. In contrast to the conventional delivery methods (oral, injections), the transdermal route allows drug administration to the circulation through the dermal vasculature and may increase drug bioavailability while avoiding painful injections. However, the presence of stratum corneum, the outermost layer of the skin, severely limits the penetration of hydrophilic and high molecular weight substances [87]. To bypass this drawback, microscale devices have been developed to enable transdermal delivery including liquid jet injectors, microneedles, and thermal ablation devices [87].

Alternatively scientists have focused on drug administration via the follicular route. The HF disrupts the stratum corneum and provides an opening to the epidermis. In certain areas such as the scalp or the face, the total area of openings can reach up to 10 % of the skin area, contributing significantly to solute permeation [88–90]. In addition the dense network of blood vessels that are associated with the HF suggests that drug release to the circulation may be feasible [91]. The heterogeneity of the harboring cell population in the HF (stem cells, gland cells, immune cells, etc.) may enable cell-specific drug targeting for treatment of skin diseases or vaccination [92–94]. Last but not least, the relatively large volume of infundibulum renders the HF a reservoir for sustained drug release to the circulation, further highlighting the importance of follicular delivery [91].

Several studies highlighted the contribution of follicular penetration during drug delivery through the skin. Mitragori and colleagues modeled the permeability of hydrophilic and hydrophobic compounds in skin, assuming that the solutes can transport through one or more of the following mechanisms: free-volume diffusion, lateral diffusion of the lipids, diffusion through pores, or diffusion through shunts (hair follicles and glands). The model predicted that high molecular weight and highly hydrophilic molecules penetrate the skin preferentially through the shunts [95]. Others suggested that there is a critical value of octanol/water partition coefficient beyond which the flux through the follicle is greatly diminished [96]. However, most studies omit the significance of sebum (a lipophilic product of sebaceous gland) during drug delivery due to lack of representative experimental models. The presence of sebum in the HF and its upward flow may hinder the delivery of hydrophilic compounds and may favor the delivery of hydrophobic compounds. Indeed, apart from molecular weight and molecular orientation, diffusion through the sebum was found to be affected by compound lipophilicity [97].

To further improve tissue targeting and drug delivery via the HF, studies have incorporated particle-based formulations. Lademann et al. demonstrated that nanoparticle-containing dye could penetrate up to 1,400 μm into the follicle of porcine skin whereas the non-particle formulation reached only 500 μm. Interestingly, the nanoparticles prolonged the storage of the dye into the follicle [98]. Nanoparticle size was shown to play critical role in follicular penetration, which was optimal for particles between 750 and 1,500 nm and decreased for larger particles [99]. In addition to this, Vogt et al. demonstrated that the size of the particles affects its uptake by the cells. They reported that only the 40 nm size nanoparticles could enter Langerhans cells that are localized around the HF. This suggests that size-specific particle formulation can be engineered to target antigen-presenting cells via the follicular route and deliver vaccines [93].

Finally, systemic delivery of a chemical through the HF has also been examined in vivo [91, 100]. Caffeine was introduced into a shampoo formulation, and its delivery into the circulation via the skin was examined in human subjects. Interestingly, the follicular route not only accelerated the delivery, but it also prolonged detection of caffeine in the blood indicating that HF may act as reservoir of chemical compounds.

5.6.7 Cell and Gene Therapy Using Hair Follicle Stem Cells

The goal of gene therapy is to restore the lost tissue function by introducing the correct gene copy at the sites where the gene is missing or is mutated [101]. Application of gene therapy for hair restoration has been attempted and showed promising results. Transduction of rat bulge-derived hair follicle stem cells with *LacZ*-encoding retrovirus showed stable expression of the transgene in the HF epithelial compartments for at least 6 months after implantation of transduced cells in an immunodeficient mouse model [102]. Retroviral gene transfer of the streptomyces tyrosinase gene was used to treat albinism. Specifically, transduction of ex vivo cultured skin from albino mice restored melanin production from the skin HF [103]. Direct gene transfer into the skin in vivo has also been reported to restore hair growth. Intradermal administration of the Sonic Hedgehog gene into C57BL/6 mice using an adenovirus resulted into anagen induction and subsequently enhanced hair growth [104]. More recently, in vivo transfection of the human telomerase reverse transcriptase DNA complexed with polyethylenimine induced telogen to anagen transition in the rat dorsal skin [105]. In addition to the treatment of hair- or skin-related disorders, gene transfer to HF could be used for delivery of proteins into the systemic circulation through the vascular plexus surrounding the follicles. To this end, it may be feasible to engineer HF that produce insulin and reverse diabetes as we have previously shown with epidermal cells using a diabetic mouse model [106].

5.6.8 Reprogramming of Hair Follicle Stem Cells

In a breakthrough study in 2006, Yamanaka and colleagues demonstrated that introduction of four transcription factors (OCT4, SOX2, KLF4, and c-Myc) into mouse embryonic fibroblasts or adult fibroblasts endowed them with enhanced proliferation capacity and potential for differentiation into all three germ layers, similar to embryonic stem cells (ESC) [107–110]. The Thomson group demonstrated that two of the transcription factors (KLF4 and c-Myc) could be replaced by NANOG and LIN28 with similar outcome [111]. The resulting cells were designated as induced pluripotent cells (iPSCs). An explosion of studies that followed demonstrated that iPSC could be generated from many human cells including blood cells [112, 113], MSC [114], fetal [114] and neonatal fibroblasts [111, 114], adipose-derived stem cells [115], adult testis [116], β-pancreatic cells [117], and T lymphocytes [118]. Interestingly, HF-derived primary keratinocytes could be reprogrammed with 100-fold higher efficiency than fibroblasts [119]. HF-derived MSC were also reprogrammed and used to understand the feedback loops that sustain self-renewal using global genomic and proteomic strategies [120]. DP cells were shown to reprogram using only two factors (Oct4, Klf4) [121], possibly suggesting the presence of endogenous factors that facilitated reprogramming. Reprogramming with fewer transcription factors or higher efficiency suggests that HF cell-derived iPSC may be useful for regenerative medicine applications as well as for development of models to study the genetics and pathophysiology of human disease.

5.7 Conclusions: Future Directions

In summary, HF stem cells have great potential for tissue engineering and regenerative medicine applications. The ease of accessibility along with the broad differentiation capacity of HF stem cells makes the HF an ideal stem cell source. However, human HF stem cells remain relatively unexplored as compared to their mouse counterparts or other human adult stem cells. As a result more studies are required to address a number of challenges that hinder application of these cells in regenerative medicine. To this end, identification of reliable HF stem cell markers is urgently needed to facilitate HF stem cell isolation. More studies are also needed to evaluate the differentiation potential of human HF stem cells and establish culture conditions for efficient differentiation. The ease of reprogramming should be further explored to identify potential small molecules that may induce reprogramming even in the absence of genetic modification [122]. Finally, more studies are necessary to establish the HF as a site for drug and gene/protein delivery, for treatment of skin diseases and wound healing, or to the blood circulation for treatment of systemic disorders.

Acknowledgments This work was supported by grants from the National Institutes of Health (R01 HL086582) and the New York State Stem Cell Science (NYSTEM Contract #C024316) to S.T. Andreadis. The authors would also like to thank Anna Mistriotis for her help with the schematic representation of the hair follicle.

References

1. Paus R, Cotsarelis G (1999) The biology of hair follicles. N Engl J Med 341:491–497
2. Schneider MR, Schmidt-Ullrich R, Paus R (2009) The hair follicle as a dynamic miniorgan. Curr Biol 19:R132–R142
3. Liu JY, Peng HF, Andreadis ST (2008) Contractile smooth muscle cells derived from hair-follicle stem cells. Cardiovasc Res 79:24–33
4. Liu JY, Peng HF, Gopinath S, Tian J, Andreadis ST (2010) Derivation of functional smooth muscle cells from multipotent human hair follicle mesenchymal stem cells. Tissue Eng Part A 16:2553–2564
5. Bajpai VK, Mistriotis P, Andreadis ST (2012) Clonal multipotency and effect of long-term in vitro expansion on differentiation potential of human hair follicle derived mesenchymal stem cells. Stem Cell Res 8:74–84
6. Goessler UR, Riedel K, Hormann K, Riedel F (2006) Perspectives of gene therapy in stem cell tissue engineering. Cells Tissues Organs 183:169–179
7. Pittenger MF, Mackay AM, Beck SC et al (1999) Multilineage potential of adult human mesenchymal stem cells. Science 284:143–147

8. Morrison SJ, Kimble J (2006) Asymmetric and symmetric stem-cell divisions in development and cancer. Nature 441:1068–1074
9. Jaks V, Kasper M, Toftgard R (2010) The hair follicle-a stem cell zoo. Exp Cell Res 316:1422–1428
10. Kligman AM (1959) The human hair cycle. J Invest Dermatol 33:307–316
11. Oliver RF (1966) Whisker growth after removal of the dermal papilla and lengths of follicle in the hooded rat. J Embryol Exp Morphol 15:331–347
12. Cotsarelis G, Sun TT, Lavker RM (1990) Label-retaining cells reside in the bulge area of pilosebaceous unit: implications for follicular stem cells, hair cycle, and skin carcinogenesis. Cell 61:1329–1337
13. Morris RJ, Potten CS (1999) Highly persistent label-retaining cells in the hair follicles of mice and their fate following induction of anagen. J Invest Dermatol 112:470–475
14. Tumbar T, Guasch G, Greco V et al (2004) Defining the epithelial stem cell niche in skin. Science 303:359–363
15. Morris RJ, Liu Y, Marles L et al (2004) Capturing and profiling adult hair follicle stem cells. Nat Biotechnol 22:411–417
16. Jaks V, Barker N, Kasper M et al (2008) Lgr5 marks cycling, yet long-lived, hair follicle stem cells. Nat Genet 40:1291–1299
17. Mobini N, Tam S, Kamino H (2005) Possible role of the bulge region in the pathogenesis of inflammatory scarring alopecia: lichen planopilaris as the prototype. J Cutan Pathol 32: 675–679
18. Taylor G, Lehrer MS, Jensen PJ, Sun TT, Lavker RM (2000) Involvement of follicular stem cells in forming not only the follicle but also the epidermis. Cell 102:451–461
19. Ito M, Liu Y, Yang Z et al (2005) Stem cells in the hair follicle bulge contribute to wound repair but not to homeostasis of the epidermis. Nat Med 11:1351–1354
20. Li L, Mignone J, Yang M et al (2003) Nestin expression in hair follicle sheath progenitor cells. Proc Natl Acad Sci USA 100:9958–9961
21. Amoh Y, Li L, Yang M et al (2004) Nascent blood vessels in the skin arise from nestin-expressing hair-follicle cells. Proc Natl Acad Sci USA 101:13291–13295
22. Amoh Y, Li L, Katsuoka K, Penman S, Hoffman RM (2005) Multipotent nestin-positive, keratin-negative hair-follicle bulge stem cells can form neurons. Proc Natl Acad Sci USA 102:5530–5534
23. Yu H, Fang D, Kumar SM et al (2006) Isolation of a novel population of multipotent adult stem cells from human hair follicles. Am J Pathol 168:1879–1888
24. Yu H, Kumar SM, Kossenkov AV, Showe L, Xu X (2010) Stem cells with neural crest characteristics derived from the bulge region of cultured human hair follicles. J Invest Dermatol 130:1227–1236
25. Ito M, Kizawa K, Hamada K, Cotsarelis G (2004) Hair follicle stem cells in the lower bulge form the secondary germ, a biochemically distinct but functionally equivalent progenitor cell population, at the termination of catagen. Differentiation 72:548–557
26. Greco V, Chen T, Rendl M et al (2009) A two-step mechanism for stem cell activation during hair regeneration. Cell Stem Cell 4:155–169
27. Nijhof JG, Braun KM, Giangreco A et al (2006) The cell-surface marker MTS24 identifies a novel population of follicular keratinocytes with characteristics of progenitor cells. Development 133:3027–3037
28. Jensen UB, Yan X, Triel C, Woo SH, Christensen R, Owens DM (2008) A distinct population of clonogenic and multipotent murine follicular keratinocytes residing in the upper isthmus. J Cell Sci 121:609–617
29. Jensen KB, Collins CA, Nascimento E et al (2009) Lrig1 expression defines a distinct multipotent stem cell population in mammalian epidermis. Cell Stem Cell 4:427–439
30. Snippert HJ, Haegebarth A, Kasper M et al (2010) Lgr6 marks stem cells in the hair follicle that generate all cell lineages of the skin. Science 327:1385–1389
31. Horsley V, O'Carroll D, Tooze R et al (2006) Blimp1 defines a progenitor population that governs cellular input to the sebaceous gland. Cell 126:597–609

32. Lako M, Armstrong L, Cairns PM, Harris S, Hole N, Jahoda CA (2002) Hair follicle dermal cells repopulate the mouse haematopoietic system. J Cell Sci 115:3967–3974
33. Jahoda CA, Whitehouse J, Reynolds AJ, Hole N (2003) Hair follicle dermal cells differentiate into adipogenic and osteogenic lineages. Exp Dermatol 12:849–859
34. Hoogduijn MJ, Gorjup E, Genever PG (2006) Comparative characterization of hair follicle dermal stem cells and bone marrow mesenchymal stem cells. Stem Cells Dev 15:49–60
35. Biernaskie J, Paris M, Morozova O et al (2009) SKPs derive from hair follicle precursors and exhibit properties of adult dermal stem cells. Cell Stem Cell 5:610–623
36. Trempus CS, Morris RJ, Bortner CD et al (2003) Enrichment for living murine keratinocytes from the hair follicle bulge with the cell surface marker CD34. J Invest Dermatol 120: 501–511
37. Merrill BJ, Gat U, DasGupta R, Fuchs E (2001) Tcf3 and Lef1 regulate lineage differentiation of multipotent stem cells in skin. Genes Dev 15:1688–1705
38. Nguyen H, Rendl M, Fuchs E (2006) Tcf3 governs stem cell features and represses cell fate determination in skin. Cell 127:171–183
39. Horsley V, Aliprantis AO, Polak L, Glimcher LH, Fuchs E (2008) NFATc1 balances quiescence and proliferation of skin stem cells. Cell 132:299–310
40. Rhee H, Polak L, Fuchs E (2006) Lhx2 maintains stem cell character in hair follicles. Science 312:1946–1949
41. Vidal VP, Chaboissier MC, Lutzkendorf S et al (2005) Sox9 is essential for outer root sheath differentiation and the formation of the hair stem cell compartment. Curr Biol 15:1340–1351
42. Jensen KB, Watt FM (2006) Single-cell expression profiling of human epidermal stem and transit-amplifying cells: Lrig1 is a regulator of stem cell quiescence. Proc Natl Acad Sci USA 103:11958–11963
43. Driskell RR, Giangreco A, Jensen KB, Mulder KW, Watt FM (2009) Sox2-positive dermal papilla cells specify hair follicle type in mammalian epidermis. Development 136:2815–2823
44. Kishimoto J, Ehama R, Wu L, Jiang S, Jiang N, Burgeson RE (1999) Selective activation of the versican promoter by epithelial-mesenchymal interactions during hair follicle development. Proc Natl Acad Sci USA 96:7336–7341
45. Iida M, Ihara S, Matsuzaki T (2007) Hair cycle-dependent changes of alkaline phosphatase activity in the mesenchyme and epithelium in mouse vibrissal follicles. Dev Growth Differ 49:185–195
46. Yu DW, Yang T, Sonoda T et al (1995) Message of nexin 1, a serine protease inhibitor, is accumulated in the follicular papilla during anagen of the hair cycle. J Cell Sci 108(Pt 12):3867–3874
47. Ito Y, Hamazaki TS, Ohnuma K, Tamaki K, Asashima M, Okochi H (2007) Isolation of murine hair-inducing cells using the cell surface marker prominin-1/CD133. J Invest Dermatol 127:1052–1060
48. Kloepper JE, Tiede S, Brinckmann J et al (2008) Immunophenotyping of the human bulge region: the quest to define useful in situ markers for human epithelial hair follicle stem cells and their niche. Exp Dermatol 17:592–609
49. Ohyama M, Terunuma A, Tock CL et al (2006) Characterization and isolation of stem cell-enriched human hair follicle bulge cells. J Clin Invest 116:249–260
50. Inoue K, Aoi N, Sato T et al (2009) Differential expression of stem-cell-associated markers in human hair follicle epithelial cells. Lab Invest 89:844–856
51. Nowak JA, Polak L, Pasolli HA, Fuchs E (2008) Hair follicle stem cells are specified and function in early skin morphogenesis. Cell Stem Cell 3:33–43
52. Tornqvist G, Sandberg A, Hagglund AC, Carlsson L (2010) Cyclic expression of lhx2 regulates hair formation. PLoS Genet 6:e1000904
53. Amoh Y, Li L, Campillo R et al (2005) Implanted hair follicle stem cells form Schwann cells that support repair of severed peripheral nerves. Proc Natl Acad Sci USA 102:17734–17738
54. Gur G, Rubin C, Katz M et al (2004) LRIG1 restricts growth factor signaling by enhancing receptor ubiquitylation and degradation. EMBO J 23:3270–3281
55. Jahoda CA, Horne KA, Oliver RF (1984) Induction of hair growth by implantation of cultured dermal papilla cells. Nature 311:560–562

56. Rendl M, Lewis L, Fuchs E (2005) Molecular dissection of mesenchymal-epithelial interactions in the hair follicle. PLoS Biol 3:e331
57. Cohen J (1961) The transplantation of individual rat and guineapig whisker papillae. J Embryol Exp Morphol 9:117–127
58. Messenger AG (1984) The culture of dermal papilla cells from human hair follicles. Br J Dermatol 110:685–689
59. Kobayashi K, Rochat A, Barrandon Y (1993) Segregation of keratinocyte colony-forming cells in the bulge of the rat vibrissa. Proc Natl Acad Sci USA 90:7391–7395
60. Magerl M, Kauser S, Paus R, Tobin DJ (2002) Simple and rapid method to isolate and culture follicular papillae from human scalp hair follicles. Exp Dermatol 11:381–385
61. Edwards RA (2007) Laser capture microdissection of mammalian tissue. J Vis Exp (8):309
62. Wu JJ, Liu RQ, Lu YG, Zhu TY, Cheng B, Men X (2005) Enzyme digestion to isolate and culture human scalp dermal papilla cells: a more efficient method. Arch Dermatol Res 297:60–67
63. Peng HF, Liu JY, Andreadis ST, Swartz DD (2011) Hair follicle-derived smooth muscle cells and small intestinal submucosa for engineering mechanically robust and vasoreactive vascular media. Tissue Eng Part A 17:981–990
64. Peng H, Schlaich EM, Row S, Andreadis ST, Swartz DD (2012) A novel ovine ex vivo arteriovenous shunt model to test vascular implantability. Cells Tissues Organs 195:108–121
65. Levy V, Lindon C, Zheng Y, Harfe BD, Morgan BA (2007) Epidermal stem cells arise from the hair follicle after wounding. FASEB J 21:1358–1366
66. Hoeller D, Huppertz B, Roos TC et al (2001) An improved and rapid method to construct skin equivalents from human hair follicles and fibroblasts. Exp Dermatol 10:264–271
67. Liu F, Luo XS, Shen HY, Dong JS, Yang J (2011) Using human hair follicle-derived keratinocytes and melanocytes for constructing pigmented tissue-engineered skin. Skin Res Technol. [Epub ahead of print]; doi: 10.1111/j.1600-0846.2011.00510.x
68. Limat A, Mauri D, Hunziker T (1996) Successful treatment of chronic leg ulcers with epidermal equivalents generated from cultured autologous outer root sheath cells. J Invest Dermatol 107:128–135
69. Limat A, Hunziker T (2002) Use of epidermal equivalents generated from follicular outer root sheath cells in vitro and for autologous grafting of chronic wounds. Cells Tissues Organs 172:79–85
70. Navsaria HA, Ojeh NO, Moiemen N, Griffiths MA, Frame JD (2004) Reepithelialization of a full-thickness burn from stem cells of hair follicles micrografted into a tissue-engineered dermal template (Integra). Plast Reconstr Surg 113:978–981
71. Gagnon V, Larouche D, Parenteau-Bareil R, Gingras M, Germain L, Berthod F (2011) Hair follicles guide nerve migration in vitro and in vivo in tissue-engineered skin. J Invest Dermatol 131:1375–1378
72. Amoh Y, Li L, Katsuoka K, Hoffman RM (2008) Multipotent hair follicle stem cells promote repair of spinal cord injury and recovery of walking function. Cell Cycle 7:1865–1869
73. Amoh Y, Kanoh M, Niiyama S et al (2009) Human hair follicle pluripotent stem (hfPS) cells promote regeneration of peripheral-nerve injury: an advantageous alternative to ES and iPS cells. J Cell Biochem 107:1016–1020
74. Amoh Y, Aki R, Hamada Y et al (2012) Nestin-positive hair follicle pluripotent stem cells can promote regeneration of impinged peripheral nerve injury. J Dermatol 39:33–38
75. Lin H, Liu F, Zhang C et al (2011) Characterization of nerve conduits seeded with neurons and Schwann cells derived from hair follicle neural crest stem cells. Tissue Eng Part A 17:1691–1698
76. Shiell RC (2001) Modern hair restoration surgery. Clin Dermatol 19:179–187
77. Kim JC, Choi YC (1995) Regrowth of grafted human scalp hair after removal of the bulb. Dermatol Surg 21:312–313
78. Jahoda CA, Oliver RF, Reynolds AJ, Forrester JC, Horne KA (1996) Human hair follicle regeneration following amputation and grafting into the nude mouse. J Invest Dermatol 107:804–807
79. Matsuzaki T, Inamatsu M, Yoshizato K (1996) The upper dermal sheath has a potential to regenerate the hair in the rat follicular epidermis. Differentiation 60:287–297

80. Reynolds AJ, Lawrence C, Cserhalmi-Friedman PB, Christiano AM, Jahoda CA (1999) Trans-gender induction of hair follicles. Nature 402:33–34
81. Jahoda CA, Oliver RF, Reynolds AJ et al (2001) Trans-species hair growth induction by human hair follicle dermal papillae. Exp Dermatol 10:229–237
82. Tang L, Madani S, Lui H, Shapiro J (2002) Regeneration of a new hair follicle from the upper half of a human hair follicle in a nude mouse. J Invest Dermatol 119:983–984
83. Inamatsu M, Matsuzaki T, Iwanari H, Yoshizato K (1998) Establishment of rat dermal papilla cell lines that sustain the potency to induce hair follicles from afollicular skin. J Invest Dermatol 111:767–775
84. McElwee KJ, Kissling S, Wenzel E, Huth A, Hoffmann R (2003) Cultured peribulbar dermal sheath cells can induce hair follicle development and contribute to the dermal sheath and dermal papilla. J Invest Dermatol 121:1267–1275
85. Blanpain C, Lowry WE, Geoghegan A, Polak L, Fuchs E (2004) Self-renewal, multipotency, and the existence of two cell populations within an epithelial stem cell niche. Cell 118: 635–648
86. Kataoka K, Medina RJ, Kageyama T et al (2003) Participation of adult mouse bone marrow cells in reconstitution of skin. Am J Pathol 163:1227–1231
87. Arora A, Prausnitz MR, Mitragotri S (2008) Micro-scale devices for transdermal drug delivery. Int J Pharm 364:227–236
88. Otberg N, Richter H, Schaefer H, Blume-Peytavi U, Sterry W, Lademann J (2004) Variations of hair follicle size and distribution in different body sites. J Invest Dermatol 122:14–19
89. Knorr F, Lademann J, Patzelt A, Sterry W, Blume-Peytavi U, Vogt A (2009) Follicular transport route – research progress and future perspectives. Eur J Pharm Biopharm 71:173–180
90. Wosicka H, Cal K (2010) Targeting to the hair follicles: current status and potential. J Dermatol Sci 57:83–89
91. Otberg N, Teichmann A, Rasuljev U, Sinkgraven R, Sterry W, Lademann J (2007) Follicular penetration of topically applied caffeine via a shampoo formulation. Skin Pharmacol Physiol 20:195–198
92. Rolland A, Wagner N, Chatelus A, Shroot B, Schaefer H (1993) Site-specific drug delivery to pilosebaceous structures using polymeric microspheres. Pharm Res 10:1738–1744
93. Vogt A, Combadiere B, Hadam S et al (2006) 40 nm, but not 750 or 1,500 nm, nanoparticles enter epidermal CD1a+ cells after transcutaneous application on human skin. J Invest Dermatol 126:1316–1322
94. Vogt A, Mahe B, Costagliola D et al (2008) Transcutaneous anti-influenza vaccination promotes both CD4 and CD8 T cell immune responses in humans. J Immunol 180:1482–1489
95. Mitragotri S (2003) Modeling skin permeability to hydrophilic and hydrophobic solutes based on four permeation pathways. J Control Release 86:69–92
96. Frum Y, Bonner MC, Eccleston GM, Meidan VM (2007) The influence of drug partition coefficient on follicular penetration: in vitro human skin studies. Eur J Pharm Sci 30: 280–287
97. Valiveti S, Lu GW (2007) Diffusion properties of model compounds in artificial sebum. Int J Pharm 345:88–94
98. Lademann J, Richter H, Teichmann A et al (2007) Nanoparticles – an efficient carrier for drug delivery into the hair follicles. Eur J Pharm Biopharm 66:159–164
99. Toll R, Jacobi U, Richter H, Lademann J, Schaefer H, Blume-Peytavi U (2004) Penetration profile of microspheres in follicular targeting of terminal hair follicles. J Invest Dermatol 123:168–176
100. Liu X, Grice JE, Lademann J et al (2011) Hair follicles contribute significantly to penetration through human skin only at times soon after application as a solvent deposited solid in man. Br J Clin Pharmacol 72:768–774
101. Cotsarelis G, Millar SE (2001) Towards a molecular understanding of hair loss and its treatment. Trends Mol Med 7:293–301
102. Sugiyama-Nakagiri Y, Akiyama M, Shimizu H (2006) Hair follicle stem cell-targeted gene transfer and reconstitution system. Gene Ther 13:732–737

103. Zhao M, Saito N, Li L et al (2000) A novel approach to gene therapy of albino hair in histoculture with a retroviral streptomyces tyrosinase gene. Pigment Cell Res 13:345–351
104. Sato N, Leopold PL, Crystal RG (1999) Induction of the hair growth phase in postnatal mice by localized transient expression of Sonic hedgehog. J Clin Invest 104:855–864
105. Jan HM, Wei MF, Peng CL, Lin SJ, Lai PS, Shieh MJ (2012) The use of polyethylenimine-DNA to topically deliver hTERT to promote hair growth. Gene Ther 19:86–93
106. Tian J, Lei P, Laychock SG, Andreadis ST (2008) Regulated insulin delivery from human epidermal cells reverses hyperglycemia. Mol Ther 16:1146–1153
107. Takahashi K, Yamanaka S (2006) Induction of pluripotent stem cells from mouse embryonic and adult fibroblast cultures by defined factors. Cell 126:663–676
108. Takahashi K, Tanabe K, Ohnuki M et al (2007) Induction of pluripotent stem cells from adult human fibroblasts by defined factors. Cell 131:861–872
109. Okita K, Ichisaka T, Yamanaka S (2007) Generation of germline-competent induced pluripotent stem cells. Nature 448:313–317
110. Okita K, Nakagawa M, Hyenjong H, Ichisaka T, Yamanaka S (2008) Generation of mouse induced pluripotent stem cells without viral vectors. Science 322:949–953
111. Yu J, Vodyanik MA, Smuga-Otto K et al (2007) Induced pluripotent stem cell lines derived from human somatic cells. Science 318:1917–1920
112. Ye Z, Zhan H, Mali P et al (2009) Human-induced pluripotent stem cells from blood cells of healthy donors and patients with acquired blood disorders. Blood 114:5473–5480
113. Loh YH, Agarwal S, Park IH et al (2009) Generation of induced pluripotent stem cells from human blood. Blood 113:5476–5479
114. Park IH, Zhao R, West JA et al (2008) Reprogramming of human somatic cells to pluripotency with defined factors. Nature 451:141–146
115. Sun N, Panetta NJ, Gupta DM et al (2009) Feeder-free derivation of induced pluripotent stem cells from adult human adipose stem cells. Proc Natl Acad Sci USA 106:15720–15725
116. Conrad S, Renninger M, Hennenlotter J et al (2008) Generation of pluripotent stem cells from adult human testis. Nature 456:344–349
117. Stadtfeld M, Brennand K, Hochedlinger K (2008) Reprogramming of pancreatic beta cells into induced pluripotent stem cells. Curr Biol 18:890–894
118. Brown ME, Rondon E, Rajesh D et al (2010) Derivation of induced pluripotent stem cells from human peripheral blood T lymphocytes. PLoS One 5:e11373
119. Aasen T, Raya A, Barrero MJ et al (2008) Efficient and rapid generation of induced pluripotent stem cells from human keratinocytes. Nat Biotechnol 26:1276–1284
120. Lu Y, Loh YH, Li H et al (2012) A self-sustaining feedback loop that regulates proteome diversity and supports self-renewal in pluripotent stem cells. In Revision
121. Tsai SY, Clavel C, Kim S et al (2010) Oct4 and klf4 reprogram dermal papilla cells into induced pluripotent stem cells. Stem Cells 28:221–228
122. Zhu S, Wei W, Ding S (2011) Chemical strategies for stem cell biology and regenerative medicine. Annu Rev Biomed Eng 13:73–90

Chapter 6
Genetically Modified Stem Cells for Transplantation

M. Ian Phillips

Abstract *Critical Challenges*: Stem cell therapies are based on a simplistic idea of harvesting stem cells from bone marrow, adipose tissue, or induced pluripotent stem cells and injecting them into tissue that requires regeneration. Such ideas are logical and appealing. The only problem is they do not work very well. Effects of stem cell therapy are modest at best and often neither effective nor long lasting. This is because injected stem cells do not survive long. These cells are taken from their comfortable niches and forced to enter a hostile environment of low oxygen, poor nutrients, attacks by immune cells, and the apoptotic agents of death.

Current Research Directions: To reach past this impasse, the emerging trend is genetic modification of stem cells for protection and facilitation. Stem cells can be modified to withstand apoptosis and inflammation and even be activated by low oxygen to switch on protective genes to make them survive longer as grafts. Stem cells can be genetically modified to deliver hormones, growth factors, and homing factors. There are multiple methods for modification from gene signaling, antisense inhibition, microRNAs, and inserting transgene switches.

Discussion of Specific Examples: Here we discuss examples of gene modification of stem cells for survival after transplantation, turning cells into insulin-producing cells, cells that could reduce plaque in Alzheimer's and at the same time repair lost neural tissue. Genetically modified stem cells could be a new step forward in stem cell therapy when designed to improve their utility in treating myocardial ischemia and heart failure, hemophilia, stroke, diabetes type 1, spinal cord injury, Alzheimer's and Parkinson's diseases, bone defects, and cancer.

M.I. Phillips, PhD, DSc
Norris Professor of Life Sciences, Center for Rare Disease Therapies,
Keck Graduate Institute, 535 Watson Drive, Claremont, CA 91711, USA
e-mail: ian_phillips@kgi.edu

M.K. Danquah, R.I. Mahato (eds.), *Emerging Trends in Cell and Gene Therapy*,
DOI 10.1007/978-1-62703-417-3_6,

Keywords Genetically modified stem cells • Transplantation • Induced pluripotent stem cells • Spinal cord injury • Stroke • Diabetes

6.1 Critical Challenges of Stem Cell Therapy

Stem cell transplantation as a new form of therapy has met with mixed results. In leukemia with bone marrow, cell transplantation has been used for over 30 years, and the method has become a standard therapy. But for newer therapies, success has been elusive. In one of the most intensely studied recent applications, stem cells for myocardial infarction and heart failure, positive results have been weak or there was no effect. The majority of transplanted stem cells (up to 90 %) do not survive beyond 48 h [1–3].

We believe that it is time to move beyond simply injecting harvested stem cells and start to genetically modify them to make them more precisely designed for functional regeneration and survival [4].

To extend the life of stem cells after their transplantation requires gene modification. Gene modification of stem cells prior to their transplantation enhances their survival and increases their function in cell therapy. Like the famous Trojan horse, the gene-modified cell has to gain entrance inside the host's walls and survive to deliver its transgene products. Using cellular, molecular, and gene manipulation techniques, the transplanted cell can be protected in a hostile environment from immune rejection, inflammation, hypoxia, and apoptosis. Genetic engineering to modify cells involves constructing functional gene sequences and inserting them into stem cells. The modifications can be simple reporter genes or complex cassettes with gene switches, cell-specific promoters, and multiple transgenes. We discuss here methods to deliver and construct gene cassettes with viral and nonviral delivery, siRNA, and conditional Cre/lox P. We review the current uses of gene-modified stem cells in various diseases.

6.1.1 Types of Stem Cells

Stem cells have three unique characteristics: they have the unique ability to renew themselves continuously, they possess the ability to differentiate into somatic cell types, and they have the ability to control their own population to keep it to a small number. When a cell is fertilized and divides a few times, those cells form a morula of 8–16 cells. Each of those cells is a totipotent stem cell, that is, they can become any type of cell. By day 5–6, a hollow ball of cells is formed and the inner mass of cells is pluripotent stem cells. These cells can turn into most types of body cells but not all. Although these cells are pre-embryonic, they have become known as "embryonic" stem cells as opposed to "adult" stem cells. Embryonic stem cells are fully pluripotent; adult stem cells are more limited in their potency. Adult stem cells generally are multipotent and can transform into tissues that are produced within the organ or tissue in which they are found. Stem cells in bone marrow can become osteocytes, blood cells, and lymph cells. Cardiac-derived stem cells can become any of the cells that are part of a functioning heart including cardiomyocytes, neurons, and endothelial cells [3, 5, 6]. It is debatable whether bone marrow cells can turn into heart cells [1] or whether they can turn into any cell which is not related to blood, bone, or lymph [7].

The problems with ESC cells are the ethical and regulatory objections to using a human cell derived from a fertilized human egg. The ethical problems have restricted the usable lines of ESCs available for research in the USA. The development of a new class of pluripotent stem cells generated without going through an embryo or pre-embryonic stage has electrified new research in this area. Induced pluripotent cells (iPS cells) are discussed below. They can be produced from adult somatic cells such as skin fibroblasts [8, 9].

Progenitor cells come from a stem cell that is set to differentiate into multipotent or lineage-specific type of cell. Tissue-specific stem cells are multipotent. These can be isolated from various tissues such as hematopoietic stem cells (HSCs), bone marrow mesenchymal stem cells (MSCs), adipose tissue-derived stem cells, amniotic fluid stem cells, and neural stem cells (NSCs). NSCs in developing or adult mammalian brain have properties of indefinite growth and multipotent potential to differentiate into three major CNS cell types, neurons, astrocytes, and oligodendrocytes [10, 11].

6.1.2 Potential of Stem Cells

Thus, stem cells are either pluripotent (ESCs and iPS) or multipotent (tissue stem cells). Progenitor cells are stem cells destined to become a certain cell type. Stem cells replicate throughout life so long as a few of them do not differentiate. The daughter cells that differentiate go on to become adult cells with specific functions in the body. Adult stem cells generally are multipotent and can transform into tissues that are produced within the organ or tissue in which they are found. Stem cells in bone marrow can become osteocytes, blood cells, and lymph cells [6]. Cardiac-derived stem cells can become any of the cells that are part of a functioning heart including cardiomyocytes, neurons, and endothelial cells [1, 12, 13]. It is debatable whether bone marrow cells can turn into heart cells or any other cell which is not related to blood, bone, or lymph [6, 14]. Cancer stem cells (CSC) are unique in that they are self-renewing like stem cells but growing into tumors, not differentiating into tissue.

6.1.3 Induced Pluripotent Stem Cells

The discovery that adult somatic cells could be induced to become pluripotent stem cells with apparently all the properties of human embryonic stem cells (hESCs), independently by Yamanaka et al. [8] and by Thompson et al. [9], was greeted with amazement and in some quarters with relief. The breakthrough, first reported in mice by S. Yamanaka in Japan, did not receive much notice, but the revelation that adult human cells could be reversed into an embryonic-like state was astounding because it was so simple. It was greeted with relief where the progress of hESCs was blocked at the federal level because producing hESCs raised ethical concerns and political consequences. Yet 5 years later, although we have discovered much about iPS cells, they are far from being the ideal solution that they seemed to promise.

When human iPS cells were first published, they appeared to have so many advantages over hESCs. Foremost was the lack of an embryo being involved which meant they could not be logically banned or targeted as unethical (although some tried). Second, the science behind making them was astonishingly simple. It only required delivery of four transcription factors found in embryos to reverse years of life as an adult cell back to an embryonic-like cell. The record for number of years as an adult cell was set by Dimos et al. [15], who induced the skin cells of an 82-year-old lady back to iPS cells.

iPS cells offer the advantages of avoiding the religious or ethical considerations that plague the use of embryonic stem cells. They also could provide autologous transplantation, for repair and regeneration of tissue without rejection. If the donor cells retain the mutation or mutations that caused the patient's disease state, it is possible to correct those mutations before implantation by gene modification with homologous recombination. Even if they cannot be corrected, the iPS cells are useful to study the mutations in lineages derived from the iPS cells. A further advantage of studying iPS cells is access to testing new drugs in those diseased human cells. At present the barriers to adopting iPS in the clinic are the limitations of methods to produce the cells and the possibility of causing teratomas.

iPS technology is expected to move health sciences forward in unique ways for diagnosis, drug screening, toxicity, repair of mutations, and treatment of human diseases. iPS cells produced from an individual are embryonic-like stem cells, and they can be regrown into any of the 200 somatic cell types. iPS cells have many similarities with ESC cells including the cell morphology, surface antigens, gene expression, telomerase activity, and the epigenetic status. iPS cells have been produced by delivering transcription factors by different types of viral vectors including retroviruses [8], lentiviruses [16], adenoviruses [17], plasmid transfections [18], transposons [19], mRNA, or recombinant proteins [20].

iPS cells are produced to be as close to human ESCs as possible to have the advantages of pluripotency that hESCs have. However, hESCs could only be transplanted allogenically into adults and rejection would always be a problem.

Despite the euphoria and literally thousands of studies, there are nagging problems with making iPS cells work in the way that was hoped. One of the reasons is that iPS cells, just like hESCs, go through a stage of producing teratomas. Several studies have found that mice produced from iPS cells are more prone to cancer. The original method for producing iPS cells was by a retrovirus to deliver the four transcription factors. Retroviruses are notorious for random insertion in the genome and being oncogenic. In Yamanaka's study [8], the method included c-Myc as a transcription factor that increased cell growth. However, the same property is the property that makes c-Myc an oncogene and therefore another prime suspect in causing cancer. Thompson's group avoided using c-Myc but also used a retrovirus and instead of c-Myc used LIN 28 as a transcription factor [9]. Eventually Yamanaka's group dispensed with c-Myc but lost efficiency. Only Oct4 and Sox2 seemed to be absolutely necessary. The other factors could be varied, with Nanog substituting for c-Myc. To avoid the dangers of retroviruses, various alternatives for non-integrating delivery have been tried. Adenovirus was used successfully. But adenovirus while not being carcinogenic has other problems. In 1999 it was prematurely used in a phase 1 gene therapy trial and caused the death of one of the participants, Jesse Gelsinger. Adenovirus produces many proteins and these induce immune reactions. It was an immune reaction to adenovirus in Jesse Gelsinger's body that made adenovirus totally unacceptable for human use [21]. There is an alternative, however, adeno-associated virus (AAV). Despite its name, AAV is unrelated to adenovirus and is proving to be a very safe and reliable vector for gene delivery. Gene therapy with rAAV for restoring sight to patients blind since birth or early childhood (Leber congenital amaurosis) has proven that in humans rAAV is very safe [22].

We have shown that AAV as the vector for the four transcription factors can be used to induce iPS in adult cells from skin or fat tissue [23]. Eventually there will come a time when iPS will be tested for therapeutic use in treating humans. Although there are now virus-free methods of making iPS, we do not know which method will have the greatest efficiency, safety, reproducibility, or efficacy. Therefore, having several different ways of producing iPS is still a viable quest, and we show how AAV can be used to produce iPS cells [23].

6.2 Current Research on Gene Modification of Stem Cells

Inducing pluripotent stem cells is a feat of genetic engineering, and considering it can reverse a skin cell of an 82-year-old back to an embryonic stem cell, so simply and elegantly, it is truly remarkable.

Obviously both embryonic and adult stem cells have great potential for treatments involving cellular repair, replacement, and regeneration. One of the limitations of cell replacement therapy is that a majority of grafted cells do not survive when grafted. Even if they are autologous or from a syngenic population, cell transplantation usually results in a loss of cells. Genetic engineering can increase survival of engrafted stem cells. The stem cells can be modified to deliver proteins to neighboring cells [4] to avoid apoptosis and inflammation or reduce graft–host rejection. In genetic modification, a gene cassette is constructed and loaded into a vector for entry into the cell. Once inside the cell, the gene construct can express or overexpress specific genes. The transgene expression can be constant leading to constitutive synthesis of specific proteins or can be controlled by a gene switch. Constitutive activation of genes is unphysiological leading to overproduction of proteins which downregulates receptors and renders the gene expression ineffective. A gene switch essentially makes the cell "intelligent" because the cell will then respond to a physiological stimulus, for example, low oxygen, high glucose levels, and hormone concentrations, or to drugs or chemical agents.

A key principle to genetic engineering for cells is to mix and match modules of functional domains that are used in nature. Thus, we can take a gene module used by yeast and a human virus module to create a chimeric regulator. Wang et al. [12] first described a gene regulatory system for gene transfer by building a gene switch that responds to increases in mifepristone, a progesterone antagonist. They fused a ligand-binding domain of a mutated human progesterone receptor to the yeast transcriptional activator GAL4 DNA-binding domain and the herpes simplex virus protein VP16-activated domain. They demonstrated that this system could be activated by the exogenous administration of mifepristone (RU 486) at low doses to activate transcription of target genes. As described below we developed a Vigilant Vector [13, 14] with a gene switch similar to this concept but built it to automatically respond to hypoxia so that no exogenous drug was required to turn the system on or off.

6.2.1 Transgenics

A very well-established gene modification of embryonic stem (ES) cells is in the production of transgenic animals. Transgenic mice with genes knocked out, or genes "knocked in" (where the number of copies of genes is increased) [24], are ubiquitous gene studies in living animals. They have been very useful for studying the role of specific genes and practical for producing specific human proteins. The method involves harvesting ES cells from the inner cell mass of the blastocyst. Using recombinant DNA (rDNA), a desired gene is made and inserted in a vector together with promoter sequences to regulate the gene expression. To replace a normal gene or knock one out, two drug-resistant genes are added to the cassette, a neo^r gene which is resistant to lethal effects of neomycin and a thymidine kinase gene (tk) which phosphorylates ganciclovir. The majority of cells fail to take the vector inside their walls. These cells can be killed by neomycin or its analogs. A few of the remaining cells allow the vector in but the gene is inserted randomly. To avoid this, these cells are killed by ganciclovir. That leaves those cells in which homologous recombination has occurred. The normal gene has been knocked out and a new, specific gene knocked in. These cells are then injected into a blastocyst, which is implanted in the uterus to produce offspring that can be bred. If the new gene is nonfunctional (i.e., a null allele), the function of the former gene may be revealed through breeding the mice with the knockout gene to homozygosity.

Ideally the function of the missing gene will be as obvious as if a limb had been cut off. In actuality several things can happen. The knocked-out gene may prevent the embryo from developing (it is embryonically lethal), or the missing gene is fully compensated by other genes, or subtle changes occur in development or in different organs so that the effect is not obvious. Nevertheless the technique has had a huge influence on revealing functional effects of proteins especially where antibodies have not been developed. The opposite of knocking in copies of a gene has been used to reveal mechanisms of diseases caused by overexpression of a protein [24]. The transgenic animal approach requires going through embryonic development. This limits the technique when a knocked-out gene is embryonically lethal. However, in a method first used by Gu et al. [25], the Cre/lox P system is able to induce the same mutation and avoid lethality.

6.2.2 Cre/lox P System

To knockout a target gene in specific cell groups or tissue in adult animals, the Cre/lox P system is a suitable technique. It is based on the viral bacteria phage P1, which produces Cre, a recombinase enzyme. Cre cuts its viral DNA into packages by cutting the DNA out between two separate lox P sites. The DNA ends, which each has a half lox P site, are then ligated by the recombinase. Gu et al. [25] used this principle with a strategy of conventional transgenic mice, in which the Cre transgene plus a promoter was inserted by homologous recombination in a cell-specific type. This mouse was crossed with a second mouse strain that had a target gene flanked by two lox P sites. In the offspring the target gene was only deleted in those specific cells that contained Cre and the lox P-"floxed" sequences. The target gene remained functional in all the other cells and the animals survived development, so the function of the targeted gene in specific cells could be studied.

More recent developments have made the technique less laborious to use [25, 26]. An example is a study by Sanniyha et al. [26] who made transgenic mice with lox P insertions flanking the gene for angiotensinogen. Angiotensinogen is a substrate for the enzyme renin and is one of the critical components for the synthesis of the peptide angiotensin. Instead of making a separate strain of Cre mice and proceeding with breeding, they simply injected Cre into the floxed mice. This had the advantage not only of being time saving but also of opening up a new way to study genes with site-directed, conditional, gene ablation in specific cells. As they were working on the brain, they were able to pinpoint anatomically a very small brain structure, the subfornical organ. By injecting Cre into the structure, they showed that angiotensin synthesis could be blocked and proved it is synthesized in the brain [27, 28].

To inhibit synthesis of proteins by inhibiting gene translation, there are two methods: antisense and RNA interference.

6.2.3 Antisense Inhibition

Antisense is based on the fact that mRNA is in the "sense" direction from 5′ to 3′. Antisense is a limited sequence of DNA in the antisense direction 3′ to 5′ designed from the known sequence of a target gene. Antisense oligonucleotides (AS-ODN) are usually built around the initiation codon of a gene (the AUG start site) and are shorter than the full-length gene. This is because the AS-ODN binds to part of the appropriate mRNA sequence and prevents the mRNA from translating the protein it would otherwise produce [29–32].

For gene modification with antisense within a cell, a viral vector can be fitted with DNA in the antisense direction. We have designed these in the adeno-associated virus and shown them to have long-lasting inhibitory effects on designated cell protein synthesis [33]. Antisense inhibition although widely used in research and approved for clinical treatment [32] is not perfect. When antisense is put into a cell, it is competing with the cell's own mRNA-copying machinery. The presence of AS-ODN may actually increase the number of cell-produced mRNA copies, thereby overcoming the endogenously administered AS-ODN. Because of this, antisense as a treatment has not proven to be a killer of cells and so not a revolutionary anticancer agent, as it was originally hoped. However, antisense has played a pivotal role in leading to the next advance in cellular gene inhibition – RNA interference – and, more recently, reemerged as antagomirs for inhibiting microRNA.

6.2.4 siRNA Gene Silencing

Fire and Mello [34] were using antisense to study behavioral effects on the primitive worm, *Caenorhabditis elegans*. They tested sense RNA and antisense RNA on the worms, but there was no effect of either. However, when they tested a combination of sense and antisense RNA, the worms started to twitch spontaneously. The gene that was holding back the twitching had been silenced. Fire and Mello had discovered gene silencing by double-stranded (ds)RNA which acted as small interfering RNA (siRNA). RNA interference has become widely recognized as a biological mechanism for the regulation of gene expression and used for intracellular inhibition. Double-stranded RNA is produced in the nucleus. In the cytoplasm, it binds to an enzyme Dicer. Dicer literally dices up the double-stranded RNA into short strands (15–20 nucleotides).

One of the strands is loaded into a protein complex, RNA-induced silencing complex (RISC). The RISC now has the single strand of short RNA as a binding site to bind to a complementary sequence on the cell's mRNA. This binding leads to cleavage of mRNA, degrading the message and stopping it from translating a specific protein; hence, it is silenced.

RNAi is a fundamental cellular process of gene regulation in the cells of animals and plants. Since both animals and plants are subject to diseases induced by viruses, RNAi may have evolved to protect cells from invasion by viruses. The genome of retroviruses is in double strands of RNA. A retrovirus, lacking cellular mechanisms and DNA, injects its genomic dsRNA into a cell to reproduce itself using the DNA of the invaded cell. RNAi protects the cell by destroying the viral RNA through the RISC mechanism.

siRNA is more powerful than antisense in silencing genes, but it has its difficulties. It is not long lasting, it may silence off-target sites, and it has not been easy to inject systemically as a therapy. We have directly compared siRNA to antisense to inhibit the beta-1 adrenergic receptor gene [35]. The effect was measured on blood pressure in hypertensive rats and on measures of heart performance, because beta-blockers have long been used for hypertension and heart failure treatments. The siRNA and AS-ODN were injected systemically in a lipofectamine vehicle. The result was a significantly better effect on lowering blood pressure and improving heart performance with the siRNA compared to the AS-ODN. Both approaches lasted about 1 week with a single injection [35].

6.2.5 microRNA

microRNAs offer completely new possibilities for gene modification, cell therapy, and drug development. They are involved in almost every biological process regulated by genes, and their absence or mutations could be the cause of many disease states from birth defects to cancer.

Although microRNAs (miRNA) were discovered over 20 years ago in *C. elegans* [36] and later found in mammals, we are still in an early stage of discovering how many there are, what they do, and how they do it. Over 500 miRNAs have been found in the human genome. A recent review in Nature Reviews suggests that miRNAs regulate one third of human genes [37]. microRNAs have become recognized as a new class of gene regulators and therefore important for gene modification of cells. miRNA are small noncoding RNAs that modify gene expression by posttranscriptional inhibition of targeted mRNA. In the nucleus miRNA is formed from introns and exons as "primary" or "pri-miRNA." But it is not a messenger RNA – it does not specify or generate a protein. The pri-miRNA, a folded-back structure of 60–70 nucleotides, is processed in the nucleus by the enzymes Drosha and Pasha. Drosha cuts out the stem-loop structure which is the "pre-miRNA." The pre-miRNA is exported out of the nucleus by exportin and into the cytoplasm where it is diced up by the enzyme Dicer RNase III. The same effect occurs. Dicer cuts the stem loop into short-length (19–25 nucleotides) inverted "mature miRNA." As with siRNA, one strand of the mature miRNA becomes part of the RISC and targets mRNA by binding to antisense complementary regions and cleaving or degrading the targeted mRNA. Multiple roles for miRNAs in gene regulation have been revealed by gene expression analysis PCR and by transgenic mice with knockouts of specific miRNA. Expression arrays are revealing specific miRNAs in different tissues and cells from invertebrates to humans. Many miRNAs (miRNAs-1, miRNAs-34, miRNAs-60, miRNAs-87, miRNAs-124a) are highly conserved between vertebrates and invertebrates [38] including the small temporal (st)RNAs discovered in *C. elegans* (e.g., let-7 RNA, lin-4) that are similar to miRNAs in humans. As these stRNAs are critical for cell differentiation and timing of neural connections, the conservation may indicate functional evolution. A survey of mouse tissues with northern blotting [38] showed that miRNAs-1 is dominant in the heart (45 %). In the liver, miRNAs-122 was 72 % of all miRNAs tested and miRNAs-124a was profound in the mouse brain.

Although the mechanism of miRNA action is principally inhibitory on target mRNA, which is essential for normal growth and differentiation in cell and tissue development, miRNAs can be involved in cancer. They can be depleted or suppressed allowing oncogenes to be overproduced. Kumar et al. [39] recently showed that global suppression of miRNAs in various cancer cell lines increased cancer cell transformation and enhanced tumorigenesis in mice. To suppress miRNA they targeted Drosha and Dicer with siRNA. Noncancerous cells did not become cancerous but did not grow.

Stem cells are regulated in their differentiation and in adult processes by microRNA. This suggests that increasing or decreasing miRNAs could be a new approach to regulate genes in, for example, retarding cancer by suppressing oncogenes or increasing developmental processes in regeneration by increasing differentiation.

6.2.6 Reporter Genes

Manipulation of genes in cells, such as stem cells, before transplantation can be done at several different levels of sophistication. If one simply wants to label cells with an internal marker so that the cells can be identified after transplantation, then a reporter gene such as a fluorescent gene like green fluorescent protein (gfp) or luciferase (Luc) or beta-galactosidase (Lac Z) gene sequence can be inserted into any of the vectors described above. Each cell marker has its own advantage or disadvantage. Fluorescent labels are visible, but not easily quantified. However, a great advantage is that they are visible using highly sensitive fluoroscopy such that the cells can be located, even under the skin in tissues and tumors. Luciferase has the advantage that it is quantifiable using luminometers, dual luciferase assays, or relative luciferase gene expression [13, 14].

6.2.7 Cell-Specific Promoters

At the next level of sophistication, a cell- or tissue-specific promoter is spliced with the selected cell marker transgene so that the transgene can be observed to be expressed in one type of cell. Selecting the promoter raises some problems. A powerful promoter like cytomegalovirus (CMV) drives a gene but is nonselective for tissue type. A more cell-specific promoter is likely to have a weaker power, and therefore, there will be less gene marker expressed.

Improving promoter power without losing cell specificity is a challenge. Also fitting a promoter into a cassette for a vector of small loading capacity, such as AAV, may require cutting the promoter into fragments and test driving for specificity. For example, we used the myosin light chain-2v promoter (MLC-2v) in the heart [13], which is 1,700 bp long. In order to fit this promoter into the AAV, we reduced the

MLC-2v to a 250 bp fragment that contained the heart-specific cis-regulatory elements [40]. To further increase power, a promoter enhancer can be added to the effective promoter fragment. SV40 and chicken beta actin or globin [41] have been tried and increase expression by severalfold. A feed-forward system can be introduced by making the product of cassette transgenes – the fusion proteins – feedback on an activating sequence to drive the promoter. Thus, more and more fusion protein is produced. If this protein is also activating an upstream activating sequence in front of a transgene TATA box, more and more transgene expression will result.

For a therapeutic approach, the gene modification needs to have a gene switch added. High-level expression powered by CMV or even the lower level of gene expression driven by a cell-specific promoter is constant. This constitutive gene expression could lead to a buildup of protein and unwanted side effects. The design of a transgene construct needs to include a regulator to control the amount of expression.

6.2.8 Gene Switches

Several different types of gene switches have been developed. Some require exogenous drugs to be applied to induce expression. These include the "Tet-on Tet-off" system using tetracycline as the switch inducer [42]. Ecdysone [43], hypoxia regulatory element (HRE) [44], and mifepristone [12] have also been used.

To make a transgene turn on and off to physiological stimulus requires genetic engineering of the cassette to include naturally occurring cellular regulatory elements. The cassette is constructed from modules which can be spliced together in a specific order. To illustrate, we have developed a "Vigilant Vector™" that is switched on by hypoxia in heart cells [45]. To develop the hypoxia switch, there were several possibilities. The natural oxygen-sensitive elements of a cell had been worked out and sequenced [46]. The hypoxia regulatory element (HRE) contains inducible factors HIF-1α and HIF-1β. When oxygen is low, the HIF-1α combines with the HIF-1β, and the fusion product acts a transcription factor in the nucleus to generate proteins in response to low oxygen, such as vascular endothelial growth factor (VEGF) and erythropoietin (EPO). By extracting the oxygen sensor in HIF-1α oxygen-dependent domain (ODD) and installing it as the oxygen sensor of a chimeric gene, we could control the genetic response to hypoxia and avoid the production of these and other proteins. The ODD module was spliced in an activator system. The DNA-binding domain is the yeast GAL4, and the activating domain is the human p65 derived from human nuclear kappa B protein. Under normal oxygen levels, the fusion protein of p65/ODD/Gal4 is ubiquitinated, and the ubiquitin tail is the signal for transport to and destruction in proteasomes. But as oxygen decreases, a threshold is reached where the fusion protein is not ubiquitinated or destroyed, and the GAL4 component of the protein binds to an inserted upstream activating sequence in front of the TATA box that activates gene expression. The lower the oxygen concentration, the greater the number of fusion proteins generated, exponentially increasing gene activity. Combined with a heart-specific promoter (MLC-2v), the whole system acts as a site-specific gene switch for hypoxia. Further the system allows for amplification of gene expression. In practical terms, when the stem cells were modified by heme oxygenase-1, an antioxidant with antiapoptotic and anti-inflammatory effects as the transgene, ischemic (mouse) hearts were protected from heart failure [47].

6.3 The Application of Genetic Modification of Stem Cells

6.3.1 Cardiology and Blood

6.3.1.1 Increase Graft Cell Survival

Adult stem cells have been proposed as a promising source for the heart repair. The success of several trials has been nothing or relatively small improvements in cardiac performance. Out of 12 studies involving 1,359 patients, the overall benefit was 4 % but 5 trials had no effect [48]. Stem cell-based therapy is confronted with the problem of poor survival in host myocardium. Graft cell survival is limited by various pathological processes such as inflammatory molecules, proapoptotic factors, lack of oxygen and/or nutrients, and the loss of cells through dispersal because they do not home to the site of myocardial infarction. The survival of engrafted stem cells requires adaptation to adverse environment in ischemic myocardium. Different strategies have been developed to increase cell survival after grafting. Pharmacologic preconditioning has been tested successfully in skeletal myoblasts and shows cytoprotective effects both in vitro and in vivo [49]. Suzuki et al. [50] reported that heat-shock treatment could improve cell tolerance to hypoxia–reoxygen insult in vitro and enhance survival when grafted into the heart. Exploiting cell growth and apoptotic regulatory factors to enhance the proliferation of viable stem cells or confer apoptosis resistance to donor cells, by gene modification, is a potential way to improve cell transplant efficiency. Akt is a powerful survival signal in many systems [51]. Mangi et al. [52] showed that Akt gene modification of MSCs improved the function of infarct rat hearts. However, as a therapy the overall application of constitutively active Akt gene may increase the risk of tumorigenesis [53]. HO-1 is the rate-limiting enzyme in the catabolism of heme, followed by production of biliverdin, free iron, and carbon monoxide (CO). All three by-products exert beneficial actions that protect the cells from oxidative damage and death [54]. Hypoxia-inducible HO-1 plasmid modification of graft mesenchymal stem cells can protect cells from subsequent hypoxia injury in vitro and improve graft cell survival in ischemic myocardium in vivo via anti-inflammatory and antiapoptosis [55]. These findings underscore the role of HO-1 for protecting grafted cells from ischemia-/inflammation-induced death.

6.3.1.2 Increase Angiogenesis in Ischemic Heart Disease

Myocardial ischemia associated with coronary artery disease is a leading cause of morbidity and mortality in the United States [56]. Although percutaneous transluminal angioplasty (PTCA) and operative coronary revascularization (CABG) procedures are effective for revascularization, there are increasing numbers of patients with extensive atherosclerotic coronary artery disease not amenable to traditional methods of revascularization. Several growth factors have appeared recently as adjuncts to regular revascularization, including vascular endothelial growth factor (VEGF) [57]. Although viruses carrying VEGF gene can maintain a therapeutic angiogenesis, VEGF expression is not under tight control and thus might cause unwanted side effects, such as angioma formation. To develop an approach for safe and long-lasting angiogenesis, we investigated neovascularization in ischemic myocardium via autologous mesenchymal stromal cells (MSCs) transplantation. Our finding suggested that bone marrow-derived MSCs play a crucial role in improving regional blood flow in ischemic myocardium and provide an optimal strategy for therapeutic angiogenesis by secreting a broad spectrum of angiogenic cytokines, including VEGF [58, 59], HGF [60], bFGF [61], and SDF-1α [61]. Increased blood supply from neovascularization would inhibit apoptosis and necrosis of hibernating and stunned myocardium in border zone. Moreover, autologous MSCs have high proliferative and self-renewal capability, which is critical for maintaining lasting effects fit for clinic treatment of patients with extensive atherosclerotic coronary disease [58]. Although autologous MSC transplantation can be administrated as “sole therapy” for neovascularization, many laboratories have developed strategies to use MSCs as vehicles for angiogenic gene therapy to enhance the benefits of neovascularization. Lei et al. [62] have reviewed improvements in angiogenic outcome via deliver of multiple growth factors with synergic effects.

6.3.1.3 Gene-Modified Stem Cells to Treat Hemophilia

Recently gene-modified bone marrow stem cell therapy approaches have been used to target life-threatening bleeding disorder, such as hemophilia. Moayeri et al. [63] used hematopoietic stem cells (HSCs) to express coagulation factor VIII (FVIII) by oncoretroviral vector. Transduced HSCs were transplanted into immunocompetent hemophilia A mice. Therapeutic levels of FVIII were detected in the serum of transplant recipient for over 6 months. More importantly, there was only minor anti-FVIII inhibitor antibody production induced following transplantation of gene-modified HSCs. In a related study, Gangadharan et al. [64] compared the therapeutic effect of achieving sustained, therapeutic levels of FVIII between gene-modified MSCs and HSCs. To test this, they used retroviral-mediated porcine FVIII vector to genetically modified bone marrow-derived MSCs and HSCs and then transplanted cells into genetically immunocompetent hemophilia A mice. They found that the FVIII activity levels drop rapidly and returned to baseline in MSC group due to the formation of anti-porcine FVIII neutralizing antibodies; however,

FVIII levels stayed high in mice treated with HSCs. They found that FVIII expression was sustained beyond 10 months because of immunologic tolerance. This investigation demonstrates that HSCs, other than MSCs, offer a sufficient and durable approach for delivering curative FVIII for treating hemophilia A.

6.3.2 Gene-Modified Stem Cells to Replenish β Cells for Treating Diabetes

Type 1 diabetes is due to the loss of pancreatic islet β cells. Therefore, the advent of stem cell technology has given rise to the hope that beta cells can be "regrown" in the pancreas by implanting stem cells. Transplantation of the pancreas is effective with the Edmonton protocol but requires rare pancreas donors and not infrequently two donors for one recipient.

Stem cells can in theory be grown in large numbers and modified to be insulin-producing cells with appropriate controls for euglycemic control. Lavon et al. [65] made hESCs constitutively expressing two different transcription factors, Foxa2 and pancreatic duodenum homeobox protein-1(Pdx1). Foxa2 is found in the early endoderm layer [66] and expressed at a very early stage in pancreas development [67]. Pdx1 is a pancreatic-specific transcription factor expressed downstream of Foxa2 and specifically involved in stem cell differentiation into β-cell progenitors [68]. Pdx1 binds and activates insulin promoter in β cells [68, 69]. But there was a limitation because expression of the insulin gene was demonstrable when the cells differentiated in vivo into teratomas.

Human ESCs carry the problem of immunological incompatibility between the cell donors and the recipients and the danger of graft vs. host disease. The levels of MHC-I expression in hESCs will increase after in vitro differentiation [70]. To eliminate the problem of immuno-incompatibility and the requirement for the classic immunosuppressive therapy employed for organ transplantation, multipotent adult stem cells are an alternative for these studies. Tang et al. [71] tested the possibility of reprogramming rat hepatic stem cells into functional insulin-producing cells by over-expression of Pdx1 via lentivirus. Their findings showed that long-term expression of Pdx1 is effective in converting hepatic stem cells into pancreatic endocrine precursor cells. When these cells transplanted into diabetic mice, they become functional insulin-producing cells and restore euglycemia. Human bone marrow-derived mesenchymal stem cells (hMSCs) may also be a source to produce insulin-producing cells as shown by Li et al. [72] who modified hMSCs with Pdx1 with a recombinant adenoviral vector. Pdx1 gene-modified hMSCs expressed multiple islet-cell genes including neurogenin3 (Ngn3), insulin, GK, Glut2, and glucagon and produced and released insulin/C-peptide in a weak glucose-regulated manner. Two weeks after injection of Pdx1-modified hMSCs in STZ-induced diabetic mice, euglycemia was observed and lasted for at least 42 days. Pancreatic ductal stem/progenitor cells have also been genetically modified into insulin-producing cells by adenovirus delivery of NeuroD [73].

We examined the potential for attracting bone marrow stem cells (BMSCs) to the pancreas using a homing factor , a chemokine, stromal cell-derived factor 1 (SDF-1) [74]. In diabetically induced rats, SDF-1 injections into the pancreas markedly increased the number of GFP-labeled BMSCs in the pancreas, but surprisingly, the majority of cells homed to the liver. The marked liver cells had typical pancreatic endocrine cell gene expression including insulin I, insulin II, Pdx1, somatostatin, and glucagon. Combined treatment with SDF-1 and labeled BMSC transplant reduced hyperglycemia to a normoglycemic range and prolonged the long-term survival of diabetic mice. One subgroup had complete normoglycemia (<150 mg/dl), restored blood insulin levels, and normal glucose tolerance. Our results suggest that a gene modification of stem cells with SDF-1 could potentially be used to improve the homing of stem cells that appears to lead to β-cell regeneration. The novel mechanism appears to involve an increase in insulin-producing cells mainly in the liver [74].

6.3.3 Gene-Modified Stem Cells to Treat Spinal Cord Injury

The adult central nervous system (CNS) has long been considered not to have the capacity to regenerate itself; thus, spinal cord injury leads to permanent loss of functions. Stem cell transplantation has been proposed as a strategy for CNS repair. Neural stem cells (NSCs) that are capable of differentiation into neurons in the brain [75] and spinal cord [76], therefore, are attractive candidates for repairing injured CNS. Three groups [77–79] demonstrated that genetically engineering NSCs with axonal growth gene or neuroprotective factor genes, such as neurotrophin-3 (NT-3), NGF, and BDNF, could exhibit better spinal cord repairing. They isolated and cultured the neural stem cells and then modified these cells with lentivirus-mediated neurotrophin-3 (NT-3). Their studies demonstrated that the NT-3-modified grafted cells could survive for a long time in vivo and migrate for long distances. Moreover, NT-3 genetically engineered NSC obviously led to a recovery of the hind limb function of the injured rats. These experiments provide a clear indication that modifying NSC with NT-3 can make NSC act as a source of neurotrophic factors and improve functional outcome in spinal cord injury via neuroregeneration.

The most promising development in spinal cord injury repair with hESCs was the Geron phase 1 study based on the work in rats of Hans Keirstead [80]. He worked on the hypothesis that glial cells act as guides to neural axons migrating down the spinal cord. Therefore, he developed neuroglia stem cells (GRNOPC1). These cells were injected into the site of spinal cord injury produced by a crush to the spinal cord. Treated animals went from paralyzed to running around on all fours within weeks of treatment. Geron which had supported the original hESC studies of J. Thompson started a phase 1 trial, the first hESC trial approved by the FDA, in July 2010. Four patients who qualified with a recent spinal cord injury were injected with very low doses of the cell GRNOPC1. No adverse effects were reported. However, it is not clear how meaningful that result was. A very low dilution could be low enough to have neither an adverse effect nor any effect. Despite being able to claim that there was no adverse effect, Geron abruptly stopped the trial in November 2011, with no plans to continue.

6.3.4 Gene-Modified Stem Cells for Stroke

About 700,000 Americans each year suffer a new or recurrent stroke. Stroke kills more than 150,000 people a year. That is about 1 of every 16 deaths. It is the number 3 cause of death behind diseases of the heart and cancer. In 2010, it is estimated that Americans have paid about $73.7 billion for stroke-related medical costs and disability [81]. Bone marrow stem cells have been demonstrated to cross the blood–brain barrier [82] and can differentiate into neurons and glia [83]. Transplantation of bone marrow stem cells in animal models of cerebral ischemia by either intracerebral or i.v. route has demonstrated therapeutic efficacy in reducing lesion size and improving functional outcome [84–87]. Although bone marrow stem cells have potential to self-renewal, these cells had reduced replicative capacity after about five cell doublings over the course of about 6 weeks in culture [88]. The limitation in life span of these cells is directly correlated with telomere shortening because of the lack of telomerase activity that is necessary for maintenance of telomere [89] and may limit clinical application of bone marrow stem cells. Overexpression of hTERT (telomerase reverse transcriptase) has been demonstrated to increase or stabilize telomere length and immortalize human cells [90, 91]. The technology of hTERT immortalization could be used to improve stem cell expansion for subsequent therapeutic cell transplantation, especially important for aging patients with stroke. Recently, hTERT-immortalized human mesenchymal stem cells have been used in rat cerebral ischemia model for brain functional repair [92]. In the experiment, human mesenchymal stem cells were isolated from healthy adult volunteers, and the primary MSCs were immortalized with hTERT-expressing retrovirus. The cell population was expanded in culture within 40 population doublings and intravenously delivered into rats 12 h after induction of transient middle cerebral artery occlusion (MCAO), to study their potential therapeutic benefit. They found that intravenous infusion of immortalized human mesenchymal stem cells 12 h after transient MCAO in the rat results in reduction in infarction volume by histological assay and magnetic resonance spectroscopy, more importantly; behavioral performance was improved in hTERT–MSC-treated group by treadmill test and Morris water maze test. Therefore, hTERT modification of mesenchymal stem cells appears beneficial to ameliorate functional deficits after stroke and enhance the efficacy of cell transplants.

MSCs were reported to promote neuronal cell survival and neurogenesis via secreting a variety of neuro-regulatory molecules, such as BDNF [93]. To further enhance this paracrine effects, Kurozumi et al. [94, 95] transfected telomerized human MSC with the BDNF gene via a fiber-mutant F/RGD adenovirus vector and investigated whether these cells contributed to improved functional recovery in a rat transient middle cerebral artery occlusion (MCAO) model. They found that BDNF production by MSC–BDNF cells was 23-fold greater than that seen in uninfected MSC. Rats that received MSC–BDNF showed significantly more functional recovery than did control rats following MCAO. Moreover, MRI analysis revealed that the rats in the MSC–BDNF group exhibited more significant recovery from ischemia after 7 and 14 days. The apoptotic cells in the ischemic boundary zone was significantly reduced in animals treated with MSC–BDNF compared to animals in the control group. Their findings suggested that BDNF gene modification of MSC may be used as a novel strategy for the treatment of stroke by promoting functional recovery and reducing infarct size in the cerebral ischemia.

6.3.5 Gene-Modified Stem Cells for Parkinson's Disease

Bone marrow mesenchymal stem cells can be used as an alternative source of cells for neural regeneration. MSCs can be genetically modified to provide sustained production of therapeutic proteins to treat neurodegenerative disorder. Parkinson's disease (PD) is a neurological disease suited for gene-modified stem cell therapy because the mechanism of substantia nigra cell degeneration is well characterized. Early study by Schwarz et al. [96] tested the efficacy of genetically modified MSCs in a rat model of Parkinson's disease. Rat MSCs were genetically engineered by transduction with retroviruses encoding tyrosine hydroxylase (TH) to convert tyrosine to L-3, 4-dihydroxyphenylalanine (L-DOPA), and GTP cyclohydrolase I, the enzyme necessary for production of the tetrahydrobiopterin cofactor for TH (BH4). Transduced cells synthesized 3, 4-dihydroxyphenylalanine (L-DOPA) in vitro and maintained their multipotentiality after retroviral transduction. In the in vivo experiment, they injected gene-modified MSCs into the striatum of 6-hydroxydopamine-lesioned rats. Their results demonstrated that L-DOPA and its metabolites can be detected in the denervated striatum of rats that received gene-modified MSCs. Most importantly, they observed a significant reduction in apomorphine-induced rotation when compared with controls. Also, they reported that the engrafted cells can survive at least 87 days, whereas the transgene expression only lasts about 9 days. Recently in the same laboratory, they have focused on the development of a new vector system to genetically engineer autologous MSC, which is a self-inactivating retrovirus (pSIR) and contains the genes for human TH and rat GTP cyclohydrolase I separated by an internal ribosome entry site (IRES) [97]. They found that transduced rMSCs can synthesize and secrete L-DOPA (89.0–283 pmol/10^6 cells/h).

Recent studies by Shen et al. [98] demonstrated that co-expression of TH and aromatic-L-amino-acid decarboxylase (AADC) which can convert L-DOPA to dopamine and GTP cyclohydrolase I (GCH1) using triple transduction with adeno-associated virus (AAV)–TH, AAV–AADC, and AAV–GCH resulted in greater dopamine production than double transduction and single transduction in denervated striatum of parkinsonian rats and improved the rotational behavior of the rats more efficiently. In a related study, Sun et al. [99] demonstrated that co-expression of four dopamine biosynthetic and transporter genes, including TH, GTPCH1, AADC, and vesicular monoamine transporter (VMAT-2, which can transport dopamine into synaptic vesicles in striatal neurons), supports efficient production of dopamine and regulated vesicular release of dopamine. The 4-gene vector improved correction of apomorphine-induced rotational behavior better than the 3-gene vector for 6 months. More importantly, only the 4-gene vector supported significant K (+)-dependent release of dopamine. Therefore, these investigations suggest that MSCs genetically modified with multiple dopamine biosynthetic and transporter genes may be the most suitable for cell therapy in patients with Parkinson's disease.

6.3.6 Gene-Modified Stem Cells to Treat Alzheimer's Disease

Alzheimer's disease (AD) is a debilitating disorder of the central nervous system which may affect up to 50 % of the population over the age of 85 years. It is a gradual loss of brain cells but most known for beginning with loss of memory, deteriorating to loss of personality, and finally death. The etiology of AD is unknown; however, research has focused on two aspects: (1) cholinergic neuron loss and (2) buildup of amyloid plaques, as the central features of Alzheimer disease.

Due to loss of cholinergic neurotransmitter systems in patients with Alzheimer's disease, early studies have focused on the development of genetically engineered cells to produce neurotrophic factors and neurotransmitters. Fisher et al. [100] developed a primary fibroblast cell line that was genetically modified to express choline acetyltransferase (ChAT). They demonstrated that in vitro these cells produced and released acetylcholine at levels that varied with the amount of choline in the culture media. In their in vivo study, they found that the ChAT-expressing fibroblasts continued to produce and release acetylcholine after transplantation into the hippocampus of rats, and the levels of acetylcholine synthesized by the cells could be regulated by the localized infusion of choline in the vicinity of the graft.

Similar results have also been obtained using ChAT human neuronal stem cells (hNSC) as demonstrated by Park et al. [101]. Although the transplantation of genetically modified cells will not cure AD, this strategy may ameliorate the progression of cognitive impairments. Nerve growth factor (NGF) can enhance cholinergic function of neurons via cell surface receptors, such as TrkA and p75 (NTR) [102]. Recently published phase 1 clinical trial [103] demonstrated that implanting autologous fibroblasts genetically modified to express human NGF into the forebrain in eight individuals with mild Alzheimer's disease showed a slower rate of cognitive decline. Also, serial PET scans showed significant ($P<0.05$) increases in cortical 18-fluorodeoxyglucose after treatment. Since both neural stem cells (NSCs) and MSCs can be integrated into brain and differentiate into neurons after transplantation, transplantation of gene-modified stem cells is a promising strategy to treat Alzheimer's disease by enhancing the NGF secretion and renewing the degenerated neuron cells.

Amyloid precursor protein (APP) produces amyloid-beta (Aβ) protein, and the accumulation of Aβ builds up plaque in the brain destroying neurons. Decreasing chronic levels of Aβ has been proposed as a possible therapy. Genetically modified neural stem cells might be useful in this therapeutic approach (a) to prevent plaques and (b) to replace lost neurons; for example, the genes for proteinases, such as neprilysin [104], have shown promise for the reduction of Aβ levels in AD brain. Lentivirus vector expressing human neprilysin intracerebral injection in transgenic mouse models of amyloidosis reduced cerebral Aβ neurodegeneration in the frontal cortex and hippocampus [105]. Hemming et al. showed that fibroblasts modified to overexpress human neprilysin gene significantly reduced amyloid plaque in Aβ transgenic mice brains [106]. Magga et al. have developed human hematopoietic stem cells (HSCM) modified with a green fluorescent protein delivered by a lentivirus [107]. These cells were effective in reducing Aβ in a model of AD.

6.3.7 Gene-Modified Stem Cells to Treat Bone Defect Disease

In United States, there are about 6.5 million fractures per year, and about 15 % of them are hard to heal. It still lacks effective therapy for these difficult cases. Bone marrow contains a population of rare progenitor cells capable of differentiating into bone, cartilage, muscle, tendon, and other connective tissues. These cells, referred to as MSCs, can be purified and culture expanded from animals and humans [108]. Bone marrow-derived stromal cells show a great promise for bone regeneration. Engineering pluripotent MSCs with BMP2s has been a recent research focus for the treatment of a variety of bone defects. Gazit et al. [109] have documented that MSC can express rhBMP2, spontaneously differentiated into osteogenic cells in vitro and enhanced segmental defect repair in a mouse model of radial segmental defect in vivo following transduction of such cells with BMP2. In a study reported by Moutsatsos et al. [110], a tetracycline-regulated expression vector encoding human BMP2 was used to transduce MSCs. Such cells were then tested in both in vitro and in vivo. Their finding showed that both bone formation and bone regeneration could be controlled by doxycycline, a tetracycline analog. Moreover, there is increased angiogenesis accompanied by the bone formation in vivo. In a related study, Hasharoni et al. [111] transduced MSCs with regulated-BMP2-expressing vector and injected engineered MSCs intramuscularly into the paraspinal muscles in mice. The nature and extent of bone formation were analyzed by microcomputerized tomography scanning and histological studies. They found that the newly formed bone fuses the spine, and a 7-day induction of vector-mediated BMP2 expression in genetically engineered MSCs was enough to form highly mineralized bone in mice injected with regulated-BMP2-transduced cells, and injected cells induced active osteogenesis at the site of implantation for up to 4 weeks post-injection. These data suggest that BMP2 vectors provide powerful gene therapy tools for bone regeneration.

6.3.8 Gene-Modified Stem Cells to Treat Cancer

Cancer gene therapy is the most promising field in gene therapy. High-efficiency, tumor-specific targeting is crucial for success in cancer gene therapy, without systemic toxicity of vector dissemination. Recent data from Hung et al. [112] provided direct evidence that bone marrow MSCs possess extensive tropism for solid tumors after systemic injection and thus can be used as delivery vehicles for cancer therapy [112]. They used micropositron emission tomography imaging with [18F]-FHBG to monitor the pTY–EFEGFP–TK vector-modified human MSCs. Micropositron emission tomography imaging revealed that tracer human MSCs could migrate to the sites of microscopic tumor lesions, engraft into these microscopic tumor lesions, and contribute to the development of a significant portion of tumor stroma. Although the mechanisms are unclear, it is believed that MSCs are likely to migrate to tumor tissues through the chemotactic effect mediated by chemokines/chemokine receptors because solid tumor tissues can express and secrete multiple chemokines such as SDF-1α hepatocyte growth factor, vascular endothelial cell growth factors (VEGF), TGFs, FGFs, platelet-derived growth factors, monocyte chemoattractant protein-1 (MCP-1), and IL-8. Therefore, MSCs can be developed to deliver genes encoding biological agents that interfere with tumor growth. Systemic delivery of genetically modified MSCs can be used as a tumor-targeting gene therapy strategy to exert antitumor effects [112].

Interferon-β (IFN-β) shows capability to anti-malignant tumor through antiproliferative and proapoptotic effects in vitro [113, 114]; however, clinical trials of IFN-β failed because the concentration of IFN-β to inhibit tumor via systemic administration is much higher than the maximally tolerated dose for human. To solve this problem, Studeny et al. [115, 116] have developed a therapeutic strategy to treat multiple tumors with IFN-β gene-engineered MSCs (MSC–IFN-β cells). They treated multiple lung metastases of human tumors in SCID mice by intravenous injection of human MSCs expressing interferon-ß and demonstrated that transplanted MSCs incorporated into the tumor architecture and MSC–IFN-β suppress the growth of pulmonary metastases and prolong the survival of mouse. Nakamizo et al. [117] extended this therapeutic strategy to the treatment of intracranial human gliomas in nude mice. They injected human MSCs expressing interferon-ß into the carotid artery of mice bearing human glioma intracranial xenografts (U87) and showed that MSC–IFN-β can track human gliomas and significantly increase animal survival. More recently, the study from Xin et al. [118] demonstrated successful inhibition of the development of lung metastases and thus prolonged the survival of these tumor-bearing mice by systemic administration of CX3CL1-expressing MSCs to the mice bearing lung metastases of C26 and B16F10 cells. In the related study, Stagg et al. [119] investigated whether MSCs can be exploited to deliver IL-2 and generate effective immune responses against the poorly immunogenic B16 melanoma in mice with normal immune systems. Their study showed that IL-2-producing MSCs mixed with B16 cells significantly delayed tumor growth in an IL-2 dose-dependent manner, while primary MSCs mixed with B16 cells and injected subcutaneously in syngenic recipients do not affect tumor growth. Moreover, they observed that matrix-embedded IL-2-producing MSCs injected in the vicinity of preestablished B16 tumors led to absence of tumor growth in 90 % of treated mice. Their study also demonstrated that tumor-bearing mice treated with IL-2-producing MSCs developed CD8-mediated tumor-specific immunity and significantly delayed tumor growth of a B16 cell challenge.

Adipose tissue-derived mesenchymal stem cells (AT-MSC) have also been shown to possess the capability to migrate actively toward tumor cells and also can be used as cellular vehicles for targeted cancer chemotherapy. Kucerova et al. [120] recently evaluated the potential of cytosine deaminase (CD)-expressing AT-MSC (CD-AT-MSC) in a human colon cancer therapy. CD is an enzyme that converts far less toxic substrate 5-fluorocytosine (5-FC) to 5-FU and the toxic metabolites production [121]. By employing retroviral vectors and G418 selection, AT-MSC transduction resulted in efficient genetic loading of AT-MSC with CD gene. CD-AT-MSC in combination with 5-fluorocytosine (5-FC) augmented the bystander effect and selective cytotoxicity on target tumor cells HT-29 in vitro. More importantly, they demonstrated that CD-AT-MSC can deliver the CD transgene to the site of tumor formation and mediate strong antitumor effect after i.v. administration of CD-AT-MSC in immunocompromised mice treated with 5-FC. Thus, AT-MSC can be used as cell vehicles to deliver prodrug-converting gene for targeted cancer gene therapy.

Acknowledgments I wish to acknowledge the expert input on parts of this chapter that I received from Henrique Cheng (Louisiana State University) and Yao Liang Tang (University of Cincinnati) and the excellent editorial help of Dilshad Contractor (Keck Graduate Institute).

References

1. Murry CE, Soonpaa MH, Reinecke H, Nakajima H, Nakajima HO, Rubart M, Pasumarthi KBS, Ismail Virag J, Bartelmez SH, Poppa V, Bradford G, Dowell JD, Williams DA, Field LJ (2004) Haematopoietic stem cells do not transdifferentiate into cardiac myocytes in myocardial infarcts. Nature 428:664–668. doi:10.1038/nature02446
2. Balsam LB, Wagers AJ, Christensen JL, Kofidis T, Weissman IL, Robbins RC (2004) Haematopoietic stem cells adopt mature haematopoietic fates in ischaemic myocardium. Nature 428:668–673. doi:10.1038/nature02460
3. Tang YL, Shen L, Qian K, Phillips MI (2007) A novel two-step procedure to expand cardiac Sca-1+ cells clonally. Biochem Biophys Res Commun 359:877–883. doi:10.1016/j.bbrc.2007.05.216
4. Phillips MI, Tang YL (2008) Genetic modification of stem cells for transplantation. Adv Drug Deliv Rev 60:160–172. doi:10.1016/j.addr.2007.08.035
5. Messina E, De Angelis L, Frati G, Morrone S, Chimenti S, Fiordaliso F, Salio M, Battaglia M, Latronico MVG, Coletta M, Vivarelli E, Frati L, Cossu G, Giacomello A (2004) Isolation and expansion of adult cardiac stem cells from human and murine heart. Circ Res 95:911–921. doi:10.1161/01.RES.0000147315.71699.51
6. Moretti A, Caron L, Nakano A, Lam JT, Bernshausen A, Chen Y, Qyang Y, Bu L, Sasaki M, Martin-Puig S, Sun Y, Evans SM, Laugwitz KL, Chien KR (2006) Multipotent embryonic

isl1+ progenitor cells lead to cardiac, smooth muscle, and endothelial cell diversification. Cell 127:1151–1165. doi:10.1016/j.cell.2006.10.029
7. Anderson DJ, Gage FH, Weissman IL (2001) Can stem cells cross lineage boundaries? Nat Med 7:393–395. doi:10.1038/86439
8. Takahashi K, Tanabe K, Ohnuki M, Narita M, Ichisaka T, Tomoda K, Yamanaka S (2007) Induction of pluripotent stem cells from adult human fibroblasts by defined factors. Cell 131:861–872. doi:10.1016/j.cell.2007.11.019
9. Yu J, Hu K, Smuga-Otto K, Tian S, Stewart R, Slukvin II, Thomson JA (2009) Human induced pluripotent stem cells free of vector and transgene sequences. Science 324:797–801. doi:10.1126/science.1172482
10. Androutsellis-Theotokis A (2010) Angiogenic factors stimulate growth of adult neural stem cells. ONE Alerts. doi:10.1371/journal.pone.0009414
11. Zhao C, Deng W, Gage FH (2008) Mechanisms and functional implications of adult neurogenesis. Cell 132:645–660. doi:10.1016/j.cell.2008.01.033
12. Wang Y, O'Malley BW, Tsai SY, O'Malley BW (1994) A regulatory system for use in gene transfer. Proc Natl Acad Sci 91:8180–8184
13. Tang Y, Jackson M, Qian K, Phillips MI (2002) Hypoxia inducible double plasmid system for myocardial ischemia gene therapy. Hypertension 39:695–698. doi:10.1161/hy0202.103784
14. Phillips MI, Tang Y, Schmidt-Ott K, Qian K, Kagiyama S (2002) Vigilant vector: heart-specific promoter in an adeno-associated virus vector for cardioprotection. Hypertension 39:651–655. doi:10.1161/hy0202.103472
15. Dimos JT, Rodolfa KT, Niakan KK, Weisenthal LM, Mitsumoto H, Chung W, Croft GF, Saphier G, Leibel R, Goland R, Wichterle H, Henderson CE, Eggan K (2008) Induced pluripotent stem cells generated from patients with ALS can be differentiated into motor neurons. Science 321:1218–1221. doi:10.1126/science.1158799
16. Chang C, Lai Y, Pawlik KM, Liu K, Sun C, Li C, Schoeb TR, Townes TM (2009) Polycistronic lentiviral vector for "Hit and Run" reprogramming of adult skin fibroblasts to induced pluripotent stem cells. Stem Cells 27:1042–1049. doi:10.1002/stem.39
17. Stadtfeld M, Nagaya M, Utikal J, Weir G, Hochedlinger K (2008) Induced pluripotent stem cells generated without viral integration. Science 322:945–949. doi:10.1126/science.1162494
18. Jia F, Wilson KD, Sun N, Gupta DM, Huang M, Li Z, Panetta NJ, Chen ZY, Robbins RC, Kay MA, Longaker MT, Wu JC (2010) A nonviral minicircle vector for deriving human iPS cells. Nat Methods 7:197–199. doi:10.1038/nmeth.1426
19. Woltjen K, Michael IP, Mohseni P, Desai R, Mileikovsky M, Hamalainen R, Cowling R, Wang W, Liu P, Gertsenstein M, Kaji K, Sung HK, Nagy A (2009) piggyBac transposition reprograms fibroblasts to induced pluripotent stem cells. Nature 458:766–770. doi:10.1038/nature07863
20. Warren L, Manos PD, Ahfeldt T, Loh YH, Li H, Lau F, Ebina W, Mandal PK, Smith ZD, Meissner A, Daley GQ, Brack AS, Collins JJ, Cowan C, Schlaeger TM, Rossi DJ (2010) Highly efficient reprogramming to pluripotency and directed differentiation of human cells with synthetic modified mRNA. Cell Stem Cell 7:618–630. doi:10.1016/j.stem.2010.08.012
21. Marshall E (1999) Gene therapy death prompts review of adenovirus vector. Science 286:2244–2245. doi:10.1126/science.286.5448.2244
22. Maguire AM, Simonelli F, Pierce EA, Pugh EN, Mingozzi F, Bennicelli J, Banfi S, Marshall KA, Testa F, Surace EM, Rossi S, Lyubarsky A, Arruda VR, Konkle B, Stone E, Sun J, Jacobs J, Dell'Osso L, Hertle R, Ma J, Redmond TM, Zhu X, Hauck B, Zelenaia O, Shindler KS, Maguire MG, Wright JF, Volpe NJ, McDonnell JW, Auricchio A, High KA, Bennett J (2008) Safety and efficacy of gene transfer for Leber's congenital amaurosis. N Engl J Med 358:2240–2248. doi:10.1056/NEJMoa0802315
23. Phillips MI, Oliveira EM (2011) Associated adeno virus vector for producing induced pluripotent stem cells (IPS) for human somatic cells. In: Gholamrezanezhad A (ed) Stem cells in clinic and research. InTech, Rijeka, pp 747–764
24. Smithies O (2005) Many little things: one geneticist's view of complex diseases. Nat Rev Genet 6:419–425. doi:10.1038/nrg1605

25. Gu H, Marth J, Orban P, Mossmann H, Rajewsky K (1994) Deletion of a DNA polymerase beta gene segment in T cells using cell type-specific gene targeting. Science 265:103–106. doi:10.1126/science.8016642
26. Sinnayah P, Lindley TE, Staber PD, Davidson BL, Cassell MD, Davisson RL (2004) Targeted viral delivery of Cre recombinase induces conditional gene deletion in cardiovascular circuits of the mouse brain. Physiol Genomics 18:25–32. doi:10.1152/physiolgenomics.00048.2004
27. Sakai K, Agassandian K, Morimoto S, Sinnayah P, Cassell MD, Davisson RL, Sigmund CD (2007) Local production of angiotensin II in the subfornical organ causes elevated drinking. J Clin Invest 117:1088–1095. doi:10.1172/JCI31242
28. Phillips MI (2004) A Cre-loxP solution for defining the brain renin-angiotensin system. Focus on "Targeted viral delivery of Cre recombinase induces conditional gene deletion in cardiovascular circuits of the mouse brain". Physiol Genomics 18:1–3. doi:10.1152/physiolgenomics.00115.2004
29. Zamecnik PC, Stephenson ML (1978) Inhibition of Rous sarcoma virus replication and cell transformation by a specific oligodeoxynucleotide. Proc Natl Acad Sci 75:280–284
30. Wahlestedt C, Pich E, Koob G, Yee F, Heilig M (1993) Modulation of anxiety and neuropeptide Y-Y1 receptors by antisense oligodeoxynucleotides. Science 259:528–531. doi:10.1126/science.8380941
31. Gyurko R, Wielbo D, Ian Phillips M (1993) Antisense inhibition of AT1 receptor mRNA and angiotensinogen mRNA in the brain of spontaneously hypertensive rats reduces hypertension of neurogenic origin. Regul Pept 49:167–174. doi:10.1016/0167-0115(93)90438-E
32. Crooke ST (2004) Progress in antisense technology. Annu Rev Med 55:61–95. doi:10.1146/annurev.med.55.091902.104408
33. Kimura B, Mohuczy D, Tang X, Phillips MI (2001) Attenuation of hypertension and heart hypertrophy by adeno-associated virus delivering angiotensinogen antisense. Hypertension 37:376–380. doi:10.1161/01.HYP.37.2.376
34. Fire A, Xu S, Montgomery MK, Kostas SA, Driver SE, Mello CC (1998) Potent and specific genetic interference by double-stranded RNA in *Caenorhabditis elegans*. Nature 391:806–811. doi:10.1038/35888
35. Arnold AS, Tang YL, Qian K, Shen L, Valencia V, Phillips MI, Zhang YC (2007) Specific beta1-adrenergic receptor silencing with small interfering RNA lowers high blood pressure and improves cardiac function in myocardial ischemia. J Hypertens 25:197–205. doi:10.1097/01.hjh.0000254374.73241.ab
36. Lee RC, Feinbaum RL, Ambros V (1993) The *C. elegans* heterochronic gene lin-4 encodes small RNAs with antisense complementarity to lin-14. Cell 75:843–854
37. Esquela-Kerscher A, Slack FJ (2006) Oncomirs – microRNAs with a role in cancer. Nat Rev Cancer 6:259–269. doi:10.1038/nrc1840
38. Lagos-Quintana M, Rauhut R, Yalcin A, Meyer J, Lendeckel W, Tuschl T (2002) Identification of tissue-specific microRNAs from mouse. Curr Biol 12:735–739
39. Kumar MS, Lu J, Mercer KL, Golub TR, Jacks T (2007) Impaired microRNA processing enhances cellular transformation and tumorigenesis. Nat Genet 39:673–677. doi:10.1038/ng2003
40. Henderson SA, Spencer M, Sen A, Kumar C, Siddiqui MA, Chien KR (1989) Structure, organization, and expression of the rat cardiac myosin light chain-2 gene. Identification of a 250-base pair fragment which confers cardiac-specific expression. J Biol Chem 264:18142–18148
41. Rincon-Arano H, Valadez-Graham V, Guerrero G, Escamilla-Del-Arenal M, Recillas-Targa F (2005) YY1 and GATA-1 interaction modulate the chicken 3′-side alpha-globin enhancer activity. J Mol Biol 349:961–975. doi:10.1016/j.jmb.2005.04.040
42. Gossen M, Freundlieb S, Bender G, Muller G, Hillen W, Bujard H (1995) Transcriptional activation by tetracyclines in mammalian cells. Science 268:1766–1769
43. No D, Yao TP, Evans RM (1996) Ecdysone-inducible gene expression in mammalian cells and transgenic mice. Proc Natl Acad Sci USA 93:3346–3351
44. Pollock R, Issner R, Zoller K, Natesan S, Rivera VM, Clackson T (2000) Delivery of a stringent dimerizer-regulated gene expression system in a single retroviral vector. Proc Natl Acad Sci USA 97:13221–13226. doi:10.1073/pnas.230446297

45. Tang Y, Schmitt-Ott K, Qian K, Kagiyama S, Phillips MI (2002) Vigilant vectors: adeno-associated virus with a biosensor to switch on amplified therapeutic genes in specific tissues in life-threatening diseases. Methods 28:259–266
46. Semenza GL (2004) O2-regulated gene expression: transcriptional control of cardiorespiratory physiology by HIF-1. J Appl Physiol 96:1173–1177. doi:10.1152/japplphysiol.00770.2003
47. Tang YL, Tang Y, Zhang YC, Agarwal A, Kasahara H, Qian K, Shen L, Phillips MI (2005) A hypoxia-inducible vigilant vector system for activating therapeutic genes in ischemia. Gene Ther 12:1163–1170. doi:10.1038/sj.gt.3302513
48. Mingliang R, Bo Z, Zhengguo W (2011) Stem cells for cardiac repair: status, mechanisms, and new strategies. Stem Cells Int 2011:310928. doi:10.4061/2011/310928
49. Tang YL, Zhu W, Cheng M, Chen L, Zhang J, Sun T, Kishore R, Phillips MI, Losordo DW, Qin G (2009) Hypoxic preconditioning enhances the benefit of cardiac progenitor cell therapy for treatment of myocardial infarction by inducing CXCR4 expression. Circ Res 104:1209–1216. doi:10.1161/CIRCRESAHA.109.197723
50. Suzuki K, Smolenski RT, Jayakumar J, Murtuza B, Brand NJ, Yacoub MH (2000) Heat shock treatment enhances graft cell survival in skeletal myoblast transplantation to the heart. Circulation 102:III216–III221
51. Datta SR, Brunet A, Greenberg ME (1999) Cellular survival: a play in three Akts. Genes Dev 13:2905–2927
52. Mangi AA, Noiseux N, Kong D, He H, Rezvani M, Ingwall JS, Dzau VJ (2003) Mesenchymal stem cells modified with Akt prevent remodeling and restore performance of infarcted hearts. Nat Med 9:1195–1201. doi:10.1038/nm912
53. Meuillet EJ, Mahadevan D, Vankayalapati H, Berggren M, Williams R, Coon A, Kozikowski AP, Powis G (2003) Specific inhibition of the Akt1 pleckstrin homology domain by D-3-deoxy-phosphatidyl-myo-inositol analogues. Mol Cancer Ther 2:389–399
54. Otterbein LE, Choi AMK (2000) Heme oxygenase: colors of defense against cellular stress. Am J Physiol Lung Cell Mol Physiol 279:L1029–L1037
55. Tang YL, Tang Y, Zhang YC, Qian K, Shen L, Phillips MI (2005) Improved graft mesenchymal stem cell survival in ischemic heart with a hypoxia-regulated heme oxygenase-1 vector. J Am Coll Cardiol 46:1339–1350. doi:10.1016/j.jacc.2005.05.079
56. Lenfant C (1998) NHLBI at 50: reflections on a half-century of research on the heart, lungs, and blood. National Heart, Lung, and Blood Institute. Interview by Charles Marwick. JAMA 280:2062–2064
57. Koransky ML, Robbins RC, Blau HM (2002) VEGF gene delivery for treatment of ischemic cardiovascular disease. Trends Cardiovasc Med 12:108–114
58. Tang YL, Zhao Q, Zhang YC, Cheng L, Liu M, Shi J, Yang YZ, Pan C, Ge J, Phillips MI (2004) Autologous mesenchymal stem cell transplantation induce VEGF and neovascularization in ischemic myocardium. Regul Pept 117:3–10
59. Tang YL, Zhao Q, Qin X, Shen L, Cheng L, Ge J, Phillips MI (2005) Paracrine action enhances the effects of autologous mesenchymal stem cell transplantation on vascular regeneration in rat model of myocardial infarction. Ann Thorac Surg 80:229–236. doi:10.1016/j.athoracsur.2005.02.072; discussion 236–237
60. Lange C, Bassler P, Lioznov MV, Bruns H, Kluth D, Zander AR, Fiegel HC (2005) Hepatocytic gene expression in cultured rat mesenchymal stem cells. Transplant Proc 37:276–279. doi:10.1016/j.transproceed.2004.11.087
61. Elmadbouh I, Haider HK, Jiang S, Idris NM, Lu G, Ashraf M (2007) Ex vivo delivered stromal cell-derived factor-1alpha promotes stem cell homing and induces angiomyogenesis in the infarcted myocardium. J Mol Cell Cardiol 42:792–803. doi:10.1016/j.yjmcc.2007.02.001
62. Lei Y, Haider HK, Shujia J, Sim ES (2004) Therapeutic angiogenesis. Devising new strategies based on past experiences. Basic Res Cardiol 99:121–132. doi:10.1007/s00395-004-0447-x
63. Moayeri M, Hawley TS, Hawley RG (2005) Correction of murine hemophilia A by hematopoietic stem cell gene therapy. Mol Ther 12:1034–1042. doi:10.1016/j.ymthe.2005.09.007

64. Gangadharan B, Parker ET, Ide LM, Spencer HT, Doering CB (2006) High-level expression of porcine factor VIII from genetically modified bone marrow–derived stem cells. Blood 107:3859–3864. doi:10.1182/blood-2005-12-4961
65. Lavon N, Yanuka O, Benvenisty N (2006) The effect of overexpression of Pdx1 and Foxa2 on the differentiation of human embryonic stem cells into pancreatic cells. Stem Cells 24:1923–1930. doi:10.1634/stemcells.2005-0397
66. Ang SL, Wierda A, Wong D, Stevens KA, Cascio S, Rossant J, Zaret KS (1993) The formation and maintenance of the definitive endoderm lineage in the mouse: involvement of HNF3/fork-head proteins. Development 119:1301–1315
67. Chakrabarti SK, Mirmira RG (2003) Transcription factors direct the development and function of pancreatic beta cells. Trends Endocrinol Metab 14:78–84
68. Ahlgren U, Jonsson J, Jonsson L, Simu K, Edlund H (1998) β-Cell-specific inactivation of the mouseIpf1/Pdx1 gene results in loss of the β-cell phenotype and maturity onset diabetes. Genes Dev 12:1763–1768. doi:10.1101/gad.12.12.1763
69. Wang J, Elghazi L, Parker SE, Kizilocak H, Asano M, Sussel L, Sosa-Pineda B (2004) The concerted activities of Pax4 and Nkx2.2 are essential to initiate pancreatic beta-cell differentiation. Dev Biol 266:178–189
70. Drukker M, Katz G, Urbach A, Schuldiner M, Markel G, Itskovitz-Eldor J, Reubinoff B, Mandelboim O, Benvenisty N (2002) Characterization of the expression of MHC proteins in human embryonic stem cells. Proc Natl Acad Sci 99:9864–9869. doi:10.1073/pnas.142298299
71. Tang DQ, Lu S, Sun YP, Rodrigues E, Chou W, Yang C, Cao LZ, Chang LJ, Yang LJ (2006) Reprogramming liver-stem WB cells into functional insulin-producing cells by persistent expression of Pdx1- and Pdx1-VP16 mediated by lentiviral vectors. Lab Invest 86:83–93. doi:10.1038/labinvest.3700368
72. Li Y, Zhang R, Qiao H, Zhang H, Wang Y, Yuan H, Liu Q, Liu D, Chen L, Pei X (2007) Generation of insulin-producing cells from PDX-1 gene-modified human mesenchymal stem cells. J Cell Physiol 211:36–44. doi:10.1002/jcp.20897
73. Noguchi H, Xu G, Matsumoto S, Kaneto H, Kobayashi N, Bonner-Weir S, Hayashi S (2006) Induction of pancreatic stem/progenitor cells into insulin-producing cells by adenoviral-mediated gene transfer technology. Cell Transplant 15:929–938
74. Cheng H, Zhang YC, Wolfe S, Valencia V, Qian K, Shen L, Tang YL, Hsu WH, Atkinson MA, Phillips MI (2011) Combinatorial treatment of bone marrow stem cells and stromal cell-derived factor 1 improves glycemia and insulin production in diabetic mice. Mol Cell Endocrinol 345:88–96. doi:10.1016/j.mce.2011.07.024
75. Dinsmore J, Ratliff J, Deacon T, Pakzaban P, Jacoby D, Galpern W, Isacson O (1996) Embryonic stem cells differentiated in vitro as a novel source of cells for transplantation. Cell Transplant 5:131–143
76. McDonald JW, Liu XZ, Qu Y, Liu S, Mickey SK, Turetsky D, Gottlieb DI, Choi DW (1999) Transplanted embryonic stem cells survive, differentiate and promote recovery in injured rat spinal cord. Nat Med 5:1410–1412. doi:10.1038/70986
77. Tang X, Cai PQ, Lin YQ, Oudega M, Blits B, Xu L, Yang YK, Zhou TH (2006) Genetic engineering neural stem cell modified by lentivirus for repair of spinal cord injury in rats. Chin Med Sci J 21:120–124
78. Cai PQ, Tang X, Lin YQ, Martin O, Sun GY, Xu L, Yang YK, Zhou TH (2006) The experimental study of genetic engineering human neural stem cells mediated by lentivirus to express multigene. Chin J Traumatol 9:43–49
79. Blits B, Kitay BM, Farahvar A, Caperton CV, Dietrich WD, Bunge MB (2005) Lentiviral vector-mediated transduction of neural progenitor cells before implantation into injured spinal cord and brain to detect their migration, deliver neurotrophic factors and repair tissue. Restor Neurol Neurosci 23:313–324
80. Coutts M, Keirstead HS (2008) Stem cells for the treatment of spinal cord injury. Exp Neurol 209:368–377. doi:10.1016/j.expneurol.2007.09.002

81. Lakshminarayan K, Schissel C, Anderson DC, Vazquez G, Jacobs DR Jr, Ezzeddine M, Luepker RV, Virnig BA (2011) Five-year rehospitalization outcomes in a cohort of patients with acute ischemic stroke: medicare linkage study. Stroke. doi:10.1161/STROKEAHA.110.605600
82. Eglitis MA, Mezey É (1997) Hematopoietic cells differentiate into both microglia and macroglia in the brains of adult mice. Proc Natl Acad Sci 94:4080–4085
83. Brazelton TR, Rossi FMV, Keshet GI, Blau HM (2000) From marrow to brain: expression of neuronal phenotypes in adult mice. Science 290:1775–1779. doi:10.1126/science.290.5497.1775
84. Chen J, Li Y, Wang L, Lu M, Zhang X, Chopp M (2001) Therapeutic benefit of intracerebral transplantation of bone marrow stromal cells after cerebral ischemia in rats. J Neurol Sci 189:49–57
85. Li Y, Chen J, Chen XG, Wang L, Gautam SC, Xu YX, Katakowski M, Zhang LJ, Lu M, Janakiraman N, Chopp M (2002) Human marrow stromal cell therapy for stroke in rat. Neurology 59:514–523
86. Iihoshi S, Honmou O, Houkin K, Hashi K, Kocsis JD (2004) A therapeutic window for intravenous administration of autologous bone marrow after cerebral ischemia in adult rats. Brain Res 1007:1–9. doi:10.1016/j.brainres.2003.09.084
87. Baker AH, Sica V, Work LM, Williams-Ignarro S, de Nigris F, Lerman LO, Casamassimi A, Lanza A, Schiano C, Rienzo M, Ignarro LJ, Napoli C (2007) Brain protection using autologous bone marrow cell, metalloproteinase inhibitors, and metabolic treatment in cerebral ischemia. Proc Natl Acad Sci USA 104:3597–3602. doi:10.1073/pnas.0611112104
88. Kobune M, Kawano Y, Ito Y, Chiba H, Nakamura K, Tsuda H, Sasaki K, Dehari H, Uchida H, Honmou O, Takahashi S, Bizen A, Takimoto R, Matsunaga T, Kato J, Kato K, Houkin K, Niitsu Y, Hamada H (2003) Telomerized human multipotent mesenchymal cells can differentiate into hematopoietic and cobblestone area-supporting cells. Exp Hematol 31:715–722
89. Harley CB (1991) Telomere loss: mitotic clock or genetic time bomb? Mutat Res 256:271–282
90. Jiang XR, Jimenez G, Chang E, Frolkis M, Kusler B, Sage M, Beeche M, Bodnar AG, Wahl GM, Tlsty TD, Chiu CP (1999) Telomerase expression in human somatic cells does not induce changes associated with a transformed phenotype. Nat Genet 21:111–114. doi:10.1038/5056
91. Bodnar AG, Ouellette M, Frolkis M, Holt SE, Chiu C, Morin GB, Harley CB, Shay JW, Lichtsteiner S, Wright WE (1998) Extension of life-span by introduction of telomerase into normal human cells. Science 279:349–352. doi:10.1126/science.279.5349.349
92. Honma T, Honmou O, Iihoshi S, Harada K, Houkin K, Hamada H, Kocsis JD (2006) Intravenous infusion of immortalized human mesenchymal stem cells protects against injury in a cerebral ischemia model in adult rat. Exp Neurol 199:56–66. doi:10.1016/j.expneurol.2005.05.004
93. Crigler L, Robey RC, Asawachaicharn A, Gaupp D, Phinney DG (2006) Human mesenchymal stem cell subpopulations express a variety of neuro-regulatory molecules and promote neuronal cell survival and neuritogenesis. Exp Neurol 198:54–64. doi:10.1016/j.expneurol.2005.10.029
94. Kurozumi K, Nakamura K, Tamiya T, Kawano Y, Kobune M, Hirai S, Uchida H, Sasaki K, Ito Y, Kato K, Honmou O, Houkin K, Date I, Hamada H (2004) BDNF gene-modified mesenchymal stem cells promote functional recovery and reduce infarct size in the rat middle cerebral artery occlusion model. Mol Ther 9:189–197. doi:10.1016/j.ymthe.2003.10.012
95. Kurozumi K, Nakamura K, Tamiya T, Kawano Y, Ishii K, Kobune M, Hirai S, Uchida H, Sasaki K, Ito Y, Kato K, Honmou O, Houkin K, Date I, Hamada H (2005) Mesenchymal stem cells that produce neurotrophic factors reduce ischemic damage in the rat middle cerebral artery occlusion model. Mol Ther 11:96–104. doi:10.1016/j.ymthe.2004.09.020
96. Schwarz EJ, Alexander GM, Prockop DJ, Azizi SA (1999) Multipotential marrow stromal cells transduced to produce L-DOPA: engraftment in a rat model of Parkinson disease. Hum Gene Ther 10:2539–2549. doi:10.1089/10430349950016870
97. Schwarz EJ, Reger RL, Alexander GM, Class R, Azizi SA, Prockop DJ (2001) Rat marrow stromal cells rapidly transduced with a self-inactivating retrovirus synthesize L-DOPA in vitro. Gene Ther 8:1214–1223. doi:10.1038/sj.gt.3301517
98. Shen Y, Muramatsu SI, Ikeguchi K, Fujimoto KI, Fan DS, Ogawa M, Mizukami H, Urabe M, Kume A, Nagatsu I, Urano F, Suzuki T, Ichinose H, Nagatsu T, Monahan J, Nakano I, Ozawa K (2000) Triple transduction with adeno-associated virus vectors expressing tyrosine hydroxylase, aromatic-L-amino-acid decarboxylase, and GTP cyclohydrolase I for gene therapy of Parkinson's disease. Hum Gene Ther 11:1509–1519. doi:10.1089/10430340050083243

99. Sun M, Kong L, Wang X, Holmes C, Gao Q, Zhang GR, Pfeilschifter J, Goldstein DS, Geller AI (2004) Coexpression of tyrosine hydroxylase, GTP cyclohydrolase I, aromatic amino acid decarboxylase, and vesicular monoamine transporter 2 from a helper virus-free herpes simplex virus type 1 vector supports high-level, long-term biochemical and behavioral correction of a rat model of Parkinson's disease. Hum Gene Ther 15:1177–1196. doi:10.1089/hum.2004.15.1177
100. Fisher LJ, Raymon HK, Gage FH (1993) Cells engineered to produce acetylcholine: therapeutic potential for Alzheimer's disease. Ann N Y Acad Sci 695:278–284
101. Park D, Lee HJ, Joo SS, Lim I, Matsumoto A, Tooyama I, Kim YB, Kim SU (2012) Human neural stem cells over-expressing choline acetyltransferase restore cognition in rat model of cognitive dysfunction. Exp Neurol. doi:10.1016/j.expneurol.2011.12.040
102. Lad SP, Neet KE, Mufson EJ (2003) Nerve growth factor: structure, function and therapeutic implications for Alzheimer's disease. Curr Drug Targets CNS Neurol Disord 2:315–334
103. Tuszynski MH, Thal L, Pay M, Salmon DP, U HS, Bakay R, Patel P, Blesch A, Vahlsing HL, Ho G, Tong G, Potkin SG, Fallon J, Hansen L, Mufson EJ, Kordower JH, Gall C, Conner J (2005) A phase 1 clinical trial of nerve growth factor gene therapy for Alzheimer disease. Nat Med 11:551–555. doi:10.1038/nm1239
104. Iwata N, Tsubuki S, Takaki Y, Shirotani K, Lu B, Gerard NP, Gerard C, Hama E, Lee H, Saido TC (2001) Metabolic regulation of brain Aβ by neprilysin. Science 292:1550–1552. doi:10.1126/science.1059946
105. Lazarov O, Marr RA (2010) Neurogenesis and Alzheimer's disease: at the crossroads. Exp Neurol 223:267–281. doi:10.1016/j.expneurol.2009.08.009
106. Hemming ML, Patterson M, Reske-Nielsen C, Lin L, Isacson O, Selkoe DJ (2007) Reducing amyloid plaque burden via ex vivo gene delivery of an Abeta-degrading protease: a novel therapeutic approach to Alzheimer disease. PLoS Med 4:e262. doi:10.1371/journal.pmed.0040262
107. Magga J, Savchenko E, Malm T, Rolova T, Pollari E, Valonen P, Lehtonen Š, Jantunen E, Aarnio J, Lehenkari P, Koistinaho M, Muona A, Koistinaho J (2011) Production of monocytic cells from bone marrow stem cells: therapeutic usage in Alzheimer's disease. J Cell Mol Med 16:1582–1838. doi:10.1111/j.1582-4934.2011.01390.x
108. Bruder SP, Kurth AA, Shea M, Hayes WC, Jaiswal N, Kadiyala S (1998) Bone regeneration by implantation of purified, culture-expanded human mesenchymal stem cells. J Orthop Res 16:155–162. doi:10.1002/jor.1100160202
109. Gazit D, Turgeman G, Kelley P, Wang E, Jalenak M, Zilberman Y, Moutsatsos I (1999) Engineered pluripotent mesenchymal cells integrate and differentiate in regenerating bone: a novel cell-mediated gene therapy. J Gene Med 1:121–133. doi:10.1002/(SICI)1521-2254(199903/04)1:2<121::AID-JGM26>3.0.CO;2-J
110. Moutsatsos IK, Turgeman G, Zhou S, Kurkalli BG, Pelled G, Tzur L, Kelley P, Stumm N, Mi S, Muller R, Zilberman Y, Gazit D (2001) Exogenously regulated stem cell-mediated gene therapy for bone regeneration. Mol Ther 3:449–461. doi:10.1006/mthe.2001.0291
111. Hasharoni A, Zilberman Y, Turgeman G, Helm GA, Liebergall M, Gazit D (2005) Murine spinal fusion induced by engineered mesenchymal stem cells that conditionally express bone morphogenetic protein-2. J Neurosurg Spine 3:47–52. doi:10.3171/spi.2005.3.1.0047
112. Hung S, Deng W, Yang WK, Liu R, Lee C, Su T, Lin R, Yang D, Chang C, Chen W, Wei H, Gelovani JG (2005) Mesenchymal stem cell targeting of microscopic tumors and tumor stroma development monitored by noninvasive in vivo positron emission tomography imaging. Clin Cancer Res 11:7749–7756. doi:10.1158/1078-0432.CCR-05-0876
113. Wong VL, Rieman DJ, Aronson L, Dalton BJ, Greig R, Anzano MA (1989) Growth-inhibitory activity of interferon-beta against human colorectal carcinoma cell lines. Int J Cancer 43:526–530
114. Lokshin A, Mayotte JE, Levitt ML (1995) Mechanism of interferon beta-induced squamous differentiation and programmed cell death in human non-small-cell lung cancer cell lines. J Natl Cancer Inst 87:206–212
115. Studeny M, Marini FC, Champlin RE, Zompetta C, Fidler IJ, Andreeff M (2002) Bone marrow-derived mesenchymal stem cells as vehicles for interferon-beta delivery into tumors. Cancer Res 62:3603–3608

116. Studeny M, Marini FC, Dembinski JL, Zompetta C, Cabreira-Hansen M, Bekele BN, Champlin RE, Andreeff M (2004) Mesenchymal stem cells: potential precursors for tumor stroma and targeted-delivery vehicles for anticancer agents. J Natl Cancer Inst 96:1593–1603. doi:10.1093/jnci/djh299
117. Nakamizo A, Marini F, Amano T, Khan A, Studeny M, Gumin J, Chen J, Hentschel S, Vecil G, Dembinski J, Andreeff M, Lang FF (2005) Human bone marrow–derived mesenchymal stem cells in the treatment of gliomas. Cancer Res 65:3307–3318. doi:10.1158/0008-5472.CAN-04-1874
118. Xin H, Kanehira M, Mizuguchi H, Hayakawa T, Kikuchi T, Nukiwa T, Saijo Y (2007) Targeted delivery of CX3CL1 to multiple lung tumors by mesenchymal stem cells. Stem Cells 25:1618–1626. doi:10.1634/stemcells.2006-0461
119. Stagg J, Lejeune L, Paquin A, Galipeau J (2004) Marrow stromal cells for interleukin-2 delivery in cancer immunotherapy. Hum Gene Ther 15:597–608. doi:10.1089/104303404323142042
120. Kucerova L, Altanerova V, Matuskova M, Tyciakova S, Altaner C (2007) Adipose tissue–derived human mesenchymal stem cells mediated prodrug cancer gene therapy. Cancer Res 67:6304–6313. doi:10.1158/0008-5472.CAN-06-4024
121. Harkin DP, Johnston PG (2003) 5-fluorouracil: mechanisms of action and clinical strategies. Nat Rev Cancer 3:330–338

Chapter 7
Induced Pluripotent Stem Cells: Basics and the Application in Disease Model and Regenerative Medicine

Zhiyong Liu and Feng Li

Abstract Since the advent of induced pluripotent stem (iPS) cells at 2006, a flood of researches have been performed to study the application of iPS-based approaches in generating various types of disease models, personalized regenerative medicine by gene therapy and tissue engineering. In this chapter, we first reviewed the comparisons between the iPS cells and the normal embryonic stem (ES) cells, followed by providing examples to highlight the advantages of using iPS cells for disease models and disease-specific and patient-specific gene repair and cell-replacement therapy. Last, using auditory organ cochlea as an example, we discussed the current status and challenges in using iPS-based approaches for regenerating auditory hair cells.

Keywords Cell-replacement therapy • Induced pluripotent stem cells • Patient-specific gene repair • Alzheimer's disease • Hutchinson–Gilford progeria syndrome

7.1 Introduction

Pluripotent stem cells during normal development (i.e., mouse and human), in a strict definition, refer to the cells that could give rise to the entire organism. Embryonic stem (ES) cells deriving from the inner cell mass of the mouse blastocysts are able to differentiate into all cell types of the three germ layers—ectoderm, endoderm, and mesoderm—and eventually form the entire mouse embryo [1–4]. In 1998, human ES

Z. Liu
Departments of Developmental Neurobiology,
St. Jude Children's Research Hospital, Memphis, TN 38105, USA

F. Li, PhD (✉)
Department of Pharmaceutical Sciences,
School of Pharmacy, Hampton University,
Kittrell Hall, Hampton, VA 23668, USA
e-mail: feng.li@hamptonu.edu

M.K. Danquah, R.I. Mahato (eds.), *Emerging Trends in Cell and Gene Therapy*,
DOI 10.1007/978-1-62703-417-3_7,

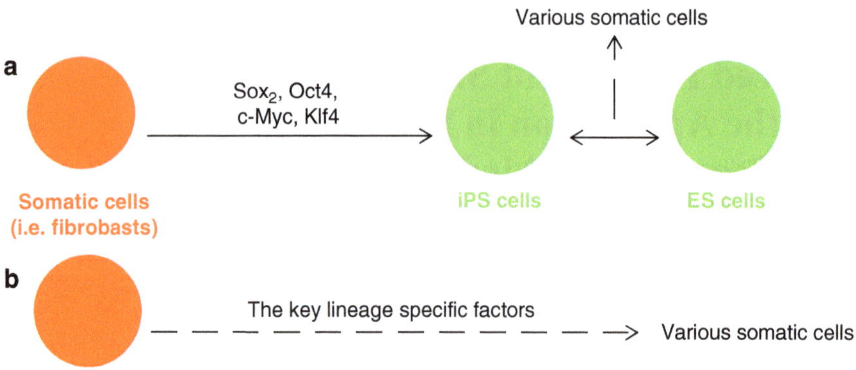

Fig. 7.1 Two approaches to generate one somatic cell from the other. (**a**) is the two-step approach. Somatic cells, such as fibroblasts, are converted into iPS cells following overexpression of the 4 iPS factors (Sox2, Oct4, c-Myc, Klf4). The iPS cells resemble the normal ES cells in various aspects. iPS cells could be further directed into various somatic cells in each permissing conditions. (**b**) is the one-step approach. One somatic cell type like fibroblast can be directly changed into any other somatic cell type by overexpressing the key lineage-specific factors (normally are transcriptional factors)

cells were derived from human blastocysts [5], after which human ES cells have been regarded as promising cell resources for cell transplantation therapeutics for various human disorders such as spinal cord injury, diabetes, and muscle dystrophy. However, using human ES cells as clinical applications has two main difficulties: The first one is the controversy in ethics involved in using human embryos as cell sources of ES cells; the second one is the tissue rejection after implantation because of the immune incompatibility between donor cells and host tissues.

To bypass these limitations, stem cell biologists keep looking for an alternative approach to generate pluripotent stem cells (Fig. 7.1). In 1960, Gurdon and his colleagues at the first time transferred the nuclei of adult frog intestinal cells into the frog enucleated oocytes and generated mature and fertile frogs [6]. This technique is referred to as somatic cell nuclear transfer (SCNT) [7]. The beauty of this study is to highlight the presence of the pluripotency-inducing factors in the cytoplasm of the oocytes. This notion is further supported by the first cloned mammal, Dolly the sheep, by using the same SCNT approach [8]. In addition, these above 2 cloning studies in nonmammal and mammals support the idea that the unknown yet pluripotency-inducing factors are very powerful and not only erase the manifestations of the differentiation/epigenetic signatures of the somatic nucleus but also reset the entire chromosome ready for generating an entire new animal [9].

With the driving hypothesis that the key transcriptional factors regulating or maintaining the pluripotency in both early embryos and ES cells are able to convert somatic cells into ES-like cells, Takahashi and Yamanaka's landmark work published in 2006 showed that overactivation of octamer-binding transcription factor 3/4 (Oct3/4), (sex-determining region Y)-box 2 (Sox2), c-Myc, and Kruppel-like factor 4 (Klf4) can reprogram a small fraction of mouse fibroblast cells into pluripotent stem cells, referred to as induced pluripotent stem (iPS) cells [10]. Although it is not clear how these 4 factors can do such a job, human iPS cells are generated in 2007 with the same 4 factors [11] or slight different combination of 4 factors (Oct4, Sox2, Nanog, and Lin 28) [12].

Following the above three pioneer studies, currently, there are multiple new ways to reprogram somatic cells into iPS cells [13]. In addition, the somatic cell types have been expanded to a wide range of other cell types such as gastric and liver cells [14], pancreatic cells [15], neuronal stem cells [16, 17], mature B cells [18], melanocytes [19], adipose cells [20], and keratinocytes [21]. These studies suggest the general capacity of these 4 iPS factors to alter the cell fate regardless of the origin of the somatic cells and extent of their differentiated state. To minimize the potential oncogenic problems caused by genomic integration of these iPS factors, DNA-free method was invented, using polyarginine (i.e., 11R) fusing version of 4 iPS factor proteins in combination with a small molecule known as valproic acid (VPA) [22].

In addition, small molecules have been shown to be able to replace Klf4 to generate iPS cells [23], which is an advancement toward the final goal of generating human iPS cells with a complete chemical approach [24, 25]. Taken together, the emergence of human iPS derived from diverse cell types makes it possible to use transplantation of iPS cells generated from patient autologous somatic tissue cells, as to be discussed later on.

7.2 Comparison Between ES Cells and iPS Cells

Before reviewing the potential application of iPS in drug screening and regenerative medicine, it is worthwhile to discuss whether iPS cells are completely identical to ES or, if not, to what extent, are similar to ES cells. The iPS and ES cells are neither identical nor distinct. They are two cell populations having their own unique characteristics (genetically and/or epigenetically) and can compensate the functions of each other, and their heterogeneities and behaviors are more complex than what was expected previously [13]. Apparently, the answer is not straightforward and needs further intensive studies. In the following sections, the properties of ES and IPS determined with various approaches will be compared and discussed.

7.2.1 Morphology

First of all, ES cells and iPS cells of the same species have similar morphologies. However, mouse ES/iPS cells are different from the human ES/iPS cells. Human ES-/iPS-cell colonies are flatter than mouse ES/iPS cells, and the latter are prone to form the dome-shaped and retractile structures (Fig. 7.2). Interestingly, human ES-/iPS-cell colonies and mouse epiblast-derived stem cells are comparable in terms of morphologies (Fig. 7.2d–f). Such disparities and similarities are proposed to reflect the distinct developmental stages of the stem cells or the “naïve state” versus “primitive state” of these two types of pluripotent stem cells [13].

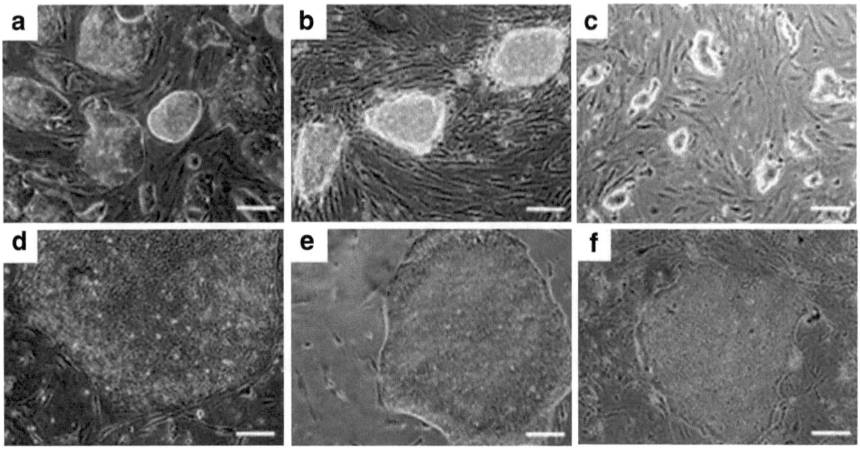

Fig. 7.2 Morphologies of mouse and human ES and iPS cells. (**a**, **b**) Mouse ES (**a**) and iPS (**b**) have similar dome-shaped colonies. (**c**) Human iPS cells that are treated with chemical inhibitors are directed into a naïve pluripotent state, and their morphologies are similar to mouse ES/iPS colonies shown in (**a**, **b**). (**d**–**f**) Human ES (**d**) and iPS (**e**), similar to mouse epiblast-derived stem cells (**f**), have the flat morphology. It might indicate a primitive pluripotent state (Reprinted from [13])

7.2.2 Gene-Expression Patterns

Both mouse and human iPS cells, to a high extent, share similar gene-expression patterns, especially the pluripotent genes that are needed to keep cells at the undifferentiated state and to differentiate into any cell types when exposed to the appropriated environmental induction signals. For an example, iPS cells express ES-specific surface antigens including stage-specific embryonic antigen 3 (SSEA-3); stage-specific embryonic antigen 4 (SSEA-4); tumor-related antigen (TRA)-1-60, TRA-1-81, and TRA-2-49/6E (alkaline phosphatase); as well as the undifferentiated ES cell-marker genes OCT3/4, SOX2, NANOG, and growth and differentiation factor 3 (GDF3) [11, 26]. In addition, genome-wide DNA microarray analyses show that the global gene-expression patterns are similar between iPS cells and ES cells [11]. Last, the epigenetic status and DNA methylation pattern or the bivalent state (H3K4 and H3K27) of genes (i.e., GATA 6, MSX2, PAX6, and hand1) are similar, but not identical, between ES and iPS cells.

7.2.3 Telomerase Activity

Pluripotent stem cells are characterized by their remarkable ability to undergo self-renewal and proliferation, which means that they have the strong telomerase activities [27]. Telomerase is a ribonucleoprotein polymerase which maintains telomere ends by addition of the telomere repeat *TTAGGG* after each cell division. Telomerase reverse transcriptase (TERT) is a catalytic subunit of the enzyme telomerase. Human iPS cells have comparable levels of TERT with the human ES cell line H9 or human embryonic carcinoma cell line, NTERA [11]. Interestingly, reprogramming process itself could increase telomerase activity of the iPS cells that are derived from human patients with a disorder of telomere maintenance (i.e., dyskeratosis congenita) by upregulating the telomerase RNA component (TERC) [28].

7.2.4 Capacity of Forming Embryonic Body

Embryonic body (EB) formation assay is a widely used method to determine the potentials of human and mouse ES cells to differentiate into all the three germ line layer-derived somatic cells [29]. Embryonic bodies are aggregates of cells derived from ES cells. Upon aggregation in vitro, EB cells begin differentiation and, to a limited extent, are able to recapitulate embryonic development. Therefore, embryonic body cells are mixed cell populations that contain a large variety of differentiated cell types. Similarly, when iPS cells are cultured in suspension culture, they form ball-shaped embryonic body and differentiate into ectoderm-, mesoderm-, and endoderm-derived lineage-specific somatic cell types [11].

7.2.5 Teratoma Formation

Teratoma formation is another routine way to determine the pluripotency of the stem cells in vivo. A teratoma is a tumor-containing tissue or organ components that resemble normal derivatives of all three germ layers. When human- and mouse-derived iPS cells are injected into immunodeficient or nude mice, they produced tumor-containing tissues of all three germ layers [10, 11]. In addition, when the mouse iPS cells are microinjected into blastocysts, iPS-derived cells are found in embryos [10].

7.2.6 Tetraploid Complementation Assay

To further determine the similar functionalities between ES cells and iPS cells, especially in terms of whether iPS cells are fully pluripotent, the most stringent or gold standard is to generate a complete embryo exclusively from iPS cells, which is referred as tetraploid complementation assay. This technique allows cells of two mammalian embryos being combined to form a new embryo [30]. Recently, mouse embryonic fibroblast-derived iPS cells are microinjected into the cavity of tetraploid blastocysts, and the tetraploid-complemented embryos are able to develop into live animals [31]. Due to the failure of the tetraploid ES cells to generate live embryos, this work can serve as a proof of principle that somatic cell-derived iPS cells indeed are fully pluripotent and can develop into all three germ layer-derived cell lineages.

7.3 Applications of iPS Cells in Human Disease Models

It is always difficult and challenging to study human diseases especially the developmental disorders. The appearance of iPS strategy makes it easy to get plenty of pluripotent stem cells which can be further directed to the interested lineage-specific cell types in vitro [32]. It provides an invaluable model to study the dynamics of the pathogenesis and for small-molecule drug screening. Here, we discussed a few recently published reports to highlight the contributions of iPS-cell-based approaches in developing human disease models and their application in drug screening.

7.3.1 Spinal Muscular Atrophy

Spinal muscular atrophy (SMA) is a genetic children disorder caused by survival of motor neuron 1 (*SMN1)* gene defect. *SMN1* gene encodes an SMN protein which is required for the survival of lower motor neurons. Therefore, the SMA patients suffer from the death of neurons in the anterior horn of spinal cord and subsequent system-wide muscle wasting (atrophy) [33]. SMN protein is found in both the cytoplasm and nuclear aggregate structures called gems, and the number of gems present is inversely correlated to disease severity [34].

Previous studies have tried to use different compounds with the aim to increase the expression levels of SMN in the various cell lines or fibroblasts derived from SMA patient. Unfortunately, the identified drugs failed in clinical trials partially because of the potential different mechanisms by which SMA protein works in cell lines and motor neurons. Given this explanation is true, it will be important to generate a motor neuron line in vitro which can resemble the pathogenesis of the SMA in vivo and further use such a neuron line for drug screening.

Recently, iPS cells derived from SMA patient (referred to as iPS–SMA) or health controls (referred to as iPS–WT) were directed to differentiate into motor neurons in vitro [35]. Both iPS–SMA- and iPS–WT-derived neurons express the motor neuron transcription factors homeobox protein Hox-B4 (HOXB4), oligodendrocyte transcription factor 2 (OLIG2), insulin gene enhancer protein ISL-1 (ISLET1, also known as ISL1), and homeobox protein 9 (HB9), all of which are important for the normal motor neuron development [36, 37], and the terminal differentiation marker of motor neurons such as SMI-32 and choline acetyltransferase (ChAT). Although similar number of motor neurons are generated from iPS–SMA and iPS–WT cells by 4 weeks in cultures, the number of iPS–SMA-derived motor neurons decreased significantly, compared with those of iPS–WT-cell-derived motor neurons by 6 weeks in vitro. Of note, the total number of the entire neuronal populations between these two groups is still comparable. It strongly supports that there is a specific effect of the SMA phenotype on motor neurons.

The iPS–SMA-derived motor neurons could also be potentially used as invaluable models for drug screening. Indeed, the two compounds valproic acid and tobramycin can significantly increase the number of gems in iPS–SMA cells [35]. Although further studies are needed to clarify whether the compounds that elevate SMN levels in SMA patient-derived iPSCs can have the same effect in motor neurons and thus rescue motor neuron loss in patients, it is a promising approach to use motor neurons derived from iPS cells of SMA patients for probing more effective drugs that can alone or synergistically rescue the SMA phenotypes in vivo [38].

7.3.2 Rett Syndrome

Rett syndrome (RTT) is another developmental neurological disease and is part of the larger group of autism spectrum disorders that are characterized by impaired social interaction and repetitive behavior. Rett syndrome is caused by mutations in X-linked gene coding methyl-CpG-binding protein 2 (MeCP2), which is a protein involved in DNA methylation and is able to bind specifically to methylated DNA and regulate a variety of different genes [39, 40].

Human RTT patient fibroblast cell-derived iPS cells are developed that carry different MeCP2 mutations [41]. RTT patient-derived iPS cells can be directed into neural progenitor cells and functional neurons expressing γ-amino butyric acid (GABA) and vesicular glutamate transpoter-1-positive (VGLUT1). Compared with neurons derived from iPS cells generated from normal control individuals, those from RTT patient-derived iPS cells have significantly decreased number of dendritic spines and synapses, smaller soma size, altered calcium signaling, and electrophysiological defects [41], which are the characteristics observed in the postmortem brains of patients with RTT.

MeCP2-targeted mutant mice are lethal and have phenotypes similar to human RTT. Previous studies indicated that reexpression of MeCP2 can partially rescue and result in a prolonged life span and delayed onset of the neurological defects in the MeCP2-null background mouse models [42]. In addition, insulin-like growth factor 1 (IGF-1) treatment can ameliorate the phenotypes of mouse RTT syndromes such as restoring spine density and synaptic amplitude, increasing postsynaptic density protein 95 (PSD-95), as well as stabilizing cortical plasticity to wild-type comparable levels [43]. When human RTT-derived neurons are treated in vitro with IGF1, an increase of the glutamatergic synapse number is observed. It suggests that human RTT iPS-cell-derived neurons are potential models for drug screening. Indeed, by using human RTT iPS-cell-derived neurons, aminoglycoside antibiotics such as gentamicin were found to increase MeCP2 expression levels [41].

7.3.3 Familial Dysautonomia

Familial dysautonomia (FD), also referred to as hereditary sensory and autonomic neuropathy III (HSAN-III) or Riley–Day syndrome, is a fatal autosomal recessive disease [44]. Patients suffering from FD have degeneration of sensory and autonomic neurons. Most, if not all, of the FD patients carry a point mutation in the *I-κ-B kinase complex-associated protein* (*IKBKAP*) gene, which leads to various levels of mRNA splicing skipping at the exon 20 of *IKBKAP* gene and decreased levels of IKAP protein, especially in the neuron cell types [45]. The molecular mechanisms of FD are poorly understood partially due to the limited access to tissues affected by FD, such as neural crest precursor cells and peripheral neurons [46].

To bypass the limitation of cell sources to study the FD, human fibroblasts obtained from FD patient were converted to iPS cells through transfection with lentiviral vectors encoding four iPS factors: Oct4, Sox2, Klf4, and c-Myc [46]. Some human fibroblasts become FD–iPS cells which resemble human ESC cells in many aspects as described in previous section but still carry the genetic deficient in the *IKBKAP* gene. These FD–iPS cells can be directed to differentiate into cell types of all three germ layers. Specifically, when FD–iPS cells were directed into neural crest precursors, they express much lower levels of *IKBKAP* transcript than those differentiated from the control human fibroblast-derived iPS cells. It means that the FD patient-derived neural crest precursor cells in the in vitro conditions mimic the pathogenesis occurring in the neural crest lineages (primary tissue affected in FD). It offers an invaluable model to perform comparative transcriptome analysis between the FD–iPS- and control iPS-derived neural crest precursors. Among the top candidate disease-related genes discovered, many of them turn out to be the genes (i.e., *ASCL1*) that play critical developmental roles in peripheral neurogenesis and neuronal differentiation [47]. The decreased expression level of these developmental genes resulted in defective peripheral neurogenesis and reduced focal adhesions which are required for cell spreading and migration [46].

As discussed, because the in vitro neural crest precursors derived from patient FD–iPS cells resemble the in vivo FD pathogenesis, they should be an ideal model for performing high-throughput screening of therapeutic agents that could potentially attenuate the defective differentiation and migration phenotypes of the peripheral neurons in FD patients. Indeed, it has been shown that long-term exposure of FD–iPS-cell-derived neural crest precursors to plant hormone kinetin [48] was able to increase the differentiation efficiency of the neural crest lineages and upregulate these aforementioned crucial developmental genes (i.e., *ASCL1*). However, the migration defective was not attenuated by kinetin treatment [46]. It suggests that combination of different drugs might rescue various neural defects. The iPS-cell-based approach is a promising model to identify more drugs in future for treatment of FD patients.

7.3.4 Alzheimer's Disease

Alzheimer's disease (AD) is a neurodegenerative disease, which is characterized by gradual memory loss and cognitive disorders in aged human beings. Its pathological features include deposition of amyloid plaques, neurofibrillary tangles, and neuronal and synaptic loss in particular brain regions [49]. Although the detailed mechanism underlying the AD pathogenesis is not clear yet, it has been hypothesized that amyloid β peptide (Aβ) accumulation triggers the complex pathogenesis pathways and eventually leads to the apparent AD phenotypes. This amyloid hypothesis was proposed primarily based on a deregulation of the balance between the production and clearance of Aβ [50].

The lack of the human neuronal cell types prevents the progress of dissecting the pathogenic cascades of AD and further probing drugs to attenuate the AD symptoms, because brain tissues are only available postmortem for studies. Recently, human dermal fibroblasts were transfected by three reprogramming factors (Oct3/4, Sox2, and Klf4) to generate human iPS cells which were further directed to differentiate into forebrain neurons expressing markers of Foxg1, Cux1, Satb2, Ctip2, and Tbr1 and especially the amyloid precursor protein, β-secretase, and γ-secretase components [51]. Therefore, those human iPS-cell-derived neuronal cells are able to synthesize Aβ. They can serve as a model to screen drugs that can block the aggregation of Aβ. These human iPS-cell-derived neuronal cells have been used to test the effects of three types of identified drugs: BSI (β-secretase inhibitor IV), GSI (γ-secretase inhibitor XXI), and sulindac sulfide (nonsteroidal anti-inflammatory drug, NSAID). The results indicated that the responsiveness of these neuronal cells to drug treatment was dependent on the differentiation state of these neurons, which might provide important preclinical information for developing new therapeutics [51].

7.3.5 Parkinson's Disease

Parkinson's disease (PD) is the second most common chronic progressive neurodegenerative disorder and is characterized primarily by major loss of nigrostriatal dopaminergic neurons [52]. Mutations of a few genes, such as PTEN-induced putative kinase 1 (PINK1), PARK2, and leucine-rich repeat kinase 2 (LRRK2), have been suggested to contribute to the pathogenesis of PD [38]. PINK1 is a mitochondrial serine/threonine protein kinase and protects cells from stress-induced mitochondrial dysfunctions. The parkin protein encoded by PARK2 gene is a component of a multiprotein E3 ubiquitin ligase that belongs to the portion of the ubiquitin–proteasome system required for global protein degradation. It has been hypothesized that parkin degrades proteins which are toxic to the dopaminergic neurons [53]. LRRK2 can interact with parkin protein and its missense mutation, *G2019S*, identified in both sporadic and familial PD cases.

In a recent report, skin fibroblasts taken from three PD patients who carry nonsense (c.1366C>T, p.Q456X) or missense (c.509T>G, p.V170G) mutations in the PINK1 gene were first reprogrammed into iPS cells (referred to as PD–iPS cells) [54]. PD–iPS cells were further directed to differentiate into dopaminergic neurons. These PD–iPS-derived dopaminergic neurons showed decreased and impaired capacity to recruit parkin to the mitochondria, increased mitochondrial copy number, and upregulation of PGC-1α, an important regulator of mitochondrial biogenesis. Intriguingly, ectopic expression of wild-type PINK1in PD–iPS-derived dopaminergic neurons can rescue these phenotypes. In another similar report, Nguyen and colleagues generated human iPS cells from PD patients carrying missense mutation *G2019S* of the LRRK2 gene [55] which were also further induced to become dopaminergic neurons in vitro. Patient-derived dopaminergic neurons displayed increased sensitivities toward hydrogen peroxide, 6-hydroxydopamine, and the proteasome inhibitor MG-132.

Taken together, it highlights not only the importance of PINK1, parkin, and LRRK2 in keeping health of dopaminergic neurons but also the similarities between PD neurons in vivo and PD–iPS-derived dopaminergic neurons in vitro. Thus, PD–iPS-derived dopaminergic neurons can be used as in vitro models to dissect the molecular mechanism of PD and probe drug targets for therapeutic purpose.

7.3.6 Hutchinson–Gilford Progeria Syndrome

Hutchinson–Gilford progeria syndrome (HGPS) is a rare but well-known congenital disease which is caused by a mutation in the lamin A (*LMNA*) gene, causing a truncated and farnesylated form of LMNA called progerin. HGPS is characterized by short stature, low body weight, early loss of hair, defect of lipid metabolism, scleroderma, decreased joint mobility, osteolysis, and premature facial features [56].

Zhang and colleagues have derived iPS cells from two HGPS patients who carry different mutations in LMNA (referred to as HGPS–iPS) and their healthy parents (referred to as WT–iPS) [57]. HGPS–iPS and WT–iPS were further induced into five different cell lineages: neural progenitors, endothelial cells, fibroblasts, vascular smooth muscle cells, and mesenchymal stem cells. Compared with the WT–iPS-derived cell types above, three primary pathological defects were observed: DNA damage, mislocalization of lamina-associated polypeptide 2, and nuclear dysmorphology. The HGPS–iPS-derived vascular smooth muscle cells which displayed the most severe phenotypes express the highest level of progerin. In addition, it was a surprise to find out calponin 1 (an actin-binding protein that is critical for cell's contraction) inclusion body was found in patient-derived vascular smooth muscle cells, suggesting that patient-derived cells have defects in handling protein load. It is consistent with the manifest symptoms of human HGPS patients. These new findings entitle HGPS–iPS-derived vascular smooth muscle cells to be good in vitro models to screen drugs that attenuate HGPS symptoms.

In summary, based on all above discussions about different human diseases, it is obvious that different lineage cell types generated in vitro from iPS cells derived from human patients carrying various mutations pretty recapitulate the symptoms observed in vivo. Therefore, iPS cells are invaluable models to study mechanisms underlying complex human disorders, especially when human samples are difficult to obtain. Last, iPS-derived various somatic cell types are also great models for drug screening.

7.4 Shortcut Approach to Generate Interested Somatic Cell Types for Modeling Human Diseases

As discussed above, the traditional way of generating interested somatic cell lineages is a two-step approach (Fig. 7.1): (1) converting fibroblasts to iPS cells by overexpressing four iPS factors (Oct3/4, Sox2, c-Myc, and Klf4) and (2) directing iPS cells to differentiate to different somatic cell types. Here, two questions incur: Do we have to go back to the pluripotent stem cell state first and then forward to terminal differentiated state? Is there a shortcut approach to obtain these somatic cell types without making stem/progenitor cells? The short answer is "yes," as suggested by recent studies, which will be discussed in the following section.

The success of generating iPS cells by the simple combination of four factors highlights that the intrinsic cell fate is primarily determined by the master transcriptional factors. If we have known the different combinations of master transcriptional factors that control each cell fate, it will be very interesting to determine whether it is possible to convert any somatic cell type (i.e., A) into our interested cell lineage (i.e., B) by just overexpressing combinations of master transcriptional factors controlling the cell fate of "B" lineage, regardless of their original distance. Indeed, it turns out to be a very efficient approach [58]. Four cases will be discussed in the following section as examples.

Ieda and colleagues overexpressed a combination of three important developmental transcriptional factors (Gata4, Mef2c, and Tbx5) in postnatal cardiac or dermal fibroblasts and finally made differentiated cardiomyocyte-like cells [59]. These induced cardiomyocyte-like cells express cardiac-specific markers, had global gene-expression files similar to wild-type cardiomyocyte, and can contract spontaneously. In addition, when fibroblasts expressing Gata4 (GATA binding protein 4), myocyte-specific enhancer factor 2C (Mef2c), and T-box transcription factor 5 (Tbx5) were transplanted into mouse hearts in vivo, they can also develop into cardiomyocytes.

When achaete–scute homolog 1 (Ascl1), POU domain, class 3, transcription factor 2 (Pou3f2, also known as Brn2), and myelin transcription factor 1-like (Myt1l) genes were overexpressed as a combination in the mouse embryonic and postnatal fibroblasts, functional glutamatergic neurons could be induced that express a variety of neuronal proteins, generate action potentials, and form functional synapses [60]. Similarly, when Ascl1, LIM homeobox transcription factor 1-alpha (Lmx1a), and nuclear receptor related 1 protein (Nurr1) were overexpressed in mouse and human fibroblasts, functional dopaminergic neurons were induced that could release dopamine and showed spontaneous electrical activity [61].

Recently, endoderm-derived terminal differentiated hepatocytes that overexpress Ascl1, Brn2, and Myt1l have been reprogrammed into functional neurons (ectoderm-derived cell types). It proved the possible cell fate conversion between lineages from different germ layers [62]. Intriguingly, single-cell and genome-wide expression analysis indicated that the combination of these three key transcriptional factors not only was sufficient to initiate the neuronal transcriptional program but also was able to inactivate the transcriptional program in the original hepatocytes.

Consistently, the remaining signatures of the donor hepatocytes gradually disappeared over time, even though a small epigenetic memory of the hepatocytes was detectable in the newly generated neuronal cells.

There are additional examples to show the sufficiency of using combinational master transcriptional factors to convert one cell type to another directly [58]. Thus, compared with the 2-step way to generate somatic cell types from iPS cells, it is a shortcut approach to generate the interested cell types for modeling the human diseases and drug screening in vitro, especially from fibroblasts dissected from human patients carrying different mutations.

7.5 Applications of iPS Cells in Gene Therapy and Cell-Based Therapy

The iPS-cell-based approach is powerful in gene therapy and tissue engineering, especially in terms of the single-gene point mutation-mediated human disorders. The principle is that (1) generating iPS cells from human patient fibroblasts in the cultured dishes, (2) in vitro correction of the mutant genes by specific gene targeting or homologous recombination, (3) directing corrected iPS cells (similar to iPS cells derived from normal healthy human) to differentiate into the interested somatic cell types, and (4) transplanting the committed somatic cell lineages in vivo to rescue the symptoms. The apparent advantage is the autologous transplantation which can bypass the immune rejection from different donors' tissues and the in vitro rapid expansion of the iPS cells to obtain enough cell sources for transplantation. Below, we will discuss two cases of iPS-based gene therapy in the hematopoietic diseases and one for diabetes.

7.5.1 Sickle Cell Disease

Sickle cell disease (SCD) is an autosomal recessive disorder which is characterized by having red blood cells with an abnormal, rigid, sickle shape. The abnormal red blood cells can occlude small capillaries and finally cause severe tissue damage. About 1/500 of the African-American population and more than 300,000 individuals all over the world suffer [63] from the SCD. The molecular basis for the pathogenesis of SCD is the A to T transversion in the sixth codon of the human β-globin gene, which changes a polar and hydrophilic glutamic acid residue to a nonpolar and hydrophobic amino acid valine in the β^S (sickle)-globin chain [64].

Hanna and colleagues generated iPS cells from fibroblasts dissected from the adult humanized knock-in mouse model of SCD [65]. In the SCD mouse model, the mouse α-globin gene was replaced with human α-globin gene, and mouse β-globin gene was replaced with human β^S (sickle)-globin gene [63]. Note that the iPS cells carry the mutant β^S (sickle)-globin gene, which is referred as to β^S-iPS cells. Then, β^S-iPS cells were electroporated with a well-designed targeting vector expressing the wild-type human β-globin gene and undergo homologous recombination to correct the mutant β^S (sickle)-globin gene. Last, hematopoietic progenitor cells derived from corrected iPS cells were transplanted into SCD mice. Excitingly, the SCD phenotypes were rescued. This promising result has motivated stem cell biologists to try the iPS-cell-based gene therapy approach in another hematopoietic disorders called β-thalassemia major (Cooley's anemia) [66].

7.5.2 β-Thalassemia

β-Thalassemias are a group of inherited blood disorders which are caused by the decreased or absent synthesis of the β-chains of hemoglobin. There are three types of β-thalassemias: thalassemia major, thalassemia intermedia, and thalassemia minor. Here we focus our discussion on the thalassemia major (also known as Cooley's anemia). Patients suffering from β-thalassemia major usually have symptoms such as severe anemia, poor growth, and skeletal abnormalities within the first 2 years of life with during infancy. β-Thalassemia major disease is prevalent in southern region of China. Without appropriate treatment, affected children will have a very short life span.

Wang and colleagues generated iPS cells from skin fibroblast cells of a 2-year-old β-41/42 homozygous patient in which the β-globin gene has a *TCTT* deletion between the 41st and 42nd amino acids [66]. Similar to the previous study [65], the iPS cells carrying the *TCTT* deletion mutation were corrected to wild-type-like iPS cells by homologous recombination. Furthermore, when hematopoietic progenitors derived from genetically corrected iPS cells were transplanted into immunodeficient mice, they were functional and could undergo normal hematopoiesis.

Taken together, the above two cases of iPS-cell-based gene therapy indicated that it might be a promising approach and is worthwhile for further human clinical trials. However, the iPS cells were obtained by virus transfection, and the remaining iPS factor genes in the genome might cause potential oncogenesis in humans. Therefore, to minimize the risk of causing tumors, the optimal approach might be generating iPS cells from small molecule drugs [67].

7.5.3 Type I Diabetes

The type 1 diabetes is characterized by the damage of the pancreatic endocrine insulin-producing beta cells due to immune attack [68]. Regenerating new insulin-producing beta cells is a promising approach for treating type I diabetes [69].

Recently, human iPS cells have been directed to differentiate into pancreatic insulin-producing beta cells in vitro [70]. Intriguingly, these cells express the terminal differentiation markers of normal pancreas beta cells in vivo, which include pancreatic and duodenal homeobox 1 (Pdx1), MafA, glucose transporter 2 (Glut2), and insulin [70]. It suggested that these new insulin-producing beta cells were fully mature and functional. Thus, human iPS cells, with the appropriate inductive signals, can develop into functional insulin-producing beta cells. It will be interesting to determine whether these insulin-producing beta cells generated in vitro are functional when they are transplanted into diabetes patients.

7.6 Auditor Hair Cell Regeneration Through the iPS-Cell-Based Approach

In the last section, we will discuss the mouse inner ear cochlear (auditory) hair cell regeneration through the iPS-cell-based or similar approaches. We will first introduce the histology and the development of the mouse cochlea, followed by the recent study of generating hair cells from iPS cells in vitro and end with the progress and challenges of in vivo hair cell regeneration in mouse and human beings.

7.6.1 Histology of Mouse Cochlea

Mammals, including mouse and human, detect sound through the mechanosensory hair cells (HCs) that reside in the cochlea. Mammalian cochlea is a spiral-like organ that resides in the ventral part of the inner ear [71]. The cochlear auditory epithelium, also referred to as the organ of Corti, has three rows of outer hair cells (OHCs) and one row of inner hair cells (IHCs) (Fig. 7.3a). The nonsensory supporting cells (SCs) surround these HCs. HCs and SCs are believed to derive from the same prosensory progenitors [72]. The basal portion of the cochlear HCs detects the high-frequency sound, and the apical part of HCs is responsible for the low-frequency sound.

7.6.2 Development of Mouse Cochlea

The primordial of the mouse inner ear, referred to as otocyst, derives from the ectoderm next to the hindbrain [73] and appears around embryonic (E) day, E8, and undergoes complex morphogenesis before the cochlear tips emerge around E12 [71]. Wnt and Notch signaling are critical in setting the boundaries between epidermis and otocyst [74, 75]. Briefly, cells with high Wnt and Notch signaling activities choose the otocyst cell fate, whereas cells with low Wnt and Notch signaling differentiate into epidermis cells. Fibroblast growth factor (Fgf) signaling pathway is another but negative one that acts to distinguish otic and non-otic cell fates, as overactivation of Fgf block development of otocyst or block expression of late otic markers [76].

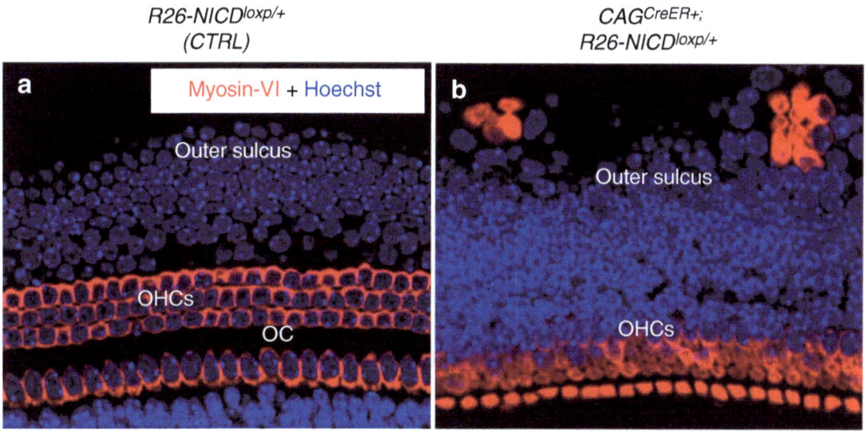

Fig. 7.3 Overactivation of Notch signaling at E10.5 generates new hair cells in cochlea. (**a**) Control cochlea contains one row of inner hair cells (IHCs) and three rows of outer hair cells (*OHCs*) inside the organ of Corti (*OC*). Hair cells are labeled in red with their specific marker myosin-VI. There are no hair cells in the outer sulcus region. (**b**) Tamoxifen injection at E10.5 causes ectopic Notch signaling in CAG^{CreER+}; $R26\text{-}NICD^{loxp/+}$ experimental cochlea and induces new hair cells in the outer sulcus regions (*arrows* in **b**) and OC and cochlear neuron regions (data not shown). Of note, the ability of Notch overactivation to induce new HCs is age dependent. Tamoxifen injection at E13 fails to generate new HCs in CAG^{CreER+}; $R26\text{-}NICD^{loxp/+}$ *cochlea* (Images modified from [99])

During the cochlear development, a variety of signaling pathways are involved. The bone morphogenetic protein 4 (BMP4) is expressed asymmetrically in the cochlear duct. Highest level of BMP4 is expressed in the abneural side, intermediate level of BMP4 is in the prosensory region (between abneural and neural side), and lowest level of BMP4 is in the neural side. In addition, when cochlear explants are cultured in vitro with intermediate level of BMP4, ectopic HCs are observed [77]. It suggests that BMP4 acts as a morphogen to determine the different cell fates across the lateral–medial axis of the cochlear duct [73].

Besides the BMP4 signals, Notch signaling is also involved in specifying the cochlear sensory regions. Cochlear development can be roughly divided into the early prosensory phase and the late cell fate determination phase [78, 79]. In the early prosensory phase between E12 and E14.5, Notch1-/Jagged1-mediated Notch signaling (primarily through) plays a critical role in specifying prosensory progenitors [80–84]. In the late cell fate determination phase (after E14.5), Notch signaling declines through "lateral inhibition" effects in progenitors that have committed to the HC fate but persists in progenitors that have committed to the SC fate [85, 86].

Atoh1, also known as Math1, is a crucial transcriptional factor required for cell fate commitment or initial differentiation of HCs. As expected, germ line Atoh1 knockout mice have no HCs [87]. Interestingly, fate mapping studies reveal that Atoh1 is initially expressed in both HCs and SCs [88, 89]. Together with the fact that higher Atoh1 expression level is present at later cochlear stages [90–92], it suggests a possibility that only the high dosage of Atoh1 is sufficient to define an HC fate.

7.6.3 Auditory HC Regeneration in Nonmammalian Vertebrates Versus Mammals

The nonmammalian vertebrates, including birds, fish, and amphibians, can regenerate HCs after damage [93]. Immediately after HC damage, SCs adjacent to the HC damage region somehow are able to sense the HC loss and become HCs directly, which process is referred to as direct trans-differentiation. Note that it occurs without involving SC proliferation. Another approach that nonmammalian vertebrates use to replace HCs is mitotic regeneration, which occurs a few days later after HC damage takes place. It means that SCs first proliferate and give rise to new daughter cells which further differentiate into new HCs. By using direct trans-differentiation and mitotic regeneration as a combination, nonmammalian vertebrates can recover their hearing capacity after HC damage.

In contrast, mammals cannot regenerate HCs after HC damage occurs, leading to permanent hearing loss. A few approaches have been tried to induce ectopic HCs. Overexpressing Atoh1 in neonatal rat cochlear explants [94], embryonic otocyst [95], and guinea pig cochlea [96] generates ectopic HCs. In addition, consistent with the roles of Notch signaling in specifying cochlear progenitor cells, overexpressing Notch activities generates ectopic HCs [97, 98]. However, the ability of Notch signaling to induce ectopic HCs is age dependent, and overexpressing Notch fails to generate new HCs at postnatal ages, as shown by our own recent study (Fig. 7.3b) [99].

7.6.4 iPS Cells Can Differentiate into New HCs In Vitro

Kazuo and colleagues have used ES cells or iPS cells to generate HCs [100].

By following the similar protocols described previously [101–103], ES or iPS cells treated with Dkk1 (Wnt signaling inhibitor), SIS3 (TGF-β signaling inhibitor), IGF1, and Fgf ligands gradually commit to endoderm, otocyst, and eventually became HCs that are reminiscent of the immature HCs in the wild-type cochlear development. Note that these immature HCs are not able to induce unless ES or iPS cells are cultured with chicken utricle stroma cells (as feeder cells). Unfortunately, it remains unclear of what contributions of chicken utricle stroma cells to the generations of HCs.

7.6.5 Challenges of Auditory HC Regeneration Using iPS Cells In Vivo

Given the appropriate induction signals present in vitro, the ability of ES cells and iPS cells to differentiate into immature HCs motivates inner ear biologists to figure out an approach to achieve HC regeneration in vivo. Currently, at least two challenges could be foreseen. The first is how to direct the cells (primarily SCs) in the damaged cochlea to behave like stem cells (like ES and iPS cells), as the case in vitro [100]. If the first question is resolved, the second is how to direct the full differentiation of the new HCs. Unfortunately, much is unknown of the signals that control the differentiation process from neonatal HCs to the adult and functional HCs in the wild-type mice in vivo.

7.7 Summary

In this chapter, we discussed the similarities between ES and iPS cells and the ease to obtain iPS cells from somatic cells by just overexpressing the four iPS transcriptional factors (Oct3/4, Sox2, c-Myc, and Klf4) and their application in modeling of human diseases. iPS-based approaches have helped us to bypass the limitations in studying various human disorders or using the patient-derived cells for efficient drug screening. Also, iPS-based gene therapy and tissue engineering provide the opportunities to offer personalized medicine in treating various diseases. However, in preclinical mouse models, autologous transplantation of iPS cells currently is only successful for the hematopoietic disorders. The main challenges for the solid human disorders are how to direct iPS-derived differentiated cells to integrate into the tissues and form functional connections to the remaining cells in situ.

Acknowledgements This work was supported by Hampton University faculty research grant of Dr. Feng Li and following grants from Dr. Jian Zuo (JZ) at St. Jude Children's Research Hospital; National Institutes of Health, DC06471, DC05168, and DC008800; the Office of Naval Research N000140911014; and the American Lebanese Syrian Associated Charities (ALSAC) of St. Jude Children's Research Hospital; JZ is a recipient of the Hartwell Individual Biomedical Research Award.

References

1. Brinster RL (1974) The effect of cells transferred into the mouse blastocyst on subsequent development. J Exp Med 140:1049–1056
2. Dewey MJ, Martin DW Jr, Martin GR, Mintz B (1977) Mosaic mice with teratocarcinoma-derived mutant cells deficient in hypoxanthine phosphoribosyltransferase. Proc Natl Acad Sci U S A 74:5564–5568
3. Evans MJ, Kaufman MH (1981) Establishment in culture of pluripotential cells from mouse embryos. Nature 292:154–156
4. Martin GR (1981) Isolation of a pluripotent cell line from early mouse embryos cultured in medium conditioned by teratocarcinoma stem cells. Proc Natl Acad Sci U S A 78:7634–7638
5. Thomson JA, Itskovitz-Eldor J, Shapiro SS, Waknitz MA, Swiergiel JJ, Marshall VS, Jones JM (1998) Embryonic stem cell lines derived from human blastocysts. Science 282:1145–1147
6. Gurdon JB (1962) Adult frogs derived from the nuclei of single somatic cells. Dev Biol 4:256–273
7. Briggs R, King TJ (1952) Transplantation of living nuclei from blastula cells into enucleated frogs' eggs. Proc Natl Acad Sci U S A 38:455–463
8. Campbell KH, McWhir J, Ritchie WA, Wilmut I (1996) Sheep cloned by nuclear transfer from a cultured cell line. Nature 380:64–66

9. Rando TA, Chang HY (2012) Aging, rejuvenation, and epigenetic reprogramming: resetting the aging clock. Cell 148:46–57
10. Takahashi K, Yamanaka S (2006) Induction of pluripotent stem cells from mouse embryonic and adult fibroblast cultures by defined factors. Cell 126:663–676
11. Takahashi K, Tanabe K, Ohnuki M, Narita M, Ichisaka T, Tomoda K, Yamanaka S (2007) Induction of pluripotent stem cells from adult human fibroblasts by defined factors. Cell 131:861–872
12. Yu J, Vodyanik MA, Smuga-Otto K, Antosiewicz-Bourget J, Frane JL, Tian S, Nie J, Jonsdottir GA, Ruotti V, Stewart R et al (2007) Induced pluripotent stem cell lines derived from human somatic cells. Science 318:1917–1920
13. Robinton DA, Daley GQ (2012) The promise of induced pluripotent stem cells in research and therapy. Nature 481:295–305
14. Aoi T, Yae K, Nakagawa M, Ichisaka T, Okita K, Takahashi K, Chiba T, Yamanaka S (2008) Generation of pluripotent stem cells from adult mouse liver and stomach cells. Science 321:699–702
15. Stadtfeld M, Brennand K, Hochedlinger K (2008) Reprogramming of pancreatic beta cells into induced pluripotent stem cells. Curr Biol 18:890–894
16. Eminli S, Utikal J, Arnold K, Jaenisch R, Hochedlinger K (2008) Reprogramming of neural progenitor cells into induced pluripotent stem cells in the absence of exogenous Sox2 expression. Stem Cells 26:2467–2474
17. Kim JB, Zaehres H, Wu G, Gentile L, Ko K, Sebastiano V, Arauzo-Bravo MJ, Ruau D, Han DW, Zenke M et al (2008) Pluripotent stem cells induced from adult neural stem cells by reprogramming with two factors. Nature 454:646–650
18. Hanna J, Markoulaki S, Schorderet P, Carey BW, Beard C, Wernig M, Creyghton MP, Steine EJ, Cassady JP, Foreman R et al (2008) Direct reprogramming of terminally differentiated mature B lymphocytes to pluripotency. Cell 133:250–264
19. Utikal J, Maherali N, Kulalert W, Hochedlinger K (2009) Sox2 is dispensable for the reprogramming of melanocytes and melanoma cells into induced pluripotent stem cells. J Cell Sci 122:3502–3510
20. Sun N, Panetta NJ, Gupta DM, Wilson KD, Lee A, Jia F, Hu S, Cherry AM, Robbins RC, Longaker MT et al (2009) Feeder-free derivation of induced pluripotent stem cells from adult human adipose stem cells. Proc Natl Acad Sci U S A 106:15720–15725
21. Maherali N, Ahfeldt T, Rigamonti A, Utikal J, Cowan C, Hochedlinger K (2008) A high-efficiency system for the generation and study of human induced pluripotent stem cells. Cell Stem Cell 3:340–345
22. Zhou H, Wu S, Joo JY, Zhu S, Han DW, Lin T, Trauger S, Bien G, Yao S, Zhu Y et al (2009) Generation of induced pluripotent stem cells using recombinant proteins. Cell Stem Cell 4:381–384
23. Lyssiotis CA, Foreman RK, Staerk J, Garcia M, Mathur D, Markoulaki S, Hanna J, Lairson LL, Charette BD, Bouchez LC et al (2009) Reprogramming of murine fibroblasts to induced pluripotent stem cells with chemical complementation of Klf4. Proc Natl Acad Sci U S A 106:8912–8917
24. Gunaseeli I, Doss MX, Antzelevitch C, Hescheler J, Sachinidis A (2010) Induced pluripotent stem cells as a model for accelerated patient- and disease-specific drug discovery. Curr Med Chem 17:759–766
25. Pei D (2008) The magic continues for the iPS strategy. Cell Res 18:221–223
26. Adewumi O, Aflatoonian B, Ahrlund-Richter L, Amit M, Andrews PW, Beighton G, Bello PA, Benvenisty N, Berry LS, Bevan S et al (2007) Characterization of human embryonic stem cell lines by the International Stem Cell Initiative. Nat Biotechnol 25:803–816
27. Cowan CA, Klimanskaya I, McMahon J, Atienza J, Witmyer J, Zucker JP, Wang S, Morton CC, McMahon AP, Powers D et al (2004) Derivation of embryonic stem-cell lines from human blastocysts. N Engl J Med 350:1353–1356
28. Agarwal S, Loh YH, McLoughlin EM, Huang J, Park IH, Miller JD, Huo H, Okuka M, Dos Reis RM, Loewer S et al (2010) Telomere elongation in induced pluripotent stem cells from dyskeratosis congenita patients. Nature 464:292–296

29. Itskovitz-Eldor J, Schuldiner M, Karsenti D, Eden A, Yanuka O, Amit M, Soreq H, Benvenisty N (2000) Differentiation of human embryonic stem cells into embryoid bodies compromising the three embryonic germ layers. Mol Med 6:88–95
30. Tam PP, Rossant J (2003) Mouse embryonic chimeras: tools for studying mammalian development. Development 130:6155–6163
31. Kang L, Wang J, Zhang Y, Kou Z, Gao S (2009) iPS cells can support full-term development of tetraploid blastocyst-complemented embryos. Cell Stem Cell 5:135–138
32. Park IH, Arora N, Huo H, Maherali N, Ahfeldt T, Shimamura A, Lensch MW, Cowan C, Hochedlinger K, Daley GQ (2008) Disease-specific induced pluripotent stem cells. Cell 134: 877–886
33. Burghes AH, Beattie CE (2009) Spinal muscular atrophy: why do low levels of survival motor neuron protein make motor neurons sick? Nat Rev Neurosci 10:597–609
34. Coovert DD, Le TT, McAndrew PE, Strasswimmer J, Crawford TO, Mendell JR, Coulson SE, Androphy EJ, Prior TW, Burghes AH (1997) The survival motor neuron protein in spinal muscular atrophy. Hum Mol Genet 6:1205–1214
35. Ebert AD, Yu J, Rose FF Jr, Mattis VB, Lorson CL, Thomson JA, Svendsen CN (2009) Induced pluripotent stem cells from a spinal muscular atrophy patient. Nature 457:277–280
36. Jessell TM (2000) Neuronal specification in the spinal cord: inductive signals and transcriptional codes. Nat Rev Genet 1:20–29
37. Wichterle H, Lieberam I, Porter JA, Jessell TM (2002) Directed differentiation of embryonic stem cells into motor neurons. Cell 110:385–397
38. Grskovic M, Javaherian A, Strulovici B, Daley GQ (2011) Induced pluripotent stem cells–opportunities for disease modelling and drug discovery. Nat Rev Drug Discov 10:915–929
39. Amir RE, Van den Veyver IB, Wan M, Tran CQ, Francke U, Zoghbi HY (1999) Rett syndrome is caused by mutations in X-linked MECP2, encoding methyl-CpG- binding protein 2. Nat Genet 23:185–188
40. Chahrour M, Jung SY, Shaw C, Zhou X, Wong ST, Qin J, Zoghbi HY (2008) MeCP2, a key contributor to neurological disease, activates and represses transcription. Science 320:1224–1229
41. Marchetto MC, Carromeu C, Acab A, Yu D, Yeo GW, Mu Y, Chen G, Gage FH, Muotri AR (2010) A model for neural development and treatment of Rett syndrome using human induced pluripotent stem cells. Cell 143:527–539
42. Giacometti E, Luikenhuis S, Beard C, Jaenisch R (2007) Partial rescue of MeCP2 deficiency by postnatal activation of MeCP2. Proc Natl Acad Sci U S A 104:1931–1936
43. Tropea D, Giacometti E, Wilson NR, Beard C, McCurry C, Fu DD, Flannery R, Jaenisch R, Sur M (2009) Partial reversal of Rett Syndrome-like symptoms in MeCP2 mutant mice. Proc Natl Acad Sci U S A 106:2029–2034
44. Slaugenhaupt SA, Blumenfeld A, Gill SP, Leyne M, Mull J, Cuajungco MP, Liebert CB, Chadwick B, Idelson M, Reznik L et al (2001) Tissue-specific expression of a splicing mutation in the IKBKAP gene causes familial Dysautonomia. Am J Hum Genet 68:598–605
45. Anderson SL, Coli R, Daly IW, Kichula EA, Rork MJ, Volpi SA, Ekstein J, Rubin BY (2001) Familial dysautonomia is caused by mutations of the IKAP gene. Am J Hum Genet 68: 753–758
46. Lee G, Papapetrou EP, Kim H, Chambers SM, Tomishima MJ, Fasano CA, Ganat YM, Menon J, Shimizu F, Viale A et al (2009) Modelling pathogenesis and treatment of familial dysautonomia using patient-specific iPSCs. Nature 461:402–406
47. Sommer L, Shah N, Rao M, Anderson DJ (1995) The cellular function of MASH1 in autonomic neurogenesis. Neuron 15:1245–1258
48. Slaugenhaupt SA, Mull J, Leyne M, Cuajungco MP, Gill SP, Hims MM, Quintero F, Axelrod FB, Gusella JF (2004) Rescue of a human mRNA splicing defect by the plant cytokinin kinetin. Hum Mol Genet 13:429–436
49. Selkoe DJ (2002) Alzheimer's disease is a synaptic failure. Science 298:789–791
50. Iwata N, Higuchi M, Saido TC (2005) Metabolism of amyloid-beta peptide and Alzheimer's disease. Pharmacol Ther 108:129–148

51. Yahata N, Asai M, Kitaoka S, Takahashi K, Asaka I, Hioki H, Kaneko T, Maruyama K, Saido TC, Nakahata T et al (2011) Anti-Abeta drug screening platform using human iPS cell-derived neurons for the treatment of Alzheimer's disease. PLoS One 6:e25788
52. Soldner F, Hockemeyer D, Beard C, Gao Q, Bell GW, Cook EG, Hargus G, Blak A, Cooper O, Mitalipova M et al (2009) Parkinson's disease patient-derived induced pluripotent stem cells free of viral reprogramming factors. Cell 136:964–977
53. Kitada T, Asakawa S, Hattori N, Matsumine H, Yamamura Y, Minoshima S, Yokochi M, Mizuno Y, Shimizu N (1998) Mutations in the parkin gene cause autosomal recessive juvenile parkinsonism. Nature 392:605–608
54. Seibler P, Graziotto J, Jeong H, Simunovic F, Klein C, Krainc D (2011) Mitochondrial Parkin recruitment is impaired in neurons derived from mutant PINK1 induced pluripotent stem cells. J Neurosci 31:5970–5976
55. Nguyen HN, Byers B, Cord B, Shcheglovitov A, Byrne J, Gujar P, Kee K, Schule B, Dolmetsch RE, Langston W et al (2011) LRRK2 mutant iPSC-derived DA neurons demonstrate increased susceptibility to oxidative stress. Cell Stem Cell 8:267–280
56. Hennekam RC (2006) Hutchinson-Gilford progeria syndrome: review of the phenotype. Am J Med Genet A 140:2603–2624
57. Zhang J, Lian Q, Zhu G, Zhou F, Sui L, Tan C, Mutalif RA, Navasankari R, Zhang Y, Tse HF et al (2010) A human iPSC model of Hutchinson Gilford Progeria reveals vascular smooth muscle and mesenchymal stem cell defects. Cell Stem Cell 8:31–45
58. Vierbuchen T, Wernig M (2011) Direct lineage conversions: unnatural but useful? Nat Biotechnol 29:892–907
59. Ieda M, Fu JD, Delgado-Olguin P, Vedantham V, Hayashi Y, Bruneau BG, Srivastava D (2010) Direct reprogramming of fibroblasts into functional cardiomyocytes by defined factors. Cell 142:375–386
60. Vierbuchen T, Ostermeier A, Pang ZP, Kokubu Y, Sudhof TC, Wernig M (2010) Direct conversion of fibroblasts to functional neurons by defined factors. Nature 463:1035–1041
61. Caiazzo M, Dell'Anno MT, Dvoretskova E, Lazarevic D, Taverna S, Leo D, Sotnikova TD, Menegon A, Roncaglia P, Colciago G et al (2011) Direct generation of functional dopaminergic neurons from mouse and human fibroblasts. Nature 476:224–227
62. Marro S, Pang ZP, Yang N, Tsai MC, Qu K, Chang HY, Sudhof TC, Wernig M (2011) Direct lineage conversion of terminally differentiated hepatocytes to functional neurons. Cell Stem Cell 9:374–382
63. Wu LC, Sun CW, Ryan TM, Pawlik KM, Ren J, Townes TM (2006) Correction of sickle cell disease by homologous recombination in embryonic stem cells. Blood 108:1183–1188
64. Ingram VM (1956) A specific chemical difference between the globins of normal human and sickle-cell anaemia haemoglobin. Nature 178:792–794
65. Hanna J, Wernig M, Markoulaki S, Sun CW, Meissner A, Cassady JP, Beard C, Brambrink T, Wu LC, Townes TM et al (2007) Treatment of sickle cell anemia mouse model with iPS cells generated from autologous skin. Science 318:1920–1923
66. Wang Y, Zheng CG, Jiang Y, Zhang J, Chen J, Yao C, Zhao Q, Liu S, Chen K, Du J et al (2012) Genetic correction of beta-thalassemia patient-specific iPS cells and its use in improving hemoglobin production in irradiated SCID mice. Cell Res 22:637–648
67. Huangfu D, Maehr R, Guo W, Eijkelenboom A, Snitow M, Chen AE, Melton DA (2008) Induction of pluripotent stem cells by defined factors is greatly improved by small- molecule compounds. Nat Biotechnol 26:795–797
68. Li F, Mahato RI (2010) RNA interference for improving the outcome of islet transplantation. Adv Drug Deliv Rev 63:47–68
69. D'Amour KA, Bang AG, Eliazer S, Kelly OG, Agulnick AD, Smart NG, Moorman MA, Kroon E, Carpenter MK, Baetge EE (2006) Production of pancreatic hormone-expressing endocrine cells from human embryonic stem cells. Nat Biotechnol 24:1392–1401
70. Zhang D, Jiang W, Liu M, Sui X, Yin X, Chen S, Shi Y, Deng H (2009) Highly efficient differentiation of human ES cells and iPS cells into mature pancreatic insulin-producing cells. Cell Res 19:429–438

71. Morsli H, Choo D, Ryan A, Johnson R, Wu DK (1998) Development of the mouse inner ear and origin of its sensory organs. J Neurosci 18:3327–3335
72. Fekete DM, Muthukumar S, Karagogeos D (1998) Hair cells and supporting cells share a common progenitor in the avian inner ear. J Neurosci 18:7811–7821
73. Groves AK, Fekete DM (2011) Shaping sound in space: the regulation of inner ear patterning. Development 139:245–257
74. Jayasena CS, Ohyama T, Segil N, Groves AK (2008) Notch signaling augments the canonical Wnt pathway to specify the size of the otic placode. Development 135:2251–2261
75. Ohyama T, Mohamed OA, Taketo MM, Dufort D, Groves AK (2006) Wnt signals mediate a fate decision between otic placode and epidermis. Development 133:865–875
76. Freter S, Muta Y, Mak SS, Rinkwitz S, Ladher RK (2008) Progressive restriction of otic fate: the role of FGF and Wnt in resolving inner ear potential. Development 135:3415–3424
77. Ohyama T, Basch ML, Mishina Y, Lyons KM, Segil N, Groves AK (2010) BMP signaling is necessary for patterning the sensory and nonsensory regions of the developing mammalian cochlea. J Neurosci 30:15044–15051
78. Bermingham-McDonogh O, Reh TA (2011) Regulated reprogramming in the regeneration of sensory receptor cells. Neuron 71:389–405
79. Hayashi T, Kokubo H, Hartman BH, Ray CA, Reh TA, Bermingham-McDonogh O (2008) Hesr1 and Hesr2 may act as early effectors of Notch signaling in the developing cochlea. Dev Biol 316:87–99
80. Brooker R, Hozumi K, Lewis J (2006) Notch ligands with contrasting functions: Jagged1 and Delta1 in the mouse inner ear. Development 133:1277–1286
81. Daudet N, Lewis J (2005) Two contrasting roles for Notch activity in chick inner ear development: specification of prosensory patches and lateral inhibition of hair-cell differentiation. Development 132:541–551
82. Eddison M, Le Roux I, Lewis J (2000) Notch signaling in the development of the inner ear: lessons from Drosophila. Proc Natl Acad Sci U S A 97:11692–11699
83. Kiernan AE, Xu J, Gridley T (2006) The Notch ligand JAG1 is required for sensory progenitor development in the mammalian inner ear. PLoS Genet 2:e4
84. Tsai H, Hardisty RE, Rhodes C, Kiernan AE, Roby P, Tymowska-Lalanne Z, Mburu P, Rastan S, Hunter AJ, Brown SD et al (2001) The mouse slalom mutant demonstrates a role for Jagged1 in neuroepithelial patterning in the organ of Corti. Hum Mol Genet 10:507–512
85. Lanford PJ, Lan Y, Jiang R, Lindsell C, Weinmaster G, Gridley T, Kelley MW (1999) Notch signalling pathway mediates hair cell development in mammalian cochlea. Nat Genet 21:289–292
86. Murata J, Tokunaga A, Okano H, Kubo T (2006) Mapping of notch activation during cochlear development in mice: implications for determination of prosensory domain and cell fate diversification. J Comp Neurol 497:502–518
87. Bermingham NA, Hassan BA, Price SD, Vollrath MA, Ben-Arie N, Eatock RA, Bellen HJ, Lysakowski A, Zoghbi HY (1999) Math1: an essential gene for the generation of inner ear hair cells. Science 284:1837–1841
88. Matei V, Pauley S, Kaing S, Rowitch D, Beisel KW, Morris K, Feng F, Jones K, Lee J, Fritzsch B (2005) Smaller inner ear sensory epithelia in Neurog 1 null mice are related to earlier hair cell cycle exit. Dev Dyn 234:633–650
89. Yang H, Xie X, Deng M, Chen X, Gan L (2010) Generation and characterization of Atoh1-Cre knock-in mouse line. Genesis 48:407–413
90. Chen P, Johnson JE, Zoghbi HY, Segil N (2002) The role of Math1 in inner ear development: uncoupling the establishment of the sensory primordium from hair cell fate determination. Development 129:2495–2505
91. Jahan I, Pan N, Kersigo J, Calisto LE, Morris KA, Kopecky B, Duncan JS, Beisel KW, Fritzsch B (2012) Expression of neurog1 instead of atoh1 can partially rescue organ of Corti cell survival. PLoS One 7:e30853
92. Pan N, Jahan I, Kersigo J, Duncan JS, Kopecky B, Fritzsch B (2012) A novel Atoh1 "self-terminating" mouse model reveals the necessity of proper Atoh1 level and duration for hair cell differentiation and viability. PLoS One 7:e30358

93. Stone JS, Cotanche DA (2007) Hair cell regeneration in the avian auditory epithelium. Int J Dev Biol 51:633–647
94. Zheng JL, Gao WQ (2000) Overexpression of Math1 induces robust production of extra hair cells in postnatal rat inner ears. Nat Neurosci 3:580–586
95. Gubbels SP, Woessner DW, Mitchell JC, Ricci AJ, Brigande JV (2008) Functional auditory hair cells produced in the mammalian cochlea by in utero gene transfer. Nature 455: 537–541
96. Izumikawa M, Minoda R, Kawamoto K, Abrashkin KA, Swiderski DL, Dolan DF, Brough DE, Raphael Y (2005) Auditory hair cell replacement and hearing improvement by Atoh1 gene therapy in deaf mammals. Nat Med 11:271–276
97. Hartman BH, Reh TA, Bermingham-McDonogh O (2010) Notch signaling specifies prosensory domains via lateral induction in the developing mammalian inner ear. Proc Natl Acad Sci U S A 107:15792–15797
98. Pan W, Jin Y, Stanger B, Kiernan AE (2010) Notch signaling is required for the generation of hair cells and supporting cells in the mammalian inner ear. Proc Natl Acad Sci U S A 107: 15798–15803
99. Liu Z, Owen T, Fang J, Zuo J (2012) Overactivation of Notch1 signaling induces ectopic hair cells in the mouse inner ear in an age-dependent manner. PLoS One 7:e34123
100. Oshima K, Shin K, Diensthuber M, Peng AW, Ricci AJ, Heller S (2010) Mechanosensitive hair cell-like cells from embryonic and induced pluripotent stem cells. Cell 141:704–716
101. Ikeda H, Osakada F, Watanabe K, Mizuseki K, Haraguchi T, Miyoshi H, Kamiya D, Honda Y, Sasai N, Yoshimura N et al (2005) Generation of Rx+/Pax6+ neural retinal precursors from embryonic stem cells. Proc Natl Acad Sci U S A 102:11331–11336
102. Lamba DA, Karl MO, Ware CB, Reh TA (2006) Efficient generation of retinal progenitor cells from human embryonic stem cells. Proc Natl Acad Sci U S A 103:12769–12774
103. Osakada F, Ikeda H, Mandai M, Wataya T, Watanabe K, Yoshimura N, Akaike A, Sasai Y, Takahashi M (2008) Toward the generation of rod and cone photoreceptors from mouse, monkey and human embryonic stem cells. Nat Biotechnol 26:215–224

Chapter 8
Gene Transfer to the Heart: Emerging Strategies for the Selection of Vectors, Delivery Techniques, and Therapeutic Targets

Michael G. Katz, Anthony S. Fargnoli, Louella A. Pritchette, and Charles R. Bridges

Abstract Heart diseases are a major cause of morbidity and mortality in contemporary society. Advances in the understanding of the molecular basis of myocardial dysfunction have placed many acquired and congenital cardiovascular diseases within the reach of gene-based therapy. Four prerequisites are required for a successful clinical application of gene therapy: (1) an effective strategy for genetic manipulation, (2) availability of vectors with enhanced myocardial tropism, (3) a clinically translatable delivery technique that will result in global or regional expression, and (4) creation of therapeutic transgenes for selected molecular targets depending on the underlying pathological state of the heart. Despite significant promise, however, several obstacles exist with gene-based therapies. These obstacles are described in detail in this chapter, along with proposed solutions. We anticipate that advances in the field will improve cardiac gene therapy in future clinical approaches.

Keywords Cardiac gene delivery • Gene overexpression • Gene blockade • Vectors for gene therapy • Physical methods of gene transfer • Cardiac molecular targets

8.1 Introduction

Heart disease remains the leading cause of mortality, morbidity, and health-care expenditure around the world. Despite progress in creating new pharmacological lines of drugs, and advances in the therapeutic and surgical treatment of cardiovascular disease, there is a significant gap between modern approaches and key

M.G. Katz • A.S. Fargnoli • L.A. Pritchette • C.R. Bridges (✉)
Department of Thoracic and Cardiovascular Surgery,
Sanger Heart and Vascular Institute, Carolinas Healthcare System,
Charlotte, NC 28203, USA
e-mail: anthony.fargnoli@carolinashealthcare.org; charles.bridges@carolinashealthcare.org

M.K. Danquah, R.I. Mahato (eds.), *Emerging Trends in Cell and Gene Therapy*,
DOI 10.1007/978-1-62703-417-3_8, © Springer Science+Business Media New York 2013

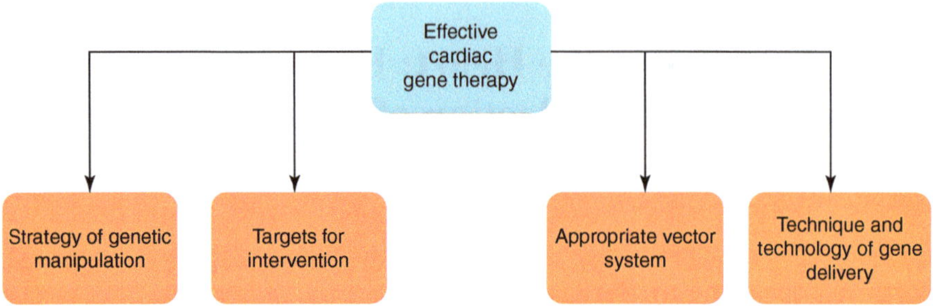

Fig. 8.1 Main prerequisites for successful cardiac gene therapy

pathological mechanisms relating to cardiac dysfunction. Therefore, there exists a need to create new methods of treatment that would have a significant impact not only on the course of the disease but that would primarily address the underlying pathological processes.

Progress in the basic science over the past two decades has improved our understanding of cardiac molecular biology and underlying genetic regulation that is involved in many clinical syndromes. These advances have indicated that gene transfer may serve as a robust management strategy for different acquired and congenital disorders.

Once a vector/gene construct has been identified, the next step is to address the delivery strategy. Successful gene delivery is associated with four basic requirements outlined in Fig. 8.1. To some extent, all of these requirements are linked and dependent upon one another; subsequently, it is not possible to consider each requirement in isolation without considering the whole treatment approach. In this chapter, we will focus on these factors and assess the critical challenges faced when determining which strategy to use for various disease mechanisms.

8.2 Strategies for Genetic Manipulation of the Cardiovascular System

Currently, there are primarily two different strategies described for cardiovascular gene therapy manipulation (Fig. 8.2). The most commonly applied strategy features overexpression of a target gene. This may involve either replacement of a missing or dysfunctional gene, for example an X-linked recessive disorder such as heart failure associated with Becker's cardiomyopathy, an autosomal recessive gene defect associated with alpha sarcoglycan deficiency in the limb girdle muscular dystrophies. More commonly, heart failure may not have a defined genetic basis (ischemic cardiomyopathy), yet certain genes are consistently downregulated. The second group of strategies relates to inactivation of dominant negative gene function involved in disease etiology or progression (other names: gene silencing or gene blockade) [1–3] (Fig. 8.3).

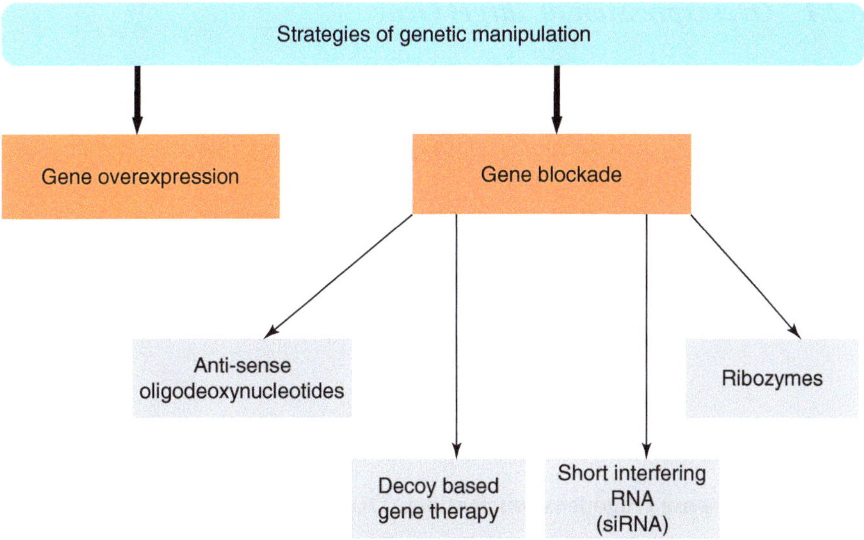

Fig. 8.2 Strategies for genetic manipulation of the cardiovascular system

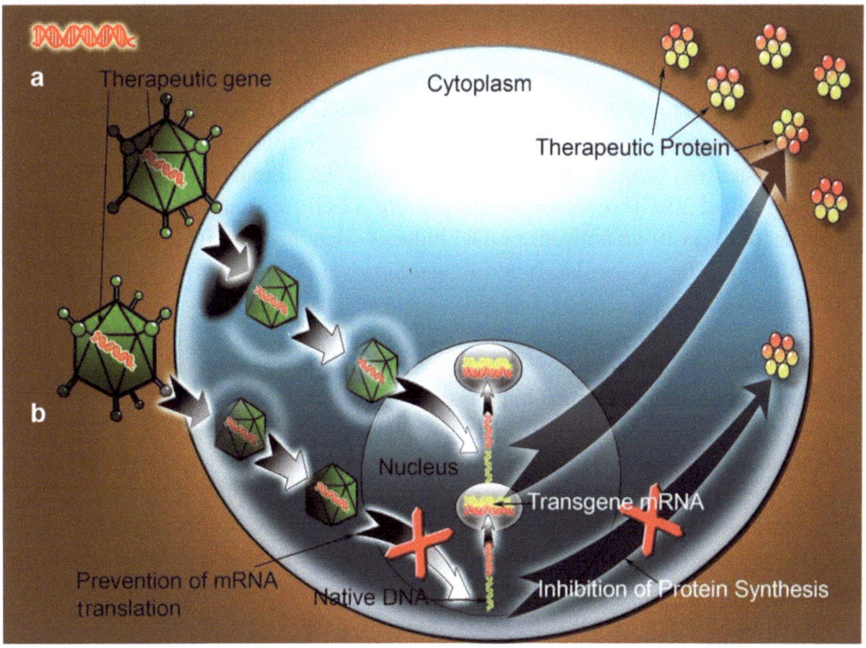

Fig. 8.3 (**a**) Gene overexpression. (**b**) Gene blockade

8.2.1 Overexpression of Target Gene

A gene's physiological function may be impaired or downregulated as a result of mutation or pathological process. Therefore, the restoration of function through exogenous delivery to replace the deficient gene seems quite logical. In this case, full-length or partial cDNA encoding the deficient gene is delivered to the target tissues using a vector system capable of expressing the therapeutic protein [1]. Several steps in the gene overexpression process may be modulated, including the transcription, RNA splicing, translation, and posttranslational modification of a protein.

8.2.2 Specific Gene Blockade

8.2.2.1 Antisense Oligodeoxynucleotides (ODN)

ODN are used as inhibitors of specific gene expression without any change in function of other genes. Single-stranded ODN may be delivered either by direct administration (as a pharmacological agent) or by transfection with a vector encoding the ODN. The ODN binds to the target mRNA transcript and prevents translation. This mechanism of action is based on the presence of two forms of ODN: the RNase H-dependent ODN, which induces the degradation of mRNA, and the steric-blocker ODN, which physically blocks the progression of mRNA translation. Concerning cardiac applications, the antisense ODN approach has been tested to prevent restenosis after balloon angioplasty [4]. Treatment with antisense ODN directed against VEGF receptors could prevent VEGF-mediated arteriogenesis [5]. Systemic delivery of an antisense ODN induces silencing of miR-208a in the myocytes, thus improving cardiac function and survival in hypertensive-induced heart failure (HF) in rats [6].

8.2.2.2 Decoy-Based Gene Therapy

Synthetic double-stranded (ds) DNA with high affinity may be introduced into target cells as a "decoy" or alternatively described as a cis-regulatory element, which binds to a sequence-specific DNA factor and changes gene transcription. Transfection of dsODN will result in the attenuation of the cis–trans interaction of cell surface receptors and remove the trans-factor from the endogenous cis-element, resulting in inhibition of gene expression [3]. dsODN containing binding sequences (decoy) for transcriptional factors are involved in the activation of pathogenic genes. Transfection of the decoy ODN prevents the binding and transactivation of the genes regulated by the target transcriptional factor [1]. The decoy strategy is very attractive for several reasons: the synthesis of the sequence-specific decoy is relatively simple and can be targeted to specific tissues; the knowledge of the exact molecular structure of the target transcription factor is unnecessary; it has been shown to be more effective than antisense technology [3]. Some restrictions of this method however are short half-life, lower uptake efficiency, and degradation by endocytosis and nucleases. Using mice and monkey models, it was found that E2F decoy transfection prevented intimal hyperplasia in cardiac allografts [7]. In a report of Yamasaki et al., the successful in vivo transfer of NFkappaB decoy ODN to inhibit vascular stenosis in balloon-injured porcine coronary arteries was demonstrated [8].

8.2.2.3 Short Interfering RNA (siRNA)

Gene silencing via siRNA technology is a novel strategy with great therapeutic potential. siRNA is a short dsRNA molecule that induces sequence-specific posttranscriptional gene modification. This mechanism is called RNA interference (RNAi). Recently, this strategy was used for the treatment of HF, and the results showed that the restoration of cardiac function was most likely through the reduction of hypertrophy [9]. Once transfected into a cell, siRNA incorporates into the nuclease complex, where it interrupts the translation of targeted genes. Successful left ventricular intracavitary delivery of DNA/siRNA complexes by means of sonoporation was demonstrated in murine hearts [10]. The incorporation of siRNA into terminally differentiated adult rat cardiac myocytes using adenovirus has also been reported [11].

8.2.2.4 Ribozymes

Another strategy used to inhibit the disease process at the transcriptional level is the use of ribozymes. Ribozyme gene therapy aims to turn off a mutated gene in a cell by targeting the mRNA transcripts copied from the gene. Therefore, protein synthesis by the target RNA may be specifically inhibited by ribozymes. This process involves three steps: (1) delivery of RNA strands engineered to function as ribozymes, (2) specific binding of the ribozyme RNA to mRNA encoded by the mutated gene, and (3) cleavage of the target mRNA, preventing it from being translated into a protein. Several studies have used ribozymes to limit neointimal hyperplasia with smooth muscle cell proliferation in response to balloon angioplasty. Ribozymes against c-myb mRNA [12] and transforming growth factor [13] prevented development of restenosis.

8.3 Cardiac Gene Delivery Vectors

Choosing the right vector for cardiovascular applications is one of the most challenging aspects of gene therapy approaches. The availability of vectors for gene transfer has improved significantly over time. The ideal vector would have the following characteristics:

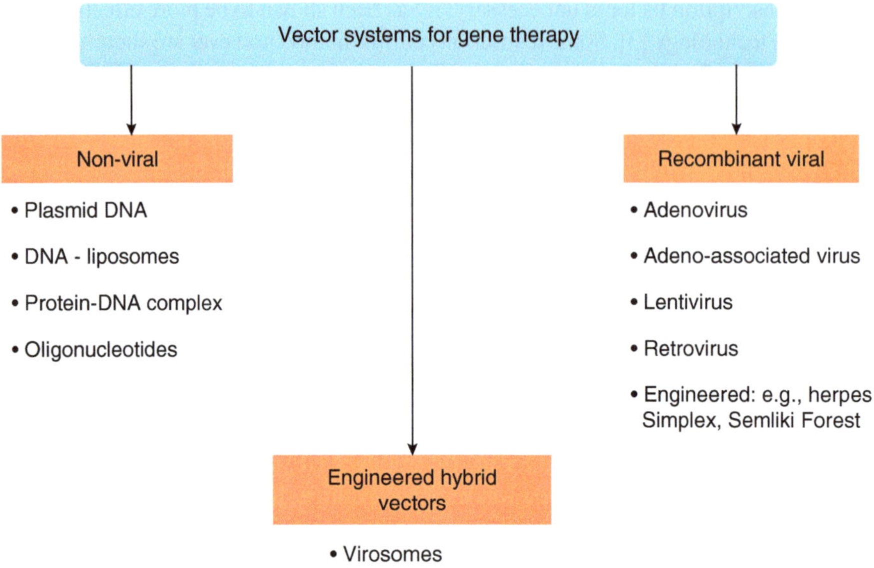

Fig. 8.4 Vector systems for gene therapy

It must be cardiotropic, result in long-term expression, minimize the risk of cellular immune response, have a large coding capacity, and have high transduction efficiency [14]. The main challenges to the vector are as follows: (1) escaping the neutralizing effects of specific antibodies and nonspecific adsorption to other blood components, (2) overcoming the endothelial barrier and penetrating the vascular wall for diffusion through the extracellular matrix, (3) uptake into the cell at the level of the plasma membrane and efficient trafficking to the nucleus, and (4) synthesis by the host of the complementary DNA strand for single-stranded delivery vectors followed by transcription and translation of the transgene [15]. A number of different vectors have been used to achieve myocardial gene transfer, these are modified or selected to enhance the probability of overcoming each of these challenges. All vectors can be classified into two main categories: the nonviral and recombinant viral (Fig. 8.4). We discuss briefly below the most commonly used vehicles.

8.3.1 Nonviral Vectors

Nonviral vectors are grouped as plasmid DNA, liposome–DNA complexes (lipoplexes), and polymer–DNA complexes (polyplexes). Oligonucleotides are also considered nonviral vectors [16]. In 1990, Lin and associates injected plasmid DNA into the left ventricle and demonstrated that the lacZ gene could be introduced and expressed in cardiac myocytes [17].

Although nonviral vectors have the major advantage of production in relatively large quantities at low cost while at the same time possess fewer toxic or immunological problems, their transfer efficiency is generally poor, independent of delivery route [18, 19]. Nevertheless, a large number of human cardiac clinical trials are based on plasmid-mediated gene transfer investigating angiogenesis in myocardial ischemia [20–23]. A major advantage of this approach is that it avoids many of the biosafety concerns associated with viral vectors. However, the level of transgene expression and the efficiency of gene transfer (percent of target cells expressing the transgene) are low, and expression is restricted to the zone of the injection site. DNA complexes are relatively more efficient [24]. There is however a major discrepancy between the data obtained in vivo and in vitro. In addition, these complexes are unstable and thus quickly removed by phagocytes when delivered, especially through intravascular delivery systems [25]. An additional shortcoming of these vectors (e.g., oligonucleotides) is their short biological half-life due to intracellular degradation and nonspecific binding [25]. The demonstration of plasmid gene transfer opened a new era of cardiovascular pharmacotherapy. Despite numerous efforts to enhance efficiency through modification, direct myocardial plasmid injection basically remains a proof-of-concept tool only [26].

8.3.2 *Viral Vectors*

As stated previously, cardiovascular gene therapy applications demand both efficient myocardial transduction initially and long-term transgene expression. Many authors strongly believe that only viral vectors appear to meet these demands in terms of performance [27, 28]. Compared to nonviral vectors, viruses have an evolutionary advantage in their interactions with the cellular surface receptors, directly leading to more efficient intracellular trafficking of packaged DNA to the nucleus. Furthermore, their protein capsid protects the message from degradation in lysosomes [15, 26, 29, 30]. Some viral vectors are able to integrate into the host genome, whereas others remain episomal. Integrating viruses result in persistent transgene expression, while viruses in episomal form lead to long-term expression in predominantly nondividing tissues (e.g., adult myocardium) but only transient expression in rapidly dividing tissues (e.g., the hematopoietic system). It should be noted that for some disorders, short-term expression in a relatively small proportion of cells would be sufficient or even desirable (e.g., angiogenesis post myocardial infarction), whereas other pathologies might require long-term expression (e.g., autosomal recessive cardiomyopathy). At present, viral vectors are the most suitable vehicles for efficient gene delivery (Table 8.1). Most of these vectors are derived from human pathogens, from which essential viral genes have been deleted.

8.3.2.1 Lentiviruses

These vectors were initially developed for HIV therapy. Lentiviral vectors can infect nondividing cells, cause long-term expression, and do not typically induce an

Table 8.1 Advantages and disadvantages of vector systems for cardiac gene therapy

Vector system	Advantages	Disadvantages
Naked plasmid DNA	Simple methodology; large DNA insert capacity and minimal safety risks	Low transduction efficiency, poor and transient expression profile
Adenovirus	Readily produced in high titers, high transduction performance, no integration in host genome	Induces inflammation and potent host immune response, short term expression, non-specific cellular tropism
Adeno-associated virus	Non-pathogenic, long term expression profile, low immunogenicity, cardiac tropism by design	Small insert capacity, complex production systems, low titer yields
Lentivirus	High transduction efficiency, long term expression profile, low immune response	Integration into the host cell genome, increased risk of oncogenesis, limited cardiac tropism

inflammatory or immune response. The major limitation is the risk for mutagenesis and oncogenesis [29]. The new generation of lentiviruses containing a mRNA and a nuclear import sequence has been used for successful myocardial transduction, although expression is usually short term [31, 32]. Fleury et al., in a study with rat cardiomyocytes in vivo, succeeded in obtaining persistent GFP transfer for up to 10 weeks [33]. In another study, the transduction efficiency of lentiviral vector-mediated SERCA2 gene transfer was about 40 %, and the positive physiological effect persisted 6 months later [34].

8.3.2.2 Adenoviruses

Adenoviral vectors have historically been the most frequently used transfer system in experimental and clinical studies. This is attributed to the vector's known advantages such as the ability to transduce nondividing cells, the ease of manufacture in very high titers, the possibility to achieve high levels of transgene expression, and a large transgene cloning capacity. However, their use is limited clinically due to transient gene expression and their inability to integrate the genome into the cellular chromosomal DNA. In addition, adenoviral vector particles are highly immunogenic and cause inflammatory and toxic reactions in the host. This is due to the fact that the adenovirus stimulates both the innate and adaptive immune systems. Using a rat model, it was confirmed that adenovirus was several orders of magnitude more efficient in transducing myocytes than plasmid DNA expressing the same construct [35, 36]. Later, it was shown that the direct intramyocardial injection of replication-deficient adenovirus can program gene expression in large animal in vivo. However, the authors noted a robust T cell-mediated immune response against the vector and limited distribution of the reporter gene [37]. Simultaneously, several groups confirmed the possibility to achieve significant cardiac gene expression after catheter-mediated delivery of adenovirus encoding phospholamban and the β2-adrenergic receptor [38, 39]. Using adenovirus to deliver selected transgenes enhanced cardiac performance several weeks after gene transfer has been demonstrated [39]. Despite sophisticated modifications in an attempt to attenuate the host immune response to the adenovirus, the risk is too high to advocate the use of this delivery vector for clinical cardiovascular applications.

8.3.2.3 Adeno-Associated Viruses

Adeno-associated virus (AAV) is a small (20 nm), non-enveloped virus that belongs to the Dependovirus genus of the parvovirus family. AAVs have a single-stranded DNA genome. The viral genome is approximately 4.7 kb in length and is composed of two major open reading frames which encode Rep (replication) and Cap (capsid) proteins [40]. For an infection to occur, AAV requires coinfection with a helper virus such as adenovirus. This allows the viral genome to replicate episomally and leads to synthesis of the AAV proteins. AAV is one of the smallest viruses, with a capsid mean diameter of 22 nm. The first AAV2 infectious clone was created in 1982 by Samulski and colleagues [41]. Several years later, it was established that AAVs can express foreign genes in mammalian cells [42]. One of the major advantages of AAV vectors is its established safety record. It has been demonstrated that after reaching a steady-state level, AAV expression may last for years with an absence of a significant immune response to the transgene [43]. Moreover, AAV vectors can be engineered to provide a wide range of cell-type tropism with the ability to transduce both dividing and nondividing cells. Due to their biological properties and advantages over other viral vector systems, AAV has gained great popularity in the last decade in many clinical trials. Seventy-five clinical trials using AAV have been initiated over the past 15 years [29] with ~10 % indicated for cardiovascular diseases [44].

AAV Endocytosis and Intracellular Trafficking

Despite the availability and diversity of AAV vectors, several biological barriers appear to limit the effectiveness of AAV-mediated gene therapy [29, 45]. Understanding the fundamental basis of these barriers has led to the establishment of methods to improve the efficiency of rAAV-mediated gene delivery [30, 46]. Clarification of the processes by which a virus first enters and traffics through a cell helps to understand the life cycle of the virus and its ability to act inside the cardiac muscle. The transport activity of AAV is mainly determined by selective receptor-mediated vesicle transcytosis [47]. This intracellular route does not appear to alter the properties of the AAV. The entry of AAV vectors into the cell involves several

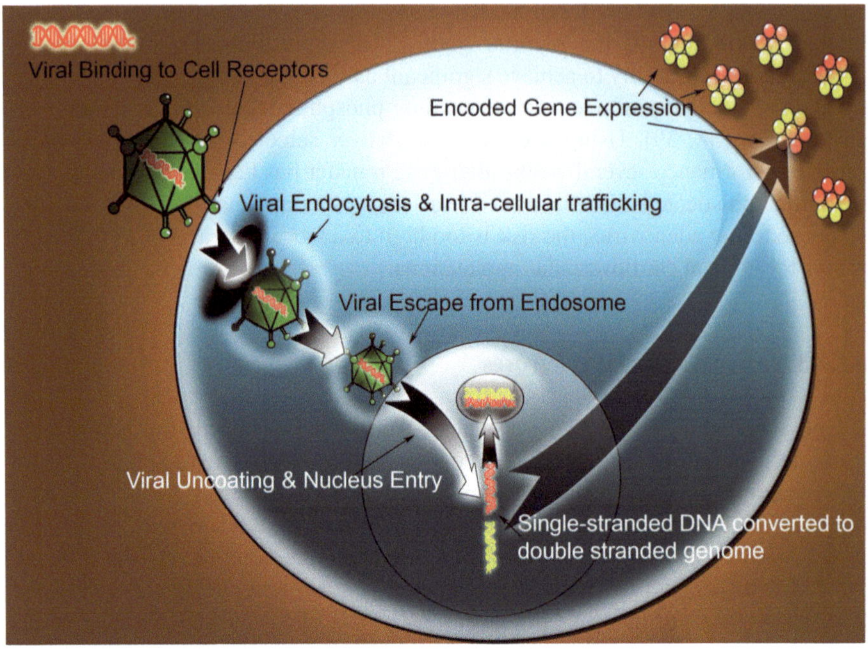

Fig. 8.5 Simplified representation of AAV endocytosis and intracellular trafficking

steps (Fig. 8.5): (1) Binding to the membrane receptor/co-receptor or attachment factors. (2) Endocytosis of the virus by the host cell occurs in distinct membrane compartments, called clathrin-coated pits, which can be internalized to form clathrin-coated vesicles. Clathrin-independent endocytosis involves the uptake in caveolae, membrane lipid rafts, and microdomains. (3) Following endocytosis, the AAV vectors are compartmentalized into early endosomes. This is the distribution station in the endocytic pathway. (4) Early endosomes then mature into late endosomes that are degraded by fusion with the lysosome secretory vesicles, and the material that will be recycled back to the plasma membrane. (5) Some separate viruses can escape lysosomal degradation via acidification of the endosome, and it is a necessary prerequisite for the release of the AAV. (6) These AAV particles are then trafficked into the nucleus, where viral uncoating leads to single-stranded DNA release. The ssDNA is then converted to dsDNA and finally to concatemers or integrated into the host genome. AAV transport can be blocked by neutralizing antibodies, temperature, and physical and chemical inhibitors through a time- and dose-dependent process. In vivo studies have noted that several serotypes of AAV are able to cross vascular endothelium with different efficiencies [48]. It is known that AAV2 has a relatively poor tropism for vascular cells, although reasonable levels of transduction have been achieved in cardiac myocytes [49]. Local delivery of AAV2 leads to transduction of underlying vascular smooth muscle cells and sequestration of AAV in the extracellular matrix around endothelial cells, thus preventing cell binding and entry. The potential of AAV6 vector for cardiac gene therapy was achieved through the use of VEGF to increase vascular permeability [50].

Challenges

Even in the setting where transient expression of a transgene may be sufficient (i.e., induction of neoangiogenesis by secreted growth factors), the results from preclinical and clinical studies have been disappointing. No single vector system is likely to be optimal for all cardiac gene therapy applications [27]. The ideal vector should be administered by the least invasive delivery route, target the desired geometric distribution of cardiac cells, express the requisite quantity of transgene product with the desired temporal regulation, be readily produced at high titers, and should avoid an immune response altogether.

8.4 Gene Delivery Techniques

Efforts to advance cardiovascular gene therapy will clearly require solutions related to delivery. The design of a delivery system consists of selectively targeting tissues of therapeutic relevance while minimizing systemic effects. This would subsequently permit the translation from the experimental phase into clinical trials once a target has been validated in preclinical models, whereby the development of safe and efficient delivery systems is a prerequisite. Several approaches to the development of myocardial-specific vectors and promoters have been explored [26, 48, 51, 52]. A complementary approach to achieve organ specificity is through the gene delivery method, which can diminish the biodistribution of vector capsids to extracardiac organs. It should be noted that a wide variety of techniques have already been designed and applied for cardiac applications [53]. Unfortunately, most of these approaches have led to limited transfection of cardiac myocytes in situ, particularly in large animals, and limited transduction efficiency with a moderate to high incidence of systemic exposure. Ideally, an optimized gene delivery system for a specific target should consider both the vector and route of administration in its design [54]. Existing methods of gene delivery can be classified by the site of injection, interventional approach, and the method of cardiac perfusion during gene delivery (Fig. 8.6).

8.4.1 Direct Gene Delivery

8.4.1.1 Intramyocardial Delivery

The majority of preclinical and clinical studies have involved intramyocardial gene delivery. This method can be performed using percutaneous catheter-based, minimally invasive surgical approaches (such as thoracoscopy) or a direct surgical route (Fig. 8.7). Moreover, there is the advantage of site selection, since multiple administrations can be carried out through the epicardium or endocardium. Anatomically, the epicardium is a serous membrane which consists of

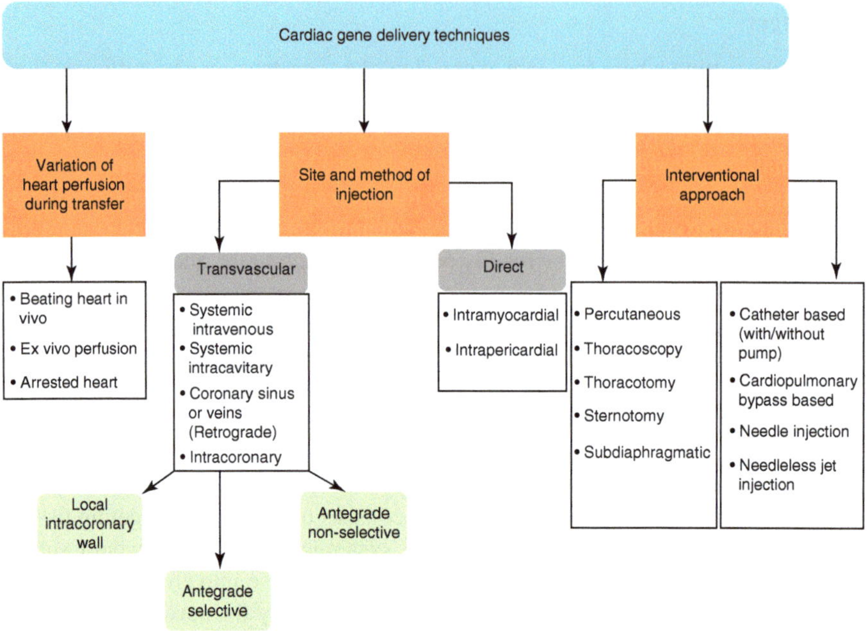

Fig. 8.6 Cardiac gene delivery techniques

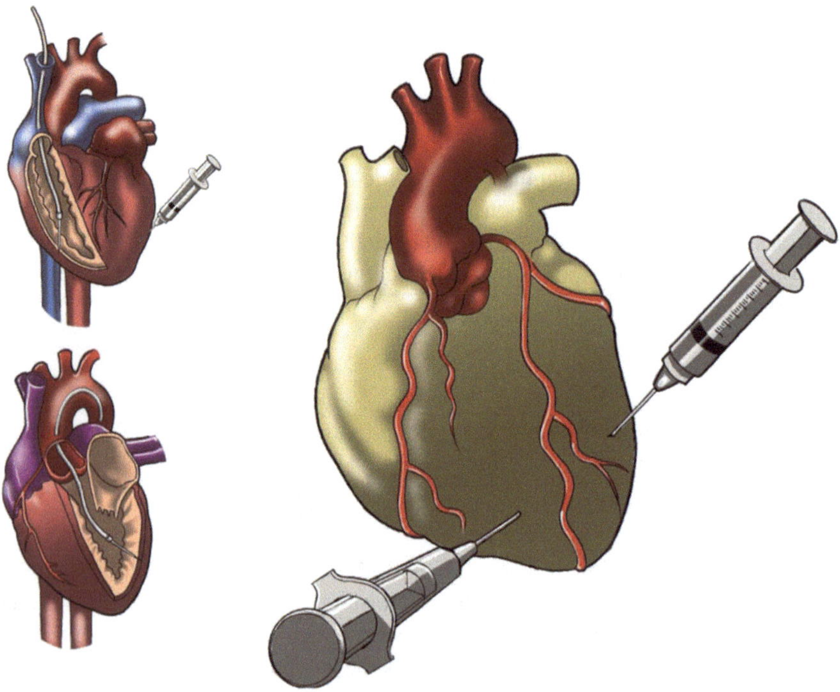

Fig. 8.7 Direct techniques of gene delivery

squamous epithelium and connective tissues. Endocardium covers the heart valves and tendons and is thus more complex in terms of its physiological function. There are three sublayers: the endothelium, subendothelium, and subendocardium.

In the early 1990s, it was demonstrated that reporter genes can be introduced and expressed in cardiac myocytes after direct injection of plasmid DNA into the LV wall [17, 55]. French et al. demonstrated several important points: the amount of recombinant protein produced increases with the amount of virus, reporter gene expression is rarely detected further than 5 mm from the injection site, the expression profile is similar in both ventricles, and the procedure itself causes minimal side effects in the heart [37]. Using a hamster model of cardiomyopathy, Tomiyasu et al. found that LV muscle injection of EBV/β2AR significantly elevated stroke volume and cardiac output [56]. In another study, robust and long-term βARKct expression was demonstrated in the rat LV after intramyocardial injection. Moreover, this technique allowed for improved cardiac contractility and normalized catecholamine levels in chronic HF animals [57]. This approach has been utilized to successfully to induce angiogenesis after VEGF transfer in several animal models of myocardial ischemia [58–60] and for the focal treatment of cardiac arrhythmias [61]. It was also demonstrated that although a substantial proportion of injected material resides in the myocardium immediately after intramuscular delivery, pervasive systemic vector leakage occurs resulting in extracardiac expression [62, 63]. Another unresolved issue associated with this technique is the acute inflammatory response secondary to injury produced at the injection site [35, 64].

Clinical Trials. Intramyocardial delivery was utilized in several clinical trials. Although positive results were not obtained [65], the safety and feasibility of this technique have been established with a reasonable risk–benefit profile [21, 22, 66].

8.4.1.2 Intrapericardial Delivery

The pericardial space is an inelastic cavity lined by a squamous serous membrane and filled with serous fluid. Several authors hypothesized that increasing the duration of vector exposure in the pericardium would result in gene expression in the myocardium (Fig. 8.8). Initially, it was shown that expression predominated in the parietal pericardium [67, 68]. Subsequently, it has been found that adding a proteolytic enzyme to a viral construct leads to a diffusion of transgene into the LV and interventricular septum [69]. However, it was also demonstrated that pericardial expression of VEGF does not improve myocardial perfusion [70] and the degree of vasculogenesis was greater in the subepicardial zone compared to subendocardial [58]. Shortcomings of intrapericardial delivery also include the high level of extracardiac expression [71] and the fact that such a delivery mode would not be suitable for patients with pericardial adhesions.

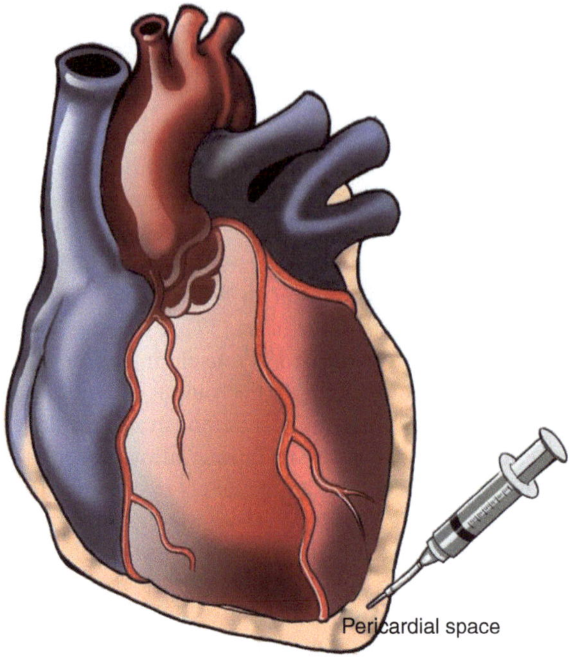

Fig. 8.8 Intrapericardial delivery

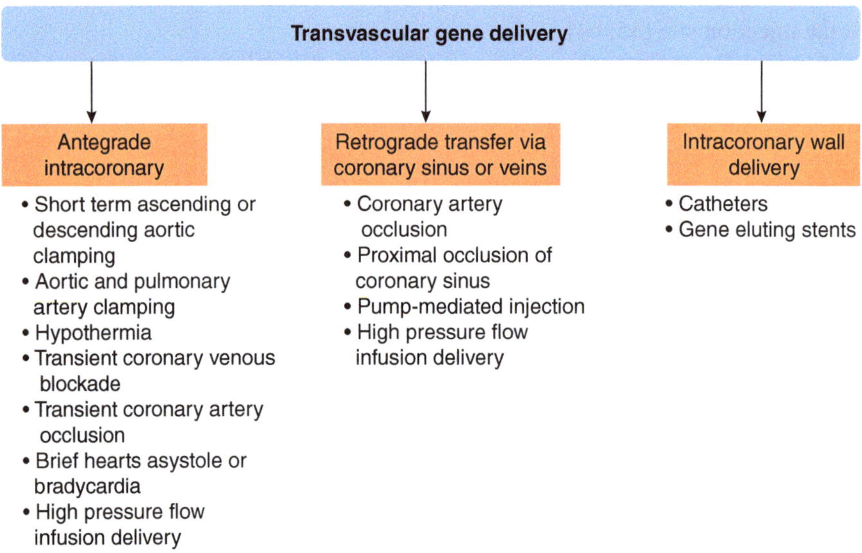

Fig. 8.9 Transvascular gene delivery

8.4.2 Transvascular Gene Delivery

A plethora of transvascular methods have been established (Fig. 8.9). However, none of these allows for the desired goal of efficient and cardiac-specific gene

Table 8.2 Advantages and disadvantages of direct and transvascular gene delivery methods

Advantages	Disadvantages
Direct cardiac gene delivery	
Safe and simple to use in the clinic	Requires guidance modalities and composite catheter systems (i.e. expensive)
Spatial and temporal control	Can cause acute inflammatory response secondary to needle stick injury
Allows for a high local concentration	Does not prevent or minimize viral escape into the system and collateral organ uptake
Good for focal treatment of regional ischemia and conduction abnormalities	Vector distribution profile is limited to injection sites
Avoids transfer across endothelial barrier	Non-homogeneous distribution is not desirable for many cardiac diseases
Was successfully used for cardiac angiogenesis	
Frequently used in clinical trials	
Limited effects (i.e. in theory independent of) neutralizing anti-bodies and T-cell activated immune response	
Transvascular cardiac gene delivery	
Efficient gene transfer	Diluted vector concentration in systemic blood circulation
Homogeneous and global distribution profile	Exposure to blood components and antibodies that neutralize or limit tropism
Extensive clinical experience with percuta-neous approaches	Systemic leakage leads to collateral organ uptake and expression
Ability to perform repeat administrations with catheter based techniques	Requires surgical manipulation to enhance transduction efficiency (e.g. aortic and pulmonary artery clamping)
Possibility to create closed loop recirculation systems to enhance transfer while minimizing collateral exposure	Vessel permeability issues which requires the use of drugs to increase vector diffusion
	Compromised vessels (e.g. atherosclerotic) impede vector distribution
	Demands complicated and expensive equipment

transfer with low systemic exposure. In Table 8.2, we summarize the benefits and limitations of direct and transvascular gene delivery methods.

8.4.2.1 Antegrade Intracoronary Gene Delivery

Percutaneous catheter-mediated intracoronary gene transfer is arguably the most clinically relevant method because of the extensive clinical experience in coronary catheterization procedures. This system features minimal invasiveness and the possibility of reaching any cardiac territory. This mode of delivery typically results in more homogenous expression compared to direct intramyocardial delivery. Most researchers initially expected that this method would be most effective for cardiac gene transfer applications. These expectations notwithstanding, studies have shown that simple antegrade intracoronary delivery results in severely limited transfection of cardiac myocytes with variable efficacy across animal species with diverse vector systems [72–74]. Upon further development of

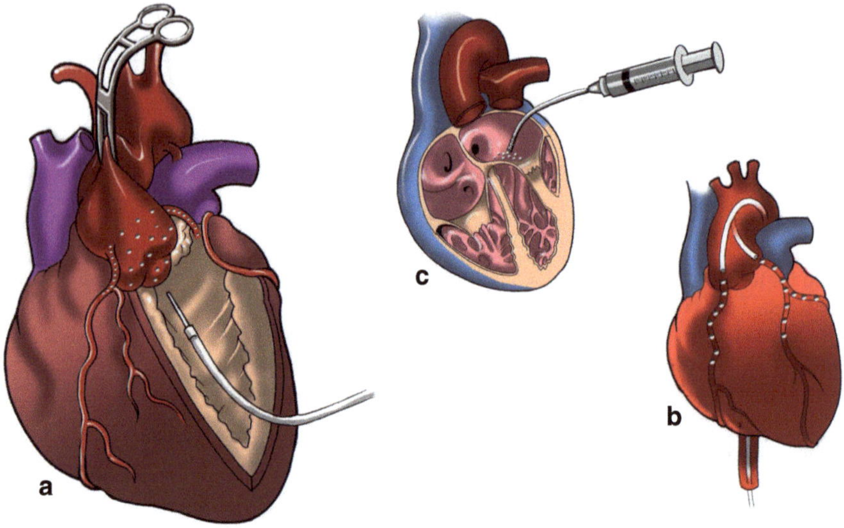

Fig. 8.10 (**a**) Nonselective antegrade intracoronary gene delivery. (**b**) Selective antegrade intracoronary gene delivery. (**c**) Intracavitary (left atrium) gene delivery

transvascular delivery systems, two different classifications have been defined: *selective* (directly into the desired artery (Fig. 8.10b) with indicated target area) and *nonselective* (indirectly into left ventricle with aortic cross clamping for distribution) (Fig. 8.10a). In an effort to achieve increased efficiency with selective catheterization, investigators began to use transient coronary occlusion [75], concomitant coronary venous blockade (Fig. 8.11a) [76], and cardiac arrest with obstruction of venous return to the heart [77]. In several studies, the dependence of transgene expression on perfusion pressure and infusion flow rate has been demonstrated [78, 79]. Donahue et al. identified parameters influencing the efficiency of intracoronary transfer. These key parameters were exposure time, high flow and pressure, virus concentration, temperature, and the use of crystalloid solution as opposed to blood [80]. The possibility of coronary antegrade perfusion without selective arterial catheterization was reported by Hajjar et al. [38], using brief aortic and pulmonary artery clamping. The improvement of homogenous vector-mediated gene expression is theoretically a result of increasing the transcoronary myocardial perfusion gradient while at the same time decreasing pulmonary blood flow, which has the net effect of limiting vector uptake by the lungs. Several laboratories have found better expression of both ventricles using this method [39, 76, 81]. Other effects included a significant decrease in cardiac contractility after aortic occlusion with elevation of afterload, both potentially improving the conditions for effective delivery [82]. Additional modifications of this method could include occlusion of descending aorta instead of ascending and the use of hypothermia [52].

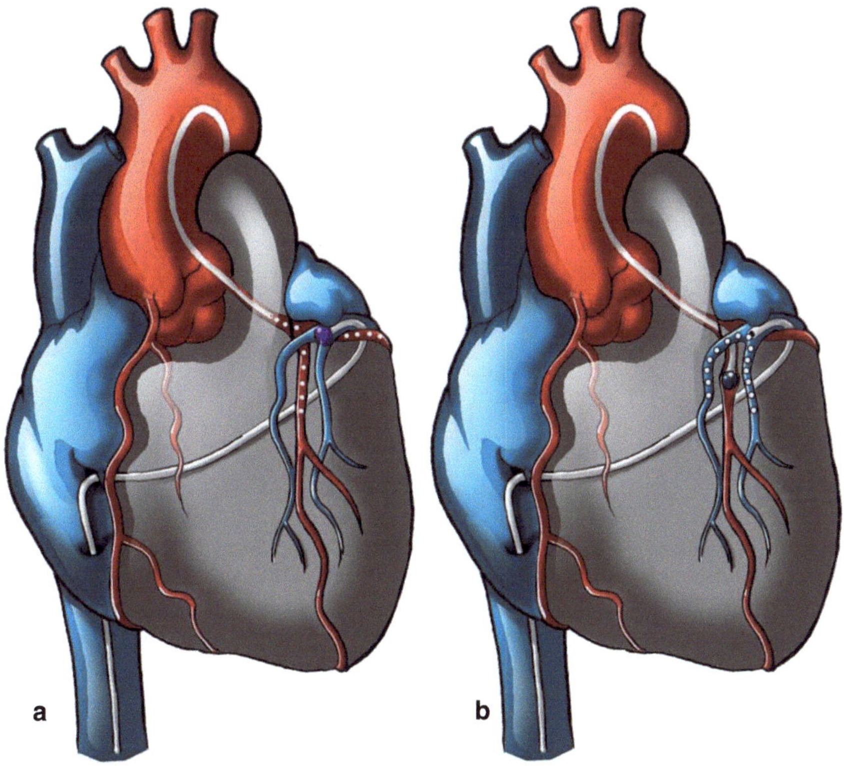

Fig. 8.11 (**a**) Selective antegrade intracoronary gene delivery with concomitant coronary venous blockade. (**b**) Selective coronary sinus or coronary venous retrograde delivery with transient coronary artery occlusion

8.4.2.2 Retrograde Intracoronary Sinus Gene Delivery

Retrograde intracoronary delivery through the coronary sinus (Fig. 8.11b), in direct comparison to antegrade, provides a more uniform distribution of agents in the presence of coronary artery disease. Another benefit of this method is that it does not have to overcome the resistance of precapillary sphincters proximally located on the arterial side of the capillary beds. Thus, less blood is shunted through the thebesian and arteriosinusoidal channels into the cardiac chambers. Many authors argue that coronary venous infusion allows for prolonged adhesion time of the vector in the cardiac endothelium. This effect directly results in both an increase in endothelial permeability and a higher pressure gradient across the interstitial capillaries and venules promoting the transfer of macromolecular particles into the interstitium of the heart [83, 84]. In support of this view, it has been reported that retrograde vector-mediated delivery of the Fibroblast Growth Factor 2 (FGF-2) gene relative to antegrade enhances collateral perfusion in pigs with chronic myocardial ischemia [85]. Also retroinfusion of VEGF reduced postischemic inflammation and myocardial reperfusion injury [86].

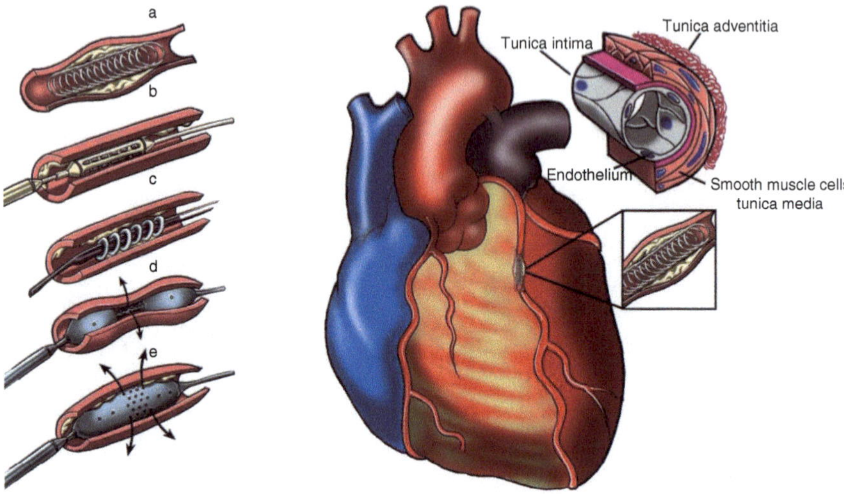

Fig. 8.12 Different catheters and stents for transvascular intracoronary wall gene delivery. (**a**) Stent. (**b**) Infiltrator nipple balloon. (**c**) Dispatch coil balloon. (**d**) Double balloon catheter. (**e**) Single balloon catheter

8.4.2.3 Transvascular Intracoronary Wall Delivery

Percutaneous coronary intervention and coronary artery bypass grafting are common treatments for patients with ischemic heart disease (IHD) requiring revascularization. The standard approach involves treatment of local coronary lesions and global management of IHD comorbidities. However, the failure rate of these procedures is relatively high [87]. Vector-mediated local gene transfer to the coronary vasculature is limited due to vessel barriers including the tunica adventitia, external and internal elastic lamina, smooth muscle cells (SMC), and tunica intima incorporating the endothelium. A clinically applicable intracoronary wall delivery device must meet these requirements: (1) provide isolation and exposition of a vascular segment, (2) maximize diffusion through the endothelium and basement membranes, (3) result in high efficiency penetration of the vascular wall while minimizing escape into the systemic circulation and perivascular space, (4) minimize or eliminate the risk of vessel dissection or perforation, and (5) permit downstream blood flow. To address these requirements, several types of balloon catheters for percutaneous gene delivery were developed [2, 14] (Fig. 8.12). *Double-balloon catheters* include two inflatable balloons separated by an intermediate space, into which the gene therapeutics can be infused through a separate lumen. After infusion, the vector remains in contact with the vessel wall between the proximal and distal occluding balloon. The efficiency of gene expression in the arterial endothelial cells using this catheter was demonstrated in a sheep model [88]. The major disadvantage of these catheters is the requirement for occlusion of vessels for significant periods of time. Other catheters have also been used which feature *porous balloons*, through which the therapy is infused under high pressure into the lumen of a single balloon that contains multiple microscopic perforations. Upon injection, the vector solution expands the balloon and exits through the pores, entering the vessel wall. The perforated balloon catheter was tested with the injection of retroviral vector containing β-galactosidase into the rabbit aorta.

Authors concluded that the practical use of this catheter is limited by the small number of cells that are actually transduced [89]. The next-generation design of the catheter type, which is named *Dispatch*, allows for maintenance of distal blood flow through a central lumen with infusion of a transgene between the artery wall and the catheter. The primary advantage of this catheter is an extended incubation time, since it can be inflated in the coronary arteries for a long period of time without inducing myocardial ischemia [90].

Unique to the *infiltrator* catheters is the attempt to enhance transfer by injecting the vector into the vessel wall via microinjection needles, which in theory decreases the chances of systemic spread of the vector while also enhancing transfer. It consists of three longitudinal polyurethane pads attached to the balloon with three linear arrays of microneedles positioned on the pads. The needle injection facilitates transgene delivery to the media and the adventitia [91]. None of these devices are ideal, but several have been shown to increase level of transfer into the coronary arterial wall. Despite these successes, however, authors have now preferred the use of *eluting stents over this catheter*. The eluting stents represent a promising platform for localized delivery to the vascular wall. The advantages of these devices include extensive clinical experience in coronary catheterization procedures, safety, permanent scaffold structure, and their function as reservoirs for viral vectors while attenuating systemic side effects. The stent coating is the main functional element, as its role is to provide a barrier between the metallic surface and the blood. It produces prolonged expression up to 28 days in neointimal SMC using phosphorylcholine stents in a rabbit model [92]. Application of bisphosphonate stents led to extensive localized Ad/GFP expression in the rat arterial wall, and an adenovirus inducible nitric oxide synthase (iNOS) attached to this stent resulted in inhibition of restenosis [93]. In addition, its function was investigated using collagen in DNA-stent coatings. Pig coronary studies comparing stents containing plasmid DNA/GFP to coated stents without DNA demonstrated that 10.8 % of neointimal cells were transduced [94]. Drug-eluting stents have been extensively used to prevent coronary restenosis in several human clinical trials [95].

8.4.2.4 Ex Vivo Gene Delivery

Heart transplantation raises a number of issues that need to be addressed. The complications associated with this modality include acute rejection, allograft vasculopathy, and a high risk for developing malignancies including skin cancer and non-Hodgkin's lymphoma. Gene delivery to the donor heart ex vivo is an active area of investigation to address these issues [96, 97] (Fig. 8.13). The possibility of intracoronary adenovirus/LacZ transfer in harvested hearts was evaluated in a piglet model. Transgene expression was detected in all cardiac areas [98]. In another study,

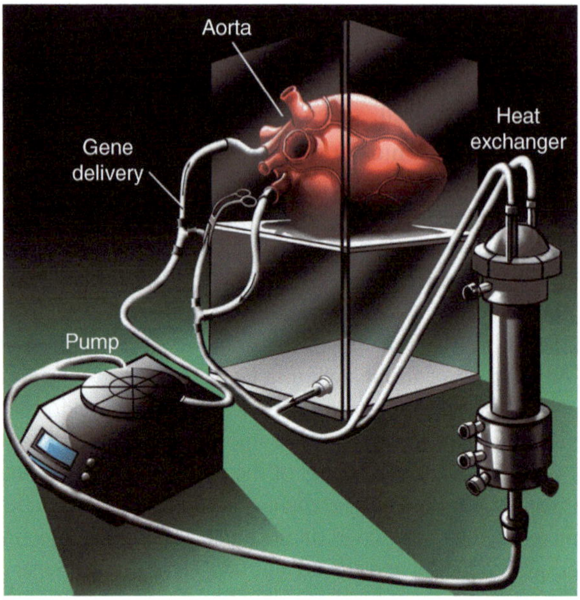

Fig. 8.13 Ex vivo gene delivery

it was also confirmed that direct plasmid DNA injection can result in detectable levels of expression in rat cardiac allografts [99]. Shah et al. administered adenoviral construct with β2-adrenergic receptor into the aortic root of a rabbit donor heart. The authors found that left ventricular performance was significantly improved after heterotopic transplantation. Thus, it was shown that it is possible to genetically modulate β-adrenergic signaling system in a transplanted heart [100].

8.4.2.5 Cardiopulmonary Bypass-Based Gene Delivery

An estimated 5.8 million Americans suffer from heart failure. In about two-thirds of those cases, the cause is ischemic heart disease (IHD). In addition, approximately 40 % of the patients with IHD have LV systolic or diastolic dysfunction. Currently, revascularization procedures such as coronary artery bypass surgery and percutaneous coronary intervention are the primary methods of treating medically refractory symptomatic IHD. In fact, more than one million cardiac procedures that require extracorporeal circulation with cardiopulmonary bypass (CPB) are carried out in the world annually. The use of CPB with cardiac arrest for the purpose of gene transfer was first demonstrated by Bridges et al. [101] and Davidson et al. [102] (Fig. 8.14b, c).

Later, it was described that cold crystalloid cardioplegia is not detrimental for transgene expression [103]. Moreover, it has been shown that there is extensive restoration of a deficient membrane protein, δ-sarcoglycan, after use of cardioplegic solution in cardiomyopathic hamsters [104]. As opposed to other researchers, Bridges and colleagues constructed the first “closed-loop” recirculating system which they called molecular cardiac surgery with recirculating delivery (MCARD). Using MCARD, they were able to achieve AAV-mediated gene expression in the

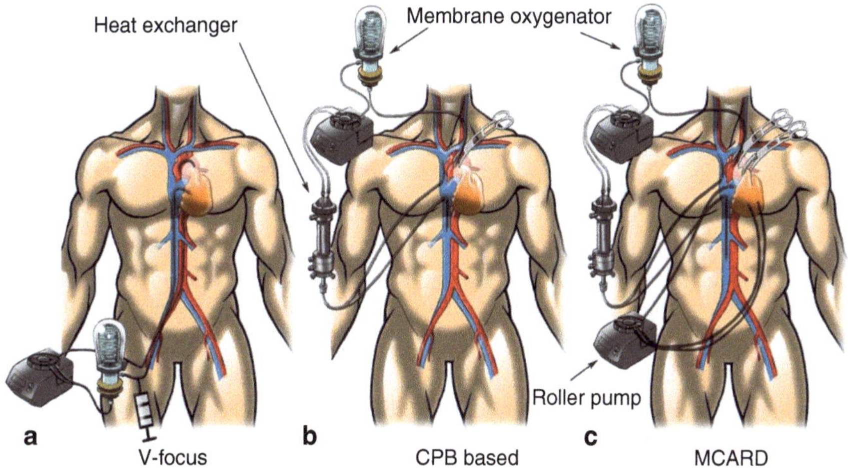

Fig. 8.14 (**a**) Catheter-based closed-loop recirculatory system (V-Focus). Coronary venous blood was drained from the coronary sinus. Following oxygenation, the blood is returned to the coronary arteries via a roller pump. Gene construction was delivered into the antegrade part the circuit. (**b**) Cardiopulmonary bypass-based gene delivery. Cardiopulmonary bypass was established via an aortic arterial cannula and a right atrial venous cannula. After cardioplegic arrest, viral particles were injected into aortic root and allowed to dwell in the myocardium. (**c**) Cardiopulmonary bypass-based closed-loop recirculatory system (MCARD). This technique integrates a separate pump circuits for the cardiac and systemic circulations, thus making it possible to achieve complete cardiac isolation. The virus/gene solution was injected into the retrograde catheter located in the coronary sinus and recirculated for 20 min. The coronary circuit then flushed to wash out residual vector

majority of myocytes in a large animal model, so the number of genome copies (gc) per cell in the left ventricle ranges from 1 gc in posterior wall to 3 gc in anterior wall [105]. In addition, they showed that more than 99 % of the gc's initial dose remains in the cardiac circuit for 20 min [106], and furthermore, T cell-mediated immune response to AAV capsid was observed only after intramyocardial injection but not in the MCARD group [105]. Later, it was established that percutaneous minimally invasive delivery system (V-Focus) reversed HF progression in a sheep model [107]. However, the V-Focus technique is not a true recirculating system since quantitative PCR reveled that there was 100 times as much gene delivery to the liver as to the heart with this technique [108] (Fig. 8.14a).

8.4.3 Physical Methods for Enhancement Gene Transfer

8.4.3.1 Sonoporation

Biocompatible microbubbles are small (1–5 μm) gas-filled microspheres. They can be administered into the circulation in various ways and can be destroyed by ultrasound (US) irradiation. This phenomenon can be applied to gene delivery. Ultrasound-targeted microbubble destruction combines low invasiveness and organ specificity. It is based upon the development of second-generation US contrast agents like perfluorocarbons or sulfur hexafluoride. Sonoporation is thought to increase the size and permeability of pores in the cell membrane allowing for enhanced gene transfer into cells. After intravenous injection, microbubbles stay stable for several minutes, can pass through the pulmonary circulation, and then can be visualized and destroyed by conventional US devices. The hearts of all rats that underwent US-mediated destruction of albumin-coated microbubbles containing adenovirus showed good myocardial expression with β-galactosidase [109]. Left ventricular injection of naked plasmid DNA and siRNA duplexes into murine heart by means of sonoporation showed much greater gene expression than intravenous administration [10]. In another study, it was demonstrated that US-mediated microbubble destruction can generate high levels of reporter gene activity restricted to the heart [110]. Interestingly, it has been noted that aortic occlusion with brief asystole increased myocardial gene expression in rats using microbubbles by 2.5 fold [111].

8.4.3.2 Electroporation

This technique involves the application of short-duration, high-intensity electric field pulses. The electrical stimulus causes membrane destabilization with subsequent opening nano-sized pores with improving permeability, allowing passage of DNA into the cells. Plasmid delivery through electroporation increased cardiac expression of VEGF in a large animal model [112]. It also provides evidence that electroporation-mediated gene delivery to the beating rat heart is an effective tool for nonviral gene transfer with a lack of toxicity and good preservation of heart function [113].

8.4.3.3 Magnetic Field-Enhanced Transfection (Magnetofection)

Magnetofection is defined as gene delivery guided and mediated by magnetic force. The basic idea is that DNA is attached to magnetic nanoparticles and the external magnetic field increases particle internalization and gene expression. Even a simple external magnet of 25 gauss can direct nanoparticles with plasmid DNA to the heart cells [114]. Marker gene expression after delivery of adenovirus/magnetic nanoparticles to the rat carotid artery was significantly greater than in nonmagnetic controls [115].

8.4.4 Guidance Systems to Identify Targeted Area

8.4.4.1 X-Ray Fluoroscopy

This method provided outstanding guidance for transcatheter therapy including coronary intervention. Catheter-based adenovirus-mediated intramyocardial gene transfer is feasible using percutaneous fluoroscopically guided coaxial catheters [116]. This procedure is not associated with hemodynamic changes or arrhythmias [117].

8.4.4.2 Real-Time MRI

Percutaneous intramyocardial gene delivery is feasible using rtMRI and permits precise 3-dimensional localization of injections involving visualization of full thickness of myocardium [118].

8.4.4.3 Electromechanical Mapping

The NOGA system is designed to display electroanatomical maps of the heart. Catheters designed to be used with this system are equipped with an electromagnetic sensor. As the catheter moves along the endocardium, local electrograms are reported. The system uses this information to construct a 3-D electroanatomical map of the left ventricle. The NOGA may be advantageous to accurately direct gene transfer to areas of myocardial ischemia where gene transfer may be potentially optimized [22]. This system was used for the injection of an adenoviral vector containing VEGF into designated ischemic sites and results in successful gene transfer and protein expression [119].

8.4.4.4 Echocardiography Guidance

Local delivery of therapeutic genes into the left ventricle was evaluated with live 3-D echo in the pig model. Accuracy defined as an injection into the target zone was 83 % [120].

Challenges

The optimal technique for cardiac gene delivery must be safe, clinically translatable, and ideally incorporate the following: (1) retrograde through the coronary venous system; (2) washout of vector after gene transfer to minimize collateral expression; (3) increased myocardial transcapillary gradient and/or enhance transendothelial transport of viral particles from the vasculature into the interstitium, using physical or pharmacological methods; and (4) a “closed-loop” for extended transgene residence time in the coronary circulation.

8.5 Cardiac Gene Therapy Molecular Targets

8.5.1 Heart Failure

Gene therapy in HF primarily targets an increase in contractility, a reduction in adverse remodeling, and inhibition of apoptosis (Fig. 8.15).

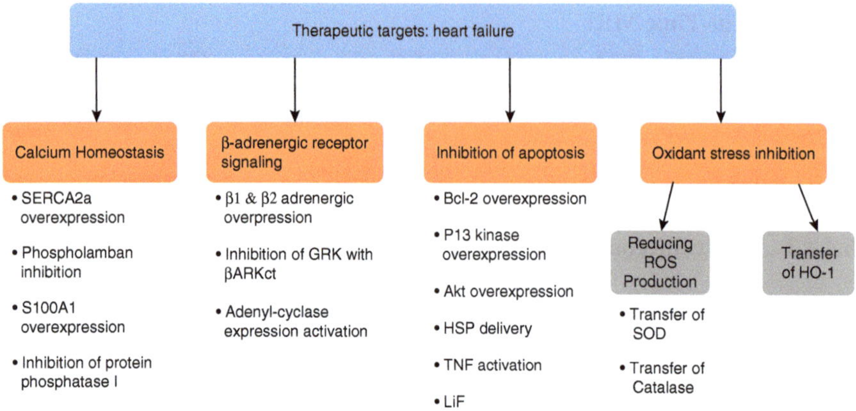

Fig. 8.15 Gene therapeutic targets: heart failure. *Abbreviations and acronyms*: *SERCA2a* sarcoplasmic reticulum calcium ATPase, *S100A1* member of S100 family of cardiac proteins, *GRK* G protein-coupled receptor kinase, *βARKct* β-adrenergic receptor kinase carboxyl-terminus, *Bcl-2* family of proteins regulated apoptosis, *P13* phosphoinositide 3-kinases, *Akt* serine/threonine kinase, *ROS* reactive oxygen species, *SOD* superoxide dismutase, *HO-1* heme oxygenase enzyme-1, *HSP* heat shock proteins, *TNF* tumor necrosis factor, *LIF* leukemia inhibitory factor

8.5.1.1 The Calcium Cycling Proteins

Ca^{2+}cycling has been found to be critically dysregulated in HF and provides an important role in excitation–contraction coupling (Fig. 8.16). To understand the Ca^{2+} handling defects in heart failure, we need to briefly describe the processes occurring in cardiac excitation–contraction coupling. During the cardiac action potential, Ca^{2+} enters the cell through depolarization-activated Ca^{2+} channels as an inward Ca^{2+} current, which contributes to the action potential plateau. Ca^{2+} entry triggers Ca^{2+} release from the sarcoplasmic reticulum (SR). This allows Ca^{2+} to bind to the myofilament protein troponin C, which then switches on the contractile process.

For relaxation to occur, there is a decline in intracellular Ca^{2+} concentration, allowing Ca^{2+} to dissociate from troponin. This requires Ca^{2+} transport out of the cytosol by pathways involving SR Ca^{2+} ATPase, sarcolemmal Na^+/Ca^{2+} exchange, sarcolemmal Ca^{2+} ATPase, or mitochondrial Ca^{2+}. Deficient SR Ca^{2+} uptake during myocyte relaxation has been identified in failing hearts from both humans and animals and is associated with a decrease in the expression and activity of sarcoplasmic reticulum calcium ATPase (SERCA2a). This protein is a Ca^{2+}ATP-dependent pump of the sarcoplasmic reticulum that has a critical role in Ca^{2+} regulation.

SERCA2a

The overexpression of SERCA2a has been demonstrated to increase contractility and normalize calcium cycling in failing human cardiomyocytes [121]. A number

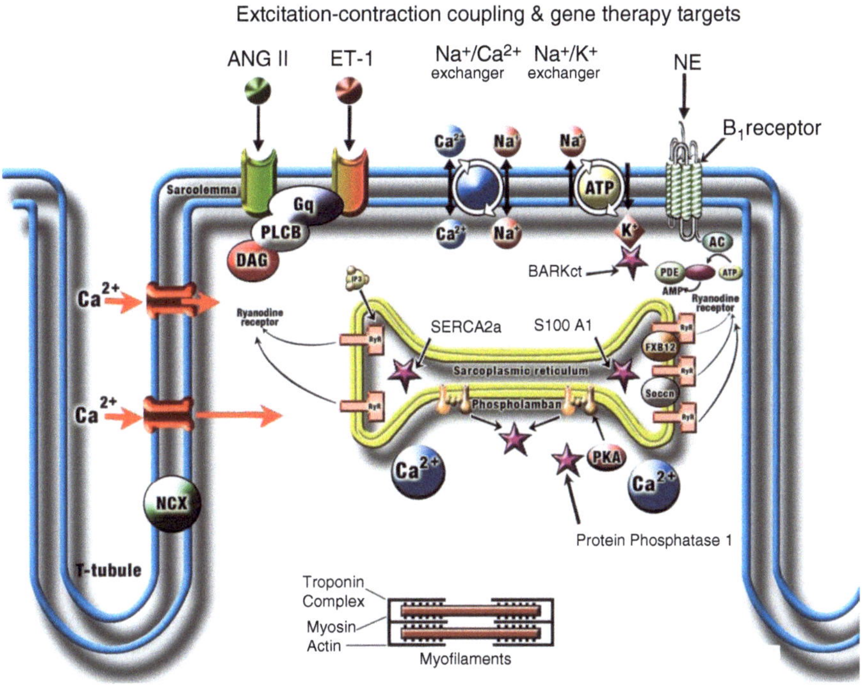

Fig. 8.16 Simplified representations of the excitation–contraction. (1) βARKct (lilac star). Molecular abnormalities associated with HF include the uncoupling of the β-adrenergic receptor system, enhanced expression and activity of the G protein-coupled receptor kinase, and loss of βAR inotropic reserve. βARKct gene delivery approach has the potential to resolve βAR downregulation and desensitization. (2) Phospholamban (lilac star) is an endogenous inhibitor of the SR Ca2+ ATPase. Phosphorylation of phospholamban by cyclic AMP- dependent or calmodulin-dependent protein kinases (PKA or CaMKII) relieves this inhibition, allowing faster muscle twitch relaxation and decline of intracellular Ca^{2+}. Because the SR Ca^{2+} ATPase competes better with Na+/Ca^{2+} exchange, phosphorylation of phospholamban also enhances Ca^{2+} content in the SR. (3) S100A1 (lilac star) plays a role in increasing SERCA2a activity, diminishing diastolic SR Ca^{2+} leak, and augmenting systolic open probability of the ryanodine receptors, causing an overall gain in SR Ca^{2+} cycling. Also, S100A1 regulates SERCA2A–phospholamban function, resulting in a balanced enhancement of SRCa^{2+} release and uptake. (4) In failing hearts, the downregulation of adrenergic receptor and cAMP-dependent protein kinase signaling leads to the inactivation of inhibitor-1 which, in turn, results in increased activity of protein phosphatase 1 (lilac star). This activation leads to the dephosphorylation of phospholamban, thus reducing calcium uptake by SERCA2a. *Abbreviations and acronyms*: *PKA* protein kinase A, *RyR* ryanodine receptors, *FKB12* calstabin 2, *Ang II* angiotensin II, *ET-1* endothelin 1, *NE* norepinephrine, *ATP* adenosine triphosphate, *cAMP* cyclic adenosine monophosphate, *AC* adenyl cyclase, *PDE* phosphodiesterase, *Gq* class of guanine nucleotide-binding proteins, *PLCB* phospholipase C beta, *DAG* diacylglycerol, *IP3* inositol trisphosphate, *Gs/Gi* stimulatory/inhibitory G protein, *L-type* long-lasting dihydropyridine receptors, *NCX* Na+/Ca^{2+} exchanger, *PDE* phosphodiesterase. Lilac stars indicate the main gene therapy targets

of animal studies with a variety of models of HF have demonstrated that overexpression of SERCA2a has a positive inotropic effect, improves oxygen utilization, attenuates the progression of HF, and prolongs survival. SERCA2a gene transfer was found to substantially decrease incidence of ventricular arrhythmias and reduce infarct size in a model of ischemia/reperfusion [122].A lentiviral vector-mediated SERCA2a intracoronary delivery after myocardial infarction in rats resulted in favorable molecular remodeling with improving systolic and diastolic function 6 months later [34]. An improvement in LV diameter, fractional shortening, and EF was also demonstrated in a tachycardia-induced HF model [123].

Clinical Trial. A first-in-human clinical trial "Calcium Upregulation by Percutaneous Administration of Gene Therapy in Cardiac Disease" (CUPID) involving gene transfer of SERCA2a cDNA via a rAAV1 vector in patients with advanced HF has been undertaken in a randomized, double-blind, placebo-controlled study. Although the results were somewhat difficult to interpret due to the absence of a clear dose response, at 12 months, SERCA2a-treated patients in the highest dose cohort demonstrated a consistent trend in clinical symptomatic improvement and in functional capacity [124].

S100A1

In cardiomyocytes, S100A1 plays an important role in increasing SERCA2a activity. This effect is achieved through diminishing diastolic SR Ca^{2+} leak and augmenting the systolic open probability of the ryanodine receptors, leading to an overall gain in SR Ca^{2+} cycling. Also, S100A1 regulates SERCA2A–phospholamban function, resulting in a balanced enhancement of SR Ca^{2+} release and uptake. S100A1 is downregulated during the development of HF [125]. Thus, in theory, S100A1 may be a promising factor in the treatment of HF. In a rat model of HF, significant cardiac recovery was demonstrated after 8 weeks in AAV6/S100A1-treated animals [126]. A study in a postinfarction pig model after 14 weeks revealed improvement in dP/dt and ejection fraction and also restoration of high-energy phosphate homeostasis in failing myocardium [127].

Phospholamban (PLN)

PLN regulates the homeostasis of SR Ca^{2+} mediating slower cytosolic Ca^{2+} decay in cardiomyocytes, which translates into diastolic relaxation. Phosphorylation of PLN suppresses its inhibitory effect. AAV-mediated overexpression of a mutant ("pseudo-phosphorylated") form of PLN improved LV function and mitigated adverse remodeling in post-MI rats [128]. Silencing of PLN expression after tachycardia-induced HF in sheep increased ejection fraction and decreased LV end-diastolic area [107]. A study in a volume-overload HF proved that adenovirus encoding antisense PLN preserved LV contractility and normalized LV mechanoenergetics [129].

8.5.1.2 The β-Adrenergic Signaling Cascade

The β-adrenergic receptor (βAR) signaling system plays an important role in the control of cardiac function, mediating the inotropic, chronotropic, and lusitropic responses to the sympathetic neurotransmitters [130, 131]. Therefore, it represents an attractive molecular target to improve heart function. Two important components of the βAR system include the β-receptors and the regulatory G protein-coupled receptor kinases (GRKs). Dysregulation of the βAR pathway, including downregulation, uncoupling of second messenger systems, and upregulation of βAR kinase (βARK1, GRK2), has been shown to be a hallmark of HF. βARs are regulated by GRK2, a member of a G protein-coupled receptor kinase family that phosphorylates and inactivate these receptors [132]. βARKct, a competitive inhibitor of GRK2, has the potential to resolve βAR downregulation and desensitization associated with HF [132, 133]. Thus, inhibiting the activity of GRK2 or lowering its expression appears to offer a novel means to enhance cardiac function.

βARKct

In a rabbit model of HF induced by myocardial infarction, it was first demonstrated that the βARKct transgene improved heart function and delayed development of HF [134]. Inhibition of myocardial βARK1 via Ad/βARKct delivery before creation of acute coronary ischemia may represent a new strategy for cardiac protection [135]. Long-term βARKct expression in the rat is by reversed LV remodeling and a normalization of the neurohumoral status of chronic HF animals [57]. The high level of βARKct expression in pressure-overload heart hypertrophy can preserve adenyl cyclase activity and βAR density and also improve cardiac function and cell morphology [136]. Based on the above results and earlier results derived from transgenic animal models, it appears that delivery of βARKct could be beneficial in the setting of IHD and HF.

8.5.2 Ischemic Heart Disease

Current gene therapy research efforts in IHD include stimulation of angiogenesis, limitations of reperfusion injury through the use of antioxidant therapy and endothelial nitric oxide synthase, and cardioprotection by using antiapoptotic proteins [137] (Fig. 8.17). Nevertheless, much of the research is devoted to the study of angiogenesis.

8.5.2.1 Stimulation of Cardiac Angiogenesis

One major focus of gene therapy for ischemic heart disease is neovascularization of fibrous postinfarct or poorly perfused (hibernating) myocardium. Therapeutic angiogenesis can be achieved by gene transfer of vascular endothelial growth factor (VEGF), hepatocyte growth factor, fibroblast growth factor, and hypoxia-induced factor 1α.

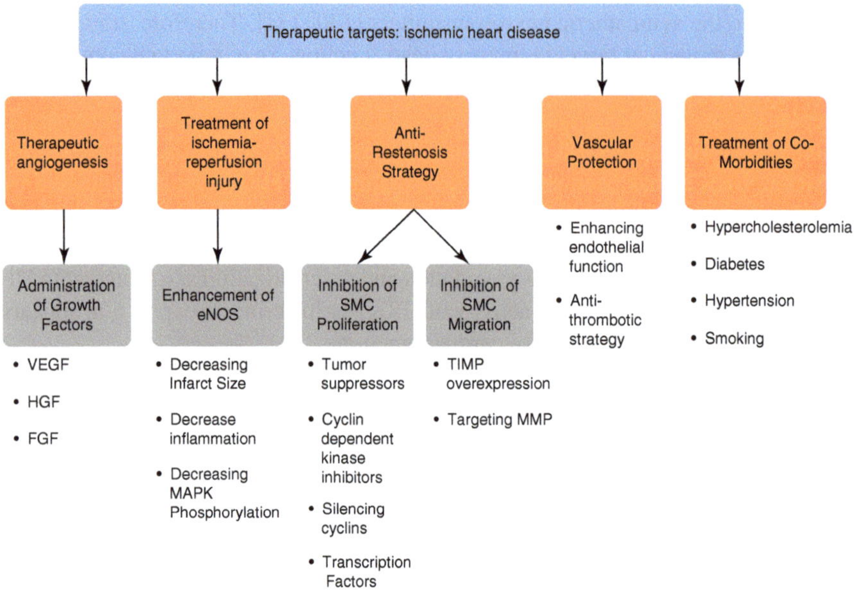

Fig. 8.17 Gene therapeutic targets: ischemic heart disease. *Abbreviations and acronyms*: *VEGF* vascular endothelial growth factor, *HGF* hepatocyte growth factor, *FGF* fibroblast growth factor, *eNOS* endothelial nitric oxide synthase, *MAPK* mitogen-activated protein kinase, *SMC* smooth muscle cells, *TIMP* tissue inhibitor matalloproteinases, *MMP* matrix metalloproteinases

VEGF

VEGF has five isoforms which act on tyrosine kinase receptors, FLK-1 and FT1. This protein factor has been shown to stimulate endothelial cell proliferation, migration, vascular permeability, and to affect fibroblast and smooth muscle growth [14, 20, 138]. Preclinical gene therapy studies with VEGF in various large animal models of myocardial ischemia have demonstrated stimulation of angiogenesis and improvement in fractional shortening [139] and reduction of infarct size and peri-infarct fibrosis [140]. In addition, it has been noted that there is an appearance of apoptosis-resistant cardiomyocytes in the border zone [141] and improvement of myocardial viability [142].

Clinical Trial. Based on the promising results of experiments of small and large animals, several clinical trials were carried out using different isoforms of VEGF mainly in patients with no other therapeutic options [14, 137, 143]. Although there has been an excellent safety record and some improvements in angina class and stress sestamibi scans, none of the randomized controlled phase 2/3 trials have shown clinically relevant positive effects [65, 143]. The most likely reason for this apparent discrepancy may be related to the placebo effect, patient selection, and ineffective gene expression [65].

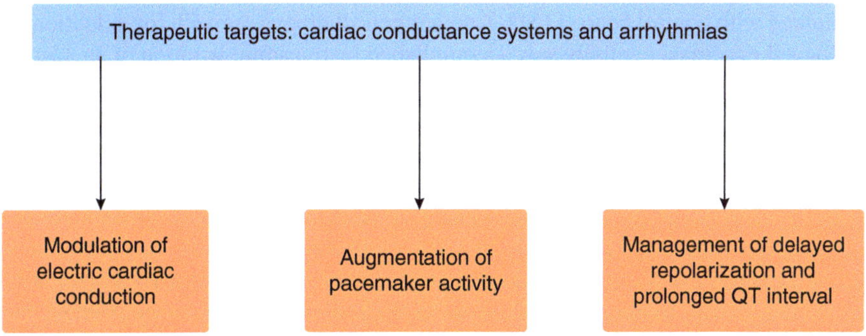

Fig. 8.18 Gene therapeutic targets: conductance system and arrhythmias

Fibroblast Growth Factor (FGF)

FGF is a heparin-binding growth factor that interacts with low-affinity cell surface receptors and high-affinity tyrosine kinase receptors. It is known that FGF stimulates endothelial cell synthesis of proteases including plasminogen activator and metalloproteinases that are necessary for angiogenesis [14]. The efficacy of FGF to promote angiogenesis has been well established in animal model of coronary ischemia [144]. Intracoronary delivery of adenovirus vector encoding FGF4 in pigs with myocardial ischemia increased regional perfusion [145]. In addition to angiogenesis, FGF5 overexpression can stimulate adaptive hypertrophy and improve wall thickening in hibernating myocardium [146].

Clinical Trial. The AGENT (angiogenic gene therapy) 3 and 4 trials of a low and high dose of adenoviral-mediated intracoronary administration of FGF4 were initiated and enrolled 532 patients. Authors found a beneficial effect on total exercise treadmill test, time to ST-segment depression and angina [147].

8.5.3 Cardiac Arrhythmias

Current approaches for the treatment of cardiac arrhythmias are limited (Fig. 8.18). Radiofrequency ablation is a strategy for focal abnormalities and still remains an experimental approach for more complex arrhythmias [148]. Implantable devices have problems such as high cost and potential risks from the invasive procedures. Thus, gene therapy targeting biological pacemaker function and conduction system is of interest. The ability to slow the heart rate during atrial fibrillation without producing heart block was demonstrated in pigs after overexpression of $G\alpha i2$ [149]. A study on the impact of atrioventricular nodal function utilized injection of fibroblasts expressing transforming growth factor-β1 [150]. Regulation of cardiac pacemaker activity was demonstrated as well through injection of plasmid DNA with a β2-adrenergic receptor. The results showed an increase in heart rate by 40 %

compared with control hearts [151]. Another gene therapy approach for induction of biological pacemaker activity was accomplished by the administration of an adenoviral construct incorporating HCN2 into the left bundle-branch system. After 48 h, all animals had sinus rhythm, and the rate was more rapid than in the controls [152]. Ventricular repolarization has also been targeted in gene transfer studies. Brunner et al. created a transgenic mouse with a long QT interval. Direct myocardial injection of adenoviral vectors expressing Kv1.5 resulted in shortening the action potential duration and the QT interval and also eliminating early afterdepolarizations [153]. In a postinfarct clinically relevant pig model of ventricular tachycardia, it was shown that gene transfer of KCNH2-G628S to the infarct border zone can eradicate cardiac ventricular arrhythmias [154].

8.5.4 Congenital Diseases

It is logical to assume that the correction of autosomal recessive genetic mutations could be treated through the exogenous delivery and expression of the normal gene. In a study of hamsters with autosomal recessive cardiomyopathy caused by a mutation in the δ-sarcoglycan (SG), it was shown that LV intramyocardial delivery of AAV/SG increased life expectancy and improved myocardial contractility and hemodynamics [155]. A similar study with intra-aortic injection AdV/SG demonstrated restoration of α-, β-, and δ-sarcoglycan in the myocyte membranes at 3 weeks after gene transfer and significantly less progression of LV dysfunction compared with controls [104]. Also it was demonstrated that genetic defects associated with inherited long QT syndrome may potentially be corrected [149, 150, 156].

8.5.4.1 Challenges

Cardiac myocytes in IHD and HF are characterized by a number of abnormalities at the molecular level. Identification of these pathological alterations and the operable mechanisms of their impact on cardiac function will allow us to better define potential targets for genetic intervention. With respect to congenital abnormalities, it should be noted that understanding the genetic mutation in heart development, which is the major challenge at present, will help to resolve cardiac repair issues through genetic reprogramming and replacement of defective genes in cardiac cells.

8.6 Conclusion

Cardiac gene therapy has advanced from the first in vitro studies in the early 1990s to current ongoing clinical trials. However, there is a need for further development in this field. Progress in vector technology is insufficient, and so far there is no vector for daily clinical practice with high transduction efficacy, long-term expression, and no immune response. Clinically reliable delivery methods also need improvement to create minimally invasive closed recirculation system that would provide extended gene residence time in coronary circulation and minimization of collateral organ expression. With regard to potential gene therapy targets, only with expanding knowledge in the key molecular mechanisms responsible for cardiovascular function in health and in disease will the discovery of novel myocardial targets be possible.

References

1. Melo LG, Pachori AS, Gnecchi M, Dzau VJ (2005) Genetic therapies for cardiovascular diseases. Trends Mol Med 11:240–250
2. Quarck R, Holvoet P (2004) Gene therapy approaches for cardiovascular diseases. Curr Gene Ther 4:207–223
3. Morishita R (2004) Perspective in progress of cardiovascular gene therapy. J Pharmacol Sci 95:1–8
4. Quarck R, De Geest B, Stengel D, Mertens A, Lox M, Theilmeier G et al (2001) Adenovirus-mediated gene transfer of human platelet-activating factor-acetylhydrolase prevents injury-induced neointima formation and reduces spontaneous atherosclerosis in apolipoprotein E-deficient mice. Circulation 103:2495–2500
5. Marchand GS, Noiseux N, Tanguay J-F, Sirois MG (2002) Blockade of in vivo VEGF-mediated angiogenesis by antisense gene therapy: role of Flk-1 and Flt-1 receptors. Am J Physiol Heart Circ Physiol 282:H194–H204
6. Montgomery RL, Hullinger TG, Semus HM, Dickinson BA, Seto AG, Lynch JM et al (2011) Therapeutic inhibition of miR-208a improves cardiac function and survival during heart failure. Circulation 124:1537–1547
7. Kawauchi M, Suzuki J, Morishita R, Wada Y, Izawa A, Tomita N et al (2000) Gene therapy for attenuating cardiac allograft arteriopathy using ex vivo E2F decoy transfection by HVJ-AVE-liposome method in mice and nonhuman primates. Circ Res 87:1063–1068
8. Yamasaki K, Asai T, Shimizu M, Aoki M, Hashiya N, Sakonjo H et al (2003) Inhibition of NFkappaB activation using cis-element 'decoy' of NFkappaB binding site reduces neointimal formation in porcine balloon-injured coronary artery model. Gene Ther 10:356–364
9. Suckau L, Fechner H, Chemaly E, Krohn S, Hadri L, Kockskämper J et al (2009) Long-term cardiac-targeted RNA interference for the treatment of heart failure restores cardiac function and reduces pathological hypertrophy. Circulation 119:1241–1252
10. Tsunoda S, Mazda O, Oda Y, Iida Y, Akabame S, Kishida T et al (2005) Sonoporation using microbubble BR14 promotes pDNA/siRNA transduction to murine heart. Biochem Biophys Res Commun 336:118–127
11. Rinne A, Littwitz C, Kienitz M-C, Gmerek A, Bösche LI, Pott L et al (2006) Gene silencing in adult rat cardiac myocytes in vitro by adenovirus-mediated RNA interference. J Muscle Res Cell Motil 27:413–421
12. Macejak DG, Lin H, Webb S, Chase J, Jensen K, Jarvis TC et al (1999) Adenovirus-mediated expression of a ribozyme to c-mybmRNA inhibits smooth muscle cell proliferation and neointima formation in vivo. J Virol 73:7745–7751
13. Yamamoto K, Morishita R, Tomita N, Shimozato T, Nakagami H, Kikuchi A et al (2000) Ribozyme oligonucleotides against TGF-b inhibited neointimal formation after vascular injury in rat model: potential application of ribozyme strategy to treat cardiovascular disease. Circulation 102:1308–1314
14. Gaffney MM, Hynes SO, Barry F, O'Brien T (2007) Cardiovascular gene therapy: current status and therapeutic potential. Br J Pharmacol 152:175–188

15. Müller OJ, Ksienzyk J, Katus HA (2008) Gene-therapy delivery strategies in cardiology. Future Cardiol 4:135–150
16. Felgner PL, Barenholz Y, Behr JP, Cheng SH, Cullis P, Huang L et al (1997) Nomenclature for synthetic gene delivery systems. Hum Gene Ther 8:511–512
17. Lin H, Parmacek MS, Morle G, Bolling S, Leiden JM (1990) Expression of recombinant genes in myocardium in vivo after direct injection of DNA. Circulation 82:2217–2221
18. Acsadi G, Jiao SS, Jani A, Duke D, Williams P, Chong W et al (1991) Direct gene transfer and expression into rat heart in vivo. New Biol 3:71–81
19. Nabel EG (1995) Gene therapy for cardiovascular disease. Circulation 91:541–548
20. Isner JM (2002) Myocardial gene therapy. Nature 415:234–239
21. Losordo DW, Vale PR, Symes JF, Dunnington CH, Esakof DD, Maysky M et al (1998) Gene therapy for myocardial angiogenesis: initial clinical results with direct myocardial injection of phVEGF 165 as sole therapy for myocardial ischemia. Circulation 98:2800–2804
22. Vale PR, Losordo DW, Milliken CE, McDonald MC, Gravelin LM, Curry CM et al (2001) Randomized, single-blind, placebo-controlled pilot study of catheter-based myocardial gene transfer for therapeutic angiogenesis using left ventricular electromechanical mapping in patients with chronic myocardial ischemia. Circulation 103:2138–2143
23. Kastrup J, Jørgensen E, Rück A, Tägil K, Glogar D, Rusyllo W et al (2005) Direct intramyocardial plasmid VEGF-A165 gene therapy in patients with stable severe angina pectoris. A randomized double-blind placebo-controlled study: the Euroinject One trial. J Am Coll Cardiol 45:982–988
24. Qin L, Pahud DR, Ding Y, Bielinska AU, Kukowska-Latallo JF, Baker JR et al (1998) Efficient transfer of genes into murine cardiac grafts by Starburst polyamidoamine dendrimers. Hum Gene Ther 9:553–560
25. Kizana E, Alexander IE (2003) Cardiac gene therapy: therapeutic potential and current progress. Curr Gene Ther 3:418–451
26. Wasala NB, Shin J-H, Duan D (2011) The evolution of heart gene delivery vectors. J Gene Med 13:557–565
27. Vinge LE, Raake PW, Koch WJ (2008) Gene therapy in heart failure. Circ Res 102: 1458–1470
28. Hinkel R, Trenkwalder T, Kupatt C (2011) Gene therapy for ischemic heart disease. Expert Opin Biol Ther 11:723–737
29. Rapti K, Chaanine AH, Hajjar RJ (2011) Targeted gene therapy for the treatment of heart failure. Can J Cardiol 27:265–283
30. Ding W, Zhang L, Yan Z, Engelhardt JF (2005) Intracellular trafficking of adeno-associated viral vectors. Gene Ther 12:873–880
31. Zhao J, Pettigrew GJ, Thomas J, Vandenberg JI, Delriviere L, Bolton EM et al (2002) Lentiviral vectors for delivery of genes into neonatal and adult ventricular cardiac myocytes in vitro and in vivo. Basic Res Cardiol 97:348–358
32. Bonci D, Cittadini A, Latronico MV, Borello U, Aycock JK, Drusco A et al (2003) 'Advanced' generation lentiviruses as efficient vectors for cardiomyocyte gene transduction in vitro and in vivo. Gene Ther 10:630–636
33. Fleury S, Simeoni E, Zuppinger C, Deglon N, von Segesser LK, Kappenberger L, Vassalli G (2003) Multiply attenuated, self-inactivating lentiviral vectors efficiency deliver and express genes for extended periods of time in adult rat cardiomyocytes in vivo. Circulation 197: 2375–2382
34. Niwano K, Arai M, Koitabashi N, Watanabe A, Ikeda Y, Miyoshi H et al (2008) Lentiviral vector-mediated SERCA2 gene transfer protects against heart failure and left ventricular remodeling after myocardial infarction in rats. Mol Ther 16:1002–1004
35. Guzman RJ, Lemarchand P, Crystal RG, Epstein SE, Finkel T (1993) Efficient gene transfer into myocardium by direct injection of adenovirus vectors. Circ Res 73:1202–1207
36. Kass-Eisler A, Falck-Pedersen E, Alvira M, Rivera J, Buttrick PM, Wittenberg BA et al (1993) Quantitative determination of adenovirus-mediated gene delivery to rat cardiac myocytes in vitro and in vivo. Proc Natl Acad Sci USA 90:11498–11502

37. French BA, Mazur W, Geske RS, Bolli R (1994) Direct in vivo gene transfer into porcine myocardium using replication-deficient adenoviral vectors. Circulation 90:2414–2424
38. Hajjar RJ, Schmidt U, Matsui T, Guerrero JL, Lee KH, Gwathmey JK et al (1998) Modulation of ventricular function through gene transfer in vivo. Proc Natl Acad Sci USA 95:5251–5256
39. Maurice JP, Hata JA, Shah AS, White DC, McDonald PH, Dolber PC et al (1999) Enhancement of cardiac function after adenoviral-mediated in vivo intracoronary β2-adrenergic receptor gene delivery. J Clin Invest 104:21–29
40. Berns KI, Giraud C (1996) Biology of adeno-associated virus. Curr Top Microbiol Immunol 218:1–23
41. Samulski RJ, Berns KI, Tan M, Muzyczka N (1982) Cloning of adeno-associated virus into pBR322: rescue of intact virus from the recombinant plasmid in human cells. Proc Natl Acad Sci USA 79:2077–2081
42. Laughlin CA, Tratschin JD, Coon H, Carter BJ (1983) Cloning of infectious adeno-associated virus genomes in bacterial plasmids. Gene 23:65–73
43. Rivera VM, Gao GP, Grant RL, Schnell MA, Zoltick PW, Rozamus LW et al (2005) Long-term pharmacologically regulated expression of erythropoietin in primates following AAV-mediated gene transfer. Blood 105:1424–1430
44. Coura Rdos S, Nardi NB (2007) The state of the art of adeno-associated virus-based vectors in gene therapy. Virol J 4:99
45. Ziello JE, Huang Y, Jovin IS (2010) Cellular endocytosis and gene delivery. Mol Med 16:222–229
46. Coura RS, Nardi NB (2008) A role for adeno-associated viral vectors in gene therapy. Gene Mol Biol 31:1–11
47. Di Pasquale G, Chiorini JA (2006) AAV transcytosis through barrier epithelia and endothelium. Mol Ther 13:506–516
48. Wang Z, Zhu T, Qiao C, Zhou L, Wang B, Zhang J et al (2005) Adeno-associated virus serotype 8 efficiently delivers genes to muscle and heart. Nat Biotechnol 23:321–328
49. Melo LG, Agraval R, Zhang L, Rezvani M, Mangi AA et al (2002) Gene therapy strategy for long-term myocardial protection using adeno-associated virus-mediated delivery of heme oxygenase gene. Circulation 105:602–607
50. Gregorovic P, Blankinship MJ, Allen JM, Crawford RW, Meuse L, Miller DG et al (2004) Systemic delivery of genes to striated muscles using adeno-associated viral vectors. Nat Med 10:828–834
51. Wang J, Faust SM, Rabinowitz JE (2011) The next step in gene delivery: molecular engineering of adeno-associated virus serotypes. J Mol Cell Cardiol 50:793–802
52. Müller OJ, Katus HA, Bekeredjian R (2007) Targeting the heart with gene therapy-optimized gene delivery methods. Cardiovasc Res 73:453–462
53. Katz MG, Swain JD, Tomasulo CE, Sumaroka M, Fargnoli A, Bridges CR (2011) Current strategies for myocardial gene delivery. J Mol Cell Cardiol 50:766–776
54. Katz MG, Fargnoli AS, Pritchette LA, Bridges CR (2012) Gene delivery technologies for cardiac applications. Gene Ther 19(6):659–669
55. Buttrick PM, Kass A, Kitsis RN, Kaplan ML, Leinwand LA (1992) Behavior of genes directly injected into the rat heart in vivo. Circ Res 70:193–198
56. Tomiyasu K, Oda Y, Nomura M, Satoh E, Fushiki S, Imanishi J et al (2000) Direct intracardiomuscular transfer of β2-adrenergic receptor gene augments cardiac output in cardiomyopathic hamsters. Gene Ther 7:2087–2093
57. Rengo G, Lymperopoulos A, Zincarelli C, Donniacuo M, Soltys S, Rabinowitz JE et al (2009) Myocardial adeno-associated virus serotype 6-bARKct gene therapy improves cardiac function and normalizes the neurohormonal axis in chronic heart failure. Circulation 119:89–98
58. Pätilä T, Ikonen T, Rutanen J, Ahonen A, Lommi J, Lappalainen K et al (2006) Vascular endothelial growth factor C-induced collateral formation in a model of myocardial ischemia. J Heart Lung Transplant 25:206–213
59. Schwarz ER, Speakman MT, Patterson M, Hale SS, Isner JM, Kedes LH et al (2000) Evaluation of the effects of intramyocardial injection of DNA expressing vascular endothelial growth

factor in a myocardial infarction model in the rat-angiogenesis and angioma formation. J Am Coll Cardiol 35:1323–1330
60. Vera Janavel GL, De Lorenzi A, Cortes C, Olea FD, Cabeza Meckert P, Bercovich A et al (2012) Effect of VEGF gene transfer on infarct size, left ventricular function and myocardial perfusion in sheep after two months of coronary artery occlusion. J Gene Med 14(4):279–287
61. Edelberg JM, Huang DT, Josephson ME, Rosenberg RD (2001) Molecular enhancement of porcine cardiac chronotropy. Heart 86:559–562
62. Grossman PM, Han Z, Palasis M, Barry JJ, Lederman RJ (2002) Incomplete retention after direct myocardial injection. Catheter Cardiovasc Interv 55:392–397
63. Bish LT, Sleeper MM, Braibard B, Cole S, Russell N, Withnall E et al (2000) Percutaneous transendocardial delivery of self-complementary AAV6 achieves global cardiac gene transfer in canines. Mol Ther 16:1953–1959
64. von Harsdorf R, Schott RJ, Shen YT, Vatner SF, Mahdavi V, Nadal-Ginard B (1993) Gene injection into canine myocardium as a useful model for studying gene expression in the heart of large mammals. Circ Res 72:688–695
65. Hedman M, Hartikainen J, Ylä-Herttuala S (2011) Progress and prospects: hurdles to cardiovascular gene therapy clinical trials. Gene Ther 18:743–749
66. Rosengart TK, Lee LY, Patel SR, Sanborn TA, Parikh M, Bergman GW et al (1999) Angiogenesis gene therapy: phase I assessment of direct intramyocardial administration of an adenovirus vector expressing VEFG 121 cDNA to individuals with clinically significant severe coronary artery disease. Circulation 100:468–474
67. Lamping KG, Rios CD, Chun JA, Ooboshi H, Davidson BL, Heistad DD (1997) Intrapericardial administration of adenovirus for gene transfer. Am J Physiol 272:H310–H317
68. March KL, Woody M, Mehdi K, Zipes DP, Brantly M, Trapnell BC (1999) Efficient in vivo catheter-based pericardial gene transfer mediated by adenoviral vectors. Clin Cardiol 22:123–129
69. Fromes Y, Salmon A, Wang X, Collin H, Rouche A, Hagege A et al (1999) Gene delivery to the myocardium by intrapericardial injection. Gene Ther 6:683–688
70. Lazarous DF, Shou M, Stiber JA, Hodge E, Thirumurti V, Goncalves L et al (1999) Adenoviral-mediated gene transfer induces sustained pericardial VEGF expression in dogs: effect on myocardial angiogenesis. Cardiovasc Res 44:294–302
71. Zhang JCL, Woo YJ, Chen JA, Swain JL, Sweeney HL (1999) Efficient transmural cardiac gene transfer by intrapericardial injection in neonatal mice. J Mol Cell Cardiol 31:721–732
72. Mühlhauser J, Jones M, Yamada I, Cirielli C, Lemarchand P, Gloe TR et al (1996) Safety and efficacy of in vivo gene transfer into the porcine heart with replication-deficient, recombinant adenovirus vectors. Gene Ther 3:145–153
73. Logeart D, Hatem SN, Heimburger M, Roux AL, Michel JB, Mercadier JJ (2001) How to optimize in vivo gene transfer to cardiac myocytes: mechanical or pharmacological procedures? Hum Gene Ther 12:1601–1610
74. Kaplitt MG, Xiao X, Samulski RJ, Li J, Ojamaa K, Klein IL et al (1996) Long-term gene transfer in porcine myocardium after coronary infusion of an adeno-associated virus vector. Ann Thorac Surg 62:1669–1676
75. Logeart D, Hatem SN, Rücker-Martin C, Chossat N, Nevo N, Haddada H et al (2000) Highly efficient adenovirus-mediated gene transfer to cardiac myocytes after single-pass coronary delivery. Hum Gene Ther 11:1015–1022
76. Hayase M, del Monte F, Kawase Y, MacNeill BD, McGregor J, Yoneyama R et al (2005) Catheter-based antegrade intracoronary viral gene delivery with coronary venous blockade. Am J Physiol Heart Circ Physiol 288:H2995–H3000
77. Ding Z, Fach C, Sasse A, G decke A, Schrader J (2004) A minimally invasive approach for efficient gene delivery to rodent hearts. Gene Ther 11:260–265
78. Wright MJ, Wightman LML, Latchman DS, Marber MS (2001) In vivo myocardial gene transfer: optimization and evaluation of intracoronary gene delivery in vivo. Gene Ther 8:1833–1839
79. Emani SM, Shah AS, Bowman MK, Emani S, Wilson K, Glower DD et al (2003) Catheter-based intracoronary myocardial adenoviral gene delivery: importance of intraluminal seal and infusion flow rate. Mol Ther 8:306–313

80. Donahue JK, Kikkawa K, Johns DC, Marban E, Lawrence JH (1997) Ultrarapid, highly efficient viral gene transfer to the heart. Proc Natl Acad Sci USA 94:4664–4668
81. Kaspar BK, Roth DM, Lai NC, Drumm JD, Erickson DA, McKirnan MD, Hammond HK (2005) Myocardial gene transfer and long-term expression following intracoronary delivery of adeno-associated virus. J Gene Med 7:316–324
82. Parsa CJ, Reed RC, Walton GB, Pascal LS, Thompsom RB, Petrofski JA et al (2005) Catheter-mediated subselective intracoronary gene delivery to the rabbit heart: introduction of a novel method. J Gene Med 7:595–603
83. Boekstegers P, Kupatt C (2004) Current concepts and applications of coronary venous retroinfusions. Basic Res Cardiol 99:373–381
84. Boekstegers P, von Degenfeld G, Giehrl W, Heinrich D, Hullin R, Kupatt C et al (2000) Myocardial gene transfer by selective pressure-regulated retroinfusion of coronary veins. Gene Ther 7:232–240
85. von Degenfeld G, Raake P, Kupatt C, Lebherz C, Hinkel R, Gildehaus FJ et al (2003) Selective pressure-regulated retroinfusion of FGF-2 into the coronary vein enhances regional myocardial blood flow and function in pigs with chronic myocardial ischemia. J Am Coll Cardiol 42:1120–1128
86. Kuppat C, Hinkel R, Vachenauer R, Horstkotte J, Raake P, Sandner T et al (2003) VEGF 165 transfection decreases postischemic NF-kappa B-dependent myocardial reperfusion injury in vivo: role eNOS phosphorylation. FASEB J 17:705–707
87. Lassaletta AD, Chu LM, Sellke FW (2011) Therapeutic neovascularization for coronary disease: current state and future prospects. Basic Res Cardiol 106:897–909
88. Rome JJ, Shayani V, Newmark KD, Farrell S, Lee SW, Virmani R et al (1994) Adenoviral vector mediated gene transfer into sheep arteries using a double balloon catheter. Hum Gene Ther 5:1249–1258
89. Flugelman MY, Jaklitsch MT, Newman KD, Casscells W, Bratthauer GL, Dichek DA (1992) Low level in vivo gene transfer into the arterial wall through a perforated balloon catheter. Circulation 85:1110–1117
90. Tahlil O, Brami M, Feldman LJ, Branellec D, Steg PG (1997) The Dispatch catheter as a delivery tool for arterial tool for arterial gene transfer. Cardiovasc Res 33:181–187
91. Pavlides GS, Barath P, Maginas A, Vasilikos V, Cokkinos DV, O'Neill WW (1997) Intramural drug delivery by direct injection within arterial wall: first clinical experience with a novel intracoronary delivery-infiltrator system. Cathet Cardiovasc Diagn 41: 287–292
92. Sharif F, Hynes SO, McMahon J, Cooney R, Conroy S, Dockery P et al (2006) Gene-eluting stents: comparison of adenoviral and adeno-associated viral gene delivery to the blood vessel wall in vivo. Hum Gene Ther 17:741–750
93. Fishbein I, Alferiev IS, Nyanguile O, Gaster R, Vohs JM, Wong GS et al (2006) Bisphosphonate-mediated gene vector delivery from the metal surfaces of stents. Proc Natl Acad Sci USA 103:159–164
94. Perstein I, Connolly JM, Cui X, Song C, Li Q, Jones PL et al (2003) DNA delivery from an intravascular stent with a denatured collagen-polylactic-polyglycolic acid-controlled release coating: mechanisms of enhanced transfection. Gene Ther 10:1420–1428
95. Lemos PA, Serruys PW, Sousa JE (2003) Drug-eluting stents: cost versus clinical benefit. Circulation 107:3003–3007
96. Lee J, Laks H, Drinkwater DC, Blitz A, Lam L, Shiraishi Y et al (1996) Cardiac gene transfer by intracoronary infusion of adenovirus vector-mediated reporter gene in the transplanted mouse heart. J Thorac Cardiovasc Surg 111:246–252
97. Kypson AP, Peppel K, Akhter SA, Lilly RE, Glower DD, Lefkowitz RJ et al (1998) Ex vivo adenovirus-mediated gene transfer to the adult rat heart. J Thorac Cardiovasc Surg 115: 623–630
98. Griscelli F, Belli E, Opolon P, Musset K, Connault E, Perricaudet M et al (2003) Adenovirus-mediated gene transfer to the transplanted piglet heart after intracoronary injection. J Gene Med 5:109–119

99. Wang J, Ma Y, Knechtle SJ (1996) Adenovirus-mediated gene transfer into rat cardiac allografts: comparison of direct injection and perfusion. Transplantation 61:1726–1729
100. Shah AS, White DC, Tai O, Hata JA, Wilson KH, Pippen A et al (2000) Adenovirus-mediated genetic manipulation of the myocardial β-adrenergic signaling system in transplanted hearts. J Thorac Cardiovasc Surg 120:581–588
101. Bridges CR, Burkman JM, Malekan R, Konig SM, Chen H, Yarnall CB et al (2002) Global cardiac-specific transgene expression using cardiopulmonary bypass with cardiac isolation. Ann Thorac Surg 73:1939–1946
102. Davidson MJ, Jones JM, Emani SM, Wilson KH, Jaggers J, Koch WJ et al (2001) Cardiac gene delivery with cardiopulmonary bypass. Circulation 104:131–133
103. Jones JM, Wilson KH, Koch WJ, Milano CA (2002) Adenoviral gene transfer to the heart during cardiopulmonary bypass: effect of myocardial protection technique on transgene expression. Eur J Cardiothorac Surg 21:847–852
104. Ikeda Y, Gu Y, Iwanada Y, Hoshijima M, Oh SS, Giordano FJ et al (2002) Restoration of deficient membrane proteins in the cardiomyopathic hamster by in vivo cardiac gene transfer. Circulation 105:502–508
105. White JD, Thesier DM, Swain JD, Katz MG, Tomasulo CE, Henderson A et al (2011) Myocardial gene delivery using molecular cardiac surgery with recombinant adeno-associated virus vectors in vivo. Gene Ther 18:546–552
106. Fargnoli AS, Katz MG, Yarnall C, Sumaroka MV, Stedman H, Rabinowitz JE et al (2011) A Pharmacokinetic analysis of molecular cardiac surgery with recirculation mediated delivery of BARKct gene therapy: developing a quantitative definition of the therapeutic window. J Card Fail 17:691–699
107. Kaye DM, Preovolos A, Marshall BS, Byrne M, Hoshijima M, Hajjar RJ et al (2007) Percutaneous cardiac recirculation mediated gene transfer of an inhibitory phospholamban peptide reverses advanced heart failure in large animals. J Am Coll Cardiol 50:253–260
108. Bridges CR (2009) Recirculating method of cardiac gene delivery should be called 'non-recirculating' method. Gene Ther 16:939–940
109. Shohet RV, Chen S, Zhou Y-T, Wang Z, Meidell RS, Unger RH, Grayburn PA (2000) Echocardiographic destruction of albumin microbubbles directs gene delivery to the myocardium. Circulation 101:2554–2556
110. Bekeredjian R, Chen S, Frenkel PA, Grayburn PA, Shohet RV (2003) Ultrasound-targeted microbubbles destruction can repeatedly direct highly specific plasmid expression to the heart. Circulation 108:1022–1026
111. Beeri R, Guerrero JL, Supple G, Sullivan S, Levine RA, Hajjar RJ (2002) New efficient catheter-based system for myocardial gene delivery. Circulation 106:1756–1759
112. Marshall WG, Boone BA, Burgos JD, Gografe SI, Baldwin MK, Danielson ML et al (2010) Electroporation-mediated delivery of a naked DNA plasmid expressing VEGF to the porcine heart enhances protein expression. Gene Ther 17:419–423
113. Ayuni EL, Gazdhar A, Giraud MN, Kadner A, Gugger M, Cecchini M et al (2010) In vivo electroporation mediated gene delivery to the beating heart. PLoS One 5:e14467
114. Kumar A, Jena PK, Bahera S, Lockey RF, Mohapatra S, Mohapatra S (2010) Multifunctional magnetic nanoparticles for targeted delivery. Nanomedicine 6:64–69
115. Polyak B, Fishbein I, Chorny M, Alferiev I, Williams D, Yellen B et al (2008) High field gradient targeting of magnetic nanoparticle-loaded endothelial cells to the surfaces of steel stents. Proc Natl Acad Sci USA 15:698–703
116. Sanborn TA, Hackett NR, Lee LY, El-Sawy T, Blanko I, Tarazona N et al (2001) Percutaneous endocardial transfer and expression of genes to the myocardium utilizing fluoroscopic guidance. Catheter Cardiovasc Interv 52:260–266
117. Gwon HC, Jeong JO, Kim HJ, Park SW, Lee SH, Park SJ et al (2001) The feasibility and safety of fluoroscopy-guided percutaneous intramyocardial gene injection in porcine heart. Int J Cardiol 79:77–88
118. Lederman RJ, Guttman MA, Peters DC, Thompson RB, Sorger JM, Dick AJ et al (2002) Catheter-based endomyocardial injection with real-time magnetic resonance imaging. Circulation 105:1282–1284

119. Kornowski R, Leon MB, Fuchs S, Vodovotz Y, Flynn MA, Gordon DA et al (2000) Electromagnetic guidance for catheter-based transendocardial injection: a platform for intramyocardial angiogenesis therapy. Results in normal and ischemic porcine models. J Am Coll Cardiol 35:1031–1039
120. Baklanov DV, de Muinck ED, Simons M, Moodie KL, Arbuckle BE, Thompson CA et al (2005) Live 3D echo guidance of catheter-based endomyocardial injection. Catheter Cardiovasc Interv 65:340–345
121. Davia K, Bernovich E, Ranu HK, del Monte F, Terracciano CM, MacLeod KT et al (2001) SERCA2a overexpression decreases the incidence of aftercontractions in adult rabbit ventricular myocytes. J Mol Cell Cardiol 33:1005–1015
122. del Monte F, Lebeche D, Guerrero JL, Tsuji T, Doye AA, Gwathmey JK et al (2004) Abrogation of ventricular arrhythmias in a model of ischemia and reperfusion by targeting myocardial calcium cycling. Proc Natl Acad Sci USA 101:5622–5627
123. Byrne MJ, Power JM, Preovolos A, Mariani JA, Hajjar RJ, Kaye DM (2008) Recirculating cardiac delivery of AAV2/1SERCA2a improves myocardial function in an experimental model of heart failure in large animals. Gene Ther 15:1550–1557
124. Jessup M, Greenberg B, Mancini D, Cappola T, Pauly DF, Jaski B et al (2011) Calcium upregulation by percutaneous administration of gene therapy in cardiac disease (CUPID). Circulation 124:304–313
125. Most P, Remppis A, Pleger ST, Katus HA, Koch WJ (2007) S100A1: a novel inotropic regulator of cardiac performance. Transition from molecular physiology to pathophysiological relevance. Am J Physiol Regul Integr Comp Physiol 293:R568–R577
126. Pleger ST, Most P, Boucher M, Soltys S, Chuprun JK, Pleger W et al (2007) Stable myocardial-specific AAV-S100A1 gene therapy results in chronic functional heart failure rescue. Circulation 115:2506–2515
127. Pleger ST, Shan C, Klienzyk J, Bekeredjian R, Boekstegers P, Hinkel R et al (2011) Cardiac AAV9-S100A1 Gene therapy rescues post-ischemic heart failure in a preclinical large animal model. Sci Transl Med 3:92ra64
128. Iwanaga Y, Hoshijima M, Gu Y, Iwatate M, Dieterle T, Ikeda Y et al (2004) Chronic phospholamban inhibition prevents progressive cardiac dysfunction and pathological remodeling after infarction in rats. J Clin Invest 113:727–736
129. Tsuji T, del Monte F, Yoshikawa Y, Abe T, Shimizu J, Nakajima-Takenaka C et al (2009) Rescue of Ca2+ overload-induced left ventricular dysfunction by targeted ablation of phospholamban. Am J Physiol Heart Circ Physiol 296:H310–H317
130. Rockman HA, Koch WJ, Lefkowitz RJ (2002) Seven-transmembrane-spanning receptors and heart function. Nature 415:206–212
131. Brodde OE (1993) Beta-adrenoreceptors in cardiac disease. Pharmacol Ther 60:405–430
132. Koch WJ, Rockman HA, Samama P, Hamilton RA, Bond RA, Milano CA et al (1995) Cardiac function in mice overexpressing the β-adrenergic receptor kinase or a βARK inhibitor. Science 268:1350–1353
133. Brinks H, Koch WJ (2010) βARKct: a therapeutic approach for improved adrenergic signaling and function in heart disease. J Cardiovasc Transl Res 3:499–506
134. White DC, Hata JA, Shah AS, Glower DD, Lefkowitz R, Koch WJ (2000) Preservation of myocardial b-adrenergic receptor signaling delays the development of heart failure after myocardial infarction. Proc Natl Acad Sci USA 97:5428–5433
135. Tevaearai HT, Walton GB, Keys JR, Koch WJ, Eckhart AD et al (2005) Acute ischemic cardiac dysfunction is attenuated via gene transfer of a peptide inhibitor of the b-adrenergic receptor kinase (βARK1). J Gene Med 7:1172–1177
136. Tachibana H, Naga Prasad SV, Lefkowitz RJ, Koch WJ, Rockman HA (2005) Level of b-adrenergic receptor kinase 1 inhibition determines degree of cardiac dysfunction after chronic pressure-overload-induced heart failure. Circulation 111:591–597
137. Lavu M, Gundewar S, Lefer DJ (2011) Gene therapy for ischemic heart disease. J Mol Cell Cardiol 50:742–750
138. Josko J, Gwozdz B, Jedrzejowska-Szypulka H, Hendryk S (2000) Vascular endothelial growth factor (VEGF) and its effect on angiogenesis. Med Sci Monit 6:1047–1052

139. Bull DA, Bailey SH, Rentz JJ, Zebrack JS, Lee M, Litwin SE et al (2003) Effect of Terplex/VEGF-165 gene therapy on left ventricular function and structure following myocardial infarction. VEGF gene therapy for myocardial infarction. J Control Release 93:175–181
140. Vera Javanel GL, Crottogini A, Cabeza Meckert P, Cuniberti L, Mele A, Papouchado M et al (2006) Plasmid-mediated VEGF gene transfer induces cardiomyogenesis and reduces myocardial infarct size in sheep. Gene Ther 13:1133–1142
141. Lähteenvuo JE, Lähteenvuo MT, Kivelä A, Rosenlew C, Falkevall A, Klar J et al (2009) Vascular endothelial growth factor-B induces myocardium-specific angiogenesis and arteriogenesis via vascular endothelial growth factor receptor-1- and neuropilin receptor-1- dependent mechanisms. Circulation 119:845–856
142. Ferrarini M, Arsic N, Recchia FA, Zentilin L, Zacchigna S, Xu X et al (2006) Adeno-associated virus-mediated transduction of VEGF 165 improves cardiac tissue viability and functional recovery after permanent coronary occlusion in conscious dogs. Circ Res 98:954–961
143. Rissanen TT, Ylä-Herttuala S (2007) Current status of cardiovascular gene therapy. Mol Ther 15:1233–1247
144. Lazarous DF, Scheinowitz M, Shou M, Hodge E, Rajanayagam S, Hunsberger S et al (1995) Effects of chronic systemic administration of basic fibroblast growth factor on collateral development in the canine heart. Circulation 91:145–153
145. Gao MH, Lai NC, McKirnan MD, Roth DA, Rubanyi GM, Roth DM, Hammond HK (2004) Increased regional function and perfusion after intracoronary delivery of adenovirus encoding FGF4: report of preclinical data. Hum Gene Ther 15:574–587
146. Suzuki G, Lee TC, Fallavollita JA, Canty JM (2005) Adenoviral gene transfer of FGF-5 to hibernating myocardium improves function and stimulates myocytes to hypertrophy and reenter the cell cycle. Circ Res 96:767–775
147. Henry TD, Grines CL, Watkins MW, Barbeau G, Moreadith R, Andrasfay T, Engler RL (2007) Effects of Ad5FGF-4 in patients with angina: an analysis of pooled data from the AGENT-3 and AGENT-4 trials. J Am Coll Cardiol 50:1038–1046
148. Donahue JK (2004) Gene therapy for cardiac arrhythmias. Ann N Y Acad Sci 1015:332–337
149. Donahue JK, Heldman AW, Fraser H, McDonald AD, Miller JM, Rade JJ et al (2000) Focal modification of electrical conduction in the heart by viral gene transfer. Nat Med 6:1395–1398
150. Bunch TJ, Mahapatra S, Bruce GK, Johnson SB, Miller DV, Horne BD et al (2006) Impact of transforming growth factor-beta1 on atrioventricular node conduction modification by injected autologous fibroblasts in the canine heart. Circulation 113:2485–2494
151. Edelberg JM, Aird WC, Rosenberg RD (1998) Enhancement of murine cardiac chronotropy by the molecular transfer of the human beta2 adrenergic receptor cDNA. J Clin Invest 101:337–343
152. Plotnikov AN, Sosunov EA, Qu J, Shlapakova IN, Anyukhovsky EP, Liu L et al (2004) Biological pacemaker implanted in canine left bundle branch provides ventricular escape rhythms that have physiologically acceptable rates. Circulation 109:506–512
153. Brunner M, Kodirov SA, Mitchell GF, Buckett PD, Shibata K, Folco EJ et al (2003) In vivo gene transfer of Kv1.5 normalized action potential duration and shortens QT interval in mice with long QT phenotype. Am J Physiol Heart Circ Physiol 285:H194–H203
154. Sasano T, McDonald AD, Kikuchi K, Donahue JK (2006) Molecular ablation of ventricular tachycardia after myocardial infarction. Nat Med 12:1256–1258
155. Kawada T, Nakazawa M, Nakauchi S, Yamazaki K, Shimamoto R, Urabe M (2002) Rescue of hereditary form of dilated cardiomyopathy by rAAV-mediated somatic gene therapy: amelioration of morphological findings, sarcolemmal permeability, cardiac performance and the prognosis of TO-2 hamsters. Proc Natl Acad Sci USA 99:901–906
156. Nuss HB, Marban E, Johns DC (1999) Overexpression of a human potassium channel suppresses cardiac hyperexcitability in rabbit ventricular myocytes. J Clin Invest 103:889–896

Chapter 9
Cell-Based Therapy for Cardiovascular Injury

Yuji Haraguchi, Tatsuya Shimizu, Masayuki Yamato, and Teruo Okano

Abstract Cardiovascular injury is a major cause of morbidity and mortality and a major public health problem especially in developed countries. Various therapies for cardiovascular injury are researched actively and have been performed clinically. Recently, a cell-based therapy appears and is focused as an alternative therapy for cardiovascular injury. Scaffold-based and cell sheet-based tissue engineering contribute to the enhancement of cell transplanting efficiency, resulting in the induction of effective therapy. These cell-based regenerative therapies have promising and enormous possibilities for curing cardiovascular injury, and the clinical trials have been started. This chapter summarizes cell-based therapies including (1) cell injection therapy and (2) scaffold-based and (3) cell sheet-based tissue engineering. In addition, cell sources are also discussed.

Keywords Cardiovascular injury • Cell injection therapy • Scaffold-based tissue engineering • Cell sheet-based tissue engineering • Organ engineering

9.1 Introduction

Although various clinical therapies for cardiovascular injuries including acute/chronic myocardial infarction and dilated cardiomyopathy are performed in the various fields for treating many patients suffering these diseases at present, many lives are still lost due to cardiovascular injury [1]. Cell-based regenerative medicine is focused as an alternative and novel therapy for curing severe cardiovascular injury. Cell therapy by the direct injection of dissociated cells has been performed

Y. Haraguchi • T. Shimizu • M. Yamato • T. Okano, PhD (✉)
Institute of Advanced Biomedical Engineering and Science,
TWIns, Tokyo Women's Medical University,
8-1 Kawada-cho, Shinjuku-ku, Tokyo, Japan
e-mail: tokano@abmes.twmu.ac.jp

M.K. Danquah, R.I. Mahato (eds.), *Emerging Trends in Cell and Gene Therapy*,
DOI 10.1007/978-1-62703-417-3_9,

clinically [2–6]. More recently, tissue engineering is developed and expected to be as a second generation cell-based regenerative therapy for cardiovascular injury [7–9]. Joseph P. Vacanti and Robert Langer are the pioneer of tissue engineering, which allows three-dimensional (3D) tissue to be fabricated by seeding living cells into 3D scaffolds, an alternative for extracellular matrix (ECM) [10]. In addition to scaffold-based tissue engineering, a scaffold-free tissue engineering, called "cell sheet engineering," has also been developed [11, 12]. These tissue engineering therapies give a good therapeutic effect for cardiovascular injury in various animal models and have been performed clinically.

Myocardial infarction is caused by the stenosis of coronary arteries providing oxygen and nutrients to the heart tissue and the necrosis of the enormous number of pulsatile cardiac cells, cardiomyocytes, found in the infarction areas [13, 14]. The enormous cell necrosis may lead to various negative heart-tissue remodeling including (1) the decrease of ventricular wall thickness, (2) ventricular dilatation, (3) the decrease of ventricular contractile function, and (4) tissue fibrosis [15, 16]. Dilated cardiomyopathy is characterized by a cardiac chamber dilation associated with impaired systolic and diastolic functions and may also lead several negative heart-tissue remodeling [14, 17]. Generally, virus infection, autoimmunity, or genetic abnormality is thought to be a trigger for dilated cardiomyopathy [14, 18]. These cardiovascular injuries may induce finally lethal cardiovascular injuries and sudden death. While the tissue regeneration may be spontaneously occurred during these events of the negative remodeling, the regeneration is insufficient. The inhibition of the negative remodeling and neovascularization in damaged tissue are generally thought to be important in the improvement of symptom.

9.2 Injection Therapy of Dissociated Cells

9.2.1 Skeletal Myoblasts

Autologous cell sources are particularly important because immunological rejection is avoided. However, at present, clinical trials using human autologous cardiomyocytes have been unaccomplished. Skeletal myoblasts are used firstly as an autologous cell source for repairing heart tissue both at experimental and clinical trials [2, 16, 19]. Skeletal myoblasts can easily and rapidly increase in vitro and have a relatively hypoxia-resistant character. Skeletal myoblast injection via epicardium for patients undergoing coronary artery bypass grafting is performed, and the phase I clinical trial shows (1) the feasibility of the cell therapy and (2) the increase of the risk of ventricular arrhythmias [2, 20, 21]. The phase II trial shows that the injection of skeletal myoblasts fails to significantly improve the cardiac function, though the clinical trial suggests an encouraging possibility that the injection of higher cell numbers may recover left ventricular (LV) dilatation [3]. On the other hand, Opie and Dib show clinically the functional efficacy of catheter-based skeletal myoblast injection via endocardium [4].

9.2.2 Cardiac Stem Cells

Cardiac stem cells (CSCs) are found in adult heart, which consequently have a renewal ability at a normal state [22, 23]. The annual rate of turning over is known to be decreased by aging (the rates at the age of 25 and 75 are approximately 1 and 0.45 %, respectively) [24]. Because (1) spontaneous heart-tissue regeneration by the expansion of CSCs after heart damage is insufficient and (2) newly formed cardiomyocytes in vivo are unable to substitute damaged myocardial tissues, the isolation and in vitro expansion of CSCs are necessary. Lee et al. have succeeded to isolate cardiospheres or cardiosphere-derived cells (CDCs) from endomyocardial biopsies and injected these cells to hearts in damaged heart model pigs autologously [25]. The intramyocardial injections of cardiospheres or CDCs provide (1) the significant improvement of LV ejection fraction (EF) and (2) the increase of LV septal wall thickness. No deaths and no tumors are found at 8 weeks after the injection of CDCs or cardiospheres. A phase I clinical trial using autologous CSCs has shown that intracoronary injection of CSCs in patients with chronic ischemic cardiomyopathy and severe heart failure is feasible, safe, and apparently highly efficacious in improving LV systolic function [6].

9.2.3 Bone Marrow- and Peripheral Blood-Derived Cells

Bone marrow- and peripheral blood-derived cells are the most used as cell sources for the clinical therapy of cardiovascular injury [5, 15, 26, 27]. Bone marrow- and peripheral blood-derived cells are consisted of several cells, namely, monocytes, hematopoietic stem cells, and endothelial progenitor cells (EPCs). EPCs can be isolated from bone marrow, peripheral blood, and umbilical cord blood [28, 29]. The transplantation of EPCs induces neovascularization, which can increase a blood perfusion rate into ischemic tissues. For example, the transplantation of EPCs induces neovascularization and the increase of blood flow in the ischemic hind limb of a rat model [29]. In vitro and in vivo human EPCs are known to differentiate into smooth muscle cells, cardiomyocytes, as well as endothelial cells (ECs) [30, 31]. However, the efficiency of cardiac differentiation is extremely low (0.4 ± 0.03 %), and the differentiation of EPCs into cardiomyocytes is unable to be confirmed by another group [32, 33]. Bone marrow cells also contain mesenchymal stem cells (MSCs). Although MSCs are a rare population (between 0.01 and 0.001 %) in bone marrow, MSCs can expand rapidly in vitro [34]. Bone marrow-derived MSCs have a multipotency including cardiac differentiation [15].

The transplantations of skeletal myoblasts, CSCs, and bone marrow- and peripheral blood-derived cells are expected to allow these cells to directly contribute to the pulsatile of damaged heart. However, there is no evidence that skeletal myoblasts can differentiate into cardiomyocytes. In addition, although CSCs, EPCs, and MSCs have the potential of cardiac differentiation in terms of RNA and protein expressions,

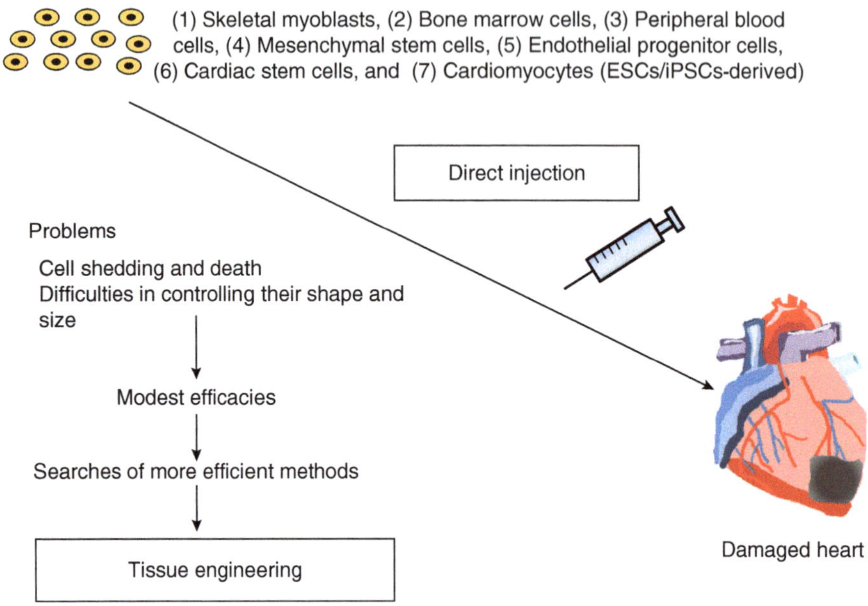

Fig. 9.1 Cell sources and a direct injection therapy for cardiovascular injury. Various cells are used for cell therapies. Although the direct injection therapy using dissociated cells shows a feasible efficacy in some cases, the therapy has some drawbacks. Tissue engineering has emerged for overcoming the drawbacks

there is no evidence clearly showing the differentiation into pulsatile cardiomyocytes. Thus, the transplantations of these cells are generally speculated to contribute to the inhibition of negative heart-tissue remodeling and neovascularization by their cytokines/chemokines, which are related to angiogenesis, anti-fibrosis, anti-apoptosis, and stem cell recruiting, produced by these cells.

Many clinical trials for cardiovascular injury using bone marrow- and peripheral blood-derived cells show a satisfactory safety, particularly with regard to arrhythmias. On the other hand, the meta-analysis studies of injection therapies using bone marrow- and peripheral blood-derived cells show the modest improvements of cardiac functions [the increases of LVEF are 3.0–3.7 %; the reductions of LV end-systolic volumes, 4.7–7.4 mL; the reductions of myocardial lesion areas, 3.5–5.6 %] [5, 26, 35, 36]. The modest efficacies may be insufficient for an accepted level which general clinicians may approve as a dependable cell-based therapy. Cell injection therapy has markedly difficulties in allowing injected cells to be delivered to target cardiac tissue and preserving the large amount of the cells on the location (Fig. 9.1). Many injected cells die and vanish after the transplantation, and only 7–10 % injected cells are found in the infarcted myocardium 3–4 days after the transplantation [37, 38]. In a clinical trial, a large percentage of injected cells are found in the liver and spleen, but not cardiac tissue, immediately after the transplantation [39]. Therefore, more effective methods are desired for spreading the cell-based regenerative therapy as a credible therapy for cardiovascular injury. Various trials to increase the therapeutic effect have been performed. Although, for example, only 1.3–2.6 % injected unselected bone marrow cells are detected in the cardiac tissue of patients with cardiovascular injury, 14–39 % CD34-positive cells purified from bone marrow-derived cells are detected in the tissue [39]. In addition, peripheral blood-derived CD34-positive cells also exhibit a superior efficacy for improving cardiac functions after myocardial infarction than unselected peripheral blood-derived cells [40]. Recently, Gavira et al. have shown that the repeated injection of skeletal myoblasts induces (1) more significant improvements in cardiac functions, (2) the increase of tissue vascularization, and (3) the decrease of fibrosis than the single injection in a porcine infarction model [41]. The most focused methodology for clearing the problem of cell loss is tissue engineering, which is summarized in details in the following chapter.

9.3 Tissue Engineering

9.3.1 Scaffold-Based Tissue Engineering

Recently, tissue engineering has been focused as a new generational cell-based therapy for cardiovascular injury [42–44]. Most popular approach of tissue engineering is based on a concept that (1) biodegradable 3D scaffolds are used as an alternative for ECM and (2) cells are seeded into the scaffolds (Fig. 9.2a). Piao et al. have fabricated 3D tissue by seeding rat bone marrow-derived mononuclear cells into a biodegradable poly-glycolide-co-caprolactone (PGCL) scaffold [45]. The transplantation of bone marrow cell-seeded PGCL scaffold effectively attenuates LV remodeling and LV systolic dysfunction in a rat infarction model via the induction of neovascularization and the differentiation of stem cells into cardiomyocytes. Zimmermann et al. have fabricated 3D tissue by a gelling mixture of cells and hydrogel solution (Fig. 9.2b) [9]. Tan et al. have used decellularized small intestinal submucosa (SIS) as a 3D scaffold (Fig. 9.2c) [46]. Though the transplantations of both SIS and MSC-seeded SIS into the heart of an infarcted rabbit model induce a significant improvement in the heart function, the MSC-seeded SIS is found to be more effective. The migration of MSCs from SIS into the infarcted area and the differentiation of MSCs into cardiomyocytes and smooth muscle cells are observed. A clinical trial by using an autologous mononuclear bone marrow cell-seeded 3D collagen type I matrix has been performed [47]. There are no lethality and no related adverse events after the transplantation. The clinical therapy shows feasible efficacies: (1) the improvement of New York Heart Association functional class (NYHA FC), from 2.3 ± 0.5 to 1.4 ± 0.3; (2) the decrease of LV end-diastolic volume, from 142 ± 24 to 117 ± 21 mL; (3) the improvement of LV filling deceleration time, from 162 ± 7 to 196 ± 8 ms; (4) the increase of scar area thickness, from 6 ± 1.4 to 9 ± 1.5 mm; and (5) the improvement of EF, from 25 ± 7 to 33 ± 5 %.

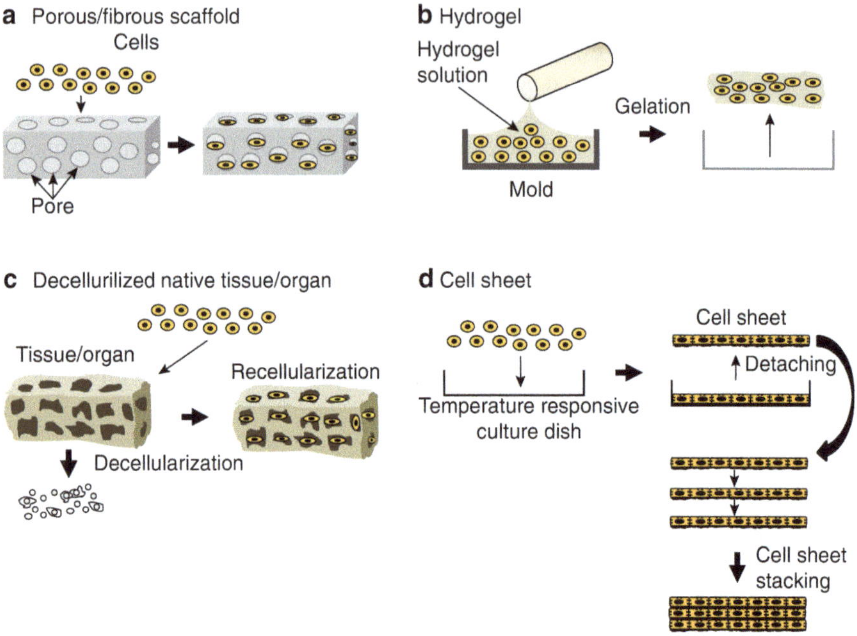

Fig. 9.2 Fabrication of three-dimensional cardiac tissue using tissue engineering. Three-dimensional (3D) tissues can be fabricated by various approaches; (**a**) cells are seeded and cultured on a porous/fibrous scaffold; (**b**) gelation of hydrogel including cells on a mold; (**c**) recellularization on a decellularized native tissues/organ; (**d**) the stacking of cell sheets without scaffolds

9.3.2 Cell Sheet-Based Tissue Engineering

9.3.2.1 Temperature-Responsive Culture Surface

Our laboratory has developed originally a temperature-responsive cell culture surface, which is covalently grafted with a temperature-responsive polymer, poly(*N*-isopropylacrylamide), and cell sheet-based scaffold-free tissue engineering using the unique culture surface (Fig. 9.2d) [11, 12, 48, 49]. Confluent cells on a temperature-responsive culture dish spontaneously detach themselves as an intact cell sheet by reducing culture temperature (Fig. 9.3). Importantly, cell sheets can conserve their cell-cell junctions, cell-surface proteins, and ECM [50–52]. Therefore, (1) 3D tissue can be easily fabricated by layering cell sheets without any scaffolds, and (2) a layered 3D tissue can adhere to host tissues without suture and other materials.

9.3.2.2 Skeletal Myoblast Sheet

Autologous skeletal myoblast sheets are already used in various damaged heart animal models. Memon et al. have compared the therapeutic effects of the transplantation of skeletal myoblast sheets with that of skeletal myoblast injection using a rat

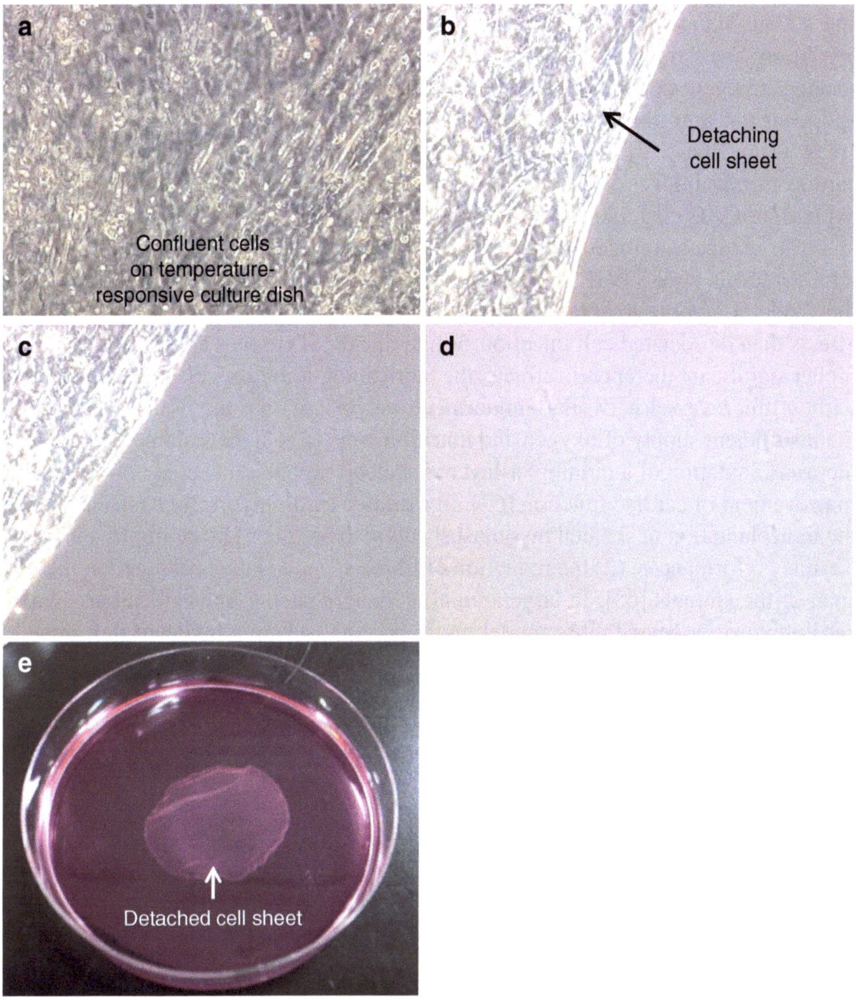

Fig. 9.3 A monolithic cell sheet detaching itself from a temperature-responsive culture dish by reducing culture temperature. Confluent cells on a temperature-responsive culture dish (**a**) are detached as an intact cell sheet (**b**, **c**) without cell residues (**d**) by decreasing culture temperature to 20°C. (**e**) A monolithic cell sheet detached from the temperature-responsive culture dish. The dish is a 100-mm culture dish

model [53]. The transplantation of skeletal myoblast sheets gives significant therapeutic effects: (1) the improvements of LVEF and fractional shortening (FS) and (2) the significant reduction of fibrosis in comparison to the injection of skeletal myoblasts, while the skeletal myoblast injection also induces the improvement of heart functions and the reduction of fibrosis in comparison to the medium-injected control. The significant recovery of anterior wall thickness is also observed in only the cell sheet-transplantation group. The productions of angiogenesis-related cytokines [vascular endothelial growth factor (VEGF) and hepatocyte growth factor (HGF)] and a chemokine [stromal cell-derived factor-1 (SDF-1)] from implanted skeletal myoblast sheets may be one of the causes of the therapeutic effects. VEGF is a strong angiogenesis factor, HGF has an anti-remodeling activity including anti-apoptosis and anti-fibrosis in infarcted heart as well as angiogenesis, and gene therapies using these cytokines have already performed clinically for ischemic cardiovascular injury [54–58]. SDF-1 recruits hematopoietic stem cells and EPCs expressing CXC chemokine receptor 4 (CXCR4), which is the receptor of SDF-1, and also induces angiogenesis in vivo via the recruitment of ECs and upregulates CXCR4 levels after VEGF stimulation [59–61]. Memon et al. have confirmed that the skeletal-myoblast-sheet transplantation gives higher enormous therapeutic effects than dissociated cell injection. While thicker 3D tissues are expected to give higher significant therapeutic effects, the fabrication of thicker cell sheet constructs, without microvessel networks, induces cell necrosis within the tissues because of the insufficient supply of oxygen and nutrition. Sekiya et al. have demonstrated that the transplantation of a quintuplet-layered skeletal myoblast sheet is optimal in the improvement of cardiac function [62]. In a dilated cardiomyopathy hamster model, the transplantation of skeletal myoblast sheets also provides (1) the improvement of cardiac performances, (2) the reduction of fibrosis, and (3) the prolongation the life span of the animals [63]. In large animal models (a pacing-induced canine dilated cardiomyopathy heart failure model and a porcine ischemic myocardium model), skeletal-myoblast-sheet transplantations give the improvements of cardiac functions and the reducing of negative cardiac remodeling [64, 65]. Based on these encouraging results in these animal models, clinical trials using autologous skeletal myoblast sheets are now in progress.

9.3.2.3 Adult Stem/Progenitor Cell Sheets

Various autologous cells including adult stem/progenitor cells are also used as the cell sources of cell therapy for cardiovascular injury in various animal models. Adipose tissue-derived stem cells, which express surface antigens similar to bone marrow-derived MSCs, have an angiogenesis activity and can also differentiate into cardiomyocytes [66, 67]. Miyahara et al. have fabricated cell sheets using adipose tissue-derived MSCs [68]. The transplantation of the MSC sheets gives (1) the improvements of cardiac performances, (2) the reversal of cardiac wall thinning, and (3) the prolongation of survival after myocardial infarction in a rat infarcted model. MSC sheets produce a large amount of angiogenesis-related cytokines (VEGF and HGF), and interestingly, in vivo a single-layer MSC sheet onto rat infarcted heart induces the formation of new and numerous blood vessels and grows to be approximately 600 μm thicker tissue. However, in vivo cardiac differentiation from the implanted MSCs is scarcely observed. Imanishi et al. have fabricated cell sheets using adipocytes differentiated from adipose tissue-derived progenitor cells and showed that the transplantation of the induced adipocyte cell sheets onto mouse acute myocardial infarction hearts gives the attenuations of (1) infarct size, (2) inflammation, and (3) negative LV remodeling [69]. Cytokines (VEGF, HGF, and adiponectin) produced from the implanted cell sheets may contribute to the therapeutic effects after the transplantation. Adiponectin is an adipose-derived plasma protein and protects cardiovascular tissues under stress conditions through several mechanisms: (1) the increase of angiogenesis and the inhibitions of (2) apoptosis, (3) inflammation, (4) fibrosis, and (5) cardiac hypertrophy [70–73]. Hida et al. have isolated some stem cells from human menstrual blood, and the stem cells also express surface antigens similar to bone marrow- and adipose tissue-derived stem cells [74]. Interestingly, cocultivating with mouse cardiac cells induces an effective differentiation from the stem cells into spontaneous beating cardiomyocytes. The transplantation of the human stem cell sheets also gives (1) the significant recovery of damaged cardiac function and (2) the decreasing of myocardial infarction area in a nude rat infarcted model. Matsuura et al. have isolated Sca-1-positive CSCs from an adult mouse using a magnetic cell sorting (MACS) system and fabricated CSC sheets [75]. The transplantation of CSC sheets improves damaged heart function through the efficient cardiomyocyte differentiations from the stem cells and its paracrine effects mediated via the secretion of soluble vascular cell adhesion molecule 1 (VCAM-1), which induces (1) the migrations of ECs and CSCs and (2) the depression of cardiomyocyte death from oxidative stress. Bone marrow-derived stem cell sheets are also successfully fabricated, and their transplantations into large animal models are now in progress in several laboratories including our laboratory.

9.3.3 Pulsatile 3D Cardiac Tissue

9.3.3.1 Fabrication of Cardiac Tissue Using Tissue Engineering

Most autologous stem/progenitor cells can hardly differentiate into beating cardiomyocytes. The therapeutic effects of these cells are generally thought to be mainly caused by the paracrine effects of various factors including cytokines/chemokines secreted from the transplanted cells as described above. The transplantation of pulsatile cardiac tissue grafts is expected to contribute to the mechanical support of damaged heart via electrical and functional couplings. Several groups fabricate 3D cardiac tissue, which can beat spontaneously in vitro, using 3D scaffolds (gelatin sponges or porous alginate scaffolds) and neonatal rat cardiac cells [7, 8]. Zimmermann et al. have fabricated pulsatile 3D myocardial tissue by a gelling mixture of neonatal rat cardiac cells and collagen solution [9, 76]. Those engineered 3D cardiac tissue grafts contract constantly and spontaneously in vitro and even after in vivo transplantation. In animal models, after being transplanted, engineered cardiac tissue shows an electrical coupling with the host myocardium without arrhythmia and survived for a long time [76]. Well-formed myofibers with typical striations, gap junctions, and a large number of blood vessels are observed within the implanted graft [8]. On the other hand, 3D scaffolds are gradually degraded in vivo. In animal infarcted models, the transplantation of the engineered myocardial tissue graft provides the improvements of damaged cardiac functions including (1) the attenuation of LV dilatation,

(2) the induction of systolic wall thickening of LV, (3) the improvement of FS, and (4) the recovery of LV contractility [8, 76]. The trials of better functional cardiac tissue fabrication by the optimization of the scaffolds are also reported [77, 78].

Three-dimensional cardiac tissue can be also easily fabricated by layering cardiac cell sheets, which are prepared from neonatal rat cardiac cells on the temperature-responsive culture surface [79]. (1) A cardiac cell sheet beats synchronously, (2) an electroconnective 3D myocardial tissue can be fabricated by layering cardiac cell sheets, and (3) the 3D tissue can adhere to host tissues without suture [79, 80]. The electrical and functional couplings of two cardiac cell sheets are established via a gap junction formed at 30–40 min after the layering [81]. The transplantation of layered cardiac cell sheets onto rat heart induces the establishments of electrical and functional connections between the implanted cardiac cells and the host heart [80]. The transplantation of layered cardiac cell sheets into damaged hearts provides (1) a significant increase in LV wall thickness, (2) decreases in cross-sectional LV area and LV end-systolic area, (3) significant improvements in LVEF and FS, and (4) the reductions of fibrosis/necrosis in scar area [82]. After cell grafting, a significantly greater numbers of mature capillaries and the decrease of the numbers of apoptotic cells are observed in the cardiac cell sheet-transplantation group compared to the injection of dissociated cardiac cells [83]. In addition, cell sheet transplantation is found to be consistently yielded a greater cell survival than cell injection by the analysis of in vivo bioluminescence imaging (more than ten times at 4 weeks after the transplantations). The significant improvements of cardiac functions—(1) the decrease of LV end-systolic diameter, (2) the improvement of FS, and (3) the improvement of end-diastolic anterior wall thickness—are also observed in the cardiac cell sheet-transplantation group compared with the injection of dissociated cardiac cells. Sekine et al. have also reported that EC cocultivation within cardiac cell sheets provides a higher therapeutic effect via prevascular networks and the production of angiogenesis-related cytokines, such as VEGF, basic fibroblast growth factor (bFGF), and HGF [84].

9.3.3.2 Human Cell Sources of Beating Cardiomyocytes

At present, clinical available beating human cardiomyocytes have been unestablished. Human embryonic stem cells (ESCs) [85] and induced pluripotent stem cells (iPSCs), which are an emerging technology for overcoming the several drawbacks of ESCs [86–88], are attractive and focused worldwide because those stem cells can differentiate into beating cardiomyocytes. Cardiac differentiation from the stem cells can be induced several methods, such as (1) embryoid body formation, (2) cultivation by media including fetal bovine serum, (3) cultivation with the supplementation of several cytokines including activin A and bone morphogenetic protein 4 (BMP-4), and (4) cocultivation with visceral endoderm-like cells, END2 cells, or using the conditioned culture medium of END2 cells [87, 89–93]. Various researches for promoting cardiac differentiation from human ESCs/iPSCs have been performed, and several factors including (1) ascorbic acid, (2) cyclosporine A, (3) p38 mitogen-activated protein kinase (MAPK) inhibitor, and (4) granulocyte colony-stimulating factor

(G-CSF) are reported to have the potential [91, 94–97]. Because the contamination of immature stem cells could lead to teratoma formation after in vivo transplantation, the purification of differentiated cardiomyocytes from a heterogeneous cell mixture and the removal of immature stem cells are important. Various researches have been performed concerning the purification and enrichment of differentiated cardiomyocytes, for example, (1) the usage of Percoll gradient centrifugation, (2) the usage of gene-modified stem cells harboring drug resistance gene in the cardiac-specific gene locus, (3) the usage of a fluorescent dye that labels mitochondria, and (4) the usage of specific cell-surface markers [activated leukocyte cell adhesion molecule (ALCAM), vascular cell adhesion molecule 1 (VCAM-1), and signal-regulatory protein alpha (SIRPA)], and some methods show a purification rate of cardiomyocytes near 100 % [95, 98–104]. After the transplantation, human stem cell-derived cardiomyocytes survive for a long term, these cardiomyocytes can integrate with the host heart tissue, and furthermore, the transplantation of these cells is focused to improve cardiac functions in damaged heart animal models [105–109]. Stevens et al. have fabricated scaffold-free 3D cardiac tissue by the self-assembly of human ESC-derived cardiomyocytes using a rotational orbital shaker, and the cardiac tissue shows a spontaneous and synchronous beating [110]. In cases of these stem cells, there are other problems, such as an immune rejection (ESCs), an ethical problem (ESCs), a tumorigenicity (iPSCs), and the chromosomal integration/insertion of exogenous genes (iPSCs), which still have to be solved before their clinical trials. However, these problems are going to be solved by the various efforts of many researchers [111–115], and clinical trials using ESCs have been started in the fields of the regenerative medicine of other tissues. In the near future, human ESC/iPSC-derived cardiomyocytes must be used in clinical application for cardiovascular injury, and these therapies using pulsatile cells are expected to provide remarkable efficacies.

9.4 Challenging Trials: From Tissue Engineering to Organ Engineering

Challenging trials for fabricating organs from tissues have been started. Our laboratory succeeds to fabricate a thicker cardiac tissue (the thickness is approximate 1 mm) by using a 1-day interval polysurgery method using cardiac cell sheets in vivo [116]. The polysurgery method can overcome the limitation of the viable size of 3D tissues due to hypoxia, nutrient insufficiency, and waste accumulation. Fabricated 3D cardiac tissue with a well-organized microvascular network can pulsate macroscopically and synchronously even after resection. Hata et al. have fabricated a cardiac tissue with a thickness of approximately 800 μm by combining cardiac cell sheets with cardiac cell-seeded decellularized porcine SIS [117]. Furthermore, the trials of in vitro fabrication of vascularized thicker cardiac tissue have been started. Kofidis et al. have fabricated fibrin gel-based cardiac tissue containing rat cardiac cells and natural vessels (rat aortas), through which culture media was perfused, and cellular viability and metabolism within the thicker tissue are improved [118]. Mixing of human

ESC-derived cardiomyocytes, human ESC-derived ECs (or human umbilical vein ECs), and embryonic fibroblasts induce the generation of significant capillary networks within engineered cardiac tissue [119]. Embryonic fibroblasts decrease EC death and increase the proliferation, and the presence of EC capillaries increases a cardiomyocyte proliferation. Our laboratory succeeds in fabricating the tubular-like structure of native microvasculature within 3D tissue containing ECs fabricated using different several co-culture systems in vitro [120–122]. Our laboratory is now trying to promote an EC tubular formation within in vitro engineered cardiac tissue and to perfuse culture media through the newly formed vessels using perfusion bioreactors.

As a further advanced therapy for cardiovascular injury, an attempt to fabricate a pulsatile tubular structure having an ability to act as an independent cardiac assist device is performing. Pulsatile cardiac tubes in vivo and in vitro are fabricated by using a novel cell sheet-wrapping device [123, 124]. When neonatal rat cardiac cell sheets are sequentially wrapped around a resected rat thoracic aorta and transplanted in the place of the abdominal aorta of athymic rats, the cardiac tubes around the abdominal aorta can produce a circulatory supportive blood pressure, which is much greater than values generated by in vitro cardiac tubes, which produced inner pressure (in vivo 5.9 ± 1.7 mmHg vs. in vitro 0.11 ± 0.01 mmHg) [123, 124]. The beating tubes are composed of cardiac tissue that resembled to native heart, namely, mature myofilaments and elongated sarcomeres [123]. The hypertrophy of functional cardiac tube is suggested to be induced by a mechanical stretching due to the host blood flow pulsation. Therefore, the application of mechanical load either in vitro or in vivo seems to be necessary for fabricating a powerful pulsatile cardiac tube. In addition, several efforts to fabricate thicker cardiac tissue should induce to create more powerful cardiac tubes that can generate independent pressures that are sufficient for the circulatory support of damaged hearts. As the next stage, our laboratory is now trying to apply the use of newer and more advanced pacing devices to synchronize the graft beatings with the host hearts to examine their effects on the host hemodynamics. Furthermore, the optimization of the pacing conditions in infarction models may be able to improve their heart failure after myocardial damage. Ott et al. have fabricated 3D cardiac tissue by reseeding neonatal rat cardiac cells into a decellularized rat whole heart by coronary perfusion with detergents [125]. When the 3D cardiac tissue survives up to 28 days by coronary perfusion in a bioreactor that simulates myocardial physiology, at day 4, the macroscopic contractions of the tissues are observed, and at day 8, with a physiological load and electrical stimulation, the tissue can generate its pumping function, which is comparable to approximately 2 % of adult or 25 % of 16-week fetal heart function.

9.5 Conclusions

As the first generation of cell therapy for cardiovascular injury, many clinical trials using the injection therapy of dissociated cells have been already performed. In addition, scaffold-based and cell sheet-based tissue engineering have now emerged as the second generation, and previous studies indicate that they have powerful potentials for improving damaged heart over the therapeutic effects of cell injection therapy. In the near future, the further development of cell sourcing and scaling-up technologies could allow us to fabricate (1) pulsatile thicker cardiac tissue using human cardiomyocytes, which can contribute to the pulsation of heart, and (2) cardiac tube, which can assist strongly original blood circulation. Finally, the accumulations and improvements of those researches and technologies may realize organ engineering, the "fabrication of bioengineered hearts."

Acknowledgement We appreciate the useful comments and technical criticism of Dr. Norio Ueno (Institute of Advanced Biomedical Engineering and Science, TWIns, Tokyo Women's Medical University). This work was supported by grants from the Global Center of Excellence Program, Multidisciplinary Education and Technology and Research Center for Regenerative Medicine (MERCREM), Innovation Center for Fusion of Advanced Technologies in the Special Coordination Funds for Promoting Science, and the High-Tech Research Center Program from the Ministry of Education, Culture, Sports Science, and Technology (MEXST), Japan, and the Japan Society for the Promotion of Science (JSPS) through the "Funding Program for World-Leading Innovative R&D on Science and Technology (FIRST Program)," initiated by the Council for Science and Technology Policy (CSTP).

References

1. Venugopal JR, Prabhakaran MP, Mukherjee S et al (2012) Biomaterial strategies for alleviation of myocardial infarction. J R Soc Interface 9:1–19
2. Menasche P, Hagege AA, Scorsin M et al (2001) Myoblast transplantation for heart failure. Lancet 357:279–280
3. Menasche P, Alfieri O, Janssens S et al (2008) The myoblast autologous grafting in ischemic cardiomyopathy (MAGIC) trial: first randomized placebo-controlled study of myoblast transplantation. Circulation 117:1189–1200
4. Opie SR, Dib N (2006) Surgical and catheter delivery of autologous myoblasts in patients with congestive heart failure. Nat Clin Pract Cardiovasc Med 3:S42–S45
5. Alaiti MA, Ishikawa M, Costa MA (2010) Bone marrow and circulating stem/progenitor cells for regenerative cardiovascular therapy. Transl Res 156:112–129
6. Bolli R, Chugh AR, D'Amario D et al (2011) Cardiac stem cells in patients with ischaemic cardiomyopathy (SCIPIO): initial results of a randomised phase 1 trial. Lancet 378: 1847–1857
7. Li RK, Jia ZQ, Weisel RD et al (1999) Survival and function of bioengineered cardiac grafts. Circulation 100:II63–II69
8. Leor J, Aboulafia-Etzion S, Dar A et al (2000) Bioengineered cardiac grafts: a new approach to repair the infarcted myocardium? Circulation 102:III56–III61
9. Zimmermann WH, Schneiderbanger K, Schubert P et al (2002) Tissue engineering of a differentiated cardiac muscle construct. Circ Res 90:223–230
10. Langer R, Vacanti JP (1993) Tissue engineering. Science 260:920–926
11. Matsuda N, Shimizu T, Yamato M, Okano T (2007) Tissue engineering based on cell sheet technology. Adv Mater 19:3089–3099
12. Masuda S, Shimizu T, Yamato M, Okano T (2008) Cell sheet engineering for heart tissue repair. Adv Drug Deliv Rev 60:277–285
13. Brueckner JK, Carmichael SW, Gest TR, Granger NA, Hansen JT, Walji AH (2006) In: Netter FH (ed) Atlas of human anatomy, Thorax 4th edn. Elsevier Science, Philadelphia

14. Ou L, Li W, Liu Y et al (2010) Animal models of cardiac disease and stem cell therapy. Open Cardiovasc Med J 4:231–239
15. Soejitno A, Wihandani DM, Kuswardhani RA (2010) Clinical applications of stem cell therapy for regenerating the heart. Acta Med Indones 42:243–257
16. Mazo M, Pelacho B, Prósper F (2010) Stem cell therapy for chronic myocardial infarction. J Cardiovasc Transl Res 3:79–88
17. del Corsso C, Campos de Carvalho AC (2011) Cell therapy in dilated cardiomyopathy: from animal models to clinical trials. Braz J Med Biol Res 44:388–393
18. Maron BJ, Towbin JA, Thiene G et al (2006) Contemporary definitions and classification of the Cardiomyopathies: an American Heart Association Scientific Statement from the Council on Clinical Cardiology, Heart Failure and Transplantation Committee; Quality of Care and Outcomes Research and Functional Genomics and Translational Biology Interdisciplinary Working Groups; and Council on Epidemiology and Prevention. Circulation 113:1807–1816
19. Taylor DA, Atkins BZ, Hungspreugs P et al (1998) Regenerating functional myocardium: improved performance after skeletal myoblast transplantation. Nat Med 4:929–933
20. Menasche P (2009) Stem cell therapy for heart failure: are arrhythmias a real safety concern? Circulation 119:2735–2740
21. Hagège AA, Marolleau JP, Vilquin JT et al (2006) Skeletal myoblast transplantation in ischemic heart failure: long-term follow-up of the first phase I cohort of patients. Circulation 114: I108–I113
22. Beltrami AP, Barlucchi L, Torella D et al (2003) Adult cardiac stem cells are multipotent and support myocardial regeneration. Cell 114:763–776
23. Oh H, Bradfute SB, Gallardo TD et al (2003) Cardiac progenitor cells from adult myocardium: homing, differentiation, and fusion after infarction. Proc Natl Acad Sci USA 100:12313–12318
24. Bergmann O, Bhardwaj RD, Bernard S et al (2009) Evidence for cardiomyocyte renewal in humans. Science 324:98–102
25. Lee ST, White AJ, Matsushita S et al (2011) Intramyocardial injection of autologous cardiospheres or cardiosphere-derived cells preserves function and minimizes adverse ventricular remodeling in pigs with heart failure post-myocardial infarction. J Am Coll Cardiol 57:455–465
26. Martin-Rendon E, Brunskill SJ, Hyde CJ et al (2008) Autologous bone marrow stem cells to treat acute myocardial infarction: a systematic review. Eur Heart J 29:1807–1818
27. Sekiguchi H, Ii M, Losordo DW (2009) The relative potency and safety of endothelial progenitor cells and unselected mononuclear cells for recovery from myocardial infarction and ischemia. J Cell Physiol 219:235–242
28. Asahara T, Murohara T, Sullivan A et al (1997) Isolation of putative progenitor endothelial cells for angiogenesis. Science 275:964–967
29. Murohara T, Ikeda H, Duan J et al (2000) Transplanted cord blood-derived endothelial precursor cells augment postnatal neovascularization. J Clin Invest 105:1527–1536
30. Yeh ET, Zhang S, Wu HD et al (2003) Transdifferentiation of human peripheral blood CD34+-enriched cell population into cardiomyocytes, endothelial cells, and smooth muscle cells in vivo. Circulation 108:2070–2073
31. Badorff C, Brandes RP, Popp R et al (2003) Transdifferentiation of blood-derived human adult endothelial progenitor cells into functionally active cardiomyocytes. Circulation 107: 1024–1032
32. Gruh I, Beilner J, Blomer U et al (2006) No evidence of transdifferentiation of human endothelial progenitor cells into cardiomyocytes after coculture with neonatal rat cardiomyocytes. Circulation 113:1326–1334
33. Koyanagi M, Bushoven P, Iwasaki M et al (2007) Notch signaling contributes to the expression of cardiac markers in human circulating progenitor cells. Circ Res 101:1139–1145
34. Pittenger MF, Martin BJ (2004) Mesenchymal stem cells and their potential as cardiac therapeutics. Circ Res 95:9–20
35. Lipinski MJ, Biondi-Zoccai GG, Abbate A et al (2007) Impact of intracoronary cell therapy on left ventricular function in the setting of acute myocardial infarction: a collaborative systematic review and meta-analysis of controlled clinical trials. J Am Coll Cardiol 50:1761–1767

36. Abdel-Latif A, Bolli R, Tleyjeh IM et al (2007) Adult bone marrow-derived cells for cardiac repair: a systematic review and meta-analysis. Arch Intern Med 167:989–997
37. Zhang M, Methot D, Poppa V et al (2001) Cardiomyocyte grafting for cardiac repair: graft cell death and anti-death strategies. J Mol Cell Cardiol 33:907–921
38. Suzuki K, Murtuza B, Beauchamp JR et al (2004) Dynamics and mediators of acute graft attrition after myoblast transplantation to the heart. FASEB J 18:1153–1155
39. Hofmann M, Wollert KC, Meyer GP et al (2005) Monitoring of bone marrow cell homing into the infarcted human myocardium. Circulation 111:2198–2202
40. Kawamoto A, Iwasaki H, Kusano K et al (2006) CD34-positive cells exhibit increased potency and safety for therapeutic neovascularization after myocardial infarction compared with total mononuclear cells. Circulation 114:2163–2169
41. Gavira JJ, Nasarre E, Abizanda G et al (2010) Repeated implantation of skeletal myoblast in a swine model of chronic myocardial infarction. Eur Heart J 31:1013–1021
42. Zimmermann WH, Cesnjevar R (2009) Cardiac tissue engineering: implications for pediatric heart surgery. Pediatr Cardiol 230:716–723
43. Vunjak-Novakovic G, Tandon N, Godier A et al (2010) Challenges in cardiac tissue engineering. Tissue Eng Part B Rev 16:169–187
44. Atala A, Lanza R, Thomson JA, Nerem R (2011) Principles of regenerative medicine. In: Radisic M, Michael VM (eds) Cardiac tissue, 2nd edn. Elsevier Science, Philadelphia
45. Piao H, Kwon JS, Piao S et al (2007) Effects of cardiac patches engineered with bone marrow-derived mononuclear cells and PGCL scaffolds in a rat myocardial infarction model. Biomaterials 28:641–649
46. Tan M, Zhi YW, Wei RQ et al (2009) Repair of infarcted myocardium using mesenchymal stem cell seeded small intestinal submucosa in rabbits. Biomaterials 30:3234–3240
47. Chachques JC, Trainini JC, Lago N et al (2007) Myocardial assistance by grafting a new bioartificial upgraded myocardium (MAGNUM clinical trial): one year follow-up. Cell Transplant 16:927–934
48. Yamada N, Okano T, Sakai H et al (1990) Thermo-responsive polymeric surface: control of attachment and detachment of cultured cells. Makromol Chem Rapid Commun 11:571–576
49. Okano T, Yamada H, Sakai H, Sakurai Y (1993) A novel recovery system for cultured cells using plasma-treated polystyrene dishes grafted with poly (*N*-isopropylacrylamide). J Biomed Mater Res 27:1243–1251
50. Kushida A, Yamato M, Konno C et al (1999) Decrease in culture temperature releases monolayer endothelial cell sheets together with deposited fibronectin matrix from temperature-responsive culture surfaces. J Biomed Mater Res 45:355–362
51. Nishida K, Yamato M, Hayashida Y et al (2004) Functional bioengineered corneal epithelial sheet grafts from corneal stem cells expanded ex vivo on a temperature-responsive cell culture surface. Transplantation 77:379–385
52. Ohashi K, Yokoyama T, Yamato M et al (2007) Engineering functional two- and three-dimensional liver systems in vivo using hepatic tissue sheets. Nat Med 13:880–885
53. Memon IA, Sawa Y, Fukushima N et al (2005) Repair of impaired myocardium by means of implantation of engineered autologous myoblast sheets. J Thorac Cardiovasc Surg 130:1333–1341
54. Liu Y, Rajur K, Tolbert E, Dworkin LD (2000) Endogenous hepatocyte growth factor ameliorates chronic renal injury by activating matrix degradation pathways. Kidney Int 58:2028–2043
55. Taniyama Y, Morishita R, Aoki M et al (2001) Therapeutic angiogenesis induced by human hepatocyte growth factor gene in rat and rabbit hindlimb ischemia models: preclinical study for treatment of peripheral arterial disease. Gene Ther 8:181–189
56. Li Y, Takemura G, Kosai K et al (2003) Postinfarction treatment with an adenoviral vector expressing hepatocyte growth factor relieves chronic left ventricular remodeling and dysfunction in mice. Circulation 107:2499–2506
57. Hinkel R, Trenkwalder T, Kupatt C (2011) Gene therapy for ischemic heart disease. Expert Opin Biol Ther 11:723–737
58. Lavu M, Gundewar S, Lefer DJ (2011) Gene therapy for ischemic heart disease. J Mol Cell Cardiol 50:742–750

59. Salcedo R, Wasserman K, Young HA (1999) Vascular endothelial growth factor and basic fibroblast growth factor induce expression of CXCR4 on human endothelial cells: in vivo neovascularization induced by stromal-derived factor-1alpha. Am J Pathol 154:1125–1135
60. Askari AT, Unzek S, Popovic ZB et al (2003) Effect of stromal-cell-derived factor 1 on stem-cell homing and tissue regeneration in ischaemic cardiomyopathy. Lancet 362:697–703
61. Yamaguchi J, Kusano KF, Masuo O et al (2003) Stromal cell-derived factor-1 effects on ex vivo expanded endothelial progenitor cell recruitment for ischemic neovascularization. Circulation 107:1322–1328
62. Sekiya N, Matsumiya G, Miyagawa S et al (2009) Layered implantation of myoblast sheets attenuates adverse cardiac remodeling of the infarcted heart. J Thorac Cardiovasc Surg 138: 985–993
63. Kondoh H, Sawa Y, Miyagawa S et al (2006) Longer preservation of cardiac performance by sheet-shaped myoblast implantation in dilated cardiomyopathic hamsters. Cardiovasc Res 69:466–475
64. Hata H, Matsumiya G, Miyagawa S et al (2006) Grafted skeletal myoblast sheets attenuate myocardial remodeling in pacing-induced canine heart failure model. J Thorac Cardiovasc Surg 132:918–924
65. Miyagawa S, Saito A, Sakaguchi T et al (2010) Impaired myocardium regeneration with skeletal cell sheets – a preclinical trial for tissue-engineered regeneration therapy. Transplantation 90:364–372
66. Planat-Bénard V, Menard C, André M et al (2004) Spontaneous cardiomyocyte differentiation from adipose tissue stroma cells. Circ Res 94:223–229
67. Gimble JM, Katz AJ, Bunnell BA (2007) Adipose-derived stem cells for regenerative medicine. Circ Res 100:1249–1260
68. Miyahara Y, Nagaya N, Kataoka M et al (2006) Monolayered mesenchymal stem cells repair scarred myocardium after myocardial infarction. Nat Med 12:459–465
69. Imanishi Y, Miyagawa S, Maeda N et al (2011) Induced adipocyte cell-sheet ameliorates cardiac dysfunction in a mouse myocardial infarction model: a novel drug delivery system for heart failure. Circulation 124:S10–S17
70. Scherer PE, Williams S, Fogliano M et al (1995) A novel serum protein similar to C1q, produced exclusively in adipocytes. J Biol Chem 270:26746–26749
71. Hu E, Liang P, Spiegelman BM (1996) AdipoQ is a novel adipose-specific gene dysregulated in obesity. J Biol Chem 271:10697–10703
72. Maeda K, Okubo K, Shimomura I et al (1996) CDNA cloning and expression of a novel adipose specific collagen-like factor, apM1 (AdiPose most abundant gene transcript 1). Biochem Biophys Res Commun 221:286–289
73. Shibata R, Ouchi N, Murohara T (2009) Adiponectin and cardiovascular disease. Circ J 73: 608–614
74. Hida N, Nishiyama N, Miyoshi S et al (2008) Novel cardiac precursor-like cells from human menstrual blood-derived mesenchymal cells. Stem Cells 26:1695–1704
75. Matsuura K, Honda A, Nagai T et al (2009) Transplantation of cardiac progenitor cells ameliorates cardiac dysfunction after myocardial infarction in mice. J Clin Invest 119:2204–2217
76. Zimmermann WH, Melnychenko I, Wasmeier G et al (2006) Engineered heart tissue grafts improve systolic and diastolic function in infarcted rat hearts. Nat Med 12:452–458
77. Zhao YS, Wang CY, Li DX et al (2005) Construction of a unidirectionally beating 3-dimensional cardiac muscle construct. J Heart Lung Transplant 24:1091–1097
78. Engelmayr GC Jr, Cheng M, Bettinger CJ et al (2008) Accordion-like honeycombs for tissue engineering of cardiac anisotropy. Nat Mater 7:1003–1010
79. Shimizu T, Yamato M, Isoi Y et al (2002) Fabrication of pulsatile cardiac tissue grafts using a novel 3-dimensional cell sheet manipulation technique and temperature-responsive cell culture surfaces. Circ Res 90:e40
80. Sekine H, Shimizu T, Kosaka S et al (2006) Cardiomyocyte bridging between hearts and bioengineered myocardial tissues with mesenchymal transition of mesothelial cells. J Heart Lung Transplant 25:324–332

81. Haraguchi Y, Shimizu T, Yamato M et al (2006) Electrical coupling of cardiomyocyte sheets occurs rapidly via functional gap junction formation. Biomaterials 27:4765–4774
82. Miyagawa S, Sawa Y, Sakakida S et al (2005) Tissue cardiomyoplasty using bioengineered contractile cardiomyocyte sheets to repair damaged myocardium: their integration with recipient myocardium. Transplantation 80:1586–1595
83. Sekine H, Shimizu T, Dobashi I et al (2011) Cardiac cell sheet transplantation improves damaged heart function via superior cell survival in comparison with dissociated cell injection. Tissue Eng Part A 17:2973–2980
84. Sekine H, Shimizu T, Hobo K et al (2008) Endothelial cell coculture within tissue-engineered cardiomyocyte sheets enhances neovascularization and improves cardiac function of ischemic hearts. Circulation 118:S145–S152
85. Thomson JA, Itskovitz-Eldor J, Shapiro SS et al (1998) Embryonic stem cell lines derived from human blastocysts. Science 282:1145–1147
86. Takahashi K, Yamanaka S (2006) Induction of pluripotent stem cells from mouse embryonic and adult fibroblast cultures by defined factors. Cell 126:663–676
87. Takahashi K, Tanabe K, Ohnuki M et al (2007) Induction of pluripotent stem cells from adult human fibroblasts by defined factors. Cell 131:861–872
88. Yu J, Vodyanik MA, Smuga-Otto K et al (2007) Induced pluripotent stem cell lines derived from human somatic cells. Science 318:1917–1920
89. Kehat I, Kenyagin-Karsenti D, Snir M et al (2001) Human embryonic stem cells can differentiate into myocytes with structural and functional properties of cardiomyocytes. J Clin Invest 108:407–414
90. Mummery C, Ward-van Oostwaard D, Doevendans P et al (2003) Differentiation of human embryonic stem cells to cardiomyocytes: role of coculture with visceral endoderm-like cells. Circulation 107:2733–2740
91. Graichen R, Xu X, Braam SR et al (2008) Enhanced cardiomyogenesis of human embryonic stem cells by a small molecular inhibitor of p38 MAPK. Differentiation 76:357–370
92. Yang L, Soonpaa MH, Adler ED et al (2008) Human cardiovascular progenitor cells develop from a KDR+ embryonic-stem-cell-derived population. Nature 453:524–528
93. Yoshida Y, Yamanaka S (2011) iPS cells: a source of cardiac regeneration. J Mol Cell Cardiol 50:327–332
94. Passier R, Oostwaard DW, Snapper J et al (2005) Increased cardiomyocyte differentiation from human embryonic stem cells in serum-free cultures. Stem Cells 23:772–780
95. Xu XQ, Graichen R, Soo SY et al (2008) Chemically defined medium supporting cardiomyocyte differentiation of human embryonic stem cells. Differentiation 76:958–970
96. Shimoji K, Yuasa S, Onizuka T et al (2010) G-CSF Promotes the proliferation of developing cardiomyocytes in vivo and in derivation from ESCs and iPSCs. Cell Stem Cell 6:227–237
97. Fujiwara M, Yan P, Otsuji TG et al (2011) Induction and enhancement of cardiac cell differentiation from mouse and human induced pluripotent stem cells with Cyclosporin-A. PLoS One 6:e16734
98. Xu C, Police S, Rao N, Carpenter MK (2002) Characterization and enrichment of cardiomyocytes derived from human embryonic stem cells. Circ Res 91:501–508
99. Laflamme MA, Gold J, Xu C et al (2005) Formation of human myocardium in the rat heart from human embryonic stem cells. Am J Pathol 167:663–671
100. Kita-Matsuo H, Barcova M, Prigozhina N et al (2009) Lentiviral vectors and protocols for creation of stable hESC lines for fluorescent tracking and drug resistance selection of cardiomyocytes. PLoS One 4:e5046
101. Rust W, Balakrishnan T, Zweigerdt R (2009) Cardiomyocyte enrichment from human embryonic stem cell cultures by selection of ALCAM surface expression. Regen Med 4:225–237
102. Hattori F, Chen H, Yamashita H et al (2010) Nongenetic method for purifying stem cell-derived cardiomyocytes. Nat Methods 7:61–66
103. Dubois NC, Craft AM, Sharma P et al (2011) SIRPA is a specific cell-surface marker for isolating cardiomyocytes derived from human pluripotent stem cells. Nat Biotechnol 29:1011–1018

104. Uosaki H, Fukushima H, Takeuchi A et al (2011) Efficient and scalable purification of cardiomyocytes from human embryonic and induced pluripotent stem cells by VCAM1 surface expression. PLoS One 6:e23657
105. Kehat I, Khimovich L, Caspi O et al (2004) Electromechanical integration of cardiomyocytes derived from human embryonic stem cells. Nat Biotechnol 22:1282–1289
106. Caspi O, Huber I, Kehat I et al (2007) Transplantation of human embryonic stem cell-derived cardiomyocytes improves myocardial performance in infarcted rat hearts. J Am Coll Cardiol 50:1884–1893
107. Dai W, Field LJ, Rubart M et al (2007) Survival and maturation of human embryonic stem cell-derived cardiomyocytes in rat hearts. J Mol Cell Cardiol 43:504–516
108. Laflamme MA, Chen KY, Naumova AV et al (2007) Cardiomyocytes derived from human embryonic stem cells in pro-survival factors enhance function of infarcted rat hearts. Nat Biotechnol 25:1015–1024
109. van Laake LW, Passier R, Monshouwer-Kloots J et al (2007) Human embryonic stem cell-derived cardiomyocytes survive and mature in the mouse heart and transiently improve function after myocardial infarction. Stem Cell Res 1:9–24
110. Stevens KR, Pabon L, Muskheli V, Murry CE (2009) Scaffold-free human cardiac tissue patch created from embryonic stem cells. Tissue Eng Part A 15:1211–1222
111. Wakayama T, Tabar V, Rodriguez I et al (2001) Differentiation of embryonic stem cell lines generated from adult somatic cells by nuclear transfer. Science 292:740–743
112. Nakagawa M, Koyanagi M, Tanabe K et al (2008) Generation of induced pluripotent stem cells without Myc from mouse and human fibroblasts. Nat Biotechnol 26:101–106
113. Nakagawa M, Takizawa N, Narita M et al (2010) Promotion of direct reprogramming by transformation-deficient Myc. Proc Natl Acad Sci USA 107:14152–14157
114. Kim D, Kim CH, Moon JI et al (2009) Generation of human induced pluripotent stem cells by direct delivery of reprogramming proteins. Cell Stem Cell 4:472–476
115. Warren L, Manos PD, Ahfeldt T et al (2010) Highly efficient reprogramming to pluripotency and directed differentiation of human cells with synthetic modified mRNA. Cell Stem Cell 7:618–630
116. Shimizu T, Sekine H, Yang J et al (2006) Polysurgery of cell sheet grafts overcomes diffusion limits to produce thick, vascularized myocardial tissues. FASEB J 20:708–710
117. Hata H, Bar A, Dorfman S et al (2010) Engineering a novel three-dimensional contractile myocardial patch with cell sheets and decellularised matrix. Eur J Cardiothorac Surg 38:450–455
118. Kofidis T, Lenz A, Boublik J et al (2003) Pulsatile perfusion and cardiomyocyte viability in a solid three-dimensional matrix. Biomaterials 24:5009–5014
119. Caspi O, Lesman A, Basevitch Y et al (2007) Tissue engineering of vascularized cardiac muscle from human embryonic stem cells. Circ Res 100:263–272
120. Sekiya S, Muraoka M, Sasagawa T et al (2010) Three-dimensional cell-dense constructs containing endothelial cell-networks are an effective tool for in vivo and in vitro vascular biology research. Microvasc Res 80:549–551
121. Sasagawa T, Shimizu T, Sekiya S et al (2010) Design of prevascularized three-dimensional cell-dense tissues using a cell sheet stacking manipulation technology. Biomaterials 31:1646–1654
122. Asakawa N, Shimizu T, Tsuda Y et al (2010) Pre-vascularization of in vitro three-dimensional tissues created by cell sheet engineering. Biomaterials 31:3903–3909
123. Sekine H, Shimizu T, Yang J et al (2006) Pulsatile myocardial tubes fabricated with cell sheet engineering. Circulation 114:I87–I93
124. Kubo H, Shimizu T, Yamato M et al (2007) Creation of myocardial tubes using cardiomyocyte sheets and an in vitro cell sheet-wrapping device. Biomaterials 28:3508–3516
125. Ott HC, Matthiesen TS, Goh SK et al (2008) Perfusion-decellularized matrix: using nature's platform to engineer a bioartificial heart. Nat Med 14:213–221

Chapter 10
Induced Pluripotent Stem Cells: New Advances in Cardiac Regenerative Medicine

Shinsuke Yuasa, Mitsushige Murata, and Keiichi Fukuda

Abstract To bring the notion of cardiac regenerative medicine to fruition, researchers have tried to determine which stem cells, embryonic stem (ES) cells or somatic stem cells, are most suitable. Thus far, there is no clear indication which is better, because both have their own advantages and disadvantages. In 2006, murine induced pluripotent stem (iPS) cells were first established. Since then, basic research into the properties of iPS cells has continued apace. Originally, human iPS cells were generated from dermal fibroblasts by retrovirus-mediated gene transfer. Although this technique is sophisticated and easy to perform, the skin biopsy is accompanied by some bleeding and pain, and there may be some damage to the host genome because of retrovirus-mediated transgene integration. However, methods of producing iPS cells have improved steadily. For clinical application in the cardiovascular field, efficient methods that produce pluripotent stem cells that can differentiate into cardiomyocytes need to be developed. Existing methods for ES cells can be applied to iPS cells to obtain cardiomyocyte differentiation. In addition, existing purification methods can be used to obtain pure cardiomyocytes from a population of mixed cells. These techniques have themselves been the subject of extensive research, and continued advances are now making the clinical application of pluripotent stem cells a reality. We are at the forefront of medical innovations in the cardiovascular field based on the use of pluripotent stem cells.

S. Yuasa, MD, PhD (✉)
Cardiology Division, Department of Internal Medicine, Keio University School of Medicine,
35-Shinanomachi Shinjuku-ku, Tokyo 160-8582, Japan

Department of Cardiology, Center for Integrated Medical Research,
Keio University School of Medicine,
35-Shinanomachi Shinjuku-ku, Tokyo 160-8582, Japan
e-mail: yuasa@a8.keio.jp

M. Murata • K. Fukuda, MD, PhD
Cardiology Division, Department of Internal Medicine, Keio University School of Medicine,
35-Shinanomachi Shinjuku-ku, Tokyo 160-8582, Japan
e-mail: kfukuda@sc.itc.keio.ac.jp

M.K. Danquah, R.I. Mahato (eds.), *Emerging Trends in Cell and Gene Therapy*,
DOI 10.1007/978-1-62703-417-3_10,

Keywords ES cells • iPS cells • Cardiomyocyte • Differentiation

10.1 Introduction

It is known that mammalian cardiomyocytes have a limited capacity to regenerate, and there have been problems achieving adult mammalian cardiomyocyte regeneration under both physiological and pathologic conditions [1–3]. Although recent evidence indicates that the adult human heart has a limited capacity for regeneration [4, 5], it remains difficult to control cardiac regeneration in diseased human heart [6, 7]. As such, cell transplantation therapy appears to be most appropriate for the diseased heart. For cardiac regenerative medicine, using cell replacement therapy to become a viable option, which of the stem cells is most suitable (i.e., embryonic stem (ES) or somatic stem cells), needs to be determined. Thus far, there is no clear consensus as to which is better, because each has its own advantages and disadvantages. Various classes of bone marrow (BM)-derived and cardiac-derived cells, including cardiac stem cells, are currently employed in clinical trials. Over the past 10 years, researchers have applied various BM-derived stem/progenitor cells for cardiac reparative therapy in animal studies. Currently, human BM-derived cells are injected into patients with proven safety and improvement of cardiac function and multiple clinical end points [8]. Resident cardiac stem cells, which are also currently in a phase I clinical trial, are alternative candidates for cardiac regenerative medicine [9, 10]. Somatic stem cell transplantation certainly has beneficial effects on cardiac functional recovery in diseased heart. However, the proliferative and differentiation abilities of somatic stem cells are not sufficient to enable complete recovery of cardiac function in severely damaged human hearts using cell replacement therapy with the current technology. This chapter therefore focuses on the potential of ES and iPS cells in the context of cardiac repair.

10.2 Potential and Challenges of iPS Cells: Comparison with ESC

Although ES cells are promising pluripotent cells with a strong proliferative capacity, there are ethical considerations that constrain the use (and destruction) of early human embryos to establish new human ES cells; furthermore, the ES cells do not display the autologous genotype of the patients in whom they are to be used [11]. In an attempt to overcome these problems and yet to maintain pluripotent stem cell characteristics, many studies have investigated various techniques, such as cell fusion and somatic nuclear transplantation; however, as yet, none has progressed to successful clinical application [12, 13]. In 2006, the first report of the establishment of murine induced pluripotent stem (iPS) cells was published [14]. In 2007, human iPS cells were generated from human adult somatic cells using the gene transfer of *OCT3/4, SOX2, KLF4, and c-MYC* [15, 16]. The ability to generate iPS cells that exhibit pluripotency and have the ability to differentiate has created an alternative to the use of ES cells, which are hampered by ethical problems and their autogenic genetic background [14–16]. The morphology, growth characteristics, and pluripotency of murine iPS cells are similar to those of ES cells. Moreover, the germline competency of iPS cells has been demonstrated using the *cis*-element of *Nanog* as a selection marker [17]. Based on promising similarities between ES cells and iPS cells, we expected that human iPS cells could provide a future cell source for cardiac regenerative medicine. Because of the similarities between iPS and ES cells, we are able to apply current knowledge regarding the differentiation of ES cells into cardiomyocytes, the purification of cardiomyocytes, and transplantation technologies for cardiac regeneration therapy to human iPS cells.

10.3 Methods Used to Generate iPS

The direct reprogramming of somatic cells to produce iPS cells represents the most significant recent advance in stem cell biology and future regenerative medicine [14–16]. In terms of the clinical use of iPS cells, there are some important issues regarding which cell types should be used for reprogramming and how these cells are reprogrammed. In order to generate human iPS cells consistently in the clinical setting, sufficient cell source material needs to be collected from patients using the least invasive procedure possible. Most common routine methods used to generate iPS cells use retrovirus- or lentivirus-mediated gene transfer into recipient cell sources (Fig. 10.1). Using these methods, it is inevitable that genomic integration of transgenes occurs, which may result in unexpected problems, such as oncogenic and functional disturbances. Because the generation of human somatic cells without genomic integration of extrinsic genes is highly desirable, we need to overcome the problems associated with the collection of cell sources and eliminate any risk of transgene integration into the host genome.

10.3.1 Methods Used to Generate iPS: Donor Cells

In mice, there are many cell types that can be reprogrammed into iPS cells, including embryonic fibroblasts [17], adult tail-tip fibroblasts (TTFs) [18], hepatocytes, gastric epithelial cells [19], pancreatic cells [20], neural stem cells [21–23], and B lymphocytes [24]. In humans, dermal fibroblasts are mainly used to derive human iPS cells [15, 16]. However, recent studies have shown that other human somatic cells, such as keratinocyte stem cells [25], adipose stem cells [26], dental stem cells [27], neural stem cells [28], and hematopoietic stem/progenitor cells [29, 30], can also be used. However, the main source remains somatic stem cells, and it is difficult

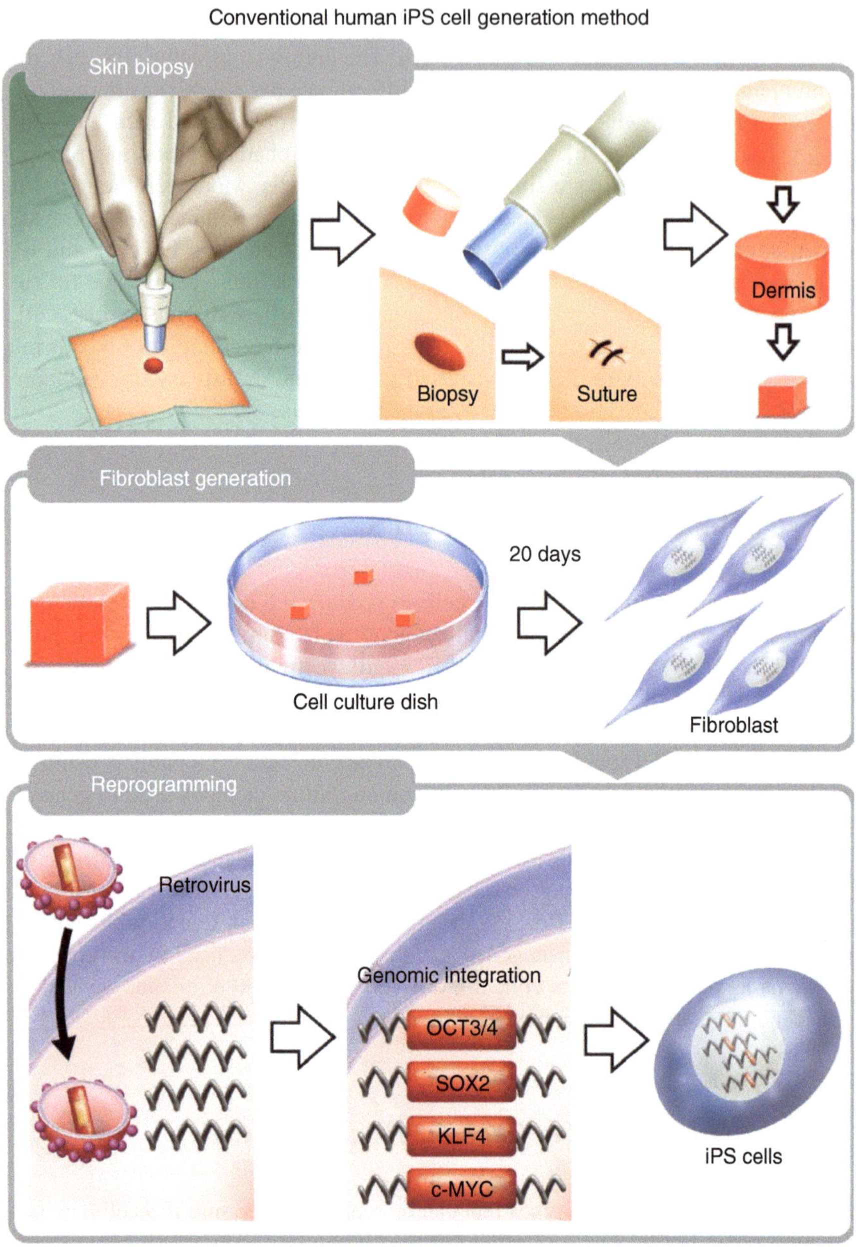

Fig. 10.1 Diagram showing conventional methods used to generate induced pluripotent stem (iPS) cells. Initially, a skin biopsy is performed and the dermis is isolated from the biopsied sample. The minced dermis is placed on a cell culture dish and fibroblast cells emerge after approximately 20 days. The reprogramming factors are introduced into the fibroblasts using retrovirus- and/or lentivirus-mediated methods. Finally, iPS cells are obtained with transgene genomic integration

to obtain human somatic stem cells using minimally invasive procedures. For use in a clinical setting, the ideal method to generate iPS cells needs to be minimally invasive, easy to perform, efficient, safe, and reliable. To achieve these goals, a novel method was developed using a combination of activated T cells in culture and Sendai virus (SeV) encoding human *OCT3/4*, *SOX2*, *KLF4*, and *c-MYC* [31]. The sampling of peripheral blood is one of the least invasive routine procedures performed in clinics, and T cells are easily cultured in vitro from the peripheral blood mononuclear cells (PBMCs) using a plate-bound anti-CD3 monoclonal antibody and recombinant interleukin (IL)-2 [32]. SeV was efficiently transfected into activated T cells [33], and the combination of activated T-cell culture and SeV-mediated gene transfer successfully generated human iPS cells [31]. Similar studies also reported that human immobilized peripheral blood cells, especially T cells, could be reprogrammed into iPS cells [34–37]. Many factors determine the efficiency of human iPS cell generation, such as transgene expression dosage and recipient cell type. Interactions between the transgene delivery system and the type or condition of recipient cells are also important. The generation of T-cell-derived iPS cell has advantages for research into stem cell reprogramming, immunological disorders, and the development of genetic markers for future applications in regenerative medicine (Fig. 10.2).

10.3.2 Methods Used to Generate iPS: Vectors

In terms of concerns regarding human reprogramming, we need to establish a method that eliminates transgene integration. To this end, there have been some developments refining existing methods, as summarized below.

1. In techniques based on the use of retroviruses and lentiviruses, it has been found that the Cre/LoxP recombination system successfully removes transgene sequences [38]. Although this system successfully removes transgene sequences, it does leave behind residual vector sequences that can still create insertional mutations, and so some risks remain.
2. Adenoviral vectors that mediate transient expression of transgenes have been used, but the efficiency of this system remains very low [39].
3. SeV is a negative-sense, single-stranded RNA virus that does not integrate into the host genome and has been used previously for generating transgene-free human iPS cells [40].
4. Nonviral methods have been used successfully to generate iPS cells. The single-vector reprogramming system combined with a *piggyBack* transposon delivery system for human somatic cells can achieve reprogramming efficiently, with any exogenous reprogramming factors that remain being completely removed from the iPS cells using subsequent *Cre* transfection [41, 42]. However, as noted above, the Cre/LoxP recombination system leaves behind residual vector sequences.
5. Derived from the Epstein–Barr virus, oriP/EBNA1 vectors are also well suited for introducing reprogramming factors into human cells [43]. The stable

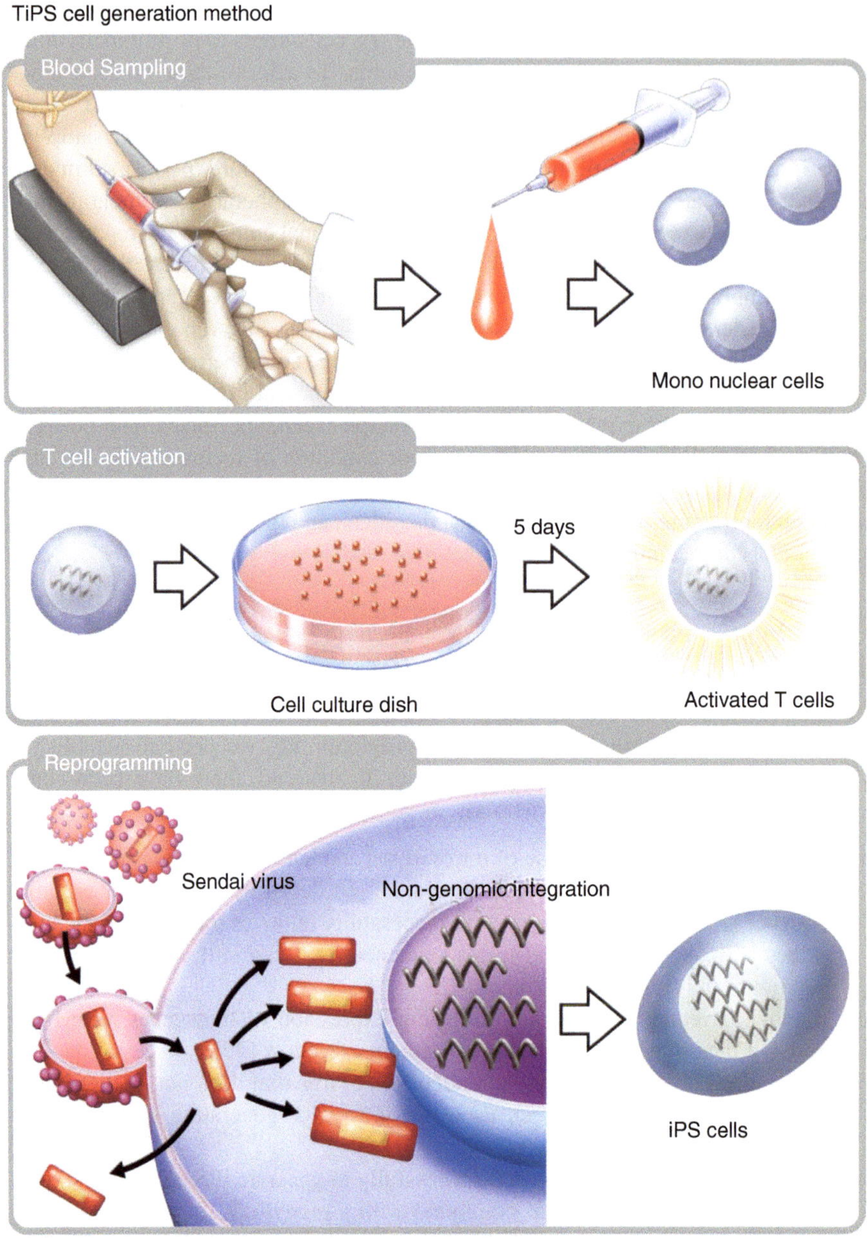

Fig. 10.2 Diagram showing methods used to obtain T-cell-derived induced pluripotent stem (TiPS) cell lines. Initially, a small blood sample is collected and mononuclear cells are separated by the Ficoll method. Mononuclear cells are cultured with anti-CD3 antibody and interleukin (IL)-2, and T cells are activated after 5 days. The reprogramming factors are introduced into activated T cells using the Sendai virus. Finally, induced pluripotent stem (iPS) cells are obtained without transgene genomic integration

extrachromosomal replication of oriP/EBNA1 vectors in mammalian cells requires only a *cis*-acting oriP element and a *trans*-acting *EBNA1* gene. These plasmids can be transfected without viral packaging and can be removed without drug selection owing to defects in plasmid synthesis and partitioning. With this system, the transgene is not integrated into the human iPS cell genome, and vector- and transgene-free human iPS cells can be isolated by subcloning without further genetic manipulation.

6. Minicircle vectors are supercoiled DNA molecules that lack a bacterial origin of replication and an antibiotic resistance gene; therefore, they are primarily composed of a eukaryotic expression cassette. Compared with plasmids, minicircle vectors benefit from higher transfection efficiencies and longer ectopic expression owing to their lower activation of exogenous silencing mechanisms. A plasmid that contains a single cassette of four reprogramming factors, namely, *OCT4*, *SOX2*, *LIN28*, and *NANOG*, each separated by sequences encoding the self-cleaving peptide 2A, can successfully generate transgene-free iPS cells from adult human adipose stem cells [44].
7. To address whether it is possible to avoid viral or DNA vectors, direct protein delivery to somatic cells has been attempted. The human immunodeficiency virus transactivator of transcription (HIV-TAT) protein contains a high proportion of basic amino acids, known as a cell-penetrating peptide (CPP), and can penetrate the cell membrane. Direct delivery of reprogramming factor proteins fused with CPP successfully generates human iPS cells from somatic cells [45, 46]. However, this method is extremely inefficient and needs to be improved prior to consideration for clinical use.
8. As a nonintegrating strategy for reprogramming, synthetic mRNA administration also accomplished the generation of human iPS cells from somatic cells [47]. The mRNAs were manufactured by using in vitro transcription reactions. A 50-guanine cap was also incorporated by inclusion of a synthetic cap analog to promote efficient translation and boost RNA half-life in the cytoplasm.
9. There are many compounds that are currently undergoing chemical screening that may prove to be viable substitutes for the reprogramming factors, with some such chemical compounds already identified [48–51]. In the future, it may be that chemical compounds alone are used to reprogram human somatic cells into iPS cells.

10.4 Tumor Formation

Mouse iPS cells can efficiently give rise to chimeric mice that are competent for germline transmission. However, the chimeras and their progenies have an increased incidence of tumor formation, primarily due to reactivation of the oncogene, *c-Myc* [17]. It has already been proven in part that iPS cells can be generated from mouse and human fibroblasts without *c-Myc*, but that the efficiency of such iPS cell generation is compromised [18]. Chimeric mice derived from mouse iPS cells generated without *c-Myc* did not demonstrate an increased incidence of tumor formation [18, 52]. The iPS cells could instead be generated by transient transgene expression without transgene integration into the host genome, which can be safer. These improved methods could decrease the risk of tumorigenicity in iPS cell-derived chimeric mice and enhance their promise for use in future cell transplantation therapy.

In terms of therapy, transplanted iPS cells are likely to carry a higher risk of tumorigenicity than ES cells, because there may be inappropriate and insufficient reprogramming of somatic cells, reactivation of exogenous genes, or other unknown reasons [53]. To circumvent the tumorigenicity of transplanted iPS cell-derived cells, a major effort has been made to identify factors that favor the acquisition of the differentiated myocyte phenotype, thereby reducing the fraction of undifferentiated iPS cells in the preparation. Although this approach may enhance the safety of iPS cell administration, it cannot be ignored that undifferentiated cells will persist in the preparation precluding the implementation of these protocols in vivo. Several attempts to reduce the risk of tumorigenicity maximally in regenerative medicine have been made using animal models. For example, studies in the nervous system minimized the risk of tumor formation from the grafted cells by separating contaminating pluripotent cells and committed neural cells using fluorescence-activated cell sorting (FACS) [54]. Another report showed that the teratoma-forming propensities of iPS cell-derived neural cells in recipient bodies depended on the iPS cells' tissue of origin such as TTF, embryonic fibroblast, hepatocyte, or gastric epithelial cells. TTF–iPS cell-derived cells showed the highest tumor-forming propensity, whereas those from MEF–iPS cells and gastric epithelial cell–iPS cells showed the lowest risk, being comparable to that from ES cells [55]. Teratoma formation by derivatives of iPS cells may be also affected by the methods used for reprogramming and differentiation, the site of transplantation, and other factors. In the heart, it remains controversial whether iPS cell-derived cardiomyocyte transplantation would form the teratoma. The ability of human ES cell-derived cardiomyocytes partially repaired myocardial infarcts and attenuated heart failure in a rodent model [56, 57]. However, transplantation of undifferentiated human ES cells resulted in the formation of teratoma [58]. In addition, rhesus ES cell-derived cardiovascular progenitor cells were engrafted in post-myocardial-infarcted nonhuman primates without tumor formation [59]. These encouraging and accumulating findings will guide the field in determining the best iPS cells with respect to cell source safety, iPS cell generation methods, differentiation methods, and transplantation cell types in animal models before embarking on human clinical trials.

10.5 Differentiation to Cardiomyocytes

Human iPS cells were generated as a substitute for human ES cells because of the ethical and immunological problems associated with the use of ES cells [14]. Because of the similarities between ES and iPS cells, the differentiation system used for ES cells can be applied to iPS cells. The first report of the establishment of a mouse ES cell line was published by Evans and Kaufman in 1981, and ES cells were used as a model of very early development as well as to generate genetically modified mice [11, 60, 61]. In the mid-1990s, ES cell research slowly moved to the development of stem cell-based cell transplantation therapy using in vivo animal models [62–64]. The differentiation of mouse ES cells into cardiomyocytes in vitro was first demonstrated in 1985 without precise characteristics [65]. Research into the use of ES cells in cardiac regenerative medicine was initiated after the mid-1990s. In 1996, Klug et al. reported that stable transfection of ES cells with the aminoglycoside phosphotransferase gene under the control of the α-cardiac myosin heavy chain promoter succeeded in purifying cardiomyocytes after differentiation in vitro [62]. That study shed light on the use of ES cells in cardiac regenerative medicine. The development of more selective and efficient methods of differentiating ES cells into cardiomyocytes progressed slowly but steadily after the late 1990s [66]. In 1998, Thomson et al. first reported on the establishment of human ES cells [67]. Although ethical issues remained, human ES cells attracted significant attention for their potential in regeneration therapy [9].

10.6 Methods Used to Differentiate iPS Cells

There have been many methods reported for the differentiation of ES cells into cardiomyocytes. The differentiation of ES cells mimics normal embryonic development, thereby providing essential information on developmental processes, including heart development. So it is generally accepted that the humoral factors that are essential for cardiomyogenesis in vivo will stimulate ES cells to differentiate into cardiomyocytes in vitro. In fact, there are generally two different strategies used to achieve the differentiation of pluripotent stem cells into cardiomyocytes, namely, the embryoid body (EB) formation system and the FACS-based system. Conventionally, the EB formation system is used only for spontaneous ES cell differentiation. In this system, ES cells are moved and cultured in floating or hanging drops to form cell aggregates and differentiate in a manner partially similar to that seen during normal early embryonic development. Differentiated EBs contains several types of differentiated cells, such as cardiomyocytes, hematopoietic cells, and neural cells. In this system, physiological factors that promote cardiomyocyte differentiation may increase the population of cardiomyocytes obtained from pluripotent stem cells. Thus far, many factors involved in early heart development have been tested in the EB formation culture system with some demonstrating increased efficiency [7]. Particular advantages of the EB formation system are that it is technically easy and suited for large-scale culture. Drawbacks include the fact that the EBs contain many types of cells, so the mechanism controlling differentiation is difficult to understand, and the fact that the selection of cardiomyocytes is technically difficult. The underlying concept of the FACS-based system depends on the collection of cardiac progenitor cells and/or mature cardiac myocytes. FACS is a relatively sophisticated technology and the resultant purified cardiac progenitor cells and mature cardiac myocytes may provide some clues as to the mechanisms involved in cardiomyocyte differentiation. However, in terms of its practical application in human cardiac regenerative medicine, it is difficult to obtain large numbers of cardiac myocytes using the FACS system, and there is also a risk of damage to the sorted cells caused by the laser emission and/or the process of single-cell sorting. Thus, considerable technical advances are required before the FACS-based system becomes a practical option for cardiac regenerative medicine.

10.6.1 Methods Used to Differentiate iPS Cells: EB

The differentiation of ES cells into any cell lineage depends, in part, on the regulatory mechanisms underlying normal early development. Information obtained in genetically modified mice displaying cardiac abnormalities has provided key information on the essential factors in embryonic heart development and cardiomyocyte differentiation. Many attempts have been made to utilize this information to increase the efficiency of ES/iPS cell differentiation into cardiomyocytes. For example, early research showed that cardiac anomalies occur in mice lacking the receptor for retinoic acid (RA), a vitamin A derivative, or when vitamin A is deficient during embryonic development, suggesting vitamin A or RA is essential for cardiac differentiation and development [68–70]. Furthermore, RA induces the differentiation of cardiomyocytes from embryonal carcinoma (EC) cells in vitro in a time- and concentration-dependent manner consistent with normal development [71]. Based on these findings, in 1997 Wobus et al. succeeded in increasing the efficiency of cardiomyocyte induction by exposing ES cells to RA under strictly controlled conditions with respect to concentration and timing [64]. Following those findings, many basic studies have sought to elucidate cardiac differentiation mechanisms and to identify cardiac differentiation promoting factors.

Although several signaling proteins, including bone morphogenetic proteins (BMPs) [72–75], Wnts [76–78], Notch [79, 80], and fibroblast growth factors (FGFs) [81] are involved in heart development, little was known as to the precise regulatory signals that mediate the differentiation of ES cells into cardiomyocytes. In mouse embryos, cardiac progenitor cells appear around embryonic day (E) 7.0 and the cardiac crescent is formed by E7.5, indicating that the growth factors expressed in these regions or in surrounding areas at the relevant developmental stage may be important for efficient cardiomyocyte induction from ES/iPS cells. Indeed, in the past decade, many studies have investigated the effects of BMPs, BMP inhibitors, Wnt, Wnt inhibitors, and Notch on the induction of cardiomyocyte from ES cells at specific developmental stages (Table 10.1). The precise and detailed temporal and spatial regulation by those molecules has made it difficult to experimentally elucidate cardiogenic programming. In other words, one molecule promotes cardiogenesis at a certain moment; however, the same molecule might inhibit cardiogenesis in a different moment or different place. Among those molecules, Wnt signals are known to play prominent and varied roles in cardiovascular development [82]. Activation of Wnt signaling downregulates the intracellular degradation of β-catenin, thereby allowing it to translocate to the nucleus and activate other transcription factors in conjunction with its cotranscription factors, the

Table 10.1 Cardiac differentiation promoting factors

Factor	Authors	Paper
Retinoic acid	Wobus, A.M. et al.	*J. Mol. Cell Cardiol.* 29, 1525–1539 (1997)
Transforming growth factor β1	Behfar, A. et al.	*FASEB J.* 16, 1558–1566 (2002)
Fibroblast growth factors	Dell'Era, P. et al.	*Circ. Res.* 93, 414–420 (2003)
Dynorphin B	Ventura, C. et al.	*Circ. Res.* 92, 623–629 (2003)
Ascorbic acid	Takahashi, T. et al.	*Circulation* 107, 1912–1916 (2003)
Nitric oxide	Kanno, S. et al.	*Proc. Natl. Acad. Sci. USA* 101, 12277–12281 (2004)
Fibroblast growth factor 2 and bone morphogenetic protein 2	Kawai, T. et al.	*Circ. J.* 68, 691–702 (2004)
Wnt11	Terami, H. et al.	*Biochem. Biophys. Res. Commun.* 325, 968–975 (2004)
Noggin	Yuasa, S. et al.	*Nat. Biotechnol.* 23(5): 607–611 (2005)
PP2 (a Src family kinase inhibitor)	Hakuno, D. et al.	*J. Biol. Chem.* 280, 39534–39544 (2005)
Wnt3a/Wnt inhibitor	Naito, A.T. et al.	*Proc. Natl. Acad. Sci. USA*. 103, 19812–19817 (2006)
Wnt3	Ueno, S. et al.	*Proc. Natl. Acad. Sci. USA*. 104, 9685–9690 (2007)
Wnt3	Kwon, C. et al.	*Proc. Natl. Acad. Sci. USA*. 104, 10894–10899 (2007)
IGFBP-4	Zhu, W. et al.	*Nature*. 454(7202):345–349 (2008)
Cyclosporin-A	Yan, P. et al.	*Biochem. Biophys. Res. Commun.* 379(1):115–120 (2009)
G-CSF	Shimoji, K. et al.	*Cell Stem Cell*. 6(3):227–237 (2010)

LEFs/TCFs; this is the canonical Wnt pathway. In contrast, the noncanonical Wnt pathway has no role in regulating β-catenin degradation, but can activate JNK and other signaling molecules. In the past studies, Wnt signaling had been implicated as an inhibitor of cardiomyocyte induction [83]. Wnt inhibitors *Crescent* and *Dkk-1* were expressed in the anterior endoderm during gastrulation and could induce the formation of beating heart muscle, while ectopic Wnt signaling repressed heart formation from the anterior mesoderm in vitro and in vivo [76, 77, 84]. However, the heart is a mesodermal organ, and the mesodermal marker *BrachyuryT* is a target of Wnt3, one of the canonical Wnt signaling ligands [85]. Therefore, it can be expected and was demonstrated that canonical Wnt signaling plays a positive role in cardiac development in vitro and in vivo [83]. In the case of ES cells, Wnt/β-catenin signaling has a biphasic role in that early treatment of differentiating cells with Wnt-3A increased cardiac differentiation through mesoderm induction and

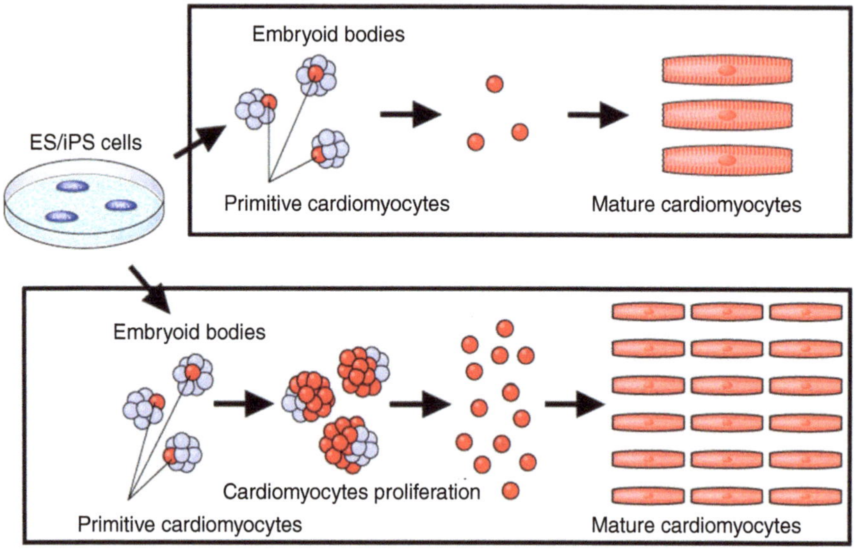

Fig. 10.3 Diagram showing the effects of granulocyte colony-stimulating factor (G-CSF) on cardiomyocyte proliferation. Embryonic stem (ES)/induced pluripotent stem (iPS) cells can differentiate into many cell types, including cardiomyocytes. Conventionally, primitive cardiomyocytes emerge spontaneously and differentiate into mature cardiomyocyte. G-CSF can promote the proliferation of primitive cardiomyocytes, boosting the yield of mature cardiomyocytes

late activation of beta-catenin signaling reduced cardiac differentiation [86–89]. Such accumulating data have therefore indicated that it is necessary to elucidate precise the regulatory signaling network at play under any given circumstance in cardiac differentiation, as a key step in maximizing the efficiency of cardiac differentiation in ES/iPS cells.

Cardiomyocyte development is a multistep process that includes initial mesodermal induction, the emergence of the cardiomyoblast, cardiomyoblast proliferation, and cardiomyocyte maturation [90]. Cardiomyocyte proliferation was therefore also investigated to improve the efficiency of cardiomyocyte acquisition, as one of the most important physiological steps in heart development that is regulated by several growth factors and cytokines during mid-gestational heart development [91–97]. These regulators act synergistically on cardiomyocyte proliferation under normal physiological conditions, making them popular targets to focus on in attempts to achieve cardiomyocyte proliferation during ES cell differentiation. In this aspect, developmental information can be useful. Both granulocyte colony-stimulating factor (G-CSF) and its receptor are expressed in the embryonic heart and involved in cardiomyocyte proliferation during development. In addition, G-CSF increased the number of cardiomyocytes derived from ES and iPS cells [98]. Thus, cardiomyocyte proliferation-promoting factors could be used to boost cardiomyocyte yield from ES and iPS cells, possibly in combination with other cardiomyocyte differentiation protocols (Fig. 10.3). Alternatively, chemical compound screening is underway to discover compounds able to promote cardiac differentiation and proliferation. For example, Takahashi et al. [99] showed that ascorbic acid promotes the induction of cardiomyocyte differentiation from ES cells.

Pivotal roles were also demonstrated for these factors in cardiomyocyte differentiation from ES cells using cardiac myocyte differentiation systems [86–89, 100–103]. However, there is no single growth factor that acts constantly throughout the entire process of organ induction during the development of multiple organ systems, suggesting that we should use a combination of the different differentiation systems with particular attention to fine spatial and temporal regulation.

10.6.2 Methods Used to Differentiate iPS Cells: Techniques Used for Cardiomyocyte Isolation

Cardiomyocytes are derived from cardiovascular progenitor cells, mesodermal progenitor cells, and pluripotent stem cells. Several marker genes have been reported for each stage, and cell surface markers are particularly useful for cell sorting without genetic manipulation. Markers that can be used for cardiomyocyte progenitor sorting are detailed below.

1. Investigations into the temporal expression of the primitive streak (PS) marker *Brachyury*, which is also a mesodermal marker, have demonstrated the sequential allocation of mesodermal cells to the hemangioblast and cardiac fates during embryonic development [104]. On the basis of these results, cell sorting for *Brachyury*-positive ES cells is likely to increase the population of cells with cardiac differentiation potential [105].
2. *ISL1*, an LIM homeodomain transcription factor, is expressed at the early stages of human cardiogenesis in a multipotent primordial progenitor and subsequently in a family of partially committed intermediate progenitors, before being down-regulated in the fully differentiated progeny. Human ES cell-derived *ISL1*-positive cardiovascular progenitors can give rise to cardiomyocyte, smooth muscle, and endothelial cell lineages [106].
3. Flk-1 (vascular endothelial growth factor receptor-2, also known as kinase insert domain protein receptor (Kdr)) is known as a lateral plate marker. An Flk-1-positive cell population appears to develop as cells exit the PS and begin to migrate to form the cardiac crescent [107, 108]. These observations support the notion that the myocardial and endothelial lineages develop through a common Flk-1-positive progenitor from ES/iPS cells and that Flk-1-positive cells may be cardiac progenitor cells [103, 109, 110].
4. During screening to identify cardiogenesis-associated genes in ES cells, *Prnp*, which encodes the cellular prion protein (PrP), is expressed in cardiomyocyte-rich EBs. PrP is expressed at the cell surface and thus serves as an effective surface marker for isolating nascent cardiomyocytes as well as cardiomyogenic progenitors [111].
5. Stage-specific embryonic antigen 1 (SSEA-1) can be used as an index of the differentiation of human ES cells as well as of the human blastocyst. Because

SSEA-1 is one of the earliest markers of human ES cell differentiation and loss of pluripotency, SSEA-1 selection enables us to isolate an early population of cardiovascular progenitor cells [59, 112].
6. Purification of human cardiomyocytes, as well as avoiding the contamination of stem cells, is an important issue for the success of cardiac regenerative medicine. TMRM (tetramethylrhodamine methyl ester perchlorate) is a fluorescent dye that labels mitochondria and which could be used to selectively mark ES/iPS cell-derived cardiomyocytes. TMRM selection would thus enable the isolation of mature cardiomyocytes [113]. Increasing the purity of cardiomyocytes derived from pluripotent stem cells will reduce the risk of tumorigenicity.

10.7 Application of iPS Cells in Cardiac Regenerative Medicine

Cardiovascular diseases are important targets for regenerative medicine because they are associated with high morbidity and mortality [114]. Most pathological processes that initiate irreversible heart dysfunction, such as myocardial infarction and cardiomyopathies, either result from or are exacerbated by a loss of heart cells. Because human heart lacks the capacity for self-repair, the prominent recovery of heart dysfunction requires the replacement of damaged cells by transplantation with large quantities of healthy cardiomyocytes. Recent studies have shown that human iPS cells can differentiate into cardiomyocyte-like cells that are similar in terms of their gene expression profiles and physiologic properties to native cardiomyocytes and ES cell-derived cardiomyocytes [115–117]. These studies highlight the potential of human iPS cells in cardiovascular regenerative medicine [118].

The first application of iPS cell technology for cardiac regenerative medicine is likely to be the transplantation of iPS cell-derived cardiomyocytes into diseased hearts to restore pump function. A considerable advantage of using human iPS cells is the possibility of creating isogenic cardiomyocytes that are genetically equivalent to the cells in the transplant recipient, thus avoiding immune rejection, which is likely to be seen with allogenic transplants (i.e., ES cell-derived cardiomyocytes may be rejected by the recipient's immune system). However, transplantation of stem cell-derived cardiomyocytes into animal models has raised many questions that need to be addressed before clinical transplantation into humans, namely, whether ES cell-derived cardiomyocyte transplantation could improve cardiac function. Laflamme et al [56]. reported an improvement in cardiac function in immunodeficient rats 4 weeks after coronary artery ligation and injection of hES cell-derived cardiomyocytes with prosurvival factors 4 days later. However, van Laake et al [57]. reported that hES cell-derived cardiomyocyte transplantation showed a significantly increased graft size, and a functional improvement was observed at 1 month, but not at 3 months [57]. Qiao et al. [119] also reported that highly enriched cardiomyocytes derived from murine ES cells transplantation improved cardiac contractile function of infarcted rat hearts at 1 and 2 months. In those experiments, there were some differences, such as transplantation timing, prosurvival factors, cardiomyocyte enrichment methods, transplantation cell number, and immunological reactions. However, many reports showed that transplanted ES cell-derived cardiomyocytes survive in host hearts in some extent, and accumulating evidence could be used to improve those methods. These data also suggest that midterm and long-term data in these kinds of experiments is essential in drawing conclusions on the long-term efficacy of cardiac cell transplantation.

Integration of grafts into host tissue requires the formation of new blood vessels to supply oxygen and nutrients from the circulation and several cytokines from vascular cells to transplanted cardiomyocytes. Vascular component cells such as endothelial cells and smooth muscle cells can be differentiated from ES and iPS cells [120, 121]. Those vascular cells can be then transplanted into an animal model, form vasculature, and connect with the host circulation [122]. Cotransplanted vascular cells with cardiomyocytes may help connect the grafts to the existing host vascular network and gain long-term benefits. Alternatively, bi-potent or tri-potent cardiac progenitor cells from human ES and iPS cells may be able to form cardiomyocytes, smooth muscle cells, and endothelial cells in situ [123]. Arrhythmic event is one of the main concerns in ES cell-derived cardiomyocyte transplantation, because of the high incidence of ventricular arrhythmias observed in the human skeletal myoblast transplantation trials [124, 125]. To assess whether hES cell-derived cardiomyocyte transplantation improves cardiac electrical activity or can be arrhythmogenic, it is necessary to observe cardiac electrical activity in vivo. Human ES cell-derived cardiomyocytes, which were demonstrated to form gap junctions with neighboring rat cardiomyocytes, integrated electrically with host cardiac tissue and did not form any significant conduction disturbances, suggesting that this approach is unlikely to cause fatal ventricular arrhythmias [126].

Another cell transplantation therapy is likely to be a biological pacemaker created from stem cell-derived cardiomyocytes. It is estimated that currently three million people have an implantable pacemaker to control cardiac rhythm disturbances, including sick sinus syndrome and atrioventricular block [127]. Although these devices treat patients successfully, there are shortcomings associated with their use, including cost, patient discomfort, cosmetic problems, and increased susceptibility to infection. These issues also have led to recent interest in the creation of a biological pacemaker. Proof-of-concept experiments using ES cell-derived cardiomyocytes have been published by two groups. Xue et al. have demonstrated the pacemaking ability of transplanted human ES cell-derived cardiomyocytes in a guinea pig preparation [128], and Kehat et al. have demonstrated that EBs injected into a swine model of complete heart block could function as an ectopic pacemaker [129]. These two studies confirmed the capacity of transplanted stem cell-derived cardiomyocytes to couple with host myocardium to function as an ectopic pacemaker, alluding to the potential of a biological pacemaker for clinical application. However, the follow-up period after transplantation in both studies was relatively short (<3 weeks) despite the use of EBs, which contain non-cardiomyocytes or undifferentiated cells. A longer follow-up period is necessary to confirm that this technique is safe without any risk of tumor formation.

10.8 Application of iPS Cells in the Genetic Analysis of Cardiac Disease

Human iPS cell-derived cardiomyocytes could also be used in pharmacologic testing. Many cardiac- and noncardiac-acting drugs prolong action potential duration (APD), giving rise to acquired long QT syndrome (LQTS), which may itself result in the life-threatening arrhythmia torsades de pointes (TdP). The most frequent reason for the removal of drugs from the market is adverse cardiac side effects [130, 131]. It is essential that any proarrhythmic risk is identified at an early stage in the drug development process, so as to define an unacceptable safety profile and to mitigate costs. Thus, a predictive, high-throughput, cell-based, in vitro QT assay system is highly desirable for cardiotoxicity screens. Although patch-clamp experiments are the accepted method for investigating action potential parameters and the precise electrophysiological properties of ion channels, the technique is time consuming and requires a skilled operator. Furthermore, even though currently available drug screening in heterologous expression systems is high throughput, these systems do not represent a native cardiac context and so may lack important accessory proteins or secondary targets that may mediate relevant adverse effects. Taking these issues into consideration, it appears that a combination of multielectrode arrays (MEAs) and stem cell-derived cardiomyocytes may represent the best system in which to measure the surface electrogenic activities of cell clusters. The MEAs may be useful for recording the electrical activity of the various derivatives of human ES and iPS cells [115, 132]. However, these systems also have issues that need to be resolved. Stem cell-derived cardiomyocytes have a similar electrophysiological phenotype to embryonic cardiomyocytes. Thus, the implications of significant phenotypic differences between these cells and adult cardiomyocytes should be considered. Moreover, hiPS cell-derived cardiomyocytes include distinct nodal and working cardiac subtypes, leading to electrophysiological heterogeneity, which may not reflect the situation in native cardiac tissues. Most cardiac ion currents undergo developmental maturation in terms of current density and properties, despite the fact that the gene expression patterns for each ion channel differ [133]. Maturation of the physiologic phenotypes of hiPS cell-derived cardiomyocytes may be critical for determining which cellular phase should be used for drug screening. Further investigations are needed to resolve these issues.

Finally, iPS cell-derived cardiomyocytes could be used as models of cardiac diseases. Genetic cardiovascular diseases include channelopathies and cardiomyopathies, which are related to abnormal electrophysiology and impaired contractility (Fig. 10.4). Genetic alterations that lead to dysfunctional cardiac ion channels are referred to as cardiac channelopathies. Common channelopathies include LQTS, Brugada syndrome, catecholaminergic polymorphic ventricular tachycardia (CPVT), and short QT syndrome, all of which are inherited arrhythmogenic diseases caused by mutations in the genes that encode ion channels or their related proteins [134, 135]. The electrical instability inherent to channelopathies (i.e., QT prolongation and triggering activity) increases the risk of fatal arrhythmias, which

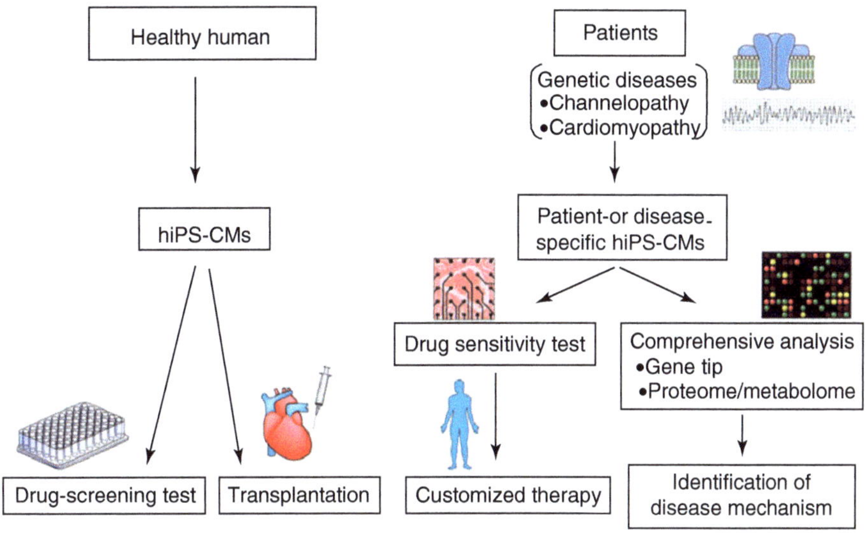

Fig. 10.4 Potential applications for human induced pluripotent stem cell-derived cardiomyocytes (hiPS-CMs). hiPS-CMs were generated from healthy or diseased human fibroblasts by transduction with defined transcription factors. The hiPS-CMs are applicable to drug-screening testing and cell transplantation therapy. In contrast, disease-specific hiPS-CMs, generated from patients with genetic diseases such as channelopathy and cardiomyopathy, can be used for drug-sensitivity testing and understanding disease mechanisms by comprehensive analysis. Drug-sensitivity testing of individual patients may lead to customized therapies, and new understanding of pathogenesis may lead to novel therapeutic approaches

may lead to sudden cardiac death. One of the merits of iPS cells is that they can be used as patient- or disease-specific stem cells, particularly in patients with genetically caused cardiac diseases. Because iPS cells derived from patients retain both their original genotype and phenotype, hiPS cells provide an excellent model for investigations into the mechanisms underlying heart disease, as well as for drug testing and toxicology (Fig. 10.1). Recently, Moretti et al. reported that patient-specific iPS cells maintained the disease genotype and recapitulated the functional features of the disorder [136]. They reprogrammed fibroblasts derived from members of a family with autosomal-dominant LQTS type 1 and created LQT1 patient-specific iPS cell-derived cardiomyocytes. These cells exhibited particular electrophysiological characteristics, including disease-specific abnormalities in APD, action potential rate adaptation, and I_{Ks}. Furthermore, precise characterization of the role of a mutation in *KCNQ1* (R190Q) revealed a dominant-negative trafficking defect associated with a reduction in the I_{Ks} current, demonstrating the benefit of using disease-specific hiPS cell-derived cardiomyocytes to investigate the pathogenesis of a genetic disease. To date, insights into the pathogenesis of the genetic arrhythmic diseases have come primarily from heterologous expression systems or genetic animal models. However, these studies often produce conflicting results depending on the cell type and animal species used. Recent experiments reported that patient-derived hiPS cell-derived cardiomyocytes recapitulated the phenotypic and mechanistic features of native cardiomyocytes [136–138]. Furthermore, it may be possible to investigate the therapeutic action of drugs to treat specific patients based on their in vitro effects on patient-specific hiPS cell-derived cardiomyocytes, resulting in the establishment of patient-specific drug-screening systems, as well as customized therapies.

Although many causative mutations in ion channels and related genes have been identified in patients with channelopathies, the genetic mutations in approximately 40 % of LQTS patients and 70 % of Brugada syndrome patients have not been identified, suggesting that there are many unknown genetic abnormalities that cause channelopathies. Furthermore, the severity of the channelopathies varies among family members, implying an association between phenotypic manifestation and epigenetic factors, in addition to genetic factors. In this regard, drug-sensitivity testing or genetic screening of disease-specific human iPS cardiomyocytes may lead to the identification of mutations that could be targeted in new medical treatments.

10.9 Conclusion

Although the adult heart may have some potential for regeneration, it remains impossible to control innate cardiac regeneration for therapeutic purposes in the diseased heart. Many clinical trials have been undertaken to investigate cardiac regeneration using adult stem cells and/or cytokines, and both clinician and patient expectations have been high. However, the results have been disappointing. We are now in the position to use newly identified stem cells, namely, iPS cells, which show considerable potential. The realization of cardiac regeneration depends on the outcome of many basic experimental investigations and the subsequent application of these results to the clinical setting. Careful clinical trials may result in gradual, yet significant, advances in this field and should be performed in tandem with further intensive investigations with the aim of realizing cardiac regeneration therapy in the clinical setting.

Acknowledgments Funding sources: This study was supported in part by research grants from the Ministry of Education, Science and Culture, Japan; by research grants from the project for realization of regenerative medicine; and by research grants from Kanrinmaru project.

Disclosures: none.

References

1. Becker RO, Chapin S, Sherry R (1974) Regeneration of the ventricular myocardium in amphibians. Nature 248:145–147
2. Fukuda K, Yuasa S (2006) Stem cells as a source of regenerative cardiomyocytes. Circ Res 98:1002–1013

3. Yuasa S, Fukuda K, Tomita Y, Fujita J, Ieda M, Tahara S, Itabashi Y, Yagi T, Kawaguchi H, Hisaka Y, Ogawa S (2004) Cardiomyocytes undergo cells division following myocardial infarction is a spatially and temporally restricted event in rats. Mol Cell Biochem 259:177–181
4. Beltrami AP, Urbanek K, Kajstura J, Yan S-M, Finato N, Bussani R, Nadal-Ginard B, Silvestri F, Leri A, Beltrami CA, Anversa P (2001) Evidence that human cardiac myocytes divide after myocardial infarction. N Engl J Med 344:1750–1757
5. Bergmann O, Bhardwaj RD, Bernard S, Zdunek S, Barnabe-Heider F, Walsh S, Zupicich J, Alkass K, Buchholz BA, Druid H, Jovinge S, Frisen J (2009) Evidence for cardiomyocyte renewal in humans. Science 324:98–102
6. Yuasa S, Fukuda K (2008) Recent advances in cardiovascular regenerative medicine: the induced pluripotent stem cell era. Expert Rev Cardiovasc Ther 6:803–810
7. Yuasa S, Fukuda K (2008) Cardiac regenerative medicine. Circ J 72:A49–A55
8. Wollert KC, Drexler H (2010) Cell therapy for the treatment of coronary heart disease: a critical appraisal. Nat Rev Cardiol 7:204–215
9. Davis DR, Zhang Y, Smith RR, Cheng K, Terrovitis J, Malliaras K, Li T-S, White A, Makkar R, Marbán E (2009) Validation of the cardiosphere method to culture cardiac progenitor cells from myocardial tissue. PLoS One 4:e7195
10. Tang X-L, Rokosh DG, Guo Y, Bolli R (2010) Cardiac progenitor cells and bone marrow-derived very small embryonic-like stem cells for cardiac repair after myocardial infarction. Circ J 74:390–404
11. Evans MJ, Kaufman MH (1981) Establishment in culture of pluripotential cells from mouse embryos. Nature 292:154–156
12. Cowan CA, Atienza J, Melton DA, Eggan K (2005) Nuclear reprogramming of somatic cells after fusion with human embryonic stem cells. Science 309:1369–1373
13. Wilmut I, Schnieke AE, McWhir J, Kind AJ, Campbell KHS (1997) Viable offspring derived from fetal and adult mammalian cells. Nature 385:810–813
14. Takahashi K, Yamanaka S (2006) Induction of pluripotent stem cells from mouse embryonic and adult fibroblast cultures by defined factors. Cell 126:663–676
15. Yu J, Vodyanik MA, Smuga-Otto K, Antosiewicz-Bourget J, Frane JL, Tian S, Nie J, Jonsdottir GA, Ruotti V, Stewart R, Slukvin II, Thomson JA (2007) Induced pluripotent stem cell lines derived from human somatic cells. Science 318:1917–1920
16. Takahashi K, Tanabe K, Ohnuki M, Narita M, Ichisaka T, Tomoda K, Yamanaka S (2007) Induction of pluripotent stem cells from adult human fibroblasts by defined factors. Cell 131:861–872
17. Okita K, Ichisaka T, Yamanaka S (2007) Generation of germline-competent induced pluripotent stem cells. Nature 448:313–317
18. Nakagawa M, Koyanagi M, Tanabe K, Takahashi K, Ichisaka T, Aoi T, Okita K, Mochiduki Y, Takizawa N, Yamanaka S (2008) Generation of induced pluripotent stem cells without Myc from mouse and human fibroblasts. Nat Biotechnol 26:101–106
19. Aoi T, Yae K, Nakagawa M, Ichisaka T, Okita K, Takahashi K, Chiba T, Yamanaka S (2008) Generation of pluripotent stem cells from adult mouse liver and stomach cells. Science 321:699–702
20. Stadtfeld M, Brennand K, Hochedlinger K (2008) Reprogramming of pancreatic [beta] cells into induced pluripotent stem cells. Curr Biol 18:890–894
21. Kim JB, Zaehres H, Wu G, Gentile L, Ko K, Sebastiano V, Arauzo-Bravo MJ, Ruau D, Han DW, Zenke M, Scholer HR (2008) Pluripotent stem cells induced from adult neural stem cells by reprogramming with two factors. Nature 454:646–650
22. Eminli S, Utikal J, Arnold K, Jaenisch R, Hochedlinger K (2008) Reprogramming of neural progenitor cells into induced pluripotent stem cells in the absence of exogenous Sox2 expression. Stem Cells 26:2467–2474
23. Silva J, Barrandon O, Nichols J, Kawaguchi J, Theunissen TW, Smith A (2008) Promotion of reprogramming to ground state pluripotency by signal inhibition. PLoS Biol 6:e253
24. Hanna J, Markoulaki S, Schorderet P, Carey BW, Beard C, Wernig M, Creyghton Menno P, Steine EJ, Cassady JP, Foreman R, Lengner CJ, Dausman Jessica A, Jaenisch R (2008) Direct

reprogramming of terminally differentiated mature b lymphocytes to pluripotency. Cell 133: 250–264
25. Aasen T, Raya A, Barrero MJ, Garreta E, Consiglio A, Gonzalez F, Vassena R, Bilic J, Pekarik V, Tiscornia G, Edel M, Boue S, Belmonte JCI (2008) Efficient and rapid generation of induced pluripotent stem cells from human keratinocytes. Nat Biotechnol 26:1276–1284
26. Sun N, Panetta NJ, Gupta DM, Wilson KD, Lee A, Jia F, Hu S, Cherry AM, Robbins RC, Longaker MT, Wu JC (2009) Feeder-free derivation of induced pluripotent stem cells from adult human adipose stem cells. Proc Nat Acad Sci 106:15720–15725
27. Yan X, Qin H, Qu C, Tuan RS, Shi S, Huang GT-J (2010) Ips cells reprogrammed from human mesenchymal-like stem/progenitor cells of dental tissue origin. Stem Cells Dev 19:469–480
28. Kim JB, Greber B, Arauzo-Bravo MJ, Meyer J, Park KI, Zaehres H, Scholer HR (2009) Direct reprogramming of human neural stem cells by OCT4. Nature 461:649–653
29. Eminli S, Foudi A, Stadtfeld M, Maherali N, Ahfeldt T, Mostoslavsky G, Hock H, Hochedlinger K (2009) Differentiation stage determines potential of hematopoietic cells for reprogramming into induced pluripotent stem cells. Nat Genet 41:968–976
30. Ye Z, Zhan H, Mali P, Dowey S, Williams DM, Jang Y-Y, Dang CV, Spivak JL, Moliterno AR, Cheng L (2009) Human-induced pluripotent stem cells from blood cells of healthy donors and patients with acquired blood disorders. Blood 114:5473–5480
31. Seki T, Yuasa S, Oda M, Egashira T, Yae K, Kusumoto D, Nakata H, Tohyama S, Hashimoto H, Kodaira M, Okada Y, Seimiya H, Fusaki N, Hasegawa M, Fukuda K (2010) Generation of induced pluripotent stem cells from human terminally differentiated circulating T cells. Cell Stem Cell 7:11–14
32. Desai-Mehta A, Lu L, Ramsey-Goldman R, Datta SK (1996) Hyperexpression of CD40 ligand by B and T cells in human lupus and its role in pathogenic autoantibody production. J Clin Invest 97:2063–2073
33. Okano S, Yonemitsu Y, Nagata S, Sata S, Onimaru M, Nakagawa K, Tomita Y, Kishihara K, Hashimoto S, Nakashima Y, Sugimachi K, Hasegawa M, Sueishi K (2003) Recombinant Sendai virus vectors for activated T lymphocytes. Gene Ther 10:1381–1391
34. Kunisato A, Wakatsuki M, Shinba H, Ota T, Ishida I, Nagao K (2011) Direct generation of induced pluripotent stem cells from human nonmobilized blood. Stem Cells Dev 20(1):159–168
35. Staerk J, Dawlaty MM, Gao Q, Maetzel D, Hanna J, Sommer CA, Mostoslavsky G, Jaenisch R (2010) Reprogramming of human peripheral blood cells to induced pluripotent stem cells. Cell Stem Cell 7:20–24
36. Loh YH, Hartung O, Li H, Guo C, Sahalie JM, Manos PD, Urbach A, Heffner GC, Grskovic M, Vigneault F, Lensch MW, Park IH, Agarwal S, Church GM, Collins JJ, Irion S, Daley GQ (2010) Reprogramming of T cells from human peripheral blood. Cell Stem Cell 7:15–19
37. Brown ME, Rondon E, Rajesh D, Mack A, Lewis R, Feng X, Zitur LJ, Learish RD, Nuwaysir EF (2010) Derivation of induced pluripotent stem cells from human peripheral blood T lymphocytes. PLoS One 5:e11373
38. Soldner F, Hockemeyer D, Beard C, Gao Q, Bell GW, Cook EG, Hargus G, Blak A, Cooper O, Mitalipova M, Isacson O, Jaenisch R (2009) Parkinson's disease patient-derived induced pluripotent stem cells free of viral reprogramming factors. Cell 136:964–977
39. Zhou W, Freed CR (2009) Adenoviral gene delivery can reprogram human fibroblasts to induced pluripotent stem cells. Stem Cells 27:2667–2674
40. Fusaki N, Ban H, Nishiyama A, Saeki K, Hasegawa M (2009) Efficient induction of transgene-free human pluripotent stem cells using a vector based on Sendai virus, an RNA virus that does not integrate into the host genome. Proc Jpn Acad Ser B 85:348–362
41. Kaji K, Norrby K, Paca A, Mileikovsky M, Mohseni P, Woltjen K (2009) Virus-free induction of pluripotency and subsequent excision of reprogramming factors. Nature 458:771–775
42. Woltjen K, Michael IP, Mohseni P, Desai R, Mileikovsky M, Hamalainen R, Cowling R, Wang W, Liu P, Gertsenstein M, Kaji K, Sung H-K, Nagy A (2009) PiggyBac transposition reprograms fibroblasts to induced pluripotent stem cells. Nature 458:766–770
43. Yu J, Hu K, Smuga-Otto K, Tian S, Stewart R, Slukvin II, Thomson JA (2009) Human induced pluripotent stem cells free of vector and transgene sequences. Science 324:797–801

44. Jia F, Wilson KD, Sun N, Gupta DM, Huang M, Li Z, Panetta NJ, Chen ZY, Robbins RC, Kay MA, Longaker MT, Wu JC (2010) A nonviral minicircle vector for deriving human iPS cells. Nat Methods 7:197–199
45. Zhou H, Wu S, Joo JY, Zhu S, Han DW, Lin T, Trauger S, Bien G, Yao S, Zhu Y, Siuzdak G, Schöler HR, Duan L, Ding S (2009) Generation of induced pluripotent stem cells using recombinant proteins. Cell Stem Cell 4:381–384
46. Kim D, Kim C-H, Moon J-I, Chung Y-G, Chang M-Y, Han B-S, Ko S, Yang E, Cha KY, Lanza R, Kim K-S (2009) Generation of human induced pluripotent stem cells by direct delivery of reprogramming proteins. Cell Stem Cell 4:472–476
47. Warren L, Manos PD, Ahfeldt T, Loh Y-H, Li H, Lau F, Ebina W, Mandal PK, Smith ZD, Meissner A, Daley GQ, Brack AS, Collins JJ, Cowan C, Schlaeger TM, Rossi DJ (2010) Highly efficient reprogramming to pluripotency and directed differentiation of human cells with synthetic modified mrna. Cell Stem Cell 7(5):618–630
48. Xu Y, Shi Y, Ding S (2008) A chemical approach to stem-cell biology and regenerative medicine. Nature 453:338–344
49. Huangfu D, Maehr R, Guo W, Eijkelenboom A, Snitow M, Chen AE, Melton DA (2008) Induction of pluripotent stem cells by defined factors is greatly improved by small-molecule compounds. Nat Biotechnol 26:795–797
50. Shi Y, Tae Do J, Desponts C, Hahm HS, Schöler HR, Ding S (2008) A combined chemical and genetic approach for the generation of induced pluripotent stem cells. Cell Stem Cell 2: 525–528
51. Cho H-J, Lee C-S, Kwon Y-W, Paek JS, Lee S-H, Hur J, Lee EJ, Roh T-Y, Chu I-S, Leem S-H, Kim Y, Kang H-J, Park Y-B, Kim H-S (2010) Induction of pluripotent stem cells from adult somatic cells by protein-based reprogramming without genetic manipulation. Blood 116: 386–395
52. Martinez-Fernandez A, Nelson TJ, Yamada S, Reyes S, Alekseev AE, Perez-Terzic C, Ikeda Y, Terzic A (2009) iPS programmed without c-MYC yield proficient cardiogenesis for functional heart chimerism. Circ Res 105:648–656
53. Yamanaka S (2007) Strategies and new developments in the generation of patient-specific pluripotent stem cells. Cell Stem Cell 1:39–49
54. Wernig M, Zhao J-P, Pruszak J, Hedlund E, Fu D, Soldner F, Broccoli V, Constantine-Paton M, Isacson O, Jaenisch R (2008) Neurons derived from reprogrammed fibroblasts functionally integrate into the fetal brain and improve symptoms of rats with parkinson's disease. Proc Natl Acad Sci 105:5856–5861
55. Miura K, Okada Y, Aoi T, Okada A, Takahashi K, Okita K, Nakagawa M, Koyanagi M, Tanabe K, Ohnuki M, Ogawa D, Ikeda E, Okano H, Yamanaka S (2009) Variation in the safety of induced pluripotent stem cell lines. Nat Biotechnol 27:743–745
56. Laflamme MA, Chen KY, Naumova AV, Muskheli V, Fugate JA, Dupras SK, Reinecke H, Xu C, Hassanipour M, Police S, O'Sullivan C, Collins L, Chen Y, Minami E, Gill EA, Ueno S, Yuan C, Gold J, Murry CE (2007) Cardiomyocytes derived from human embryonic stem cells in pro-survival factors enhance function of infarcted rat hearts. Nat Biotechnol 25:1015–1024
57. van Laake LW, Passier R, Doevendans PA, Mummery CL (2008) Human embryonic stem cell-derived cardiomyocytes and cardiac repair in rodents. Circ Res 102:1008–1010
58. Caspi O, Huber I, Kehat I, Habib M, Arbel G, Gepstein A, Yankelson L, Aronson D, Beyar R, Gepstein L (2007) Transplantation of human embryonic stem cell-derived cardiomyocytes improves myocardial performance in infarcted rat hearts. J Am Coll Cardiol 50: 1884–1893
59. Blin G, Nury D, Stefanovic S, Neri T, Guillevic O, Brinon B, Bellamy V, Rücker-Martin C, Barbry P, Bel A, Bruneval P, Cowan C, Pouly J, Mitalipov S, Gouadon E, Binder P, Hagège A, Desnos M, Renaud J-F, Menasché P, Pucéat M (2010) A purified population of multipotent cardiovascular progenitors derived from primate pluripotent stem cells engrafts in postmyocardial infarcted nonhuman primates. J Clin Invest 120:1125–1139
60. Murry CE, Keller G (2008) Differentiation of embryonic stem cells to clinically relevant populations: lessons from embryonic development. Cell 132:661–680

61. Manis JP (2007) Knock out, knock in, knock down – genetically manipulated mice and the nobel prize. N Engl J Med 357:2426–2429
62. Klug MG, Soonpaa MH, Koh GY, Field LJ (1996) Genetically selected cardiomyocytes from differentiating embronic stem cells form stable intracardiac grafts. J Clin Invest 98: 216–224
63. Kolossov E, Fleischmann BK, Liu Q, Bloch W, Viatchenko-Karpinski S, Manzke O, Ji GJ, Bohlen H, Addicks K, Hescheler J (1998) Functional characteristics of ES cell-derived cardiac precursor cells identified by tissue-specific expression of the green fluorescent protein. J Cell Biol 143:2045–2056
64. Wobus AM, Kaomei G, Shan J, Wellner M-C, Rohwedel J, Guanju J, Fleischmann B, Katus HA, Hescheler J, Franz W-M (1997) Retinoic acid accelerates embryonic stem cell-derived cardiac differentiation and enhances development of ventricular cardiomyocytes. J Mol Cell Cardiol 29:1525–1539
65. Doetschman TC, Eistetter H, Katz M, Schmidt W, Kemler R (1985) The in vitro development of blastocyst-derived embryonic stem cell lines: formation of visceral yolk sac, blood islands and myocardium. J Embryol Exp Morphol 87:27–45
66. Hattan N, Kawaguchi H, Ando K, Kuwabara E, Fujita J, Murata M, Suematsu M, Mori H, Fukuda K (2005) Purified cardiomyocytes from bone marrow mesenchymal stem cells produce stable intracardiac grafts in mice. Cardiovasc Res 65:334–344
67. Thomson JA, Itskovitz-Eldor J, Shapiro SS, Waknitz MA, Swiergiel JJ, Marshall VS, Jones JM (1998) Embryonic stem cell lines derived from human blastocysts. Science 282: 1145–1147
68. Dyson E, Sucov HM, Kubalak SW, Schmid-Schonbein GW, DeLano FA, Evans DM, Ross J Jr, Chien KR (1995) Atrial-like phenotype is associated with embryonic ventricular failure in retinoid X receptor alpha -/- mice. Proc Natl Acad Sci 92:7386–7390
69. Kastner P, Grondona JM, Mark M, Gansmuller A, LeMeur M, Decimo D, Vonesch J-L, Dolle P, Chambon P (1994) Genetic analysis of RXR[alpha] developmental function: convergence of RXR and RAR signaling pathways in heart and eye morphogenesis. Cell 78:987–1003
70. Osmond MK, Butler AJ, Voon FC, Bellairs R (1991) The effects of retinoic acid on heart formation in the early chick embryo. Development 113:1405–1417
71. Edwards MK, Harris JF, McBurney MW (1983) Induced muscle differentiation in an embryonal carcinoma cell line. Mol Cell Biol 3:2280–2286
72. Winnier G, Blessing M, Labosky PA, Hogan BL (1995) Bone morphogenetic protein-4 is required for mesoderm formation and patterning in the mouse. Genes Dev 9:2105–2116
73. Zhang H, Bradley A (1996) Mice deficient for BMP2 are nonviable and have defects in amnion/chorion and cardiac development. Development 122:2977–2986
74. Schlange T, Andree B, Arnold H-H, Brand T (2000) BMP2 is required for early heart development during a distinct time period. Mech Dev 91:259–270
75. Gaussin V, Van de Putte T, Mishina Y, Hanks MC, Zwijsen A, Huylebroeck D, Behringer RR, Schneider MD (2002) Endocardial cushion and myocardial defects after cardiac myocyte-specific conditional deletion of the bone morphogenetic protein receptor alk3. Proc Natl Acad Sci 99:2878–2883
76. Marvin MJ, Di Rocco G, Gardiner A, Bush SM, Lassar AB (2001) Inhibition of Wnt activity induces heart formation from posterior mesoderm. Genes Dev 15:316–327
77. Schneider VA, Mercola M (2001) Wnt antagonism initiates cardiogenesis in xenopus laevis. Genes Dev 15:304–315
78. Foley AC, Mercola M (2005) Heart induction by Wnt antagonists depends on the homeodomain transcription factor hex. Genes Dev 19:387–396
79. Timmerman LA, Grego-Bessa J, Raya A, Bertran E, Perez-Pomares JM, Diez J, Aranda S, Palomo S, McCormick F, Izpisua-Belmonte JC, de la Pompa JL (2004) Notch promotes epithelial-mesenchymal transition during cardiac development and oncogenic transformation. Genes Dev 18:99–115
80. Maillard I, Fang T, Pear WS (2005) Regulation of lymphoid development, differentiation, and function by the notch pathway. Annu Rev Immunol 23:945–974

81. Mima T, Ueno H, Fischman DA, Williams LT, Mikawa T (1995) Fibroblast growth factor receptor is required for in vivo cardiac myocyte proliferation at early embryonic stages of heart development. Proc Natl Acad Sci 92:467–471
82. Gessert S, Kuhl M (2010) The multiple phases and faces of Wnt signaling during cardiac differentiation and development. Circ Res 107:186–199
83. Tzahor E (2007) Wnt/[beta]-catenin signaling and cardiogenesis: timing does matter. Dev Cell 13:10–13
84. Tzahor E, Lassar AB (2001) Wnt signals from the neural tube block ectopic cardiogenesis. Genes Dev 15:255–260
85. Yamaguchi TP, Takada S, Yoshikawa Y, Wu N, McMahon AP (1999) T (Brachyury) is a direct target of Wnt3a during paraxial mesoderm specification. Genes Dev 13:3185–3190
86. Ueno S, Weidinger G, Osugi T, Kohn AD, Golob JL, Pabon L, Reinecke H, Moon RT, Murry CE (2007) From the cover: biphasic role for Wnt/beta-catenin signaling in cardiac specification in zebrafish and embryonic stem cells. Proc Natl Acad Sci 104:9685–9690
87. Naito AT, Shiojima I, Akazawa H, Hidaka K, Morisaki T, Kikuchi A, Komuro I (2006) Developmental stage-specific biphasic roles of Wnt/beta-catenin signaling in cardiomyogenesis and hematopoiesis. Proc Natl Acad Sci 103:19812–19817
88. Lin L, Cui L, Zhou W, Dufort D, Zhang X, Cai C-L, Bu L, Yang L, Martin J, Kemler R, Rosenfeld MG, Chen J, Evans SM (2007) Beta-catenin directly regulates islet1 expression in cardiovascular progenitors and is required for multiple aspects of cardiogenesis. Proc Natl Acad Sci 104:9313–9318
89. Klaus A, Saga Y, Taketo MM, Tzahor E, Birchmeier W (2007) Distinct roles of Wnt/beta-catenin and bmp signaling during early cardiogenesis. Proc Natl Acad Sci 104:18531–18536
90. Srivastava D (2006) Making or breaking the heart: from lineage determination to morphogenesis. Cell 126:1037–1048
91. Olson EN, Schneider MD (2003) Sizing up the heart: development redux in disease. Genes Dev 17:1937–1956
92. Kajstura J, Cheng W, Reiss K, Anversa P (1994) The IGF-1-IGF-1 receptor system modulates myocyte proliferation but not myocyte cellular hypertrophy in vitro. Exp Cell Res 215:273–283
93. Reiss K, Cheng W, Ferber A, Kajstura J, Li P, Li B, Olivetti G, Homcy CJ, Baserga R, Anversa P (1996) Overexpression of insulin-like growth factor-1 in the heart is coupled with myocyte proliferation in transgenic mice. PNAS 93:8630–8635
94. Lavine KJ, Yu K, White AC, Zhang X, Smith C, Partanen J, Ornitz DM (2005) Endocardial and epicardial derived FGF signals regulate myocardial proliferation and differentiation in vivo. Dev Cell 8:85–95
95. Sucov HM, Dyson E, Gumeringer CL, Price J, Chien KR, Evans RM (1994) RXR alpha mutant mice establish a genetic basis for vitamin a signaling in heart morphogenesis. Genes Dev 8:1007–1018
96. Kastner P, Messaddeq N, Mark M, Wendling O, Grondona JM, Ward S, Ghyselinck N, Chambon P (1997) Vitamin A deficiency and mutations of RXRalpha, RXRbeta and RARalpha lead to early differentiation of embryonic ventricular cardiomyocytes. Development 124:4749–4758
97. Chen H, Shi S, Acosta L, Li W, Lu J, Bao S, Chen Z, Yang Z, Schneider MD, Chien KR, Conway SJ, Yoder MC, Haneline LS, Franco D, Shou W (2004) Bmp10 is essential for maintaining cardiac growth during murine cardiogenesis. Development 131:2219–2231
98. Shimoji K, Yuasa S, Onizuka T, Hattori F, Tanaka T, Hara M, Ohno Y, Chen H, Egasgira T, Seki T, Yae K, Koshimizu U, Ogawa S, Fukuda K (2010) G-CSF promotes the proliferation of developing cardiomyocytes in vivo and in derivation from ESCs and iPSCs. Cell Stem Cell 6:227–237
99. Takahashi T, Lord B, Schulze PC, Fryer RM, Sarang SS, Gullans SR, Lee RT (2003) Ascorbic acid enhances differentiation of embryonic stem cells into cardiac myocytes. Circulation 107:1912–1916
100. Yuasa S, Itabashi Y, Koshimizu U, Tanaka T, Sugimura K, Kinoshita M, Hattori F, Fukami S, Shimazaki T, Ogawa S, Okano H, Fukuda K (2005) Transient inhibition of BMP signaling by noggin induces cardiomyocyte differentiation of mouse embryonic stem cells. Nat Biotechnol 23:607–611

101. Chen VC, Stull R, Joo D, Cheng X, Keller G (2008) Notch signaling respecifies the hemangioblast to a cardiac fate. Nat Biotechnol 26:1169–1178
102. Behfar A, Zingman LV, Hodgson DM, Rauzier J-M, Kane GC, Terzic A, Puceat M (2002) Stem cell differentiation requires a paracrine pathway in the heart. FASEB J 16:1558–1566
103. Yang L, Soonpaa MH, Adler ED, Roepke TK, Kattman SJ, Kennedy M, Henckaerts E, Bonham K, Abbott GW, Linden RM, Field LJ, Keller GM (2008) Human cardiovascular progenitor cells develop from a KDR+ embryonic-stem-cell-derived population. Nature 453:524–528
104. Kinder SJ, Tsang TE, Quinlan GA, Hadjantonakis AK, Nagy A, Tam PP (1999) The orderly allocation of mesodermal cells to the extraembryonic structures and the anteroposterior axis during gastrulation of the mouse embryo. Development 126:4691–4701
105. Kattman SJ, Huber TL, Keller GM (2006) Multipotent flk-1+ cardiovascular progenitor cells give rise to the cardiomyocyte, endothelial, and vascular smooth muscle lineages. Dev Cell 11:723–732
106. Bu L, Jiang X, Martin-Puig S, Caron L, Zhu S, Shao Y, Roberts DJ, Huang PL, Domian IJ, Chien KR (2009) Human ISL1 heart progenitors generate diverse multipotent cardiovascular cell lineages. Nature 460:113–117
107. Kataoka H, Takakura N, Nishikawa S, Tsuchida K, Kodama H, Kunisada T, Risau W, Kita T, Nishikawa SI (1997) Expressions of PDGF receptor alpha, c-Kit and Flk1 genes clustering in mouse chromosome 5 define distinct subsets of nascent mesodermal cells. Dev Growth Differ 39:729–740
108. Ema M, Takahashi S, Rossant J (2006) Deletion of the selection cassette, but not cis-acting elements, in targeted Flk1-lacZ allele reveals Flk1 expression in multipotent mesodermal progenitors. Blood 107:111–117
109. Yamashita JK, Takano M, Hiraoka-Kanie M, Shimazu C, Peishi Y, Yanagi K, Nakano A, Inoue E, Kita F, Nishikawa S-I (2005) Prospective identification of cardiac progenitors by a novel single cell-based cardiomyocyte induction. FASEB J 19(11):1534–1536, 2005:04-3540fje
110. Nelson TJ, Faustino RS, Chiriac A, Crespo-Diaz R, Behfar A, Terzic A (2008) Cxcr4+/Flk-1+ biomarkers select a cardiopoietic lineage from embryonic stem cells. Stem Cells 26:1464–1473
111. Hidaka K, Shirai M, Lee J-K, Wakayama T, Kodama I, Schneider MD, Morisaki T (2010) The cellular prion protein identifies bipotential cardiomyogenic progenitors. Circ Res 106:111–119
112. Leschik J, Stefanovic S, Brinon B, Puceat M (2008) Cardiac commitment of primate embryonic stem cells. Nat Protoc 3:1381–1387
113. Hattori F, Chen H, Yamashita H, Tohyama S, Satoh Y-S, Yuasa S, Li W, Yamakawa H, Tanaka T, Onitsuka T, Shimoji K, Ohno Y, Egashira T, Kaneda R, Murata M, Hidaka K, Morisaki T, Sasaki E, Suzuki T, Sano M, Makino S, Oikawa S, Fukuda K (2010) Nongenetic method for purifying stem cell-derived cardiomyocytes. Nat Methods 7:61–66
114. Passier R, van Laake LW, Mummery CL (2008) Stem-cell-based therapy and lessons from the heart. Nature 453:322–329
115. Tanaka T, Tohyama S, Murata M, Nomura F, Kaneko T, Chen H, Hattori F, Egashira T, Seki T, Ohno Y, Koshimizu U, Yuasa S, Ogawa S, Yamanaka S, Yasuda K, Fukuda K (2009) In vitro pharmacologic testing using human induced pluripotent stem cell-derived cardiomyocytes. Biochem Biophys Res Commun 385:497–502
116. Yokoo N, Baba S, Kaichi S, Niwa A, Mima T, Doi H, Yamanaka S, Nakahata T, Heike T (2009) The effects of cardioactive drugs on cardiomyocytes derived from human induced pluripotent stem cells. Biochem Biophys Res Commun 387:482–488
117. Zhang J, Wilson GF, Soerens AG, Koonce CH, Yu J, Palecek SP, Thomson JA, Kamp TJ (2009) Functional cardiomyocytes derived from human induced pluripotent stem cells. Circ Res 104:e30–e41
118. Murata M, Tohyama S, Fukuda K (2010) Impacts of recent advances in cardiovascular regenerative medicine on clinical therapies and drug discovery. Pharmacol Ther 126:109–118
119. Qiao H, Zhang H, Yamanaka S, Patel VV, Petrenko NB, Huang B, Muenz LR, Ferrari VA, Boheler KR, Zhou R (2011) Long-term improvement in postinfarct left ventricular global and

regional contractile function is mediated by embryonic stem cell – derived cardiomyocytes/ clinical perspective. Circ Cardiovasc Imaging 4:33–41
120. Levenberg S, Golub JS, Amit M, Itskovitz-Eldor J, Langer R (2002) Endothelial cells derived from human embryonic stem cells. Proc Natl Acad Sci USA 99:4391–4396
121. Homma K, Sone M, Taura D, Yamahara K, Suzuki Y, Takahashi K, Sonoyama T, Inuzuka M, Fukunaga Y, Tamura N, Itoh H, Yamanaka S, Nakao K (2010) Sirt1 plays an important role in mediating greater functionality of human ES/iPS-derived vascular endothelial cells. Atherosclerosis 212:42–47
122. Ferreira LS, Gerecht S, Shieh HF, Watson N, Rupnick MA, Dallabrida SM, Vunjak-Novakovic G, Langer R (2007) Vascular progenitor cells isolated from human embryonic stem cells give rise to endothelial and smooth muscle like cells and form vascular networks in vivo. Circ Res 101:286–294
123. Wu SM, Chien KR, Mummery C (2008) Origins and fates of cardiovascular progenitor cells. Cell 132:537–543
124. Menasché P, Hagège AA, Vilquin J-T, Desnos M, Abergel E, Pouzet B, Bel A, Sarateanu S, Scorsin M, Schwartz K, Bruneval P, Benbunan M, Marolleau J-P, Duboc D (2003) Autologous skeletal myoblast transplantation for severe postinfarction left ventricular dysfunction. J Am Coll Cardiol 41:1078–1083
125. Eisen HJ (2008) Skeletal myoblast transplantation: no magic bullet for ischemic cardiomyopathy. Nat Clin Pract Cardiovasc Med 5:520–521
126. Gepstein L, Ding C, Rehemedula D, Wilson EE, Yankelson L, Caspi O, Gepstein A, Huber I, Olgin JE (2010) In vivo assessment of the electrophysiological integration and arrhythmogenic risk of myocardial cell transplantation strategies. Stem Cells 28:2151–2161
127. Wood MA, Ellenbogen KA (2002) Cardiology patient pages. Cardiac pacemakers from the patient's perspective. Circulation 105:2136–2138
128. Xue T, Cho HC, Akar FG, Tsang SY, Jones SP, Marban E, Tomaselli GF, Li RA (2005) Functional integration of electrically active cardiac derivatives from genetically engineered human embryonic stem cells with quiescent recipient ventricular cardiomyocytes: insights into the development of cell-based pacemakers. Circulation 111:11–20
129. Kehat I, Khimovich L, Caspi O, Gepstein A, Shofti R, Arbel G, Huber I, Satin J, Itskovitz-Eldor J, Gepstein L (2004) Electromechanical integration of cardiomyocytes derived from human embryonic stem cells. Nat Biotechnol 22:1282–1289
130. Gussak I, Litwin J, Kleiman R, Grisanti S, Morganroth J (2004) Drug-induced cardiac toxicity: emphasizing the role of electrocardiography in clinical research and drug development. J Electrocardiol 37:19–24
131. Roden DM (2004) Drug-induced prolongation of the qt interval. N Engl J Med 350: 1013–1022
132. Harding SE, Ali NN, Brito-Martins M, Gorelik J (2007) The human embryonic stem cell-derived cardiomyocyte as a pharmacological model. Pharmacol Ther 113:341–353
133. Sartiani L, Bettiol E, Stillitano F, Mugelli A, Cerbai E, Jaconi ME (2007) Developmental changes in cardiomyocytes differentiated from human embryonic stem cells: a molecular and electrophysiological approach. Stem Cells 25:1136–1144
134. Marban E (2002) Cardiac channelopathies. Nature 415:213–218
135. Priori SG, Bloise R, Crotti L (2001) The long QT syndrome. Europace 3:16–27
136. Moretti A, Bellin M, Welling A, Jung CB, Lam JT, Bott-Flügel L, Dorn T, Goedel A, Höhnke C, Hofmann F, Seyfarth M, Sinnecker D, Schömig A, Laugwitz K-L (2010) Patient-specific induced pluripotent stem-cell models for long-QT syndrome. N Engl J Med 363:1397–1409
137. Itzhaki I, Maizels L, Huber I, Zwi-Dantsis L, Caspi O, Winterstern A, Feldman O, Gepstein A, Arbel G, Hammerman H, Boulos M, Gepstein L (2011) Modelling the long QT syndrome with induced pluripotent stem cells. Nature 471(7337):225–229
138. Carvajal-Vergara X, Sevilla A, D'Souza SL, Ang Y-S, Schaniel C, Lee D-F, Yang L, Kaplan AD, Adler ED, Rozov R, Ge Y, Cohen N, Edelmann LJ, Chang B, Waghray A, Su J, Pardo S, Lichtenbelt KD, Tartaglia M, Gelb BD, Lemischka IR (2010) Patient-specific induced pluripotent stem-cell-derived models of leopard syndrome. Nature 465:808–812

Chapter 11
Dendritic Cells for Cancer Immunotherapy

Nicolas Larmonier and Emmanuel Katsanis

Abstract The mainstay of cancer treatment remains surgery, radiotherapy, and cytotoxic chemotherapy, which are associated with significant side effects. Cancer immunotherapy, the manipulation of the immune system to eliminate tumor cells, has been considered for several decades as an alternative to these therapies. Among immunotherapeutic modalities, the perspective of using dendritic cell vaccines to stimulate antitumor immunity has shown some promises but also limitations. Dendritic cells are the most potent antigen-presenting cells of the immune system, playing a pivotal role in the initiation and regulation of tumor-specific immune responses as they are endowed with the unique ability to take up, process, and present tumor antigens to $CD4^+$ or $CD8^+$ T lymphocytes. Dendritic cells also contribute to the activation of natural killer cells and to the orchestration of humoral immunity. This unique capability has been widely exploited in cancer vaccination approaches against a variety of malignancies. However, tumors commonly develop so-called "immune escape" mechanisms including the secretion of immunosuppressive molecules and/or the promotion of immunosuppressive cells such as regulatory T cells that impair dendritic cell functions and therefore compromise the success of dendritic cell vaccination. Specific radio- or chemoimmunotherapeutic manipulations can blunt tolerogenic cells and revert the cancer-induced immunosuppressive environment into a pro-inflammatory context that can enhance dendritic cell capability to effectively prime and sustain antitumor immune responses.

Keywords Cancer • Immunotherapy • Dendritic cells • Immunosurveillance • Antigen-presenting cells

N. Larmonier (✉) • E. Katsanis (✉)
Department of Pediatrics, University of Arizona,
1501 N Campbell Ave, Tucson, AZ 85724, USA
e-mail: nrlarmon@email.arizona.edu; katsanis@peds.arizona.edu

M.K. Danquah, R.I. Mahato (eds.), *Emerging Trends in Cell and Gene Therapy*,
DOI 10.1007/978-1-62703-417-3_11, © Springer Science+Business Media New York 2013

11.1 Dendritic Cells as Orchestrators of Cancer Immunosurveillance

Immunity against tumors depends on a finely tuned balance that involves complex and highly orchestrated innate and adaptive immune responses. Cancer immunotherapy aims at promoting tumor eradication through the activation of these immune responses. The major advantages of immunotherapy over conventional therapies include relatively limited side effects, the specific targeting of tumor cells, and the generation of a long-lasting memory response against tumor-specific antigens [1–4]. The unique ability of dendritic cells (DCs) to function as professional antigen-presenting cells (APC) has positioned them as key players in the organization and control of antitumor immunity. This central role has also been the basis for the development of these cells as promising tools in cancer immunotherapeutic approaches [5, 6].

DCs consist of a heterogeneous population of cells made of specific subsets defined by their anatomic distribution, phenotype, mode of antigen presentation, and cytokine production profile [7–9]. DCs serve as sentinels that continuously take up antigens in peripheral tissues and migrate to the secondary lymphoid organs (lymph nodes, spleen) where they present processed antigenic proteins or lipids on major histocompatibility complexes (MHC) class I or class II or on CD1d antigen-presenting molecules [2, 8]. In an appropriate pro-inflammatory environment, antigen presentation by DC results in T cell proliferation and differentiation into $CD8^+$ cytotoxic (CTL) or $CD4^+$ helper (Th) effector T lymphocytes. This clonal expansion and activation of CTL and Th cells eventually leads to the elimination of target cells expressing the specific antigens. The differentiation of effector T lymphocytes toward a defined subset (Th-1, Th-2, Th-17, Treg) depends on the DC subset, on the level of activation of DC, and on the nature of the cytokines they secrete [2, 8–11]. DCs are also endowed with the capacity of modulating the function of other effector immune cells such as NK, B, or NKT cells [12–15]. However, although essential for the generation of adaptive immune response, DC can also participate in the mechanisms of immune tolerance, thereby playing a central role in the control of autoimmunity [16–20]. These “tolerogenic” DCs may anergize effector T lymphocytes [17, 21, 22], promote $FoxP3^+$ regulatory T cells (Treg), or drive the differentiation of anergic IL-10-secreting immunosuppressive Tr-1 cells [23–26].

The cancer immunosurveillance theory, now widely accepted, envisions that the immune system can recognize newly arising malignant cells before they become clinically apparent [27–33]. Tumor immunosurveillance relies on a multistage process tightly regulated by DC [6, 34, 35]. The main initial source of available antigens is provided in the form of apoptotic or necrotic cancer cell debris that result from tumor cell killing by macrophages, NK, NKT, or other cytotoxic innate immune cells or by chemotherapeutic agents or radiotherapy [33, 36–38]. Immature DCs, attracted to the tumor site, take up the released tumor-derived antigens and subsequently enter a maturation and activation phase. The acquired antigens are processed into peptides and presented to T lymphocytes on MHC class I or class II molecules. Activated tumor-specific $CD8^+$ CTL (the primary effector cytotoxic cells of the immune system) express a specific profile of chemokine receptors responsible for their migration from the secondary lymphoid organs to the tumor beds where they eliminate tumor cells using several killing mechanisms [34]. CTL activation, proliferation, and survival are further supported by $CD4^+$ T helper lymphocytes primed by DC. Activated $CD4^+$ T cells also support the activation of cytotoxic NK or macrophages [31, 38]. Activated NKT cells recognizing tumor-derived glycolipids associated with CD1d expressed by DC may also participate to tumor cell destruction [39]. DCs can therefore orchestrate and control an immune attack against cancer at virtually all of its stages (initiation, maintenance and regulation, activation of diverse cytotoxic effectors). They thus theoretically represent strategic targets for immune intervention strategies and have successfully been used in animals and humans to induce specific anticancer immunity after loading with tumor antigens [6, 40–47].

However, cancers commonly avoid immune detection and elimination using multiple strategies [4, 31, 48–50]. Considerable advances have been made in the past decade in our understanding of the mechanisms underlying the escape of tumor cells from destruction by immune responses [51]. For instance, the ability of cancer cells to downregulate the expression of major histocompatibility complex (MHC) class I molecules prevents their recognition and thus killing by cytotoxic T lymphocytes [51–55]. Cancer cells have also evolved multiple mechanisms of resistance to cell death (anti-apoptotic molecule expression, deficiencies in key factors involved in the apoptosis molecular cascade, lack or downregulation of receptors for death ligands such as Fas ligand or TRAIL, resistance to the perforin/granzyme system) inflicted by CTL [31, 56–60]. Tumors also produce immunosuppressive factors that negatively affect the function of DC, T, and natural killer (NK) cells, such as nitric oxide (NO), IL-6, IL-10, tumor growth factor beta (TGF-β), indoleamine 2,3-dioxygensase (IDO), arginase-1, prostaglandin E2 (PGE_2), vascular endothelial growth factor (VEGF), and cyclooxygenase-2 (COX-2) [31, 56, 57, 61]. This immunosuppressive tumor environment may also foster the generation and/or promotion of immunosuppressive cells such as Treg, type 2 macrophages (M2), myeloid-derived suppressor cells (MDSC), and as mentioned above immature/tolerogenic DC which can block antitumoral T or NK cell activation and/or induce lymphocyte anergy or apoptosis [17, 62–70]. While the mechanisms responsible for the production and accumulation of tolerogenic DC or the impairment of DC-based anticancer vaccine efficacy are not fully understood, extensive research has focused on developing strategies to restore the functions of DC in tumors, to design more effective DC vaccines that resist inhibition by the tumor microenvironment, and to develop combinatorial DC-based chemoimmunotherapeutic approaches.

11.2 Dendritic Cell-Based Vaccines in Cancer Immunotherapy

Growing knowledge about the immunobiology of DC has led to the rapid development of these cells for tumor immunotherapy and has prompted extensive research and clinical trials to evaluate the therapeutic efficacy of DC-based cancer vaccines.

Although multiple human studies have established that DC vaccines are safe and in most patients lead to the generation of immunological responses, complete cancer remissions have been limited, underscoring the need for further advances in the field [40–44, 71–74]. The design of effective DC-based cancer vaccines should follow a certain number of criteria and characteristics, but widely accepted standardized conditions are still to be defined. DC-based cancer vaccines should be able to access the lymph nodes after injection, display a mature pro-inflammatory phenotype, have the capacity to polarize Th-1 responses, and maintain their ability to present antigen for a sufficient period of time so a productive antitumor T lymphocyte-mediated response can be generated. The development of DC-based vaccines is associated with specific questions and challenges including but not limited to the following: (1) the choice of the DC subset to be used and the DC generation/expansion method, (2) the types and source of tumor antigens and the technique for loading antigen onto DC, (3) the choice of optimal adjuvants to induce DC maturation, and (4) the route, dose, and timing of DC vaccine delivery.

11.2.1 Generation of Dendritic Cells for Cancer Immunotherapy

DCs originate from bone marrow hematopoietic progenitors and represent a heterogenous cell population as it pertains to their lineage origin, phenotype, function, and localization [8, 9]. Two major subsets of blood DC have been identified: myeloid DC ($CD123^-CD14^-CD11c^+$) and plasmacytoid (lymphoid) DC ($CD123^+CD14^-CD11c^+$). Although it is now accepted that both subsets have the potential when appropriately stimulated to prime type 1 immune responses and induce antitumoral CTLs, the primary source of DC currently used in clinical trials and in most animal studies consists of myeloid DC generated from monocytes. Combination of multiple DC subsets has also been evaluated. The scarcity of DC in vivo has prompted the development of ex vivo differentiation and expansion procedures that allow DC generation in large number as starting material for vaccine preparation. These techniques have evolved over the last decade. The most common approach for preparing DC for clinical use consists in the differentiation of monocytes obtained from peripheral blood mononuclear cells (PBMC) purified by density centrifugation. The PBMC are cultured in the presence of granulocyte-macrophage colony-stimulating factor (GM-CSF) and interleukin-4 (IL-4). Immature DCs, characterized by high antigen uptake potential but low T cell stimulatory potential, are obtained after a 4–5-day culture. After loading with tumor antigens, DC are activated, typically for 24–48 h with various molecules such as TNF-α, IFN-γ, LPS, CpG, IL-1β, or CD40L as detailed hereunder, which leads to the generation of mature DC characterized by low endocytosis potential and significantly increased ability to activate T lymphocytes. These mature DCs are characterized by the upregulation of MHC class II-antigenic peptide complexes; increase expression of co-stimulatory molecules CD40, CD80, CD86, CD83, and members of the TNF receptor superfamily such as OX40L and 4-1 BBL; and production of pro-inflammatory cytokines such as TNF-a and IL-12

[2, 8]. Activated DCs also express the CCR7 chemokine receptor responsible for their homing to the lymph nodes (in response to the chemokine MIP-3β) where they encounter and activate naïve T cells. The combinations of GM-CSF and IL-15 or Flt3-ligand (Flt3-L) have also been used to generate DC ex vivo. Other approaches include the culture of unselected bone marrow cells [41, 73, 75–77] or bone marrow or blood-derived CD34[+] hematopoietic stem cells [78–80] in the presence of differentiation cytokines (GM-CSF, Flt3-L, TNF-α with or without IL-4) [81]. An additional strategy consists in the isolation of native blood immature DC after in vivo infusion of donors with granulocyte colony-stimulating factor (G-CSF) and Flt3-L to expand the DC pool in vivo [82], but the low number of isolated DC represents a major drawback of this approach. Difference between these various populations of DC as it relates to their ability to prime optimal antitumor immune responses remains however elusive, and further investigation is needed to develop standardized functional grade DC generation protocols.

11.2.2 Tumor Antigen Selection and Antigen Loading Techniques

The activation and expansion of specific antitumor T lymphocyte clones requires the presentation of defined tumor antigens on MHC class I and class II molecules expressed by DC. The identification and selection of optimal tumor antigen(s) for loading of DC has been the subject of extensive discussions. The choice of the type of tumor antigens impacts the specificity and quality of the immune response that will be induced by the vaccine [83–91]. Tumor antigens should be as specific of the malignancy as possible to prevent generation of immune responses against normal tissues (autoimmunity) and to reduce the risk of tolerance induced by self-antigen presentation. The ideal tumor antigens should lead to the induction of a wide repertoire of tumor-specific CD4[+] helper T cells and CD8[+] CTL displaying high affinity for the antigenic structures. A broad range of antigen pulsing techniques have been developed and include DC incubation with defined tumor peptides, whole-tumor cell lysates, apoptotic or necrotic tumor cells, exosomes, tumor-derived heat shock proteins, or DC transfection with RNA purified from tumors or encoding specific tumor antigens [83–91]. α-Galactosylceramide-loaded DCs have also been used to trigger NKT antitumoral activity [92]. However, an optimal strategy that most efficiently stimulates DC antigen processing and presentation has not yet been widely accepted.

Frequently used in clinical trials, the loading of DC with peptides derived from defined tumor antigens requires to identify specific antigens expressed by tumors, which is a major limitation. In addition, the restriction to a relatively small number of defined MHC class I and II binding peptides results in the induction of a limited repertoire of tumor-specific T cell clones, which may foster the emergence of tumor escape variants inasmuch that tumor cells loose expression of the particular epitopes contained in the vaccine. Transfection of DC with RNA or DNA encoding of one or a few characterized specific tumor antigen peptides or proteins presents similar limitations. A variety of immunogenic peptides have however been identified and utilized for known tumor-associated antigens such as tyrosinase, MAGE, Melan-A/MART, MUC1 CEA, Her-2/Neu, and survivin, which are restricted to specific HLA types in humans [93–95]. Alternatively, the full-length antigenic protein can be used to load DC which may result in both $CD4^{+}$ and $CD8^{+}$ T cell responses [93, 96].

DC loading with autologous whole dead tumor cells has the theoretical advantage that essentially all of the antigenic components of the tumor could be represented (including peptides, proteins, lipids, carbohydrates), leading therefore to the stimulation of a wider repertoire of tumor-specific $CD8^{+}$ and $CD4^{+}$ T lymphocytes as well as NKT cells (induction of polyclonal immune responses). However, there is still debate about the most advantageous form of dead tumor cells (apoptotic vs. necrotic vs. whole lysates) to be used to load DC [47, 97]. It has been reported that the uptake of apoptotic bodies efficiently induces the maturation of DCs that become more adept at eliciting specific cytotoxic T cells (CTL) [98–100]. However, others have proposed that necrotic cells or whole-tumor lysates are a superior source of tumor antigens for DC loading [84, 87, 101, 102]. Some concerns have also been raised about the lack of immunogenicity of lysates that may contain immunosuppressive factors [103, 104]. A main limitation inherent to the use of autologous tumor cells as a source of antigens to pulse DC is their relative difficulty of production. Indeed, the small size of many tumor specimens obtained by biopsy makes it difficult to obtain enough material for therapy, especially when multiple immunizations are required. This issue may however be partially overcome by the utilization of amplified tumor-derived total RNA [105, 106]. Nonetheless, an efficient antitumor immunity as well as clinical responses to DC vaccination has been reported in various clinical trials using whole-tumor cell preparations [107, 108].

The demonstration that cell lysate or necrotic tumor cell immunogenicity is associated with members of the chaperone protein family (heat shock proteins, HSP) has led to the development of tumor-derived HSP-based vaccines [109–117]. The specific immunogenic properties of these molecules are related to the repertoire of tumor-derived antigenic peptides that are carried by the chaperones [110]. They also constitute natural adjuvants per se capable of activating DC, enhancing their potential to process and present antigens and to stimulate T cell responses [111]. Instead of purifying chaperone proteins from tumor cell lysate, our group has developed a novel anticancer vaccine named chaperone-rich cell lysate (CRCL). CRCL is generated by a free solution-isoelectric focusing technique (FS-IEF), using tumor lysates, which results in an enrichment for chaperone proteins rather than a purification of them [118–124]. CRCL contains HSP90, HSP70 family members, the endoplasmic reticulum chaperone glucose-regulated protein (GRP) 94/glycoprotein (gp)96, and calreticulin. CRCL preserves its antigenic components, while excluding some presumed immunosuppressive factors present in unfractionated lysates [120]. CRCL combines the relative simplicity of lysate preparations, along with a high-yield and extensive antigen repertoire of chaperone proteins. CRCL provides tumor antigens to DC and by virtue of its adjuvant effects triggers DC activation (expression of CD40, CD80/86, CD70 fundamental for T cell activation) [121–124]. Importantly,

DCs loaded with tumor-derived CRCL resist regulatory T cell and TGF-β-mediated suppression [45], and tumor-derived CRCL-loaded DCs trigger a strong protective immune response and improve survival of tumor-bearing animals [104, 121]. Thus, the enhanced immunogenicity arising from CRCL-pulsed DC as a vaccine indicates that CRCL may represent an antigen source of choice for DC-based personalized anticancer immunotherapies [104].

Additional procedures have been developed for DC loading, including DC-tumor cell fusion [86, 126] and genetic approaches such as vaccinia viruses and lentivirus [127, 128]. This multiplicity of the aforementioned DC-loading techniques warrants further comparative large-scale studies to establish standardized procedures.

11.2.3 DC Activation and Maturation

The efficient activation of antitumor-specific T lymphocytes by loaded DC requires the presentation of tumor-derived epitopes on MHC class I and II molecules in the context of a second signal displayed by DC co-stimulatory molecules (including CD80, CD86, or CD40) and of pro-inflammatory cytokines (IL-12, TNF-α) secreted by activated DC. The generation of fully activated DC is obtained by the addition of maturation signals in the culture. Several DC-activation agents have been used including cytokines (such as interferons, TNF-α, GM-CSF, PGE2, or IL-1β), ligands of the TNF receptor family such as CD40-L, or adjuvants such as TLR ligands (LPS, CpG, poly-I:C) [1, 41, 73]. However, no consensus has been reached as to whether DC maturation should be induced in vitro or in vivo following their administration. Additionally, if immature DC are considered to be less potent in inducing T cell activation compared to mature DC, there is concern that activated DC may become exhausted during the culture stage. Alternative strategies to produce activated DC include the genetic modification of these cells to make them express co-stimulatory molecules or to secrete pro-inflammatory cytokines [129].

11.2.4 Route, Dose, and Timing of DC-Based Vaccine Administration

The route of DC vaccine delivery critically influences vaccine efficacy since DC must migrate to the secondary lymphoid organs where tumor antigen presentation to T cells takes place. In many clinical trials and in animal tumor models, DC-based vaccines have been administered intravenously, intradermally, subcutaneously, intratumorally, or into the lymphatic system or lymph nodes [73]. Intradermal and subcutaneous delivery may substantially limit the vaccine efficiency since only 5–10 % of the cells reach the draining lymph nodes. Intranodal DC administration has been associated with better responses and enhanced Th-1 helper lymphocyte function compared to intradermal or intravenous injection [130]. Intratumoral delivery has also been considered, but this approach remains limited by the accessibility of the tumor site [131, 132]. DC migratory ability may be fostered by pre-administration of TLR ligands or pro-inflammatory cytokines increasing CCL21, the ligand of CCR7 expressed by activated DC and that direct their homing to the draining lymph nodes [73, 133]. Depending on the choice of the delivery route, T lymphocytes with different homing properties and function may be induced, suggesting therefore that combining different administration sites may confer therapeutic benefits.

Different doses and frequency of DC vaccine inoculations have been investigated. In clinical trials, doses ranging from 10^6 to 10^8 cells are usually used with no significant improvement with higher doses. The optimal timing between injections is still debatable, but repeated injections usually separated by a 2-week interval has been a general scheme in many immunotherapy protocols [134].

11.2.5 In Situ Manipulation of DC

Because the ex vivo generation of DC is complex and expensive, the possibility of loading and activating DC directly in situ has been considered. Numerous techniques have thus been explored to provide to DC in vivo both tumor antigens and activation signals [135]. One approach consists in the inoculation of tumor antigens coupled with antibodies that recognize DC-specific markers such as DEC205 or cancer antigens conjugated to molecules that specifically bind to DC receptors. The simultaneous delivery of pro-inflammatory signals such as CD40-activating antibodies or TLR ligands can promote the full maturation of DC [5, 41, 73, 135].

11.3 Harnessing the Non-conventional Cytotoxic Function of Dendritic Cells in Cancer Immunotherapy

In cancer immunotherapeutic strategies, the interest in DC-based vaccines has centered on the antigen-presenting and immunostimulatory function of these cells. However, although the direct elimination of tumor cells has primarily been attributed to highly specialized killer cells such as CTL, NKT, NK, or macrophages, many studies conducted in rodents and humans have highlighted the possibility that several DC subsets can exert direct cytotoxic activity against cancer cells in vitro and in vivo [34, 35, 136–138]. This less conventional aspect of DC biology has however received limited attention, and controversy has arisen as it relates to the mode of induction and the mechanism(s) underlying their killing activity [136, 139–141]. How the cytotoxic function of DC may influence their antigen-presenting function and ability to activate effector lymphocytes is still the subject to investigation

[34, 125, 136, 138, 139, 142]. Different independent studies have reported that the killing activity of diverse DC subpopulations (native DC differentiating in vivo or DC generated in vitro from dedicated precursors) may be innate or triggered by distinct signals including Toll-like receptor (TLR) agonists such as LPS or CpG, CD40L, or IFN-γ [142]. A variety of cytotoxic mechanisms responsible for DC-mediated tumor cell killing have been described which include the perforin/granzyme system, death receptor ligands (FasL, TRAIL, and other TNF family members), ROS, and/or NO [139]. In human, peripheral blood [143–147], umbilical cord blood $CD34^+$ [148], or monocyte-derived [148–152] DCs exhibiting spontaneous or induced tumoricidal abilities have been described. The cytotoxic mechanisms employed by these killer DC (KDC) are various and may involve the death receptor ligand tumor necrosis factor-related apoptosis-inducing ligand (TRAIL), the perforin/granzyme, the Fas/FasL, or the CD40/CD40L systems [139].

Different aspects of KDC properties should be considered with regard to their potential advantage for clinical use. (1) These cells may directly participate in the cytotoxic mechanisms leading to tumor cell elimination. Tumors can adapt to resist killing by NK or CTL through conventional pathways; therefore, the variety of the mechanisms used by KDC to kill their targets may provide a significant advantage as it diversifies cytotoxic effector responses. (2) Importantly, from an immunologic perspective, by allowing for the rapid uptake of released tumor antigens, before their clearance by scavenger neutrophils or macrophages, KDC-mediated tumor cell killing is of considerable relevance for the acquisition of tumor-derived material in a more efficient manner. (3) Following killing and capture of cancer cell debris, KDC are capable of switching their function from killers to messengers capable of processing and presenting or cross-presenting acquired tumor antigens to $CD4^+$ or $CD8^+$ T lymphocytes. The unification of these properties therefore makes KDC highly desirable for the induction of specific antitumoral immunity. (4) Finally, the fundamental observation that KDC cytotoxic activity is mainly directed toward tumor cells implies their specific recognition through cell surface receptors (such as NKG2D or other unidentified molecules) and importantly provides these cells with the ability to spare nonmalignant cells. One may therefore logically expect relatively limited side effects associated with the exploitation of the killing potential of KDC in clinic.

Different applications for KDC in human cancer immunotherapy can be envisioned. One approach may consist in the administration of KDC generated in vitro that are allowed to kill, capture, and process tumor cells in culture. A second approach may entail the systemic or intra-/peri-tumoral injection of KDC generated in vitro. Encouraging results of this approach have been reported [153]. Additional therapies may be designed to promote the tumoricidal activity of DC in vivo and/or the recruitment of these KDC to the tumor site, as reported by Stary et al. using the TLR-7 ligand imiquimod [154]. In all these approaches, the choice of the type of DC activation signal(s) is critical since it may determine the nature of the killing mechanism. This is an important point to consider since tumor cells may develop resistance to specific death pathways. It may, therefore, be advantageous to promote simultaneously the tumoricidal activities of multiple KDC subsets, capable of inducing tumor cell killing by different mechanisms, to overcome the emergence of resistant tumor variants.

Therefore, the concept of KDC as a "multitasking" cell population that can act at virtually all levels of antitumor immune response opens new perspectives for the development of DC-based cancer vaccines. However, additional studies are required to determine whether KDC may promote a more immunogenic type of cancer cell death which may foster tumor antigen uptake, processing, and presentation. It will also be essential to clearly delineate the potential advantages of using KDC over conventional tumor antigen-loaded DC as cancer vaccines.

11.4 Current Challenges in Dendritic Cell-Based Cancer Immunotherapy and Combination Therapies

Even if proven clinically safe and efficient to prime and sustain immune responses, conventional DC-based immunotherapy has not yielded the enthusiasm initially expected because of the relatively limited objective clinical responses that have been observed in cancer patients [5, 73, 74, 155, 156]. This disappointing lack of clinical effects may be partly attributed to the end-stage nature of the patients included in DC vaccination trials. These cancer patients are usually heavily immunocompromised by the suppressive environment created in the course of tumor progression. As underlined in section 1, cancer cells can, by multiple mechanisms, alter the development of antitumor immunity and exploit several immune regulatory mechanisms to their advantage, leading to the impairment of DC function. Therefore, even if optimal DC-based cancer vaccines are generated, the suppressive tumor environment may significantly avert their efficiency in vivo. A number of molecules, such as TGF-β, IL-10, IL-13, VEGF, IDO, or PGE_2, produced by tumor or stromal cells may exert inhibitory effects on the immune system [7, 17, 157–162]. These tumor-derived factors inhibit DC differentiation and promote accumulation of immature DC (iDC), plasmacytoid DC (pDC), immunosuppressive regulatory DC, as well as myeloid-derived suppressor cells (MDSC) [17, 56, 163–167]. Accumulation of several populations of regulatory DC in the spleen and the lymph nodes of tumor-bearing hosts inhibits CTL responses [46, 168–172]. An additional major obstacle for successful cancer immunotherapy is the expansion of $CD4^+CD25^+FoxP3^+$ Treg induced by tumors [46, 172–176]. Tumor-induced Treg compromise the function of antitumor effector $CD8^+$ CTL, curtail $CD4^+$ T cell help, and impede the maturation, activation, and antigen-presenting capability of DC [45, 66, 172]. Studies in humans and in animal models have demonstrated that Treg elimination and/or Treg functional inactivation using different approaches (chemotherapeutic drugs or specific antibodies) significantly enhances antitumoral immunity [46, 174, 177, 178]. Therefore, associating DC-based therapy with Treg elimination or inactivation strategies, and more generally with approaches aimed at overcoming tumor-induced tolerance (inhibition of immunosuppressive molecules or cells such as TGF-β or myeloid-derived suppressor cells), may enhance the clinical efficiency of DC-based cancer vaccines. Several studies have indicated that the efficacy of antigen-loaded DC can be efficiently promoted by TGF-β antagonists and TGF-β receptor kinase inhibitors [179–181]. Similarly, immunotoxins such as the recombinant IL-2 diphtheria toxin conjugate (ONTAK) and LMB-2, which target Treg, have been shown to enhance the immunostimulatory effect of tumor antigen-pulsed DC, leading to the stimulation of helper and cytotoxic T cell responses [178, 182]. The benefit of Treg depletion upon treatment with immunotoxins has also been observed in patients with metastatic renal carcinoma and melanoma [178, 182]. Cyclophosphamide facilitates adoptive immunotherapy of established tumor through the elimination/inactivation of immunosuppressive Treg [45, 174, 183]. Our group has reported that in an established lymphoma model, the efficiency of DC pulsed with total tumor cell lysates is significantly enhanced by imatinib mesylate, a chemotherapeutic drug used to treat BCR-ABL$^+$ leukemia [46]. In addition, specific chemotherapeutic agents such as cisplatin or doxorubicin may promote DC vaccination efficacy by making tumor cells more susceptible to cytotoxic effects of CTL or by inducing an "immunogenic" type of cell death that fosters antigen uptake and activation of DC [31, 184, 185]. This synergistic effect is associated with the inhibition of Treg and the efficient activation of effector CD4$^+$ and CD8$^+$ T lymphocytes. Therefore, chemoimmunotherapy approaches may be of interest to enhance the efficacy of DC-based vaccination.

11.5 Conclusions and Perspectives

The potential of DC-based vaccines for cancer immunotherapy has been evaluated for decades. With the improved understanding of DC immunobiology, significant progresses have been made in the design, optimization, and translation to human of DC vaccines. Importantly DC-based therapy is usually associated with no or minimal side effects. However, although specific T lymphocyte immune responses induced by tumor antigen-loaded DC were detected in many cancer patients, the absence of objective tumor regression has eroded the initial enthusiasm for DC-based immunotherapy. One of the major challenges in the field lies on the establishment of immune tolerance by developing tumor, which explains the limited clinical benefits provided by DC-based vaccination, especially in patients with terminal-stage disease. The use of DC loaded with total dead tumor cells or with multiple antigens may partly overcome these immunosuppressive phenomena as they virtually contain all the antigens harbored by the tumor and can therefore trigger the activation of a much wider repertoire of tumor-specific T lymphocytes. The generated polyclonal immune responses may prevent the outgrowth of tumor escape variants. However, a limitation of this approach is the amount of tumor needed for vaccine preparation of multiple injections.

It is likely that the future of DC-based cancer immunotherapy may consist of combination strategies associating tumor antigen-loaded DC vaccination with chemotherapeutic agents that may not only directly target tumor cells but may also eliminate or avert the function of immunosuppressive cells or block the production of suppressive molecules. The judicious choice of chemotherapeutic molecules that would simultaneous promote the function of injected DC vaccine (chemoimmunodulatory agents) may also enhance the therapeutic efficacy of these cells. Cytokines and activation agents may also be administered to promote the survival, activation, and antigen-presenting function of DC in vivo. An optimal condition for the application of DC vaccines would be the instance of minimum residual disease. Because the elimination of primary and highly suppressive tumors is unlikely to be achieved with DC-based vaccination alone as a frontline therapy, initial surgery, radiation, or chemotherapy may reduce tumor burden to the point that DC vaccination can be performed more successfully with the objective of inducing specific and durable memory T lymphocyte responses that can prevent tumor recurrence and eliminate arising metastases. Further evaluation and standardization of optimal approaches are still needed to significantly improve the outcome of cancer immunotherapy.

References

1. Banchereau J, Palucka AK (2005) Dendritic cells as therapeutic vaccines against cancer. Nat Rev Immunol 5:296–306
2. Banchereau J, Steinman RM (1998) Dendritic cells and the control of immunity. Nature 392:245–252
3. Palucka AK, Ueno H, Fay J et al (2008) Dendritic cells: a critical player in cancer therapy? J Immunother 31:793–805
4. Smyth MJ, Godfrey DI, Trapani JA (2001) A fresh look at tumor immunosurveillance and immunotherapy. Nat Immunol 2:293–299
5. Adema GJ (2009) Dendritic cells from bench to bedside and back. Immunol Lett 122: 128–130
6. Melief CJ (2008) Cancer immunotherapy by dendritic cells. Immunity 29:372–383
7. Ueno H, Klechevsky E, Morita R et al (2007) Dendritic cell subsets in health and disease. Immunol Rev 219:118–142
8. Banchereau J, Briere F, Caux C et al (2000) Immunobiology of dendritic cells. Annu Rev Immunol 18:767–811
9. Dubsky P, Ueno H, Piqueras B et al (2005) Human dendritic cell subsets for vaccination. J Clin Immunol 25:551–572
10. Pulendran B, Smith JL, Caspary G et al (1999) Distinct dendritic cell subsets differentially regulate the class of immune response in vivo. Proc Natl Acad Sci USA 96:1036–1041
11. Kimura A, Naka T, Kishimoto T (2007) IL-6-dependent and -independent pathways in the development of interleukin 17-producing T helper cells. Proc Natl Acad Sci USA 104:12099–12104
12. Jego G, Palucka AK, Blanck JP et al (2003) Plasmacytoid dendritic cells induce plasma cell differentiation through type I interferon and interleukin 6. Immunity 19:225–234
13. Batista FD, Harwood NE (2009) The who, how and where of antigen presentation to B cells. Nat Rev Immunol 9:15–27
14. Walzer T, Dalod M, Robbins SH et al (2005) Natural-killer cells and dendritic cells: "l'union fait la force". Blood 106:2252–2258
15. Fujii S, Shimizu K, Hemmi H et al (2007) Innate Valpha14(+) natural killer T cells mature dendritic cells, leading to strong adaptive immunity. Immunol Rev 220:183–198

16. Steinbrink K, Jonuleit H, Muller G et al (1999) Interleukin-10-treated human dendritic cells induce a melanoma-antigen-specific anergy in CD8(+) T cells resulting in a failure to lyse tumor cells. Blood 93:1634–1642
17. Gabrilovich D (2004) Mechanisms and functional significance of tumour-induced dendritic-cell defects. Nat Rev Immunol 4:941–952
18. Monti P, Leone BE, Zerbi A et al (2004) Tumor-derived MUC1 mucins interact with differentiating monocytes and induce IL-10highIL-12low regulatory dendritic cell. J Immunol 172: 7341–7349
19. Ohnmacht C, Pullner A, King SB et al (2009) Constitutive ablation of dendritic cells breaks self-tolerance of CD4 T cells and results in spontaneous fatal autoimmunity. J Exp Med 206:549–559
20. Jonuleit H, Schmitt E, Schuler G et al (2000) Induction of interleukin 10-producing, nonproliferating CD4(+) T cells with regulatory properties by repetitive stimulation with allogeneic immature human dendritic cells. J Exp Med 192:1213–1222
21. Steinman RM, Hawiger D, Liu K et al (2003) Dendritic cell function in vivo during the steady state: a role in peripheral tolerance. Ann N Y Acad Sci 987:15–25
22. Tuettenberg A, Huter E, Hubo M et al (2009) The role of ICOS in directing T cell responses: ICOS-dependent induction of T cell anergy by tolerogenic dendritic cells. J Immunol 182:3349–3356
23. Banerjee DK, Dhodapkar MV, Matayeva E et al (2006) Expansion of FOXP3high regulatory T cells by human dendritic cells (DCs) in vitro and after injection of cytokine-matured DCs in myeloma patients. Blood 108:2655–2661
24. Yamazaki S, Steinman RM (2009) Dendritic cells as controllers of antigen-specific Foxp3+ regulatory T cells. J Dermatol Sci 54:69–75
25. Belkaid Y, Oldenhove G (2008) Tuning microenvironments: induction of regulatory T cells by dendritic cells. Immunity 29:362–371
26. Roncarolo MG, Gregori S, Battaglia M et al (2006) Interleukin-10-secreting type 1 regulatory T cells in rodents and humans. Immunol Rev 212:28–50
27. Stutman O (1974) Tumor development after 3-methylcholanthrene in immunologically deficient athymic-nude mice. Science 183:534–536
28. Kaplan DH, Shankaran V, Dighe AS et al (1998) Demonstration of an interferon gamma-dependent tumor surveillance system in immunocompetent mice. Proc Natl Acad Sci USA 95:7556–7561
29. Smyth MJ, Thia KY, Street SE et al (2000) Perforin-mediated cytotoxicity is critical for surveillance of spontaneous lymphoma. J Exp Med 192:755–760
30. Smyth MJ, Thia KY, Street SE et al (2000) Differential tumor surveillance by natural killer (NK) and NKT cells. J Exp Med 191:661–668
31. Zitvogel L, Tesniere A, Kroemer G (2006) Cancer despite immunosurveillance: immunoselection and immunosubversion. Nat Rev Immunol 6:715–727
32. Pure E, Allison JP, Schreiber RD (2005) Breaking down the barriers to cancer immunotherapy. Nat Immunol 6:1207–1210
33. Dunn GP, Old LJ, Schreiber RD (2004) The immunobiology of cancer immunosurveillance and immunoediting. Immunity 21:137–148
34. Chan CW, Housseau F (2008) The 'kiss of death' by dendritic cells to cancer cells. Cell Death Differ 15:58–69
35. Ullrich E, Chaput N, Zitvogel L (2008) Killer dendritic cells and their potential role in immunotherapy. Horm Metab Res 40:75–81
36. Larmonier N, Billerey C, Rebe C et al (2002) An atypical caspase-independent death pathway for an immunogenic cancer cell line. Oncogene 21:6091–6100
37. Kepp O, Tesniere A, Zitvogel L et al (2009) The immunogenicity of tumor cell death. Curr Opin Oncol 21:71–76
38. Bonnotte B, Larmonier N, Favre N et al (2001) Identification of tumor-infiltrating macrophages as the killers of tumor cells after immunization in a rat model system. J Immunol 167: 5077–5083

39. Shimizu K, Kurosawa Y, Taniguchi M et al (2007) Cross-presentation of glycolipid from tumor cells loaded with alpha-galactosylceramide leads to potent and long-lived T cell mediated immunity via dendritic cells. J Exp Med 204:2641–2653
40. Gilboa E, Nair SK, Lyerly HK (1998) Immunotherapy of cancer with dendritic-cell-based vaccines. Cancer Immunol Immunother 46:82–87
41. Nestle FO, Farkas A, Conrad C (2005) Dendritic-cell-based therapeutic vaccination against cancer. Curr Opin Immunol 17:163–169
42. Palucka AK, Laupeze B, Aspord C et al (2005) Immunotherapy via dendritic cells. Adv Exp Med Biol 560:105–114
43. Mayordomo JI, Zorina T, Storkus WJ et al (1995) Bone marrow-derived dendritic cells pulsed with synthetic tumour peptides elicit protective and therapeutic antitumour immunity. Nat Med 1:1297–1302
44. Shimizu J, Suda T, Yoshioka T et al (1989) Induction of tumor-specific in vivo protective immunity by immunization with tumor antigen-pulsed antigen-presenting cells. J Immunol 142:1053–1059
45. Larmonier N, Cantrell J, Lacasse C et al (2008) Chaperone-rich tumor cell lysate-mediated activation of antigen-presenting cells resists regulatory T cell suppression. J Leukoc Biol 83:1049–1059
46. Larmonier N, Janikashvili N, LaCasse CJ et al (2008) Imatinib mesylate inhibits CD4+ CD25+ regulatory T cell activity and enhances active immunotherapy against BCR-ABL- tumors. J Immunol 181:6955–6963
47. Larmonier N, Merino D, Nicolas A et al (2006) Apoptotic, necrotic, or fused tumor cells: an equivalent source of antigen for dendritic cell loading. Apoptosis 11:1513–1524
48. Staveley-O'Carroll K, Sotomayor E, Montgomery J et al (1998) Induction of antigen-specific T cell energy: an early event in the course of tumor progression. Proc Natl Acad Sci USA 95:1178–1183
49. Feinberg MB, Silvestri G (2002) T(S) cells and immune tolerance induction: a regulatory renaissance? Nat Immunol 3:215–217
50. Pardoll D (2003) Does the immune system see tumors as foreign or self? Annu Rev Immunol 21:807–839
51. Khong HT, Restifo NP (2002) Natural selection of tumor variants in the generation of "tumor escape" phenotypes. Nat Immunol 3:999–1005
52. Restifo NP, Esquivel F, Kawakami Y et al (1993) Identification of human cancers deficient in antigen processing. J Exp Med 177:265–272
53. Hicklin DJ, Wang Z, Arienti F et al (1998) beta2-Microglobulin mutations, HLA class I antigen loss, and tumor progression in melanoma. J Clin Invest 101:2720–2729
54. Restifo NP, Marincola FM, Kawakami Y et al (1996) Loss of functional beta 2-microglobulin in metastatic melanomas from five patients receiving immunotherapy. J Natl Cancer Inst 88:100–108
55. Garrido F, Ruiz-Cabello F, Cabrera T et al (1997) Implications for immunosurveillance of altered HLA class I phenotypes in human tumours. Immunol Today 18:89–95
56. Rabinovich GA, Gabrilovich D, Sotomayor EM (2007) Immunosuppressive strategies that are mediated by tumor cells. Annu Rev Immunol 25:267–296
57. Lu B, Finn OJ (2008) T-cell death and cancer immune tolerance. Cell Death Differ 15:70–79
58. Vucic D, Fairbrother WJ (2007) The inhibitor of apoptosis proteins as therapeutic targets in cancer. Clin Cancer Res 13:5995–6000
59. Rampino N, Yamamoto H, Ionov Y et al (1997) Somatic frameshift mutations in the BAX gene in colon cancers of the microsatellite mutator phenotype. Science 275:967–969
60. Medema JP, de Jong J, Peltenburg LT et al (2001) Blockade of the granzyme B/perforin pathway through overexpression of the serine protease inhibitor PI-9/SPI-6 constitutes a mechanism for immune escape by tumors. Proc Natl Acad Sci USA 98:11515–11520
61. Vicari AP, Caux C, Trinchieri G (2002) Tumour escape from immune surveillance through dendritic cell inactivation. Semin Cancer Biol 12:33–42

62. Fricke I, Gabrilovich DI (2006) Dendritic cells and tumor microenvironment: a dangerous liaison. Immunol Invest 35:459–483
63. Kusmartsev S, Gabrilovich DI (2006) Role of immature myeloid cells in mechanisms of immune evasion in cancer. Cancer Immunol Immunother 55:237–245
64. Terme M, Chaput N, Combadiere B et al (2008) Regulatory T cells control dendritic cell/NK cell cross-talk in lymph nodes at the steady state by inhibiting CD4+ self-reactive T cells. J Immunol 180:4679–4686
65. Shevach EM (2009) Mechanisms of foxp3+ T regulatory cell-mediated suppression. Immunity 30:636–645
66. Antony PA, Piccirillo CA, Akpinarli A et al (2005) CD8+ T cell immunity against a tumor/self-antigen is augmented by CD4+ T helper cells and hindered by naturally occurring T regulatory cells. J Immunol 174:2591–2601
67. Piccirillo CA, Shevach EM (2001) Cutting edge: control of CD8+ T cell activation by CD4+ CD25+ immunoregulatory cells. J Immunol 167:1137–1140
68. Almand B, Clark JI, Nikitina E et al (2001) Increased production of immature myeloid cells in cancer patients: a mechanism of immunosuppression in cancer. J Immunol 166:678–689
69. Nagaraj S, Gabrilovich DI (2008) Tumor escape mechanism governed by myeloid-derived suppressor cells. Cancer Res 68:2561–2563
70. Zou W (2006) Regulatory T cells, tumour immunity and immunotherapy. Nat Rev Immunol 6:295–307
71. Nagaraj S, Ziske C, Strehl J et al (2006) Dendritic cells pulsed with alpha-galactosylceramide induce anti-tumor immunity against pancreatic cancer in vivo. Int Immunol 18:1279–1283
72. Paczesny S, Ueno H, Fay J et al (2003) Dendritic cells as vectors for immunotherapy of cancer. Semin Cancer Biol 13:439–447
73. Figdor CG, de Vries IJ, Lesterhuis WJ et al (2004) Dendritic cell immunotherapy: mapping the way. Nat Med 10:475–480
74. Steinman RM, Banchereau J (2007) Taking dendritic cells into medicine. Nature 449: 419–426
75. Inaba K, Inaba M, Romani N et al (1992) Generation of large numbers of dendritic cells from mouse bone marrow cultures supplemented with granulocyte/macrophage colony-stimulating factor. J Exp Med 176:1693–1702
76. Schreurs MW, Eggert AA, de Boer AJ et al (1999) Generation and functional characterization of mouse monocyte-derived dendritic cells. Eur J Immunol 29:2835–2841
77. Romani N, Gruner S, Brang D et al (1994) Proliferating dendritic cell progenitors in human blood. J Exp Med 180:83–93
78. Caux C, Vanbervliet B, Massacrier C et al (1996) CD34+ hematopoietic progenitors from human cord blood differentiate along two independent dendritic cell pathways in response to GM-CSF+ TNF alpha. J Exp Med 184:695–706
79. Fay JW, Palucka AK, Paczesny S et al (2006) Long-term outcomes in patients with metastatic melanoma vaccinated with melanoma peptide-pulsed CD34(+) progenitor-derived dendritic cells. Cancer Immunol Immunother 55:1209–1218
80. Paczesny S, Banchereau J, Wittkowski KM et al (2004) Expansion of melanoma-specific cytolytic CD8+ T cell precursors in patients with metastatic melanoma vaccinated with CD34+ progenitor-derived dendritic cells. J Exp Med 199:1503–1511
81. Banchereau J, Ueno H, Dhodapkar M et al (2005) Immune and clinical outcomes in patients with stage IV melanoma vaccinated with peptide-pulsed dendritic cells derived from CD34+ progenitors and activated with type I interferon. J Immunother 28:505–516
82. Pulendran B, Banchereau J, Burkeholder S et al (2000) Flt3-ligand and granulocyte colony-stimulating factor mobilize distinct human dendritic cell subsets in vivo. J Immunol 165: 566–572
83. Ashley DM, Faiola B, Nair S et al (1997) Bone marrow-generated dendritic cells pulsed with tumor extracts or tumor RNA induce antitumor immunity against central nervous system tumors. J Exp Med 186:1177–1182

84. Fields RC, Shimizu K, Mule JJ (1998) Murine dendritic cells pulsed with whole tumor lysates mediate potent antitumor immune responses in vitro and in vivo. Proc Natl Acad Sci USA 95:9482–9487
85. Geiger C, Regn S, Weinzierl A et al (2005) A generic RNA-pulsed dendritic cell vaccine strategy for renal cell carcinoma. J Transl Med 3:29
86. Phan V, Errington F, Cheong SC et al (2003) A new genetic method to generate and isolate small, short-lived but highly potent dendritic cell-tumor cell hybrid vaccines. Nat Med 9: 1215–1219
87. Sauter B, Albert ML, Francisco L et al (2000) Consequences of cell death: exposure to necrotic tumor cells, but not primary tissue cells or apoptotic cells, induces the maturation of immunostimulatory dendritic cells. J Exp Med 191:423–434
88. Wolfers J, Lozier A, Raposo G et al (2001) Tumor-derived exosomes are a source of shared tumor rejection antigens for CTL cross-priming. Nat Med 7:297–303
89. Ueda G, Tamura Y, Hirai I et al (2004) Tumor-derived heat shock protein 70-pulsed dendritic cells elicit tumor-specific cytotoxic T lymphocytes (CTLs) and tumor immunity. Cancer Sci 95:248–253
90. Wang XH, Qin Y, Hu MH et al (2005) Dendritic cells pulsed with gp96-peptide complexes derived from human hepatocellular carcinoma (HCC) induce specific cytotoxic T lymphocytes. Cancer Immunol Immunother 54:971–980
91. Andre F, Schartz NE, Movassagh M et al (2002) Malignant effusions and immunogenic tumour-derived exosomes. Lancet 360:295–305
92. Fujii S, Shimizu K, Hemmi H et al (2006) Glycolipid alpha-C-galactosylceramide is a distinct inducer of dendritic cell function during innate and adaptive immune responses of mice. Proc Natl Acad Sci USA 103:11252–11257
93. Nencioni A, Grunebach F, Schmidt SM et al (2008) The use of dendritic cells in cancer immunotherapy. Crit Rev Oncol Hematol 65:191–199
94. Grunebach F, Erndt S, Hantschel M et al (2008) Generation of antigen-specific CTL responses using RGS1 mRNA transfected dendritic cells. Cancer Immunol Immunother 57:1483–1491
95. Waldhauer I, Goehlsdorf D, Gieseke F et al (2008) Tumor-associated MICA is shed by ADAM proteases. Cancer Res 68:6368–6376
96. Tuyaerts S, Aerts JL, Corthals J et al (2007) Current approaches in dendritic cell generation and future implications for cancer immunotherapy. Cancer Immunol Immunother 56:1513–1537
97. Melero I, Vile RG, Colombo MP (2000) Feeding dendritic cells with tumor antigens: self-service buffet or a la carte? Gene Ther 7:1167–1170
98. Albert ML, Sauter B, Bhardwaj N (1998) Dendritic cells acquire antigen from apoptotic cells and induce class I-restricted CTLs. Nature 392:86–89
99. Schnurr M, Scholz C, Rothenfusser S et al (2002) Apoptotic pancreatic tumor cells are superior to cell lysates in promoting cross-priming of cytotoxic T cells and activate NK and gammadelta T cells. Cancer Res 62:2347–2352
100. Jarnjak-Jankovic S, Pettersen RD, Saeboe-Larssen S et al (2005) Preclinical evaluation of autologous dendritic cells transfected with mRNA or loaded with apoptotic cells for immunotherapy of high-risk neuroblastoma. Cancer Gene Ther 12:699–707
101. Gallucci S, Lolkema M, Matzinger P (1999) Natural adjuvants: endogenous activators of dendritic cells. Nat Med 5:1249–1255
102. Basu S, Binder RJ, Suto R et al (2000) Necrotic but not apoptotic cell death releases heat shock proteins, which deliver a partial maturation signal to dendritic cells and activate the NF-kappa B pathway. Int Immunol 12:1539–1546
103. Kalos M (2003) Tumor antigen-specific T cells and cancer immunotherapy: current issues and future prospects. Vaccine 21:781–786
104. Zeng Y, Graner MW, Katsanis E (2006) Chaperone-rich cell lysates, immune activation and tumor vaccination. Cancer Immunol Immunother 55:329–338
105. Nencioni A, Muller MR, Grunebach F et al (2003) Dendritic cells transfected with tumor RNA for the induction of antitumor CTL in colorectal cancer. Cancer Gene Ther 10: 209–214

106. Muller MR, Grunebach F, Nencioni A et al (2003) Transfection of dendritic cells with RNA induces CD4- and CD8-mediated T cell immunity against breast carcinomas and reveals the immunodominance of presented T cell epitopes. J Immunol 170:5892–5896
107. O'Rourke MG, Johnson M, Lanagan C et al (2003) Durable complete clinical responses in a phase I/II trial using an autologous melanoma cell/dendritic cell vaccine. Cancer Immunol Immunother 52:387–395
108. Holtl L, Zelle-Rieser C, Gander H et al (2002) Immunotherapy of metastatic renal cell carcinoma with tumor lysate-pulsed autologous dendritic cells. Clin Cancer Res 8:3369–3376
109. Blachere NE, Srivastava PK (1995) Heat shock protein-based cancer vaccines and related thoughts on immunogenicity of human tumors. Semin Cancer Biol 6:349–355
110. Suto R, Srivastava PK (1995) A mechanism for the specific immunogenicity of heat shock protein-chaperoned peptides. Science 269:1585–1588
111. Asea A, Kraeft SK, Kurt-Jones EA et al (2000) HSP70 stimulates cytokine production through a CD14-dependant pathway, demonstrating its dual role as a chaperone and cytokine. Nat Med 6:435–442
112. Udono H, Srivastava PK (1993) Heat shock protein 70-associated peptides elicit specific cancer immunity. J Exp Med 178:1391–1396
113. Nair S, Wearsch PA, Mitchell DA et al (1999) Calreticulin displays in vivo peptide-binding activity and can elicit CTL responses against bound peptides. J Immunol 162:6426–6432
114. Arnold D, Faath S, Rammensee H et al (1995) Cross-priming of minor histocompatibility antigen-specific cytotoxic T cells upon immunization with the heat shock protein gp96. J Exp Med 182:885–889
115. Srivastava PK, Udono H (1994) Heat shock protein-peptide complexes in cancer immunotherapy. Curr Opin Immunol 6:728–732
116. Srivastava PK, Menoret A, Basu S et al (1998) Heat shock proteins come of age: primitive functions acquire new roles in an adaptive world. Immunity 8:657–665
117. Ishii T, Udono H, Yamano T et al (1999) Isolation of MHC class I-restricted tumor antigen peptide and its precursors associated with heat shock proteins hsp70, hsp90, and gp96. J Immunol 162:1303–1309
118. Zeng Y, Chen X, Larmonier N et al (2006) Natural killer cells play a key role in the antitumor immunity generated by chaperone-rich cell lysate vaccination. Int J Cancer 119:2624–2631
119. Graner M, Raymond A, Akporiaye E et al (2000) Tumor-derived multiple chaperone enrichment by free-solution isoelectric focusing yields potent antitumor vaccines. Cancer Immunol Immunother 49:476–484
120. Graner MW, Likhacheva A, Davis J et al (2004) Cargo from tumor-expressed albumin inhibits T-cell activation and responses. Cancer Res 64:8085–8092
121. Zeng Y, Feng H, Graner MW et al (2003) Tumor-derived, chaperone-rich cell lysate activates dendritic cells and elicits potent antitumor immunity. Blood 101:4485–4491
122. Feng H, Zeng Y, Graner MW et al (2003) Exogenous stress proteins enhance the immunogenicity of apoptotic tumor cells and stimulate antitumor immunity. Blood 101:245–252
123. Zeng Y, Graner MW, Thompson S et al (2005) Induction of BCR-ABL-specific immunity following vaccination with chaperone-rich cell lysates derived from BCR-ABL+tumor cells. Blood 105:2016–2022
124. Li G, Zeng Y, Chen X et al (2007) Human ovarian tumour-derived chaperone-rich cell lysate (CRCL) elicits T cell responses in vitro. Clin Exp Immunol 148:136–145
125. Chauvin C, Philippeau JM, Hemont C et al (2008) Killer dendritic cells link innate and adaptive immunity against established osteosarcoma in rats. Cancer Res 68:9433–9440
126. Yasuda T, Kamigaki T, Kawasaki K et al (2007) Superior anti-tumor protection and therapeutic efficacy of vaccination with allogeneic and semiallogeneic dendritic cell/tumor cell fusion hybrids for murine colon adenocarcinoma. Cancer Immunol Immunother 56:1025–1036
127. Lopes L, Fletcher K, Ikeda Y et al (2006) Lentiviral vector expression of tumour antigens in dendritic cells as an immunotherapeutic strategy. Cancer Immunol Immunother 55:1011–1016
128. Wang B, He J, Liu C et al (2006) An effective cancer vaccine modality: lentiviral modification of dendritic cells expressing multiple cancer-specific antigens. Vaccine 24:3477–3489

129. Hodge JW, Rad AN, Grosenbach DW et al (2000) Enhanced activation of T cells by dendritic cells engineered to hyperexpress a triad of costimulatory molecules. J Natl Cancer Inst 92:1228–1239
130. Nestle FO, Alijagic S, Gilliet M et al (1998) Vaccination of melanoma patients with peptide- or tumor lysate-pulsed dendritic cells. Nat Med 4:328–332
131. Ehtesham M, Kabos P, Gutierrez MA et al (2003) Intratumoral dendritic cell vaccination elicits potent tumoricidal immunity against malignant glioma in rats. J Immunother 26:107–116
132. Guo J, Zhu J, Sheng X et al (2007) Intratumoral injection of dendritic cells in combination with local hyperthermia induces systemic antitumor effect in patients with advanced melanoma. Int J Cancer 120:2418–2425
133. MartIn-Fontecha A, Sebastiani S, Hopken UE et al (2003) Regulation of dendritic cell migration to the draining lymph node: impact on T lymphocyte traffic and priming. J Exp Med 198: 615–621
134. Palucka K, Banchereau J, Mellman I (2010) Designing vaccines based on biology of human dendritic cell subsets. Immunity 33:464–478
135. den Brok MH, Nierkens S, Figdor CG et al (2005) Dendritic cells: tools and targets for antitumor vaccination. Expert Rev Vaccines 4:699–710
136. Bonmort M, Dalod M, Mignot G et al (2008) Killer dendritic cells: IKDC and the others. Curr Opin Immunol 20:558–565
137. Wesa AK, Storkus WJ (2008) Killer dendritic cells: mechanisms of action and therapeutic implications for cancer. Cell Death Differ 15:51–57
138. Chauvin C, Josien R (2008) Dendritic cells as killers: mechanistic aspects and potential roles. J Immunol 181:11–16
139. Larmonier N, Fraszczak J, Lakomy D et al (2010) Killer dendritic cells and their potential for cancer immunotherapy. Cancer Immunol Immunother: CII 59:1–11
140. Chan CW, Crafton E, Fan HN et al (2006) Interferon-producing killer dendritic cells provide a link between innate and adaptive immunity. Nat Med 12:207–213
141. Blasius AL, Barchet W, Cella M et al (2007) Development and function of murine B220+CD11c+NK1.1+ cells identify them as a subset of NK cells. J Exp Med 204: 2561–2568
142. Fraszczak J, Trad M, Janikashvili N et al (2010) Peroxynitrite-dependent killing of cancer cells and presentation of released tumor antigens by activated dendritic cells. J Immunol 184: 1876–1884
143. Fanger NA, Maliszewski CR, Schooley K et al (1999) Human dendritic cells mediate cellular apoptosis via tumor necrosis factor-related apoptosis-inducing ligand (TRAIL). J Exp Med 190:1155–1164
144. Janjic BM, Lu G, Pimenov A et al (2002) Innate direct anticancer effector function of human immature dendritic cells. I. Involvement of an apoptosis-inducing pathway. J Immunol 168: 1823–1830
145. Lu G, Janjic BM, Janjic J et al (2002) Innate direct anticancer effector function of human immature dendritic cells. II. Role of TNF, lymphotoxin-alpha(1)beta(2), Fas ligand, and TNF-related apoptosis-inducing ligand. J Immunol 168:1831–1839
146. Manna PP, Mohanakumar T (2002) Human dendritic cell mediated cytotoxicity against breast carcinoma cells in vitro. J Leukoc Biol 72:312–320
147. Schmitz M, Zhao S, Deuse Y et al (2005) Tumoricidal potential of native blood dendritic cells: direct tumor cell killing and activation of NK cell-mediated cytotoxicity. J Immunol 174:4127–4134
148. Shi J, Ikeda K, Fujii N et al (2005) Activated human umbilical cord blood dendritic cells kill tumor cells without damaging normal hematological progenitor cells. Cancer Sci 96:127–133
149. Chapoval AI, Tamada K, Chen L (2000) In vitro growth inhibition of a broad spectrum of tumor cell lines by activated human dendritic cells. Blood 95:2346–2351
150. Liu S, Yu Y, Zhang M et al (2001) The involvement of TNF-alpha-related apoptosis-inducing ligand in the enhanced cytotoxicity of IFN-beta-stimulated human dendritic cells to tumor cells. J Immunol 166:5407–5415

151. Vanderheyde N, Aksoy E, Amraoui Z et al (2001) Tumoricidal activity of monocyte-derived dendritic cells: evidence for a caspase-8-dependent, Fas-associated death domain-independent mechanism. J Immunol 167:3565–3569
152. Yang R, Xu D, Zhang A et al (2001) Immature dendritic cells kill ovarian carcinoma cells by a FAS/FASL pathway, enabling them to sensitize tumor-specific CTLs. Int J Cancer 94:407–413
153. Triozzi PL, Khurram R, Aldrich WA et al (2000) Intratumoral injection of dendritic cells derived in vitro in patients with metastatic cancer. Cancer 89:2646–2654
154. Stary G, Bangert C, Tauber M et al (2007) Tumoricidal activity of TLR7/8-activated inflammatory dendritic cells. J Exp Med 204:1441–1451
155. De Vries IJ, Krooshoop DJ, Scharenborg NM et al (2003) Effective migration of antigen-pulsed dendritic cells to lymph nodes in melanoma patients is determined by their maturation state. Cancer Res 63:12–17
156. Rosenberg SA, Yang JC, Restifo NP (2004) Cancer immunotherapy: moving beyond current vaccines. Nat Med 10:909–915
157. Cheng F, Wang HW, Cuenca A et al (2003) A critical role for Stat3 signaling in immune tolerance. Immunity 19:425–436
158. Bromberg JF, Wrzeszczynska MH, Devgan G et al (1999) Stat3 as an oncogene. Cell 98:295–303
159. Wang T, Niu G, Kortylewski M et al (2004) Regulation of the innate and adaptive immune responses by Stat-3 signaling in tumor cells. Nat Med 10:48–54
160. Kortylewski M, Kujawski M, Wang T et al (2005) Inhibiting Stat3 signaling in the hematopoietic system elicits multicomponent antitumor immunity. Nat Med 11:1314–1321
161. Burdelya L, Kujawski M, Niu G et al (2005) Stat3 activity in melanoma cells affects migration of immune effector cells and nitric oxide-mediated antitumor effects. J Immunol 174:3925–3931
162. Evel-Kabler K, Song XT, Aldrich M et al (2006) SOCS1 restricts dendritic cells' ability to break self tolerance and induce antitumor immunity by regulating IL-12 production and signaling. J Clin Invest 116:90–100
163. Melief CJ (2003) Mini-review: regulation of cytotoxic T lymphocyte responses by dendritic cells: peaceful coexistence of cross-priming and direct priming? Eur J Immunol 33:2645–2654
164. Vermi W, Bonecchi R, Facchetti F et al (2003) Recruitment of immature plasmacytoid dendritic cells (plasmacytoid monocytes) and myeloid dendritic cells in primary cutaneous melanomas. J Pathol 200:255–268
165. Salio M, Cella M, Vermi W et al (2003) Plasmacytoid dendritic cells prime IFN-gamma-secreting melanoma-specific CD8 lymphocytes and are found in primary melanoma lesions. Eur J Immunol 33:1052–1062
166. Munn DH, Sharma MD, Hou D et al (2004) Expression of indoleamine 2,3-dioxygenase by plasmacytoid dendritic cells in tumor-draining lymph nodes. J Clin Invest 114:280–290
167. Zhang M, Tang H, Guo Z et al (2004) Splenic stroma drives mature dendritic cells to differentiate into regulatory dendritic cells. Nat Immunol 5:1124–1133
168. Gabrilovich DI, Velders MP, Sotomayor EM et al (2001) Mechanism of immune dysfunction in cancer mediated by immature Gr-1+ myeloid cells. J Immunol 166:5398–5406
169. Li Q, Pan PY, Gu P et al (2004) Role of immature myeloid Gr-1+ cells in the development of antitumor immunity. Cancer Res 64:1130–1139
170. Huang B, Pan PY, Li Q et al (2006) Gr-1+CD115+ immature myeloid suppressor cells mediate the development of tumor-induced T regulatory cells and T-cell anergy in tumor-bearing host. Cancer Res 66:1123–1131
171. Curiel TJ, Coukos G, Zou L et al (2004) Specific recruitment of regulatory T cells in ovarian carcinoma fosters immune privilege and predicts reduced survival. Nat Med 10:942–949
172. Larmonier N, Marron M, Zeng Y et al (2007) Tumor-derived CD4(+)CD25 (+) regulatory T cell suppression of dendritic cell function involves TGF-beta and IL-10. Cancer Immunol Immunother 56:48–59

173. Sakaguchi S (2000) Regulatory T cells: key controllers of immunologic self-tolerance. Cell 101:455–458
174. Ghiringhelli F, Larmonier N, Schmitt E et al (2004) CD4+CD25+ regulatory T cells suppress tumor immunity but are sensitive to cyclophosphamide which allows immunotherapy of established tumors to be curative. Eur J Immunol 34:336–344
175. Liyanage UK, Moore TT, Joo HG et al (2002) Prevalence of regulatory T cells is increased in peripheral blood and tumor microenvironment of patients with pancreas or breast adenocarcinoma. J Immunol 169:2756–2761
176. Viguier M, Lemaitre F, Verola O et al (2004) Foxp3 expressing CD4+CD25(high) regulatory T cells are overrepresented in human metastatic melanoma lymph nodes and inhibit the function of infiltrating T cells. J Immunol 173:1444–1453
177. Prasad SJ, Farrand KJ, Matthews SA et al (2005) Dendritic cells loaded with stressed tumor cells elicit long-lasting protective tumor immunity in mice depleted of CD4+CD25+ regulatory T cells. J Immunol 174:90–98
178. Dannull J, Su Z, Rizzieri D et al (2005) Enhancement of vaccine-mediated antitumor immunity in cancer patients after depletion of regulatory T cells. J Clin Invest 115:3623–3633
179. Yang YA, Dukhanina O, Tang B et al (2002) Lifetime exposure to a soluble TGF-beta antagonist protects mice against metastasis without adverse side effects. J Clin Invest 109: 1607–1615
180. Muraoka RS, Dumont N, Ritter CA et al (2002) Blockade of TGF-beta inhibits mammary tumor cell viability, migration, and metastases. J Clin Invest 109:1551–1559
181. Uhl M, Aulwurm S, Wischhusen J et al (2004) SD-208, a novel transforming growth factor beta receptor I kinase inhibitor, inhibits growth and invasiveness and enhances immunogenicity of murine and human glioma cells in vitro and in vivo. Cancer Res 64:7954–7961
182. Attia P, Powell DJ Jr, Maker AV et al (2006) Selective elimination of human regulatory T lymphocytes in vitro with the recombinant immunotoxin LMB-2. J Immunother 29:208–214
183. North RJ (1982) Cyclophosphamide-facilitated adoptive immunotherapy of an established tumor depends on elimination of tumor-induced suppressor T cells. J Exp Med 155: 1063–1074
184. Ramakrishnan R, Assudani D, Nagaraj S et al (2010) Chemotherapy enhances tumor cell susceptibility to CTL-mediated killing during cancer immunotherapy in mice. J Clin Invest 120:1111–1124
185. Taieb J, Chaput N, Schartz N et al (2006) Chemoimmunotherapy of tumors: cyclophosphamide synergizes with exosome based vaccines. J Immunol 176:2722–2729

Chapter 12
Mesenchymal Stem Cells: Prospects for Cancer Therapy

Long-Jun Dai, Xu-Yong Sun, Jie Luo, and Garth L. Warnock

Abstract Cancer remains one of the leading causes of mortality and morbidity throughout the world. To a significant extent, current conventional cancer therapies are symptomatic and passive in nature. The major obstacle for the development of effective cancer therapy is believed to be the lack of sufficient specificity. Since the discovery of tumor-oriented homing capacity of mesenchymal stem cells (MSCs), the application of specific anticancer gene-engineered MSCs has held great potential for cancer therapies. The MSC-based multiple-targeted anticancer strategy is based on MSCs' capacity of tumor-directed migration and incorporation and in situ expression of tumor-specific anticancer genes. Aimed at translating the benchwork to meaningful clinical applications, we will describe MSCs' tumor tropism and their use as therapeutic vehicles, the multiple-targeted anticancer potential of engineered MSCs and a personalized strategy for cancer therapy.

Keywords Mesenchymal stem cells • Gene therapy • Cancer therapy • Cytotherapy

L.-J. Dai, Ph.D. (✉) • G.L. Warnock
Department of Surgery, University of British Columbia,
400-828 West 10th Avenue, Vancouver, BC V5Z 1L8, Canada
e-mail: ljdai@mail.ubc.ca, longjundai@gmail.com

X.-Y. Sun
Department of Organ Transplantation,
Institute of Transplant Medicine,
303 Hospital of PLA,
Nanning, China

J. Luo
Department of Neurosurgery,
Taihe Hospital, Hubei University of Medicine,
Shiyan, China

M.K. Danquah, R.I. Mahato (eds.), *Emerging Trends in Cell and Gene Therapy*,
DOI 10.1007/978-1-62703-417-3_12,

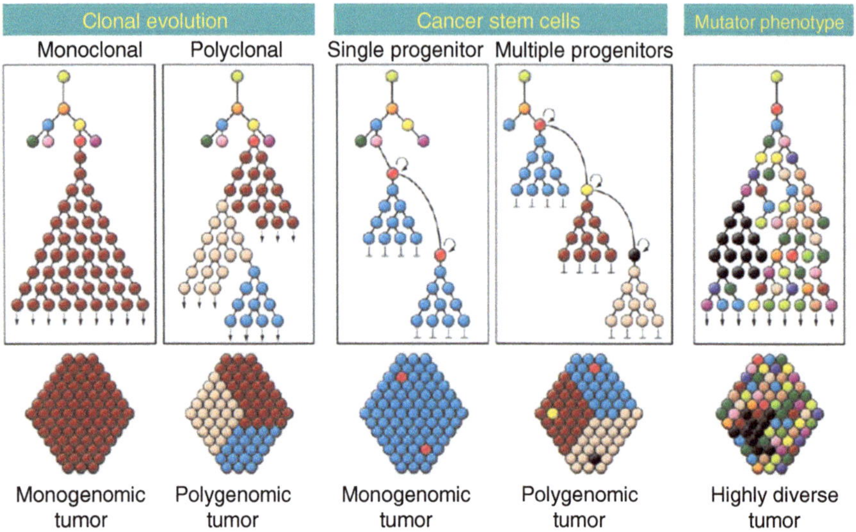

Fig. 12.1 Hypothetical models explaining intratumor heterogeneity. Different models of tumor progression can give rise to distinct types of intratumor heterogeneity. Shown here are the clonal evolution, cancer stem cells, and the mutator phenotype. The different models can result in distinct spatial distributions of subpopulation (Reproduced from Ref. [8] with permission)

12.1 Introduction

Cancer is one of the top life-threatening diseases accounting for an estimated one in four human deaths. About 7.6 million people died from cancer in 2008 worldwide, and this number is projected to rise to 13.2 million deaths in 2030 due to the aging and growth of the population [1]. According to recent cancer statistics, a total of 1,638,910 new cancer cases and 577,190 deaths from cancer are projected to occur in the United States in 2012 [2]. The dramatic systemic effects of tumors cause the majority of cancer deaths, rather than the effects of the direct overgrowth of the primary tumor or even the metastases. When fatalities occur, the majority of cancer patients die from vital organ failure, cachexia, or therapy-related life-threatening complications [3, 4]. Current conventional cancer therapies (surgery, chemotherapy, and radiotherapy) are, to a significant extent, symptomatic and passive in nature. Despite improved treatment models, many cancers remain unresponsive to traditional therapy. The futility of current therapies is primarily attributed to the lack of therapeutic specificity. Therefore, it is critical to explore efficient remedial strategies specifically targeting neoplasms. Since the discovery of the tumor-oriented homing capacity of mesenchymal stem cells (MSCs), the application of specific anticancer gene-engineered MSCs has held great potential for cancer therapies.

The profound heterogeneity of cancer is the essential cause for the futility and resistance of conventional anticancer therapies. In nearly 50 % of all cancer cases, resistance to chemotherapy already exists before drug treatment starts (intrinsic resistance), and in a large proportion of the remaining 50 %, resistance develops during treatment (acquired resistance) [5]. All efforts to overcome resistance to chemotherapy so far have failed, owing to the enormous heterogeneity and complex biology of cancer cells, with wide individual variations [6]. Tumor heterogeneity has been classified on varied levels, including molecular heterogeneity, genomic heterogeneity, intertumoral heterogeneity, and intratumoral heterogeneity [7]. Figure 12.1 represents the hypothetical models explaining the heterogeneity of breast cancer. There is a growing concern that the extent and nature of heterogeneity within cancers may simply defy rational explanation and explication [9]. Finding common vulnerabilities to target for the treatment of multiple cancer types would make for an extremely attractive and alternative way forward. Theoretically, it is impossible for any given drug to efficiently act on any given cancer. The ideal anticancer strategy should meet following requirements: (1) Anticancer actions are specifically confined to the tumor site; (2) the sensitivity of anticancer agents can be predetermined for personalizing cancer treatment; (3) diversified anticancer mechanisms are capable of acting concurrently; and (4) the anticancer agents can be adaptively replaced. The implementation of MSC-mediated anticancer synopsis meets most, if not all, of the above criteria, thereby holding great promise for the development of an efficient strategy to treat cancer. In this chapter, we present a general description of MSCs, the interactions of MSCs with cancer, and a proposed putative personalized strategy with anticancer gene-engineered MSCs.

12.2 Overview of MSCs

12.2.1 What Are Mesenchymal Stem Cells?

Mesenchymal stem cells (MSCs) are a group of adult stem cells naturally found in the body and are the first type of stem cells to be utilized in clinical regenerative medicine. MSCs were first identified in the stromal compartment of bone marrow by Friedenstein and colleagues in the 1960s [10, 11]. In recent years, MSCs have gained popularity among stem cell researchers due to their capability of self-renewal and differentiation into many different cell types, particularly cells of mesodermal origin such as osteoblasts, chondrocytes, and adipocytes in culture [12]. As a result of their supposed capacity for self-renewal and differentiation, these cells were first considered as stem cells by Caplan and named mesenchymal stem cells in 1991 [13]. Furthermore, MSCs have generated considerable biomedical interest since their multi-lineage potential was first identified by Pittenger et al. in 1999 [14].

In addition to the bone marrow, MSCs have been found to reside in marrow-distant mesenchymal tissues such as skeletal muscles, adipose tissue, pancreas, placenta, synovial fluid, dental tissue, parathyroid gland, fallopian tube, umbilical cord blood, and circulating blood. It has been assumed that basically all organs containing connective tissue also contain MSCs [15]. Despite the wide distribution of MSCs in the body, the bone marrow remains the principal source for most MSC-based preclinical and clinical studies where MSCs have mainly been characterized after isolation. Actually, MSCs are a rare population in bone marrow aspirates, representing at a frequency of 0.1–5/10^5 cells in rodents and 1–20/10^5 cells in humans [16]. The total number of MSCs is inversely proportional with the age of individuals. When grouped by decade, a dramatic decrease in MSCs per nucleated marrow cell could be observed, with a tenfold decrease from birth to teens and another tenfold decrease from teens to the elderly [17].

MSCs demonstrate heterogeneity in their morphology and have various appearances such as fibroblast-like, spindle shaped, and very small circular-shaped [18]. The morphology of these cells also varies greatly with their seeding density, changing dramatically especially when confluence is reached in cell culture condition [19]. The relation between the morphology and their cell functions remains unclear. MSCs express a number of markers phenotypically. However, none of them are specific to these cells. According to the International Society for Cellular Therapy (ISCT), human MSCs under standard culture conditions must satisfy at least three criteria: (1) They must be plastic adherent; (2) they must express CD105, CD73, and CD90 and not CD45, CD34, CD14, CD11b, CD79, or CD19 and HLA-DR surface molecules by flow cytometry; (3) they must be capable of differentiating into osteoblasts, adipocytes, and chondroblasts [20]. Other markers that are generally accepted include CD44, CD71, Stro-1, and adhesion molecules such as CD106, CD166, and CD29 [21]. In ongoing and future studies, it is important that investigators continue to gather new information regarding modification of, or additions to, these characteristics. This information will be instrumental in comparing data from different laboratories and in the clinical translation of MSCs for cellular therapy. Thorough evaluation of those criteria does not rule out the heterogeneity of MSCs from different sources; however, detailed descriptions of cell markers and behavior allow us to decide which cell source to use for a specific therapy in a certain individual.

12.2.2 Why Choose MSCs for Cancer Therapy?

The most prominent reason to use MSCs for cancer therapy is attributed to their capacity for tumor-directed migration and incorporation. As described in the following sections, the tropism of MSCs for tumor microenvironment is independent of tumor type, immunocompetence, and the route of MSC delivery. MSCs are immune-privileged cells. The immune phenotype of MSCs is generally described as major histocompatibility complex I (MHC I) positive and MHC II negative. They also lack the co-stimulatory molecules CD80, CD86, and CD40. Although expressing low levels of MHC I antigens can activate T cells, the absence of co-stimulatory molecules cannot initiate secondary signals, thus leaving the T cells anergic. Therefore, MSCs possess a reduced risk of allogeneic transplant rejection [22]. This has significant clinical implications, whereby engineered MSCs could be used in patients as a cell therapy without the considerations and complications surrounding immunomodulation associated with their use. This property could theoretically allow for the development of an MSC bank where allogeneic cells could be stored and used for patients. Other benefits of choosing MSCs for cancer therapy include easy acquisition, fast *ex vivo* expansion, and the feasibility of autologous transplantation.

12.3 The Interactions of MSC and Cancer

12.3.1 Tumor Tropism Properties of MSCs

The specific tumor-oriented migration and incorporation of MSCs have been demonstrated in various preclinical models, exhibiting the potential for MSCs to be used as ideal carriers for anticancer agents [23]. The first evidence of this tropism of MSCs to tumors was demonstrated when rat MSCs were implanted into rats bearing syngeneic gliomas [24]. Since then, an increasing number of studies have verified MSC tropism toward primary and metastatic tumor locations. Tumor-directed migration and incorporation of MSCs were evidenced in a number of preclinical studies in vitro using transwell migration assays and in vivo using animal tumor models. The homing capacity of MSCs has been demonstrated with almost all tested human cancer cell lines including lung cancer [25], malignant gliomas [26–28], Kaposi's sarcomas [29], breast cancer [30, 31], colon carcinoma [32], melanoma [33], ovarian cancer [30], and pancreatic cancer [34, 35]. The high frequency of MSC migration and incorporation was observed in in vitro co-culture and in vivo xenograft tumors, respectively. These consistent findings are independent of tumor type, immunocompetence, and the route of MSC delivery. It has recently been shown that MSCs are able to migrate specifically to and incorporate within tumors, and this property can be used to deliver targeted anticancer therapies.

Although the precise molecular mechanism by which MSCs are able to migrate and home into tumor sites are not yet fully understood, the complex multistep process by which leukocytes migrate to peripheral sites of inflammation has been proposed as a paradigm [36]. Tumors can be characterized as "wounds that never heal," serving as a continuous source of cytokines, chemokines, and other inflammatory mediators [37]. These signals are capable of recruiting respondent cell types including MSCs. The preconditions for this phenomenon are the production of chemoattractant molecules from tumor tissue and the expression of corresponding receptors in MSCs. The possible pathways and prospective models were summarized in recent reviews [36, 38].

12.3.2 Dually Characterized Roles of Integrated MSCs with Regard to Tumorigenesis

While research has established that MSCs migrate and integrate toward tumor tissues, their fate and function inside the tumor appears ambiguous and sometimes paradoxical. Native MSCs have been shown to suppress tumor growth in models of glioma [24], Kaposi's sarcoma [29], malignant melanoma [39], Lewis lung carcinoma [39], and colon carcinoma [40]. The release of soluble factors by MSCs has also been shown to reduce tumor growth and progression of glioma [24], melanoma, and lung carcinoma models [39]. Conditioned media from MSCs exhibit the ability to downregulate NFκB in hepatoma and breast cancer cells resulting in a decrease in their in vitro proliferation [41]. While the precise mechanism underlying intrinsic antitumor properties of MSCs has not been fully investigated, it is presumably related to the downregulation of Akt, NFκB, and Wnt signaling pathways [23]. On the other hand, several studies have demonstrated that MSCs can augment tumor growth [42–44]. MSC-mediated promotion of tumor growth is possibly initiated by MSC-derived immunosuppressive factors and by the contribution of MSCs to tumor stroma and tumor vascularization. It is not an overstatement to describe MSCs as a "double-edged sword" in their interaction with tumors. If MSCs are properly engineered with anticancer genes, they could be employed as a "single-edged sword" against cancers. This viewpoint was partially revealed by Luetzkendorf et al.'s recent study [45]. In mixed subcutaneous xenografts, lentiviral TRAIL-transgenic MSCs inhibited colorectal carcinoma, but wild-type MSCs exerted a colorectal carcinoma growth-supporting effect under the same experimental circumstance. The detailed pro- and anti-tumorigenic effects of MSCs were described in our recent review [38].

12.3.3 MSCs as Therapeutic Vehicles for Cancer Therapy

MSCs have been considered ideal vehicles to deliver anticancer agents since the discovery of their tumor-directed homing capacity. In addition to tumor-homing properties, MSCs are also easily transduced with integrating vectors due to having high levels of amphotropic receptors [46] which allow long-term gene expression without affecting their phenotypes [47, 48]. To date, a number of anticancer genes have been engineered into MSCs and have successfully caused anticancer effects in various carcinoma models. MSCs can also be utilized to deliver prodrug-converting enzymes, such as HSV-tk and cytosine deaminase, which can convert systemically administrated and non-active forms of prodrugs into active forms in the tumor microenvironment [4]. This therapeutic regimen has been successfully investigated in experimental models of glioma [49], melanoma [50], colon carcinoma [51], prostate cancer [52], and pancreatic cancer [35]. The methods of MSC administration has been classified as directional, semi-directional, and systemic deliveries [53]. The selection of delivery route of MSCs is based on considering all potential factors, such as the type, location, and stage of cancer, and the feasibility of surgical interventions.

12.4 Multiple-Targeted Anticancer Effects of Engineered MSCs

12.4.1 Engineered MSCs Targeting Tumor Cells Through Multiple Mechanisms

The major obstacle limiting the effectiveness of conventional therapies for cancer treatment is their tumor specificity. Advanced drug targeting of tumor cells is often impossible when treating highly invasive and infiltrative tumors, because of tumor cells' high migration and invasiveness. Uncontrolled drug distribution in the body, i.e., insufficient concentration at the tumor site and toxic concentration on normal cells, is attributed to anticancer inefficacy and is often the direct cause of side effects and sometimes life-threatening complications. Targeting solid tumors with antitumor gene therapy has also been hindered by systemic toxicity, low efficiency of delivery, and nominal temporal expression. However, MSC-mediated anticancer scenario can overcome these limitations, mainly through preferentially homing to sites of primary and metastatic tumors and delivering antitumor agents. Anticancer gene-engineered MSCs are capable of specifically targeting and acting on tumors through multiple selections.

12.4.1.1 The First Selection Rests with MSCs' Tumor-Directed Migration and Incorporation

In addition to the intrinsic anticancer effects of MSCs, the presence of MSCs in the tumor microenvironment allows the agents that are delivered by MSCs to exert their anticancer function locally and efficiently. Therefore, the systemic and organ-specific side effects of anticancer agents can be greatly minimized by using this cell-based vector system.

12.4.1.2 The Second Selection Lies in the Anticancer Genes Carried by MSCs

Research using MSCs as a vehicle for agents to treat cancers has been greatly motivated by advances in the study of specific anticancer genes. The products of specific anticancer genes can selectively induce apoptosis in cancer cells without affecting healthy cells. A number of anticancer genes have been engineered into MSCs and successfully caused anticancer effects in various carcinoma models [4, 23]. In the tumor microenvironment, engineered MSCs could serve as a constant source of anticancer agent production and locally release anticancer agents acting on adjacent tumor cells, thereby inducing tumor growth inhibition or apoptosis.

12.4.1.3 Additional Selections Can Be Made by Modifying the Vector Construction

Taking the advantage of unique protein expression in individual organs, the organ-specific expression of MSC-carried anticancer genes can be achieved through the modification of vector construction. For example, pancreas- or insulinoma-specific anticancer gene-bearing vectors can be made by employing an insulin promoter. For the same purpose, the unique expression of albumin by hepatocytes, neurotransmitter expression by neurons, and surfactant from pulmonary alveoli can also be used to construct organ-specific expression vectors. MSCs engineered with organ-specific vectors express anticancer proteins only when they home to the tumor located in the specific corresponding organ or to the metastatic sites with the same cell type.

12.4.2 Synergism of TRAIL and PTEN with Regard to Cancer Cell Apoptosis

MSC-based cancer therapy is capable of providing multiple anticancer agents simultaneously, which may potentiate therapeutic efficiency through synergistic effect on the induction of cancer cell apoptosis. There are two major signaling pathways that lead to apoptosis in mammalian cells: the intrinsic pathway and the extrinsic pathway. The extrinsic death pathway (also known as type I apoptosis) is mitochondrially independent. It is initiated through apoptotic signal transduction cascades mediated by the members of TNF receptor superfamily, such as TNF-related apoptosis-inducing ligand (TRAIL). By contrast, the intrinsic pathway (also known as type II apoptosis) is mitochondrially dependent and controlled by pro- and anti-apoptotic Bcl2 family proteins in the mitochondria. The intrinsic pathway has a substantial role in chemotherapy- and radiation-induced cell death. As discussed in next section, it is also the mechanism underlying PTEN (phosphatase and tensin homolog)-mediated cell death. MSC-mediated therapeutic spectrum can be dramatically broadened by using multiple anticancer gene-engineered MSCs, and theoretically, a synergistic effect can be achieved through the application of multiple anticancer agents simultaneously.

12.4.2.1 TRAIL Induces Tumor Cell Death Through the Extrinsic Pathway of Apoptosis

TNF-related apoptosis-inducing ligand (TRAIL) is one of few anticancer proteins which selectively causes apoptosis of tumor cells through the activation of death receptors, with no effects on healthy cells [54]. It is known that there are five TRAIL receptors, i.e., TRAIL receptor 1 (death receptor 4, DR4), TRAIL receptor 2 (death receptor 5, DR5), TRAIL receptor 3 (decoy receptor 1, DcR1), TRAIL receptor 4 (decoy receptor 2, DcR2), and a soluble receptor, osteoprotegerin (OPG) [55]. There is a death domain in the intracellular region of DR4 or DR5, which can recruit death-inducing signaling complex (DISC) upon TRAIL stimulation, and therefore, activate a downstream caspase cascade leading to cell apoptosis. There is no intact death domain in the intracellular region of DcR1, DcR2, and OPG, so they are unable to induce apoptosis, even though they can compete with DR4 or DR5 for binding with TRAIL [56] and overexpression of DcR1 and/or DcR2 blocks TRAIL-mediated apoptosis in some cell types [57]. In our recent liver cancer studies, the high expression of DR5 was verified on human hepatoma cells (HepG2) [58]. As shown in Fig. 12.2a, *TRAIL*-engineered MSCs-induced HepG2 cell death is proportionally related with the MSC content on direct co-cultures, even though native

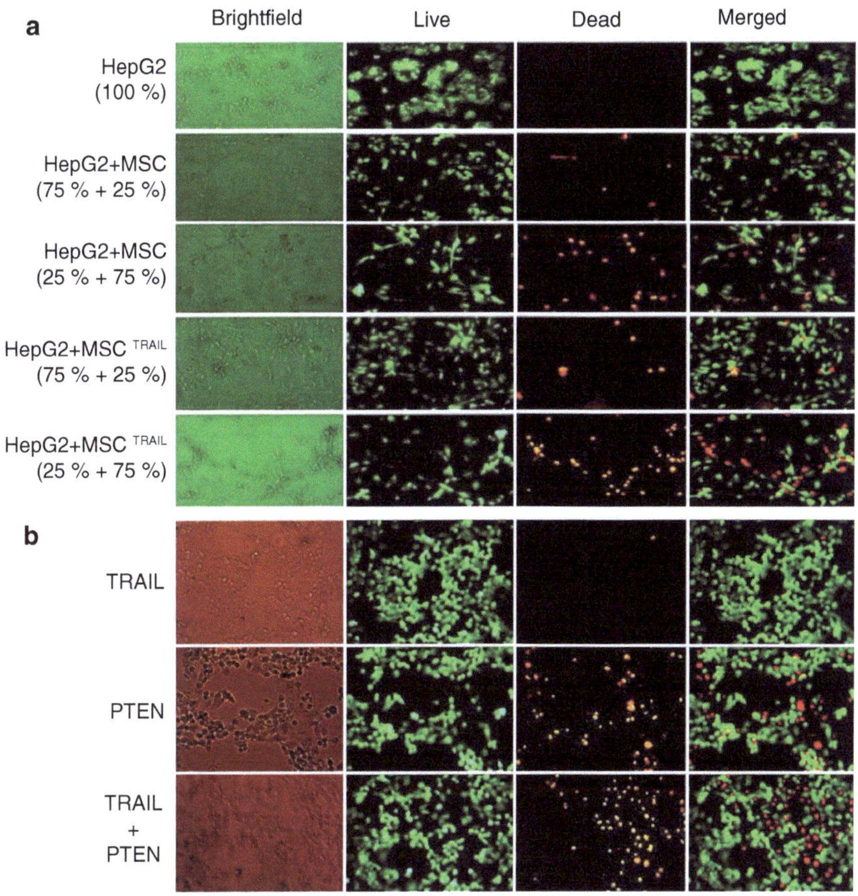

Fig. 12.2 (**a**) Cell viability of DR5⁺HepG2 cells directly co-cultured with MSCs. Live cells stained with calcein and dead cells stained with EthD-1 are shown in *green* and *red*, respectively. The whole population of cells and dead cell distribution are presented in the bright-field and merged images. MSCs exhibit intrinsic inhibition on HepG2 which is potentiated by TRAIL transfection (Modified from [58]). (**b**) TRAIL and PTEN-induced Panc-1 cell death. Panc-1 cells were pre-detected for death receptors and showed $DR4^-$ and $DR5^-$. The *top* two rows represent the cells transfected with TRAIL or PTEN individually, and *bottom* row shows the cells with co-transfection of the combination. The most serious cell death was observed in PTEN and TRAIL/PTEN transfected Panc-1 cells (Reproduced from [4])

MSCs exert intrinsic inhibition to HepG2 cells. The tumor specificity of TRAIL-induced apoptosis is determined by the death receptor expression in tumor cells. However, the expression of death receptors is varied with the alteration of tumor heterogeneity, and the tumor cell can also be desensitized to TRAIL through death receptor internalization. The MSC-mediated therapeutic spectrum can be dramatically broadened by using multiple anticancer gene-engineered MSCs, and theoretically, a synergistic effect can be achieved via the simultaneous application of multiple anticancer agents. For example, a pancreatic cancer cell line (Panc-1) lacks in death receptors and therefore shows no response to TRAIL (Fig. 12.2b). However, remarkable cell death was induced by PTEN or the combination of TRAIL and PTEN in these particular cells (Fig. 12.2b) [4].

12.4.2.2 PTEN Antagonizes PI3K-AKT-mTOR Pathway in Tumorigenesis

PI3K-AKT-mTOR signaling pathway is the most frequently activated pathway in human cancers, because it promotes cell growth, survival, and proliferation. It contributes to the evasion of apoptosis, loss of cell-cycle control, and genomic instability during tumorigenesis through numerous mechanisms [59]. In response to ligand binding to the receptor tyrosine kinase (RTK) or G-protein coupled receptor, PI3K (phosphoinositide 3-kinase) is activated and converts phosphatidylinositol 4,5 phosphate (PIP_2) to phosphatidylinositol 3,4,5-trisphosphate (PIP_3), a critical second messenger in cellular signaling. PIP_3 transduces activating signals by binding to the pleckstrin homology (PH) domains of proteins, including phosphatidylinositide-dependent kinase 1 (PDK1) and the serine-threonine kinase AKT, thereby recruiting them to the membrane. AKT is a centrally important downstream effector of PIP_3. After membrane enrichment, AKT is fully activated following phosphorylation by PDK1 together with mammalian target of rapamycin (mTOR). The activated AKT promotes cellular survival, cell-cycle progression, and growth through mTOR, thereby contributing to carcinogenesis [60]. PI3K-AKT-mTOR survival pathway is also known as an anti-apoptotic pathway. Apoptosis, induced by a number of stimuli, including growth factor withdrawal, UV irradiation, cell-cycle discordance, and activation of FAS signaling, is suppressed by the activation of PI3K-AKT-mTOR pathway. PTEN dephosphorylates PIP_3 to PIP_2, thereby directly opposes the activity of PI3K. Thus, PTEN functions as the central negative regulator of the PI3K-AKT-mTOR pathway in controlling apoptosis. PTEN activity is lost by mutations, deletions, or promoter methylation at very high frequency in many primary and metastatic human cancers [61]. Compared to other classical tumor suppressor genes, PTEN is haploinsufficient because a single copy is unable to prevent cancer. Loss of its heterozygosity or partial inhibition of its expression/activity is sufficient to promote carcinogenesis [62]. Thus, restoring PTEN function in cancer cells would break down the *PTEN* mutation-dependent cancer cell growth (oncogene addiction) and holds great promise for cancer therapy.

12.4.2.3 Synergistic Effect of TRAIL and PTEN

TRAIL and PTEN induce cancer cell apoptosis through extrinsic pathway and intrinsic pathway, respectively. The complementary nature or synergistic effect between these two anticancer genes can be explained in following three aspects. Firstly, in cancer cells, apoptosis induced by the extrinsic pathway complements that which is induced by the intrinsic pathway, so targeting death receptors is considered a useful therapeutic approach [63]. Secondly, as shown in Fig. 12.2, PTEN is able to induce apoptosis on tumor cells that are insensitive to TRAIL due to the lack of death receptors. In addition to its direct induction of apoptosis through the intrinsic pathway, PTEN has also been demonstrated to sensitize tumor cells to death receptor-mediated apoptosis induced by TRAIL [64, 65]. A number of possible mechanisms have been postulated for this synergism, including the upregulation of TRAIL receptors [66, 67], the clustering of TRAIL receptors into lipid rafts [68], the downregulation of apoptotic pathway inhibitors [69], or the enhanced cleavage of caspases [70]. Finally, PTEN plays a critical role in regulating the apoptotic threshold to multiple stimuli, including death ligands and chemotherapeutic agents [64].

12.5 Putative Personalized Strategy for Cancer Therapy

The ultimate goal of this area of research is to develop a cellular therapy for humans. MSC-mediated anticancer treatment has definite potential for translation to clinical medicine. MSCs can be acquired from patients' own body, quickly expanded in vitro and easily transfected with expression vectors. The exhibition of the powerful tumor-directed migration capacity of MSCs makes them suitable for use in anticancer therapies. Anticancer gene-engineered MSCs could be eventually used as an alternative treatment for cancer patients without the concern of rejection or other ethical problems. Since there exists a great deal of variation among the cancer patient population with respect to degree of carcinogenic differentiation and preparation of human MSCs, it is unlikely to expect a singular fixed therapeutic model that would successfully perform on different types of cancers. In order to translate the benchwork to the real clinical application, it is necessary to develop a specifically personalized treatment for each individual patient. Figure 12.3 illustrated putative personalized strategy with anticancer gene-engineered MSCs.

12.6 Summary and Clinical Perspectives

There is a pressing clinical demand for new efficient remedies to replace existing symptomatic anticancer therapies. The extensive achievements of MSC and anticancer agent studies have laid the foundation for the exploitation of MSC-based cancer therapies. MSCs possess powerful capabilities of tumor-directed migration and incorporation, acting as optimal vehicles to deliver anticancer agents. Although MSCs have both positive and negative effects on tumor progression, profound anticancer effects have been demonstrated by using felicitously engineered MSCs. MSC-mediated anticancer therapy relies on tumor-specific selections provided by MSCs and MSC-carried

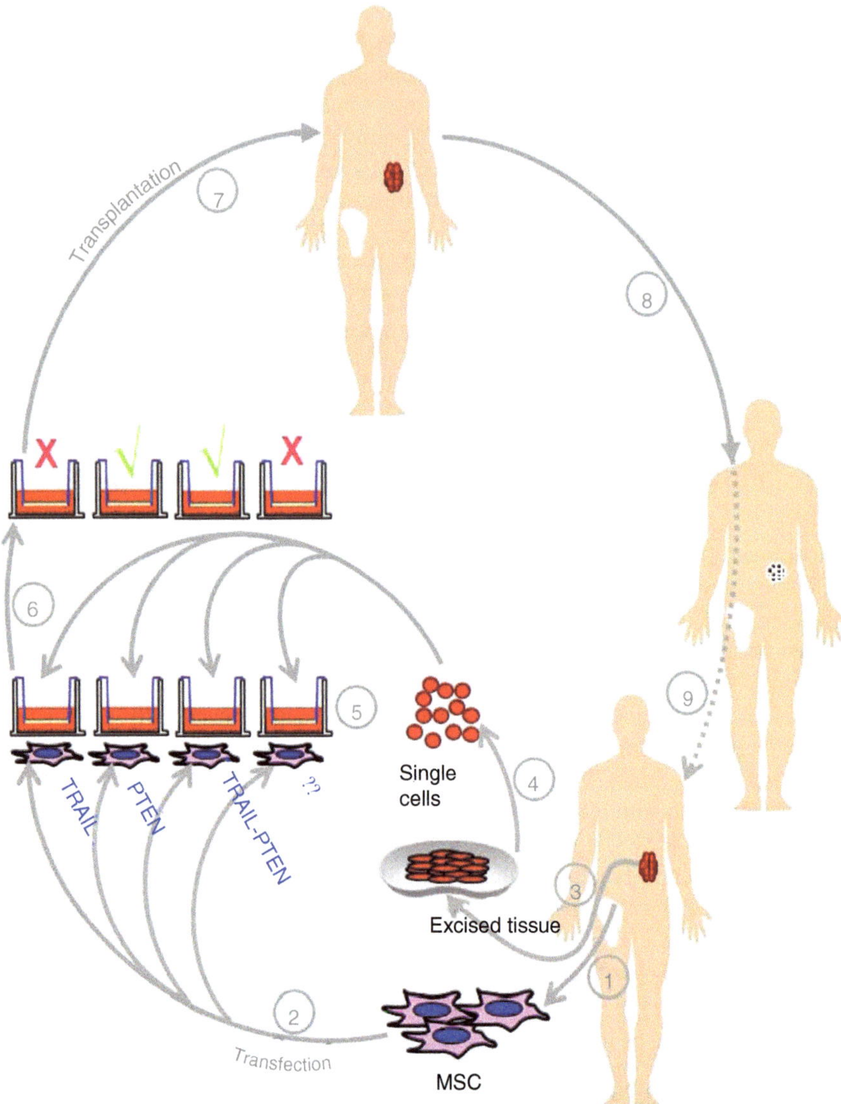

Fig. 12.3 Putative personalized cancer therapy with engineered MSCs. Operable patients provide direct access to tumor tissue, so that a customized therapeutic strategy can be arranged. (*1*) MSC isolation and *ex vivo* expansion. (*2*) Anticancer gene transfection on a test scale. (*3*) Excised tissue from cancer patient. (*4*) Primary tumor cell preparation. (*5*) Primary cell co-culture with anticancer gene-engineered MSCs. (*6*) Sensitivity determination by monitoring cell viability. (*7*) Selected gene transfection on a large scale followed by cell transplantation. Cell number, cell delivery route, and frequency are determined individually. (*8*) Achievement of successful treatment. (*9*) With regard to the patients not responding to the treatment or not receiving surgical intervention, MSCs engineered with multiple anticancer genes may be considered as a replacement

anticancer agents. Homed directly at the tumor microenvironment, engineered MSCs are able to express and/or release anticancer agents constantly acting on the adjacent tumor cells. However, almost all of the findings are confined to cell culture and/or animal cancer models, and more well-designed preclinical studies are definitely required before applying this strategy to real clinical settings. The following precautions may need to be taken when planning engineered MCS-related clinical studies: (1) Ensure that all standby MSCs are not contaminated with cancerous stem cells, in addition to the regular quality control; (2) upon reaching adequate number of required MSCs, restrict in vitro manipulation on MSCs as much as possible; (3) ensure the homing capacity of MSCs is not altered by gene transfection; (4) ensure the cytogenetic stability of MSCs remains after gene transfection; (5) ensure the delivery route, cell number, and frequency of MSC transplantation are determined by overall considerations on all multifaceted factors, such as type, location, and stage of cancer as well as the consideration of surgical intervention; and (6) allogeneic MSCs can be considered when suitable autologous MSCs are not approachable.

In conclusion, the intense progresses in both stem cell and anticancer gene studies have built up great potential for exploiting new efficient cancer therapies. The combination of human MSCs and specific anticancer genes can selectively act upon targeted tumor cells. Further translational studies could lead to novel and effective treatments for cancer. Hopefully, the mesenchymal stem cell-based and multiple-targeted anticancer strategy can benefit future cancer patients by at least providing options to patients in their terminal stages.

Acknowledgments This study was supported by VGH and UBC Hospital Foundation and Guangxi Ministry of Science and Technology. The authors are grateful to Crystal Robertson for her assistance in preparing the manuscript.

References

1. Brawley OW (2011) Avoidable cancer deaths globally. CA Cancer J Clin 61:67–68
2. Siegel R, Naishadham D, Jemal A (2012) Cancer statistics, 2012. CA Cancer J Clin 62: 10–29
3. Egeblad M, Nakasone ES, Werb Z (2010) Tumors as organs: complex tissues that interface with the entire organism. Dev Cell 18:884–901
4. Sun XY, Nong J, Qin K, Warnock GL, Dai LJ (2011) Mesenchymal stem cell-mediated cancer therapy: a dual-targeted strategy of personalized medicine. World J Stem Cells 3:96–103
5. Lippert TH, Ruoff HJ, Volm M (2011) Current status of methods to assess cancer drug resistance. Int J Med Sci 8:245–253
6. Lippert TH, Ruoff HJ, Volm M (2008) Intrinsic and acquired drug resistance in malignant tumors. The main reason for therapeutic failure. Arzneimittelforschung 58:261–264
7. Bertos NR, Park M (2011) Breast cancer – one term, many entities? J Clin Invest 121: 3789–3796
8. Russnes HG, Navin N, Hicks J, Borresen-Dale AL (2011) Insight into the heterogeneity of breast cancer through next-generation sequencing. J Clin Invest 121:3810–3818
9. Cao Y, DePinho RA, Ernst M, Vousden K (2011) Cancer research: past, present and future. Nat Rev Cancer 11:749–754

10. Friedenstein AJ, Piatetzky S II, Petrakova KV (1966) Osteogenesis in transplants of bone marrow cells. J Embryol Exp Morphol 16:381–390
11. Friedenstein AJ, Petrakova KV, Kurolesova AI, Frolova GP (1968) Heterotopic of bone marrow. Analysis of precursor cells for osteogenic and hematopoietic tissues. Transplantation 6:230–247
12. Bruder SP, Jaiswal N, Haynesworth SE (1997) Growth kinetics, self-renewal, and the osteogenic potential of purified human mesenchymal stem cells during extensive subcultivation and following cryopreservation. J Cell Biochem 64:278–294
13. Caplan AI (1991) Mesenchymal stem cells. J Orthop Res 9:641–650
14. Pittenger MF, Mackay AM, Beck SC, Jaiswal RK, Douglas R, Mosca JD et al (1999) Multilineage potential of adult human mesenchymal stem cells. Science 284:143–147
15. Vaananen HK (2005) Mesenchymal stem cells. Ann Med 37:469–479
16. Short B, Brouard N, Occhiodoro-Scott T, Ramakrishnan A, Simmons PJ (2003) Mesenchymal stem cells. Arch Med Res 34:565–571
17. Caplan AI (2009) Why are MSCs therapeutic? New data: new insight. J Pathol 217:318–324
18. Pevsner-Fischer M, Levin S, Zipori D (2011) The origins of mesenchymal stromal cell heterogeneity. Stem Cell Rev 7:560–568
19. Wong RS (2011) Mesenchymal stem cells: angels or demons? J Biomed Biotechnol 2011: 459510
20. Dominici M, Le Blanc K, Mueller I, Slaper-Cortenbach I, Marini F, Krause D et al (2006) Minimal criteria for defining multipotent mesenchymal stromal cells. The International Society for Cellular Therapy position statement. Cytotherapy 8:315–317
21. Chamberlain G, Fox J, Ashton B, Middleton J (2007) Concise review: mesenchymal stem cells: their phenotype, differentiation capacity, immunological features, and potential for homing. Stem Cells 25:2739–2749
22. Javazon EH, Beggs KJ, Flake AW (2004) Mesenchymal stem cells: paradoxes of passaging. Exp Hematol 32:414–425
23. Loebinger MR, Janes SM (2010) Stem cells as vectors for antitumour therapy. Thorax 65: 362–369
24. Nakamura K, Ito Y, Kawano Y, Kurozumi K, Kobune M, Tsuda H et al (2004) Antitumor effect of genetically engineered mesenchymal stem cells in a rat glioma model. Gene Ther 11: 1155–1164
25. Loebinger MR, Kyrtatos PG, Turmaine M, Price AN, Pankhurst Q, Lythgoe MF et al (2009) Magnetic resonance imaging of mesenchymal stem cells homing to pulmonary metastases using biocompatible magnetic nanoparticles. Cancer Res 69:8862–8867
26. Sasportas LS, Kasmieh R, Wakimoto H, Hingtgen S, van de Water JA, Mohapatra G et al (2009) Assessment of therapeutic efficacy and fate of engineered human mesenchymal stem cells for cancer therapy. Proc Natl Acad Sci USA 106:4822–4827
27. Sonabend AM, Ulasov IV, Tyler MA, Rivera AA, Mathis JM, Lesniak MS (2008) Mesenchymal stem cells effectively deliver an oncolytic adenovirus to intracranial glioma. Stem Cells 26: 831–841
28. Yang B, Wu X, Mao Y, Bao W, Gao L, Zhou P et al (2009) Dual-targeted antitumor effects against brainstem glioma by intravenous delivery of tumor necrosis factor-related, apoptosis-inducing, ligand-engineered human mesenchymal stem cells. Neurosurgery 65:610–624; discussion 24
29. Khakoo AY, Pati S, Anderson SA, Reid W, Elshal MF, Rovira II et al (2006) Human mesenchymal stem cells exert potent antitumorigenic effects in a model of Kaposi's sarcoma. J Exp Med 203:1235–1247
30. Kidd S, Spaeth E, Dembinski JL, Dietrich M, Watson K, Klopp A et al (2009) Direct evidence of mesenchymal stem cell tropism for tumor and wounding microenvironments using in vivo bioluminescent imaging. Stem Cells 27:2614–2623
31. Patel SA, Meyer JR, Greco SJ, Corcoran KE, Bryan M, Rameshwar P (2010) Mesenchymal stem cells protect breast cancer cells through regulatory T cells: role of mesenchymal stem cell-derived TGF-beta. J Immunol 184:5885–5894

32. Menon LG, Picinich S, Koneru R, Gao H, Lin SY, Koneru M et al (2007) Differential gene expression associated with migration of mesenchymal stem cells to conditioned medium from tumor cells or bone marrow cells. Stem Cells 25:520–528
33. Studeny M, Marini FC, Champlin RE, Zompetta C, Fidler IJ, Andreeff M (2002) Bone marrow-derived mesenchymal stem cells as vehicles for interferon-beta delivery into tumors. Cancer Res 62:3603–3608
34. Kidd S, Caldwell L, Dietrich M, Samudio I, Spaeth EL, Watson K et al (2010) Mesenchymal stromal cells alone or expressing interferon-beta suppress pancreatic tumors in vivo, an effect countered by anti-inflammatory treatment. Cytotherapy 12:615–625
35. Zischek C, Niess H, Ischenko I, Conrad C, Huss R, Jauch KW et al (2009) Targeting tumor stroma using engineered mesenchymal stem cells reduces the growth of pancreatic carcinoma. Ann Surg 250:747–753
36. Salem HK, Thiemermann C (2010) Mesenchymal stromal cells: current understanding and clinical status. Stem Cells 28:585–596
37. Dvorak HF (1986) Tumors: wounds that do not heal. Similarities between tumor stroma generation and wound healing. N Engl J Med 315:1650–1659
38. Dai LJ, Moniri MR, Zeng ZR, Zhou JX, Rayat J, Warnock GL (2011) Potential implications of mesenchymal stem cells in cancer therapy. Cancer Lett 305:8–20
39. Maestroni GJ, Hertens E, Galli P (1999) Factor(s) from nonmacrophage bone marrow stromal cells inhibit Lewis lung carcinoma and B16 melanoma growth in mice. Cell Mol Life Sci 55:663–667
40. Ohlsson LB, Varas L, Kjellman C, Edvardsen K, Lindvall M (2003) Mesenchymal progenitor cell-mediated inhibition of tumor growth in vivo and in vitro in gelatin matrix. Exp Mol Pathol 75:248–255
41. Qiao L, Zhao TJ, Wang FZ, Shan CL, Ye LH, Zhang XD (2008) NF-kappaB downregulation may be involved the depression of tumor cell proliferation mediated by human mesenchymal stem cells. Acta Pharmacol Sin 29:333–340
42. Karnoub AE, Dash AB, Vo AP, Sullivan A, Brooks MW, Bell GW et al (2007) Mesenchymal stem cells within tumour stroma promote breast cancer metastasis. Nature 449:557–563
43. Coffelt SB, Marini FC, Watson K, Zwezdaryk KJ, Dembinski JL, LaMarca HL et al (2009) The pro-inflammatory peptide LL-37 promotes ovarian tumor progression through recruitment of multipotent mesenchymal stromal cells. Proc Natl Acad Sci USA 106:3806–3811
44. Djouad F, Plence P, Bony C, Tropel P, Apparailly F, Sany J et al (2003) Immunosuppressive effect of mesenchymal stem cells favors tumor growth in allogeneic animals. Blood 102:3837–3844
45. Luetzkendorf J, Mueller LP, Mueller T, Caysa H, Nerger K, Schmoll HJ (2010) Growth inhibition of colorectal carcinoma by lentiviral TRAIL-transgenic human mesenchymal stem cells requires their substantial intratumoral presence. J Cell Mol Med 14:2292–2304
46. Ehtesham M, Kabos P, Kabosova A, Neuman T, Black KL, Yu JS (2002) The use of interleukin 12-secreting neural stem cells for the treatment of intracranial glioma. Cancer Res 62:5657–5663
47. Pisati F, Belicchi M, Acerbi F, Marchesi C, Giussani C, Gavina M et al (2007) Effect of human skin-derived stem cells on vessel architecture, tumor growth, and tumor invasion in brain tumor animal models. Cancer Res 67:3054–3063
48. Moore XL, Lu J, Sun L, Zhu CJ, Tan P, Wong MC (2004) Endothelial progenitor cells' "homing" specificity to brain tumors. Gene Ther 11:811–818
49. Uchibori R, Okada T, Ito T, Urabe M, Mizukami H, Kume A et al (2009) Retroviral vector-producing mesenchymal stem cells for targeted suicide cancer gene therapy. J Gene Med 11:373–381
50. Kucerova L, Matuskova M, Pastorakova A, Tyciakova S, Jakubikova J, Bohovic R et al (2008) Cytosine deaminase expressing human mesenchymal stem cells mediated tumour regression in melanoma bearing mice. J Gene Med 10:1071–1082
51. Kucerova L, Altanerova V, Matuskova M, Tyciakova S, Altaner C (2007) Adipose tissue-derived human mesenchymal stem cells mediated prodrug cancer gene therapy. Cancer Res 67:6304–6313

52. Cavarretta IT, Altanerova V, Matuskova M, Kucerova L, Culig Z, Altaner C (2010) Adipose tissue-derived mesenchymal stem cells expressing prodrug-converting enzyme inhibit human prostate tumor growth. Mol Ther 18:223–231
53. Zhang ZX, Guan LX, Zhang K, Zhang Q, Dai LJ (2008) A combined procedure to deliver autologous mesenchymal stromal cells to patients with traumatic brain injury. Cytotherapy 10:134–139
54. Wu GS (2009) TRAIL as a target in anti-cancer therapy. Cancer Lett 285:1–5
55. Zhu DM, Shi J, Liu S, Liu Y, Zheng D (2011) HIV infection enhances TRAIL-induced cell death in macrophage by down-regulating decoy receptor expression and generation of reactive oxygen species. PLoS One 6:e18291
56. Pan G, O'Rourke K, Chinnaiyan AM, Gentz R, Ebner R, Ni J et al (1997) The receptor for the cytotoxic ligand TRAIL. Science 276:111–113
57. Sheridan JP, Marsters SA, Pitti RM, Gurney A, Skubatch M, Baldwin D et al (1997) Control of TRAIL-induced apoptosis by a family of signaling and decoy receptors. Science 277:818–821
58. Sun XY, Nong J, Qin K, Lu H, Moniri MR, Dai LJ et al (2011) MSC(TRAIL)-mediated HepG2 cell death in direct and indirect co-cultures. Anticancer Res 31:3705–3712
59. Liu W, Zhou Y, Reske SN, Shen C (2008) PTEN mutation: many birds with one stone in tumorigenesis. Anticancer Res 28:3613–3619
60. Fine B, Hodakoski C, Koujak S, Su T, Saal LH, Maurer M et al (2009) Activation of the PI3K pathway in cancer through inhibition of PTEN by exchange factor P-REX2a. Science 325:1261–1265
61. Chalhoub N, Baker SJ (2009) PTEN and the PI3-kinase pathway in cancer. Annu Rev Pathol 4:127–150
62. Peyrou M, Bourgoin L, Foti M (2010) PTEN in liver diseases and cancer. World J Gastroenterol 16:4627–4633
63. Johnstone RW, Frew AJ, Smyth MJ (2008) The TRAIL apoptotic pathway in cancer onset, progression and therapy. Nat Rev Cancer 8:782–798
64. Yuan XJ, Whang YE (2002) PTEN sensitizes prostate cancer cells to death receptor-mediated and drug-induced apoptosis through a FADD-dependent pathway. Oncogene 21:319–327
65. Opel D, Westhoff MA, Bender A, Braun V, Debatin KM, Fulda S (2008) Phosphatidylinositol 3-kinase inhibition broadly sensitizes glioblastoma cells to death receptor- and drug-induced apoptosis. Cancer Res 68:6271–6280
66. Shankar S, Singh TR, Srivastava RK (2004) Ionizing radiation enhances the therapeutic potential of TRAIL in prostate cancer in vitro and in vivo: intracellular mechanisms. Prostate 61:35–49
67. Shankar S, Chen X, Srivastava RK (2005) Effects of sequential treatments with chemotherapeutic drugs followed by TRAIL on prostate cancer in vitro and in vivo. Prostate 62:165–186
68. Dumitru CA, Carpinteiro A, Trarbach T, Hengge UR, Gulbins E (2007) Doxorubicin enhances TRAIL-induced cell death via ceramide-enriched membrane platforms. Apoptosis 12: 1533–1541
69. El-Zawahry A, McKillop J, Voelkel-Johnson C (2005) Doxorubicin increases the effectiveness of Apo2L/TRAIL for tumor growth inhibition of prostate cancer xenografts. BMC Cancer 5:2
70. Belyanskaya LL, Marti TM, Hopkins-Donaldson S, Kurtz S, Felley-Bosco E, Stahel RA (2007) Human agonistic TRAIL receptor antibodies Mapatumumab and Lexatumumab induce apoptosis in malignant mesothelioma and act synergistically with cisplatin. Mol Cancer 6:66

Chapter 13
RNA Interference for Oncology: Clinical Prospects Beyond the Hype

Eishi Ashihara and Taira Maekawa

Abstract RNA interference (RNAi) is a process of sequence-specific posttranscriptional gene silencing induced by double-strand RNA, and this phenomenon has been shown to function in higher organisms including mammals, and methods that exploit RNAi mechanisms have been developing. Recently, RNAi induced by short interfering siRNAs has been experimentally introduced as a cancer therapy and is expected to be developed as a nucleic acid-based medicine. Moreover, RNAi technology is used in biomarker-based screening, which is a new screening method based on transcriptional profiling to identify the specific transcriptional activities altered by the compounds of interest. In this chapter, we briefly review the mechanism of RNAi and discuss in detail some of the most recent findings concerning the administration of potential nucleic acid-based drugs. We next discuss several current clinical trials of RNAi therapies against cancers. Finally, we introduce a new high-throughput screening method based on transcriptional profiling for drug discovery. Current studies and clinical trials demonstrate that RNAi technology could establish a novel and promising therapeutic tool against cancers.

Keywords RNA interference • siRNA • microRNA • Cancer • Cancer biomarker • Cancer therapy

---

E. Ashihara (✉)
Department of Clinical and Translational Physiology,
Kyoto Pharmaceutical University,
Misasagi-Nakauchicho 5, Kyoto 607-8414, Japan
e-mail: ash@mb.kyoto-phu.ac.jp

T. Maekawa
Department of Transfusion Medicine and Cell Therapy,
Kyoto University Hospital,
Kyoto, Japan

M.K. Danquah, R.I. Mahato (eds.), *Emerging Trends in Cell and Gene Therapy*,
DOI 10.1007/978-1-62703-417-3_13,

13.1 Introduction

RNA interference (RNAi) is a process of sequence-specific posttranscriptional gene silencing induced by double-strand RNA (dsRNA), and this phenomenon was discovered in *Caenorhabditis elegans* (*C. elegans*) [1]. RNAi has been shown to function in higher organisms including mammals, and methods that exploit RNAi mechanisms have been developing. Aberrant expression of endogenous normal or mutant genes occurs in pathological conditions, resulting in alterations in signal pathways, cellular proliferation, and apoptosis. Posttranscriptional gene regulation by RNAi controls these alterations positively or negatively, and consequently RNAi has now been well established as a method for experimental analyses of gene function in vitro. Recently, short interfering RNA (siRNA), which induces RNAi, has been experimentally introduced as a cancer therapy and is expected to be developed as a nucleic acid-based medicine, and several clinical trials of RNAi therapies against cancers are ongoing. To develop nuclear medicine against cancers, we have two important issues to overcome: one is to select suitable gene targets and another is to develop effective drug delivery systems (DDSs). DDSs are divided into two categories: viral vector-based carriers and nonviral-based carriers. Although viral vectors are the most powerful tools for transfection so far, especially retroviral and lentiviral vectors randomly integrate into host cells' DNA and those might induce insertional mutagenesis [2–4]. The use of nonviral DDSs including cationic liposomes [5, 6] and atelocollagen [7, 8] is preferred because it offers greater safety for clinical application than does the use of viral DDSs.

In addition to the development of a nucleic acid-based medicine, RNAi is put to practical use for a high-throughput screening for development of molecular targeting agents. The alternation of the related gene transcripts which are investigated after the knockdown of the targeted gene transcript by RNAi is compared with that of gene transcripts treated by compounds with unknown functions. The compounds which demonstrate the resemble alternation are recognized as molecular target compounds for the interested gene [9–11]. In this chapter, we discuss the application of RNAi for the development of medicine against cancers.

13.2 Mechanisms of RNAi

RNAi processes can be roughly divided into the initiation phase and the effector phase. In the initiation phase, following introduction of dsRNA into a target cell, dsRNA encounters a dsDNA-specific RNAse III family ribonuclease Dicer. Dicer is a modular enzyme and is composed of an N-terminal helicase domain, an RNA-binding Piwi/Argonaute/Zwille (PAZ) domain, two tandem RNAse III domains, and a dsRNA-binding domain [12]. Dicer acts to produce both siRNAs and microRNAs (miRNAs) [13–16]. dsRNA is processed into shorter lengths of 21–23 nucleotides (nts) dsRNAs, termed siRNAs by the ribonuclease activity of Dicer. dsRNA precursors are sequentially processed by the two RNAse III domains of Dicer and

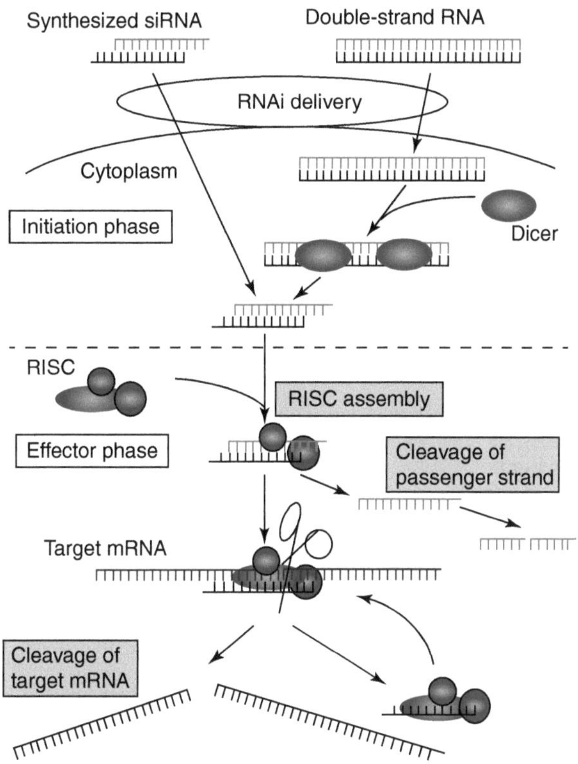

Fig. 13.1 Mechanisms of RNA interference. Synthesized short interference RNA (*siRNA*) or double-strand (ds) RNA is introduced into a target cell. The dsRNA is processed into siRNA length of 21–23 nucleotides by Dicer (initiation phase). siRNA then enters an RNA-induced silencing complex (*RISC*) assembly pathway. The dsRNA unwinds to form two single strands of RNA. The passenger strand rapidly degrades and the guide strand binds and cleaves the target mRNA, resulting in mRNA degradation (effector phase)

cleaved into smaller dsRNAs with 3′ dinucleotide overhangs [12]. In the biogenesis of miRNA, pre-miRNA is also processed into a miRNA duplex (Biogenesis of miRNA is discussed below).

In the second effector phase, smaller dsRNAs enter into an RNA-induced silencing complex (RISC) assembly pathway [17]. RISC is ribonucleoprotein complex that contains Argonaute (Ago) proteins, siRNAs or miRNAs, and complementary mRNAs. Ago is a family of proteins characterized by the presence of a PAZ domain and a PIWI domain [18]. The PAZ domain of Ago protein is likely to engage siRNA or miRNA, and the PIWI domain adopts an RNAse H-like structure that can catalyze the cleavage of the guide strand. The dsRNA is unwound by ATP-dependent RNA helicase activity to form two single strands of RNA. dsRNA is unwounded by ATP-dependent RNA helicase activity to form two single strands of RNA. The guide (antisense) strand directs silencing targeted mRNA, and the other strand is called the passenger (sense) strand. Ago2 protein binds the guide strand and cleaves its targeted RNA at the phosphodiester bond which is positioned between nucleotides 10 and 11. The cleaved products are rapidly degraded because of its unprotected ends, and the passenger strand is also degraded. After dissociation of cleaved mRNAs from siRNA, the RISC encounters and cleaves mRNA, resulting in decrease of expression of the target gene (Fig. 13.1).

Table 13.1 Target genes for experimental RNA interference cancer therapies

Target genes	Cancers
1. Proliferation/anti-apoptosis	
BCL-2	Lung cancer, prostate cancer, fibrosarcoma
VEGF	Ewing's sarcoma, prostate cancer
PLK-1	Urinary bladder cancer, lung cancer (liver metastasis)
Survivin	Glioblastoma, rhabdomyosarcoma
CDC25B	Hepatocellular carcinoma
EGFR	Glioblastoma
Telomerase	Malignant melanoma
EZH2	Prostate cancer (bone metastasis)
FGF-4	Germinoma
2. Signal transduction	
ERK1/2	Hepatocellular carcinoma
STAT3	Colon cancer, prostate cancer, breast cancer
β-catenin	Colon cancer, multiple myeloma
BCR-ABL	Chronic myelogenous leukemia
LYN	Chronic myelogenous leukemia
3. Drug resistance	
MDR1	Colon cancer
MRP7/ABCC10	Non-small cell lung cancer
RPN2	Breast cancer
ABCG10	Gastric cancer
FGFR1	Breast cancer
4. Metastasis/angiogenesis	
VEGF/VEGFR	Ewing's sarcoma, breast cancer, colon cancer, prostate cancer
u-PA/u-PAR	Squamous carcinoma
CCR7	Colon cancer
LYN	Ewing's sarcoma
RhoC	Hepatocellular carcinoma

13.3 Target Genes for Cancer Therapy

The RNAi technology in the clinical setting has relied on localized drug delivery first. This reason is that the localized administration could maintain higher concentrations of siRNAs in the targeted diseases. However, thanks to the development of DDSs (see Refs. [19, 20]), RNAi has recently been evaluated as a therapeutic strategy for cancer treatment. To develop nuclear medicine against cancers, suitable gene targets should be selected (Table 13.1). The definition of cancers is cell proliferation without normal regulation, and one of the most important characteristics of cancers is to bereave the host's life with their malignant behaviors. Such targets include anti-apoptotic proteins, cell cycle regulators, transcription factors, signal transduction proteins, and factors associated with malignant biological behaviors of cancer cells, all of these genes are associated with the poor prognosis of cancer patients.

Among such suitable genes, BCL2 protein is one of the anti-apoptotic members of BCL family proteins and contributes to resistance to apoptosis against external stimuli, including cytotoxic agents. BCL2 participates in tumorigenesis and progression and its overexpression in tumor cells correlates with the poor prognosis of the cancer patients [21–24]. Many studies have demonstrated that siRNA treatment against BCL2 inhibited the proliferation of tumor cells [5, 25–27]. Intravenous administration of synthetic BCL2 siRNA, using a cationic or pegylated cationic liposome, suppressed tumor progression in a xenograft mouse model, and BCL2 siRNA treatment significantly elongated the survival of cancer-bearing mice [5, 27]. Oblimersen sodium is a 18-mer phosphorothioate antisense oligonucleotide designed to bind to the first six codons of the human BCL2 mRNA [28]. Though this nucleic acid medicine is an antisense oligonucleotide, it has been also used in a substantial number of clinical trials against several types of cancers [29–33]. These observations indicate that BCL2 is a suitable target for cancer therapy.

Signal transduction molecules are other candidates for RNAi. Member of the signal transduces and activator of transcription (STAT) family act as key components of cytokine signaling pathways that regulate gene expression. Among STAT family, STAT3 is most strongly implicated in carcinogenesis. Its constitutively active form is detected in variety of cancers and dysregulates the downstream target genes of cell proliferation [34] and survival [35, 36]. RNAi therapy against STAT3 demonstrates the inhibition of tumor progression as well as invasion [37–40].

Bcr-Abl fusion protein, which is created by the molecular consequence of the *t*(9; 22) translocation, is a constitutively active tyrosine kinase that causes Philadelphia (Ph)-positive leukemias [41]. Imatinib mesylate (IM; Gleevec™, Glivec™) was developed as a first-generation tyrosine kinase inhibitor (TKI), and its emergence has dramatically changed the outcomes of therapies against Ph-positive leukemia, especially chronic myelogenous leukemia (CML) [42–45]. Moreover, several second generation TKIs developed to overcome resistance to IM have yielded excellent outcomes [46–49]. These clinical observations demonstrated that targeting Bcr-Abl protein is a promising strategy to eliminate Bcr-Abl-positive leukemic cells. The approach to downregulate the expression of Bcr-Abl mRNA by RNAi was investigated in vitro [50–53]. Koldehoff et al. reported the in vivo administration of synthetic Bcr-Abl siRNA with cationic liposomes in a patient with recurrent Ph-positive CML resistant to IM [54]. This patient had a high level of Bcr-Abl transcripts and subcutaneous nodule, and she was treated with 10 μg/kg of Bcr-Abl siRNA intravenously by a bolus injection and 300 μg iRNA was directly injected into CML node. The level of Bcr-Abl mRNA transcript was drastically decreased; however, no obvious effects were observed after the second and third courses. Although this report was not constructed as a clinical trial, these observations are worth noting for developing nuclear medicine against CML.

β-catenin is a downstream protein of the canonical Wnt signaling pathway that has been shown to play an important role in the process of development, proliferation, and differentiation [55]. In the absence of Wnt signals, adenomatous polyposis coli (APC), Axin, glycogen synthase kinase-3β (GSK3β), and casein kinase 1α

(CK1α) form a complex called the "β-catenin destruction complex." GSK3β and CK1α target serine/threonine residues at the N terminus of β-catenin for phosphorylation [56]. Phosphorylated β-catenin is recognized and polyubiquitinated by β-transducin repeat-containing protein (β-TrCP), a component of a ubiquitin ligase complex, targeting β-catenin for degradation by the 26S proteasome [57, 58]. On the other hand, the binding of Wnt ligands to Frizzled (Fz) receptors and the low-density lipoprotein receptor-related protein 5/6 (LRP5/6) co-receptors induces the phosphorylation of Disheveled (Dvl) and prevents GSK3β-dependent phosphorylation of β-catenin. Stabilized β-catenin translocates into the nucleus and interacts with T cell factor (TCF)/lymphocyte enhancer factor (LEF). In the absence of β-catenin, TCF/LEF, which interacts with Groucho and histone deacetylase (HDAC), acts as a repressor of the transcription [59]. The β-catenin/TCF complex regulates the transcription of a number of genes associated with cell proliferation and apoptosis, as well as the expression of growth factors. Typical β-catenin/TCF target genes that are associated with cell proliferation are c-myc and cyclin D1. The c-myc oncogene regulates cell cycle progression and apoptosis. Cyclin D1 activates cyclin-dependent kinases leading to cell cycle progression. Recently, this pathway has been focused on as it is involved in cancer development. Aberrant activation of Wnt/β-catenin signaling is observed in many human cancers. Genetic mutations of Wnt signaling pathway components are primarily responsible for this aberrant activation and cause β-catenin to escape the degradation process and lead to nuclear stabilized β-catenin accumulation [60]. Treatment of siRNAs against β-catenin successfully suppressed the proliferation of colon cancer cells and myeloma cells by inducing caspase-dependent apoptosis [61–63]. Thus, β-catenin represents a suitable target for RNAi therapy.

Molecules controlling cell division are also useful targets for cancer therapy. Polo-like kinases (PLKs) belong to the family of serine/threonine kinases. PLK family has identified PLK-1, PLK-2 (SNK), PLK-3 (FNK), and PLK-4 (SAK) in mammalians so far and PLKs function as regulators of both cell cycle progression and cellular response to DNA damage. PLK-1 is the best characterized among them to date. PLK-1 regulates cell division at several points in the mitotic phase: mitotic entry through CDK1 activation, bipolar spindle formation, chromosome alignment, segregation of chromosomes, and cytokinesis [64]. Whereas PLK-1 is scarcely detectable in most adult tissues [65, 66], PLK-1 is overexpressed in cancerous tissues [65], and many reports have described that PLK-1 is overexpressed in cancerous tissues and that PLK-1 expression levels were tightly correlated with histological grades of tumors, clinical stages, and prognosis of the patients.

Inhibition of PLK-1 activity in cancer cells induces mitotic arrest and tumor cell apoptosis. Depletion of PLK-1 mRNA also inhibits the functions of PLK-1 protein in DNA damages and spindle formation and causes the inhibition of the cell proliferation in a time- and a dose-dependent manner. PLK-1 siRNA treatment induces an arrest at the G2/M phase in the cell cycle with the increase of CDC2/Cyclin B1 and the transfected cells had dumbbell-like and misaligned nuclei. Moreover, the caspase activation was induced in these cells [6, 67, 68]. These observations indicate that PLK-1 could be an excellent target for cancer therapy.

Other candidate siRNA targets are molecules that define the malignant behavior of cancerous cells. The vascular endothelial growth factor (VEGF)/VEGF receptor (VEGFR) axis plays an important role in angio- and lymphangiogenesis. VEGF family has seven members. Among them, VEGF-A stimulates angiogenesis in tumor masses, enhances the permeability of the blood vessels, and promotes the motility of cancer cells, which results in metastases [69, 70]. The previous investigations reveal that VEGF-A depletion successfully prevents metastasis of cancers [71, 72]. In contrast to VEGF-A, VEGF-C and VEGF-D are associated with tumor lymphangiogenesis and lymph node metastasis. Depletion of VEGF-C/D inhibits metastasis of cancers [73, 74]. Another example of the molecule associated with metastasis is the urokinase-type plasminogen activator (u-PA). u-PA binds to u-PA receptor (u-PAR), and this molecule activates plasminogen and matrix metalloproteases, which enhances the degradation of basement membranes and extracellular matrices and promotes metastases [75, 76]. Data using a mouse model demonstrated that the administration of u-PAR inhibited metastasis and progression of oral squamous cell carcinoma [77]. These molecules associated with metastasis will also be attractive targets of RNAi therapy.

13.4 microRNAs

microRNAs (miRNAs), as the name suggests, are very short RNAs consisted of 21 nts. Those short RNAs regulate target gene expression through translation repression or mRNA degradation, and consequently miRNAs involve diverse biological processes in eukaryocytes. miRNAs are derived from stem-loop-structured primary miRNAs (pri-miRNAs) by the cleavage activity of Drocha, a nuclear-localized member of the RNAse III family, to yield short precursor miRNAs called pre-miRNAs. Pre-miRNAs comprising 70–90 nts exhibit a hairpin structure with a 5′-phosphate and a 3′-2 nts overhang. After translocation from the nucleus to the cytoplasm by Exprtin-5 pre-miRNAs are processed by Dicer into miRNAs of 21 nts. miRNAs as well as siRNAs enter into RISC assembly pathway. Unlike siRNAs, the mature miRNAs often have a partially complementary sequence to the target mRNAs, and a single miRNA might bind to numerous target genes. Therefore, a single miRNA has diverse functions including proliferation, differentiation, and apoptosis [78].

One of the mechanisms of carcinogenesis is the imbalance of oncogenes and tumor suppressor genes caused by several factors including carcinogen. miRNAs affect gene expression by regulating the translation of mRNAs into proteins. In many cancers, some kinds of miRNAs negatively regulate tumor suppressor. miRs-15/16 are downregulated in chronic lymphocytic leukemia (CLL). miR-15a and mir16-2 recognize target sites on the 3′UTR of BCL-2, an anti-apoptotic oncogene [79]. These miRNAs regulate BCL-2 expression in normal cells. However, these are deleted in patients with CLL. On the contrary, other kinds of miRNAs regulate carcinogenesis and tumor progression. Mir-17-92 cluster is overexpressed in lung cancer tissues [80] and its target genes are PTEN and RB2 [81]. These observations indicate that the overexpression of this miR-17-93 cluster induces the carcinogenesis in lung tissues. Anti-miRNA oligonucleotides (AMOs) can suppress the miRNA activity [82], and recently MAOs are developed as nucleic acid medicines [83–86]. miRNAs regulating anti-apoptosis and cell proliferation are also suitable target molecules against cancers.

13.5 Preclinical Application of RNAi

Before the clinical trials for RNAi therapy, preclinical studies are performed. We introduce two applications of PLK-1 siRNA for cancer therapy. One application is an intravesical treatment against urinary bladder cancers. PLK-1 protein is overexpressed in urinary bladder tumors, and moreover PLK-1 expression levels are correlated with histological grades of tumors, clinical stages, and prognosis of the patients [6]. Superficial urinary bladder cancers are approximately 70 % of urinary bladder cancers at initial diagnosis. After resected transurethrally, Bacillus Calmette-Guerin (BCG), mitomycin C, and Adriamycin are administered intravesically to prevent the recurrence of or diminish the residual cancers [87]. However, half of superficial cancers recur, and consequently novel intravesical treatment should be developed. Clinical trials of RNAi therapy often rely on localized drug delivery because maintenance of higher siRNAs concentrations is necessary for efficacy against the targeted diseases. The urinary bladder which is closed to the urethra is considered as a "putative" in vitro space. In accordance with the unique idea, the efficacy of intravesical therapy of PLK-1 siRNA against urinary bladder cancers was investigated. Bladder cancer-bearing mice were established by the implantation of luciferase (Luc)-labeled UM-UC-3 bladder cancer cells into the murine bladder cavity through the urethra. After the engraftment of cancer cells in the bladder was evaluated by using the in vivo imaging system (IVIS) of bioluminescence imaging (BLI) [88], cancer-bearing mice were treated with PLK-1 siRNA/cationic liposome complexes. Tumor progression was significantly suppressed by the intravesical treatment of PLK-1 siRNA [6].

Another application is a systemic administration of siRNAs against liver metastatic tumors of lung cancers. Distant metastasis is one of the life-threatening factors in lung cancer patients. Despite the development of new molecular targeting agents [89, 90], current therapies are not sufficient to cure or manage the patients with distant metastasis [91, 92]. Therefore, novel therapies should be developed. Kawata et al. investigated the effects of PLK-1 siRNA on the liver metastasis of lung cancers in an orthotopic liver metastatic mouse model. Spleens were exposed to allow direct intrasplenic injections of Luc-labeled A549 non-small cell lung cancer cells. After the removal of spleens, the Luc-labeled A549 cell engraftment was confirmed by using IVIS, and then PLK-1 siRNA/atelocollagen complexes were administered by intravenous injection for 10 days. On day 35, mice treated with PLK-1 siRNA/atelocollagen complex showed the significant suppression of tumor growth compared to mice treated with nonsense siRNA/atelocollagen complex or PBS/atelocollagen complex which showed extensive metastases in the livers. These findings indicate that PLK-1 siRNA/atelocollagen complex is an attractive therapeutic tool for further development as a treatment against liver metastasis of lung cancer [8].

13.6 Adverse Effects of RNAi

Although RNAi shows excellent specificity in gene silencing, several adverse effects are brought in in vivo application. One probable adverse effect is activation of immune reaction. Mammalian immune cells express family of Toll-like receptors (TLRs), which play an essential role in innate immune responses. TLRs recognize microbial ligands including bacterial lipopolysaccharide, lipopeptides, or viral and bacterial RNA and DNA. Among 13 TLRs, TLR7 and TLR8 recognize ssRNA sequence-dependently and produce interferons (IFNs) and inflammatory cytokines such as IL-12 and TNF-α through the activation of NF-κB and IFN regulatory factor (IRF)-7. For this immune response, the length of single-strand RNA (ssRNA) is important and 16–19 nt ssRNA induces IFN production although 12 nt ssRNAs contains the immunostimulatory motif (GUCCUUAA) [93]. The administration of siRNAs into mammalian cells activates the immune systems also sequence-independently. siRNAs induce dsRNA-activated protein kinase (PKR) autophosphorylation and PKR produces IFNs through the activation of NK-κB and IRF-3. TLR3 recognizes unmethylated CpG DNA but not ssRNA. dsRNA directly binds to TLR3 and this signaling pathway is activated sequence-independently [94]. Interestingly, although the receptors recognizing a ssRNA containing a CpG motif and a 6 nt poly-(G) run at the 3′ end are still unknown, a ssRNA activates monocytes [95]. TLR 9, which expresses in endosomes, recognizes CpG oligodeoxynucleotides (ODNs). Purified recombinant TLR 9 binds CpG ODNs directly in a sequence- and pH-dependent manner [96]. Thus, the activation of immune response by siRNAs is dependent on their sequence and chemical nature, implying that chemical modifications of siRNAs might prevent the immune activation. The 2′ position of nucleotides is within TLR-7-interacting sequences and 2′ O-methyl or 2′ fluoro modification abrogate immune response. Furthermore, the uridine or guanosine modification is most effective [97]. Locked nucleic acid modifications of the 3′ of 5′ termini of the sense strand of siRNAs can reduce the immunostimulatory effects [93]. siRNAs conjugated to cholesterol have no significant activation of immune system and improve the distribution of siRNA to the targeted organ including the liver. Systemic administration of cholesterol-conjugated apolipoprotein B siRNAs induces a decrease of apolipoprotein B expression in liver and jejunum of mice, resulting in a decrease in cholesterol levels without the activation of immune systems [98].

Besides perfect complementarity of siRNAs in target RNA sequence, partially complementary sequences in unintended RNAs induce gene silencing (off-target effect). This effect is induced by the sequence complementarity in the seed region of siRNAs or short-hairpin RNAs (shRNAs) [99]. Moreover, the 7 nt motif complementary to 2–8 nt at the 5′ end of antisense strands of siRNAs has been shown to be a key determinant in directing off-target effects [100]. There are several ways to control the off-target effects. The in silico screening of siRNA constructs are useful for optimization to prevent the off-target effects, and several groups have been developing algorithm [101, 102]. Chemical modification is also useful. For example, the O-methyl modification of the 2′-position of the ribose within the seed region of siRNAs reduces the off-target effect [103]. Asymmetrically designed siRNAs reduce off-target effects compared to symmetric siRNAs. Sun et al. designed asymmetric RNA duplexes of various lengths with overhangs at the 3′ and 5′ ends of the antisense strand to target genes. All siRNAs against target genes were designed to match the same 19 nt sequence. The asymmetric siRNAs effectively induced gene silencing of targeted genes without silencing of nontargeted genes [104].

shRNAs can also induce stable gene silencing. Consequently, it is possible that long-term silencing by shRAN overexpression causes fatal adverse effects. Because shRNA is processed through the miRNA pathway, the miRNA maturation is blocked in response to shRNA concentration. Grimm et al. demonstrated that the sustained high-level shRNA expression in the liver of mice by AAV vector downregulated liver-derived miRNAs, resulting in hepatic injury and death. Morbidity was associated with the downregulation of liver-derived miRNAs [105]. They speculated that saturation of Exportin-5 whose function is nuclear transport inhibited the miRNA maturation pathway. On the contrary, Constein et al. demonstrated that the administration of synthesized siRNAs induced acute and long-term gene silencing without interrupting the endogenous miRNA biogenesis [106]. As mentioned by Grimm et al. [105], higher expression of shRNAs by viral vector might influence the miRNA biogenesis. Considering these findings, careful modification and formulation of siRNAs could avoid the competition between siRNA and miRNA.

13.7 Clinical Trials of RNAi Towards Cancer Therapies

siRNA cancer therapies have been conducted in clinical settings, but few clinical trials for cancer therapy are ongoing (Table 13.2). Alnylam Pharmaceuticals is developing ALN-VSP01 targeting kinase spindle protein and VEGF, and conducting a Phase I study in patients with advanced tumors with liver involvement. Calando Pharmaceuticals is conducting a Phase I study of CALAA-01 in patients with solid tumors refractory to standard-of-care therapies. CALAA-01 is composed of RRM2 siRNA and CDP nanoparticles called Rondel™, and CALAA-01 has been proven safe and effective in mice and nonhuman primates' studies. Clinical studies using LNAs are also ongoing. Santaris Pharma has developed LNA against Bcl-2, SPC2996, for use in an ongoing Phase I/II study in patients with relapsed or refractory chronic lymphocytic leukemia is ongoing. Enzon Pharmaceuticals has developed a LNA against hypoxia-inducible factor-1α and a Phase I/II study in patients with advanced solid tumors or lymphoma is ongoing. National Cancer Institute and

Table 13.2 Clinical trials of RNAi towards cancer therapies

Sponsor	siRNA	Target genes	Disease	Root	Phase	Year
Santaris	SPC2996[a]	Bcl-2	Chronic lymphocytic leukemia	i.v.	Phase I/II	2005
Santaris&Enzon	ENZ-2968[a]	HIF-1α	Metastatic liver tumors	i.v.	Phase I/II	2007
Calando	CALLA-01	RRM2	Relapsed or refractory solid cancers	i.v.	Phase I	2008
Alnylam	ALN-VSP01	KSP+VEGF	Metastatic liver tumors	i.v.	Phase I	2009
Silence Therapeutics	Atu027	PKN-3	Advanced solid cancers	i.v.	Phase I	2009
Silenseed	siG12D LODER	KRAS[b]	Locally advanced adenocarcinoma of pancreas	Local[c]	Phase 0	2010
National Cancer Institute	TKM080301	PLK-1	Metastatic liver tumors	i.v.	Phase I	2012
Tekmira Pharmaceuticals	TKM080301	PLK-1	Solid tumors or lymphomas	i.v.	Phase I	2012

HIF-1α hypoxia-inducible factor-1 α, *RRM2* ribonucleotide reductase M2 subunit, *KSP* kinase spindle protein, *PKN-3* protein kinase N3, *PLK-1* polo-like kinase 1

[a]ENZ-2968 is LNA (locked nucleic acid), KRAS

[b]KRAS with G12D mutation

[c]siRNA is administered using an endoscopic ultrasound biopsy needle

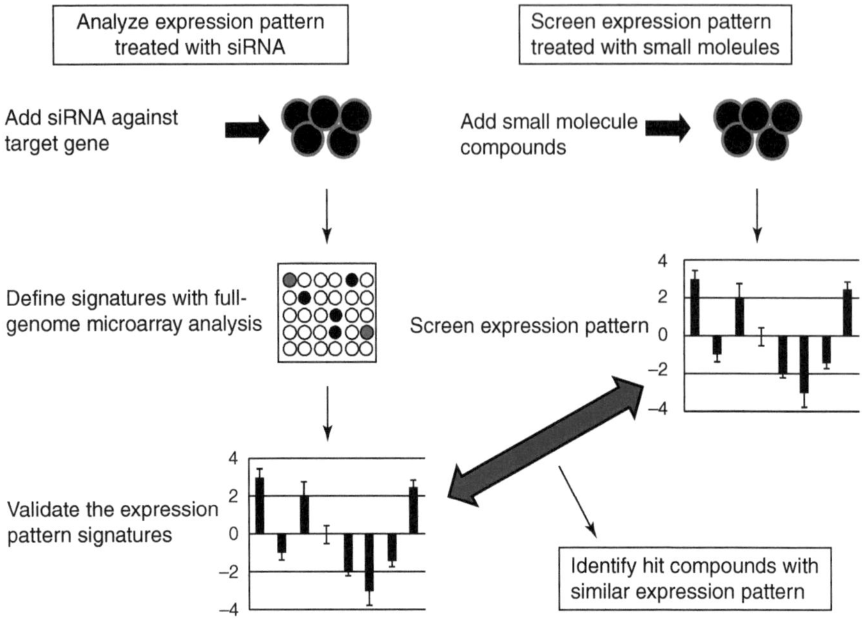

Fig. 13.2 Biomarker-based screening using RNA interference. This assay proceeds in two steps: the first step consists of setting up the signature of siRNA against target gene. The second step involves screening for compounds with the similar expression patterns. Consequently, hit compounds that inhibit the downstream signal of the target gene

Tekmira Pharmaceuticals are conducting clinical trials on PLK-1 RNA interference against solid tumor or lymphoma. As clinical trials of cancer therapies have just started, their outcomes are expected.

13.8 Biomarker-Based Screening

RNA interference technology is also used in the field of drug discovery. The biomarker-based screening is a new high-throughput screening method based on transcriptional profiling and identifies the specific transcriptional activities altered by the compounds of interest. PGX Health, A division of Clinical Data Inc. (formerly Avalon Pharmaceuticals, MD, USA) assessed the transcriptional response of a colon cancer cell line to treatment with β-catenin siRNA using full-genome microarray analysis [9]. Nine biomarkers were selected for their potential as indicators for cancer therapy. A library of 90,000 individual compounds was screened to identify compounds that showed a similar expression pattern to the siRNA (Fig. 13.2). Finally, the compound LC-363 was detected based on its ability to mimic the effect of β-catenin knockdown. The effect of AV-65, one of LC-363 compound series, on MM cells and CML cells was investigated. AV-65 inhibited the proliferation of MM and CML cells by promoting the degradation of β-catenin and inhibiting β-catenin/TCF transcriptional activity. AV-65 decreased the expression of c-myc, cyclin D1, and survivin, which resulted in the inhibition of tumor cell proliferation through the apoptotic pathway [10, 11]. Moreover, AV-65 treatment prolonged the survival of orthotopic MM-bearing mice [11]. A clinical study with this compound series in solid and hematopoietic malignancies will be carried out in the future.

13.9 Conclusion

RNAi therapy against cancers has just started and the outcomes are expected. However, it should be warranted to establish the pharmacokinetics and pharmacodynamics of siRNAs on the administration for the potential approval of siRNA as a tool for cancer therapy. Moreover, to maximize efficacy and to minimize adverse effects of RNAi, it should be determined whether siRNAs are best administered alone or in combination with chemotherapeutic agents [107], and whether it is better to administer a single specific siRNA or multiple specific siRNAs [108–110].

In conclusion, RNAi therapy represents a powerful strategy against cancers and may offer a novel and attractive therapeutic option. The success of RNAi depends on the suitable selection of target genes. Besides developing nucleic acid-based medicine, RNAi technology is applied into the field of drug discovery. We anticipate that RNAi technology could establish a novel and promising therapeutic tool against cancers.

References

1. Fire A, Xu S, Montgomery MK, Kostas SA, Driver SE, Mello CC (1998) Potent and specific genetic interference by double-stranded RNA in Caenorhabditis elegans. Nature 391(6669):806–811
2. Check E (2002) A tragic setback. Nature 420(6912):116–118
3. Hacein-Bey-Abina S, von Kalle C, Schmidt M, Le Deist F, Wulffraat N, McIntyre E, Radford I, Villeval JL, Fraser CC, Cavazzana-Calvo M, Fischer A (2003) A serious adverse event after successful gene therapy for X-linked severe combined immunodeficiency. N Engl J Med 348(3):255–256
4. Nguyen T, Menocal EM, Harborth J, Fruehauf JH (2008) RNAi therapeutics: an update on delivery. Curr Opin Mol Ther 10(2):158–167
5. Yano J, Hirabayashi K, Nakagawa S, Yamaguchi T, Nogawa M, Kashimori I, Naito H, Kitagawa H, Ishiyama K, Ohgi T, Irimura T (2004) Antitumor activity of small interfering RNA/cationic liposome complex in mouse models of cancer. Clin Cancer Res 10(22):7721–7726
6. Nogawa M, Yuasa T, Kimura S, Tanaka M, Kuroda J, Sato K, Yokota A, Segawa H, Toda Y, Kageyama S, Yoshiki T, Okada Y, Maekawa T (2005) Intravesical administration of small interfering RNA targeting PLK-1 successfully prevents the growth of bladder cancer. J Clin Invest 115(4):978–985. doi:10.1172/JCI23043
7. Sano A, Maeda M, Nagahara S, Ochiya T, Honma K, Itoh H, Miyata T, Fujioka K (2003) Atelocollagen for protein and gene delivery. Adv Drug Deliv Rev 55(12):1651–1677

8. Kawata E, Ashihara E, Kimura S, Takenaka K, Sato K, Tanaka R, Yokota A, Kamitsuji Y, Takeuchi M, Kuroda J, Tanaka F, Yoshikawa T, Maekawa T (2008) Administration of PLK-1 small interfering RNA with atelocollagen prevents the growth of liver metastases of lung cancer. Mol Cancer Ther 7(9):2904–2912. doi:10.1158/1535-7163.MCT-08-0473
9. Bol D, Ebner R (2006) Gene expression profiling in the discovery, optimization and development of novel drugs: one universal screening platform. Pharmacogenomics 7(2):227–235. doi:10.2217/14622416.7.2.227
10. Nagao R, Ashihara E, Kimura S, Strovel JW, Yao H, Takeuchi M, Tanaka R, Hayashi Y, Hirai H, Padia J, Strand K, Maekawa T (2011) Growth inhibition of imatinib-resistant CML cells with the T315I mutation and hypoxia-adaptation by AV65 – a novel Wnt/beta-catenin signaling inhibitor. Cancer Lett 312(1):91–100. doi:10.1016/j.canlet.2011.08.002
11. Yao H, Ashihara E, Strovel JW, Nakagawa Y, Kuroda J, Nagao R, Tanaka R, Yokota A, Takeuchi M, Hayashi Y, Shimazaki C, Taniwaki M, Strand K, Padia J, Hirai H, Kimura S, Maekawa T (2011) AV-65, a novel Wnt/beta-catenin signal inhibitor, successfully suppresses progression of multiple myeloma in a mouse model. Blood Cancer J 1:e43
12. Bernstein E, Caudy AA, Hammond SM, Hannon GJ (2001) Role for a bidentate ribonuclease in the initiation step of RNA interference. Nature 409(6818):363–366
13. Grishok A, Pasquinelli AE, Conte D, Li N, Parrish S, Ha I, Baillie DL, Fire A, Ruvkun G, Mello CC (2001) Genes and mechanisms related to RNA interference regulate expression of the small temporal RNAs that control C. elegans developmental timing. Cell 106(1):23–34
14. Hutvagner G, Zamore PD (2002) A microRNA in a multiple-turnover RNAi enzyme complex. Science 297(5589):2056–2060
15. Kolb FA, Zhang H, Jaronczyk K, Tahbaz N, Hobman TC, Filipowicz W (2005) Human dicer: purification, properties, and interaction with PAZ PIWI domain proteins. Methods Enzymol 392:316–336
16. Murchison EP, Hannon GJ (2004) MiRNAs on the move: miRNA biogenesis and the RNAi machinery. Curr Opin Cell Biol 16(3):223–229
17. Agrawal N, Dasaradhi PV, Mohmmed A, Malhotra P, Bhatnagar RK, Mukherjee SK (2003) RNA interference: biology, mechanism, and applications. Microbiol Mol Biol Rev 67(4):657–685
18. Parker JS, Barford D (2006) Argonaute: a scaffold for the function of short regulatory RNAs. Trends Biochem Sci 31(11):622–630
19. Whitehead KA, Langer R, Anderson DG (2009) Knocking down barriers: advances in siRNA delivery. Nat Rev Drug Discov 8(2):129–138. doi:10.1038/nrd2742
20. Oh YK, Park TG (2009) siRNA delivery systems for cancer treatment. Adv Drug Deliv Rev 61(10):850–862. doi:10.1016/j.addr.2009.04.018
21. Iqbal J, Neppalli VT, Wright G, Dave BJ, Horsman DE, Rosenwald A, Lynch J, Hans CP, Weisenburger DD, Greiner TC, Gascoyne RD, Campo E, Ott G, Muller-Hermelink HK, Delabie J, Jaffe ES, Grogan TM, Connors JM, Vose JM, Armitage JO, Staudt LM, Chan WC (2006) BCL2 expression is a prognostic marker for the activated B-cell-like type of diffuse large B-cell lymphoma. J Clin Oncol 24(6):961–968
22. Lauwers GY, Scott GV, Karpeh MS (1995) Immunohistochemical evaluation of bcl-2 protein expression in gastric adenocarcinomas. Cancer 75(9):2209–2213
23. Pezzella F, Turley H, Kuzu I, Tungekar MF, Dunnill MS, Pierce CB, Harris A, Gatter KC, Mason DY (1993) bcl-2 protein in non-small-cell lung carcinoma. N Engl J Med 329(10):690–694
24. Sinicrope FA, Hart J, Michelassi F, Lee JJ (1995) Prognostic value of bcl-2 oncoprotein expression in stage II colon carcinoma. Clin Cancer Res 1(10):1103–1110
25. Fu GF, Lin XH, Han QW, Fan YR, Xu YF, Guo D, Xu GX, Hou YY (2005) RNA interference remarkably suppresses bcl-2 gene expression in cancer cells in vitro and in vivo. Cancer Biol Ther 4(8):822–829
26. Ruckert F, Samm N, Lehner AK, Saeger HD, Grutzmann R, Pilarsky C (2010) Simultaneous gene silencing of Bcl-2, XIAP and survivin re-sensitizes pancreatic cancer cells towards apoptosis. BMC Cancer 10:379

27. Sonoke S, Ueda T, Fujiwara K, Sato Y, Takagaki K, Hirabayashi K, Ohgi T, Yano J (2008) Tumor regression in mice by delivery of Bcl-2 small interfering RNA with pegylated cationic liposomes. Cancer Res 68(21):8843–8851
28. Klasa RJ, Gillum AM, Klem RE, Frankel SR (2002) Oblimersen Bcl-2 antisense: facilitating apoptosis in anticancer treatment. Antisense Nucleic Acid Drug Dev 12(3):193–213
29. Advani PP, Paulus A, Masood A, Sher T, Chanan-Khan A (2011) Pharmacokinetic evaluation of oblimersen sodium for the treatment of chronic lymphocytic leukemia. Expert Opin Drug Metab Toxicol 7(6):765–774. doi:10.1517/17425255.2011.579105
30. Chanan-Khan AA, Niesvizky R, Hohl RJ, Zimmerman TM, Christiansen NP, Schiller GJ, Callander N, Lister J, Oken M, Jagannath S (2009) Phase III randomised study of dexamethasone with or without oblimersen sodium for patients with advanced multiple myeloma. Leuk Lymphoma 50(4):559–565
31. Galatin PS, Advani RH, Fisher GA, Francisco B, Julian T, Losa R, Sierra MI, Sikic BI (2011) Phase I trial of oblimersen (Genasense(R)) and gemcitabine in refractory and advanced malignancies. Invest New Drugs 29(5):971–977
32. Rudin CM, Salgia R, Wang X, Hodgson LD, Masters GA, Green M, Vokes EE (2008) Randomized phase II study of carboplatin and etoposide with or without the bcl-2 antisense oligonucleotide oblimersen for extensive-stage small-cell lung cancer: CALGB 30103. J Clin Oncol 26(6):870–876
33. Sternberg CN, Dumez H, Van Poppel H, Skoneczna I, Sella A, Daugaard G, Gil T, Graham J, Carpentier P, Calabro F, Collette L, Lacombe D (2009) Docetaxel plus oblimersen sodium (Bcl-2 antisense oligonucleotide): an EORTC multicenter, randomized phase II study in patients with castration-resistant prostate cancer. Ann Oncol 20(7):1264–1269
34. Bromberg J (2002) Stat proteins and oncogenesis. J Clin Invest 109(9):1139–1142. doi:10.1172/JCI15617
35. Konnikova L, Simeone MC, Kruger MM, Kotecki M, Cochran BH (2005) Signal transducer and activator of transcription 3 (STAT3) regulates human telomerase reverse transcriptase (hTERT) expression in human cancer and primary cells. Cancer Res 65(15):6516–6520. doi:10.1158/0008-5472.CAN-05-0924
36. Masuda M, Suzui M, Yasumatu R, Nakashima T, Kuratomi Y, Azuma K, Tomita K, Komiyama S, Weinstein IB (2002) Constitutive activation of signal transducers and activators of transcription 3 correlates with cyclin D1 overexpression and may provide a novel prognostic marker in head and neck squamous cell carcinoma. Cancer Res 62(12):3351–3355
37. Xu Y, Li X, Zhang S, Shen D, Li H, Wu Y, Qiu Y, Ji Y, Chen F (2012) Targeting Stat3 suppresses growth of U251 cell-derived tumours in nude mice. J Clin Neurosci 19(3):443–446. doi:10.1016/j.jocn.2011.04.017
38. Gao L, Zhang L, Hu J, Li F, Shao Y, Zhao D, Kalvakolanu DV, Kopecko DJ, Zhao X, Xu DQ (2005) Down-regulation of signal transducer and activator of transcription 3 expression using vector-based small interfering RNAs suppresses growth of human prostate tumor in vivo. Clin Cancer Res 11(17):6333–6341. doi:10.1158/1078-0432.CCR-05-0148
39. Fan Y, Zhang YL, Wu Y, Zhang W, Wang YH, Cheng ZM, Li H (2008) Inhibition of signal transducer and activator of transcription 3 expression by RNA interference suppresses invasion through inducing anoikis in human colon cancer cells. World J Gastroenterol 14(3):428–434
40. Ling X, Arlinghaus RB (2005) Knockdown of STAT3 expression by RNA interference inhibits the induction of breast tumors in immunocompetent mice. Cancer Res 65(7):2532–2536. doi:10.1158/0008-5472.CAN-04-2425
41. Sawyers CL (1999) Chronic myeloid leukemia. N Engl J Med 340(17):1330–1340. doi:10.1056/NEJM199904293401706
42. O'Brien SG, Guilhot F, Larson RA, Gathmann I, Baccarani M, Cervantes F, Cornelissen JJ, Fischer T, Hochhaus A, Hughes T, Lechner K, Nielsen JL, Rousselot P, Reiffers J, Saglio G, Shepherd J, Simonsson B, Gratwohl A, Goldman JM, Kantarjian H, Taylor K, Verhoef G, Bolton AE, Capdeville R, Druker BJ, Investigators I (2003) Imatinib compared with interferon and low-dose cytarabine for newly diagnosed chronic-phase chronic myeloid leukemia. N Engl J Med 348(11):994–1004. doi:10.1056/NEJMoa022457

43. Sawyers CL, Hochhaus A, Feldman E, Goldman JM, Miller CB, Ottmann OG, Schiffer CA, Talpaz M, Guilhot F, Deininger MW, Fischer T, O'Brien SG, Stone RM, Gambacorti-Passerini CB, Russell NH, Reiffers JJ, Shea TC, Chapuis B, Coutre S, Tura S, Morra E, Larson RA, Saven A, Peschel C, Gratwohl A, Mandelli F, Ben-Am M, Gathmann I, Capdeville R, Paquette RL, Druker BJ (2002) Imatinib induces hematologic and cytogenetic responses in patients with chronic myelogenous leukemia in myeloid blast crisis: results of a phase II study. Blood 99(10):3530–3539
44. Talpaz M, Silver RT, Druker BJ, Goldman JM, Gambacorti-Passerini C, Guilhot F, Schiffer CA, Fischer T, Deininger MW, Lennard AL, Hochhaus A, Ottmann OG, Gratwohl A, Baccarani M, Stone R, Tura S, Mahon FX, Fernandes-Reese S, Gathmann I, Capdeville R, Kantarjian HM, Sawyers CL (2002) Imatinib induces durable hematologic and cytogenetic responses in patients with accelerated phase chronic myeloid leukemia: results of a phase 2 study. Blood 99(6):1928–1937
45. Druker BJ, Guilhot F, O'Brien SG, Gathmann I, Kantarjian H, Gattermann N, Deininger MW, Silver RT, Goldman JM, Stone RM, Cervantes F, Hochhaus A, Powell BL, Gabrilove JL, Rousselot P, Reiffers J, Cornelissen JJ, Hughes T, Agis H, Fischer T, Verhoef G, Shepherd J, Saglio G, Gratwohl A, Nielsen JL, Radich JP, Simonsson B, Taylor K, Baccarani M, So C, Letvak L, Larson RA, Investigators I (2006) Five-year follow-up of patients receiving imatinib for chronic myeloid leukemia. N Engl J Med 355(23):2408–2417. doi:10.1056/NEJMoa062867
46. Kantarjian H, Giles F, Wunderle L, Bhalla K, O'Brien S, Wassmann B, Tanaka C, Manley P, Rae P, Mietlowski W, Bochinski K, Hochhaus A, Griffin JD, Hoelzer D, Albitar M, Dugan M, Cortes J, Alland L, Ottmann OG (2006) Nilotinib in imatinib-resistant CML and Philadelphia chromosome-positive ALL. N Engl J Med 354(24):2542–2551. doi:10.1056/NEJMoa055104
47. Talpaz M, Shah NP, Kantarjian H, Donato N, Nicoll J, Paquette R, Cortes J, O'Brien S, Nicaise C, Bleickardt E, Blackwood-Chirchir MA, Iyer V, Chen TT, Huang F, Decillis AP, Sawyers CL (2006) Dasatinib in imatinib-resistant Philadelphia chromosome-positive leukemias. N Engl J Med 354(24):2531–2541. doi:10.1056/NEJMoa055229
48. Kimura S, Ashihara E, Maekawa T (2006) New tyrosine kinase inhibitors in the treatment of chronic myeloid leukemia. Curr Pharm Biotechnol 7(5):371–379
49. Khoury HJ, Cortes JE, Kantarjian HM, Gambacorti-Passerini C, Baccarani M, Kim DW, Zaritskey A, Countouriotis A, Besson N, Leip E, Kelly V, Brummendorf TH (2012) Bosutinib is active in chronic phase chronic myeloid leukemia after imatinib and dasatinib and/or nilotinib therapy failure. Blood 119(15):3403–3412. doi:10.1182/blood-2011-11-390120
50. Li MJ, McMahon R, Snyder DS, Yee JK, Rossi JJ (2003) Specific killing of Ph+ chronic myeloid leukemia cells by a lentiviral vector-delivered anti-bcr/abl small hairpin RNA. Oligonucleotides 13(5):401–409. doi:10.1089/154545703322617087
51. Scherr M, Battmer K, Winkler T, Heidenreich O, Ganser A, Eder M (2003) Specific inhibition of bcr-abl gene expression by small interfering RNA. Blood 101(4):1566–1569. doi:10.1182/blood-2002-06-1685
52. Rangatia J, Bonnet D (2006) Transient or long-term silencing of BCR-ABL alone induces cell cycle and proliferation arrest, apoptosis and differentiation. Leukemia 20(1):68–76. doi:10.1038/sj.leu.2403999
53. Arthanari Y, Pluen A, Rajendran R, Aojula H, Demonacos C (2010) Delivery of therapeutic shRNA and siRNA by Tat fusion peptide targeting BCR-ABL fusion gene in Chronic Myeloid Leukemia cells. J Control Release 145(3):272–280. doi:10.1016/j.jconrel.2010.04.011
54. Koldehoff M, Steckel NK, Beelen DW, Elmaagacli AH (2007) Therapeutic application of small interfering RNA directed against bcr-abl transcripts to a patient with imatinib-resistant chronic myeloid leukaemia. Clin Exp Med 7(2):47–55. doi:10.1007/s10238-007-0125-z
55. Wodarz A, Nusse R (1998) Mechanisms of Wnt signaling in development. Annu Rev Cell Dev Biol 14:59–88. doi:10.1146/annurev.cellbio.14.1.59
56. Liu C, Li Y, Semenov M, Han C, Baeg GH, Tan Y, Zhang Z, Lin X, He X (2002) Control of beta-catenin phosphorylation/degradation by a dual-kinase mechanism. Cell 108(6):837–847
57. Latres E, Chiaur DS, Pagano M (1999) The human F box protein beta-Trcp associates with the Cul1/Skp1 complex and regulates the stability of beta-catenin. Oncogene 18(4):849–854. doi:10.1038/sj.onc.1202653

58. Kitagawa M, Hatakeyama S, Shirane M, Matsumoto M, Ishida N, Hattori K, Nakamichi I, Kikuchi A, Nakayama K, Nakayama K (1999) An F-box protein, FWD1, mediates ubiquitin-dependent proteolysis of beta-catenin. EMBO J 18(9):2401–2410. doi:10.1093/emboj/18.9.2401
59. Roose J, Molenaar M, Peterson J, Hurenkamp J, Brantjes H, Moerer P, van de Wetering M, Destree O, Clevers H (1998) The Xenopus Wnt effector XTcf-3 interacts with Groucho-related transcriptional repressors. Nature 395(6702):608–612. doi:10.1038/26989
60. Moon RT, Kohn AD, De Ferrari GV, Kaykas A (2004) WNT and beta-catenin signalling: diseases and therapies. Nat Rev Genet 5(9):691–701. doi:10.1038/nrg1427
61. Ashihara E, Kawata E, Nakagawa Y, Shimazaski C, Kuroda J, Taniguchi K, Uchiyama H, Tanaka R, Yokota A, Takeuchi M, Kamitsuji Y, Inaba T, Taniwaki M, Kimura S, Maekawa T (2009) beta-catenin small interfering RNA successfully suppressed progression of multiple myeloma in a mouse model. Clin Cancer Res 15(8):2731–2738. doi:10.1158/1078-0432.CCR-08-1350
62. Verma UN, Surabhi RM, Schmaltieg A, Becerra C, Gaynor RB (2003) Small interfering RNAs directed against beta-catenin inhibit the in vitro and in vivo growth of colon cancer cells. Clin Cancer Res 9(4):1291–1300
63. Liyan W, Xun S, Xiangwei M (2011) Effect of beta -catenin siRNA on proliferation and apoptosis of hepatoma cell line SMMC-7721 and HepG-2. Hepatogastroenterology 58(110–111):1757–1764. doi:10.5754/hge11017
64. Barr FA, Sillje HH, Nigg EA (2004) Polo-like kinases and the orchestration of cell division. Nat Rev Mol Cell Biol 5(6):429–440. doi:10.1038/nrm1401
65. Lake RJ, Jelinek WR (1993) Cell cycle- and terminal differentiation-associated regulation of the mouse mRNA encoding a conserved mitotic protein kinase. Mol Cell Biol 13(12):7793–7801
66. Hamanaka R, Maloid S, Smith MR, O'Connell CD, Longo DL, Ferris DK (1994) Cloning and characterization of human and murine homologues of the Drosophila polo serine-threonine kinase. Cell Growth Differ 5(3):249–257
67. Liu X, Erikson RL (2003) Polo-like kinase (Plk)1 depletion induces apoptosis in cancer cells. Proc Natl Acad Sci U S A 100(10):5789–5794. doi:10.1073/pnas.1031523100
68. Kawata E, Ashihara E, Maekawa T (2011) RNA interference against polo-like kinase-1 in advanced non-small cell lung cancers. J Clin Bioinform 1(1):6. doi:10.1186/2043-9113-1-6
69. Yancopoulos GD, Davis S, Gale NW, Rudge JS, Wiegand SJ, Holash J (2000) Vascular-specific growth factors and blood vessel formation. Nature 407(6801):242–248. doi:10.1038/35025215
70. Ellis LM (2004) Angiogenesis and its role in colorectal tumor and metastasis formation. Semin Oncol 31(6 Suppl 17):3–9. doi:10.1053/j.seminoncol.2004.11.028
71. Wang S, Liu H, Ren L, Pan Y, Zhang Y (2008) Inhibiting colorectal carcinoma growth and metastasis by blocking the expression of VEGF using RNA interference. Neoplasia 10(4):399–407
72. Salva E, Kabasakal L, Eren F, Ozkan N, Cakalagaoglu F, Akbuga J (2012) Local delivery of chitosan/VEGF siRNA nanoplexes reduces angiogenesis and growth of breast cancer in vivo. Nucleic Acid Ther 22(1):40–48. doi:10.1089/nat.2011.0312
73. Chen Z, Varney ML, Backora MW, Cowan K, Solheim JC, Talmadge JE, Singh RK (2005) Down-regulation of vascular endothelial cell growth factor-C expression using small interfering RNA vectors in mammary tumors inhibits tumor lymphangiogenesis and spontaneous metastasis and enhances survival. Cancer Res 65(19):9004–9011. doi:10.1158/0008-5472.CAN-05-0885
74. Lin W, Jiang L, Chen Y, She F, Han S, Zhu J, Zhou L, Tang N, Wang X, Li X (2012) Vascular endothelial growth factor-D promotes growth, lymphangiogenesis and lymphatic metastasis in gallbladder cancer. Cancer Lett 314(2):127–136. doi:10.1016/j.canlet.2011.09.004
75. Andreasen PA, Kjoller L, Christensen L, Duffy MJ (1997) The urokinase-type plasminogen activator system in cancer metastasis: a review. Int J Cancer 72(1):1–22
76. Dass K, Ahmad A, Azmi AS, Sarkar SH, Sarkar FH (2008) Evolving role of uPA/uPAR system in human cancers. Cancer Treat Rev 34(2):122–136
77. Zhou H, Tang Y, Liang X, Yang X, Yang J, Zhu G, Zheng M, Zhang C (2009) RNAi targeting urokinase-type plasminogen activator receptor inhibits metastasis and progression of oral squamous cell carcinoma in vivo. Int J Cancer 125(2):453–462

78. Cho WC (2010) MicroRNAs in cancer – from research to therapy. Biochim Biophys Acta 1805(2):209–217. doi:10.1016/j.bbcan.2009.11.003
79. Cimmino A, Calin GA, Fabbri M, Iorio MV, Ferracin M, Shimizu M, Wojcik SE, Aqeilan RI, Zupo S, Dono M, Rassenti L, Alder H, Volinia S, Liu CG, Kipps TJ, Negrini M, Croce CM (2005) miR-15 and miR-16 induce apoptosis by targeting BCL2. Proc Natl Acad Sci USA 102(39):13944–13949. doi:10.1073/pnas.0506654102
80. Hayashita Y, Osada H, Tatematsu Y, Yamada H, Yanagisawa K, Tomida S, Yatabe Y, Kawahara K, Sekido Y, Takahashi T (2005) A polycistronic microRNA cluster, miR-17-92, is overexpressed in human lung cancers and enhances cell proliferation. Cancer Res 65(21):9628–9632. doi:10.1158/0008-5472.CAN-05-2352
81. Lewis BP, Shih IH, Jones-Rhoades MW, Bartel DP, Burge CB (2003) Prediction of mammalian microRNA targets. Cell 115(7):787–798
82. Krutzfeldt J, Kuwajima S, Braich R, Rajeev KG, Pena J, Tuschl T, Manoharan M, Stoffel M (2007) Specificity, duplex degradation and subcellular localization of antagomirs. Nucleic Acids Res 35(9):2885–2892. doi:10.1093/nar/gkm024
83. Park JK, Kogure T, Nuovo GJ, Jiang J, He L, Kim JH, Phelps MA, Papenfuss TL, Croce CM, Patel T, Schmittgen TD (2011) miR-221 silencing blocks hepatocellular carcinoma and promotes survival. Cancer Res 71(24):7608–7616. doi:10.1158/0008-5472.CAN-11-1144
84. Sun L, Yan W, Wang Y, Sun G, Luo H, Zhang J, Wang X, You Y, Yang Z, Liu N (2011) MicroRNA-10b induces glioma cell invasion by modulating MMP-14 and uPAR expression via HOXD10. Brain Res 1389:9–18. doi:10.1016/j.brainres.2011.03.013
85. Shi SJ, Zhong ZR, Liu J, Zhang ZR, Sun X, Gong T (2012) Solid lipid nanoparticles loaded with anti-microRNA oligonucleotides (AMOs) for suppression of microRNA-21 functions in human lung cancer cells. Pharm Res 29(1):97–109. doi:10.1007/s11095-011-0514-6
86. Meng W, Jiang L, Lu L, Hu H, Yu H, Ding D, Xiao K, Zheng W, Guo H, Ma W (2012) Anti-miR-155 oligonucleotide enhances chemosensitivity of U251 cell to taxol by inducing apoptosis. Cell Biol Int. doi:10.1042/CBI20100918
87. Gee J, Sabichi AL, Grossman HB (2002) Chemoprevention of superficial bladder cancer. Crit Rev Oncol Hematol 43(3):277–286
88. Nogawa M, Yuasa T, Kimura S, Kuroda J, Sato K, Segawa H, Yokota A, Maekawa T (2005) Monitoring luciferase-labeled cancer cell growth and metastasis in different in vivo models. Cancer Lett 217(2):243–253. doi:10.1016/j.canlet.2004.07.010
89. Sharma SV, Bell DW, Settleman J, Haber DA (2007) Epidermal growth factor receptor mutations in lung cancer. Nat Rev Cancer 7(3):169–181. doi:10.1038/nrc2088
90. Oxnard GR, Miller VA (2010) Use of erlotinib or gefitinib as initial therapy in advanced NSCLC. Oncology 24(5):392–399
91. Bremnes RM, Sundstrom S, Aasebo U, Kaasa S, Hatlevoll R, Aamdal S, Norweigian Lung Cancer Study G (2003) The value of prognostic factors in small cell lung cancer: results from a randomised multicenter study with minimum 5 year follow-up. Lung Cancer 39(3):303–313
92. Hoang T, Xu R, Schiller JH, Bonomi P, Johnson DH (2005) Clinical model to predict survival in chemonaive patients with advanced non-small-cell lung cancer treated with third-generation chemotherapy regimens based on eastern cooperative oncology group data. J Clin Oncol 23(1):175–183. doi:10.1200/JCO.2005.04.177
93. Hornung V, Guenthner-Biller M, Bourquin C, Ablasser A, Schlee M, Uematsu S, Noronha A, Manoharan M, Akira S, de Fougerolles A, Endres S, Hartmann G (2005) Sequence-specific potent induction of IFN-alpha by short interfering RNA in plasmacytoid dendritic cells through TLR7. Nat Med 11(3):263–270. doi:10.1038/nm1191
94. Trinchieri G, Sher A (2007) Cooperation of Toll-like receptor signals in innate immune defence. Nat Rev Immunol 7(3):179–190. doi:10.1038/nri2038
95. Marques JT, Williams BR (2005) Activation of the mammalian immune system by siRNAs. Nat Biotechnol 23(11):1399–1405. doi:10.1038/nbt1161
96. Rutz M, Metzger J, Gellert T, Luppa P, Lipford GB, Wagner H, Bauer S (2004) Toll-like receptor 9 binds single-stranded CpG-DNA in a sequence- and pH-dependent manner. Eur J Immunol 34(9):2541–2550. doi:10.1002/eji.200425218

97. Judge AD, Bola G, Lee AC, MacLachlan I (2006) Design of noninflammatory synthetic siRNA mediating potent gene silencing in vivo. Mol Ther 13(3):494–505. doi:10.1016/j.ymthe.2005.11.002
98. Soutschek J, Akinc A, Bramlage B, Charisse K, Constien R, Donoghue M, Elbashir S, Geick A, Hadwiger P, Harborth J, John M, Kesavan V, Lavine G, Pandey RK, Racie T, Rajeev KG, Rohl I, Toudjarska I, Wang G, Wuschko S, Bumcrot D, Koteliansky V, Limmer S, Manoharan M, Vornlocher HP (2004) Therapeutic silencing of an endogenous gene by systemic administration of modified siRNAs. Nature 432(7014):173–178. doi:10.1038/nature03121
99. Jackson AL, Burchard J, Schelter J, Chau BN, Cleary M, Lim L, Linsley PS (2006) Widespread siRNA "off-target" transcript silencing mediated by seed region sequence complementarity. RNA 12(7):1179–1187. doi:10.1261/rna.25706
100. Lin X, Ruan X, Anderson MG, McDowell JA, Kroeger PE, Fesik SW, Shen Y (2005) siRNA-mediated off-target gene silencing triggered by a 7 nt complementation. Nucleic Acids Res 33(14):4527–4535. doi:10.1093/nar/gki762
101. Yamada T, Morishita S (2005) Accelerated off-target search algorithm for siRNA. Bioinformatics 21(8):1316–1324. doi:10.1093/bioinformatics/bti155
102. Park YK, Park SM, Choi YC, Lee D, Won M, Kim YJ (2008) AsiDesigner: exon-based siRNA design server considering alternative splicing. Nucleic Acids Res 36(Web Server issue):W97–W103. doi:10.1093/nar/gkn280
103. Jackson AL, Burchard J, Leake D, Reynolds A, Schelter J, Guo J, Johnson JM, Lim L, Karpilow J, Nichols K, Marshall W, Khvorova A, Linsley PS (2006) Position-specific chemical modification of siRNAs reduces "off-target" transcript silencing. RNA 12(7): 1197–1205. doi:10.1261/rna.30706
104. Sun X, Rogoff HA, Li CJ (2008) Asymmetric RNA duplexes mediate RNA interference in mammalian cells. Nat Biotechnol 26(12):1379–1382. doi:10.1038/nbt.1512
105. Grimm D, Streetz KL, Jopling CL, Storm TA, Pandey K, Davis CR, Marion P, Salazar F, Kay MA (2006) Fatality in mice due to oversaturation of cellular microRNA/short hairpin RNA pathways. Nature 441(7092):537–541. doi:10.1038/nature04791
106. John M, Constien R, Akinc A, Goldberg M, Moon YA, Spranger M, Hadwiger P, Soutschek J, Vornlocher HP, Manoharan M, Stoffel M, Langer R, Anderson DG, Horton JD, Koteliansky V, Bumcrot D (2007) Effective RNAi-mediated gene silencing without interruption of the endogenous microRNA pathway. Nature 449(7163):745–747. doi:10.1038/nature06179
107. von Bueren AO, Shalaby T, Oehler-Janne C, Arnold L, Stearns D, Eberhart CG, Arcaro A, Pruschy M, Grotzer MA (2009) RNA interference-mediated c-MYC inhibition prevents cell growth and decreases sensitivity to radio- and chemotherapy in childhood medulloblastoma cells. BMC Cancer 9:10. doi:10.1186/1471-2407-9-10
108. Shi XH, Liang ZY, Ren XY, Liu TH (2009) Combined silencing of K-ras and Akt2 oncogenes achieves synergistic effects in inhibiting pancreatic cancer cell growth in vitro and in vivo. Cancer Gene Ther 16(3):227–236. doi:10.1038/cgt.2008.82
109. Honma K, Iwao-Koizumi K, Takeshita F, Yamamoto Y, Yoshida T, Nishio K, Nagahara S, Kato K, Ochiya T (2008) RPN2 gene confers docetaxel resistance in breast cancer. Nat Med 14(9):939–948. doi:10.1038/nm.1858
110. Mu P, Nagahara S, Makita N, Tarumi Y, Kadomatsu K, Takei Y (2009) Systemic delivery of siRNA specific to tumor mediated by atelocollagen: combined therapy using siRNA targeting Bcl-xL and cisplatin against prostate cancer. Int J Cancer 125(12):2978–2990. doi:10.1002/ijc.24382

Chapter 14
Cell Therapy Using Adult Stem Cells in Osteonecrosis and Nonunion Fractures

Jean-Philippe Hauzeur

Abstract Nonunion fractures and aseptic bone necrosis are both characterized by some impairment of the cellular part of bone repair: a reduction of the mesenchymal stem cell (MSC) number and an impairment of the osteoblastic activation. Both seem to be good candidates for cell-based therapies using stem cells, especially MSC. Many animal studies, together with a few human trials, have been published. In this chapter, a review of the human trials is discussed.

The majority of the trials used autologous bone marrow aspirate to implant MSC. Only one tested culture to expand MSC before local implantation.

In nonunion fractures, a direct injection – 15 to 150 ml – was made in four case studies. In another, the bone marrow aspirate was concentrated before injection. The results were encouraging.

In bone necrosis, only two level II studies were published. The results at 24 months were positive in terms of reduction of the necrosis and appearance of collapse. These results were confirmed at 60 months. In three case studies, treatment with concentrated bone marrow aspirates was deemed useful with good results in 76–96 %.

These results are interesting but need confirmation by controlled studies.

Keywords Nonunion fractures • Osteonecrosis • Mesenchymal stem cells • BM aspirate

14.1 Introduction

The physiological bone repair process is impaired in delayed or nonunion (NU) fractures [1] and aseptic bone necrosis (ON) [2].

J.-P. Hauzeur, MD, PhD
Department of Rheumatology, University of Liège,
CHU Sart Tilman, 156, Av J F. Debecker, B 1200, Brussels, Belgium
e-mail: jean-philippe@hauzeur.org

M.K. Danquah, R.I. Mahato (eds.), *Emerging Trends in Cell and Gene Therapy*,
DOI 10.1007/978-1-62703-417-3_14, © Springer Science+Business Media New York 2013

Although, in both diseases, the physiopathological factors are different, in both diseases, bone lesions are neither repaired at the right time nor in the right manner.

Bone healing is produced by a cellular mechanism including mesenchymal stem cells (MSC). The MSC are normally recruited in the pathological area. These non-hematopoietic progenitor cells are able to be differentiated in osteoblasts under the influence of growth factors such as bone morphogenetic proteins (BMP), platelet-derived growth factor, transforming growth factor beta, insulin-like growth factor, fibroblast growth factor, and PTH.

MSC can be found mainly in bone marrow but also in fat tissue, synovium, periosteum, skeletal muscles, and umbilical cord. Some recent data suggests that the osteogenic differentiation capability of MSC from bone marrow and from periosteum is higher than MSC from adipose tissues [10].

In nonunion, the etiology is not clearly understood. Excessive mechanical instability of the fracture, a reduction of bone vascularity, and smoking are cited. Furthermore, some genetic predisposition could exist. In atrophic NU sites, osteoblast progenitor cells are significantly reduced [3]. In bone marrow from the iliac crest of atrophic NU, bone marrow-derived mesenchymal stem cells are in smaller number and have a reduced proliferation capacity [4].

In nontraumatic ON, apoptosis of osteocytes and cancerous bone lining cells in the necrotic lesion, as well as those at some distance from the lesion, in the proximal femur are increased [5]. The replicative capacities of osteoblastic cells obtained from the intertrochanteric area of the femur are reduced in patients with ON [6]. The number and the activity of fibroblast colony-forming units, reflecting the number of mesenchymal stem cells that could potentially give rise to mature osteoblasts, have been shown to be decreased in ON [7, 8]. Moreover, the capillaries serving as a conduit for the stem cells and bone cells needed in bone repair in addition to providing blood supply could be altered by emboli or thrombosis in ON [9].

So, in both pathological conditions, some impairment of the cellular part of the repair could exist, thereby provoking a reduction of MSC and of osteoblastic activation.

Several methods could be used to increase MSC population and its osteogenic differentiation in the pathological area:

- A local injection of bone marrow aspirates
- A preliminary culture of the bone marrow aspirate to increase the number of MSC cells
- A preliminary culture of the bone marrow aspirate to produce an expansion and an osteogenic differentiation of the MSC
- A genetic modification of the injected MSC to increase the secretion of growth factors like BMP and VEGF [11, 12]

The best treatment remains to be found for both conditions. Among the approaches developed so far, cell-based therapies to improve bone repair seem to be the most promising. These are based on the concept of the regenerative medicine and aim to recover an optimal bone repair process. This chapter summarizes the data published so far.

14.2 Clinical Trials in Nonunion Fractures

A recent review of the current technologies in bone healing and repair in human studies did not find any level I evidence concerning bone marrow aspirates or gene therapy [13]. Only a few studies support the therapeutic use of bone marrow transplantation in human [2].

A systematic review was conducted using PubMed, Medline. This research was completed checking references cited in listed articles. The key words were "bone marrow," "stem cells," "MSC," "nonunion fractures," and "cell-based treatment."

Unlike animals, in humans, only bone marrow (BM) aspirate implantations were, until now, used. The BM aspirates were, in some studies, concentrated before implantation (Table 14.1).

14.2.1 BM Aspirate

Connolly and coauthors seem to be the first to report results in a case of infected NU of the tibia in 1986 [14]. In a further report on the use of marrow graft for osteogenesis from 1986 to 1995 including 100 patients having a tibial NU, a good response was found in 80 % [15]. No complications were reported. The method used was made under general anesthesia. The patient was placed in a prone position and the marrow was aspirated in 3–5-ml aliquots. Simultaneously with the marrow aspiration, a second marrow needle was inserted into the site of the NU to directly inject the BM aspirate. The total injected volume was 100–150 ml. In two cases a second injection was performed. The authors gave no reason for this second injection. The healing time ranged from 6 to 10 months.

In 1990, Healey et al. published good results in 7/8 cases of NU after BM aspirate injection in situ [16]. These cases were all failures of osseous reconstruction (autologous iliac crest bone grafting) after lower-extremity resections for sarcoma affecting bone. The bone marrow, 5–6 ml at the beginning of the series to 3 ml at the end, was aspirated from the iliac crest under general anesthesia and directly injected in NU, until a total of 50 ml had been grafted. No heparin was used to avoid potential impairment of bone healing associated with heparin [17, 18]. In four cases a second injection was made when no healing process was observed on review of serial roentgenogram. The healing time ranged from 4 to 36 weeks (mean 18).

In 1993 Garg et al. applied a technique they had tested earlier on rabbits [19]. They grafted bone marrow percutaneously in 20 ununited long bone fractures (15 in the tibia, 3 in the humerus, and 2 in the ulna). Under general anesthesia, 15–20 ml of bone marrow aspirates (3–4 aspirations of 5 ml) from the posterior iliac crest was directly injected into the NU sites twice, at an interval of 3 weeks. All cases were immobilized in a plaster cast. In 17/20 cases, bone fusion was observed after 5 (3–7) months.

In 2005, Goel et al. presented results of BM grafting in tibial NU [20]. Under local anesthesia, 3–5 ml of marrow was aspirated from the anterior iliac crest and

Table 14.1 Nonunion fractures

Reference	Authors	Year	Cases	Aspirated vol (ml)	Concentrated vol (ml)	Repetition (week)	Healing time (month)	Positive results (%)	Safety
[14]	Connolly	1998	100	100–150	ND	2 cases × 2	6–10	80	Ok
[16]	Healey	1990	8	50	ND	5 cases × 2	4–36	88	Ok
[19]	Garg	1993	20	15–20	ND	× 2/3 weeks	3–7	85	Ok
[20]	Goel	2005	20	15	ND	× 2–3/4–6 weeks if no callus (mean 2.3 weeks)	2.5 (mean 3.3)	75	Ok
[21]	Hernigou	2005	60	300	50	No	1–4 (mean 3)	88	Ok

injected immediately percutaneously into and about the nonunion site. Subsequent aspirations were performed 1 cm posterior to the previous site until a maximum of 15 ml of marrow was injected. Injections were repeated at 4–6 weeks if there was no radiological evidence of callus formation. The procedure was considered a failure if there was no clinical and radiological union at 6 weeks following the third injection. The results revealed clinical and radiological bone union in 15 out of 20 patients (75 %), with an average time to union following the first injection of 14 weeks. Four patients (20 %) showed no evidence of union and were considered a failure. There were no cases of infection following the injection and no complications at the donor site.

14.2.2 Concentrated BM Aspirate

Only one trial using a concentration of the BM aspirate was published. In 2005, Hernigou et al. reported the results of a retrospective study including 60 tibial NU [21]. Under general anesthesia, 300-ml BM was aspirated from both anterior iliac crests, then filtered and concentrated by centrifugation on a cell separator. The 50-ml concentrated bone marrow was injected in NU. Weight bearing was not allowed during a minimum of 1 month and until a callus had appeared. Failure was considered when no healing existed after 6 months. In 53/60 patients, bone union was obtained in a mean of 12 weeks (range 4–16 week). They quantified the number of injected MSC and found a significantly lower count of MSC in the negative cases.

14.2.3 Others

Until now, there have been no human studies using gene-modified MSC, expanded MSC, or differentiated MSC in osteoblasts. Only a recent publication concerns the effect of autologous osteoblast (OB) to improve the fracture healing [22]. The autologous OB cells were obtained from a 4-week culture of 3–5-ml bone marrow aspirate. A mixture with 0.4 ml (12×10^6 cells) and fibrin was prepared and injected under local anesthesia into the fracture area. In this randomized study, a significant fracture healing acceleration was shown.

14.3 Clinical Trials in Osteonecrosis

A systematic review was also conducted using PubMed, Medline. This research was completed checking references cited in listed articles. The key words were "bone marrow," "stem cells," "MSC," "osteonecrosis," "bone necrosis," "avascular bone necrosis," and "cell-based treatment."

In 2002, Hernigou et al. reported the results of a non-controlled study of femoral head osteonecrosis [23]. The patients were followed up from between 5 to 11 years with a mean of 7 years. When patients were treated before collapse, hip replacement was done in 9 of the 145 hips. Total hip replacement was necessary in 25 of the 44 hips operated after collapse. The authors classified this study as level III evidence. However, the study did not have any control. The evaluation was only based on a comparison with the estimated natural evolution of cases as published in other studies. The correct level of evidence seems to be level IV. The method for implanting the bone marrow aspirate in the necrotic area was the same as that described by the same author in NU. The volume of BM aspiration made under general anesthesia was 300 ml. A filtration and a concentration by cell separator were performed. The final injected volume to inject into the necrotic area was 50 ml.

In 2004, Gangji et al. published a controlled, double-blind, prospective study including 18 femoral head ON before collapse [24]. These authors used core decompression with a 5-mm trephine with or without concentrated BM aspirate [25]. The method used to obtain and to prepare BM was Hernigou's method. After a 24-month follow-up period, there was a significant reduction in pain and joint symptoms within the BM graft group ($P = .021$). At 24 months, 5 of the 8 hips in the control group had deteriorated with an appearance of a collapse of the femoral head, whereas only 1 of the 10 hips in the BM graft group had progressed to this stage ($P = .016$). Survival analysis showed a significant difference in the time of collapse between the two groups. In addition, in the BM graft group, the volume of the necrotic lesion decreased by 35 %.

In 2011, the same group published a 5-year follow-up result of these cohorts [26]. Bone marrow implantation afforded a significant reduction in pain and joint symptoms and reduced the incidence of fractural stages. At 60 months, 8 of the 11 hips in the control group had deteriorated to the fractural stage, whereas only 3 of the 13 hips in the bone marrow graft group had progressed to this stage. At 60 months, survival analysis showed a significant difference in the failure time between the two groups. Patients suffered only minor side effects after treatment.

In 2008, a publication in Chinese presented a retrospective study using a different method of treatment [27]. A 3-tunnel core decompression was performed in the femoral head to allow implantation of bone marrow MSC and decalcified bone matrix. Among the 87 patients (103 hips), the average rate of excellent to good results (based on clinical and radiological evaluation) was deemed to be 75.7 % after a follow-up of a mean of 26 months. No further details were provided.

In 2010, Wang et al. reported the results of 59 ON of the femoral head (before or after collapse) in a prospective non-controlled study [28]. The 100–180-ml BM aspirate was concentrated to 30–50 ml. The implantation into the necrotic area was done through two to three holes made using a trocar with a 3.5-mm outer diameter. The follow-up was a mean of 27 months (range: 12–40). Clinically, the overall success was deemed in 80 % and hip replacement was made in 7/59 hips (11.9 %).

In 2011, Yoshioka et al. published a retrospective non-controlled study of six patients (nine hips) suffering from corticoid-induced ON with SLE [29]. The X-ray stage was non-collapsed in six hips and collapsed in three hips. The BM aspirate from iliac crest (336 ± 88 ml) was concentrated to 31 ± 7.6 ml.

In a follow-up report of a minimum of 3 years, significant improvement in pain and Harris Hip score was observed. Only one hip required replacement.

Finally, in 2012, a Chinese publication detailed a prospective randomized controlled study using MSC expansion obtained after a 2-week culture of BM aspirates in 100 ON patients (104 hips) [30]. The etiological factors were trauma in 20 patients, corticosteroid use in 24, alcohol abuse in 19, Caisson disease in 11, and idiopathic in 30. All had non-collapsed ON stages. A volume of 10 ml BM was aspirated, not in the iliac crest as in the other studies but in the subtrochanteric area where the core decompression has been made. The decompression was an original technique including the removal of the necrotic tissue by a custom-made trephine with a collapsible scraping end, a plugging of the bored bone core into the decompression tunnel, and a scaling of the outlet of this tunnel with bone wax. The MSC present in BM was expanded by a 2-week culture. In each case, 2.10^6 MSC was prepared and injected into the necrotic area. The MSC was injected through the bored plug. Postoperative cares included bed rest with skin traction for 3 weeks and non-weight bearing for 6 weeks.

At 60 months, only 2/53 MSC treated hips compared with 10/44 control hips needed surgical treatment. MSC treatment significantly reduced the volume of necrosis and improved the Harris Hip score. No complications were observed in either group. A summary of some key studies is presented in Table 14.2.

14.4 Concluding Remarks

In NU, this review shows that the therapeutic effect of MSC is only supported by some studies using BM aspirate, concentrated or not, of evidence level IV. Several differences between these studies must be noted. The type of NU and the therapeutic methods were not the same. Different methods to harvest and to inject bone marrow were used. The volume and the number of injected MSC (when evaluated) were quite variable.

Good results were found in all studies. With small volume (15–20 ml) and without any concentration, they were 83 % [19] and 75 %, respectively [20]. With larger volumes (300 ml), and after concentration, the positive results increased slightly to 88 % [21]. Clearly, the question of the best method, and the interest of larger BM aspirate volumes, remained to be resolved.

An additional question is the interest of an injection of volume larger than the volume of the lesions. What are the effects on homing and proliferation of injected MSC? In addition, it remains to clarify if the bone repair is boosted by the injected MSC or by other components of the BM aspirate like endothelial cells and growth factors. Trials using BMP have proven their efficacy in seven studies with level I evidence [13].

Table 14.2 Osteonecrosis

Reference	Authors	Year	Study design	Non-collapsed/collapsed hips	Aspirated vol (ml)	Concentrated vol (ml)	Follow-up (year)	Positive results (%)	Remarks
[23]	Hernigou	2002	Retrospective	145/44	300	50	5–11 (7.7)	94/43	
[24]	Gangji	2004	Controlled prospective	18/0	400	50	2	90	
[26]	Ji	2008	Retrospective	103/0	?	?	2.1	76	Three-tunnel core
[27]	Wang	2010	Open prospective	50/9	100–180	30–50	2.5	88/44	Two to three tunnel core
[25]	Gangji	2011	Controlled prospective	24/0	400	50	5	77	Including cases Ref. [24]
[28]	Yoshida	2011	Retrospective	6/3	336	31	3	83/66	Corticoids

In ON, the effect of BM implantation was tested in two trials, with level II evidence [24, 25], and four trials, with level IV evidence [23]. The method of harvesting and concentrating the injected volume of bone marrow, as well as the method of implantation, was the same in three, whereas two other studies used two or three tunnel cores. All the results are very promising but need to be confirmed in larger randomized control studies. The same answer concerning the relationship between the injected volume and the lesion volume also needs further research.

A first randomized clinical trial, using MSC expansion before implantation into the necrotic area, was recently published [30]. The results were very positive with a good safety level, an absence of collapse in 95 %, and a significant reduction of the necrotic area. The comparison with the other studies is however not easy because the core decompression was made following a new technique. This factor could interfere with the cell-based part of the trial.

Finally, we have found no data to confirm that the therapeutic effect of BM aspirate is due to its cellular part, especially MSC, rather than to growth factors. In conclusion, these reviews confirm that BM aspirate could induce bone repair in NU and ON. However, the data is very preliminary and many questions remain to be answered.

Bone reconstruction is a long process. Radiological tools give late imaging of the bone repair, together with a late response concerning the efficacy of any therapeutical approach of bone healing impairment. It is important to detect and assess the osteogenic process due to the treatment as soon as possible. This early evaluation could allow us to confirm that the bone healing has started and that the evolution is positive.

Another objective of such quantification is to monitor the bone healing process: to detect any delay, to compare the osteogenesis rate for different therapeutical programs, such as different cell populations and levels of differentiation, as well as the dose-response rate, the relationships between different cells, growth factors and non-biological factors. Finally, tracking and following the homing of the injected cells postinjection is a key point in understanding and developing the therapeutical success of bone repair.

References

1. Gaston MS, Simpson AH (2007) Inhibition of fracture healing. J Bone Joint Surg Br 89(12):1553–1560, Review
2. Gangji V, Hauzeur JP (2009) Cellular-based therapy for osteonecrosis. Orthop Clin North Am 40(2):213–221
3. Hernigou P, Beaujean F (1997) Bone marrow in patients with pseudarthrosis. A study of progenitor cells by in vitro cloning. Rev Chir Orthop Reparatrice Appar Mot 83(1):33–40
4. Seebach C, Henrich D, Tewksbury R, Wilhelm K, Marzi I (2007) Number and proliferative capacity of human mesenchymal stem cells are modulated positively in multiple trauma patients and negatively in atrophic nonunions. Calcif Tissue Int 80(4):294–300
5. Calder JD, Pearse MF, Revell PA (2001) The extent of osteocyte death in the proximal femur of patients with osteonecrosis of the femoral head. J Bone Joint Surg Br 83(3):419–422

6. Gangji V, Hauzeur JP, Schoutens A, Hinsenkamp M, Appelboom T, Egrise D (2003) Abnormalities in the replicative capacity of osteoblastic cells in the proximal femur of patients with osteonecrosis of the femoral head. J Rheumatol 30(2):348–351
7. Hernigou P, Beaujean F (1997) Abnormalities in the bone marrow of the iliac crest in patients who have osteonecrosis secondary to corticosteroid therapy or alcohol abuse. J Bone Joint Surg Am 79(7):1047–1053
8. Hernigou P, Beaujean F, Lambotte JC (1999) Decrease in the mesenchymal stem-cell pool in the proximal femur in corticosteroid-induced osteonecrosis. J Bone Joint Surg Br 81(2):349–355
9. Simkin PA, Downey DJ (1987) Hypothesis: retrograde embolization of marrow fat may cause osteonecrosis. J Rheumatol 14(5):870–872
10. Hayashi O, Katsube Y, Hirose M, Ohgushi H, Ito H (2008) Comparison of osteogenic ability of rat mesenchymal stem cells from bone marrow, periosteum, and adipose tissue. Calcif Tissue Int 82(3):238–247
11. Tang Y, Tang W, Lin Y, Long J, Wang H, Liu L, Tian W (2008) Combination of bone tissue engineering and BMP-2 gene transfection promotes bone healing in osteoporotic rats. Cell Biol Int 32(9):1150–1157
12. Liu BY, Zhao DW (2009) Treatment for osteonecrosis of femoral head by hVEGF-165 gene modified marrow stromal stem cells under arthroscope. Zhonghua Yi Xue Za Zhi 89(37):2629–2633
13. Novicoff WM, Manaswi A, Hogan MV, Brubaker SM, Mihalko WM, Saleh KJ (2008) Critical analysis of the evidence for current technologies in bone-healing and repair. J Bone Joint Surg Am 90(Suppl 1):85–91
14. Connolly JF, Shindell R (1986) Percutaneous marrow injection for an ununited tibia. Nebr Med J 71(4):105–107
15. Connolly JF (1998) Clinical use of marrow osteoprogenitor cells to stimulate osteogenesis. Clin Orthop Relat Res 355S:S257–S266
16. Healey JH, Zimmerman PA, McDonnell JM, Lane JM (1990) Percutaneous bone marrow grafting of delayed union and nonunion in cancer patients. Clin Orthop Relat Res 256:280–285
17. Connolly JF, Guse R, Tiedeman J, Dehne R (1989) Autologous marrow injection for delayed unions of the tibia: a preliminary report. J Orthop Trauma 3(4):276–282
18. Stinchfield FE, Sankaran B, Samilson R (1956) The effect of anticoagulant therapy on bone repair. J Bone Joint Surg Am 38-A(2):270–282
19. Garg NK, Gaur S, Sharma S (1993) Percutaneous autogenous bone marrow grafting in 20 cases of ununited fracture. Acta Orthop Scand 64(6):671–672
20. Goel A, Sangwan SS, Siwach RC, Ali AM (2005) Percutaneous bone marrow grafting for the treatment of tibial non-union. Injury 36:203–206
21. Hernigou P, Poignard A, Beaujean F, Rouard H (2005) Percutaneous autologous bone-marrow grafting for nonunions. Influence of the number and concentration of progenitor cells. J Bone Joint Surg Am 87(7):1430–1437
22. Kim SJ, Shin YW, Yang KH, Kim SB, Yoo MJ, Han SK, Im SA, Won YD, Sung YB, Jeon TS, Chang CH, Jang JD, Lee SB, Kim HC, Lee SY (2009) A multi-center, randomized, clinical study to compare the effect and safety of autologous cultured osteoblast(Ossron) injection to treat fractures. BMC Musculoskelet Disord 12:10–20
23. Hernigou P, Beaujean F (2002) Treatment of osteonecrosis with autologous bone marrow grafting. Clin Orthop Relat Res 405:14–23
24. Gangji V, Hauzeur JP, Matos C et al (2004) Treatment of osteonecrosis of the femoral head with implantation of autologous bone marrow cells: a pilot study. J Bone Joint Surg Am 86-A:1153–1160
25. Hauzeur JP, Orloff S, Taverne-Verbanck J, Pasteels JL (1986) Diagnosis of aseptic osteonecrosis of the femoral head by percutaneous transtrochanterian needle biopsy. Clin Rheumatol 5(3):346–358
26. Gangji V, De Maertelaer V, Hauzeur JP (2011) Autologous bone marrow cell implantation in the treatment of non-traumatic osteonecrosis of the femoral head: five-year follow-up of a prospective controlled study. Bone 49(5):1005–1009

27. Ji WF, Ding WH, Ma ZC, Li J, Tong PJ (2008) Three-tunnels core decompression with implantation of bone marrow stromal cells (bMSCs) and decalcified bone matrix (DBM) for the treatment of early femoral head necrosis. Zhongguo Gu Shang 21(10):776–778
28. Wang BL, Sun W, Shi ZC, Zhang NF, Yue DB, Guo WS, Xu SQ, Lou JN, Li ZR (2010) Treatment of nontraumatic osteonecrosis of the femoral head with the implantation of core decompression and concentrated autologous bone marrow containing mononuclear cells. Arch Orthop Trauma Surg 130(7):859–865
29. Yoshioka T, Mishima H, Akaogi H, Sakai S, Li M, Ochiai N (2011) Concentrated autologous bone marrow aspirate transplantation treatment for corticosteroid-induced osteonecrosis of the femoral head in systemic lupus erythematosus. Int Orthop 35(6):823–829
30. Zhao D, Cui D, Wang B, Tian F, Guo L, Yang L, Liu B, Yu X (2012) Treatment of early stage osteonecrosis of the femoral head with autologous implantation of bone marrow-derived and cultured mesenchymal stem cells. Bone 50(1):325–330

Chapter 15
Stem Cell Therapies for the Treatment of Spinal Cord Injuries: Current Progress in Basic Science and Clinical Research

Hilary North and John Kessler

Abstract Spinal cord injury (SCI) is a debilitating condition affecting an estimated 1,275,000 Americans at a cost of over 40 billion dollars each year. The main causes of SCI are automobile accidents, falls, other accidents, and violence such as gunshot or stab wounds. Depending on the precise location and severity of the insult, patients experience a range of motor, sensory, and autonomic impairments resulting either from disruption of ascending and descending axonal tracts or damage to the local neuronal circuitry at the injury site. Although much effort has been dedicated to the development of treatments and cures for this condition, to date, there is no effective way to reinstate motor, sensory, or autonomic functions. The burgeoning field of stem cell research has offered exciting new possibilities for the treatment of SCI, but little success has been realized in the limited clinical trials that have been performed thus far. The following chapter will review the cellular consequences of SCI, the efforts made to counteract these consequences by non-stem cell approaches, the stem cell-based strategies currently being investigated in preclinical studies, and the current state of clinical stem cell trials on patients suffering from SCI.

Keywords Spinal cord injuries • Cell replacement • Neuroprotection • Trophic support

H. North
Department of Neurology, Feinberg School of Medicine,
Northwestern University,
Chicago, IL 60611, USA

J. Kessler (✉)
Feinberg School of Medicine, Northwestern University,
Chicago, IL 60611, USA

Departments of Neurology and Molecular Pharmacology and Biological Chemistry,
Feinberg School of Medicine, Northwestern University,
675 N. St. Clair, Galter 20-100, Chicago, IL 60611, USA
e-mail: jakessler@northwestern.edu

M.K. Danquah, R.I. Mahato (eds.), *Emerging Trends in Cell and Gene Therapy*,
DOI 10.1007/978-1-62703-417-3_15,

15.1 Epidemiology and Heterogeneity of Spinal Cord Injuries (SCIs)

It is difficult to know precisely how many individuals worldwide suffer from SCI, but estimates range from 11.5 to 58 millions. Spinal cord injuries occur in numerous different ways resulting in heterogeneous clinical presentations [1, 2]. This variability poses a challenge to both clinicians and researchers. The most common causes of SCIs are car accidents, which account for up to 58 % of incidents in developed nations, falls, sports, or violence such as stab and gunshot wounds. The incidence of SCI is highest in persons aged 20–40 years and disproportionately afflicts males three to four times more frequently than females. The spinal cord is subject to four vectors of force: extension, flexion, rotation, and compression, and most injuries are a combination of these, resulting in a crush injury. However SCI is occasionally caused by penetration or maceration of the spinal column [3, 4]. Despite the prevalence of contusion-type injuries, transection lesions have been used as a model in many animal studies because of their reproducibility; their relevance to the clinical state is questionable, however, and contusion injury models are being used increasingly in animal studies since they provide a histological response profile that is more relevant to human cases [5]. While the cervical spine is the most common location of SCI (40–70 % of cases), and 34–60 % of injuries are incomplete, the injuries that occur in humans under disparate conditions, at various levels of the spinal cord, and with varying degrees of severity are heterogeneous. It is therefore difficult to assemble a cohort of patient subjects suffering from sufficiently similar injuries to properly test therapeutic approaches, and this has been a major roadblock in the clinical assessment of potential SCI treatments (Fig. 15.1).

15.1.1 Critical Challenges Facing the Treatment of SCI

15.1.1.1 Description of Pathological Responses Post-SCI

Demyelination

The myelin sheath surrounding the axons of neurons is essential for proper transmission of action potentials and communication of signals between cells. Demyelination of spared axons after SCI and death of the oligodendroglia (OL) responsible for producing myelin may contribute to the functional deficits experienced by patients. Contusive injury to the spinal cord results in both immediate and secondary damages [5]. Immediately, there is damage to axons and cell bodies at the site of injury and disruption of the blood-brain barrier (BBB). Within the first day following injury, spinal neurons die of exocytotic damage and necrosis. At the same time, OL undergo their first of two waves of apoptotic death. The second, subacute wave of OL apoptosis lasts for several weeks post-injury. As a consequence

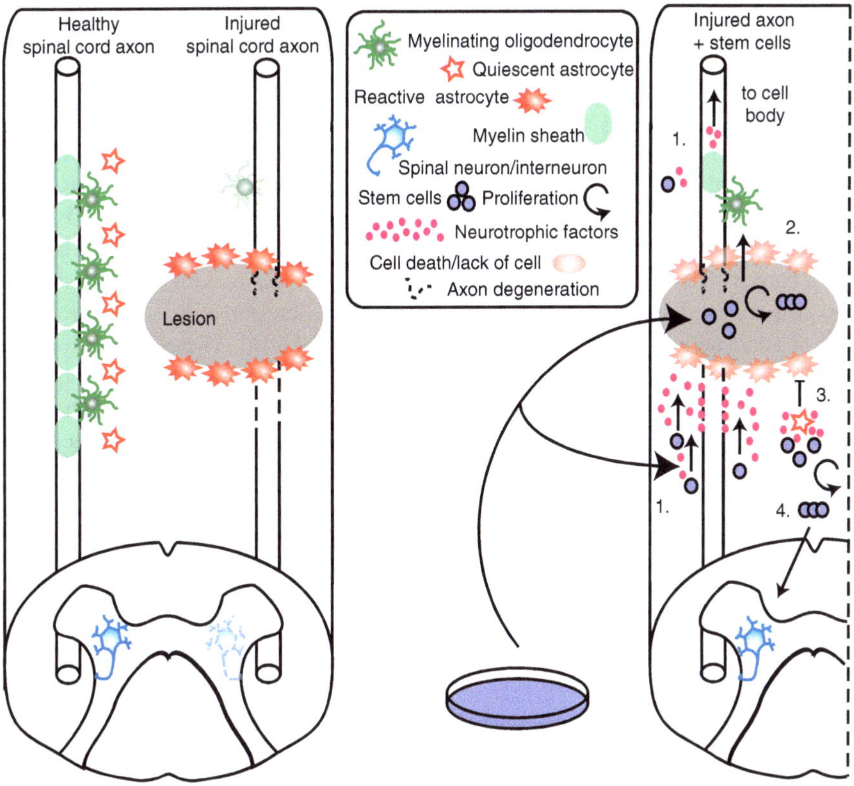

Fig. 15.1 Spinal cord injury challenges and stem cell therapy approaches. In the healthy spinal cord, oligodendrocytes myelinate axons and quiescent astrocytes exist throughout the tissue. After an injury, astrocytes become enlarged, more plentiful and migrate to the lesion area to form a scar that impedes axon regrowth. Injured axons degenerate away from the lesion, and oligodendrocytes die, resulting in demyelination of spared fibers. Stem cells have been implanted into the injured human spinal cord in attempts to (*1*) increase levels of neurotrophic factors that protect injured neurons from dying and promote outgrowth, (*2*) replace the myelinating oligodendrocyte population, (*3*) to alleviate glial scar formation, and (*4*) to replace local neuron populations

of this extensive OL death, any spared axonal fibers that may have survived the initial injury begin to lose their myelination and become unable to transmit action potentials effectively. This demyelination persists; demyelinated axons have been observed even 10 years after SCI in humans [6]. It is estimated that just 10–15 % of white matter is required for retention of significant functionality [7], so preventing demyelination of axons spared after injury could have significant clinical benefits. Additionally, any successful axonal sprouting resulting in restoration of neuronal circuits during recovery from injury will be less efficiently myelinated and similarly compromised in signal transmission capability. Substantial efforts have therefore focused on stem cell transplantation and other strategies to replace or spare OL after SCI.

Glial Scar Formation

While neurons and OL die in the days and weeks following injury, astrocytes instead change their morphology and proliferate in response to the assault. These changes also occur in two waves: the first, reactive astrogliosis is marked by a hypertrophic morphological change and upregulation of intermediate filaments such as glial fibrillary acidic protein (GFAP). The primary function of reactive astrogliosis is to protect the central nervous system (CNS) by helping to reseal the breeched blood-brain barrier (BBB). The second phase of gliosis, hyperplasia, is an accumulation of new astrocytes, thought to be generated from the proliferation of existing astrocytes and of adult neural stem/progenitor cells in the spinal cord. These new astrocytes join with infiltrating and proliferating microglia and macrophages that together surround the lesion site and form the glial scar. While the initial hypertrophic phase of glial scar formation has beneficial effects on wound closure and functional recovery after SCI, the later, hyperplasic phase is thought to be inhibitory to axonal regeneration [8–10]. In addition to posing a physical barrier to axon regeneration, the glial scar contains a number of molecules such as chondroitin sulfate proteoglycan (CSPG), Nogo, and myelin-associated glycoprotein (MAG) that inhibits axon regrowth. Mitigation of glial scar formation has therefore been a major focus of research into potential SCI therapies.

Immune Reaction

Immune cells play important roles in the cellular responses to SCI beyond their incorporation in the glial scar. The breech in the BBB allows infiltration of immune cells such as lymphocytes, macrophages, T cells, and neutrophils, contributing to secondary damage by upregulating necrosis and apoptosis. Though such immune components may have beneficial effects after injury [11], microglia/macrophages in particular are thought to contribute to secondary damage. After SCI, microglia become activated, undergo morphological changes, and secrete factors that induce apoptosis of OL, contributing to the demyelination of spared axons. Thus, control of the immune response is a strategy being investigated for the minimization of secondary damage to the injured spinal cord.

Endogenous Reparative Processes

Although severe neural injuries rarely recover in a meaningful way, some endogenous cellular and molecular responses indicate that the body makes an inherent attempt at regaining neural connectivity and functionality. Understanding which of these changes may improve functional recovery and designing strategies to harness and enhance the natural ability of the body to repair damaged tissue may lead to effective therapies. Fundamentally, CNS axons have the ability to regenerate. After injury, axons revert from their functional "transmission" state to their developmental "growth" mode. Regrowth potentially occurs at the slow rate of 1 mm per day, but it is actively inhibited by a number of factors such as the glial scar and passively inhibited by the absence of growth factors that are required for the robust outgrowth observed during development [12, 13]. Progenitor cells that are quiescent during the normal, healthy adult state begin to proliferate, possibly in an attempt to replenish the diminishing OL population [14–17]. Various forms of plasticity occur including the sprouting of injured axons onto spared axons to improve connectivity [18, 19], functional reorganization of sensorimotor cortical areas [18, 20], and subcortical rubrospinal compensation for some lost corticospinal function [20].

Alas, these endogenous responses to injury and apparent attempts at functional recovery after SCI fall far short of what is necessary for the majority of patients to regain sensory and motor control. Many strategies investigated at the laboratory and clinical trial levels have attempted to foster a growth-permissive environment so that the damaged spinal cord can successfully repair itself by enhancement of the inherent regenerative responses that occur after injury.

Consequences to the Patient

Taken together, these primary and secondary injuries to the spinal cord result in a loss of synaptic transmission both between neurons within a circuit and between neurons and their terminal innervation sites in the musculature and sensory receptors. Depending on the spinal column level at which the injury occurs, patients experience a range of functional deficiencies that produce varying effects on lifestyle. Injuries are classified as either complete or incomplete; a complete injury leaves a patient unable to exert voluntary muscle movements or detect conscious sensory information from all parts of the body situated below the level of injury. The level of injury to the spinal cord dictates the areas of lost function. For example, complete injuries between cervical vertebral levels two and four (C2–C4) result in loss of sensory and motor function in all limbs, loss of autonomic functions (bowel, bladder, sexual), and loss of spontaneous respiration; complete lesions at C5 to T1 lead to loss of all function in the trunk and legs with varying effects on the arms; thoracic lesions spare the arms but affect everything below the chest, back, and abdomen; lumbar and sacral lesions lead to varying effects on the legs and on autonomic functions; L1 and L2 control flexion at the hips which is important for standing and walking, and lesions below L2 spare hip flexion; and L3 through S5 innervate the remainder of the legs and mediate autonomic functions [21].

15.1.1.2 Treatment Approaches to SCI

There are currently no FDA-approved therapies for spinal cord injury, and no interventions that have been proven to alter clinical outcome. A widely accepted and implemented treatment for acute spinal cord injury involves surgical decompression of the injury site. Past studies have not shown that this intervention alters clinical outcome, but a large multicenter trial is in progress to define the potential benefits. Infusion of methylprednisolone is also used in the acute setting of SCI. This treatment was widely adopted following the report of the Second National Acute Spinal Cord Injury Study (NASCIS II) in 1990 and became an implied standard of care. However, subsequent clinical studies and critical reviews suggest that methylprednisolone has little or no positive impact on clinical outcome [22–26]. Furthermore, methylprednisolone use may lead to complications for the patient that may outweigh any potential benefits [27].

A variety of other approaches have been utilized clinically and experimentally with varying but marginal degrees of success. Peripheral nerves, which are known to regenerate after injury, have been grafted into central spinal cord injuries in an attempt to harness the regenerative capacity of the peripheral nerves. These studies yielded limited success, and spinal axon regeneration was limited to short distances [28, 29]. When a similar approach was taken in a nonhuman primate study, no functional recovery was observed [30]. Although one human study reports some functional improvement in one patient after a peripheral nerve graft, proper control patients were not included [31], and, anecdotally, this approach seems unsuccessful after complete SCI [32]. With remyelination of spared axons in mind, Schwann cells, the cellular population responsible for myelinating peripheral nerve axons, have been implanted into rodent models of SCI including contusion injury, lateral hemisection, and complete spinal cord transection [33–35]. Although injured axons successfully grew into the Schwann cell grafts, remyelinated, and were able to conduct electrical impulses [33, 35, 36], the axons failed to grow out of the graft and innervate the host tissue. The feasibility of this approach for human use was investigated when human Schwann cells were implanted into immune-compromised injured rodents [37]. While some functional improvement was reported in the human Schwann cell-injected rodents, it was not sufficient to allow weight support by the hind limbs in more than one animal. Schwann cell implantation has also been used in combination with other strategies such as the upregulation of neurotrophic factors [38, 39] or other compounds or cell types [4, 40, 41].

Because various neurotrophic factors are known to be crucial to the ability of young axons to grow during development, one approach for treating injury has been to artificially increase the levels of these factors in the adult nervous system, where they are substantially diminished. This has been done both in isolation [42] and in combination with other strategies [38, 39] with some limited success. For example, in one study, injection of glial cell line-derived neurotrophic factor (GDNF) into injured rodent spinal cords significantly improved behavioral functionality compared to saline-injected control animals [42]. Recovery was, however, limited, and the mechanism of action remains unclear. Other molecular pathways involved in axon elongation have also been targeted for upregulation after injury. For example, the second messenger cyclic adenosine monophosphate (cAMP) and rho-family GTPases have been targeted for their ability to promote axon outgrowth and to overcome the negative effects of inhibitory molecules present in the injured spinal cord including Nogo, MAG, and oligodendrocyte myelin glycoprotein (OMgp) [43, 44].

A host of alternative substrates has been investigated for their potential to facilitate central axon regeneration. Nitrocellulose treated with compounds known to promote growth such as laminin and poly-L-lysine [45], collagen matrices [46, 47], carbon filaments [48], ionic synthetic hydrogels [49], and self-assembling peptide amphiphiles (PA) designed to mimic the extracellular matrix [50] has been implanted into the lesion sites of rodents to facilitate axon regeneration and produced limited favorable outcomes. Experiments investigating the effects of the PA, which presents a high density of laminin-mimicking epitopes to the cells its surrounds, revealed that this approach facilitates not only axon outgrowth but also beneficial effects on the glial scar as described below.

Finally, some approaches are designed to mitigate the inhibitory effects of the glial scar. The glial scar formed after injury expresses a number of molecular factors, most notably proteoglycans, which are inhibitory to axon outgrowth. Application of chondroitinase, an enzyme that largely removes the sugar chain from the chondroitin family of proteoglycans (CSPGs), has successfully relieved inhibition in the scar and improved growth of injured axons towards their original targets [51]. Other inhibitory factors found in the glial scar such as semaphorin3 and members of the Eph/ephrin receptor tyrosine kinase family, and the myelin-associated protein Nogo, have also been inhibited with a limited degree of success [52]. Other approaches have attempted to limit the amount of gliosis as a comprehensive way to diminish expression of these inhibitory molecular factors as well as the physical barrier that the scar presents. Molecular targeting of the signaling cascades that mediate reactivity and proliferation of astrocytes has had promising effects on axon regeneration. For example, BMP signaling attenuation can promote lesion closure without inhibiting axon elongation [10], and a laminin-mimicking PA inhibits glial scar formation in this manner [50]. The glial scar remains a main focus of SCI research but attempts to limit its inhibitory capacity appear to be insufficient to restoring axon integrity on their own.

While some of these approaches have yielded moderate regeneration into the lesion, rarely have axons been shown to grow through the lesion.

15.1.1.3 Stem Cell Strategies

There are several different strategies for using stem cells to treat SCI including (a) replacement of dead or dysfunctional cells, both neurons and OL; (b) creation of a more growth- friendly environment to encourage the regeneration or survival, via trophic support, of existing, damaged axons or improving the integrity of the growth substrate by filling cavities; and (c) relief of detrimental, inhibitory glial scarring. Before addressing the major challenge of improving SCI in patients using stem cells, however, researchers and clinicians face a host of roadblocks involving the general use of stem cells for therapy including immune reaction to implants [53], potential development of cancers from the stem cells which are highly proliferative [54], and other dangers posed by stem cell implantation including pain [55]. Finally, it is difficult to conclude from the current stem cell clinical research which approach is most promising, as each approach has been evaluated separately and in a range of SCI types rather than compared directly under controlled conditions [53].

The potential immune response after transplantation of stem cells is a major issue, and a number of the clinical trials that have occurred to date have primarily tested strategies to avoid immune rejection. An ideal cell implantation approach would utilize autologous stem cells (i.e., a stem cell population isolated from patient) and therefore not require the complicated and risk- filled process of immunosuppression [53]. Several stem cell populations fulfill this criterion and have been tested in various clinical trials: bone marrow stromal cells [56–58], hematopoietic (blood-derived) stem cells [59], olfactory ensheathing cells [60], and umbilical cord cells [61, 62], which may be available autologously if the patient had them cryoprotected at birth [53]. However, procedures for utilizing neural stem cells, to date, require the use of heterologous populations, and clinical applications using them may require immunosuppression of the patient [53]. Alternatively, the field of induced pluripotent stem cell (iPSC) generation has been rapidly expanding and may one day provide reliable sources of neural stem cells for use in treating SCI [63, 64].

15.1.2 Current Research into Stem Cell Strategies

15.1.2.1 Cell Replacement

Although SCI is generally viewed as a problem involving interruptions in descending motor and ascending sensory tracts, local cell death near the site of injury also poses a major hurdle to functional recovery. The gray matter contains local neuronal circuits and interneurons, and cell bodies of myelinating OL. These populations die of exocytotic damage and necrosis, or apoptosis, respectively, in the days and weeks following SCI, and the damaged adult spinal cord cannot replace these cells from endogenous stem/progenitor cells [65, 66]. A logical approach to replacing these depleted local cell populations is implantation of pluripotent embryonic stem (ES) cells or other stem cells with the potential to generate neurons or OL.

A major challenge to all stem or progenitor cell implantation strategies is determining precisely what type of cell to implant. Theoretically, ES cells are an ideal population because of their pluripotency and their expansive capacity. However, since implantation of unmanipulated ES cells leads to teratomas [67], substantial effort has been devoted to devising protocols for differentiating ES cells into specific cell types, or into more restricted stem and progenitor states, prior to implantation. McDonald et al. coaxed rodent ES cells towards a neural identity by treatment in culture with retinoic acid [68, 69]. Once transplanted into the injured spinal cord, the treated ES cells differentiated into oligodendrocytes and astrocytes and, to a lesser extent, neurons. They survived more than a month, migrated 8 mm from the site of injection, did not form tumors, and enhanced functional recovery as measured by open field testing (BBB scale).

A more focused strategy involves pre-differentiating cells to an OL phenotype to potentially remyelinate spared axons after an incomplete injury [70, 71]. Keirstead et al. differentiated human ES cells (hESCs) into an OL-committed precursor state in vitro and implanted the resulting oligodendrocyte precursor cells (OPCs) into the spinal cord 1 week after SCI. Many cells migrated to the lesion site and differentiated into mature oligodendrocytes. The stem cell treatment enhanced functional recovery on open field testing even though there was no notable difference in remyelination between the implanted group and the control group in which remyelination was also robust. There were, however, fewer demyelinated axons following injury and fewer improperly myelinated axons in the experimental group. The authors propose that the OPC implants remyelinate axons more effectively than endogenous remyelination mechanisms leading to the observed functional recovery. This approach is ineffective in models of complete SCI [5], presumably because there are few spared axons to remyelinate.

ES cells differentiated into even more restricted lineages have also been used to produce neurons after implantation. These neuronal lineage-restricted precursor cells (NRPs) are generated by the differentiation of the stem cells in culture and isolation by FACS sorting according to the expression of the cell-surface antigen embryonic neural cell adhesion molecule (E-NCAM) [72]. When implanted into the spinal cord, these NRPs can survive at least a month, differentiate into neurons, and extend processes into the gray and white matter [72]. Some groups have pre-differentiated pluripotent stem cells into specific types of neurons before implantation. For example, Harper et al. exposed ES cells to retinoic acid and sonic hedgehog [73] in culture to differentiate them into neurons with a transcriptional profile akin to that of spinal motoneurons. These motoneurons survived for over 1 month when implanted into injured rodent spinal cords, but their growth was hindered, presumably by the same factors that inhibit endogenous axon growth in the injured spinal cord [74].

Other studies have attempted to introduce new neurons into injured spinal cords by means of fetal grafts. Anderson et al. implanted solid or suspended populations of neural cells from fetal spinal cord, brainstem, and neocortex into injured cat spinal cords and found integration into the host tissue as well as differentiation of graft cells into specific types of neurons, but functional results were limited and variable [75]. Similar experiments performed in rodent models of SCI have generated equivalent results: injured axons grew into, but not out of, the fetal tissue grafts [76–79] with statistically significant but only minimally improved functional recovery. The authors hypothesize that fetal grafts enhanced function by acting as a relay capable of transmitting neuronal impulses over the lesion site [4]. Such grafts might also contribute growth factors and trophic support to compromised host axons, a targeted approach taken by other groups and discussed below. Indeed, this approach has been more successful when performed in conjunction with neurotrophin delivery [80, 81].

Finally, there have been attempts to replace damaged neurons by implanting stem cells derived from non-neural lineages. Some evidence suggests that certain non-neural stem cell lines have the ability to "transdifferentiate" into neural cells. This approach was attractive because it opened up the possibility that grafts could be generated from a patient's own tissue, thereby eliminating concerns about immune rejection. Stem cells derived from bone marrow, umbilical cord, blood, and skin have been implanted into the injured rodent spinal cord with varying degrees of improvement in functional recovery [82]. However, there is little if any evidence of transdifferentiation of such cells into neural phenotypes. Indeed, clinical trials of bone marrow implantation into injured human spinal cords have revealed no significant improvements in patient condition, and the mechanism underlying the results of the preclinical studies remains unclear [57, 83].

15.1.2.2 Neuroprotection and Trophic Support

Another approach is to try to condition the environment at and around the lesion site to encourage axonal outgrowth. Stem cells are an intriguing tool for this task since they secrete a variety of growth factors and cytokines. Thus, their presence alone, regardless of whether they incorporate into neural circuits in the injured spinal cord, could potentiate the regrowth and remyelination of injured neurons and facilitate the survival of cells at risk of death from secondary injury processes.

Facilitation of Axon Regeneration and Local Neuron Regrowth

The possibility that the unidentified properties governing the ability of PNS axons to regenerate might hypothetically assist CNS axons to regenerate was first examined in the early 1900s [84]. Tello noted that rabbit-denervated cortical fibers were able to grow into peripheral nerve grafts. This work was not widely accepted, however, and it was not until the 1980s that the concept was revisited in the context of SCI research [29]. Researchers at McGill University in Canada reported sprouting of injured rat spinal cord axons into peripheral grafts containing Schwann cells. This work has since been applied successfully to chronic SCI in rats [85] and acute SCI in nonhuman primates [30].

These successful animal studies and similar others prompted a search for the specific factor(s) of PNS cells that can promote CNS regeneration. Schwann cells, the myelinating glia of the PNS, are thought to be largely responsible for the effects of PNS grafts on injured CNS axons [29]. A number of animal model and clinical studies have therefore investigated the potential for Schwann cell grafts to improve axon regeneration after SCI. Paino and Bunge filled the cystic cavities of adult rat injured spinal cords with collagen rolls containing Schwann cells and noted that the axons growing into the graft were usually associated with the transplanted cells, underscoring the potential of Schwann cells to provide unique facilitation of spinal axon outgrowth. Similar approaches were tested in conjunction with methylprednisolone treatment [40] and with the myelinating cell population of olfactory ensheathing cells [41]. Olfactory ensheathing cells are glial cells that are found around sensory axons in the olfactory mucosa of the nose. They are the only type of glial cell found in both the central and peripheral nervous system, making them an attractive type of cell to use for their accessibility and potential compatibility within the spinal cord. They have been shown to promote axonal growth in a manner similar to Schwann cells [86], and results from several dozen preclinical studies in animal models of SCI demonstrated limited recovery of motor, sensory, and bladder function, possibly by assisting the myelination of regenerating axons after injury [87, 88]. These preclinical studies formed the foundation for several clinical trials investigating cell replacement therapies [60, 89], reviewed in Sect. 15.3 of this chapter.

The marginally promising results of Schwann cell and olfactory ensheathing cell preclinical and clinical trials prompted studies of the potential benefits of combinational approaches. Taylor et al. combined the use of stromal cell implants to provide a permissive substrate for axon growth with lentiviral upregulation of NT-3 to stimulate axon sprouting [90]. Like many approaches, this succeeded in allowing axons to grow into but not beyond the lesion site. A combination of NT-3 and cAMP stimulation, but neither manipulation individually, generated slightly more success and allowed axons to grow beyond the lesion site [91], but this required a preconditioning of the neurons with cAMP prior to injury, a clinically unrealistic constraint. Combinations of transplants and neurotrophic factors have also shown limited success in animal models of chronic SCI [80, 92, 93].

Neuroprotection

Implantation of fetal tissue grafts into the injured spinal cord of newborns improves motor recovery, and this is also true, albeit to a less robust extent, in adults.

One mechanism contributing to this phenomenon is the protection of axotomized neurons from dieback and retrograde-induced cell death. Mori et al. reported in 1997 [94] that this applies to the projection neurons of the red nucleus (RN) or the rubrospinal tract. The authors performed left-sided hemisections on adult rat spinal cords to axotomize the cells of the right red nucleus and, at the same time, filled the resulting cavity with embryonic day (E) 14 (of the 22 total days of rat gestation) spinal cord tissue. Using tracing dyes that migrate along intact axons, it is possible to determine which cells were successfully axotomized and whether or not subsequent cell death occurred. The fetal tissue transplant rescued roughly 50 % of the axotomized cells from death, although the surviving RN projections did not traverse the fetal graft and lesion site, and the full mechanism of functional improvement remained unclear.

One way fetal tissue or stem cell grafts may contribute to neuroprotection of cells that survive SCI is by secretion of neurotrophic factors abundant in developing, but not adult tissues [95–98]. Some research groups have genetically enhanced the ability of cells to produce neurotrophic factors to improve the ability of implanted cells to support neuronal and glial survival and function after experimental SCI [99–102]. Liu et al. in 1999 reported enhanced regeneration of rubrospinal axons and improvement in forelimb function after implantation of fibroblasts modified to secrete brain-derived neurotrophic factor (BDNF) [102]. The same group reported a similar effect on supraspinal neurons in chronic SCI [101]. Implantation of fibroblasts secreting both BDNF and neurotrophin-3 (NT-3) improved rubrospinal regrowth [100]. Another approach involved engineering neural precursor cells (NPCs) to secrete Noggin, an inhibitor of bone morphogenetic protein (BMP) signaling, to inhibit BMP-mediated differentiation of the endogenous NPCs into the astrocytic component of the glial scar [99].

15.1.2.3 Endogenous Stem Cells

Although the precise complement of stem cells present in the adult spinal cord and their proliferative capacity, differentiation potential, and molecular profiles remain unclear, it is easy to imagine the potential benefits of orchestrating repair by directing them to replace myelinating oligodendrocytes, replace neurons comprising local circuitry, attenuate glial scar formation, and secrete trophic factors to facilitate recovery. The obvious advantage of this approach is the amelioration of problems and issues related to stem cell transplantation and immune rejection.

Replacement of Myelinating Oligodendrocytes

After SCI, endogenous stem and progenitor cells in the adult spinal cord proliferate and then differentiate into mature glia. However, these cells have a propensity to produce astrocytes rather than the functional, myelinating OL. In order to specifically target the oligodendrocyte lineage, some studies have attempted to increase generation of NG2-expressing progenitor cells that are oligodendrocyte precursors. Administration of glial growth factor 2 (GGF2) and fibroblast growth factor 2 (FGF2) to rats 24 h after SCI increased NG2+ progenitor cells, myelinated axons, and functional recovery [103]. Since BMP signaling promotes astroglial rather than OL differentiation of these cells [104, 105], it may also be possible to increase oligodendrocyte lineage commitment by endogenous progenitor cells by inhibiting BMP signaling.

Reduction of Glial Scarring

A major roadblock for axons attempting to regenerate after SCI is the glial scar. In addition to acting as a physical blockade to axon outgrowth, the glial scar contains a number of extracellular matrix proteins that are inhibitory for axon outgrowth including CGSP, Nogo, MAG, and others. Efforts to chemically mitigate the presence of inhibitory factors in the glial scar have had some limited successes, but inhibition of scar formation from injury onset may prove to be a more effective strategy [52, 106, 107]. Because endogenous neural stem or progenitor cells contribute to scar formation, it may be possible to attenuate the process by manipulating the propensity of the stem/progenitor cells to differentiate into astrocytes [5, 10].

15.2 Importance of Nonhuman Primate Studies

When translating the outcomes of rodent studies to clinical applications, the differences between rodent and human spinal cords, both anatomical and physiological, should be kept in mind [108]. Nonhuman primate studies may lend additional insight to the likelihood of success in human trials since their spinal cord anatomy more closely resembles ours. Although embryonic stem cells have been isolated from nonhuman primates [109], most nonhuman primate SCI stem cell studies have focused on the implantation of human stem cells into injured animals. Iwanami et al. [110] inflicted C5 contusion injuries to adult marmosets and injected, directly into the lesion 9 weeks after injury, human neural stem progenitor cells that had been cultured from 8-week-old fetuses and propagated in vitro as neurospheres. The marmosets that received cell implants reportedly experienced a significant improvement in grip strength compared to controls. Histological analysis after completion of the behavioral portion of the study revealed that implanted cells became both neurons and glia. However, the variability in outcomes between treatment groups was so large that the significance of the functional improvement has been questioned [53]. Indeed, the procedure failed to qualify for a clinical trial [111].

Although it has been suggested that primate SCI may be more relevant to the human condition than rodents injuries, new differences between nonhuman primate and human SCI are being uncovered. A recent study reported spontaneous recovery of function after SCI in nonhuman primates, making it difficult to determine whether improvements after treatment in these animal models are truly a result of the therapy [112].

15.2.1 Case Studies: Clinical Trials of Stem Cells for the Treatment of SCI

Despite the limited success in treatment of experimental SCI, a number of approaches have been translated into clinical trials. It is important for patients and families to manage expectations of outcomes of clinical trials [113]. But the lack of effective treatments in the mainstream medical arena has left a void that can be filled by procedures that are often offered without legitimate expectations of success [114]. This section details the protocols used and outcomes for the patients undergoing these treatments (Table 15.1).

15.2.1.1 Autologous Bone Marrow Stem Cell Therapy

Intravenous administration of bone marrow stem cells (BMSCs) was tested by several groups for the treatment of SCI [70, 110, 115, 116]. While this approach was demonstrated to be safe and without major side effects, no functional improvements were observed. Sykova et al. (2006) observed modest improvements in SCI condition by administration of unmanipulated, autologously generated cultures of the cells intravenously or intra-arterially to 20 patients with both acute and chronic SCIs. However, only 5 of the 20 patients demonstrated functional improvement of any kind [57]. Park and colleagues investigated the potential of these cells when injected directly into the spinal cord lesion site. They coadministered the cells with granulocyte macrophage colony-stimulating factor (GM-CSF), a cytokine that stimulates

Table 15.1 Summary of clinical trials

Cell type	Administration route	Outcomes	Reference
BMSCs	Intravenous (IV)	Modest functional improvements in 5 of 20 patients	[57]
BMSCs	Intravenous (IV)	Safe; effectiveness was not determined	[70, 110, 115, 116]
BMSCs plus macrophage factor	Lesion site injection	Safe; effectiveness was not determined	[117]
MSCs plus T cells	Lesion site injection plus IV	Some sensory improvement in 2 of 2 patients; effectiveness inconclusive	[118]
BMSCs plus macrophage factor	Lesion site injection	Significant improvement in patients with acute SCI; no improvement in patients with chronic SCI	[58]
BMSCs selected for CD34 expression	Varied	Safe; morphological changes detectable by MRI; motor improvement in most patients; sensory improvement in two patients; quality-of-life improvement in all patients	[56]
Mononuclear cells from bone marrow	Lumbar puncture (LP)	Safe; functional improvement in nearly one third of patients; acute SCI more responsive than chronic	[119]
Mononuclear cells from bone marrow	LP	Safe; effectiveness inconclusive	[120–122]
MSCs from bone marrow	Lesion site and intradural injections; lumbar tap after 2 months	Decrease in spinal cavity size detectable on MRI; electrophysiological improvements; motor improvement in six of ten patients	[117]
MSCs from bone marrow	Lesion site injection	Minimal improvement in some patients	[123]
Olfactory ensheathing cells	Intraspinal implantation	Safe; no apparent clinical benefits or improvements visible by MRI	[60]
Olfactory ensheathing cells	Intraspinal implantation	Some potential clinical benefit reported	[124–126]
Schwann cells	Lesion site injection	Safe; functional improvement in one of four patients; no changes observed by MRI	[89]
Autologous macrophages	Lesion site injection	ProNeuron technology; safe; functional improvement in three of eight patients	[59]
MSCs from adipose tissue	IV	Safe; no tumor formation; effectiveness not yet determined	[127]

stem cells to become monocytes that differentiate into macrophages during the immune response [128] and reported in 2005 upon the safety of this procedure. Though safe, its effectiveness as a SCI treatment was not determined. The following year, Moviglia et al. combined T cell and marrow stromal cell (MSC) therapy to treat two paraplegic patients suffering from cervical or thoracic SCI [118]. Autologous T cells were cocultured with autologous MSCs. The T cells were administered intravenously and the cocultured MSCs were transplanted directly to the lesion site by a feeding artery. The goal of the coculture of T cells and MSCs was to induce a controlled inflammatory response that could, as in the 2005 macrophage study described above, boost reparative processes and potentially condition the lesion site for a more successful stem cell transplant. Although some sensory improvement was observed in both of the subjects, the low number of patients and the marginal benefits that were observed limit the conclusiveness of the results [5, 118].

Yoon et al. (2007) reported on a phase I/II clinical trial of autologous BMSC implantation along with GM-CSF administration into acutely and chronically injured spinal cords [58]. While there was significant functional recovery in the acute SCI patient cohort, no improvement was observed in the chronic (>8-week-old) SCI group. Geffner et al. (2008) reported on eight case studies of patients with SCI (four acute and four chronic) who received autologous BMSC implants through a variety of administration routes [56]. BMSCs were isolated from bone marrow samples by FACS sorting of the CD34+ population. A laminectomy was then performed to expose the injured portion of the spinal cord, and the scar tissue was resected and the cord detethered. A series of micropunctures were introduced using a 21-gauge needle and a total of 20 ml of stem cell suspension was delivered into various sites and cavities in and around the lesion epicenter. After suturing closed the dura, an additional 30 ml of cell suspension was delivered to the spinal canal and a final 30 ml delivered intravenously. This amounted to a total of 1.2×10^6 cells/kg of body weight. Subjects were evaluated for motor and sensory abilities before the procedure and then at 6, 12, and 24 months after implantation. In addition to the American Spinal Injury Association (ASIA) scale evaluation for motor and sensory function, Frankel scores were also collected to chart the severity of impairment, and Ashworth scores to track changes in muscle spasticity. Quality-of-life improvements were measured using the Barthel index [129]. No safety concerns or side effects were observed. The authors reported "noticeable" morphological changes to the spinal cord as detected by MRI. Motor scores improved in most of the patients, and to a statistically significant degree, after 24 months, in two of these. Sensory ability improved in the same two patients who experienced the most motor recovery. Quality of life improved in all eight patients, when aspects such as bladder function were taken into account using the Barthel method and the authors' own bladder function scoring method. While this trial has been reported as a success, the benefits that were reported were small.

Mononuclear cells have also been isolated from the bone marrow and tested clinically for therapeutic effect after SCI. Kumar et al. published in 2009 the safety assessment of isolating such cells autologously and injecting into patients by way of lumbar puncture (LP) [119]. Bone marrow was collected from patients' iliac crests, and mononuclear cells were isolated by centrifugation and FACS separation of the CD43+ subpopulation. These cells were then injected, unmanipulated, intrathecally into 215 patients with paraplegia, 49 patients with quadriplegia, and 33 patients with nontraumatic spinal cord myelopathy. Follow-ups including ASIA scoring occurred every 3 months for 21 months. The treatment was determined to be safe, as no serious side effects or tumor formation occurred. ASIA data indicated that approximately one third of the patients being scored showed sensory and motor improvement. These improvements correlated with improvement in bladder function, as well. The authors noted that recent SCIs were more responsive to the treatment than chronic SCIs, and lower SCIs were likewise more responsive that upper-level injuries such as those in the cervical region of the spine. Notably, the number of CD34+ cells that were successfully delivered to the patient had a direct correlation with the reported success of the treatment, suggesting that this cell population's therapeutic potential might be worthy of follow-up investigation.

A number of other groups have investigated the effects of intrathecal administration of cells derived from bone marrow. Callera et al. determined that intrathecal introduction of autologous mononuclear BMCs is safe, but the trial did not reveal whether this method is useful for treating SCI [120]. Saito and colleagues reported on a single patient who had received bone marrow stem cells intrathecally and experienced slight but steady improvement. This confirms the safeness of this procedure, but again, its usefulness for SCI therapy could not be determined from the small sample size and lack of control group [121]. Pal et al. isolated bone marrow cells autologously and expanded them in culture before administering them intrathecally to patients [122]. In this study, 30 patients were divided between acute and chronic SCI groups. Despite the relatively large number of participants, at the time the 2009 report was published, only the safety and not the effectiveness of this therapy could be determined. Diverse methods of isolation, cell population selection, and delivery were tested in an animal study by Paul et al. [130]. The authors concluded that the intrathecal route is the most promising method of bone marrow cell administration, as it is minimally invasive yet more effective than IV delivery.

Two very recent studies examined mesenchymal stem cell (MSC) therapy. Park et al. isolated autologous MSCs from iliac crest bone marrow of each patient and expanded the cell population in culture for 4 weeks [117]. Cells were injected both into the spinal cord directly and into the intradural space. One and two months later, additional cells were administered to each patient by lumbar tap. The ten patients, each of whom had cervical level SCIs, were assessed for changes in motor power in the extremities and by MRI and electrophysiology. After 6 months, six of the ten patients demonstrated improvement in the motor assessment and three in the quality-of-life index (ADL). In three patients, MRI revealed a decrease in cavity size concomitant with improvements in elecrophysiological activity. None of the ten had adverse reactions to the transplantation. In a similar fashion, Bhanot et al. examined MSC used to treat chronic SCI [123]. Autologous MSCs were administered directly to the lesion area after laminectomy. While one patient experienced improvement in motor control and two others experienced limited improvement in sensation capacity, none of three different doses of cell administration was able to improve clinical outcome in a meaningful way. This study reinforced the safety but limited therapeutic usefulness of autologous MSCs [127]. Overall, the clinical experience with BMSC transplantation has been disappointing.

Bone marrow-derived cells, both MSCs and bone marrow-derived progenitor cells (BMPCs), are also the focus of two clinical studies currently recruiting participants at the time this text went to press. A Chinese group at Guangzhou General Hospital intends to combine intravenous and intrathecal administration of autologous bone marrow-derived MSCs. This study will determine the safety and efficacy of a less-invasive means of delivering MSCs to the injured spinal cord than the direct implantation. Similarly, a trial at the Baylor College of Medicine will test the safety and efficacy of autologous BMPCs administered intravenously to children aged 1–15 years with SCI.

15.2.1.2 Autologous Olfactory Ensheathing Cell Therapy

Feron et al. reported in 2005 on a phase I clinical trial that examined the effects of intraspinal implantation of olfactory ensheathing cells into three male paraplegics with chronic injuries to the thoracic spinal cord [60]. The study participants underwent nasal biopsy for the excision of olfactory mucosa samples. Cells were dissociated from this tissue sample and grown in culture, undergoing passages every 3 days until enough cells were generated, 4 weeks later. The subjects underwent multiple vertebral laminectomies to remove the vertebrae directly above as well as just rostral and caudal to the injury site. The dura was opened and cells were injected directly into the injury as well as into the uninjured cord rostral and caudal to the epicenter. The three patients received different amounts of cells (12, 24, and 28 millions, respectively). Three additional patients were used as control subjects. By 1 year after implantation surgery, no adverse physical or psychological side effects were noted. MRI studies showed no change in spinal cord structure by 1 year after surgery, and there were no apparent clinical benefits. This study builds on the authors' previous publications reporting on the safety of the olfactory mucosa biopsy itself as well as procedures for growing the olfactory ensheathing cells in culture. In 2008, a 3-year follow-up of these same patients was published supporting the safety and feasibility of this technique [131]. Although other less well-controlled studies have suggested some clinical benefits from implantation of olfactory ensheathing cells [124–126, 132], in toto these studies have shown little if any therapeutic benefits.

15.2.1.3 Autologous Schwann Cell Therapy

Saberi et al. (2008) examined the effects of autologous Schwann cell transplantation in four patients with chronic thoracic spinal cord injuries [89]. The authors subjected the patients to 6 months of physical therapy prior to implantation surgery to ensure that their SCI condition was stable and not improving on its own. To prepare a culture of autologous Schwann cells for implantation, the sural nerve was surgically excised (12–15 cm) and the surrounding cells dissociated in culture. The cells were not passaged, but the media was frequently changed. Cell purity, which was found to be 95–99 %, was confirmed immunohistochemically and by cell morphology. Patients underwent a laminectomy to expose the spinal cord at the level of injury where the dura was opened at the midline. The autologous Schwann cells, resuspended in serum prepared from the patient's blood, were injected into 5 or 6 positions within the injured area by hand using a 30.5-gauge needle.

The subjects were evaluated according to MRI, the ASIA scale, and reporting of sphincter and sexual function up to 1 year after surgery. Only one of the four subjects showed signs of motor or sensory improvement. However, this patient had an incomplete SCI, rather than complete, and had also undergone extensive physical therapy during the year after surgery, so little conclusion can be drawn from this improvement. Further, all four subjects developed transient paresthesias or increased muscle spasms following the procedure. MRI data showed no signs of change, pathological or otherwise, in the spinal cords 1 year after surgery. Taken together, the findings of Saberi et al. indicate that intraspinal autologous Schwann cell transplantation is safe but likely ineffective for the treatment of SCI. Additional trials of Schwann cell transplantation by other groups are in early stages.

15.2.1.4 Autologous Macrophage Therapy

In one of the more rigorous clinical trials in the field of cell replacement therapy for acute SCI, Knoller et al. administered incubated autologous macrophages (developed commercially by the ProCord division of ProNeuron Biotechnologies) to eight patients [59]. This trial was based on animal models of SCI that suggested that artificially boosting local immune responses at the site of the injury by injecting activated macrophages enhanced functional recovery. Eight patients with acute (<14-day-old) complete SCIs between C5 and T11 donated their own blood (200 ml) and a 10 × 3 cm swatch of their own skin for the preparation of the activated autologous macrophages. Monocytes were isolated from the blood by centrifugation and coincubated on the skin explant for 24 h. Purity of the cell cultures was ascertained by morphological phenotype and flow cytometry targeting activated macrophage-specific antigens. 4 × 106 CD14+ cells were resuspended in medium and administered over four injections into the caudal edge of the lesion site by a 30-gauge syringe. Recovery was measured according to the ASIA scale as well as motor and sensory scales. By 1 year after surgery, function had improved in three of the eight patients. No safety concerns were encountered that could be traced to the therapy itself, and based on the functional recovery seen in nearly half of the subjects, a phase II trial was planned. However, it was later abandoned due to financial concerns and lack of a convincing benefit to the patients [5].

15.2.1.5 Adipose Tissue-Derived Mesenchymal Stem Cell Therapy

Also, recently tested in clinical trial were mesenchymal stem cells isolated from adipose tissue (AdMSCs). AdMSCs were expanded in culture and displayed morphology, immunoreactivity, and differentiation capacity consistent with other MSC populations and were stable for a dozen passages [127]. Eight patients suffering from chronic (more than 1-year-old) SCIs received one IV injection of 4×10^8 hAdMSCs. This procedure was reported to be safe, including a lack of tumor development, after a 3-month analysis. Treatment efficacy has not yet been reported.

15.2.1.6 Stem Cells from Umbilical Cord

Ongoing efforts to utilize stem cells found in umbilical cord blood (UCB) to treat SCI are currently underway in China and elsewhere. The large Chinese study will also coadminister lithium to increase the proliferative nature of UCB stem cells and their propensity to differentiate neutrally. Though results of the clinical trial have not yet been published, these researchers are testing injection of these treated stem cells into the entry sites of the dorsal roots to the injured spinal cord [61]. Likewise, the Spinal Cord Injury Network USA is planning to treat human SCI with UCB stem cells conditioned by lithium [62].

15.2.1.7 Human Embryonic Stem Cells

Perhaps the most controversial trial of stem cells for the treatment of SCI has been a study organized by Geron Inc. to utilize human embryonic stem cells (hESCs) in acute SCI. Preclinical studies discussed above suggested that transplantation of OL precursor cells derived from hESCs enhanced recovery in rats subjected to moderate injuries of the spinal cord. Geron generated and extensively tested a human embryonic stem cell line named GRNOPC1 for potential use in humans. After the publication of initial, promising results with these cells in experimental SCI, subsequent work revealed a more complicated picture [71, 133]. Similar work on the cervical spinal cord of rats likewise demonstrated functional improvement in OPC-injected animals over control groups as well as a decrease in cavities present in the injured spinal cord [71]. But in this case, histological analyses revealed that the implanted rats actually experienced similar [71] or even *less* [53] remyelination of axons, calling into question the mechanism underlying the previously observed functional improvements. The authors credited neuroprotection resulting from the implantation of the OPC line as an explanation for the discrepancy. A phase 1 clinical trial began with its first patient receiving implantation in October of 2010, and a total of four patients with complete T3-T10 SCI were treated. GRNOPC1 cells were administered at one time point between 7 and 14 days post-injury, and the study planned to evaluate safety as well as improvements in neurological function as measured by ASIA and sensory scales monthly over the following year. Specifically, lower extremity motor function, bladder and bowel function, and the donor-specific immune responsiveness would be evaluated. At the time this text was sent to press, no results from the clinical cases had been reported at clinicaltrials.gov [134]. Despite having spent 12 years [135] and many millions of dollars [136], possibly as many as 170 million [53], developing this therapy, Geron unexpectedly halted their clinical trial. This was reportedly a business-driven decision unrelated to the scientific progress of the clinical trial. While no significant adverse events have been reported, there has also been no evidence of significant recovery in the four patients who were treated.

Stem Cells Isolated from the Human Central Nervous System

Stem cells isolated from the human central nervous system will also be tested in a clinical trial of SCI sponsored by Stem Cells, Inc. Patients with thoracic spinal cord injuries will receive intramedullary transplantations of human central nervous system stem cells (HuCNS-SCs). The study, planned to last until 2016, will examine the safety and efficacy of this approach.

The Maryland-based company Neuralstem Inc. plans to begin a clinical trial to test the safety and viability of its human spinal cord stem cells (HSSCs) for the treatment of SCI. The company filed for investigational new drug (IND) approval from the US FDA in August of 2010 and plans to do so in India, as well. Clinical trials will be conducted at multiple sites and will enroll patients with chronic (1–2 years old) SCIs. This trial is based upon preclinical work performed on a rat model of SCI that showed improvement after transplantation of the Neuralstem HSSCs [137].

15.3 Concerns for Patients Considering a Clinical Trial for SCI

SCI patients considering joining a clinical trial of an experimental treatment should seek the advice of a number of sources before committing. The International Campaign for Cures of spinal cord injury Paralysis (ICCP) comprises an assortment of organizations dedicated to supporting individuals with SCI. Potential study participants should consult their guide entitled "Experimental treatments for spinal cord injury: what you should know if you are considering participation in a clinical trial" [138]. For more in-depth learning about the guidelines under which clinical trials for SCI must be conducted, consult reviews by Fawcett and colleagues [139–142]. If possible, the potential participant should learn from a medical professional whether the treatment offered has been rigorously tested in animal models, with the understanding that many treatments effective in rodents and nonhuman primates have yielded no significant improvements in human patients. The ICCP website and aforementioned downloadable pamphlet provide a list of questions a SCI patient considering clinical trial participation should ask, as well as contact information for a number of organizations to help in finding answers.

15.4 Conclusion

Spinal cord injuries produce a number of biological responses, none of which is easily addressed using the therapies developed to date. Adding to the challenge of developing treatment strategies are the imperfection of the animal models available and the heterogeneous nature of the injuries. Successful treatment of human SCI will likely involve a combination of approaches to treat the multitude of biological consequences of the injury, and stem cells may be able to address a number of the biological issues. Results from animal and human studies are promising in the sense that transplantation has largely been safe, and some patients have experienced improvements. However, no therapy has yet been shown to alter the outcome of SCI significantly. Moreover, the biotechnology/pharmaceutical industry has been abandoning development of therapies for SCI because the cost and challenge are great and the market is relatively small compared to many other medical problems. Thus, the challenges facing the SCI community are not just biological.

In sum, stem cells studies have increased our understanding of SCI and the potential for CNS repair in humans and have generated some exciting results in a few animal models and clinical trials. But achievement of an accepted, reliable treatment will likely depend on new approaches combined with those already underway.

References

1. Bunge RP, Puckett WR, Becerra JL, Marcillo A, Quencer RM (1993) Observations on the pathology of human spinal cord injury. A review and classification of 22 new cases with details from a case of chronic cord compression with extensive focal demyelination. Adv Neurol 59:75–89
2. Kakulas BA (1999) A review of the neuropathology of human spinal cord injury with emphasis on special features. J Spinal Cord Med 22(2):119–124
3. Schwab ME, Bartholdi D (1996) Degeneration and regeneration of axons in the lesioned spinal cord. Physiol Rev 76(2):319–370
4. Thuret S, Moon LD, Gage FH (2006) Therapeutic interventions after spinal cord injury. Nat Rev Neurosci 7(8):628–643
5. Sahni V, Kessler JA (2010) Stem cell therapies for spinal cord injury. Nat Rev Neurol 6(7): 363–372
6. Guest JD, Hiester ED, Bunge RP (2005) Demyelination and Schwann cell responses adjacent to injury epicenter cavities following chronic human spinal cord injury. Exp Neurol 192(2):384–393
7. Reier PJ (2004) Cellular transplantation strategies for spinal cord injury and translational neurobiology. NeuroRx 1(4):424–451, PMCID: 534951
8. Fawcett JW, Asher RA (1999) The glial scar and central nervous system repair. Brain Res Bull 49(6):377–391
9. Barnabe-Heider F, Frisen J (2008) Stem cells for spinal cord repair. Cell Stem Cell 3(1):16–24
10. Sahni V, Mukhopadhyay A, Tysseling V, Hebert A, Birch D, McGuire TL et al (2010) BMPR1a and BMPR1b signaling exert opposing effects on gliosis after spinal cord injury. J Neurosci 30(5):1839–1855
11. Jones TB, McDaniel EE, Popovich PG (2005) Inflammatory-mediated injury and repair in the traumatically injured spinal cord. Curr Pharm Des 11(10):1223–1236

12. Fawcett JW (2006) Overcoming inhibition in the damaged spinal cord. J Neurotrauma 23(3–4):371–383
13. Pettigrew DB, Shockley KP, Crutcher KA (2001) Disruption of spinal cord white matter and sciatic nerve geometry inhibits axonal growth in vitro in the absence of glial scarring. BMC Neurosci 2:8, PMCID: 32296
14. Horky LL, Galimi F, Gage FH, Horner PJ (2006) Fate of endogenous stem/progenitor cells following spinal cord injury. J Comp Neurol 498(4):525–538
15. Yang H, Lu P, McKay HM, Bernot T, Keirstead H, Steward O et al (2006) Endogenous neurogenesis replaces oligodendrocytes and astrocytes after primate spinal cord injury. J Neurosci 26(8):2157–2166
16. Yamamoto S, Yamamoto N, Kitamura T, Nakamura K, Nakafuku M (2001) Proliferation of parenchymal neural progenitors in response to injury in the adult rat spinal cord. Exp Neurol 172(1):115–127
17. Azari MF, Profyris C, Zang DW, Petratos S, Cheema SS (2005) Induction of endogenous neural precursors in mouse models of spinal cord injury and disease. Eur J Neurol 12(8): 638–648
18. Bareyre FM, Kerschensteiner M, Raineteau O, Mettenleiter TC, Weinmann O, Schwab ME (2004) The injured spinal cord spontaneously forms a new intraspinal circuit in adult rats. Nat Neurosci 7(3):269–277
19. Raineteau O, Fouad K, Bareyre FM, Schwab ME (2002) Reorganization of descending motor tracts in the rat spinal cord. Eur J Neurosci 16(9):1761–1771
20. Raineteau O, Schwab ME (2001) Plasticity of motor systems after incomplete spinal cord injury. Nat Rev Neurosci 2(4):263–273
21. SCI-Info-Pages [cited]. http://www.sci-info-pages.com. Accessed on Feb 2012
22. Bracken MB, Shepard MJ, Collins WF, Holford TR, Young W, Baskin DS et al (1990) A randomized, controlled trial of methylprednisolone or naloxone in the treatment of acute spinal-cord injury. Results of the Second National Acute Spinal Cord Injury Study. N Engl J Med 322(20):1405–1411
23. Bracken MB, Shepard MJ, Hellenbrand KG, Collins WF, Leo LS, Freeman DF et al (1985) Methylprednisolone and neurological function 1 year after spinal cord injury. Results of the National Acute Spinal Cord Injury Study. J Neurosurg 63(5):704–713
24. Bracken MB, Shepard MJ, Holford TR, Leo-Summers L, Aldrich EF, Fazl M et al (1997) Administration of methylprednisolone for 24 or 48 hours or tirilazad mesylate for 48 hours in the treatment of acute spinal cord injury. Results of the Third National Acute Spinal Cord Injury Randomized Controlled Trial. National Acute Spinal Cord Injury Study. JAMA 277(20):1597–1604
25. Tator CH, Duncan EG, Edmonds VE, Lapczak LI, Andrews DF (1987) Comparison of surgical and conservative management in 208 patients with acute spinal cord injury. Can J Neurol Sci 14(1):60–69
26. Vaccaro AR, Daugherty RJ, Sheehan TP, Dante SJ, Cotler JM, Balderston RA et al (1997) Neurologic outcome of early versus late surgery for cervical spinal cord injury. Spine (Phila Pa 1976) 22(22):2609–2613
27. Hugenholtz H (2003) Methylprednisolone for acute spinal cord injury: not a standard of care. CMAJ 168(9):1145–1146, PMCID: 153684
28. Richardson PM, Issa VM, Aguayo AJ (1984) Regeneration of long spinal axons in the rat. J Neurocytol 13(1):165–182
29. Richardson PM, McGuinness UM, Aguayo AJ (1980) Axons from CNS neurons regenerate into PNS grafts. Nature 284(5753):264–265
30. Levi AD, Dancausse H, Li X, Duncan S, Horkey L, Oliviera M (2002) Peripheral nerve grafts promoting central nervous system regeneration after spinal cord injury in the primate. J Neurosurg 96(2 Suppl):197–205
31. Cheng H, Liao KK, Liao SF, Chuang TY, Shih YH (2004) Spinal cord repair with acidic fibroblast growth factor as a treatment for a patient with chronic paraplegia. Spine (Phila Pa 1976) 29(14):E284–E288

32. Amador MJ, Guest JD (2005) An appraisal of ongoing experimental procedures in human spinal cord injury. J Neurol Phys Ther 29(2):70–86
33. Takami T, Oudega M, Bates ML, Wood PM, Kleitman N, Bunge MB (2002) Schwann cell but not olfactory ensheathing glia transplants improve hindlimb locomotor performance in the moderately contused adult rat thoracic spinal cord. J Neurosci 22(15):6670–6681
34. Bamber NI, Li H, Lu X, Oudega M, Aebischer P, Xu XM (2001) Neurotrophins BDNF and NT-3 promote axonal re-entry into the distal host spinal cord through Schwann cell-seeded mini-channels. Eur J Neurosci 13(2):257–268
35. Xu XM, Guenard V, Kleitman N, Bunge MB (1995) Axonal regeneration into Schwann cell-seeded guidance channels grafted into transected adult rat spinal cord. J Comp Neurol 351(1): 145–160
36. Pinzon A, Calancie B, Oudega M, Noga BR (2001) Conduction of impulses by axons regenerated in a Schwann cell graft in the transected adult rat thoracic spinal cord. J Neurosci Res 64(5):533–541
37. Guest JD, Rao A, Olson L, Bunge MB, Bunge RP (1997) The ability of human Schwann cell grafts to promote regeneration in the transected nude rat spinal cord. Exp Neurol 148(2): 502–522
38. Menei P, Montero-Menei C, Whittemore SR, Bunge RP, Bunge MB (1998) Schwann cells genetically modified to secrete human BDNF promote enhanced axonal regrowth across transected adult rat spinal cord. Eur J Neurosci 10(2):607–621
39. Xu XM, Guenard V, Kleitman N, Aebischer P, Bunge MB (1995) A combination of BDNF and NT-3 promotes supraspinal axonal regeneration into Schwann cell grafts in adult rat thoracic spinal cord. Exp Neurol 134(2):261–272
40. Chen A, Xu XM, Kleitman N, Bunge MB (1996) Methylprednisolone administration improves axonal regeneration into Schwann cell grafts in transected adult rat thoracic spinal cord. Exp Neurol 138(2):261–276
41. Ramon-Cueto A, Plant GW, Avila J, Bunge MB (1998) Long-distance axonal regeneration in the transected adult rat spinal cord is promoted by olfactory ensheathing glia transplants. J Neurosci 18(10):3803–3815
42. Cheng H, Wu JP, Tzeng SF (2002) Neuroprotection of glial cell line-derived neurotrophic factor in damaged spinal cords following contusive injury. J Neurosci Res 69(3):397–405
43. Nikulina E, Tidwell JL, Dai HN, Bregman BS, Filbin MT (2004) The phosphodiesterase inhibitor rolipram delivered after a spinal cord lesion promotes axonal regeneration and functional recovery. Proc Natl Acad Sci U S A 101(23):8786–8790, PMCID: 423273
44. Pearse DD, Pereira FC, Marcillo AE, Bates ML, Berrocal YA, Filbin MT et al (2004) CAMP and Schwann cells promote axonal growth and functional recovery after spinal cord injury. Nat Med 10(6):610–616
45. Schreyer DJ, Jones EG (1987) Growth of corticospinal axons on prosthetic substrates introduced into the spinal cord of neonatal rats. Brain Res 432(2):291–299
46. Marchand R, Woerly S (1990) Transected spinal cords grafted with in situ self-assembled collagen matrices. Neuroscience 36(1):45–60
47. Goldsmith HS, de la Torre JC (1992) Axonal regeneration after spinal cord transection and reconstruction. Brain Res 589(2):217–224
48. Khan T, Dauzvardis M, Sayers S (1991) Carbon filament implants promote axonal growth across the transected rat spinal cord. Brain Res 541(1):139–145
49. Woerly S, Lavallee C, Marchand R (1992) Intracerebral implantation of ionic synthetic hydrogels: effect of polar substrata on astrocytosis and axons. J Neural Transplant Plast 3(1):21–34, PMCID: 2565137
50. Tysseling-Mattiace VM, Sahni V, Niece KL, Birch D, Czeisler C, Fehlings MG et al (2008) Self-assembling nanofibers inhibit glial scar formation and promote axon elongation after spinal cord injury. J Neurosci 28(14):3814–3823, PMCID: 2752951
51. Moon LD, Asher RA, Rhodes KE, Fawcett JW (2001) Regeneration of CNS axons back to their target following treatment of adult rat brain with chondroitinase ABC. Nat Neurosci 4(5):465–466

52. Silver J, Miller JH (2004) Regeneration beyond the glial scar. Nat Rev Neurosci 5(2): 146–156
53. Thomas KE, Moon LD (2011) Will stem cell therapies be safe and effective for treating spinal cord injuries? Br Med Bull 98:127–142
54. Amariglio N, Hirshberg A, Scheithauer BW, Cohen Y, Loewenthal R, Trakhtenbrot L et al (2009) Donor-derived brain tumor following neural stem cell transplantation in an ataxia telangiectasia patient. PLoS Med 6(2):e1000029, PMCID: 2642879
55. Hofstetter CP, Holmstrom NA, Lilja JA, Schweinhardt P, Hao J, Spenger C et al (2005) Allodynia limits the usefulness of intraspinal neural stem cell grafts; directed differentiation improves outcome. Nat Neurosci 8(3):346–353
56. Geffner LF, Santacruz P, Izurieta M, Flor L, Maldonado B, Auad AH et al (2008) Administration of autologous bone marrow stem cells into spinal cord injury patients via multiple routes is safe and improves their quality of life: comprehensive case studies. Cell Transplant 17(12):1277–1293
57. Sykova E, Homola A, Mazanec R, Lachmann H, Konradova SL, Kobylka P et al (2006) Autologous bone marrow transplantation in patients with subacute and chronic spinal cord injury. Cell Transplant 15(8–9):675–687
58. Yoon SH, Shim YS, Park YH, Chung JK, Nam JH, Kim MO et al (2007) Complete spinal cord injury treatment using autologous bone marrow cell transplantation and bone marrow stimulation with granulocyte macrophage-colony stimulating factor: phase I/II clinical trial. Stem Cells 25(8):2066–2073
59. Knoller N, Auerbach G, Fulga V, Zelig G, Attias J, Bakimer R et al (2005) Clinical experience using incubated autologous macrophages as a treatment for complete spinal cord injury: phase I study results. J Neurosurg Spine 3(3):173–181
60. Feron F, Perry C, Cochrane J, Licina P, Nowitzke A, Urquhart S et al (2005) Autologous olfactory ensheathing cell transplantation in human spinal cord injury. Brain 128(Pt 12): 2951–2960
61. Chinese Spinal Cord Injury Network (Cited 2012) Lithium and cord blood cell for spinal cord injury. http://www.chinascinet.org/index.php?option=com_content&task=view&id=31&Itemid=57. Accessed on Feb 2012
62. SciNetUSA. http://www.spinal-injury.net/spinal-cord-injury-reseach-cure.htm. Accessed on Feb 2012
63. Kim K, Doi A, Wen B, Ng K, Zhao R, Cahan P et al (2010) Epigenetic memory in induced pluripotent stem cells. Nature 467(7313):285–290, PMCID: 3150836
64. Miura K, Okada Y, Aoi T, Okada A, Takahashi K, Okita K et al (2009) Variation in the safety of induced pluripotent stem cell lines. Nat Biotechnol 27(8):743–745
65. Johansson CB, Momma S, Clarke DL, Risling M, Lendahl U, Frisen J (1999) Identification of a neural stem cell in the adult mammalian central nervous system. Cell 96(1):25–34
66. Meletis K, Barnabe-Heider F, Carlen M, Evergren E, Tomilin N, Shupliakov O et al (2008) Spinal cord injury reveals multilineage differentiation of ependymal cells. PLoS Biol 6(7):e182, PMCID: 2475541
67. Prokhorova TA, Harkness LM, Frandsen U, Ditzel N, Schroder HD, Burns JS et al (2009) Teratoma formation by human embryonic stem cells is site dependent and enhanced by the presence of Matrigel. Stem Cells Dev 18(1):47–54
68. McDonald JW, Liu XZ, Qu Y, Liu S, Mickey SK, Turetsky D et al (1999) Transplanted embryonic stem cells survive, differentiate and promote recovery in injured rat spinal cord. Nat Med 5(12):1410–1412
69. Bain G, Kitchens D, Yao M, Huettner JE, Gottlieb DI (1995) Embryonic stem cells express neuronal properties in vitro. Dev Biol 168(2):342–357
70. Keirstead HS, Nistor G, Bernal G, Totoiu M, Cloutier F, Sharp K et al (2005) Human embryonic stem cell-derived oligodendrocyte progenitor cell transplants remyelinate and restore locomotion after spinal cord injury. J Neurosci 25(19):4694–4705

71. Sharp J, Frame J, Siegenthaler M, Nistor G, Keirstead HS (2010) Human embryonic stem cell-derived oligodendrocyte progenitor cell transplants improve recovery after cervical spinal cord injury. Stem Cells 28(1):152–163
72. Han SS, Kang DY, Mujtaba T, Rao MS, Fischer I (2002) Grafted lineage-restricted precursors differentiate exclusively into neurons in the adult spinal cord. Exp Neurol 177(2):360–375
73. Wichterle H, Lieberam I, Porter JA, Jessell TM (2002) Directed differentiation of embryonic stem cells into motor neurons. Cell 110(3):385–397
74. Harper JM, Krishnan C, Darman JS, Deshpande DM, Peck S, Shats I et al (2004) Axonal growth of embryonic stem cell-derived motoneurons in vitro and in motoneuron-injured adult rats. Proc Natl Acad Sci U S A 101(18):7123–7128, PMCID: 406476
75. Anderson DK, Howland DR, Reier PJ (1995) Fetal neural grafts and repair of the injured spinal cord. Brain Pathol 5(4):451–457
76. Bregman BS, McAtee M, Dai HN, Kuhn PL (1997) Neurotrophic factors increase axonal growth after spinal cord injury and transplantation in the adult rat. Exp Neurol 148(2):475–494
77. Jakeman LB, Reier PJ (1991) Axonal projections between fetal spinal cord transplants and the adult rat spinal cord: a neuroanatomical tracing study of local interactions. J Comp Neurol 307(2):311–334
78. Kunkel-Bagden E, Bregman BS (1990) Spinal cord transplants enhance the recovery of locomotor function after spinal cord injury at birth. Exp Brain Res 81(1):25–34
79. Bregman BS, Kunkel-Bagden E, Reier PJ, Dai HN, McAtee M, Gao D (1993) Recovery of function after spinal cord injury: mechanisms underlying transplant-mediated recovery of function differ after spinal cord injury in newborn and adult rats. Exp Neurol 123(1):3–16
80. Bregman BS, Coumans JV, Dai HN, Kuhn PL, Lynskey J, McAtee M et al (2002) Transplants and neurotrophic factors increase regeneration and recovery of function after spinal cord injury. Prog Brain Res 137:257–273
81. Coumans JV, Lin TT, Dai HN, MacArthur L, McAtee M, Nash C et al (2001) Axonal regeneration and functional recovery after complete spinal cord transection in rats by delayed treatment with transplants and neurotrophins. J Neurosci 21(23):9334–9344
82. Parr AM, Tator CH, Keating A (2007) Bone marrow-derived mesenchymal stromal cells for the repair of central nervous system injury. Bone Marrow Transplant 40(7):609–619
83. Liu Y, Rao MS (2003) Transdifferentiation–fact or artifact. J Cell Biochem 88(1):29–40
84. Tello F (1911) La influencia de1 neurotropismo en la regeneration de 10s centros nerviosos. Trab Lab Invest Biol 9:123–159
85. Houle JD (1991) Demonstration of the potential for chronically injured neurons to regenerate axons into intraspinal peripheral nerve grafts. Exp Neurol 113(1):1–9
86. Doucette R (1995) Olfactory ensheathing cells: potential for glial cell transplantation into areas of CNS injury. Histol Histopathol 10(2):503–507
87. Radtke C, Sasaki M, Lankford KL, Vogt PM, Kocsis JD (2008) Potential of olfactory ensheathing cells for cell-based therapy in spinal cord injury. J Rehabil Res Dev 45(1): 141–151
88. Richter MW, Roskams AJ (2008) Olfactory ensheathing cell transplantation following spinal cord injury: hype or hope? Exp Neurol 209(2):353–367
89. Saberi H, Moshayedi P, Aghayan HR, Arjmand B, Hosseini SK, Emami-Razavi SH et al (2008) Treatment of chronic thoracic spinal cord injury patients with autologous Schwann cell transplantation: an interim report on safety considerations and possible outcomes. Neurosci Lett 443(1):46–50
90. Taylor L, Jones L, Tuszynski MH, Blesch A (2006) Neurotrophin-3 gradients established by lentiviral gene delivery promote short-distance axonal bridging beyond cellular grafts in the injured spinal cord. J Neurosci 26(38):9713–9721
91. Lu P, Yang H, Jones LL, Filbin MT, Tuszynski MH (2004) Combinatorial therapy with neurotrophins and cAMP promotes axonal regeneration beyond sites of spinal cord injury. J Neurosci 24(28):6402–6409

92. Kadoya K, Tsukada S, Lu P, Coppola G, Geschwind D, Filbin MT et al (2009) Combined intrinsic and extrinsic neuronal mechanisms facilitate bridging axonal regeneration one year after spinal cord injury. Neuron 64(2):165–172, PMCID: 2773653
93. Hofstetter CP, Schwarz EJ, Hess D, Widenfalk J, El Manira A, Prockop DJ et al (2002) Marrow stromal cells form guiding strands in the injured spinal cord and promote recovery. Proc Natl Acad Sci U S A 99(4):2199–2204, PMCID: 122342
94. Mori F, Himes BT, Kowada M, Murray M, Tessler A (1997) Fetal spinal cord transplants rescue some axotomized rubrospinal neurons from retrograde cell death in adult rats. Exp Neurol 143(1):45–60
95. Bregman BS, Reier PJ (1986) Neural tissue transplants rescue axotomized rubrospinal cells from retrograde death. J Comp Neurol 244(1):86–95
96. Eagleson KL, Cunningham TJ, Haun F (1992) Rescue of both rapidly and slowly degenerating neurons in the dorsal lateral geniculate nucleus of adult rats by a cortically derived neuron survival factor. Exp Neurol 116(2):156–162
97. Haun F, Cunningham TJ (1984) Cortical transplants reveal CNS trophic interactions in situ. Brain Res 317(2):290–294
98. Himes BT, Goldberger ME, Tessler A (1994) Grafts of fetal central nervous system tissue rescue axotomized Clarke's nucleus neurons in adult and neonatal operates. J Comp Neurol 339(1):117–131
99. Tang BL, Low CB (2007) Genetic manipulation of neural stem cells for transplantation into the injured spinal cord. Cell Mol Neurobiol 27(1):75–85
100. Tobias CA, Shumsky JS, Shibata M, Tuszynski MH, Fischer I, Tessler A et al (2003) Delayed grafting of BDNF and NT-3 producing fibroblasts into the injured spinal cord stimulates sprouting, partially rescues axotomized red nucleus neurons from loss and atrophy, and provides limited regeneration. Exp Neurol 184(1):97–113
101. Jin Y, Fischer I, Tessler A, Houle JD (2002) Transplants of fibroblasts genetically modified to express BDNF promote axonal regeneration from supraspinal neurons following chronic spinal cord injury. Exp Neurol 177(1):265–275
102. Liu Y, Kim D, Himes BT, Chow SY, Schallert T, Murray M et al (1999) Transplants of fibroblasts genetically modified to express BDNF promote regeneration of adult rat rubrospinal axons and recovery of forelimb function. J Neurosci 19(11):4370–4387
103. Whittaker MT, Zai LJ, Lee HJ, Pajoohesh-Ganji A, Wu J, Sharp A et al (2012) GGF2 (Nrg1-beta3) treatment enhances NG2(+) cell response and improves functional recovery after spinal cord injury. Glia 60(2):281–294
104. Mabie PC, Mehler MF, Marmur R, Papavasiliou A, Song Q, Kessler JA (1997) Bone morphogenetic proteins induce astroglial differentiation of oligodendroglial-astroglial progenitor cells. J Neurosci 17(11):4112–4120
105. Wang Y, Cheng X, He Q, Zheng Y, Kim DH, Whittemore SR et al (2011) Astrocytes from the contused spinal cord inhibit oligodendrocyte differentiation of adult oligodendrocyte precursor cells by increasing the expression of bone morphogenetic proteins. J Neurosci 31(16): 6053–6058, PMCID: 3081104
106. Busch SA, Silver J (2007) The role of extracellular matrix in CNS regeneration. Curr Opin Neurobiol 17(1):120–127
107. Filbin MT (2003) Myelin-associated inhibitors of axonal regeneration in the adult mammalian CNS. Nat Rev Neurosci 4(9):703–713
108. Lemon RN (2008) Descending pathways in motor control. Annu Rev Neurosci 31:195–218
109. Wolf DP (2008) Nonhuman primate embryonic stem cells: an underutilized resource. Regen Med 3(2):129–131
110. Iwanami A, Kaneko S, Nakamura M, Kanemura Y, Mori H, Kobayashi S et al (2005) Transplantation of human neural stem cells for spinal cord injury in primates. J Neurosci Res 80(2):182–190
111. Yamane J, Nakamura M, Iwanami A, Sakaguchi M, Katoh H, Yamada M et al (2010) Transplantation of galectin-1-expressing human neural stem cells into the injured spinal cord of adult common marmosets. J Neurosci Res 88(7):1394–1405

112. Rosenzweig ES, Courtine G, Jindrich DL, Brock JH, Ferguson AR, Strand SC et al (2010) Extensive spontaneous plasticity of corticospinal projections after primate spinal cord injury. Nat Neurosci 13(12):1505–1510, PMCID: 3144760
113. Magnus T, Liu Y, Parker GC, Rao MS (2008) Stem cell myths. Philos Trans R Soc Lond B Biol Sci 363(1489):9–22, PMCID: 2605483
114. Blight A, Curt A, Ditunno JF, Dobkin B, Ellaway P, Fawcett J et al (2009) Position statement on the sale of unproven cellular therapies for spinal cord injury: the international campaign for cures of spinal cord injury paralysis. Spinal Cord 47(9):713–714
115. Boyd JG, Doucette R, Kawaja MD (2005) Defining the role of olfactory ensheathing cells in facilitating axon remyelination following damage to the spinal cord. FASEB J 19(7): 694–703
116. Mitsui T, Shumsky JS, Lepore AC, Murray M, Fischer I (2005) Transplantation of neuronal and glial restricted precursors into contused spinal cord improves bladder and motor functions, decreases thermal hypersensitivity, and modifies intraspinal circuitry. J Neurosci 25(42): 9624–9636
117. Park JH, Kim DY, Sung IY, Choi GH, Jeon MH, Kim KK, et al (2011) Long-term results of spinal cord injury therapy using mesenchymal stem cells derived from bone marrow in humans. Neurosurgery 70(5):1238–1247
118. Moviglia GA, Fernandez Vina R, Brizuela JA, Saslavsky J, Vrsalovic F, Varela G et al (2006) Combined protocol of cell therapy for chronic spinal cord injury. Report on the electrical and functional recovery of two patients. Cytotherapy 8(3):202–209
119. Kumar AA, Kumar SR, Narayanan R, Arul K, Baskaran M (2009) Autologous bone marrow derived mononuclear cell therapy for spinal cord injury: a phase I/II clinical safety and primary efficacy data. Exp Clin Transplant 7(4):241–248
120. Callera F, do Nascimento RX (2006) Delivery of autologous bone marrow precursor cells into the spinal cord via lumbar puncture technique in patients with spinal cord injury: a preliminary safety study. Exp Hematol 34(2):130–131
121. Saito F, Nakatani T, Iwase M, Maeda Y, Hirakawa A, Murao Y et al (2008) Spinal cord injury treatment with intrathecal autologous bone marrow stromal cell transplantation: the first clinical trial case report. J Trauma 64(1):53–59
122. Pal R, Venkataramana NK, Bansal A, Balaraju S, Jan M, Chandra R et al (2009) Ex vivo-expanded autologous bone marrow-derived mesenchymal stromal cells in human spinal cord injury/paraplegia: a pilot clinical study. Cytotherapy 11(7):897–911
123. Bhanot Y, Rao S, Ghosh D, Balaraju S, Radhika CR, Satish Kumar KV (2011) Autologous mesenchymal stem cells in chronic spinal cord injury. Br J Neurosurg 25(4):516–522
124. Lima C, Escada P, Pratas-Vital J, Branco C, Arcangeli CA, Lazzeri G et al (2010) Olfactory mucosal autografts and rehabilitation for chronic traumatic spinal cord injury. Neurorehabil Neural Repair 24(1):10–22
125. Chhabra HS, Lima C, Sachdeva S, Mittal A, Nigam V, Chaturvedi D et al (2009) Autologous olfactory [corrected] mucosal transplant in chronic spinal cord injury: an Indian Pilot Study. Spinal Cord 47(12):887–895
126. Lima C, Pratas-Vital J, Escada P, Hasse-Ferreira A, Capucho C, Peduzzi JD (2006) Olfactory mucosa autografts in human spinal cord injury: a pilot clinical study. J Spinal Cord Med 29(3):191–203; discussion 4–6. PMCID: 1864811
127. Ra JC, Shin IS, Kim SH, Kang SK, Kang BC, Lee HY et al (2011) Safety of intravenous infusion of human adipose tissue-derived mesenchymal stem cells in animals and humans. Stem Cells Dev 20(8):1297–1308
128. Park HC, Shim YS, Ha Y, Yoon SH, Park SR, Choi BH et al (2005) Treatment of complete spinal cord injury patients by autologous bone marrow cell transplantation and administration of granulocyte-macrophage colony stimulating factor. Tissue Eng 11(5–6):913–922
129. Stroke Center [cited]. http://www.strokecenter.org. Accessed on Feb 2012
130. Paul C, Samdani AF, Betz RR, Fischer I, Neuhuber B (2009) Grafting of human bone marrow stromal cells into spinal cord injury: a comparison of delivery methods. Spine (Phila Pa 1976) 34(4):328–334, PMCID: 3073497

131. Mackay-Sim A, Feron F, Cochrane J, Bassingthwaighte L, Bayliss C, Davies W et al (2008) Autologous olfactory ensheathing cell transplantation in human paraplegia: a 3-year clinical trial. Brain 131(Pt 9):2376–2386, PMCID: 2525447
132. Iwatsuki K, Yoshimine T, Kishima H, Aoki M, Yoshimura K, Ishihara M et al (2008) Transplantation of olfactory mucosa following spinal cord injury promotes recovery in rats. Neuroreport 19(13):1249–1252
133. Cloutier F, Siegenthaler MM, Nistor G, Keirstead HS (2006) Transplantation of human embryonic stem cell-derived oligodendrocyte progenitors into rat spinal cord injuries does not cause harm. Regen Med 1(4):469–479
134. ClinicalTrials [cited]. http://www.clinicaltrials.gov. Accessed on Feb 2012
135. Hooshmand MJ, Sontag CJ, Uchida N, Tamaki S, Anderson AJ, Cummings BJ (2009) Analysis of host-mediated repair mechanisms after human CNS-stem cell transplantation for spinal cord injury: correlation of engraftment with recovery. PLoS One 4(6):e5871, PMCID: 2690693
136. Geron Clinical Trial Halted (2011) [updated 2011; cited 2012]. http://www.reeve.uci.edu/news-geron-clinical-trial-halted.html. Accessed on Feb 2012
137. Cizkova D, Kakinohana O, Kucharova K, Marsala S, Johe K, Hazel T et al (2007) Functional recovery in rats with ischemic paraplegia after spinal grafting of human spinal stem cells. Neuroscience 147(2):546–560
138. Steeves J, Fawcett J, Tuszynski MH, Lammertse D et al (2007) Experimental treatments for spinal cord injury: what you should know if you are considering participating in a clinical trial. International Campaign for Cures of Spinal Cord Injury Paralysis. http://www.christopherreeve.org. Accessed on Feb 2012
139. Fawcett JW, Curt A, Steeves JD, Coleman WP, Tuszynski MH, Lammertse D et al (2007) Guidelines for the conduct of clinical trials for spinal cord injury as developed by the ICCP panel: spontaneous recovery after spinal cord injury and statistical power needed for therapeutic clinical trials. Spinal Cord 45(3):190–205
140. Steeves JD, Lammertse D, Curt A, Fawcett JW, Tuszynski MH, Ditunno JF et al (2007) Guidelines for the conduct of clinical trials for spinal cord injury (SCI) as developed by the ICCP panel: clinical trial outcome measures. Spinal Cord 45(3):206–221
141. Tuszynski MH, Steeves JD, Fawcett JW, Lammertse D, Kalichman M, Rask C et al (2007) Guidelines for the conduct of clinical trials for spinal cord injury as developed by the ICCP panel: clinical trial inclusion/exclusion criteria and ethics. Spinal Cord 45(3):222–231
142. Lammertse D, Tuszynski MH, Steeves JD, Curt A, Fawcett JW, Rask C et al (2007) Guidelines for the conduct of clinical trials for spinal cord injury as developed by the ICCP panel: clinical trial design. Spinal Cord 45(3):232–242

Chapter 16
Pluripotent Stem Cells for Neural Regeneration

Syed Ameer Basha Paspala, Aleem Ahmed Khan, Sandeep Vishwakarma, and T.V. Ramakrishna Murthy

Abstract Neurodegenerative disorders remain challenging to treat using traditional pharmacological or neurosurgical approaches. In contrast, cell therapy is a promising strategy for ameliorating irreparable brain tissue damage during the process of neurogenesis. Currently, more efficient methodologies for isolating neural stem cells from a plethora of sources including hematopoietic stem cells and mesenchymal stem cells are continually being developed. The availability of neural stem cells would ensure that damaged neural tissues can be regenerated and fast-track translation from bed to bedside. In this chapter, we discuss various sources of neural stem cells, strategies for their isolation and characterization, and application of stem cells in the treatment of neurological disorders. Historically, clinical application of cell therapy for treating neurological diseases has been hindered due to numerous technical difficulties. Therefore, these barriers and potential ways of addressing them are also discussed.

Keywords Adult stem cells • Neural stem cells • Neural regeneration • Human umbilical cord blood

---

S.A.B. Paspala (✉)
PAN Research Foundation, Hyderabad, AP, India

Department of Neurology, CARE, The Institute of Medical Sciences,
Hyderabad, AP, India
e-mail: syedpaspala@gmail.com

A.A. Khan • T.V.R. Murthy
Habeebullah's Memorial Centre for Stem Cell Sciences,
Centre for Liver Research and Diagnostics, Hyderabad, AP, India

S. Vishwakarma
PAN Research Foundation, Hyderabad, AP, India

Centre for Liver Research and Diagnostics, Deccan College of Medical Sciences,
Kanchanbagh, Hyderabad, AP, India

M.K. Danquah, R.I. Mahato (eds.), *Emerging Trends in Cell and Gene Therapy*,
DOI 10.1007/978-1-62703-417-3_16,

16.1 Introduction

Pharmacological or neurosurgical therapies currently in practice to treat the neurological damage in various neurodegenerative disorders are not efficient in preventing or reverting these progressive neurodegenerative processes. Recently, a new approach in the form of cell therapy using stem cell has been introduced. Their use in the treatment of neurological diseases is highly restricted owing to several practical implications in isolating the pure neural progenitors and selecting the ideal source for harvesting these neural progenitors. The strategy of isolation and characterization of neural stem cells from various sources will therefore provide major impetus and open up interesting therapeutic modalities for treating several neurodegenerative disorders. The high regenerative potential of pluripotent stem cells in neurological damages suggests that various embryonic/adult sources serve as a proxy for neural stem cells for cell-based therapy.

The concept of regeneration is a new term in the treatment of nervous system disorders which is now well accepted and demonstrated. In terms of neural regeneration, subventricular zone (SVZ) and the dentate region are the most happening and fertile zones in the brain. These zones are rich in pluripotent stem cells. There is a good amount of preclinical data supporting the beneficial role of pluripotent stem cells in various neurological disorders. Safety and efficacy of these cells is well documented in the limited clinical material available till now. The discovery of the stem cells in the central and peripheral nervous system is a relatively recent event. Neural stem cells (NSCs) are generated lifelong by the process of neurogenesis in a specific area of central nervous system (CNS). New neurons are generated from the SVZ of the lateral ventricles and the sub-granular layer of dentate gyrus of the hippocampus. Even in adult brain, the SVZ is the highest neurogenic region in the brain. This is the first region with highest number of NSCs, characterized for their capacity to give rise to neural and glial cells (astrocytes and oligodendrocytes) as well. Cell therapy using these cells has imbibed a ray of hope in the effective management of several neurological disorders.

16.2 Types of Stem Cells

According to their developmental status, stem cells generally are divided into embryonic, fetal, and adult stem cells.

16.2.1 Embryonic Stem Cells (ES Cells)

ES cells are capable of giving rise to all derivatives of the three primary germ layers: ectoderm, endoderm, and mesoderm (Fig. 16.1). Human ES cells are derived from discarded, non-transferred human embryos, from the inner cell mass of blastocyst

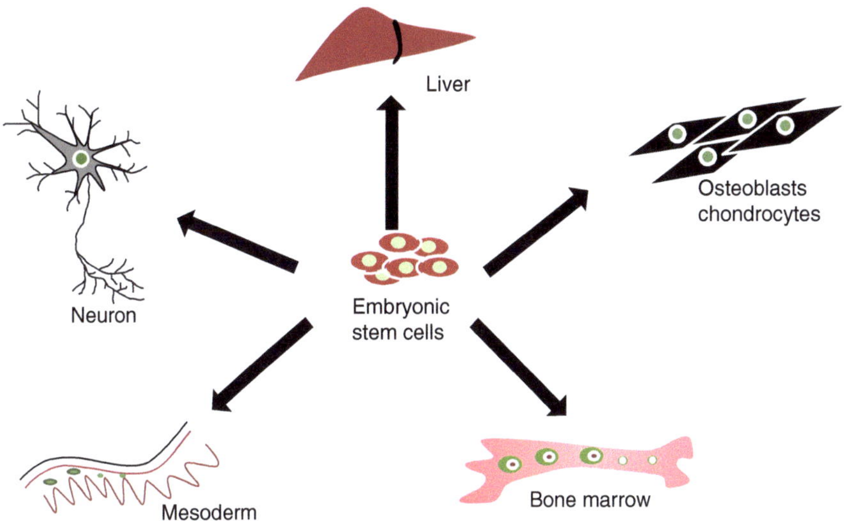

Fig. 16.1 Differentiation potential of embryonic stem cells

using an immunosurgical technique. NSCs derived from embryo can differentiate into neurons, astrocytes, and oligodendrocytes and are capable of forming mature progeny [1] and dopamine neurons [2] both in vitro and in vivo. Recently NSCs derived from embryo are identified by CD133 biomarker. In comparison to the adult stem cells, ES cells are clinically very effective for the treatment of neurological disorders. However, ES cells have some demerits such as in somatic cell nuclear transfer (SCNT), not all of the donor cell's genetic information is transferred and the resulting hybrid cells retain those mitochondrial structures which originally belonged to the egg (e.g., Dolly sheep had hybrid DNA). Previously more than 120 ES cell lines have been reported worldwide. ES cell lines are gradually degraded and will soon be useless for research [3].

Some of the cell lines are genetically identical to others; hence, it has ethical issue thereby restricting the number of available cell lines to 11 which are used for research. Moreover, these cell lines are grown on mouse feeder layers and are not suitable for clinical applications as there is always an associated risk of viral transmission at the time of stem cell transplantation. In clinical therapy, a large number of eggs required for SCNT and the human embryos required as a source of ES cells are extremely sensitive ethical issues. Currently, in India, the spare human embryos from in vitro fertilization (IVF) programmers are permitted for research activity [3].

16.2.2 Adult Stem Cells

Specialized cells are found in many body tissues and functions in tissue homeostasis and repair. They have been propagated from bone marrow, liver, brain, dental pulp, hair follicles, skin, skeletal muscle, adipose tissue, and blood (Fig. 16.2). In vitro,

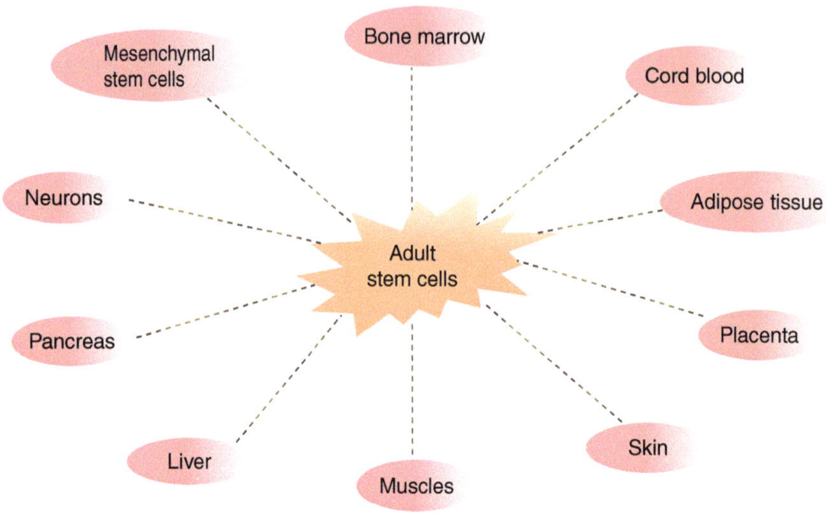

Fig. 16.2 Sources of adult stem cells

they have been shown to differentiate into a wide variety of cell [4–7] types such as osteoblasts, chondrocytes, endothelial cells, skeletal myocytes, glia, neurons, and cardiac myocytes. Unlike ES cells, the use of adult stem cells in research and therapy is not controversial. They were mainly studied in humans and model organisms such as mice and rats.

16.2.3 Neural Stem Cells (NSCs)

NSCs have been isolated and characterized from various regions of the adult CNS from biopsies and postmortem tissues. In several studies, adult tissue-derived NSCs have been transplanted in animal models and studied as functional engraftment, supporting their potential use for a possible cell therapy [7].

16.2.3.1 Origin of Neural Stem Cells

The generation of new neurons in adult mammalian brain occurs throughout the life within two- to three-layered cortical regions, the hippocampus, and olfactory bulb (OB), where it is sustained by endogenous stem cells. The SVZ is the most active NSCs compartment which represents a remnant of the embryonic germinal neuroepithelium (Fig. 16.3). This region persists throughout life as an active mitotic layer in the wall of the telencephalic lateral ventricles and along its rostral extension toward the OB. A complete turnover of the resident proliferating cell population occurs every 12–28 days in the SVZ; about 30,000 new neuronal precursors (neuroblasts) are

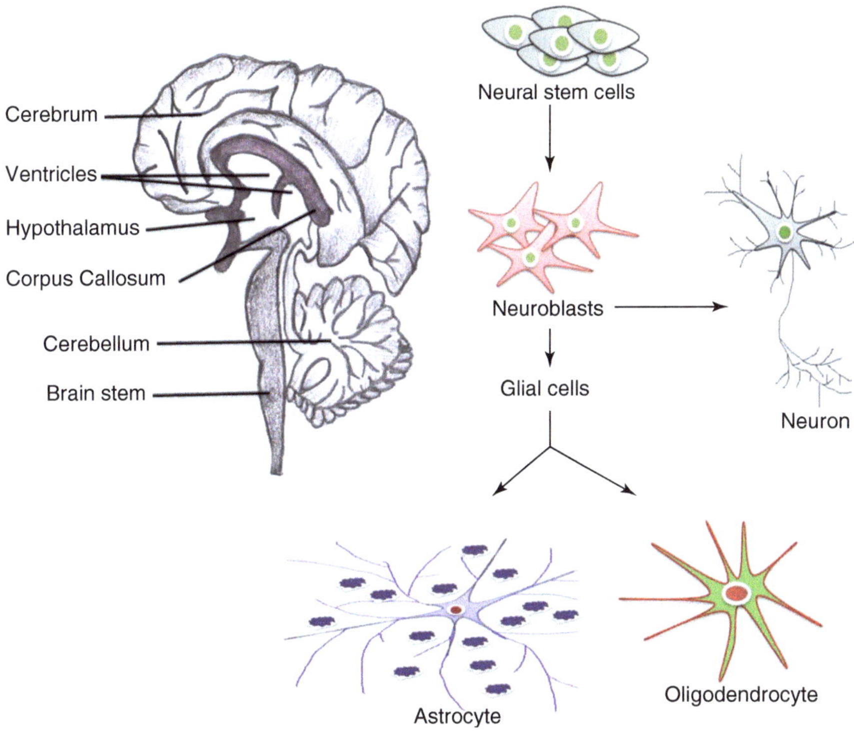

Fig. 16.3 Neural stem cells arise in the SVZ which can differentiate into progenitor cells and then into neurons and glial cells (astrocyte and oligodendrocyte)

produced every day and migrate to the OB [8]. Migrating, proliferating neuroblasts and astrocytes are the main types of cells in SVZ region. They reach the more superficial OB layers and terminally differentiate into granular and periglomerular neurons. Glial tubes are composed of a special type of astroglia that expresses the marker of mature CNS astrocytes (glial fibrillary acidic protein) but also contain the cytoskeletal proteins vimentin and nestin. Astroglial tubes and NSCs do not coexist solely within the periventricular aspect of the SVZ but also within the rostral migratory stream that extends into the OB, with the former perhaps contributing to create an appropriate stem cell "niche" for the maintenance of NSCs all along the pathway.

Adult neurogenesis is a spatially confined process, constrained within the boundaries of the SVZ. Astrocytes and ependymal cells of the SVZ may act as "stromal" elements of the CNS by producing molecules that affect the neuronal versus glial fate of the stem/progenitor cells. Furthermore, the extracellular matrix of the SVZ contains tenascin and proteoglycans, molecules that are important in the formation of developmental compartments and in the control of cell adhesion, migration, and differentiation. The proximity of the SVZ with the cerebrospinal fluid (CSF), the enlarged intercellular spaces, the reduced cell-cell contacts, and the presence of molecules linked to water cotransport contribute to create in the SVZ a cytoarchitectural/biochemical niche, which is very different from the environment of the mature CNS parenchyma [8]. In recent years, neurogenesis has been reported to occur in other regions of the adult brain under normal conditions, such as neocortex, amygdala, and substantia nigra. However, other research groups were not able to replicate some of these reports [9].

However, the organization of the adult SVZ in human is different from that in other mammalian species. The lateral ventricular wall consists of four layers with various thickness and cell densities:

1. A monolayer of ependymal cells (layer I)
2. A hypocellular gap containing astrocytic processes (layer II)
3. A ribbon of cells composed of astrocytes (layer III)
4. A transitional zone into the brain parenchyma (layer IV)

Astrocytes proliferate in vivo and behave as multipotent progenitors in vitro, but no chain migration has been observed in the human SVZ. However, newborn cells that express cell cycle proteins (Ki-67 and proliferating cell nuclear antigen) have been detected in the granular and glomerular layers of the human OB, but no clear evidence of the presence of a migratory pathway from the SVZ has been demonstrated. Therefore, it has been suggested that individual cells might migrate separately to the OB. These results indicate that in comparison with rodents, precursor cells in the human OB are rare but not completely absent [2]. However, these endogenous NSCs are very difficult to isolate and are used for immediate cell therapy.

16.2.3.2 Isolation and Culturing of Neural Stem Cells

The neural stem cells were first isolated and expanded from the embryonic and adult mouse striatum in the early 1990s in a culture system referred to as the neurosphere assay [10]. Later, it was found that not only embryonic CNS but also adult CNS in vitro possesses the ability to generate neurosphere-forming cells in vitro including neural epithelial progenitor cells, radial glial cells, SVZ cells, and ependymal cells [11]. Most studies have shown that the NSCs derived from the brain respond to either basic fibroblast growth factor (bFGF) or EGF and NSCs cultured as neurospheres from the early embryonic forebrain do not respond to EGF until they acquire EGF receptors at later stages of development in vitro or in vivo. However, NSCs cultures from the adult murine hippocampus form as monolayer in the presence of bFGF [12].

These neurospheres on repeated passages produce self-renewing, proliferating, and differentiating cells, typically presenting prominin-1 cell surface antigen which is also known as CD133. These cells are uniquely separated directly by magnetic beads conjugated with antibodies (MACS) or fluorescence-activated cell sorting (FACS) by negative selection of $CD34^-$ and $CD45^-$ antigen marker cells ($CD133^+CD34^-CD45^-$). These cells, upon transplantation into the brain of an immunodeficient neonatal mice (the sorted/expanded $CD133^+$) showed potent engraftment, proliferation, migration, and neural differentiation [13]. However previously, stemlike cells have been purified from various organs as side population (SP) cells, based on their property to exclude Hoechst 33342 [14].

16.2.3.3 Characteristics of Neural Stem Cells

Proliferating cell population in the adult CNS shares the expression of number of stem cell markers such as nestin, Notch1, and SOX2. Notch pathway appears to play an essential role in the maintenance of stem/progenitor cell pool as well as in regulating asymmetric versus symmetric division, both during embryogenesis and in adulthood. Expression of Notch1 or one of its downstream regulators such as HES-1 inhibits neural differentiation and results in the maintenance of a progenitor state [11]. Numerous specific genes/pathways have been identified as important regulators of neural stem cell proliferation, many of which are important for several other cell types, including other stem cells. Some of these are Bmi-I, P21, nucleostemin, maternal embryonic leucine zipper kinase, P53, Rb, and Akt among others [12].

16.2.4 Alternative Sources of Neural Stem Cells/Progenitor Cells for Cell Therapy

16.2.4.1 Olfactory Ensheathed Cells/Olfactory Mucosa Cells

The nose contains neurons that send signals to the brain when triggered by odor molecules. Because olfactory tissue is exposed to the external environment, it contains cells with considerable regeneration potential, including renewable neurons and progenitor/stem cells. Through a relatively innocuous biopsy procedure, olfactory tissue can be obtained from the nasal cavity. It can also be retrieved from the olfactory bulb, but this requires an invasive penetration of the cranial cavity that although unsuitable for human patients has been the procedure for most of the supporting animal research.

Problems of rejection, overgrowth, disease transmission, and ethical issues can be avoided because a person's own olfactory mucosa can be used. OECs theoretically promote axonal regeneration by producing insulating myelin sheaths around growing and damaged axons, secreting growth factors, and generating structural and matrix macromolecules that lay the tracks for axonal elongation. These properties have led to an increasing use of olfactory ensheathing cells (OECs) in preclinical models of transplantation for spinal cord repair including complete transection, hemisection, tract lesion, and contusion with over 50 published studies in the last 10 years.

Nasal olfactory ensheathing cells transplants assist recovery in spinal cord injury (SCI), including complete transaction [15], and there is evidence that adult olfactory tissue is effective when transplanted 1 month after spinal cord transaction in the rat [16]. According to the promising results obtained from animal experiments, several clinical trials were started in a large series which recruited more than 400 patients for transplantation of fetal olfactory bulb-derived cells. The results of 171 operations have been published [17]. In addition, a single-blinded controlled study also demonstrated the safety and feasibility of intraspinal transplantation of autologous OECs in human SCI [18].

Whereas in Huang's procedure the fetal tissue's undifferentiated nature minimized immunological rejection [19], Feron et al. tested the feasibility and safety of transplantation of autologous OECs into the injured spinal cord in human paraplegia [18]. OECs were grown and purified in vitro from nasal biopsies and injected by microinjection. Twelve to twenty-eight million cells were injected into the region of damaged spinal cord. Posttransplantation follow-up demonstrated the procedure to be safe as no significant medical, surgical, or other complications developed even after 1 year of cell implantation. There was no evidence of spinal cord damage or of cyst, syrinx, or tumor formation. In this clinical trial, there was no naturopathic pain reported by the participants, no change in psychosocial status, and no evidence of deterioration in neurological status. This indicates that the OECs transplantation may be a safe method by in vitro propagation before transplantation [18]. However, in a recent report of OECs transplantation, some adverse effects were seen [20]. Of the 327 patients recruited, 16 (4.9 %) patients experienced various complications including headache, short-term fever, seizure, central nervous system infection, pneumonia, respiratory failure, urinary tract infection, heart failure, and possible pulmonary embolism along with four deaths (1.2 %). In another report of Chew et al., it was reported that in a woman who received an injection into each frontal lobe in Beijing, China, amyotrophic lateral sclerosis (ALS) progressed at a more rapid rate after the procedure and she suffered disabling side effects [21]. In a phase I/II clinical study designed to test the feasibility and safety of transplantation of autologous OECs into the injured spinal cord in human paraplegia, no adverse findings were seen even 3 years after transplantation into spinal cords injured at least 2 years prior to transplantation. The magnetic resonance images (MRIs) at 3 years showed no change from preoperative MRIs or intervening MRIs at 1 and 2 years, with no evidence of any tumor of introduced cells and no development of posttraumatic syringomyelia or other adverse radiological findings. There were no significant functional changes in any patient and no neuropathic pain. In one transplant recipient, there was an improvement over three segments in light touch and pinprick sensitivity bilaterally, anteriorly, and posteriorly. This report concluded that transplantation of autologous OECs into the injured spinal cord is feasible and safe up to 3 years postimplantation; however, this conclusion should be considered preliminary because of the small number of trial patients [22].

16.2.4.2 Bone Marrow (BM)

The bone marrow stroma contains mesenchymal stem cells (also called marrow stromal cells). These multipotent cells have the ability to differentiate into a variety of cell types.

Recently, Larson et al. [23] described the isolation and expansion of human mesenchymal stem cells (MSCs), isolated from 1 to 4 ml of bone marrow aspirates from the iliac crest of normal adult.

The potential of bone marrow cells to differentiate into myelin-forming cells and to repair the demyelinated rat spinal cord in vivo was studied using cell transplantation techniques [24]. Transplantation of the hematopoietic and non-hematopoietic precursor cells and lymphocytes in demyelinated dorsal column showed remyelination to occur in the immunocompromised rats. These transplanted bone marrow cells showed a characteristic Schwann cell pattern of myelination. Transplantation of CD34+ve hematopoietic progenitor cells failed to form myelin. These results therefore underscore that bone marrow cells have the potential of in vivo differentiation into myelinated cells which can repair the demyelinated CNS [24]. From the above findings, it is clearly evident that non-hematopoietic cells do have the ability of transdifferentiation into cells of neural lineage upon exposure to appropriate differentiation media. These cells develop electrophysiological characteristics of neurons, or neuron-like MSCs fail to generate action potential owing to the lack of voltage-gated ion channels. Transplantation of MSCs 1 week after the injury demonstrated better results in comparison to cells transplanted immediately. Histological assessment of the site of injury 5 weeks after transplantation showed regenerative changes which included longitudinal arrangement of immature astrocytes along with formation of bundles bridging the site of injury. In addition, bundles of neurofilament and some 5-hydroxytryptamine-positive fibers were seen at the interface between graft and scar tissue. MSCs constitute an easily accessible and expandable cell source for the repair of SCI [25]. MSCs cocultured with fetal spinal cord-derived neurosphere cells stimulate the development of extensive processes. These cells expedite the process of tissue repair leaving apparently smaller cavities than in controls. Although the number of grafted MSCs gradually decreased, some treated animals showed remarkable functional recovery [26].

To assess the migration and transdifferentiation of the bone marrow stem cells (BMSCs) upon transplantation, mice with cerebral infarct were subjected to BMSCs transplantation. BMSCs from mice were harvested, characterized, and cultured. The cultured cells were CD45+ve (low expression), CD90+ve, and Sca-1 (high expression) after flow cytometric analysis. After 4 weeks of transplantation, a large subset of the cells was found to survive in normal brain with many cells located in the close proximity to the transplanted site. Fluorescent immunohistochemistry of the transplanted cells demonstrated these cells to express neuronal specific markers such as NeuN, MAP2, and doublecortin [27]. Similar results were obtained when BMSCs were transplanted in mice with SCI. These transplanted cells were also found to express astrocytic GFAP protein at the site of SCI. In the similar fashion, Zurita et al. demonstrated progressive recovery of the paraplegic animals after bone marrow stem cell transplantation [28]. The transplanted cells were found to express neuronal and astroglial markers along with marked ependymal proliferation as demonstrated by nestin positivity. These transplanted cells survived well in the spinal cord tissue with the formation of cell bridges within the centromedullary cavity. These findings therefore support the clinical feasibility of using BMSCs in chronic paraplegia.

In another study using combinatorial approaches (cAMP/NT-3) that induce both the neuronal soma and the axonal regeneration, axons across the SCI sites were demonstrated by Lu et al. [29]. This approach may be a novel strategy which can be further explored to regenerate injured spinal cord. MSC transplants occupied the lesion cavity and were associated with preservation of host tissue and white matter (myelin), demonstrating that these cells exert neuroprotective effects. The tissue matrix formed by MSC grafts supported greater axonal growth than that found in specimens without grafts. Uniform random sampling of axon profiles revealed that the majority of neurites in MSC grafts were oriented with their long axis parallel to that of the spinal cord, suggesting longitudinally directed growth [30]. Later it was demonstrated that BDNF-expressing marrow stromal cells support extensive axonal growth at sites of spinal cord injury [31].

Similarly, grafting of human MSCs derived from aspirates of four different donors into a subtotal cervical hemisection in adult female rats showed cell integration at the injury site with little migration away from the graft [32]. Immunocytochemical (ICC) analysis demonstrated robust axonal growth through the grafts of animals treated with MSCs, suggesting that MSCs support axonal growth after spinal cord injury. However, the amount of axon growth through the graft site varied considerably between groups of animals treated with different MSC lots, suggesting that efficacy may be donor dependent. Similarly, a battery of behavioral tests showed partial recovery in some treatment groups but not in others. Using enzyme-linked immunosorbent assay (ELISA), variations were found in secretion patterns of selected growth factors and cytokines between different MSC lots. In a dorsal root ganglion explant culture system analyzing the efficacy of conditioned medium from three donors, it was found that average axon lengths increased for all groups compared to control. These results suggest that human MSCs produce factors important for mediating axon outgrowth and recovery after SCI but that MSC lots from different donors vary considerably. To qualify MSC lots for future clinical application, such notable differences in donor or lot-lot efficacy highlight the need for establishing adequate characterization, including the development of relevant efficacy assays [32]. Results demonstrated that transplantation of Schwann cells derived from bone marrow stromal cells (BMSC-SCs) promotes axonal regeneration of lesioned spinal cord resulting in recovery of hind limb function in rats [33].

Transplantation of bone marrow cells into the SCI has been found to improve neurologic activities in experimental animal studies. However, it is unclear whether bone marrow cells can similarly improve the neurologic functions of complete SCI in human patients. To address this issue, clinical outcome of autologous cell transplantation in conjunction with the administration of granulocyte macrophage-colony stimulating factor (GM-CSF) was analyzed in complete SCI patients [34]. Sensory improvements were noted immediately along with significant motor improvements 3–7 months after the procedure. Four patients showed neurologic improvements in their AIS grades (from A to C), and one patient improved to AIS grade B from A. No immediate worsening of neurological status was found. Side effects of GM-CSF treatment such as fever and myalgia were observed. Serious complications and increasing mortality and morbidity were not found. The follow-up study with MRI after injury showed slight improvement within the zone of cell transplantation. BMT and GM-CSF administration was demonstrated to be a safe protocol to manage SCI patients especially those with acute complete SCI [34]. However, the beneficial effects of unmanipulated autologous bone marrow cells in patients with spinal cord injury are yet to be confirmed. Since last few years, macroporous polymer combined hydrogels based on the derivatives of pHEMA has been considered the most suitable material for bridging cavities at SCI lesions [35]. Subarachnoid injection has been recently reported as a minimally invasive method for the transplantation of bone marrow stromal cells in spinal cord injury. It may be, however, less effective than direct injection into the spinal cord in terms of cell delivery [36]. In another study, Yoshihara et al. indicated that combination of beneficial effects of rat MSCs and exercise protocol were not sufficient to enhance behavioral recovery [37]. Similarly, Cao et al. [38] demonstrated transdifferentiation of transplanted marrow stromal cells and reactive changes of glial cells in a completely transected rat spinal cord. In another experiment, the safety, therapeutic time window, implantation strategy, method of administration, and functional improvement of transversal spinal cord injury patients receiving unmanipulated autologous bone marrow was investigated. The results of the study demonstrated the implantation of autologous bone marrow cells to be safe as no complications were observed following implantation even after 2 years of follow-up. However, longer follow-ups are required to determine the safety and confirm the observed beneficial effects that are due to the cell therapy [35]. From the above findings, it is evident that stem cell transplantation within a therapeutic window of 3–4 weeks following injury will play an important role in any type of SCI. In addition, clinical trials involving a larger population of patients and different cell types are needed before further conclusions could be drawn.

Cellular and extracellular inhibitors are thought to restrict axon growth after chronic SCI, confronting the axon with a combination of chronic astrocytosis and extracellular matrix-associated inhibitors that collectively constitute the chronic "scar." However, SCI does not create impenetrable boundaries that inhibit the balance of local and diffusible signals that appear to generate robust axonal growth even without resecting chronic scar tissue [39]. In a rabbit model of spinal cord ischemia, transplantation of MSCs was found to enhance angiogenesis and improve functional recovery. This study also supported the perspective that the therapeutic time window is critical for the therapeutic effect of MSCs [40]. Another study by Koda et al. [41] observed that a number of double-positive cells for GFP and glial markers are larger in the G-CSF-treated mice than in the control mice after bone marrow cells were transplanted into lethally irradiated C57BL/6 mice. Staining with Luxol fast blue (LFB) also revealed that G-CSF promoted white matter sparing.

Recently, a phase I/II open-label nonrandomized trial was done to assess the safety and efficacy of autologous bone marrow cell transplantation coupled with the administration of granulocyte macrophage-colony stimulating factor. The study included 35 patients divided into 4 groups all with complete SCI and all received BMCs by injection into the surrounding area of injury site:

- The first group ($n=17$) received BMC transplantation within 14 days after injury.
- The second group ($n=6$) received within 14 days to 8 weeks.
- The third group ($n=12$) received at >8 weeks after injury.
- The last group ($n=13$) were conservatively treated with conventional decompression and fusion surgery without receiving BMC transplantation.

All the patients were followed up for a mean period of 10.4 months after injury and underwent preoperative and follow-up neurological assessments as per the American Spinal Injury Association impairment scale, magnetic resonance imaging, and electrophysiological monitoring. At 4 months follow-up, MRI showed enlargement of spinal cords and the small enhancement of the cell implantation sites; furthermore, patients that received BMC transplantation and GM-CSF were not associated with any serious adverse events increasing morbidities. The AIS grade increased in 30.4 % of the acute and subacute treated patients (AIS A to B or C), whereas no significant improvement was observed in the chronic treatment group. Increasing neuropathic pain during the treatment and tumor formation at the site of transplantation are still to be investigated. Long-term and large-population multicenter clinical studies are therefore required to determine its precise therapeutic effect.

16.2.4.3 Cord Blood

Umbilical cord blood/cord blood which is a rich source of stem cells (CD $34^{+}/45^{-}$) is being increasingly used on an experimental basis as an alternative to bone marrow. To date, more than 70 different diseases have been treated using cord blood transplants. Cord blood contains multiple populations of pluripotent stem cells and can be considered the best alternative to ES cells. In addition to this, cord blood stem cells are capable of giving rise to hematopoietic, epithelial, endothelial, and neural tissues both in vitro and in vivo. Thus, cord blood stem cells have the potential to treat a wide variety of diseases including cardiovascular, ophthalmic, orthopedic, neurologic, and endocrine diseases. However, owing to few ethical considerations, their applications in humans is prohibited in some countries.

16.2.4.4 The Use of Human Umbilical Cord Blood (hUCB): A Rich Source of Nonembryonic or Adult Stem Cells

Saporta et al. [42] experimented with the umbilical cord blood cells in SCI in in vitro model. He reported that the cord blood cells can be used to regenerate the behavioral effects of spinal cord and also reported that cells of cord blood were found in only at the site of injury in the spinal cord of the rat. This report supports the hypothesis that the cord blood stem cells have the capacity to migrate to and heal the neurological damage caused by traumatic assault. Functional score assessment in SCI rat models at day 7 and 14 after $CD34^{+}$ cord blood (CB) cells transplantation showed better improvement as compared to BMCs. Histological evaluation revealed that bromodeoxyuridine (BrdU)-labeled CD34+ CB and BMS cells survived and migrated into the injured area. Some of these cells expressed glial fibrillary acidic protein or neuronal nuclear antigen (NeuN). CD34-positive cells of cord blood showed recovery in functional behavior when they were intraspinally transplanted in hemisectioned spinal cord rats. The results of the study suggested that cord blood CD34-positive stem cells may be employed for routine allogenic and autologous transplantations as a treatment modality for human SCI [43]. Another report by Li et al. [44] reported that intraspinal transplantation of human cord blood CD34+ cells resulted in the improvement of neurological function after SCI in rats compared with the control group ($p < 0.05$). Moreover, the results showed that intraspinally administered human cord blood CD34+ cells survived, differentiated, and expedited the process of functional recovery after SCI in rats. In another study, transplanted hUCB differentiated into various neural cells and improved the motor function in cord-injured rat model. Similarly, Nishio et al. [45] suggested that transplanted CD34+ fraction cells from hUCB may have therapeutic effects for SCI. These hematopoietic stem cells (CD34+ cells) promoted restoration of spinal cord tissue and recovery of hind limb function in adult rats. The results of this study provided important preclinical data regarding HUCB stem cell-based therapy for SCI. Recently Dasari et al. [46] showed that hUCB cells differentiated into several neural phenotypes including neurons, oligodendrocytes, and astrocytes. Ultrastructural analysis of axons revealed that hUCB cells formed morphologically normal-appearing myelin sheaths around axons in the injured areas of spinal cord. These findings demonstrated that hUCB cells when transplanted into the spinal cord 7 days after weight-drop injury survived for at least 2 weeks, differentiated into oligodendrocytes and neurons, and improved locomotor function. Due to difficulties in maintaining graft in the aging rat CNS, recent study selected the NOD SCID mouse for in vivo characterization of hUCB cells. Stereotaxically transplanted hUCB cells survived and differentiated into neuronal cells at either 5 or 30 days after transplantation. At early time points, many differentiated hUCB cells expressing characteristic neuronal proteins were detected. However, at 1 month postgrafting, hUCB cells were no longer detected [47].

Recent findings in SCI rats treated with neurally induced progenitor cells of hUCB showed recovery of somatosensory potentials. The grafted cells especially exhibited oligodendrocytic phenotype around the necrotic cavity. These results thereby demonstrate that neurally induced progenitor cells of hUCB might be a therapeutic resource to repair damaged spinal cords [48]. Cord blood contains a mixture of different types of stem cells in numbers not seen in any other location including embryonic-like stem cells, hematopoietic stem cells, endothelial stem cells, epithelial stem cells, and MSCs. Extensive published work from multiple investigators has demonstrated that CB stem cells are amenable to neurological applications including as evidenced by in vitro studies, preclinical animal models of disease, and more recently by patient clinical trials. Therefore, umbilical CB stem cells are unique in their ability to be used for stem cell transplantation in the treatment of blood disorders, as well as in regenerative medicine to treat patients with neurological disease [49].

16.2.4.5 Skin

Skin contains epidermal and dermal layers. The outermost epidermis consists of stratified squamous epithelium with an underlying connective tissue section, or dermis, and a hypodermis, or basement membrane. Mammalian skin is composed of two primary layers; the epidermis and the dermis, can be generated in adult skin [50]. Skin-derived stem cells (SKPs) can generate both neural and mesodermal cell types and that most of the neural cells generated by SKPs have characteristics of peripheral neurons and Schwann cells, consistent with a potential neural crest origin. Recently a subpopulation of nestin– vimentin+ phenotype of fibroblasts cells appeared multipotent which showed neural cell differentiation characters [20].

Recently, a milestone was achieved when induced pluripotent stem cells (iPS) were produced from adult human cells [51–53]. With the same principle used earlier in mouse models, Yamanaka had successfully transformed human fibroblasts into pluripotent stem cells using the same four pivotal genes, Oct-3/4, SOX2, Klf4, and c-Myc, with a retroviral system. Thomson and colleagues used Oct-4, SOX2, Nanog, and a different gene LIN28 using a lentiviral system. Further, Daley and his team derived iPS cells (using Oct-4, SOX2, Klf4, and Myc transcriptional factors) from fetal, neonatal, and adult human primary cells, including dermal fibroblasts isolated from a skin biopsy of a healthy research subject. This data demonstrated that defined factors can reprogram human cells to pluripotency and can be established a method whereby patient-specific cells might be established in culture [34].

16.2.4.6 Hair Follicles

Hair follicles are known to contain a well-characterized niche for adult stem cells: the bulge, which contains epithelial and melanocytic stem cells. Nestin-positive cells were identified in the bulge area in mouse and were found to give rise to neurons, smooth muscle cells, and melanocytes [54]. Neural-crest-like stem cells have been identified in mouse whisker hair follicles, and bulge cells from mouse whisker hairs grow as adherent monolayer cells and appear to be multipotent [55].

16.2.4.7 Adipose Tissue

The adipose tissue is a highly complex tissue and consists of mature adipocytes, preadipocytes, fibroblasts, vascular smooth muscle cells, endothelial cells, resident monocytes/macrophages [56], and lymphocytes [57]. Since this tissue provides a rich source of pluripotent adipose tissue-derived stromal cells, it has been considered as another alternative source for the isolation of stromal cells in larger quantities using less invasive procedures [58]. Adipose tissue (AT) is another alternative source that can be obtained by a less invasive method and in larger quantities than BM. It has been demonstrated that AT contains stem cells similar to BM-MSCs, which are termed processed lipoaspirated cells [57]. These cells can be isolated from cosmetic liposuctions in large numbers and grown easily under standard tissue culture conditions. These processed lipoaspirated (PLA) cells and clones can be further differentiated into putative neurogenic cells, exhibiting a neuronal-like morphology and expressing several proteins consistent with the neuronal phenotype. The multilineage differentiation capacity of PLA cells has been confirmed. Before application to human neurological diseases, additional in vitro experiments and preclinical trials are necessary.

16.2.4.8 Wharton's Jelly

Wharton's jelly cells are the cells isolated from the gelatinous connective tissue of the umbilical cord. Wharton's jelly as a source of primitive cell types was established based on the low levels of collagen expressed in gelatinous connective tissue and the fact that, during embryogenesis, totipotent cells such as primordial germ cells and hematopoietic stem cells migrate from the yolk sac through this region to populate target tissues in the embryo and fetus. Wharton's jelly cells have been cultured for more than 80 population doublings with no indications of senescence, changes in morphology, increased growth rate, or change in ability to differentiate into neurons. Thus, Wharton's jelly cells possess one of the defining characteristics of stem cells, the ability to self-renew. Wharton's jelly cells have telomerase activity, which is usually characteristic of human embryonic stem cells. Colonies of Wharton's jelly cells were also found to express NSE, c-kit, and, even more intriguing, TH, a marker for catecholaminergic neurons. However, the expression of markers for non-neuronal cell lineages by these cells remains to be determined [59].

16.2.4.9 Amniotic Placental Fluid

Amniotic fluid contains a heterogeneous population of cells which are contributed mainly from the fetal skin; the fetal digestive, respiratory, and urinary tract; and the placental membranes [60–63]. Recent discoveries of stem cell populations in amniotic fluid have postulated that amniotic fluid is a promising alternative source of fetal stem cells for cellular therapies [64–67].

16.2.4.10 Macrophages

Recruitment of macrophages is limited in CNS and the resident microglia cells are the main immune cells that are activated after SCI [68]. It has been shown that controlled boosting of local immune response by delivering of autologous macrophages, which alternatively activates wound-healing phenotype, can promote recovery from the spinal cord injury. Implantation of macrophages activated by preincubation with peripheral nerve fragments led to partial recovery of paraplegic rats [69]. The possible mechanisms are activation of infiltrating T cells and increased production of trophic factors and brain-derived neurotrophic factor [70, 71] that leads to removal of inhibitory myelin debris [69]. The results of phase I studies show that out of eight patients in the study, three recovered clinically significant neurological motor and sensory function. In other clinical studies this therapy has been found to be well tolerated in acute SCI patients [72].

16.2.4.11 Dendritic Cells

In animal model studies, transplantation of dendritic cells into the injured spinal cord of mice led to better functional recovery as compared to controls [73]. The implanted dendritic cells induced proliferation of endogenous neural stem/progenitor cells (NSPCs) and led to de novo neurogenesis. This observation was attributed to the action of secreted neurotrophic factors, such as neurotrophin-3, cell-attached plasma membrane molecules, and possible activation of microglia/macrophages by implanted dendritic cell. Dendritic cells pulsed (incubated) with encephalitogenic or non-encephalitogenic peptides derived from myelin basic protein when administered intravenously or locally to the site of injury promoted recovery from SCI [74].

16.2.4.12 Schwann Cells (SCs)

Schwann cells are the supporting cells of the peripheral nervous system. Like oligodendrocytes, Schwann cells wrap themselves around nerve axons, but the difference is that a single Schwann cell makes up a single segment of an axon's myelin sheath. Schwann cells originating from dorsal and ventral roots are one of the cellular components that migrate to the site of tissue damage after spinal SCI [75–77]. The remyelinating capability of Schwann cells has been demonstrated in a number of studies [76, 78], and the functioning status of this myelin in conduction of neural impulses has been confirmed [79, 80].

16.2.4.13 Human Fetus

Fetal-derived multipotent fetal stem cells (FSCs) are generally more tissue-specific than ES cells. Therefore, FSCs are able to generate a more limited number of progenitor types. One of the particular therapeutic advantages of FSCs as compared with ES cells is the fact that FSCs do not form teratomas in vivo. Moreover, the FSCs obtained up to week 12 offer the possibility of transplanting these primitive stem cells without frequent rejection reactions in contrast to UCB and BM stem cell transplants. Recent work has revealed the possibility of using FSCs or their progenitors, isolated from particular tissues, for multiple therapeutic applications involving tissue regeneration [81–84]. FSCs can cross both the placental and blood–brain barrier. These cells can be administered intravenously and hence can be employed as a therapeutic application for repair of diverse brain disorders. The two areas of fetal CNS, SVZ and hippocampus have been considered as the richest source for the isolation of neural progenitor cells.

16.3 Regeneration of Central Nervous Tissue

In our current investigation which aimed to explore application of stem cells in the treatment of neurological disorders, literature reported by Stocum et al. [85] documents that neural stem cells were extensively found in two areas of adult mammalian brain (anterior part of the SVZ of the lateral ventricle, from where the immature neurons migrate through the rostral migratory stream into the OB). These stem cells differentiate into granule neurons and periglomerular interneurons. The other is the sub-granular zone of the dentate gyrus (from where the stem cells migrate into the hippocampus). These cells develop into mature granular neurons. Cells from these two regions (subventricular and sub-granular zone) could be a primitive source for progenitor population which can be widely employed in the treatment of all neurological problems such as the neurodegenerative diseases and spinal cord injury [43]. This cell therapy may also support quiescent neurogenesis in brain leading to stimulation and proliferation of progenitors from those two regions. These progenitor cells are very few in number to recover or substitute the degenerated neurons in the brain.

Another important aspect is that these cells sometimes survive or differentiate with difficulty. Stocum et al. [85] demonstrated that immature neurons originating from the SVZ migrate to the damaged striatal area. Although these cells start to express markers for striatal medium-size spiny neurons with longer survival period (>1 year) of some progenitor cells, others died within few weeks. Due to the above experimental example, investigators have put much effort for developing new methods and clinical procedures for neural cell/stem cell-based therapies for spinal cord and other degenerative neural diseases. Further experiments have given some hope for cell-based therapies in which allogenic NSCs transplantation to spinal cord injuries showed partial recovery from paralysis. However, slight improvements were possibly seen, owing to the effect of transplanted cells on host cells, but the recovery may not be attributable to the transplanted cells. However, there are evidences which demonstrate that when NSCs are transplanted in spinal cord injuries, the progenitor cells differentiate into glial cells and also oligodendrocytes. Intravenous transplantation of fluorescein isothiocyanate-labeled (FITC) human umbilical cord blood cells in rats showed recovery of locomotory behavior after 5 days. In histological examination, less than 1,000 labeled cells survived after transplantation; however, these cells were not found to differentiate into neurons and/or glia.

Although when bone marrow cells were transplanted into a group of 32 patients of 2–12 years after complete spinal cord injury in a study at the University of Sao Paulo, a modest improvement in the lower extremity function was seen in 15 patients. The same study showed improvements were likely due to axon remyelination by grafted cells and/or juxtacrine effects or paracrine of the transplanted cells on host neurons. He also showed direct evidence for paracrine/juxtacrine effects of transplanted cells on host neural tissue. Mikami et al. [73] reported that transplanted splenic dendritic cells supported proliferating and differentiating host NSCs and induced axon sprouting, accompanied by partial recovery from hind limb paralysis into lesioned mouse spinal cords. However, there was a confirmation that coculture of spinal cord NSCs with dendritic cells significantly enhanced the survival and proliferation of the NSCs. Mikami et al. [73] also reported that transplantation of human MSCs into the dentate gyrus of the mouse hippocampus promoted neurogenesis by endogenous NSCs and astrocytes derived from embryonic glial-restricted precursors. The rat spinal cord developed axon regrowth and inhibited initial scarring that was associated with significant improvement of locomotor function.

Stocum et al. reported that transplantation of oligodendrocyte precursors generated from adipose tissue-derived stromal cells in SCI model promotes functional recovery by both remyelination and induction of proliferation and differentiation of host NSCs [85].

Based on the current knowledge of NSCs from different sources, there is a need to standardize methodology for the isolation and characterization of NSCs from potential sources. This will enable the researchers to harvest high number of viable and well-characterized cells which can be further employed to treat patients with neurological disorders [86].

16.4 Induced Pluripotent Stem Cells in Neural Regeneration

The increasing availability of iPSCs derived from adult human somatic cells provides new prospects for cell-replacement strategies and disease-related basic research in neurological diseases. iPSCs are artificially derived from a non-pluripotent cell, typically an adult somatic cell through reprogramming. iPSCs were initially derived from mouse embryonic and adult fibroblasts by overexpression of particular transcription factors, which have become famous as the "Yamanaka factors," which includes 24 candidate genes known to be pluripotency associated. After elimination of irrelevant factors, a minimum of four factors remained that were minimally required to generate mouse iPSCs. These factors are octamer 3/4 (OCT3/4), SRY-box-containing gene 2 (SOX2), cytoplasmic Myc protein (c-MYC), and Kruppel-like factor 4 (KLF4). Yamanaka's group used retroviral vectors encoding OCT4 (also known as Pou5F1), SOX2, KLF4, and c-MYC, while the group of James Thomson used lentiviral vectors encoding OCT4, SOX2, NANOG, and Lin-28 to reprogram human fibroblasts in to iPSCs. Cell types that have been used for iPSCs derivation include keratinocytes, pancreatic β cells, neural cells, mature B and T cells, melanocytes, hepatocytes, amniotic cells, and cells derived from adipose tissue. However, so far only fibroblasts have been used to generate iPSCs from patients suffering from neurological diseases. Retroviral and lentiviral vectors have been widely used for the delivery of reprogramming factors. New strategies have been suggested to generate safe and less tumorogenic iPSCs by using nonviral methods or by omitting the oncogenic factors c-MYC and KLF4. Therefore, attempts have been made to derive iPSCs by using plasmids rather than viruses. The use of iPSCs in the treatment of neurological disorders requires that iPSCs should differentiate into the relevant neuronal subtypes. The earliest recognizable cell type in the neural lineage is the neural ectoderm. Yamanaka's group also showed that human iPSCs can differentiate into β-III-tubulin-positive neurons as well as GFAP-positive astrocytes. Park et al. in 2008 [34], for the first time, created patient as well as disease-specific iPSCs from skin fibroblasts of patients that suffered from a variety of genetic diseases, including adenosine deaminase deficiency-related severe combined immunodeficiency, Gaucher disease type III, Duchenne (DMD) and Becker muscular dystrophy (BMD), Parkinson's disease (PD), Huntington's disease (HD), juvenile-onset type 1 diabetes mellitus, Down syndrome (DS)/trisomy 21, and the carrier state of Lesch-Nyhan syndrome.

16.5 Hurdles and Future Prospects

Knowledge of basic biology of development is extensive. Though it is explored and grossly understood to a major extent, the fine-tuning controls are yet to be understood. The cell therapy is being affected by various factors involved in cell proliferation, migration, engraftment and differentiation within the host in diseased condition (Table 16.1), which play a definite role in cell therapy. Though the researchers have tried to understand the cells and applied them for beneficial effects on animal models of different diseases like Parkinson's disease, multiple sclerosis, Huntington's disease, ALS, and ischemic stroke (Table 16.2). The major problem was that these animal models are definitely different from humans, who are supposedly the targets of cell therapy benefits. Further, the behavior of these cells in a diseased environment is not fully understood, and the correct clues of their survival and engraftment into the system are also to be understood in depth, so that the results can be controlled, repeated, and compared. There are several mechanisms playing different roles in cell therapy depending upon the pathophysiology, pathology of the disease, and native environment (Table 16.3). With this review, it's evident that to get stem cell therapy accepted as a standard therapy, we need to get the clinicians as well as basic science experts to put their heads together and standardize the protocols at all levels.

Table 16.1 Cell therapy hitches

1. Type of disease
2. Source of cells
3. Quality of cells
4. No. of cells to be transplanted
5. Route of transplantation
6. Time of transplantation
7. Cell migration
8. Cell survival
9. Cell proliferation and differentiation
10. Functional integration and clinical improvement
11. Unwanted effects

Table 16.2 Transplantation studies of hNSCs in animal models of neurological diseases

Disease	Disease model	Administration route	Source of NSCs	Therapeutic effect	Result	Reference
Parkinson's disease	MPTP-treated monkeys	Direct transplantation into injury site	Human fetal telencephalon	Rescue of substantia nigra neurons	Not assessed	[87]
	MPTP-treated monkeys	Direct transplantation into injury site	Human fetal telencephalon	Migration through nigrostriatal pathways	Not assessed	[88]
	MPTP-treated monkeys	Direct transplantation into injury site	Human fetal cortex	GDNF delivery	Functional recovery	[89]
Multiple sclerosis	Lysolecithin-injected rat	Direct transplantation into injury site	Adult human brain tissue	Dispersion and differentiation to oligodendrocyte	Not assessed	[90, 91]
Huntington's disease	Striatal QA injection rat model	Striatal injection	Human fetal NSCs	Migration to basal ganglia and differentiation to neurons and astrocytes	Functional improvement	[92])
ALS	SOD1 G93A rat model	Intraspinal injection	Human fetal spinal NSCs	Synapse formation of NSCs with host motoneurons	Not assessed	[93]
	SOD1 G93A rat model	Intraspinal injection	Human fetal brain	GDNF delivery	No functional improvement	[94]
	SOD1 G93A rat model	Intraspinal injection	Human fetal brain	GDNF delivery and migration to degenerating areas	No functional improvement	[95]
Ischemic stroke	MCAO rat model	Injection into ischemic striatum	Human fetal brain	Differentiation to neurons	Not assessed	[96]

Table 16.3 Mechanisms of cell therapy benefits

1. Modulation of inflammation
2. Instillation of trophic factors
3. Stem cell population replacement
4. Activation of native stem cells
5. Repair of anatomical/pathological abnormalities

References

1. Reubinoff BE, Pera MF, Vajta G, Trounson AO (2001) Effective cryopreservation of human embryonic stem cells by the open pulled straw vitrification method. Hum Reprod 16(10): 2187–2194
2. Geraerts M, Krylyshkina O, Debyser Z, Baekelandt V (2007) Concise review: therapeutic strategies for Parkinson disease based on the modulation of adult neurogenesis. Stem Cells 25(2):263–270
3. ISSCF procs. Embryonic stem cells and factors of rejection. Mumbai, 2005
4. Cotsarelis G, Cheng SZ, Dong G, Park JH, Lee JE, Eom YW et al (1989) Existence of slow-cycling limbal epithelial basal cells that can be preferentially stimulated to proliferate: implications on epithelial stem cells. Cell 57:201–209
5. Dennis JE, Charbord P (2002) Origin and differentiation of human and murine stroma. Stem Cells 20:205–214
6. Eglitis MA, Mezey E (1997) Hematopoietic cells differentiate into both microglia and macroglia in the brains of adult mice. Proc Natl Acad Sci U S A 94(8):4080–4085
7. Graziadei PP, Graziadei MGA (1980) Neurogenesis and neuron regeneration in the olfactory system of mammals. III. Deafferentiation and reinnervation of the olfactory bulb following section of the fila olfactoria in rat. J Neurocytol 9:145–162
8. Galli R, Gritti A, Bonfanti L, Vescovi AL (2003) Neural stem cells an overview. Circ Res 92: 598–608
9. Hermann A, Maisel M, Wegner F, Liebau S, Kim DW, Gerlach M et al (2006) Multipotent neural stem cells from the adult tegmentum with dopaminergic potential develop essential properties of functional neurons. Stem Cells 24(4):949–964
10. Reynolds BA, Tetzlaff W, Weiss S (1992) A multipotent EGF-responsive striatal embryonic progenitor cell produces neurons and astrocytes. J Neurosci 12:4565–4574
11. Rao M (2006) Neural development and neural stem cell, 2nd edn. Humana Press, Totowa
12. Kornblum HI, Hussain R, Wiesen J, Miettinen P, Zurcher SD, Chow K et al (1998) Abnormal astrocyte development and neuronal death in mice lacking the epidermal growth factor receptor. J Neurosci Res 53(6):697–717
13. Uchida N, Buck DW, He D, Reitsma MJ, Masek M, Phan TV et al (2000) Direct isolation of human central nervous system stem cells. Proc Natl Acad Sci U S A 97(26):14720–14725
14. Goodell MA, Rosenzweig M, Kim H, Marks DF, DeMaria M, Paradis G et al (1997) Dye efflux studies suggest that hematopoietic stem cells expressing low or undetectable levels of CD34 antigen exist in multiple species. Nat Med 3(12):1337–1345
15. Lu J, Féron F, Ho SM, Mackay-Sim A, Waite PM (2001) Transplantation of nasal olfactory tissue promotes partial recovery in paraplegic adult rats. Brain Res 889(1–2):344–357
16. Lu J, Ashwell K (2002) Olfactory ensheathing cells: their potential use for repairing the injured spinal cord. Spine 27(8):887–892
17. Huang H, Chen L, Wang H, Xiu B, Li B, Wang R et al (2003) Influence of patients' age on functional recovery after transplantation of olfactory ensheathing cells into injured spinal cord injury. Chin Med J (Engl) 116(10):1488–1491
18. Féron F, Perry C, Cochrane J, Licina P, Nowitzke A, Urquhart S et al (2005) Autologous olfactory ensheathing cell transplantation in human spinal cord injury. Brain 128(12):2951–2960
19. Huang HC (2006) Young Tai-Horng, differences in the effect on neural stem cells of fetal bovine serum in substrate-coated and soluble form. Biomaterials 27:5901–5908

20. Chen L, Huang H, Zhang J, Zhang F, Liu Y, Xi H et al (2007) Short-term outcome of olfactory ensheathing cells transplantation for treatment of amyotrophic lateral sclerosis. Zhongguo Xiu Fu Chong Jian Wai Ke Za Zhi 21(9):961–966
21. Chew S, Khandji AG, Montes J, Mitsumoto H, Gordon PH (2007) Olfactory ensheathing glia injections in Beijing: misleading patients with ALS. Amyotroph Lateral Scler 8(5):314–316
22. Mackay SA, Feron F, Cochrane J, Bassingthwaighte L, Bayliss C, Davies W et al (2008) Autologous olfactory ensheathing cell transplantation in human paraplegia: a 3-year clinical trial. Brain 131:2376–2386
23. Larson BL, Ylostalo J, Prockop DJ (2008) Human multipotent stromal cells undergo sharp transition from division to development in culture. Stem Cells 26:193–201
24. Sasaki M, Honmou O, Akiyama Y, Uede T, Hashi K, Kocsis JD (2003) Transplantation of an acutely isolated bone marrow fraction repairs demyelinated adult rat spinal cord axons. Glia 35(1):26–34
25. Hofstetter CP, Schwarz EJ, Hess D, Widenfalk J, El Manira A, Prockop DJ et al (2002) Marrow stromal cells form guiding strands in the injured spinal cord and promote recovery. Proc Natl Acad Sci U S A 99(4):2199–2204
26. Wu S, Suzuki Y, Ejiri Y, Noda T, Bai H, Kitada M (2003) Bone marrow stromal cells enhance differentiation of cocultured neurosphere cells and promote regeneration of injured spinal cord. J Neurosci Res 72:343–351
27. Lee J, Kuroda S, Shichinohe H, Ikeda J, Seki T, Hida K et al (2003) Migration and differentiation of nuclear fluorescence-labeled bone marrow stromal cells after transplantation into cerebral infarct and spinal cord injury in mice. Neuropathology 23(3):169–180
28. Zurita M, Vaquero J (2004) Functional recovery in chronic paraplegia after bone marrow stromal cells transplantation. Neuroreport 15(7):1105–1108
29. Lu P, Yang H, Jones LL, Filbin MT, Tuszynski MH (2004) Combinatorial therapy with neurotrophins and cAMP promotes axonal regeneration beyond sites of spinal cord injury. J Neurosci 24(28):6402–6409
30. Ankeny DP, McTigue DM, Jakeman LB (2004) Bone marrow transplants provide tissue protection and directional guidance for axons after contusive spinal cord injury in rats. Exp Neurol 190(1):17–31
31. Lu P, Jones LL, Tuszynski MH (2005) BDNF-expressing marrow stromal cells support extensive axonal growth at sites of spinal cord injury. Exp Neurol 191(2):344–360
32. Neuhuber B, Himes TB, Shumsky JS, Gallo G, Fischer I (2005) Axon growth and recovery of function supported by human bone marrow stromal cells in the injured spinal cord exhibit donor variations. Brain Res 1035(1):73–85
33. Kamada T, Koda M, Dezawa M, Anahara R, Toyama Y, Yoshinaga K et al (2005) Transplantation of bone marrow stromal cell-derived Schwann cells promotes axonal regeneration and functional recovery after complete transection of adult rat spinal cord. J Neuropathol Exp Neurol 64(1):37–45
34. Park IH, Zhao R, West JA, Yabuuchi A, Huo H, Ince TA et al (2008) Reprogramming of human somatic cells to pluripotency with defined factors. Nature 451(7175):141–146
35. Sykova E, Homola A, Mazanec R, Lachmann H, Konrádová SL, Kobylka P et al (2006) Autologous bone marrow transplantation in patients with subacute and chronic spinal cord injury. Cell Transplant 15:1–100
36. Nishida K, Tanaka N, Nakanishi K, Kamei N, Hamasaki T, Yanada S et al (2006) Magnetic targeting of bone marrow stromal cells into spinal cord: through cerebrospinal fluid. Neuroreport 17(12):1269–1272
37. Yoshihara H, Shumsky JS, Neuhuber B, Otsuka T, Fischer I, Murray M (2006) Combining motor training with transplantation of rat bone marrow stromal cells does not improve repair or recovery in rats with thoracic contusion injuries. Brain Res 1119(1):65–75
38. Cao Q, Ding P, Lu J et al (2007) 2′, 3′-Cyclic nucleotide 3′-phosphodiesterase cells derived from transplanted marrow stromal cells and host tissue contribute to perineurial compartment formation in injured rat spinal cord. J Neurosci Res 85(1):116–130

39. Lu P, Jones LL, Mark TH (2007) Axon regeneration through scars and into sites of chronic spinal cord injury. Exp Neurol 203(1):8–21
40. Shi E, Kazui T, Jiang X, Washiyama N, Yamashita K, Terada H et al (2007) Therapeutic benefit of intrathecal injection of marrow stromal cells on ischemia-injured spinal cord. Ann Thorac Surg 83(4):1484–1490
41. Koda M, Nishio Y, Kamada T, Someya Y, Okawa A, Mori C et al (2007) Granulocyte colony-stimulating factor (G-CSF) mobilizes bone marrow-derived cells into injured spinal cord and promotes functional recovery after compression-induced spinal cord injury in mice. Brain Res 1149:223–231
42. Saporta S, Kim JJ, Willing AE, Fu ES, Davis CD, Sanberg PR (2003) Human umbilical cord blood stem cells infusion in spinal cord injury: engraftment and beneficial influence on behavior. J Hematother Stem Cell Res 12:271–278
43. Zhao ZM, Li HJ, Liu HY, Lu SH, Yang RC, Zhang QJ et al (2004) Intraspinal transplantation of CD34+ human umbilical cord blood cells after spinal cord hemisection injury improves functional recovery in adult rats. Cell Transplant 13(2):113–122
44. Li HJ, Liu HY, Zhao ZM, Lu SH, Yang RC, Zhu HF et al (2004) Transplantation of human umbilical cord stem cells improves neurological function recovery after spinal cord injury in rats. Zhongguo Yi Xue Ke Xue Yuan Xue Bao 26(1):38–42
45. Nishio Y, Koda M, Kamada T, Someya Y, Yoshinaga K, Okada S et al (2006) The use of hemopoietic stem cells derived from human umbilical cord blood to promote restoration of spinal cord tissue and recovery of hindlimb function in adult rats. J Neurosurg Spine 5(5):424–433
46. Dasari VR, Spomar DG, Li L, Gujrati M, Rao JS, Dinh DH (2008) Umbilical cord blood stem cell mediated downregulation of fas improves functional recovery of rats after spinal cord injury. Neurochem Res 33(1):134–149
47. Walczak P, Chen N, Eve D, Hudson J, Zigova T, Sanchez-Ramos J et al (2007) Long-term cultured human umbilical cord neural-like cells transplanted into the striatum of NOD SCID mice. Brain Res Bull 74(1–3):155–163
48. Cho SR, Yang MS, Yim SH et al (2008) Neurally induced umbilical cord blood cells modestly repair injured spinal cords. Neuroreport 19(13):1259–1263
49. David HN, Haelewyn B, Rouillon C, Lecoq M, Chazalviel L, Apiou G et al (2008) Neuroprotective effects of xenon: a therapeutic window of opportunity in rats subjected to transient cerebral ischemia. FASEB J 22(4):1275–1286
50. Nurse CA, Macintyre L, Diamond J (1984) Reinnervation of the rat touch dome restores the Merkel cell population reduced after denervation. Neuroscience 13:563–571
51. Kolata G (2007) Scientists bypass need for embryo to get stem cells, The New York Times. Retrieved on 11 Dec 2007
52. Vogel G, Holden C (2007) Developmental biology. Field leaps forward with new stem cell advances. Science 318(5854):1224–1225
53. Takahashi K, Tanabe K, Ohnuki M, Narita M, Ichisaka T, Tomoda K et al (2007) Induction of pluripotent stem cells from adult human fibroblasts by defined factors. Cell 131(5):861–872
54. Amoh Y, Li L, Katsuoka K, Penman S, Hoffman RM (2005) Multipotent nestin-positive, keratin-negative hair-follicle bulge stem cells can form neurons. Proc Natl Acad Sci U S A 102:5530–5534
55. Sieber-Blum M, Grim M (2004) The adult hair follicle: cradle for pluripotent neural crest stem cells. Birth Defects Res C Embryo Today 72:162–172
56. Weisberg SP, McCann D, Desai M, Rosenbaum M, Leibel RL, Ferrante AW Jr (2003) Obesity is associated with macrophage accumulation in adipose tissue. J Clin Invest 112(12):1796–1808
57. Caspar-Bauguil S, Cousin B, Galinier A, Segafredo C, Nibbelink M, André M et al (2005) Adipose tissues as an ancestral immune organ: site-specific change in obesity. FEBS Lett 579(17):3487–3492
58. Katz AJ, Tholpady A, Tholpady SS, Shang H, Ogle RC (2005) Cell surface and transcriptional characterization of human adipose-derived adherent stromal (hADAS) cells. Stem Cells 23(3):412–423

59. Mitchell KE, Weiss ML, Mitchell BM, Martin P, Davis D, Morales L et al (2003) Matrix cells from Wharton's jelly form neurons and glia. Stem Cells 21:50–60
60. Priest RE, Marimuthu KM, Priest JH (1978) Origin of cells in human amniotic fluid cultures. Lab Invest 39:106–109
61. Prusa AR, Hengstschläger M (2002) Amniotic fluid cells and human stem cell research: a new connection. Med Sci Monit 8:253–257
62. Milunsky A (1998) Genetic disorders and the fetus: diagnosis, prevention, and treatment. The Johns Hopkins University Press, Baltimore, pp 128–149
63. Tyden O, Bergstrom S, Nilsson BA (1981) Origin of amniotic fluid cells in mid-trimester pregnancies. Br J Obstet Gynaecol 88:278–286
64. Prusa AR, Marton E, Rosner M, Bernaschek G, Hengstschlager M (2003) Oct-4-expressing cells in human amniotic fluid: a new source for stem cell research? Hum Reprod 18:1489–1493
65. Prusa AR, Marton E, Rosner M, Bettelheim D, Lubec G, Pollack A et al (2004) Neurogenic cells in human amniotic fluid. Am J Obstet Gynecol 191:309–314
66. Mitka M (2001) Amniotic cells show promise for fetal tissue engineering. JAMA 286:2083
67. Kaviani A, Perry TE, Dzakovic A, Jennings RW, Ziegler MM, Fauza DO (2001) The amniotic fluid as a source of cells for fetal tissue engineering. J Pediatr Surg 36:1662–1665
68. Barami K, Diaz FG (2000) Cellular transplantation and spinal cord injury. Neurosurgery 47:691–700
69. Rapalino O, Lazarov-Spiegler O, Agranov E, Velan GJ, Yoles E, Fraidakis M et al (1998) Implantation of stimulated homologous macrophages results in partial recovery of paraplegic rats. Nat Med 4:814–821
70. Bomstein Y, Marder JB, Vitner K, Smirnov I, Lisaey G, Butovsky O et al (2003) Features of skin-coincubated macrophages that promote recovery from spinal cord injury. J Neuroimmunol 142:10–16
71. Franzen R, Schoenen J, Leprince P, Joosten E, Moonen G, Martin D (1998) Effects of macrophage transplantation in the injured adult rat spinal cord: a combined immunocytochemical and biochemical study. J Neurosci Res 51(3):316–327
72. Knoller N, Auerbach G, Fulga V, Zelig G, Attias J, Bakimer R et al (2005) Clinical experience using incubated autologous macrophages as a treatment for complete spinal cord injury: phase I study results. J Neurosurg Spine 3:173–181
73. Mikami Y, Okano H, Sakaguchi M, Nakamura M, Shimazaki T, Okano HJ et al (2004) Implantation of dendritic cells in injured adult spinal cord results in activation of endogenous neural stem/progenitor cells leading to de novo neurogenesis and functional recovery. J Neurosci Res 76:453–465
74. Hauben E, Gothilf A, Cohen A, Butovsky O, Nevo U, Smirnov I et al (2003) Vaccination with dendritic cells pulsed with peptides of myelin basic protein promotes functional recovery from spinal cord injury. J Neurosci 23:8808–8819
75. Brook GA, Houweling DA, Gieling RG, Hermanns T, Joosten EA, Bär DP et al (2000) Attempted endogenous tissue repair following experimental spinal cord injury in the rat: involvement of cell adhesion molecules L1 and NCAM? Eur J Neurosci 12:3224–3238
76. Jasmin L, Janni G, Moallem TM, Lappi DA, Ohara PT (2000) Schwann cells are removed from the spinal cord after effecting recovery from paraplegia. J Neurosci 20:9215–9223
77. Von Euler M, Janson AM, Larsen JO, Seiger A, Forno L, Bunge MB et al (2002) Spontaneous axonal regeneration in rodent spinal cord after ischemic injury. J Neuropathol Exp Neurol 61:64–75
78. Guest JD, Hiester ED, Bunge RP (2005) Demyelination and Schwann cell responses adjacent to injury epicenter cavities following chronic human spinal cord injury. Exp Neurol 192:384–393
79. Pennon A, Calancie B, Oudega M, Noga BR (2001) Conduction of impulses by axons regenerated in a Schwann cell graft in the transected adult rat thoracic spinal cord. J Neurosci Res 64:533–541
80. Kohama I, Lankford KL, Preiningerova J, White FA, Vollmer TL, Kocsis JD (2001) Transplantation of cryopreserved adult human Schwann cells enhances axonal conduction in demyelinated spinal cord. J Neurosci 21:944–950

81. Guettier C (2005) Which stem cells for adult liver? Ann Pathol 25:33–44
82. Rollini P, Kaiser S, Faes-van't Hull E, Kapp U, Leyvraz S (2004) Long-term expansion of transplantable human fetal liver hematopoietic stem cells. Blood 103:1166–1170
83. Suen PM, Leung PS (2005) Pancreatic stem cells: a glimmer of hope for diabetes? J Pancreas 6:422–424
84. Tarasenko YI, Yu Y, Jordan PM, Bottenstein J, Wu P (2004) Effect of growth factors on proliferation and phenotypic differentiation of human fetal neural stem cells. J Neurosci Res 78(5):625–636
85. Stocum DL, Zupanc GKH (2008) Stretching the limits: stem cells in regeneration science. Dev Dyn 237(12):3648–3671
86. Grandel H, Kaslin J, Ganz J, Wenzel I, Brand M (2006) Neural stem cells and neurogenesis in the adult zebrafish brain: origin, proliferation dynamics, migration and cell fate. Dev Biol 295:263–277
87. Bjugstad KB, Redmond DE, Teng YD et al (2005) Neural stem cells implanted into MPTP-treated monkeys increase the size of endogenous dopamine neurons found in the striatum: a return to control measure. Cell Transplant 14:183–192
88. Bjugstad KB, Teng YD, Redmond DE, Elsworth JD, Roth RH, Cornelius SK et al (2008) Human neural stem cells migrate along the nigrostriatal pathway in a primate model of Parkinson's disease. Exp Neurol 211:362–369
89. Emborg ME, Ebert AD, Moirano J, Peng S, Suzuki M, Capowski E et al (2008) GDNF-secreting human neural progenitor cells increase tyrosine hydroxylase and VMAT2 expression in MPTP-treated cynomolgus monkeys. Cell Transplant 17:383–395
90. Windrem MS, Roy NS, Wang J, Nunes M, Benraiss A, Goodman R et al (2002) Progenitor cells derived from the adult human subcortical white matter disperse and differentiate as oligodendrocytes within demyelinated lesions of the rat brain. J Neurosci Res 69:966–975
91. Windrem MS, Schanz SJ, Guo M, Tian GF, Washco V, Stanwood N et al (2008) Neonatal chimerization with human glial progenitor cells can both remyelinate and rescue the otherwise lethally hypomyelinated shiverer mouse. Cell Stem Cell 2:553–565
92. McBride JL, Behrstock SP, Chen EY, Jakel RJ, Siegel I, Svendsen CN et al (2004) Human neural stem cell transplants improve motor function in a rat model of Huntington's disease. J Comp Neurol 475:211–219
93. Xu L, Ryugo DK, Pongstaporn T, Johe K, Koliatsos VE (2009) Human neural stem cell grafts in the spinal cord of SOD1 transgenic rats: Differentiation and structural integration into the segmental motor circuitry. J Comp Neurol 514:297–309
94. Klein SM, Behrstock S, McHugh J, Hoffmann K, Wallace K, Suzuki M et al (2005) GDNF delivery using human neural progenitor cells in a rat model of ALS. Hum Gene Ther 16:509–521
95. Suzuki M, McHugh J, Tork C, Shelley B, Klein SM, Aebischer P et al (2007) GDNF-secreting human neural progenitor cells protect dyingmotor neurons, but not their projection to muscle, in a rat model of familial ALS. PLoS One 2:e689
96. Darsalia V, Kallur T, Kokaia Z (2007) Survival, migration and neuronal differentiation of human fetal striatal and cortical neural stem cells grafted in stroke-damaged rat striatum. Eur J Neurosci 26:605–614

Chapter 17
Synovial Mesenchymal Stem Cells and Their Applications in Musculoskeletal Regeneration

Yvonne Peck and Dong-An Wang

Abstract Stem cell therapy represents the cutting edge of regenerative medicine. The prospect of engineering whole functional tissue has long been sought after, and stem cells hold the key to this. Recently, a class of stem cells known as synovial mesenchymal stem cells (SMSCs) has come to the forefront of musculoskeletal stem cell research. They were first extracted from the synovium of the knee joints and exhibit the common MSC trait of multipotency. SMSCs however display high *in vitro* expandability and superior chondrogenic potential as compared to their other MSC counterparts. This chapter starts with a review on the extraction, isolation, and culture techniques of SMSCs, followed by an overview of SMSC characteristic and properties. This chapter will then delve into SMSC applications for musculoskeletal regeneration, including that for cartilage, bones, tendons, ligaments, and muscles. For each of the tissues, some of the current challenges and the differing strategies adopted to overcome them such as transgenic enhancement, growth factor induction, and scaffold-based tissue engineering will be discussed in detail.

Keywords Synovial mesenchymal stem cells • Bone regeneration • Gene therapy • Bioreactor • Meniscal engineering

Y. Peck • D.-A. Wang, PhD (✉)
Division of BioEngineering, School of Chemical and Biomedical Engineering,
Nanyang Technological University,
70 Nanyang Drive N1.3-B2-13, Singapore 637457, Republic of Singapore
e-mail: dawang@ntu.edu.sg

M.K. Danquah, R.I. Mahato (eds.), *Emerging Trends in Cell and Gene Therapy*,
DOI 10.1007/978-1-62703-417-3_17,

17.1 Introduction

In the past decade, the field of regenerative medicine has been recognized as a new thrust in the therapies for various musculoskeletal diseases and injuries. It holds great promise in healing damaged tissue by addressing the problem of transplanted organ shortages. Current techniques in tissue regeneration primarily center on the delivery of therapeutic cells either by direct transplantation or by seeding the cells onto scaffolds prior to transplantation. However, the availability of therapeutic cells has been a concern as existing treatments largely rely on autologous cell sources. Overdependence on autologous cells raises a host of issues such as limited cell sources and donor site morbidity.

In order to overcome these limitations, many efforts have been made to discover an alternative cell source to replace the autologous cells. Among the various possible cell sources, mesenchymal stem cells (MSCs) are regarded as promising candidates for use in cell-based therapy to regenerate lost tissue due to them possessing high proliferative capacity while retaining their multipotency [1]. Furthermore, they also possess anti-inflammatory and immunomodulatory effects which help in the process of tissue repair [2, 3].

The term 'Mesenchymal Stem Cell' was originally employed to describe the cells isolated from the bone marrow as they were the first to be identified and characterized. However, in recent years, MSCs have also been found in other tissues including the adipose tissue [4, 5], synovium [6], periosteum [7], and skeletal muscle [8]. Among these various sources of MSCs, the MSCs isolated from synovium have recently gained popularity as the reparative cell source for musculoskeletal regeneration, especially for cartilage. Many studies have proven that these cells have superior *in vitro* expandability and chondrogenic potential over other mesenchymal tissue stem cells [9, 10]. This unique population of cells was first discovered and characterized by De Bari and colleagues in year 2001 [6].

To date, many different names have been used to define both the stem and progenitor cells that originate from the synovium. These different names include synovium-derived stem cells, synovium-derived mesenchymal stem cells, or synovial progenitor cells [2]. However, for consistency, the term synovial mesenchymal stem cells (SMSCs) will be used in this book chapter to describe the mesenchymal stem cells isolated from the synovial membrane, the subsynovial connective tissue, as well as the synovial fluid.

SMSCs possess a few traits that give them their chondrogenic superiority, namely, the fact that the cells express higher numbers of hyaluronan receptors (CD44) than other MSCs. Furthermore, they are also capable of producing uridine diphosphoglucose dehydrogenase (UDPGD), an enzyme that is important in hyaluronan synthesis [11, 12]. Further evidence also hints at the chondrogenic potential of SMSCs, albeit indirectly. In a disease known as synovial chondromatosis, cartilaginous bodies are formed in the synovial membrane. Interestingly, SMSCs were observed to undergo a process comparable to human chondromatosis during *in vitro* chondrogenesis. In addition, chondrocyte-like cells have been discovered in pannus formed in joints affected by rheumatoid arthritis (RA). Lastly, the articular cartilage and synovium share a common origin as both tissues are derived from the interzone cells. All these supporting evidences suggest that SMSCs may potentially emerge as a powerful therapeutic tool in the repair and regeneration of the musculoskeletal system.

Since SMSCs represent a relatively new cell source, researchers have only begun to examine its characteristics. Many studies have attempted to characterize the SMSCs behavior under *in vitro* and *in vivo* environments, with varying types of scaffolds and combinations of biochemical factors. The properties of these cells are also frequently compared with some of the better-characterized stem cells derived from other mesenchymal tissues, such as bone marrow-derived mesenchymal stem cells (BMSCs) [9, 10, 13–15]. Thus, in the last decade, a myriad of information regarding the phenotypic and functional properties of SMSCs has been obtained through the analysis of their epitope profiles, transcriptional profiles, proliferation potential, as well as their differentiation capabilities [9, 10, 16].

Therefore, this chapter first discusses the extraction, isolation, and culture techniques of SMSCs and then continues with a review of SMSC characteristics and properties. More importantly, this chapter will highlight the various SMSC applications for the repair and regeneration of bone, cartilage, muscle, tendon, and ligament. However, a large portion of this chapter will be devoted to the cartilage itself. The different challenges faced for each tissue will be explored, as well as their potential solutions such as transgenic enhancements, growth factor induction, and scaffold-based tissue engineering.

17.2 Harvest of Synovial Tissue and Extraction of SMSCs

The synovium is a thin membrane consisting of two to three layers of specialized cells called synoviocytes which line the joint space, thus forming a synovial fluid-filled cavity surrounding the cartilage and the surfaces of tendon [12, 17]. The synoviocytes can be further divided into two distinct types: type A cells and type B cells. Type A cells originate from the bone marrow and display macrophage markers such as CD68 and CD14. In contrast, type B cells are fibroblast-like cells with the special ability to show the expression of UDPGD [18]. Apart from being able to synthesize a matrix that is abundant in type III, V, and VI collagen, several adhesion molecules such as vascular cell adhesion molecule 1 (VCAM-1) and CD44 are also expressed by the type B cells [18, 19]. The major function of synovium is to provide lubrication for the joint tissues, as well as to supply them with oxygen, nutrients, and proteins [20].

One of the advantages of synovial tissue from which the SMSCs are mainly harvested is that it has a high regenerative capacity which ensures an abundant supply of cells. This is supported by the fact that it is able to heal completely after surgical [21, 22] and chemical [23] synovectomy in rabbits and horses [24]. This tissue can be readily harvested by arthroscopy which is minimally invasive, thus causing no donor site morbidity [11]. The first successful isolation of SMSCs from human synovial lining was done by De Bari and colleagues based on the selective plastic adherent property of MSCs where this selectivity excludes nonadherent type A cells. Subsequent investigations have proven that these adherent cells displayed multilineage differentiation capacity and were stable for at least ten passages. Notably, a very small amount of synovial tissue is enough for the derivation of these cells [6]. One recent study has further confirmed these findings by showing that each milligram of synovial tissue collected was able to give rise to about 21,000 cells after 2 weeks of culture at optimal density and the cells proliferative ability was preserved even at passage 10 [9].

Synovial tissue can be harvested from different locations in the knee joint. Some studies have been directed to probe the difference in chondrogenic potential between SMSCs derived from different harvest sites. Fibrous synovium located at the interior of the lateral joint capsule and adipose synovium derived from the infrapatellar fat pad have been reported to show no difference in their capacity for proliferation and differentiation [25]. A similar result was shown by Nagase and colleagues as they compared the chondrogenic ability of SMSCs isolated from four different sites in the knee joint including the infrapatellar fat pad, the suprapatellar pouch, and both the medial inner and outer regions. Although there was no apparent difference in their growth and differentiation ability, the degree of vascularity was reported to differ between the different harvest sites. However, the medial outer region of the osteoarthritis (OA) synovium was revealed to give rise to SMSCs with higher colony-forming potential than the other regions, and this potential was correlated with the presence of higher numbers of α-smooth muscle actin-positive vessels and CD31+ endothelial cells [26].

Apart from harvesting SMSCs from healthy synovial tissues, some studies have shown that these cells can also be isolated from pathological synovium of patients with rheumatoid arthritis (RA) or OA [25–27]. A recent study has set out to investigate if the inflammatory environment in the RA synovium would affect the functions of SMSCs. The results demonstrated that the SMSC chondrogenic and clonogenic potential diminished with each increase in the extent of synovitis in RA [28]. Therefore, joint inflammation needs to be suppressed before any effective treatment of RA utilizing SMSCs can take place.

Besides synovium, some works have reported that SMSCs can also be isolated from the subsynovium tissue comprised of fibrous and adipose connective tissues, and these cells have been proven to retain their multipotency. The adipose synovium is more commonly referred to as the infrapatellar fat pad, from which many studies have shown successful extraction of SMSCs [25]. For instance, a recent study set out to investigate the effect of patient profiles on the growth rate and cell surface marker expression of the infrapatellar fat pad-derived SMSCs. Interestingly, the results shown that the age and gender of patient do affect the rate of cell proliferation and cell surface marker expression. Cells from female patients were shown to proliferate faster and express higher cell surface marker consistently. Nonetheless, the study also determined that seeding densities can affect cell proliferation rate to a greater extent as compared to the patient profiles [29].

A study by Jones et al. [14] reported that MSCs can also be found in the synovial fluid (SF) of patients suffering from OA and RA, with a significantly higher number found in the latter case. This indicates that these MSCs may play a role in the pathophysiology of arthritis. A later study by Morito et al. attempted to characterize these SF-derived MSCs by investigating their *in vitro* differentiation potential, surface markers, as well as gene expression profiles. They reported that these SF-derived MSCs have shown multipotentiality and have epitope profiles similar to that of SMSCs and BMSCs. Furthermore, gene expression profiles analysis revealed that these MSCs have higher similarity to SMSCs than to BMSCs, suggesting they are derived from the synovium [14].

17.3 Isolation, Cultivation, and Purification of SMSCs

Generally, isolation of SMSCs starts with the rinsing of the freshly harvested synovial tissue from the knee joints in sterilized phosphate buffer saline (PBS). After thorough rinsing, the synovial tissue is cut into fine pieces which are then digested in a solution containing an appropriate concentration of collagenase in the culture medium [6]. After the SMSCs are released from the synovial tissues, they are then collected and washed before being seeded onto a 60 cm^2 dish with a cell density of 10^3–10^5 cells in each plate. These seeded cells are cultured for 2 weeks as passage 0. The colony-forming units-fibroblast assay can be used to specify the number of SMSCs in the culture. The two culture media that are most suitable to support the robust *in vitro* proliferation of SMSCs are comprised of either high-glucose Dulbecco's Modified Eagle's Medium (DMEM) or α-minimum essential medium (α-MEM) supplemented with 10 % of fetal bovine serum (FBS) and antibiotics [6, 9, 14, 26]. Although this conventional enzymatic method has been widely used, a recent study suggested that a newly developed nonenzymatic direct explant technique could be better for isolating SMSCs, especially for OA patients since it is simpler and less invasive [30].

Following the isolation of cells, the SMSCs can be expanded by culturing in monolayer in either of the aforementioned culture media types at 37°C in a humidified atmosphere of 5 % CO_2. The medium should be replaced every 3 days. The period of the primary culture of SMSCs should be around 10–25 days for them to reach confluency of 80~90 %. For subculture, the cells can be first harvested by digestion using trypsin-EDTA and then replated with a 1:4 dilution [6].

Although the initial culturing technique used by De Bari and colleagues was assumed to produce a homogeneous population of cells, later studies suggested that this adherent cell population derived from the synovium was actually a lot more heterogeneous than progenitor cells obtained from other sources like bone marrow. This heterogeneity is shown in their heterogeneous growth and differentiation capabilities [31, 32]. One disadvantage of a mixed population of cells is that these cells are not suitable for clinical use since they are likely to produce unpredictable biological activities. In order to overcome this problem, there are several purification methods that are currently in development with the aim of yielding cell populations that are more homogeneous in their functions [33].

Previously, the two different methods that were used to purify MSCs were the long passaging culture method and limited diluted method [9]. However, more recently, more advanced techniques such as magnetic-activated cell sorting (MACS) and fluorescence-activated cell sorting (FACS) have been utilized for MSC purification [34, 35]. There are two alternative ways in which the purification procedure can be done: either subjecting the cells to selection immediately after being freshly harvested or after selecting them based on their plastic adherent property [34]. The MACS was shown to be able to purify MSCs based on the presence of certain unique cell surface markers, for instance, CD14 for macrophages [35]. A recent study by Bilgen et al. reported that a more homogeneous population of synovial fibroblasts, showing enhanced chondrogenic potential, was obtained after subjecting the original mixed population of synoviocytes to CD14-negative isolation. Typically, the negative selection is done by coating the monodiperse supramagnetic polystyrene beads with primary monoclonal antibody targeting CD14, and this enables subsequent separation of macrophages from SMSCs. Jun Qi et al. also demonstrated that MACS can be used to enrich CD105+ SMSCs, resulting in a more homogeneous cell population to be used as a potential cell source for cartilage engineering [36].

The other technique, FACS, was shown to be utilized for the detection and isolation of another subpopulation of cells known as the side population (SP) cells from the synovium-derived cells [37, 38]. These cells have been isolated from many other adults' tissues through staining with the DNA dye Hoechst 33342, subsequently separating them from the rest of the cells by FACS based on their strong dye exclusion property. These SP cells were shown to behave like stem cells as they expressed cell markers such as CD34, Flk-1, c-Kit, and Abcg-2. Furthermore, they were able to differentiate into multiple lineages like the chondrogenic, osteogenic, and myogenic lineages [37]. Although the SP cells and SMSCs are both derived from the synovial tissues, their actual relationship still remains unclear [12].

17.4 SMSC Proliferative Ability and Multipotency

Since the discovery of SMSCs, these cells have been known to possess high proliferative and multilineage differentiation potential. The high proliferative potential of SMSCS is displayed through their ability to undergo extensive *in vitro* expansion in monolayer for up to ten passages with similar growth kinetics [6]. This remarkable self-renewal capacity was reported to be comparable to that of BMSCs [9]. Notably, the SMSCs manage to maintain a linear growth curve for at least 30 population doublings. Despite possessing such a high ability to proliferate, the activity of telomerase, which is an enzyme that elongates the telomeres to enable numerous rounds of cell divisions, remains undetectable in these cells. Thus, cell senescence is still being observed at later passages (after passage 10). However, this problem does not become a limiting factor for cell expansion as a sufficiently large number of cells can be obtained only after a few passages (roughly a billion cells after six passages, extracted from 10 to 50 mg synovial tissue of the knee joint) [6]. As for the multilineage potential of SMSCs, it has been proven that these cells are able to differentiate into osteoblasts, chondrocytes, adipocytes, and, to a lesser extent, myocytes when they are cultured in lineage-specific medium. Among these different lineages, SMSCs have been proven to be particularly efficient in undergoing chondrogenesis. This potential was reported to be unaffected by cell passage number, donor age, or cryopreservation [6, 9, 10].

Generally, the use of serum as a nutritional supplement in the cell culture medium is essential for the *in vitro* expansion of MSCs. The selection of the type of serum used is crucial not only in supporting cell proliferation and differentiation but also in affecting MSC gene expression and transcriptional profiles. Typically, the fetal bovine serum (FBS) has been used in most studies to supplement the cell culture medium for *in vitro* culture of MSCs [39]. The use of FBS is not favorable, especially when the cells are cultured for therapeutic purposes. This is because there exists a certain degree of risk of prion diseases and zoonoses transmission from the FBS [40] but, more importantly is the possibility that the xenogenic proteins will evoke an adverse immune response [41]. Therefore, the use of autologous or allogenic human serum is preferable for *in vitro* expansion of MSCs.

Many studies have been directed to compare the effect of using different sera on the proliferative and differentiation ability of MSCs, yielding various results and conclusions. Some investigators reported that the FBS was better than human serum [42, 43], while some demonstrated that the FBS and human serum have comparable proliferative effects [41, 44]. There were also studies that concluded that MSCs have higher rate of proliferation in human serum than in FBS [39, 45–47].

In a recent work, the effect of autologous human serum on SMSCs and BMSCs was compared. The results showed that SMSCs expanded more in human serum as compared to in FBS, while BMSCs showed otherwise. This phenomenon is related to the platelet-derived growth factor (PDGF) signaling due to the fact that the human serum contains large amounts of these growth factors. The SMSCs have higher expressions of receptors for PDGF as compared to BMSCs, resulting in their higher proliferative ability in human serum [48]. This compatibility of SMSCs with human serum has also been shown by another work, in which higher expandability was achieved in human serum without compromising their differentiation potential [49].

17.5 SMSC Characteristics and Properties

17.5.1 SMSC Genotypic, Phenotypic, and Functional Characterization

It is widely known that SMSCs and BMSCs share many similarities in terms of their phenotypic and functional properties. This is easily confirmed by phenotypic characterization using the flow cytometric immunophenotyping technique. With this technique, the expression of various cell surface markers can be analyzed, and many studies have revealed that both SMSCs and BMSCs have shown negative expression for hematopoietic stem cell markers (including CD34 and CD117), lymphocytic cell markers (LFA-1, CD11a, CD20), endothelial cell markers (CD31, CD62e), monocytic or macrophages cell markers (CD14, CD68), a leukocytic cell marker (CD45), and an MHC class II cell surface receptor (HLA-DR). On the contrary, both cell sources positively expressed markers such as CD90, CD105, CD166, CD44, CD73, CD10, CD13, CD49a, and CD147, which are typical markers characteristic of MSCs [12, 16, 27, 28, 34, 35, 50, 51].

The immunophenotype of SMSCs was shown to be affected by many factors such as cell passage number, cell culture medium, cell preculture time, the different harvest sites, and also the condition of synovial membrane (isolated from healthy joints, OA or RA) [12, 27, 34, 51, 52]. For example, SMSCs were shown to have a higher expression of PDGF receptor α when cultured in the presence of human serum as compared to being cultured in FBS [48]. Besides that, the amount of cells expressing mesenchymal markers such as CD90, CD44, and CD105 in the knee joints of OA patients was found to exceed those in healthy joints [53]. One study also revealed that the cell surface marker expression profiles were different for the freshly isolated cells and the cells after the first passage. Initial expression of CD14, CD34, CD45, CD62e, and HLA-DR vanished following the first passage while the expression of CD105 and CD166 appeared. Furthermore, this was accompanied by an upregulation for CD10, CD13, CD44, CD49a, and CD73 [51]. Similarly, another study reported that the SMSCs have shown a higher expression level of CD9, CD44, CD54, CD90, and CD166 after *in vitro* expansion [34].

Interestingly, the expression of certain phenotypic markers may be a good indicator of the differentiation potential of SMSCs. For example, SMSCs showing triplicate positive expression for CD9, CD90, and CD166 have been proven to be multipotent [34]. Besides that, the expression of CD90 or Thy-1 may also be related to the chondrogenic potential of SMSCs [26]. Despite the use of all these markers for their identification, SMSCs still cannot be completely characterized. This is because the immunotyping technique lacks specificity where the characterization of SMSCs is simply based on the absence of hematopoietic and endothelial markers. Furthermore, it also relies on detecting the expression of molecules which can also be found on more mature stromal cells [54].

As for genotypic characterization, some investigators have analyzed the gene expression profiles of the various mesenchymal tissues using hierarchical clustering analysis and principal component analysis (PCA). The results revealed that SMSCs, together with chondrocytes and MSCs derived from the meniscus and the intra-articular ligament, belong to a single cluster distinct from adipose-, muscle-, and bone marrow-derived MSCs. Furthermore, this study also showed that SMSCs were able to express proline-arginine-rich end leucine-rich repeat protein (PRELP), an intra-articular tissue MSC-specific gene consistently, whereas BMSCs upregulated the expression of this protein only during *in vitro* chondrogenesis [55]. In another study, it was shown that activin A, which is an important cell-signaling protein and recently found to play a vital role in the maintenance of multipotency of mesenchymal progenitor cells, was being expressed at a higher level in BMSCs as compared to SMSCs

[16, 56]. These differences between the two cell sources may be attributed to genotypic variation. According to the DNA microarray analysis done in a study, the transcription profiles of BMSCs and SMSCs differ by 46 genes from a set of 268 genes.

17.5.2 Immunosuppressive Potential of SMSCs

In recent years, MSCs have gained more interest after they have demonstrated the ability to exert immunosuppressive and immunoregulatory effects on cells of the immune system. This ability is a result of an immune tolerant phenotype displayed by the cells. Under normal culture conditions, BMSCs only express MHC class I molecules but not MHC class II molecules. However, both classes of molecules have been shown to be upregulated upon induction by interferon-gamma (IFN-γ). Besides low expressions of HLA molecules, they also express low levels of co-stimulatory molecules such as B7-1, B7-2, CD40, or CD40L [1, 57].

SMSCs do not actively display immunomodulatory activities; instead these immunosuppressive functions are activated following exposure to the inflammatory environment. This conclusion is rooted in an important observation that these immunosuppressive functions of SMSCs can be inhibited by the use of anti-IFN-γ receptor antibodies. Thus, exposure to inflammatory cytokines such as IFN-γ, TNF-α, or IL-1β is crucial in activating the MSC-mediated immunosuppression [54]. It has been reported in various studies that MSCs are capable of suppressing and regulating the functions of T cells [58, 59], B cells [60, 61], and natural killer (NK) cells [62, 63]. To date, the immunosuppressive mechanisms of MSCs have yet to be elucidated. However, there are two main conditions that need to be fulfilled: direct cell-to-cell contact and the presence of soluble immune modulators such as indoleamine 2,3-dioxygenase (IDO), prostaglandin E2 (PGE-2), or nitric oxide (NO) [12, 64].

The immunosuppression function of SMSCs was demonstrated to be comparable to that of BMSCs. An upregulation of MHC I and II molecules similar to BMSCs was observed in SMSCs following autocrine stimulation by IFN-γ. Besides that, SMSCs were also capable of inhibiting T cells proliferation in a mixed lymphocyte reaction (MLR). This suppression of T cell proliferative activity is believed to be linked to IDO activity, which is induced by SMSCs to the same extent as BMSCs [12, 16].

Although the immunosuppressive capabilities of SMSCs are valuable, a few potential problems still have to be investigated. There still exists a possibility that SMSCs may elicit an immune response from the body. Some investigators reported that BMSCs exhibit contradictory properties at low concentrations of IFN-γ. Under such a condition, the levels of MHC class II molecules expression increased and the cells become phagocytic. They also act as antigen-presenting cells to support CD4+ T cells proliferation [65]. It is thus imperative to devise a detailed characterization of the SMSC phenotype so that their *in vivo* biological properties can be understood through comprehensive preclinical studies, with the aim to widen their clinical applications.

17.6 SMSC Applications in Musculoskeletal Regeneration

17.6.1 SMSCs in Cartilage Regeneration

Articular cartilage has a very limited self-healing capacity due to its avascular nature. Typically, the defects in the articular cartilage can be divided into two different types: full- and partial-thickness defects. Full-thickness defects that reach the subchondral bone are capable of self-healing. However, the self-healing process is still limited and dependent upon the size and location of the defects [66]. In general, small full-thickness defects are able to heal spontaneously with the regeneration of hyaline cartilage, whereas bigger defects are only able to heal with fibrous tissue [67]. In contrast to full-thickness defects, partial-thickness defects do not reach the subchondral bone; thus, no spontaneous repair mechanism is triggered [68].

Despite the many efforts that have been made to regenerate defects on the articular cartilage, there is still no one method to date that is truly effective and successful. Most of the methods used thus far have some shortcomings. For example, treatment of lesions by microfracturing results in defect repair by fibrocartilage that is both mechanically and hydroelastically inferior as compared to the native hyaline cartilage [69]. The efficiency of another repair method known as mosaicplasty [70] is also questionable since this technique is limited by poor graft integration and donor site morbidity, hence rendering it ineffective for the repair of larger defects [67]. However, a recently developed technique known as autologous matrix-induced chondrogenesis (AMIC) has yielded satisfactory early [71] and midterm [72] results for the repair of larger cartilage defects, though further follow-up results are not yet available [73]. Untreated focal cartilage injuries may potentially develop into more extensive defects which eventually require invasive treatments like joint replacement surgery. Total knee replacement has been proven to be useful for older and less active patients, even if they only regain partial function of the knee [74]. Therefore, it is imperative to develop an alternative treatment method for the younger and more active patients.

Cell-based therapies have gained much interest as they can potentially provide biological replacements for injured cartilage. Autologous chondrocyte implantation (ACI) has been employed in treating focal cartilage defects, but this method is largely limited by the availability of autologous articular chondrocytes. The human articular chondrocytes are difficult to extract and have low *in vitro* expandability. Additionally, chondrocytes are unable to sustain prolonged expansion as they will lose their proliferative ability and also undergo dedifferentiation [75]. To date, there are still no conclusive results regarding the mid- and long-term efficiency of ACI.

A more promising approach, which is worth exploring, is the use of stem cells in tissue engineering for regeneration and repair of defective cartilage [73]. Among the various types of stem cells, adult MSCs are the most favorable for use. This is due to some advantages that MSCs have over their embryonic counterparts which include lesser tendency for tumor formation and lower risks for rejection and disease transmission [76]. Furthermore, MSCs are also known to possess some immunosuppressive potential [59]. BMSCs are the earliest discovered MSCs and have been extensively studied. There is a wealth of information in published works relating to their use in cartilage regeneration. However, the more recently discovered SMSCs display superior chondrogenic potential over BMSCs [10].

In one study, rat SMSCs surpassed other MSCs from bone marrow, periosteum, adipose, and muscle in *in vitro* chondrogenesis, as evidenced by heavier pellet formation. The greater pellet weight was a direct result of a higher rate of cartilage matrix formation [10]. Similar results were shown in another study utilizing human MSCs, where cartilage pellets comprised of SMSCs were considerably larger than those from BMSCs in patient-matched comparisons [10]. A study by Sakaguchi et al. also emphasized the superior chondrogenic potential of SMSCs over other MSCs by directly comparing yield, expandability, differentiation potential, and epitope profiles [9]. More recently, the development of more efficient cell-sorting techniques such as MACS and FACS has allowed specific subpopulations of cells from SMSCs to be isolated, thus enabling the assessment of their respective chondrogenic potential. Some researchers have reported that the CD105±enriched subpopulation of cells from the rat SMSCs has shown greater chondrogenic potential as compared to the non-sorted SMSCs, stemming from the fact that the CD105+ group expresses higher levels of type II collagen and SOX-9 [36]. In a separate study, the CD105+ subpopulation of SMSCs extracted from both OA and normal human synovial membranes were found to have comparable chondrogenic capacity despite the difference in health status of the patients [77].

Besides that, subpopulations of SMSCs enriched for CD73, CD106, and CD271 markers were also investigated for their ability to undergo chondrogenesis. The results indicated that the CD271+ SMSC subpopulation maintained the highest chondrogenic potential as compared to the other two subpopulations of cells. The CD106+ SMSC subpopulation appeared to be the least differentiated due to the highest level of MSC marker expression. Another study demonstrated CD14-negative selection of cells with the aim to enhance chondrogenesis by excluding monocytes and macrophages from a mixed SMSC population [78]. The efficiency of FACS and MACS enables these cell surface markers to be distinguished from one another, thus providing a means to select a specific subpopulation of cells for differentiation toward a defined cell lineage. This is because different subpopulations were shown to have different multipotentiality [79].

Besides various comparative studies on different MSC sources and subpopulations, some investigators also compared the chondrogenic ability of SMSCs extracted from different harvest sites. A study by Mochizuki et al. compared SMSCs harvested from human fibrous synovium, adipose synovium, and subcutaneous fat. The results revealed that both fibrous- and adipose-derived SMSCs outperformed MSCs isolated from subcutaneous fat in chondrogenesis, with the adipose SMSCs in particular showing the highest chondrogenic ability. Therefore, these two groups of cells can potentially be used as cell sources for cartilage tissue engineering. In addition to the different harvest sites, the chondrogenic potential of SMSCs was also found to be affected by preculture conditions. A shorter culture period was found to be more favorable in order to preserve the chondrogenic differentiation ability of SMSCs [26].

Apart from investigating *in vitro* chondrogenesis, it is also vital that the *in vivo* performance of SMSCs be evaluated. A recent work by Koga et al. has attempted to investigate the *in vivo* chondrogenic potential of rabbit MSCs. They reported that both SMSCs and BMSCs have greater *in vivo* chondrogenic potential as compared to the MSCs derived from the adipose and muscle tissue. Furthermore, SMSCs were found to have the advantage of possessing a higher proliferative ability. In order to stimulate higher production of cartilage matrix, higher density of SMSCs was coupled with the use of a periosteal patch during transplantation [80].

17.6.1.1 Growth Factors Induction

Over the years, the directed chondrogenic differentiation of SMSCs has been extensively investigated by utilizing various cytokines and hormones. Although it is likely that no single cocktail of growth factors will ever become a universal solution for optimizing chondrogenesis of SMSCs, there are a few of them which show great efficacy [81]. Currently, the most commonly used chondrogenic medium to induce SMSC chondrogenesis contains dexamethasone (DEX), ascorbate, insulin-transferring-selenium mixture (ITS+ premix), proline, and sodium pyruvate, along with a growth factor from the transforming growth factor-beta (TGF-β) superfamily [6, 9, 26].

Members of the transforming growth factor-beta (TGF-β) superfamily are well recognized for their ability to mediate chondrogenic differentiation in several MSC lines [81, 82]. The mechanism of TGF-β signaling has been elucidated. It starts with the binding of TGF-β to specific cell surface receptors, which then triggers a number of intracellular kinase pathways, and eventually, transcription factors such as SOX-9 are activated to induce the expression of chondrogenic genes [82].

Accordingly, many studies have been directed to investigate the efficacy of the various members of the TGF-β superfamily such as TGF-β1, 2, and 3, bone morphogenetic protein 2 (BMP-2), and bone morphogenetic protein 7 (BMP-7) in mediating SMSC chondrogenic differentiation [9, 10, 26, 36, 79, 83]. Among these, only TGF-β1, 2, and 3 are well proven to be able to fully induce chondrogenesis by increased production of proteoglycan and type II collagen, even when used individually [84, 85].

Thus far, TGF-β1 has only produced generally mixed results regarding its role as an inducer of SMSC chondrogenesis. A number of studies reported successful induction of chondrogenesis when SMSC culture was supplemented with this growth factor [86, 87]. One of the studies showed that the expression level of type II collagen, an essential chondrogenic marker, has been elevated in both rabbit SMSC pellets and synovial explants following culture in the presence of TGF-β1 [88]. Conversely, some studies reported that although TGF-β1 is an essential mediator for chondrogenesis, using it alone may not be sufficient to stimulate differentiation of SMSCs into chondrocytes. This was confirmed by some researchers using human and bovine SMSCs for chondrogenic differentiation in 3D culture using alginate hydrogel [89, 90]. Treatment by TGF-β1 was reported to downregulate the expression of type II collagen in the bovine SMSCs [90]. These discrepancies indicate that factors such as cell species and the culture system adopted should also be taken into consideration when selecting the growth factors to be used.

For better performance, other growth factors have been investigated for use in combination with TGF-β1. It was thought that the use of insulin-like growth factor (IGF-I) could enhance the chondrogenic induction effect of TGF-β1 [87, 91], since they are both anabolic growth factors present during chondrogenesis [92, 93]. The efficacy of this combination has been investigated by Sakimura et al. where they reported higher glycosaminoglycan (GAG) production by the SMSCs seeded on a polyglycolic acid (PGA) scaffold when these two growth factors were used together in cultures [87]. Similar results were shown by Pei et al., where simultaneous use of TGF-β1 and IGF-I yielded the best results as compared to sequential use of these two factors or any other combinations of chondrogenic growth factors for SMSCs [91]. However, Bilgen et al. reported that no such enhancement effect was observed [78].

It was also suggested by some studies that basic fibroblast growth factor (bFGF) can be used for pretreatment of SMSCs prior to induction of chondrogenesis since it has been shown to be effective both as a mitotic stimulator for MSC *ex vivo* expansion and also in subsequently regulating their ability to differentiate. Furthermore, bFGF also has the ability to induce both the prechondrogenic cells and chondrocytes to synthesize more cartilage matrix [94]. Kim et al. demonstrated that supplementation of bFGF in SMSC monolayer cultures caused cells to shrink and express a greater amount of actin, as well as to proliferate at a higher rate. They also showed that for micromass pellet cultures, SMSCs supplemented with bFGF were greater in size, weight, and GAG accumulation. The results suggested that this growth factor can potentially be used to enhance proliferation and chondrogenesis of SMSCs [95]. Besides that, there are also results from another study that reported a higher chondrogenic potential for the bFGF-expanded SMSCs derived from infrapatellar fat pad. The SMSCs that were pretreated with bFGF during *in vitro* expansion showed significantly higher accumulation of matrix during subsequent chondrogenesis as compared to untreated cells [96]. However, in a separate study, it was found that the bFGF could suppress SMSC chondrogenesis if used in combination with TGF-β1 or TGF-β1 + BMP-2 in a pellet culture [94].

Another growth factor that is also frequently used is TGF-β3. Some studies revealed that TGF-β3 and DEX were inadequate to mediate chondrogenesis in human SMSC pellet culture. However, the addition of BMP-2 to TGF-β3 and DEX significantly enhances SMSC chondrogenic potential [10]. This condition has been used in many recent studies since it has been proven to be an effective stimulus for chondrogenesis [9, 10, 36, 67]. This enhancement effect of BMP-2 on TGF-β3-induced chondrogenesis was shown by Rui and colleagues using human SMSCs extracted from OA patients in a pellet culture system. The group of cells that was treated with TGF-β3 and BMP-2 generally showed enhanced chondrogenesis as evidenced by the increased pellet size and weight, higher matrix production, and chondrogenic markers expression than the group treated with only TGF-β3 [97]. The need for supplementation with exogenous BMPs in SMSC cultures is further shown by a recent study where successful chondrogenesis of SMSCs was only achieved when both TGF-β3 and BMP-6 were applied concurrently [98].

Apart from being used in conjunction with TGF-β3, the BMPs can also be used separately as chondrogenic stimulators for SMSCs. Generally, BMP-2 and BMP-7 performed better than TGF-β1 under serum- and DEX-free conditions [83, 89]. In a study by Miyamoto et al., they reported that simultaneous addition of BMP-7 and TGF-β1 was able to enhance *in vitro* chondrogenesis of SMSCs, particularly when a high dose of BMP-7 was used [86]. Interestingly, the gene expression levels of some chondrogenic markers such as aggrecan and type II collagen were found to be affected by the dosage of BMP-2 and BMP-7 in the range of 50–200 ng/ml or higher [89]. The chondrogenic-promoting effect on human SMSCs of another BMP member, BMP-14, was also investigated in a recent study using cells from RA patients. BMP-14 was found to have little effect on SMSC proliferation but was able to promote the TGF-β3-induced chondrogenesis to a great extent [99].

Despite their high chondrogenic efficacies, BMP-2 and BMP-7 may not be suitable for use in long-term culturing since the presence of either of these factors seemed to be associated with SMSC hypertrophic differentiation [83, 86, 89, 90]. In addition, supplementation with BMP-2 alone did not produce similar gene expression levels in SMSCs as that of chondrocytes *in vitro* [89, 90]. Moreover, the quality of the cartilage tissue produced was also questionable due to undesirable, sustained expression of type I collagen by BMP-2-treated SMSCs [89].

DEX, which is a synthetic glucocorticoid, is also believed to be an important chondrogenic inducer of SMSCs. It has been reported that DEX has a supportive role in TGF-β1- and TGF-β3-induced chondrogenesis [9, 89, 100]. However, under certain conditions, the use of DEX can adversely affect the chondrogenic potential of SMSCs. DEX is known to suppress the chondrogenic induction effect of BMP-2, inhibiting the gene expression of several important chondrogenic markers such as SOX-9, type II collagen, and aggrecan in SMSCs [89, 90].

In order to be able to induce full differentiation of SMSCs into articular chondrocytes, more extensive research is required to refine the stimulation conditions for chondrogenesis. In addition to selecting and formulating the optimal cocktail of growth factors, other factors such as culture system, timing, and dosing also need to be determined.

17.6.1.2 Gene Therapy

As discussed in the previous section on growth factor induction of SMSC chondrogenesis, many locally administered recombinant proteins such as isoforms of TGF-β, IGF-1, bFGF, BMP-2, or BMP-7 have been widely used in augmenting the various aspects of cartilage tissue repair. Nonetheless, one recurring problem is the transient residence time due to the short half-lives of many proteins in the *in vivo* environment, making it difficult to deliver these factors to cartilage repair sites at therapeutic concentrations [101, 102]. This can be remedied by using gene therapy to lengthen the period of growth factors synthesis. This is achieved by the insertion of coding sequences which allow sustained synthesis of bioactive anabolic agents, both locally and regionally [103]. Thus, if trials are favorable, gene therapy may be the key innovation that allows this treatment to be effective at repairing and regenerating articular cartilage for patients.

In gene therapy, delivery of the desired coding sequence into the cell nucleus is inherently difficult due to the many defense mechanisms of the cell. Apart from extracellular defenses such as phagocytes and T cells, the cell membrane itself repels foreign DNA fragments due to electrostatic repulsion, as both are negatively charged. Furthermore, foreign DNA fragments must also bypass endonuclease degradation and endosomal engulfment once they are in the cell [103]. Thus, in order to ensure efficient gene delivery into the target cell nucleus, it is crucial to employ vectors that are capable of overcoming all these challenges. Viral vectors can fulfill these criteria as they are particularly efficient at penetrating cell membranes and delivering the transgene into the cell nucleus. To date, no other known nonviral delivery methods approach the efficiency of viral vectors. The commonly used viral vectors in gene therapy include those from adenoviruses, adeno-associated viruses, retroviruses, and lentiviruses. Basically, the performances of viral vectors are gauged based on their transduction efficiency and, more importantly, their safety profile [104, 105]. Each type of the viral vector has its own advantages and disadvantages. For example, the ability of lentiviruses and retroviruses to integrate foreign genes into the host genome prolongs gene-induced repair. However, this same ability can cause insertional mutagenesis, which may lead to undesirable consequences. Adenoviruses on the other hand do not integrate their transgenes into the host, making them safer to be used for gene delivery [103].

Prior works using recombinant adenoviral vectors have been successful in delivering the gene of TGF-β3 to cells such as rat BMSCs and articular chondrocytes for *in vitro* chondrogenesis [106]. However, the pioneering work utilizing this technique on SMSCs was done by Zhang et al. [107]. In their preliminary study, they successfully constructed a dual-functioning adenoviral vector (Ad-dual) consisting of transgene encoding TGF-β3 and short hairpin RNA (shRNA) targeting type I collagen. They reasoned that this Ad-dual not only promotes sustained synthesis of TGF-β3 but at the same time also suppresses the production of undesirable type I collagen which compromises the quality of the repaired articular cartilage. The suppression of type I collagen is done by employing the RNA interference (RNAi) strategy. Their initial hypothesis was later confirmed when the Ad-dual vector was shown to work as intended in SMSCs, as supported by the concurrent suppression of type I collagen and promotion of TGF-β3 synthesis [107]. In their follow-up study, they investigated the transduction effect on SMSC *in vitro* chondrogenic differentiation using several different combinations of recombinant adenoviruses in a 3D alginate hydrogel culture system. The various combinations included infecting SMSCs with recombinant adenoviruses encoding TGF-β3 (Ad-TGF-β3) or anti-Col I shRNA (Ad-shRNA) independently, concurrently (Ad-combination), or conjugately (Ad-double), which is the Ad-dual vector described earlier. The results showed that the expression of cartilage-specific genes such as type II collagen, aggrecan, and cartilage oligomeric matrix protein (COMP) were elevated in Ad-TGF-β3, Ad-combination, and the Ad-double infected groups. An undesirable increase in type I collagen expression was observed in the Ad-TGF-β3 infected group, but this did not happen in the other two groups. Despite being able to induce chondrogenesis and suppress hypertrophic differentiation of SMSCs, the dual-functioning adenoviral vectors are still in need of refinement since this group of vectors cannot compete with the AD-TGF-β3 group in promoting chondrogenesis. Furthermore, the expression of transgene by cells transducted with adenoviral vectors is not permanent due to the lack of integration of the viral DNA into host genome [108].

In a recent study by Varshney et al., adenoviral vectors were also employed for transduction of gene encoding TGF-β3 into rabbit articular chondrocytes. These transfected chondrocytes were later cocultured with SMSCs to supply them with sustained, localized, and overexpressed growth factors. They reported that the SMSCs were able to undergo chondrogenesis to a much greater extent under the influence of the transgenic growth factors. Interestingly, the transient expression of the transgenic TGF-β3 allows timely termination of its supply to SMSCs in the later stage to avoid hypertrophic differentiation [109]. Besides transduction of growth factor genes, another study also probed the gene transduction effect of an enzyme known as histone deacetylase 4 (HDAC4) on TGF-β1-induced chondrogenesis in SMSCs. Adenoviral vectors were utilized to facilitate the transfer of gene encoding HDAC4 into SMSCs. This enzyme is believed to possess the ability to suppress chondrocyte hypertrophy. The results showed that HDAC4-transduced SMSCs underwent rapid and extensive chondrogenesis in the presence of TGF-β1. At the same time, the cells also showed low levels of hypertrophic differentiation as evidenced by the downregulation of a chondrocyte hypertrophy marker, type X collagen [110].

In order to have better performance in cartilage repair, the sustained expression of TGF-β3 and permanent suppression of type I collagen may be necessary. In another study by Zhang et al., they aimed to address the shortcomings in their previous work by switching to dual-functioning lentiviral vectors (LV), which are integrative in nature and thus support prolonged gene expression. They devised four different arrangements of the two expression cassettes and tested separately for their role in TGF-β3 secretion and type I collagen suppression. The different arrangements were constructed to determine the best configuration that displayed the highest expression level of TGF-β3 while maintaining the lowest level of type I collagen. All the constructed LVs were used in SMSCs transduction and the cells were then encapsulated in alginate hydrogel for further culturing. The results showed that the four vectors differed in their ability in inhibiting type I collagen and also their varying inductive efficacies in mediating upregulation of chondrocytic markers. They concluded that LV-1 which has two expression cassettes arranged in a distant and reverse order displayed the greatest potential in promoting Col I-suppressed chondrogenesis. Even though LV-1 showed the greatest potential, its current performance is only modest at best. Thus, further work is needed to find an optimal arrangement that exhibits both desirable features [111].

As a whole, current preclinical studies have returned promising results regarding the effectiveness of gene delivery techniques. Viral vectors are able to give the desired results because they ensure strong expression of therapeutic genes, but uncertainties remain over the possibility of side effects of long-term applications. One example would be that the overexpression of transgene products may have adverse effects on non-targeted organs. Thus, it is critical that future gene delivery systems include components that allow the duration of *in vivo* expression to be controlled and timed accurately.

17.6.1.3 Scaffolds

After the selection of an appropriate therapeutic cell source, the local delivery of the cells to repair sites becomes an important issue to be addressed in cartilage tissue engineering. One effective way of cell delivery is by utilizing the various types of scaffolds available. Generally, a scaffold in tissue engineering has to serve at least two purposes: it has to deliver cells to the target sites efficiently, and it also needs to support the subsequent development of new tissue by providing a favorable microenvironment for cells to reside, proliferate, and differentiate.

It has been established that chondrocytes tend to undergo dedifferentiation after an extended culture period under 2D conditions. Interestingly, the dedifferentiated chondrocytes can regain their lost phenotype and metabolic activity once they are transferred to 3D culture systems [112]. Therefore, it is thought to be the same for SMSCs where 3D culture systems are preferable over 2D culture systems in engineered chondrogenesis. This is supported by the results from various studies using the pellet culture system, which is one of the earliest 3D culture system used [9, 91, 95, 97]. Another similar culture system known as the micromass pellet culture system was also used in some studies. Although these two types of 3D culture system have been shown to support chondrogenic differentiation of SMSCs, the cell pellets produced are too small in size to be clinically useful, especially for the repair of larger defects [113].

In order to overcome this size limitation, other more complicated 3D scaffolds have been extensively developed to ensure that the engineered tissue constructs are large enough for therapeutic purposes. Among the various 3D scaffolds available, hydrogels are especially popular for cartilage engineering. This is mainly due to them having high water content which is comparable to that of the native cartilage. Besides that, they are also known to have excellent cell compatibility and are easily implanted. Hydrogels that are currently in use for cartilage engineering range from purely natural materials to purely synthetic materials [114].

Many studies have utilized alginate hydrogels for SMSC encapsulation before subjecting the stem cells to chondrogenic differentiation [89, 90, 109, 111, 115]. A recent study by Park et al. observed that the fibroblast-like SMSCs assumed a rounded shape following encapsulation. The encapsulated SMSCs were later reported to show higher expression of chondrocytic markers [90]. Besides alginate, collagen gels are also frequently used. In an earlier study, cartilaginous tissues were successfully formed by embedding SMSCs in collagen gel followed by *in vitro* culture. Phenotypic change was observed in the embedded SMSCs where they became more like the chondrocytes. Furthermore, they also showed increased chondroitin sulphate secretion, an important cartilage matrix component [116]. Also, in a study by Buckley et al., the use of agarose hydrogels in making porcine SMSC embedded 3D constructs was proven to be feasible. They showed that chondrogenesis was successfully induced in the cell-laden agarose contructs by pretreatment with bFGF [96].

A notable development in hydrogel scaffolds was elucidated in a study by Fan et al. where a novel injectable gellan hydrogel was utilized as a cell carrier to deliver rabbit SMSCs for *in vitro* engineered cartilage. The results showed that this hydrogel has high cell compatibility as indicated by the high cell viability. In addition, under the influence of appropriate chondrogenic inducers such as TGF-β1 and 3, the constructs were able to form cartilaginous tissue after 3 weeks of *in vitro* culture [117]. In a subsequent work by Fan et al., they explored SMSC chondrogenesis in both biodegradable and nonbiodegradable photopolymerized synthetic hydrogels. Both the biodegradable phosphoester-poly(ethylene glycol) (PhosPEG)-based hydrogel and nondegradable poly(ethylene glycol) diacrylate (PEGDA)-based hydrogel were shown to support the proliferation and chondrogenic differentiation of SMSCs in the presence of appropriate growth factors. This suggests that they are suitable as cell vehicles for SMSCs chondrogenesis [118].

Apart from using hydrogels, some studies have also probed the use of synthetic polymeric scaffolds. Some of the more commonly used materials include the polyglycolic acid (PGA) and poly(lactic-co-glycolic) acid (PLGA) [119–121]. For example, a study by Pei et al. showed that SMSCs that were premixed with fibrin gel before seeded onto PGA scaffolds were able to undergo chondrogenesis in the presence of various important growth factors coupled with biomechanical stimulation [122].

However, the recent trend in scaffold research points to the development of hybrid scaffolds which combine the various advantages of different materials. One recent study utilized a three-dimensional hybrid scaffold made of chitosan and alginate composite to support the proliferation and chondrogenic differentiation of SMSCs. The results showed that the SMSCs were able to attach and proliferate well on the porous chitosan-alginate composite scaffolds. The cells were later shown to undergo chondrogenesis as evidenced by the elevation in chondrocyte-related gene expression [36]. In another study by Gong et al., they investigated the use of a hybrid scaffold constructed using chitosan and collagen type I in directing SMSC chondrogenesis. The scaffolds were fabricated using freeze-drying and cross-linking techniques resulting in their porous structures. Results indicated that the hybrid scaffolds were able to support both *in vitro* and *in vivo* chondrogenic differentiation of SMSCs when treated with appropriate growth factors [123].

Currently, a number of scaffolds have already been approved for clinical use, but there are still some concerns that remain about safety issues arising from long-term usage of these materials. The use of synthetic polymers may raise problems concerning *in situ* retention and degradation [124, 125]. On the other hand, the use of natural materials may pose risks of pathogen transmission and precipitating immunological reactions [126, 127]. Based on these reasons, a group of researchers have devised a scaffold-free delivery system called the tissue-engineered construct (TEC) using porcine SMSCs. This special culture condition yielded pure tissue constructs which were later used for *in vivo* repair of chondral defects in a porcine model. The TEC showed satisfactory results during *in vitro* chondrogenesis when placed in a chondrogenic medium. As for *in vivo* repair, the implanted TECs were shown to form cartilaginous tissue in defects and appeared to integrate well with the repair site. Moreover, when TECs were subjected to static compression and friction tests, their mechanical properties were reported to be comparable to that of native porcine cartilage [128]. In another study, the TECs were formed using human SMSCs. The results suggested that the mechanical properties of the constructs can be improved by extending the culture period in ascorbate 2-phosphate (Asc-2P)-supplemented growth medium. Under proper chondrogenic induction, these TECs from human SMSCs were also able to differentiate into chondrocytic cells [129].

Besides TECs, other scaffold-free techniques have also been investigated. For example, an earlier study by Koga et al. used what is called a local adherent technique to transplant rabbit SMSCs into chondral defects under *ex vivo* and *in vivo* conditions. They reported that over 60 % of the SMSCs in suspension were found to adhere to the defects after placing them there for 10 min, which later helped in cartilage regeneration [130]. Another technique was demonstrated by a recent study with the use of an intra-articular magnet coupled with magnetically labelled SMSCs to repair osteochondral defects in a rat model. An intra-articular magnet was implanted at the bottom of the defect prior to injection of the magnetically labelled SMSCs into the knee. The results showed that regeneration of the articular cartilage occurred after 4, 8, and 12 weeks of treatment. Although the initial results from this study are encouraging, a few shortcomings still need to be addressed. This includes finding an absorbable magnet so that invasive surgery for removal is not needed after repair of cartilage is complete [131].

All these studies showed that cell delivery systems or scaffolds can exert varying effects on directing the chondrogenesis of SMSCs. It is however difficult to elucidate an optimal cell delivery system or scaffold solely based on existing literature since not many studies have made direct comparisons between the different delivery methods.

17.6.1.4 Bioreactor

The *in vitro* engineering of cartilaginous tissue has been extensively studied for years with many researchers striving to produce engineered tissues which are clinically useful. Despite all of these efforts, there have been no successful attempts at *in vivo* implantation of cultured articular cartilage in patients. A possible reason for this is the lack of replication of physiological conditions during the *in vitro* culturing process of the tissue. Previously neglected environmental factors such as mechanical stimulation [132] and hypoxia [133] may be crucial in promoting *in vitro* chondrogenesis of MSCs. A recent study by Li et al. has investigated the effect of oxygen tension on SMSC *ex vivo* expansion and *in vitro* chondrogenesis. The SMSCs were cultured in either hypoxia (5 % O_2) or normoxia (21 % O_2) environment. They concluded that SMSCs showed the highest expansion rate in a hypoxic environment coupled with supplementation of FGF-2 and plated on ECM derived from SMSCs.

Furthermore, chondrogenic hypertrophic markers were downregulated under hypoxic conditions [134].

The native articular cartilage has to bear both constant loads and mechanical stresses produced by the movement of joints. In view of that, many different bioreactor systems have been used to provide mechanical stimulation to tissue explants or cell-scaffold constructs. This effort in mimicking the mechanical load experienced *in vivo* is believed to help in *in vitro* cartilaginous tissue formation. A study by Sakao et al. applied intermittent hydrostatic pressure (IHP) to SMSCs encapsulated in alginate beads. The results showed that chondrogenic differentiation was induced in the pressure-loaded SMSCs to a much greater extent than those in the control group. This is supported by the increased expression of proteoglycan core protein, type II collagen, and SOX-9. They concluded that application of IHP as high as 5.0 MPa on SMSCs could induce chondrogenesis, more specifically by the MAP kinase/JNK pathway [115]. In a few other studies, rotating bioreactor systems were used to culture SMSC-seeded PGA scaffolds, and results showed successful induction of chondrogenesis in these cell-laden scaffolds [35, 122].

17.6.2 SMSCs in Meniscal Engineering

Apart from the repair and regeneration of articular cartilage, another important structure of the knee that deserves equal attention is the meniscus. The knee meniscus is a fibrocartilaginous structure which can be found between the femoral condyle and the tibia plateau [135]. It plays an important role in maintaining the normal function of the knee, one of which includes preventing the degeneration of articular cartilage. It can be divided into a medial and a lateral component, both of which are semilunar in shape and consist of cells surrounded by specialized extracellular matrix molecules.

One special characteristic of the meniscus is the regional variation of cell phenotype and ECM composition. The outer region is made up of fibroblast-like cells surrounded by matrix containing high levels of type I collagen together with a small amount of glycoproteins and type III and V collagen [136, 137]. As for the inner region, its cells appear to be more rounded and surrounded by matrix having higher levels of type II collagen and a smaller but not negligible amount of type I collagen. In addition, this inner region also has a higher amount of GAGs than the outer zone [137, 138].

Besides cell phenotype and ECM composition, vascularization of the meniscus is also subjected to regional variation. Interestingly, vascularization decreases as this tissue matures. At maturity, only the peripheral region which is about 10–25 % of the meniscus has blood vessels and nerves [139, 140]. Consequently, this delineates two different regions of the meniscus which are red-red zone (vascular or neural outer region) and the white-white zone (avascular or aneural inner region). In actual fact, blood circulation directly affects the healing capacity of each area. This causes the inner region to be far more susceptible to long-lasting lesions since this region is incapable of healing [141].

Consequently, avascular meniscal injuries are generally more complicated and need to be handled more effectively. The most common way of treating these injuries is by a surgical procedure known as partial meniscectomy. However, this partial removal of the menisci will inevitably lead to the development of osteoarthritis, the severity of which is determined to a certain extent by the amount of resected tissue [142–144]. Thus, this necessitates an alternative therapeutic method. Similar to articular cartilage, meniscal tissue engineering is seen as a promising novel treatment strategy as both the structural and functional properties of menisci could potentially be restored.

A suitable cell source is often crucial in the field of tissue engineering even though there are some meniscal engineering techniques that are based on acellular scaffolds. The appropriate cell type to be used is controversial, but Hoben and Athanasiou have recently reported the criteria for an ideal cell source for cell-based meniscal engineering. According to them, an ideal cell source consists of cells that are preferably autologous, easily and abundantly available, capable of *in vitro* expansion, and also able to synthesize fibrocartilaginous matrix [145].

SMSCs are known for their high proliferation rate and superior chondrogenic potential and can be readily harvested from the synovial tissue through noninvasive arthroscopy. Therefore, SMSCs are slowly gaining attention for use in meniscal regeneration. Moreover, some studies have shown that the gene expression profiles of the meniscal cells and SMSCs are reasonably similar to each other [55, 146]. This is also supported by the findings of one study which reported that meniscal cells have a gene expression profile that shows more resemblance to that of the SMSCs as compared to BMSCS [146]. This suggests that the cells derived from the synovium hold more promise than BMSC as a cell source for meniscal regeneration.

Currently, there are two distinct strategies being adopted for the repair and regeneration of meniscal defects using SMSCs: the first of which is a scaffold-free strategy that involves direct infusion of SMSCs into the injured sites with the intention to generate a reparative response [146, 147]. The second strategy involves the construction of a whole physical tissue prior to implantation [140]. The first strategy has been investigated by some studies where they have reported that the SMSCs were able to differentiate directly into meniscal or cartilage cells by intra-articular injection into a rat model [146, 147]. The injected SMSCs were shown to adhere to the defects and underwent differentiation into meniscal cells, subsequently enhancing meniscal regeneration locally [146]. The latter strategy of constructing a physical tissue was attempted via two different approaches. Some studies demonstrated the formation of an implantable, cell-seeded scaffold [142, 148] while one other study has shown the use of pure cell-based tissue constructs formed by specialized culturing techniques [128].

The studies that utilized implantable cell-seeded scaffolds yielded mixed results. Synthetic scaffolds comprised of PGA and PLLA used in one of the studies showed poor performance due to suboptimal ECM production [142], while some other studies have reported encouraging results using nonwoven PGA scaffolds. The discrepancies in results may be due to the application of varying biochemical stimuli. In another recent study, an interesting outcome was observed when SMSCs were cocultured with meniscal cells on a small intestine submucosa (SIS) scaffold. The cell-laden SIS constructs were shown to have higher cell survival rates, with higher glycosaminoglycan (GAG) content and elevated chondrogenic gene expression as compared to techniques using only synoviocytes [148].

In order to achieve better results in tissue regeneration, the effects of various growth factors and hormones on SMSCs have also been widely investigated. Several studies have reported successful chondrogenic induction of SMSCs for meniscal regeneration by using either TGF-β1 only or in combination with other growth factors like FGF-2 and IGF-1 [87, 142, 148, 149].

Although it is known that SMSCs carry the potential to achieve successful meniscal repair and regeneration, the specific culture environment and stimuli needed for fibrochondrogenesis have yet to be established. Moreover, the optimal amounts of type I and type II collagen in the extracellular matrix for meniscal applications have also yet to be defined. Many more detailed studies and experiments are thus needed to address these current challenges.

17.6.3 SMSCs in Bone Regeneration

SMSCs have a much higher tendency to differentiate into the chondrocytic cells rather than into osteoblastic cells [12]. Thus, as of now, there are relatively fewer studies that focus on osteogenic differentiation of SMSCs into bony tissue for repair and regeneration [12]. However, some studies have shown that SMSCs harvested either from healthy or diseased knee joints can undergo *in vitro* osteogenesis when cultured in osteogenic medium. The commonly used osteogenic medium contains complete medium supplemented with appropriate concentrations of dexamethasone, β-glycerol phosphate, and ascorbate-2-phosphate [6, 9, 150].

In a study by Sakaguchi et al., the osteogenic potential of SMSCs was compared against MSCs derived from other mesenchymal tissues. According to them, SMSCs showed a greater extent of calcification as compared to MSCs harvested from skeletal muscle and adipose tissue, indicating their higher osteogenic potential [9]. Mochizuki et al. also reported that SMSCs derived from both fibrous and adipose synovium have higher osteogenic differentiation ability than the MSCs derived from subcutaneous fat as evidenced by the higher number of alizarin red-positive colonies observed in the two cell groups [25]. Further studies also confirmed that the osteogenic potential of SMSCs varies among specific subpopulations of cells. A study by Arufe et al. has investigated the osteogenic potential of three different subpopulations of cells denoted as CD73+, CD106+, and CD271+ isolated from the human synovial membranes. Among these three groups that were tested, the CD73+ subpopulation of SMSCs showed the greatest osteogenic potential where more than half the cells were differentiated into osteoblastic cells, followed by CD271+ and CD106+ [79].

Although the aforementioned studies have indicated the capability of SMSCs to differentiate into osteoblastic cells, their osteogenic potential still may not be as high as BMSCs and periosteal mesenchymal stem cells (PMSCs). A study compared the *in vitro* osteogenic differentiation capacities of SMSCs and BMSCs by quantifying the expression of osteogenic markers such as alkaline phosphatase (ALP) and osteocalcin. It was reported that the SMSCs showed much lower expression for the osteogenic markers tested as compared to BMSCs [16]. The osteogenic potency of human SMSCs and PMSCs was compared in another study by De Bari et al. They investigated both *in vitro* and *in vivo* osteogenic potential of these two types of cells. For *in vitro* osteogenesis, SMSCs were observed to have a much lower level of ALP activity and calcium deposits than PMSCs. Similar results were observed for the *in vivo* bone formation capacities of both types of cells [31].

The specific factors that are needed to direct SMSCs to commit into the osteoblastic lineage and become terminally differentiated bone cells have not been elucidated. However, there are a few general parameters that must be controlled to ensure successful bone tissue regeneration such as the manipulation of bone-forming cells, the choice of scaffold biomaterials, and the various chemical and physical cues needed to induce osteogenesis. Therefore, after selecting the appropriate cell source, the subsequent step is to make use of a suitable scaffold for efficient delivery of cells to the bone repair sites. The scaffold must also be able to provide some osteogenic induction signals, either chemically, physically, or both. De Bari et al. utilized the osteoinductive Collagraft scaffolds which are commonly employed for use in orthopedic clinical practice. These scaffolds are made up of hydroxyapatite, tricalcium phosphate, and type I collagen. The compositions of these scaffolds mimic the nature of bone matrix, thus providing useful physical cues for SMSCs to differentiate. The SMSCs were seeded onto these scaffolds prior to implantation into nude mice. Successful *in vivo* bone formation was observed after a period of 8 weeks [31].

A study by Shi et al. has generated much interest lately as they investigated the use of a multicomponent scaffold which is capable of controlled release of a nitrogenous bisphosphonate additive called alendronate (AL) for SMSC osteogenesis [151]. AL is capable of inhibiting osteoclastic bone resorption, and it is also commonly used in treating metabolic bone disease. In addition, this compound was also shown to be effective in promoting proliferation and maturation of osteoblasts [152], as well as enhancing osteogenic differentiation of BMSCs [153]. The AL-releasing scaffolds were synthesized in three steps starting with the hybridization of AL with hydroxyapatite (HA) nanoparticles, followed by self-assembly into mesoporous silica (MS) particles. Lastly the HA-AL-loaded MS constructs were incorporated into a bulk of poly(lactic-co-glycolic acid) (PLGA) microspheres. This multilevel structure not only significantly increased the encapsulation efficiency of the strongly hydrophilic AL but also allowed sustained release of the compound. The SMSCs seeded on these scaffolds were induced to undergo *in vitro* osteogenic differentiation to a great extent. This was evidenced by the strong expression of several important osteogenic markers such as ALP, type I collagen, osteocalcin, Runx-2, and BMP-2 [151]. The group furthered their work by using another similar multicomponent scaffolding system composed of PLGA/HA sintered microspherical scaffolds. These newly designed scaffolds were shown to be capable of releasing both AL and DEX in a controlled manner to enhance SMSC osteogenesis. DEX is another osteoinductive agent known to stimulate the initiation of BMSC differentiation and promote terminal maturation of osteoprogenitor cells at the late stages of differentiation [154]. The results reported that *in vitro* SMSC osteogenesis was greatly enhanced when seeded on these scaffolds as indicated by the significant amount of ALP produced and the high level of bone calcification [155].

Although many bioscaffolds have been employed for the construction of tissue-engineered bones, a group had recently attempted the use of scaffold-free constructs to repair bone defects. Rabbit SMSCs and the ECM secreted by them were used to form pure cell-based constructs called tissue-engineered constructs (TECs). The TECs were further processed by alternating soaking processes to have the hydroxyapatite (HAp) crystals formed on top of them. These crystals were used to enhance the osteoinductive properties of basic TECs. As indicated by their preliminary *in vivo* results, these TEC-HAp composites demonstrated improved osteoinduction when implanted into the osteochondral defects of a rabbit model, suggesting their potential use in the repair and regeneration of bone [156].

Many of the previously cited studies have shown the possibility of directing the SMSCs to commit into a specific linage by providing the necessary chemical or physical cues. In view of the abundant supply and high proliferative profile of SMSCs, the development of a more efficient delivery system that provides the necessary osteoinductive signals will help to harness these advantages of SMSCs to be applied in bone tissue engineering.

17.6.4 SMSCs in Tendon and Ligament Regeneration

Tendons and ligaments are frequently injured as a consequence of trauma or disease which can cause significant morbidity. Moreover, the healing process of the tendons and ligaments is comparatively slower and inefficient as compared to other connective tissues due to their poor vascularization [157, 158]. In the last few decades, many researchers have focused on developing different techniques to improve the healing of these tissues.

The repair and regeneration of the tendon and ligament is particularly challenging as these tissues are highly specialized in their functions. For example, the ECM of these structures needs to withstand high and constant mechanical loading within the human body. Current treatment methods remain inefficient as each of them is associated with some disadvantages. For example, the use of autografts, which is a common option for the reconstruction of injured tendon and ligament, may lead to donor site morbidity, and the use of allografts may pose the risk of disease transmission from the donor [159]. All these shortcomings have hastened the development of better strategies such as tissue engineering for the repair of these tissues. One of the more popular research areas involves the utilization of adult MSCs to regenerate functional tendons and ligaments.

The potential role of SMSCs in mediating tendon and ligament repair was reported by Jones et al. when they found that SMSCs in the synovial fluid (SF) were 20-fold higher in the knee of OA than RA patients [160]. Morito et al. later showed that the number of SMSCs was much higher (100-fold) in the SF of patients with ligament injury as compared to healthy patients. Furthermore, they also reported that the SMSCs injected into a partial ligament injury model of rabbit were able to adhere to the injured ligament, suggesting that the SMSCs may contribute to ligament repair [161].

The anterior cruciate ligament (ACL) has been the focus for ligament repair and regeneration since it plays an important role in maintaining stability and the function of the knee joint. It is also more prone to injuries. Thus, it makes a good illustration for ligament repair or regeneration [158, 162, 163]. In recent years, autologous hamstring tendon grafts have been preferably used to reconstruct the ACL since this method causes less donor site morbidity and anterior knee pain as compared to other methods such as the use of the patellar tendon autografts [164, 165]. One major problem associated with this method of reconstruction is the slow tendon-bone healing process. A study investigated the use of SMSCs in enhancing tendon-bone healing in a rat model. The results showed that the implantation of SMSCs hastened the early remodeling of tendon-bone healing by synthesizing larger amounts of collagen fibers after 1 week. Furthermore, the application of SMSCs also helped in forming more oblique collagen fibers which resemble Sharpey's fibers, linking the bone to tendon after 2 weeks post-surgery [130].

In another study, Ozturk et al. have investigated the use of SMSCs to engineer a tendon synovial cell biomembrane which functions to prevent adhesions after tendon surgery in the hand [166]. Adhesions are a common complication of the hand tendon surgery, which may cause severe disability if not carefully treated [166–168]. The SMSCs were chosen because of their ability to produce uridine diphosphoglucose dehydrogenase (UDPGD), an enzyme that is important in HA synthesis [11, 12, 169] which in turn is commonly used to create anti-adhesion barriers [170–172]. In their study, they infused SMSCs into a type I collagen matrix to form a barrier to separate the repaired tendon from its surrounding tissue. The results showed that the infused SMSCs were able to grow and form a surface layer *in vitro* after 14 days. Additionally, the actual presence of HA was observed by histological staining. This indicates that this SMSC collagen membrane is capable of producing endogenous HA [166]. In a recent study, a group of researchers have also hypothesized that SMSCs overexpressing hyaluronic acid synthase 2 (*has2*) may be effectively used as anti-adhesion therapeutic cells after surgery of the digital flexor tendons [173].

Although SMSCs have attracted much interest because of their multipotentiality and high proliferative ability, their applications in the repair and regeneration of tendons and ligaments are still limited. To date, most studies have only reported the use of SMSCs in enhancing the healing process of these tissues after surgery. Therefore, more extensive research is needed to explore the direct applications of SMSCs in engineering functional tendons and ligaments.

17.6.5 MSCs in Skeletal Muscle Regeneration

Skeletal muscle is the most abundant tissue in the human body, and despite being primarily a post-mitotic tissue, the skeletal muscle still retains the ability to regenerate with the help of the satellite cells which are present between the plasma membrane and the adjacent basal lamina of muscle fibers [174–177]. Although skeletal muscle loss can occur either due to trauma or various myopathies, cell-based therapies thus far have emphasized on treating the latter cause.

Muscular dystrophies (MD) are a category of myopathies affecting the skeletal muscle and are caused by mutations in the genes encoding for various muscular proteins. This group of diseases not only restricts the mobility of patients but in the worst case, such as for Duchenne muscular dystrophy (DMD), it can cause full paralysis and premature death [178, 179]. For patients suffering from DMD, muscle regeneration is progressively suppressed due to the exhaustion of the satellite cell pool resulting from recurring cycles of degeneration and regeneration. This condition ultimately causes total muscle function loss [176, 177].

To date, most research on cell-based therapies for MD has focused on the use of the satellite cells in skeletal muscle engineering. However, there are certain shortcomings that are associated with the application of these cells. For instance, the *in vitro* expandability of the satellite cells is very limited as they undergo rapid dedifferentiation following a small number of cell cycles [180]. In addition, their inability to cross the endothelial cell wall into target tissues hinders systemic delivery [181]. Therefore, MSCs have recently emerged as an alternative and preferable cell source since they possess some clear advantages over the satellite cells for use in skeletal muscle tissue engineering. The basis of stem cell therapy for treating MD comes from the differentiation potential of stem cells which can be harnessed to regenerate lost muscle fibers and also their self-renewal capacity which can help replenish the satellite cell pool [177].

From the onset of their discovery, SMSCs have been reported to be able to form sporadic atypical myotubes *in vitro* in proper inductive medium as evidenced by the positive immunostaining of skeletal muscle-specific myosin heavy chain (SKM-MHC) [6]. In a follow-up work by De Bari et al., they investigated the *in vivo* myogenic potential of SMSCs by injecting culture-expanded human SMSCs into cardiotoxin-damaged muscle in a nude mice model. The results suggested that the injected human SMSCs contributed to the repair and regeneration of the damaged muscle. This is supported by their integration with muscle fibers of the host and the expression of human muscle markers, myosin heavy chain type IIx/d (MyHC-IIx/d). Furthermore, they also reported that the human SMSCs replenished the satellite cell pool by forming sustainable functional satellite cells [182].

In the same study, they also explored the capacity of human SMSCs in correcting genetic muscle disorders such as MDs by transplanting the cells into the dystrophic muscles of an mdx mouse model. They found that the application of human SMSCs to the damage site was not only able to restore the expression of dystrophin but was also able to restore the expression of another crucial molecule, mouse mechano growth factor (MGF), which is responsible for the maintenance and repair of skeletal muscle [182].

In a recent study by Meng et al., they attempted to repeat and extend the previous work of De Bari et al. [182]. They investigated the contribution of human SMSCs in skeletal muscle regeneration by assessing *in vitro* myogenic potential of human SMSCs and also their *in vivo* muscular regeneration capacity. For *in vivo* muscular regeneration, the human SMSCs were injected intramuscularly into cryo-damaged muscles of Rag2-/c chain-/C5-mice. The results of SMSC *in vitro* myogenesis correspond with the study done by De Bari et al., both showing markedly limited myogenic potential [6]. However, the *in vivo* results differ greatly as this current study failed to observe muscle fiber formation by human SMSCs in the mice model. Furthermore, the donor human SMSCs also did not seem to give rise to satellite cells as reported by De Bari et al. since they were not found under the basal lamina of the muscle fiber but instead resided in the interstitial space of the grafted muscle. These discrepancies in the results of the two studies may be due to differences in experimental design [177].

In order to overcome the lack of *in vitro* and *in vivo* myogenic potential of human SMSCs, Meng et al. genetically modified the human SMSCs by transducing the cells with lentiviral vectors carrying the human MyoD gene to probe the effects of overexpression of this gene on their myogenic potential. They showed that human SMSCs can be directed to differentiate along the myogenic lineage by the ectopic expression of hMyoD1. After lentiviral transduction, the overexpression of MyoD by transduced SMSCs caused the cells to express higher levels of muscle-specific markers such as desmin and myogenin, augmenting the *in vitro* myotube formation. Besides that, *in vivo* regeneration of muscle was also enhanced by these genetically modified cells [177].

It has been previously reported that MSCs can exert a paracrine effect on differentiation via the secretion of various cytokines [183]. It is notable that the human SMSCs transplanted into skeletal muscles were able to secrete a few important extracellular matrix components such as laminin $\alpha 2$ and collagen VI. A deficiency of laminin $\alpha 2$ is associated with one form of congenital muscular dystrophy called the merosin-deficient CMD or MDC1A, while the deficiency of collagen VI leads to Bethlem myopathy (BM) and Ullrich congenital muscular dystrophy (UCMD). This indicates that aside from possessing myogenic potential, SMSCs also have a wide trophic function.

Thus far, there are relatively fewer studies on the applications of SMSCs in skeletal muscle engineering as compared to other MSC sources due to their limited myogenic potential. However, it has been discovered that the myogenic potential can be augmented through genetic modifications. Additionally, the SMSCs have also been found to be able to secrete important ECM molecules. All these necessitate more extensive studies and research in this field so that the SMSCs may be utilized to treat muscular dystrophies either directly by generating new muscle fibers or indirectly by secreting ECM molecules to minimize muscle pathological changes.

17.7 Conclusion

As a whole, SMSC research has returned encouraging results in the area of musculoskeletal regeneration. In the near future, it is hoped that much progress will be made to overcome the various challenges faced so that SMSCs can become a truly effective therapeutic cell source. A much more in-depth understanding of the science of manipulating SMSCs will prove useful in hastening clinical trials in humans. Thus far, a number of studies have revealed the possibility of engineering SMSCs to suit specific applications, and this helps in bridging the vastly different requirements of regenerating the various musculoskeletal tissues.

References

1. Pittenger MF, Mackay AM, Beck SC, Jaiswal RK, Douglas R, Mosca JD, Moorman MA, Simonetti DW, Craig S, Marshak DR (1999) Multilineage potential of adult human mesenchymal stem cells. Science 284(5411):143–147. doi:10.1126/science.284.5411.143
2. Steinhoff G (2011) Regenerative medicine [electronic resource]. Springer: Dordrecht
3. Yoo KH, Jang IK, Lee MW, Kim HE, Yang MS, Eom Y, Lee JE, Kim YJ, Yang SK, Jung HL et al (2009) Comparison of immunomodulatory properties of mesenchymal stem cells derived from adult human tissues. Cell Immunol 259:150–156
4. Zuk PA, Zhu M, Mizuno H, Huang J, Futrell JW, Katz AJ, Benhaim P, Lorenz HP, Hedrick MH (2001) Multilineage cells from human adipose tissue: implications for cell-based therapies. Tissue Eng 7:211–228
5. Zuk PA, Zhu M, Ashjian P, De Ugarte DA, Huang JI, Mizuno H, Alfonso ZC, Fraser JK, Benhaim P, Hedrick MH (2002) Human adipose tissue is a source of multipotent stem cells. Mol Biol Cell 13(12):4279–4295
6. De Bari C, Dell'Accio F, Tylzanowski P, Luyten FP (2001) Multipotent mesenchymal stem cells from adult human synovial membrane. Arthritis Rheum 44(8):1928–1942. doi:10.1002/1529-0131(200108)44:8<1928::aid-art331>3.0.co;2-p
7. De Bari C, Dell'Accio F, Luyten FP (2001) Human periosteum-derived cells maintain phenotypic stability and chondrogenic potential throughout expansion regardless of donor age. Arthritis Rheum 44(1):85–95
8. Asakura A, Komaki M, Rudnicki MA (2001) Muscle satellite cells are multipotential stem cells that exhibit myogenic, osteogenic, and adipogenic differentiation. Differentiation 68(4–5):245–253
9. Sakaguchi Y, Sekiya I, Yagishita K, Muneta T (2005) Comparison of human stem cells derived from various mesenchymal tissues: superiority of synovium as a cell source. Arthritis Rheum 52:2521–2529
10. Shirasawa S, Sekiya I, Sakaguchi Y, Yagishita K, Ichinose S, Muneta T (2006) In vitro chondrogenesis of human synovium-derived mesenchymal stem cells: optimal condition and comparison with bone marrow-derived cells. J Cell Biochem 97(1):84–97. doi:10.1002/jcb.20546
11. Archer CW, Dowthwaite GP, Francis-West P (2003) Development of synovial joints. Birth Defects Res C Embryo Today 69(2):144–155. doi:10.1002/bdrc.10015
12. Fan JB, Varshney RR, Ren L, Cai DZ, Wang DA (2009) Synovium-derived mesenchymal stem cells: a new cell source for musculoskeletal regeneration. Tissue Eng Part B Rev 15(1):75–86. doi:10.1089/ten.teb.2008.0586
13. Ichinose S, Muneta T, Koga H, Segawa Y, Tagami M, Tsuji K, Sekiya I (2009) Morphological differences during in vitro chondrogenesis of bone marrow-, synovium-MSCs, and chondrocytes. Lab Invest 90(2):210–221
14. Jones EA, English A, Henshaw K, Kinsey SE, Markham AF, Emery P, McGonagle D (2004) Enumeration and phenotypic characterization of synovial fluid multipotential mesenchymal progenitor cells in inflammatory and degenerative arthritis. Arthritis Rheum 50(3):817–827. doi:10.1002/art.20203
15. Lee SY, Nakagawa T, Reddi AH (2010) Mesenchymal progenitor cells derived from synovium and infrapatellar fat pad as a source for superficial zone cartilage tissue engineering: analysis of superficial zone protein/lubricin expression. Tissue Eng Part A 16(1):317–325

16. Djouad F, Bony C, Häupl T, Uzé G, Lahlou N, Louis-Plence P, Apparailly F, Canovas F, Rème T, Sany J, Jorgensen C, Noël D (2005) Transcriptional profiles discriminate bone marrow-derived and synovium-derived mesenchymal stem cells. Arthritis Res Ther 7(6):1–12. doi:10.1186/ar1827
17. O'Connell JX (2000) Pathology of the synovium. Am J Clin Pathol 114(5):773–784
18. FitzGerald O, Bresnihan B (1995) Synovial membrane cellularity and vascularity. Ann Rheum Dis 54(6):511–515. doi:10.1136/ard.54.6.511
19. Edwards JCW (1994) The nature and origins of synovium – experimental approaches to the study of synoviocyte differentiation. J Anat 184:493–501
20. Vigorita VJ, Ghelman B, Mintz D (2008) Orthopaedic pathology. Lippincott Williams and Wilkins, Philadelphia
21. Bentley G, Kreutner A, Ferguson AB (1975) Synovial regeneration and articular cartilage changes after synovectomy in normal and steroid treated rabbits. J Bone Joint Surg Br 57(4):454–462
22. Mitchell N, Blackwell P (1968) The electron microscopy of regenerating synovium after subtotal synovectomy in rabbits. J Bone Joint Surg Am 50(4):675–686
23. Campbell WG Jr, Callahan BC (1971) Regeneration of synovium of rabbit knees after total chemical synovectomy by ingrowth of connective tissue-forming elements from adjacent bone. A light and electron microscopic study. Lab Invest 24(5):404–422
24. Theoret CL, Barber SM, Moyana T, Townsend HGG, Archer JF (1996) Repair and function of synovium after arthroscopic synovectomy of the dorsal compartment of the equine antebrachiocarpal joint. Vet Surg 25(2):142–153
25. Mochizuki T, Muneta T, Sakaguchi Y, Nimura A, Yokoyama A, Koga H, Sekiya I (2006) Higher chondrogenic potential of fibrous synovium– and adipose synovium–derived cells compared with subcutaneous fat–derived cells: distinguishing properties of mesenchymal stem cells in humans. Arthritis Rheum 54(3):843–853. doi:10.1002/art.21651
26. Nagase T, Muneta T, Ju YJ, Hara K, Morito T, Koga H, Nimura A, Mochizuki T, Sekiya I (2008) Analysis of the chondrogenic potential of human synovial stem cells according to harvest site and culture parameters in knees with medial compartment osteoarthritis. Arthritis Rheum 58(5):1389–1398
27. Zimmermann T, Kunisch E, Pfeiffer R, Hirth A, Stahl HD, Sack U, Laube A, Liesaus E, Roth A, Palombo-Kinne E, Emmrich F, Kinne RW (2001) Isolation and characterization of rheumatoid arthritis synovial fibroblasts from primary culture – primary culture cells markedly differ from fourth-passage cells. Arthritis Res 3(1):72–76
28. Jones E, Churchman SM, English A, Buch MH, Horner EA, Burgoyne CH, Reece R, Kinsey S, Emery P, McGonagle D, Ponchel F (2010) Mesenchymal stem cells in rheumatoid synovium: enumeration and functional assessment in relation to synovial inflammation level. Ann Rheum Dis 69(2):450–457. doi:10.1136/ard.2008.106435
29. Fossett E, Khan WS, Longo UG, Smitham PJ (2012) Effect of age and gender on cell proliferation and cell surface characterization of synovial fat pad derived mesenchymal stem cells. J Orthop Res 30(7):1013–1018
30. Lee D-H, Joo S-D, Han S-B, Im J, Lee S-H, Sonn CH, Lee K-M (2011) Isolation and expansion of synovial CD34– CD44+ CD90+ mesenchymal stem cells: comparison of an enzymatic method and a direct explant technique. Connect Tissue Res 52(3):226–234. doi:10.3109/03008207.2010.516850
31. De Bari C, Dell'Accio F, Karystinou A, Guillot PV, Fisk NM, Jones EA, McGonagle D, Khan IM, Archer CW, Mitsiadis TA, Donaldson AN, Luyten FP, Pitzalis C (2008) A biomarker-based mathematical model to predict bone-forming potency of human synovial and periosteal mesenchymal stem cells. Arthritis Rheum 58(1):240–250. doi:10.1002/art.23143
32. Ho AD, Wagner W, Franke W (2008) Heterogeneity of mesenchymal stromal cell preparations. Cytotherapy 10(4):320–330
33. Karystinou A, Dell'Accio F, Kurth TB, Wackerhage H, Khan IM, Archer CW, Jones EA, Mitsiadis TA, de Bari C (2009) Distinct mesenchymal progenitor cell subsets in the adult human synovium. Rheumatology 48(9):1057–1064

34. Fickert S, Fiedler J, Brenner RE (2003) Identification, quantification and isolation of mesenchymal progenitor cell from osteoarthritic synovium by fluorescence automated cell sorting. Osteoarthritis Cartilage 11(11):790–800
35. Pei M, He F, Kish VL, Vunjak-Novakovic G (2008) Engineering of functional cartilage tissue using stem cells from synovial lining: a preliminary study. Clin Orthop Relat Res 466(8):1880–1889
36. Qi J, Chen A, You H, Li K, Zhang D, Guo F (2011) Proliferation and chondrogenic differentiation of CD105-positive enriched rat synovium-derived mesenchymal stem cells in three-dimensional porous scaffolds. Biomed Mater 6(1)
37. Teramura T, Fukuda K, Kurashimo S, Hosoi Y, Miki Y, Asada S, Hamanishi C (2008) Isolation and characterization of side population stem cells in articular synovial tissue. BMC Musculoskelet Disord 9(1):86
38. Yamane S, Reddi AH (2008) Induction of chondrogenesis and superficial zone protein accumulation in synovial side population cells by BMP-7 and TGF-b1. J Orthop Res 26(4):485–492
39. Shahdadfar A, Frønsdal K, Haug T, Reinholt FP, Brinchmann JE (2005) In vitro expansion of human mesenchymal stem cells: choice of serum is a determinant of cell proliferation, differentiation, gene expression, and transcriptome stability. Stem Cells 23(9):1357–1366. doi:10.1634/stemcells.2005-0094
40. Doerr HW, Cínatl J, Stürmer M, Rabenau HF (2003) Prions and orthopedic surgery. Infection 31(3):163–171
41. Spees JL, Gregory CA, Singh H, Tucker HA, Peister A, Lynch PJ, Hsu SH, Smith J, Prockop DJ (2004) Internalized antigens must be removed to prepare hypoimmunogenic mesenchymal stem cells for cell and gene therapy. Mol Ther 9(5):747–756
42. Kuznetsov SA, Mankani MH, Robey PG (2000) Effect of serum on human bone marrow stromal cells: ex vivo expansion and in vivo bone formation. Transplantation 70(12):1780–1787
43. Yamaguchi M, Hirayama F, Wakamoto S, Fujihara M, Murahashi H, Sato N, Ikebuchi K, Sawada K-I, Koike T, Kuwabara M, Azuma H, Ikeda H (2002) Bone marrow stromal cells prepared using AB serum and bFGF for hematopoietic stem cells expansion. Transfusion 42(7):921–927. doi:10.1046/j.1537-2995.2002.00149.x
44. Yamamoto N, Isobe M, Negishi A, Yoshimasu H, Shimokawa H, Ohya K, Amagasa T, Kasugai S (2003) Effects of autologous serum on osteoblastic differentiation in human bone marrow cells. J Med Dent Sci 50(1):63–69
45. Stute N, Holtz K, Bubenheim M, Lange C, Blake F, Zander AR (2004) Autologous serum for isolation and expansion of human mesenchymal stem cells for clinical use. Exp Hematol 32(12):1212–1225
46. Mizuno N, Shiba H, Ozeki Y, Mouri Y, Niitani M, Inui T, Hayashi H, Suzuki K, Tanaka S, Kawaguchi H, Kurihara H (2006) Human autologous serum obtained using a completely closed bag system as a substitute for foetal calf serum in human mesenchymal stem cell cultures. Cell Biol Int 30(6):521–524
47. Kobayashi T, Watanabe H, Yanagawa T, Tsutsumi S, Kayakabe M, Shinozaki T, Higuchi H, Takagishi K (2005) Motility and growth of human bone-marrow mesenchymal stem cells during ex vivo expansion in autologous serum. J Bone Joint Surg Br 87-B(10):1426–1433. doi:10.1302/0301-620x.87b10.16160
48. Nimura A, Muneta T, Koga H, Mochizuki T, Suzuki K, Makino H, Umezawa A, Sekiya I (2008) Increased proliferation of human synovial mesenchymal stem cells with autologous human serum: comparisons with bone marrow mesenchymal stem cells and with fetal bovine serum. Arthritis Rheum 58(2):501–510. doi:10.1002/art.23219
49. Tateishi K, Ando W, Higuchi C, Hart DA, Hashimoto J, Nakata K, Yoshikawa H, Nakamura N (2008) Comparison of human serum with fetal bovine serum for expansion and differentiation of human synovial MSC: potential feasibility for clinical applications. Cell Transplant 17(5):549–557

50. Djouad F, Noel D, Uze G, Haupl T, Plence P, Bony C, Apparailly F, Jorgensen C (2005) Phenotypic, genotypic and functional characterization of mesenchymal stem cells from synovial membrane compared with bone marrow. Arthritis Res Ther 7:S28–S29. doi:10.1186/ar1594
51. Jo CH, Ahn HJ, Kim HJ, Seong SC, Lee MC (2007) Surface characterization and chondrogenic differentiation of mesenchymal stromal cells derived from synovium. Cytotherapy 9(4):316–327
52. Galligan CL, Baig E, Bykerk V, Keystone EC, Fish EN (2007) Distinctive gene expression signatures in rheumatoid arthritis synovial tissue fibroblast cells: correlates with disease activity. Genes Immun 8(6):480–491
53. Hermida-Gómez T, Fuentes-Boquete I, Gimeno-Longas MJ, Muiños-López E, Díaz-Prado S, de Toro FJ, Blanco FJ (2011) Quantification of cells expressing mesenchymal stem cell markers in healthy and osteoarthritic synovial membranes. J Rheumatol 38(2):339–349. doi:10.3899/jrheum.100614
54. Marigo I, Dazzi F (2011) The immunomodulatory properties of mesenchymal stem cells. Semin Immunopathol 33(6):593–602. doi:10.1007/s00281-011-0267-7
55. Segawa Y, Muneta T, Makino H, Nimura A, Mochizuki T, Ju YJ, Ezura Y, Umezawa A, Sekiya I (2009) Mesenchymal stem cells derived from synovium, meniscus, anterior cruciate ligament, and articular chondrocytes share similar gene expression profiles. J Orthop Res 27:435–441
56. Djouad F, Jackson WM, Bobick BE, Janjanin S, Song Y, Huang GTJ, Tuan RS (2010) Activin A expression regulates multipotency of mesenchymal progenitor cells. Stem Cell Res Ther 1(2):11
57. Deans RJ, Moseley AB (2000) Mesenchymal stem cells: biology and potential clinical uses. Exp Hematol 28(8):875–884
58. Glennie S, Soeiro I, Dyson PJ, Lam EWF, Dazzi F (2005) Bone marrow mesenchymal stem cells induce division arrest anergy of activated T cells. Blood 105(7):2821–2827
59. Krampera M, Glennie S, Dyson J, Scott D, Laylor R, Simpson E, Dazzi F (2003) Bone marrow mesenchymal stem cells inhibit the response of naive and memory antigen-specific T cells to their cognate peptide. Blood 101(9):3722–3729
60. Deng W, Han Q, Liao L, You S, Deng H, Zhao RCH (2005) Effects of allogeneic bone marrow-derived mesenchymal stem cells on T and B lymphocytes from BXSB mice. DNA Cell Biol 24(7):458–463
61. Corcione A, Benvenuto F, Ferretti E, Giunti D, Cappiello V, Cazzanti F, Risso M, Gualandi F, Mancardi GL, Pistoia V, Uccelli A (2006) Human mesenchymal stem cells modulate B-cell functions. Blood 107(1):367–372
62. Sotiropoulou PA, Perez SA, Gritzapis AD, Baxevanis CN, Papamichail M (2006) Interactions between human mesenchymal stem cells and natural killer cells. Stem Cells 24(1):74–85
63. Spaggiari GM, Capobianco A, Becchetti S, Mingari MC, Moretta L (2006) Mesenchymal stem cell-natural killer cell interactions: evidence that activated NK cells are capable of killing MSCs, whereas MSCs can inhibit IL-2-induced NK-cell proliferation. Blood 107(4):1484–1490
64. Gebler A, Zabel O, Seliger B (2012) The immunomodulatory capacity of mesenchymal stem cells. Trends Mol Med 18(2):128–134
65. Chan JL, Tang KC, Patel AP, Bonilla LM, Pierobon N, Ponzio NM, Rameshwar P (2006) Antigen-presenting property of mesenchymal stem cells occurs during a narrow window at low levels of interferon-g. Blood 107(12):4817–4824
66. Convery FR, Akeson WH, Keown GH (1972) The repair of large osteochondral defects. An experimental study in horses. Clin Orthop Relat Res 82:253–262
67. Koga H, Muneta T, Ju YJ, Nagase T, Nimura A, Mochizuki T, Ichinose S, von der Mark K, Sekiya I (2007) Synovial stem cells are regionally specified according to local microenvironments after implantation for cartilage regeneration. Stem Cells 25:689–696
68. Fuller JA, Ghadially FN (1972) Ultrastructural observations on surgically produced partial-thickness defects in articular cartilage. Clin Orthop Relat Res 86:193–205

69. Richter W (2009) Mesenchymal stem cells and cartilage in situ regeneration. J Intern Med 266(4):390–405. doi:10.1111/j.1365-2796.2009.02153.x
70. Matsusue YaY T, Hama H (1993) Arthroscopic multiple osteochondral transplantation to the chondral defect in the knee associated with anterior cruciate ligament disruption. Arthroscopy 9:318–321
71. Steinwachs MR, Guggi T, Kreuz PC (2008) Marrow stimulation techniques. Injury 39(1 suppl):26–31
72. Gille J, Schuseil E, Wimmer J, Gellissen J, Schulz A, Behrens P (2010) Mid-term results of Autologous Matrix-Induced Chondrogenesis for treatment of focal cartilage defects in the knee. Knee Surg Sports Traumatol Arthrosc 18(11):1456–1464. doi:10.1007/s00167-010-1042-3
73. Khan WS, Johnson DS, Hardingham TE (2010) The potential of stem cells in the treatment of knee cartilage defects. Knee 17(6):369–374
74. Yarashi T, Rutherford J, Kapoor A, Anand S, Johnson D (2009) Knee outcome scores – do we get people back to normal? J Bone Joint Surg Br 91-B(SUPP III):427
75. Cancedda R, Dozin B, Giannoni P, Quarto R (2003) Tissue engineering and cell therapy of cartilage and bone. Matrix Biol 22(1):81–91
76. Raghunath J, Salacinski HJ, Sales KM, Butler PE, Seifalian AM (2005) Advancing cartilage tissue engineering: the application of stem cell technology. Curr Opin Biotechnol 16(5):503–509
77. Arufe MC, De la Fuente A, Fuentes-Boquete I, De Toro FJ, Blanco FJ (2009) Differentiation of synovial CD-105+ human mesenchymal stem cells into chondrocyte-like cells through spheroid formation. J Cell Biochem 108(1):145–155. doi:10.1002/jcb.22238
78. Bilgen B, Ren YX, Pei M, Aaron RK, Ciombor DM (2009) CD14-negative isolation enhances chondrogenesis in synovial fibroblasts. Tissue Eng Part A 15(11):3261–3270. doi:10.1089/ten.tea.2008.0273
79. Arufe MC, De la Fuente A, Fuentes I, de Toro FJ, Blanco FJ (2010) Chondrogenic potential of subpopulations of cells expressing mesenchymal stem cell markers derived from human synovial membranes. J Cell Biochem 111(4):834–845. doi:10.1002/jcb.22768
80. Koga H, Muneta T, Nagase T, Nimura A, Ju Y-J, Mochizuki T, Sekiya I (2008) Comparison of mesenchymal tissues-derived stem cells for in vivo chondrogenesis: suitable conditions for cell therapy of cartilage defects in rabbit. Cell Tissue Res 333(2):207–215. doi:10.1007/s00441-008-0633-5
81. Fox DB, Warnock JJ (2011) Cell-based meniscal tissue engineering: a case for synoviocytes. Clin Orthop Relat Res 469(10):2806–2816
82. Shi Y, Massagué J (2003) Mechanisms of TGF-2 signaling from cell membrane to the nucleus. Cell 113(6):685–700
83. Shintani N, Hunziker EB (2007) Chondrogenic differentiation of bovine synovium: bone morphogenetic proteins 2 and 7 and transforming growth factor b1 induce the formation of different types of cartilaginous tissue. Arthritis Rheum 56(6):1869–1879
84. Mackay AM, Beck SC, Murphy JM, Barry FP, Chichester CO, Pittenger MF (1998) Chondrogenic differentiation of cultured human mesenchymal stem cells from marrow. Tissue Eng 4(4):415–428
85. Barry F, Boynton RE, Liu B, Murphy JM (2001) Chondrogenic differentiation of mesenchymal stem cells from bone marrow: differentiation-dependent gene expression of matrix components. Exp Cell Res 268(2):189–200
86. Miyamoto C, Matsumoto T, Sakimura K, Shindo H (2007) Osteogenic protein-1 with transforming growth factor-b1: potent inducer of chondrogenesis of synovial mesenchymal stem cells in vitro. J Orthop Sci 12(6):555–561
87. Sakimura K, Matsumoto T, Miyamoto C, Osaki M, Shindo H (2006) Effects of insulin-like growth factor I on transforming growth factor beta(1) induced chondrogenesis of synovium-derived mesenchymal stem cells cultured in a polyglycolic acid scaffold. Cells Tissues Organs 183(2):55–61. doi:10.1159/000095509
88. Nishimura K, Solchaga LA, Caplan AI, Yoo JU, Goldberg VM, Johnstone B (1999) Chondroprogenitor cells of synovial tissue. Arthritis Rheum 42(12):2631–2637

89. Kurth T, Hedbom E, Shintani N, Sugimoto M, Chen FH, Haspl M, Martinovic S, Hunziker EB (2007) Chondrogenic potential of human synovial mesenchymal stem cells in alginate. Osteoarthritis Cartilage 15(10):1178–1189
90. Park Y, Sugimoto M, Watrin A, Chiquet M, Hunziker EB (2005) BMP-2 induces the expression of chondrocyte-specific genes in bovine synovium-derived progenitor cells cultured in three-dimensional alginate hydrogel. Osteoarthritis Cartilage 13(6):527–536
91. Pei M, He F, Vunjak-Novakovic G (2008) Synovium-derived stem cell-based chondrogenesis. Differentiation 76(10):1044–1056
92. Laviola L, Natalicchio A, Giorgino F (2007) The IGF-I signaling pathway. Curr Pharm Des 13(7):663–669
93. Longobardi L, O'Rear L, Aakula S, Johnstone B, Shimer K, Chytil A, Horton WA, Moses HL, Spagnoli A (2006) Effect of IGF-I in the chondrogenesis of bone marrow mesenchymal stem cells in the presence or absence of TGF-b signaling. J Bone Miner Res 21(4):626–636
94. Lee S, Kim JH, Jo CH, Seong SC, Lee JC, Lee MC (2009) Effect of serum and growth factors on chondrogenic differentiation of synovium-derived stromal cells. Tissue Eng Part A 15(11):3401–3415
95. Kim JH, Lee MC, Seong SC, Park KH, Lee S (2011) Enhanced proliferation and chondrogenic differentiation of human synovium-derived stem cells expanded with basic fibroblast growth factor. Tissue Eng Part A 17(7–8):991–1002
96. Buckley CT, Kelly DJ (2012) Expansion in the presence of FGF-2 enhances the functional development of cartilaginous tissues engineered using infrapatellar fat pad derived MSCs. J Mech Behav Biomed Mater 11:102–111
97. Rui YF, Du L, Wang Y, Lui PPY, Tang TT, Chan KM, Dai KR (2010) Bone morphogenetic protein 2 promotes transforming growth factor b3-induced chondrogenesis of human osteoarthritic synovium-derived stem cells. Chin Med J 123(21):3040–3048
98. Dickhut A, Pelttari K, Janicki P, Wagner W, Eckstein V, Egermann M, Richter W (2009) Calcification or dedifferentiation: requirement to lock mesenchymal stem cells in a desired differentiation stage. J Cell Physiol 219(1):219–226
99. Liu FL, Lin LH, Sytwu HK, Chang DM (2010) GDF-5 is suppressed by IL-1b and enhances TGF-b3-mediated chondrogenic differentiation in human rheumatoid fibroblast-like synoviocytes. Exp Mol Pathol 88(1):163–170
100. Yoshimura H, Muneta T, Nimura A, Yokoyama A, Koga H, Sekiya I (2007) Comparison of rat mesenchymal stem cells derived from bone marrow, synovium, periosteum, adipose tissue, and muscle. Cell Tissue Res 327(3):449–462. doi:10.1007/s00441-006-0308-z
101. Trippel SB, Ghivizzani SC, Nixon AJ (2004) Gene-based approaches for the repair of articular cartilage. Gene Ther 11(4):351–359
102. Steinert AF, Nöth U, Tuan RS (2008) Concepts in gene therapy for cartilage repair. Injury 39(1 suppl):97–113
103. Nixon AJ, Goodrich LR, Scimeca MS, Witte TH, Schnabel LV, Watts AE, Robbins PD (2007) Gene therapy in musculoskeletal repair. Ann N Y Acad Sci 1117(1):310–327. doi:10.1196/annals.1402.065
104. Marshall E (1999) Gene therapy death prompts review of adenovirus vector. Science 286(5448):2244–2245
105. Marshall E (2000) Gene therapy on trial. Science 288(5468):951–957
106. Hao J, Yao Y, Varshney RR, Wang L, Prakash C, Li H, Wang DA (2008) Gene transfer and living release of transforming growth factor-b3 for cartilage tissue engineering applications. Tissue Eng Part C Methods 14(4):273–280
107. Zhang F, Yao Y, Hao J, Zhou R, Liu C, Gong Y, Wang DA (2010) A dual-functioning adenoviral vector encoding both transforming growth factor-b3 and shRNA silencing type I collagen: construction and controlled release for chondrogenesis. J Control Release 142(1):70–77
108. Yao Y, Zhang F, Zhou R, Su K, Fan J, Wang DA (2010) Effects of combinational adenoviral vector-mediated TGFb3 transgene and shRNA silencing type I collagen on articular chondrogenesis of synovium-derived mesenchymal stem cells. Biotechnol Bioeng 106(5):818–828

109. Varshney RR, Zhou R, Hao J, Yeo SS, Chooi WH, Fan J, Wang D-A (2010) Chondrogenesis of synovium-derived mesenchymal stem cells in gene-transferred co-culture system. Biomaterials 31(26):6876–6891. doi:10.1016/j.biomaterials.2010.05.038
110. Pei M, Chen D, Li J, Wei L (2009) Histone deacetylase 4 promotes TGF-b1-induced synovium-derived stem cell chondrogenesis but inhibits chondrogenically differentiated stem cell hypertrophy. Differentiation 78(5):260–268
111. Zhang F, Yao Y, Zhou R, Su K, Citra F, Wang DA (2011) Optimal construction and delivery of dual-functioning lentiviral vectors for type i collagen-suppressed chondrogenesis in synovium-derived mesenchymal stem cells. Pharm Res 28(6):1338–1348
112. Chung C, Burdick JA (2008) Engineering cartilage tissue. Adv Drug Deliv Rev 60(2):243–262
113. De Bari C, Dell'Accio F, Luyten FP (2004) Failure of in vitro–differentiated mesenchymal stem cells from the synovial membrane to form ectopic stable cartilage in vivo. Arthritis Rheum 50(1):142–150. doi:10.1002/art.11450
114. Tibbitt MW, Anseth KS (2009) Hydrogels as extracellular matrix mimics for 3D cell culture. Biotechnol Bioeng 103(4):655–663
115. Sakao K, Takahashi KA, Arai Y, Inoue A, Tonomura H, Saito M, Yamamoto T, Kanamura N, Imanishi J, Mazda O, Kubo T (2008) Induction of chondrogenic phenotype in synovium-derived progenitor cells by intermittent hydrostatic pressure. Osteoarthritis Cartilage 16(7):805–814
116. Yokoyama A, Sekiya I, Miyazaki K, Ichinose S, Hata Y, Muneta T (2005) In vitro cartilage formation of composites of synovium-derived mesenchymal stem cells with collagen gel. Cell Tissue Res 322(2):289–298
117. Fan J, Gong Y, Ren L, Varshney RR, Cai D, Wang DA (2010) In vitro engineered cartilage using synovium-derived mesenchymal stem cells with injectable gellan hydrogels. Acta Biomater 6(3):1178–1185
118. Fan J, Ren L, Liang R, Gong Y, Cai D, Wang D-A (2010) Chondrogenesis of synovium-derived mesenchymal stem cells in photopolymerizing hydrogel scaffolds. J Biomater Sci Polym Ed 21(12):1653–1667. doi:10.1163/092050609x12531835454314
119. Mahmoudifar N, Doran PM (2010) Chondrogenic differentiation of human adipose-derived stem cells in polyglycolic acid mesh scaffolds under dynamic culture conditions. Biomaterials 31(14):3858–3867
120. Xin X, Hussain M, Mao JJ (2007) Continuing differentiation of human mesenchymal stem cells and induced chondrogenic and osteogenic lineages in electrospun PLGA nanofiber scaffold. Biomaterials 28(2):316–325
121. Yang DJ, Huang DQ, Xu FY (2007) Feasibility of constructing artificial cartilage with rabbit mesenchymal stem cells and polyglycolic acid scaffold. J Clin Rehabilitative Tissue Eng Res 11(14):2761–2764
122. Pei M, He F, Boyce BM, Kish VL (2009) Repair of full-thickness femoral condyle cartilage defects using allogeneic synovial cell-engineered tissue constructs. Osteoarthritis Cartilage 17(6):714–722
123. Gong Z, Xiong H, Long X, Wei L, Li J, Wu Y, Lin Z (2010) Use of synovium-derived stromal cells and chitosan/collagen type I scaffolds for cartilage tissue engineering. Biomed Mater 5(5):055005
124. Daniels AU, Andriano KP, Smutz WP, Chang MKO, Heller J (1994) Evaluation of absorbable poly(ortho esters) far use in surgical implants. J Appl Biomater 5(1):51–64. doi:10.1002/jab.770050108
125. van der Elst M, Klein C, de Blieck-Hogervorst JM, Patka P, Haarman H (1999) Bone tissue response to biodegradable polymers used for intramedullary fracture fixation: a long-term in vivo study in sheep femora. Biomaterials 20(2):121–128. doi:10.1016/s0142-9612(98)00117-3
126. Yang CL, Hillas PJ, Baez JA, Nokelainen M, Balan J, Tang J, Spiro R, Polarek JW (2004) The application of recombinant human collagen in tissue engineering. BioDrugs 18(2):103–119. doi:10.2165/00063030-200418020-00004

127. Martin MJ, Muotri A, Gage F, Varki A (2005) Human embryonic stem cells express an immunogenic nonhuman sialic acid. Nat Med 11(2):228–232
128. Ando W, Tateishi K, Hart DA, Katakai D, Tanaka Y, Nakata K, Hashimoto J, Fujie H, Shino K, Yoshikawa H, Nakamura N (2007) Cartilage repair using an in vitro generated scaffold-free tissue-engineered construct derived from porcine synovial mesenchymal stem cells. Biomaterials 28(36):5462–5470
129. Ando W, Tateishi K, Katakai D, Hart DA, Higuchi C, Nakata K, Hashimoto J, Fujie H, Shino K, Yoshikawa H, Nakamura N (2008) In vitro generation of a scaffold-free tissue-engineered construct (TEC) derived from human synovial mesenchymal stem cells: biological and mechanical properties and further chondrogenic potential. Tissue Eng Part A 14(12):2041–2049
130. Ju YJ, Muneta T, Yoshimura H, Koga H, Sekiya I (2008) Synovial mesenchymal stem cells accelerate early remodeling of tendon-bone healing. Cell Tissue Res 332(3):469–478
131. Hori J, Deie M, Kobayashi T, Yasunaga Y, Kawamata S, Ochi M (2011) Articular cartilage repair using an intra-articular magnet and synovium-derived cells. J Orthop Res 29(4):531–538
132. Potier E, Noailly J, Ito K (2010) Directing bone marrow-derived stromal cell function with mechanics. J Biomech 43(5):807–817
133. Araldi E, Schipani E (2010) Hypoxia, HIFs and bone development. Bone 47(2):190–196
134. Li J, Pei M (2011) Optimization of an in vitro three-dimensional microenvironment to reprogram synovium-derived stem cells for cartilage tissue engineering. Tissue Eng Part A 17(5–6):703–712
135. Kohn D, Moreno B (1995) Meniscus insertion anatomy as a basis for meniscus replacement: a morphological cadaveric study. Arthroscopy 11(1):96–103
136. McDevitt CA, Mukherjee S, Kambic H, Parker R (2002) Emerging concepts of the cell biology of the meniscus. Curr Opin Orthop 13(5):345–350
137. Melrose J, Smith S, Cake M, Read R, Whitelock J (2005) Comparative spatial and temporal localisation of perlecan, aggrecan and type I, II and IV collagen in the ovine meniscus: an ageing study. Histochem Cell Biol 124(3–4):225–235
138. Le Graverand MPH, Ou Y, Schield-Yee T, Barclay L, Hart D, Natsume T, Rattner JB (2001) The cells of the rabbit meniscus: their arrangement, interrelationship, morphological variations and cytoarchitecture. J Anat 198(5):525–535
139. Clark CR, Ogden JA (1983) Development of the menisci of the human knee joint. Morphological changes and their potential role in childhood meniscal injury. J Bone Joint Surg Am 65(4):538–547
140. Makris EA, Hadidi P, Athanasiou KA (2011) The knee meniscus: structure-function, pathophysiology, current repair techniques, and prospects for regeneration. Biomaterials 32(30):7411–7431
141. Arnoczky SP, Warren RF (1982) Microvasculature of the human meniscus. Am J Sports Med 10(2):90–95
142. Fox DB, Warnock JJ, Stoker AM, Luther JK, Cockrell M (2010) Effects of growth factors on equine synovial fibroblasts seeded on synthetic scaffolds for avascular meniscal tissue engineering. Res Vet Sci 88(2):326–332
143. Cox JS, Cordell LD (1977) The degenerative effects of medial meniscus tears in dogs' knees. Clin Orthop Relat Res 125:236–242
144. Higuchi H, Kimura M, Shirakura K, Terauchi M, Takagishi K (2000) Factors affecting long-term results after arthroscopic partial meniscectomy. Clin Orthop Relat Res 377:161–168
145. Hoben GM, Athanasiou KA (2006) Meniscal repair with fibrocartilage engineering. Sports Med Arthrosc Rev 14(3):129–137
146. Horie M, Sekiya I, Muneta T, Ichinose S, Matsumoto K, Saito H, Murakami T, Kobayashi E (2009) Intra-articular injected synovial stem cells differentiate into meniscal cells directly and promote meniscal regeneration without mobilization to distant organs in rat massive meniscal defect. Stem Cells 27(4):878–887

147. Mizuno K, Muneta T, Morito T, Ichinose S, Koga H, Nimura A, Mochizuki T, Sekiya I (2008) Exogenous synovial stem cells adhere to defect of meniscus and differentiate into cartilage cells. J Med Dent Sci 55(1):101–111
148. Tan Y, Zhang Y, Pei M (2010) Meniscus reconstruction through coculturing meniscus cells with synovium-derived stem cells on small intestine submucosa-a pilot study to engineer meniscus tissue constructs. Tissue Eng Part A 16(1):67–79
149. Warnock JJ, Fox DB, Stoker AM, Cook JL (2011) Evaluation of in vitro growth factor treatments on fibrochondrogenesis by synovial membrane cells from osteoarthritic and nonosteoarthritic joints of dogs. Am J Vet Res 72(4):500–511. doi:10.2460/ajvr.72.4.500
150. Wang Y, Rui Y, Du L, Tang T, Dai K (2009) In vitro study on multiple differentiation potential of swine synovium-derived MSCs. Zhongguo Xiu Fu Chong Jian Wai Ke Za Zhi 23(6):737–741
151. Shi X, Wang Y, Varshney RR, Ren L, Zhang F, Wang DA (2009) In-vitro osteogenesis of synovium stem cells induced by controlled release of bisphosphate additives from microspherical mesoporous silica composite. Biomaterials 30(23–24):3996–4005
152. Im GI, Qureshi SA, Kenney J, Rubash HE, Shanbhag AS (2004) Osteoblast proliferation and maturation by bisphosphonates. Biomaterials 25(18):4105–4115
153. Von Knoch F, Jaquiery C, Kowalsky M, Schaeren S, Alabre C, Martin I, Rubash HE, Shanbhag AS (2005) Effects of bisphosphonates on proliferation and osteoblast differentiation of human bone marrow stromal cells. Biomaterials 26(34):6941–6949
154. Porter RM, Huckle WR, Goldstein AS (2003) Effect of dexamethasone withdrawal on osteoblastic differentiation of bone marrow stromal cells. J Cell Biochem 90(1):13–22
155. Wang Y, Shi X, Ren L, Yao Y, Wang DA (2010) In vitro osteogenesis of synovium mesenchymal cells induced by controlled release of alendronate and dexamethasone from a sintered microspherical scaffold. J Biomater Sci Polym Ed 21(8–9):1227–1238
156. Matsusaki M, Kadowaki K, Tateishi K, Higuchi C, Ando W, Hart DA, Tanaka Y, Take Y, Akashi M, Yoshikawa H, Nakamura N (2009) Scaffold-free tissue-engineered construct-hydroxyapatite composites generated by an alternate soaking process: potential for repair of bone defects. Tissue Eng Part A 15(1):55–63
157. Longo UG, Lamberti A, Maffulli N, Denaro V (2011) Tissue engineered biological augmentation for tendon healing: a systematic review. Br Med Bull 98(1):31–59
158. Yates EW, Rupani A, Foley GT, Khan WS, Cartmell S, Anand SJ (2012) Ligament tissue engineering and its potential role in anterior cruciate ligament reconstruction. Stem Cells Int. 2012. doi:10.1155/2012/438125
159. Kew SJ, Gwynne JH, Enea D, Abu-Rub M, Pandit A, Zeugolis D, Brooks RA, Rushton N, Best SM, Cameron RE (2011) Regeneration and repair of tendon and ligament tissue using collagen fibre biomaterials. Acta Biomater 7(9):3237–3247
160. Jones EA, English A, Henshaw K, Kinsey SE, Markham AF, Emery P, McGonagle D (2004) Enumeration and phenotypic characterization of synovial fluid multipotential mesenchymal progenitor cells in inflammatory and degenerative arthritis. Arthritis Rheum 50(3):817–827
161. Morito T, Muneta T, Hara K, Ju YJ, Mochizuki T, Makino H, Umezawa A, Sekiya I (2008) Synovial fluid-derived mesenchymal stem cells increase after intra-articular ligament injury in humans. Rheumatology 47(8):1137–1143. doi:10.1093/rheumatology/ken114
162. Ge Z, Goh JCH, Lee EH (2005) Selection of cell source for ligament tissue engineering. Cell Transplant 14(8):573–583
163. Zhao MH, Lu QC, Li HQ, Yue YL (2007) Research and progress in the application of ligament tissue engineering in the reconstruction of anterior cruciate ligament. J Clin Rehabil Tissue Eng Res 11(2):333–335
164. Kartus J, Movin T, Karlsson J (2001) Donor-site morbidity and anterior knee problems after anterior cruciate ligament reconstruction using autografts. Arthroscopy 17(9):971–980
165. Goldblatt JP, Fitzsimmons SE, Balk E, Richmond JC (2005) Reconstruction of the anterior cruciate ligament: meta-analysis of patellar tendon versus hamstring tendon autograft. Arthroscopy 21(7):791–803

166. Ozturk AM, Yam A, Chin SI, Heong TS, Helvacioglu F, Tan A (2008) Synovial cell culture and tissue engineering of a tendon synovial cell biomembrane. J Biomed Mater Res A 84A(4):1120–1126. doi:10.1002/jbm.a.31738
167. Komurcu M, Akkus O, Basbozkurt M, Gur E, Akkas N (1997) Reduction of restrictive adhesions by local aprotinin application and primary sheath repair in surgically traumatized flexor tendons of the rabbit. J Hand Surg 22(5):826–832
168. Tang JB (2006) Tendon injuries across the world: treatment. Injury 37(11):1036–1042
169. Wilkinson LS, Pitsillides AA, Worrall JG, Edwards JCW (1992) Light microscopic characterization of the fibroblast-like synovial intimal cell (synoviocyte). Arthritis Rheum 35(10):1179–1184. doi:10.1002/art.1780351010
170. Liu Y, Skardal A, Shu XZ, Prestwich GD (2008) Prevention of peritendinous adhesions using a hyaluronan-derived hydrogel film following partial-thickness flexor tendon injury. J Orthop Res 26(4):562–569
171. Işik S, Öztürk S, Gürses S, Yetmez M, Güler MM, Selmanpakoglu N, Günhan O (1999) Prevention of restrictive adhesions in primary tendon repair by HA-membrane: experimental research in chickens. Br J Plast Surg 52(5):373–379
172. Kitamura M, Ito K, Nishiya T, Fukushima S, Ishii S (2002) Preventing adhesions using high molecular weight hyaluronan in chicken flexor tendon injuries. Hokkaido J Orthop Traumatol 44(1):29–33
173. Chen G, Zhang S, Zhang Z (2011) Over-expression of has2 in synovium-derived mesenchymal stem cells may prevent adhesions following surgery of the digital flexor tendons. Med Hypotheses 76(3):314–316
174. Aziz A, Sebastian S, Dilworth FJ (2012) The origin and fate of muscle satellite cells. Stem Cell Rev Rep 8(2):609–622. doi:10.1007/s12015-012-9352-0
175. Ehrhardt J, Morgan J (2005) Regenerative capacity of skeletal muscle. Curr Opin Neurol 18(5):548–553
176. Wagers AJ, Conboy IM (2005) Cellular and molecular signatures of muscle regeneration: current concepts and controversies in adult myogenesis. Cell 122(5):659–667
177. Meng J, Adkin CF, Arechavala-Gomeza V, Boldrin L, Muntoni F, Morgan JE (2010) The contribution of human synovial stem cells to skeletal muscle regeneration. Neuromuscul Disord 20(1):6–15
178. Cossu G, Sampaolesi M (2007) New therapies for Duchenne muscular dystrophy: challenges, prospects and clinical trials. Trends Mol Med 13(12):520–526
179. Turner N, Badylak S (2012) Regeneration of skeletal muscle. Cell Tissue Res 347(3):759–774. doi:10.1007/s00441-011-1185-7
180. Carlson ME, Conboy IM (2007) Loss of stem cell regenerative capacity within aged niches. Aging Cell 6(3):371–382
181. Sampaolesi M, Torrente Y, Innocenzi A, Tonlorenzi R, D'Antona G, Pellegrino MA, Barresi R, Bresolin N, De Angelis MGC, Campbell KP, Bottinelli R, Cossu G (2003) Cell therapy of a-sarcoglycan null dystrophic mice through intra-arterial delivery of mesoangioblasts. Science 301(5632):487–492
182. De Bari C, Dell'Accio F, Vandenabeele F, Vermeesch JR, Raymackers JM, Luyten FP (2003) Skeletal muscle repair by adult human mesenchymal stem cells from synovial membrane. J Cell Biol 160(6):909–918
183. Da Silva Meirelles L, Nardi NB (2009) Methodology, biology and clinical applications of mesenchymal stem cells. Front Biosci 14(11):4281–4298

Chapter 18
Stem Cell Therapy Aided Liver Regeneration

Ningning Yang and Manrong Jiang

Abstract Liver transplantation is the only lifesaving therapeutic strategy currently available for patients with end-stage liver disease. Due to the limited numbers of liver donors, an alternative therapy is required for liver disease. Stem cell therapy has high potential to provide a promising treatment, and more than 400 clinical trials are underway using this new strategy for liver disease. Stem cells are converted to hepatocytes through two possible mechanisms: transdifferentiation and cell fusion. Endogenous hepatic stem cells or exogenous stem cells are delivered to the bodies using systemic infusion and liver-related or extrahepatic injection. Current clinical trials and some challenges of stem cell therapy in liver disease are also discussed.

Keywords Liver disease • Endogenous hepatic stem cells • Extrahepatic stem cells Bone marrow stem cells • Stem cell therapy • Liver regeneration

18.1 Introduction

The liver which is the largest internal organ in the human body functions as a "processing plant," aiding in food digestion, detoxifying harmful substances, clearing waste products from the bloodstream, and storing vitamins, minerals, and nutrients, besides synthesizing numerous blood factors [1]. Due to its multiple roles in the body, the liver is prone to many diseases that cause liver dysfunction. In normal physiological situation, the liver does not require any external cell source to repair

N. Yang, PhD (✉)
Department of Pharmaceutical Sciences, Manchester University College of Pharmacy, 10627 Diebold Road, Fort Wayne, IN 46845, USA
e-mail: nyang@manchester.edu

M. Jiang
Department of Anatomy and Cell Biology, Indiana University School of Medicine-Fort Wayne, 2101 East Coliseum Boulevard, Fort Wayne, IN 46805, USA

M.K. Danquah, R.I. Mahato (eds.), *Emerging Trends in Cell and Gene Therapy*,
DOI 10.1007/978-1-62703-417-3_18,

the injury because of self-regeneration [2]. Nevertheless, this self-regenerative ability cannot compensate for massive or chronic liver injury. Termination of the sources of injury is a way to prevent further liver injury [3]. For example, alcohol consumption should be forbidden in the patients with alcohol-related hepatic fibrosis. But for liver cirrhosis or other liver damage at the end stage, the only way to treat may be liver transplantation. Liver transplantation is not always applicable because of limited availability of compatible donors [4].

Stem cells are cells with capability of self-renewing and potential of multilineage differentiation. Recently, stem cell therapy has been considered as a possible alternative to whole liver transplantation [5]. Some stem cells used in current research are adult cells, while some of them are fetal or embryonic cells. Adult stem cells can be found in bone marrow, blood, skin, adipose tissue, liver, spleen, brain, and at many other locations. Fetal stem cells can be found in placental and umbilical cord blood. Embryonic cells are usually generated from the inner cell mass of embryos. All of these cell types may have great potential application in therapeutics to treat liver diseases.

It has already been reported by many investigators that stem cells may differentiate into functional mature hepatocytes, which can produce albumin and metabolize urea, in vitro and in vivo [6–9]. These results are extremely promising and, therefore, have encouraged further research to bring stem cell therapy to the clinic. In fact, over 400 clinical trials are underway in stem cell therapy for hepatic diseases. Part of them focus on facilitation of liver regeneration and repair. Although stem cells therapy may require another 5–10 years to go to real clinical application, it is the time to investigate the current progress on stem cells therapy-aided hepatic regeneration.

In this chapter, the mechanisms, sources, and delivery methods of hepatic stem cells in liver disease treatment are discussed. Current clinical trials and challenges of stem cell therapy in liver disease are also addressed.

18.2 Liver Diseases, Regeneration, and Repair

The liver is a vital organ in the human body and has a wide range of functions, including food digestion, nutrients absorption, protein synthesis, and toxin removal. Because of its important role, liver is frequently invaded by many kinds of pathogens that result in liver problems or diseases. Currently, liver transplantation is the only real treatment for patients with end-stage liver disease. Due to the shortage of liver donors and other technical problems, alternative treatment strategies are needed for patients. Stem cell therapy in this arena appears to be the most provocative approach.

18.2.1 Liver Diseases

There are nine clinically distinct forms of hepatotoxicity. They are (1) necrosis, (2) hepatitis, (3) cholestasis, (4) fibrosis, (5) fatty liver, (6) cirrhosis, (7) granulomatous hepatitis, (8) autoimmune hepatitis, and (9) venoocclusive diseases. The most common type of liver disease (or hepatic diseases) is hepatitis or inflammation of the liver, which is caused by various viruses (hepatitis A–E types). Inflammation and temporary scarring (fibrosis) are the early-stage liver diseases [10]. Liver disease usually develops from inflammation to fibrosis, and then to the next stage that is permanent scarring (cirrhosis), and finally liver failure or liver cancer. Liver cancers include primary liver cancer which arises in the liver itself and secondary liver cancer which derives from other parts of the body and metastasizes to the liver. Liver failure is also called end-stage liver disease. The signs and symptoms of liver disease may vary greatly by individuals. About half of patients show no symptoms at all until at the very late stage of the disease. The most common signs or symptoms include fatigue, itching, jaundice (yellowing of the skin and eyes), abdominal pain, dark urine, pale stools, mental confusion, and fluid retention. For diagnosis of liver diseases, blood tests help to determine the reason and severity of inflammation. Ultrasound and other types of liver scans are usually used to obtain more details. A liver biopsy is sometimes recommended to get the pathophysiological information [11].

18.2.2 Mechanism of Liver Regeneration

The liver is one of the organs in the body capable of repair, renovation, and regeneration after damage and regulating its own size and growth. Only one fourth of a liver can regrow into a whole organ. Liver regeneration occurs after surgery or loss of mass induced by toxins and viruses. However, liver regeneration is actually compensatory growth because the replaced lobes do not regrow. Liver regeneration is triggered by various biochemical signals and metabolic demands generated from other tissues; however, internal signals generated inside the liver determine the extent of the response. The process is achieved by coordination of hepatocytes, various non-parenchymal cells, and the extracellular matrix, involving in numerous hormones, cytokines, growth factors, and metabolic factors to stimulate quiescent hepatocytes to reenter cell cycle and proliferate [10].

During the process of liver regeneration, hepatocytes first respond to the regenerative signals and reenter the cell cycle. Mitogenic growth factors such as EGF and HGF induce hepatocytes in G0 phase to enter G1 phase of the cell cycle by activating cyclin D, NF-κB, and STAT3. Stellate cells and Kupffer cells reenter cell cycle shortly after the proliferation of hepatocytes. Later, endothelial cells are undergoing proliferation to reestablish sinusoidal architecture [10, 11]. Hepatocytes also transduce and receive mitogenic factors from stellate cells, Kupffer cells, and endothelial cells (Fig. 18.1).

18.2.3 Current Treatment Strategies and Their Limitations

Due to the capacity of liver regeneration, current treatments for patients with early stages of liver diseases include removing the injury-causing stimulus such as

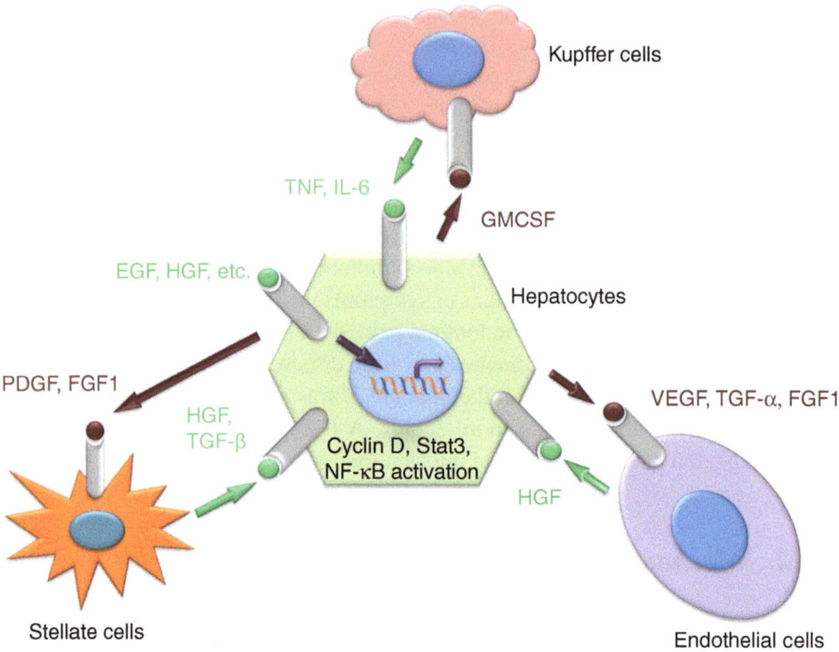

Fig. 18.1 Molecular mechanism of liver regeneration. Binding of the growth factors such as EGF and HGF to their receptors in quiescent hepatocytes leads to the activation of cyclin D, Stat3, NF-κB, etc. The cells reenter cell cycle and divide. After that, stellate cells, Kupffer cells, and endothelial cells begin to proliferate. The mitogenic factors are further interchanged during these cell types as indicated. *EGF* epidermal growth factor, *HGF*, hepatocyte growth factor, *VEGF* vascular endothelial growth factor, *FGF1* fibroblast growth factor 1, *GMCSF*, granulocyte-macrophage colony-stimulating factor, *TNF* tumor necrosis factor, *TGF* transforming growth factor, *IL-6* interleukin-6, *NF-κB* nuclear factor-κB, *STAT* signal transducer and activator of transcription

avoiding alcohol and a specially formulated liver disease diet, antiviral therapy, and surgical resection. However, liver transplantation is the only option currently available for patients with end-stage liver disease because of irreversible loss of liver function.

The first human liver transplant was performed by Dr. Thomas Starzl on March 1, 1963, in a child, and the child unfortunately died intraoperatively [12]. Due to the poor results of transplantation in the 1960s, physicians realized that liver transplantation should be further studied, and this technique remained experimental during the 1970s [13–16]. The introduction of the immunosuppressive medicine, cyclosporine, into clinical practice dramatically increased patient outcomes and made liver transplantation a standard clinical protocol in the 1980s [17, 18]. By now, liver transplant is a successful treatment for patients with acute or chronic hepatic failure. About 80–85 % patients survive after 1 year, and the outcome has been increasingly promising [19–21].

Despite of the success of liver transplantation, there are several obstacles to be considered, including a serious shortage of liver donors, risks involved during the surgery, possibility of graft rejection, recurrence of liver disease, and high cost. The surgical risk may include bleeding, infection, blood clots in the liver, and damage to the bile ducts. The average cost during the first year of the transplant is more than $300,000. Acute rejection which occurs within days or weeks after the surgery is the most common mode of graft rejection, and current immunosuppressive agents have shown to be effective minimizing rejection rates. Occasional chronic rejection occurs after 1 year and may require the lifelong immunosuppressive treatment. Recurrent hepatitis C virus (HCV) infection after transplantation is universal and becomes a large diagnostic and clinical burden because more than half of all adult liver transplantations are carried out for the patients with chronic hepatitis C [22–24]. As indicated earlier, the biggest limitation for the transplant is the lack of usable donor organs. According to the data from the Organ Procurement and Transplantation Network (OPTN), in the United States, there were about 16,000 patients on the waiting list in 2008, but only 6,138 patients received liver transplantation at the same year [20]. About one tenth of the patients die annually while on the waiting list. Therefore, an alternative treatment is urgently needed to compensate for or replace liver transplantation. Stem cell therapy came into the sight of the researchers due to its profound ability to modulate liver regeneration. This emerging technique appears to be the most promising and effective way of treating liver diseases.

18.3 Mechanisms of Stem Cells in Liver Regeneration

The mechanisms underlying the process of conversion of stem cells to hepatocytes remain unclear. As shown in Fig. 18.2, there are two possible mechanisms with current experimental evidences: transdifferentiation of stem cells and fusion of donor stem cells with resident hepatocytes [25–27]. The evidences indicating that bone marrow stem cells can differentiate into hepatocytes in vitro and in vivo suggest that transdifferentiation is a possible mechanism to explain the transformation of stem cell into hepatocytes [28, 29]. However, transdifferentiation cannot explain the study that FAH-deficient mice were rescued by transplantation of BM stem cells [30]. Later research demonstrated that cell fusion is another mechanism to produce bone marrow-derived hepatocytes [31, 32]. Some key representative studies investigating mechanisms of stem cells in liver regeneration are summarized in Table 18.1. All these approaches appear to be extremely provocative at this time.

18.3.1 Transdifferentiation

Numerous laboratories have investigated the liver-regenerating capacity of bone marrow-derived stem cells or adult hepatic stem/progenitor cells [20, 28, 33, 34].

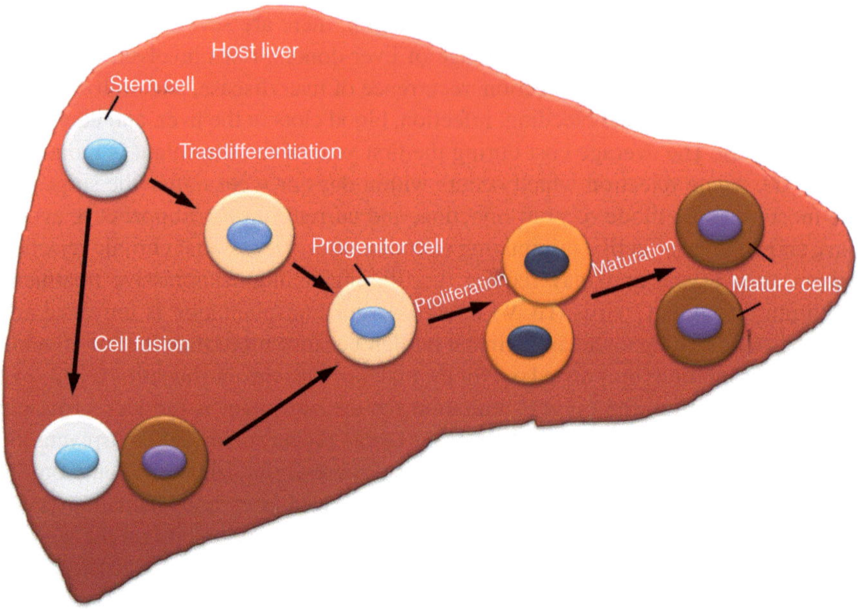

Fig. 18.2 Conversion of stem cells to hepatocytes. Endogenous hepatic stem cells and exogenous stem cells by circulating or injection give rise to progenitor cells via two possible mechanisms: transdifferentiation and cell fusion. After the proliferation of progenitors, cells become mature hepatic cells to repair the injured host liver

These studies found that both adult human and animal liver cells can be differentiated from endogenous hepatic or extrahepatic stem cells. The differentiated liver cells expressed specific hepatic biomarkers, including hepatocyte nuclear factor-3beta (HNF-3beta), transthyretin, and cytokeratin 8, 18, and 19 (CK8, 18, 19), and showed functionally hepatic characteristics such as urea and albumin secretion and glycogen storage [29, 35, 36].

The assumption that the hepatocytes are derived from stem cells via transdifferentiation was supported by the evidence that there was a range of percentage of Y-chromosome-positive liver cells (4–43 %) in female patients who received bone marrow transplantations from male donors [34, 36]. More direct evidence of transdifferentiation came from the experiments of transplanting human bone marrow cells into rat livers [37]. Sato et al. separated human bone marrow cells into mesenchymal stem cells (MSCs), CD34-positive (CD34+) cells, and non-MSCs/CD34- cells and checked the function of these cells by transplanting into variously injured rat livers. Hepatocyte-like cells with human-specific markers were only observed in MSC-transplanted rat livers. Transdifferentiation rather than cell fusion was involved because the analysis of fluorescence in situ hybridization (FISH) indicated that both human and rat chromosomes were independently present in liver cells [37]. Another study also examined the response and functional changes of a highly enriched

Table 18.1 Summary of representative studies investigating mechanisms of stem cells in liver regeneration

Mechanisms	Cell sources	Experimental models	Evidences	References
Transdifferentiation	Rat bone marrow cells	Rats	Y-chromosome percentage; expression of hepatic biomarkers	[28]
		In vitro cell culture	Expression of CK8/18	[29]
	Human bone marrow cells	Human patients	Y-chromosome percentage	[33]
	Human, mouse, rat bone marrow cells	In vitro cell culture	Expression of HNF-3beta, CK18	[36]
	Human mesenchymal stem cells	Rats	Expression of CK18/19; albumin secretion	[37]
	Mouse hematopoietic stem cells	Mice	Chromosome analysis; expression of CK18, albumin	[38]
Cell fusion	Mouse bone marrow cells	Mice	Rescue of FAH deficiency	[30]
			Karyotype; X- , Y-chromosome numbers	[31]
			FAH expression	[32]
	Mouse hematopoietic myelomonocytic cells	Mice	Cre/lox-based strategy	[41, 42]

population of hematopoietic stem cells (HSCs) in vitro and in vivo when exposed to damaged liver tissues. The authors found that HSCs were converted into functional liver cells without cell fusion [38].

18.3.2 Cell Fusion

The mechanism of transdifferentiation could not explain the liver regeneration in $FAH^{-/-}$ mouse bone marrow transplantation models. Lagasse et al. injected adult bone marrow cells into the fumarylacetoacetate hydrolase (FAH) knockout mouse, an animal model of hereditary type I tyrosinemia. The transplantation of BM cells rescued this potentially fatal enzyme deficiency and restored the biochemical function of mouse liver. Therefore, the authors questioned about previous assumption of transdifferentiation in the process of stem cell conversion [30]. However, all these approaches remain experimentally to date.

To find the underlying mechanism for liver regeneration in $FAH^{-/-}$ mouse model, two groups showed that the cells in healthy liver tissue after bone marrow transplantation of $FAH^{-/-}$ mice contained both donor and host genetic markers, suggesting that external stem cells could fuse with host liver cells and repair injured liver tissue in FAH-deficient mice [32, 39]. Usually, cell fusion leads to a hybrid cell containing both donor and host chromosomes and reprogramming of gene expression. Wang et al. showed that hybrid cells could also divide into daughter cells of a normal chromosome set, probably through a reduction division [31]. Another study further confirmed and extended the mechanism of cell fusion. The authors used a Cre/lox recombination system to detect cell fusion events. Cre-expressing bone marrow cells were injected into mouse tissues which contained lacZ reporter gene downstream of a loxP-flanked (floxed) stop cassette. By examining the expression of reporter gene, it was shown that bone marrow-derived cells (BMDCs) fused with Purkinje cells in the brain, cardiomyocytes in the heart, and hepatocytes in liver [40]. Later, two independent studies demonstrated that hematopoietic myelomonocytic cells in bone marrow are the major source for hepatocyte fusion in $FAH^{-/-}$ mouse models [41, 42].

18.4 Sources of Liver Stem Cells

Besides stem cells, mature hepatocytes are used as another important cell source for transplantation, but the number needed can be quite large because of their poor engraftment [43]. Although stem cells have less specialized function than mature hepatocytes, they are highly proliferative with possible long survival term [26, 44]. According to the origins, there are endogenous hepatic stem cells and extrahepatic stem cells. The origins of these stem cells are listed in Fig. 18.3. All the stem cells have various fates after transplanted to the livers. Endogenous hepatic stem cells can finally develop to mature hepatocytes, while extrahepatic stem cells usually can only differentiate to hepatocyte-like cells and play similar function as hepatocytes [45, 46].

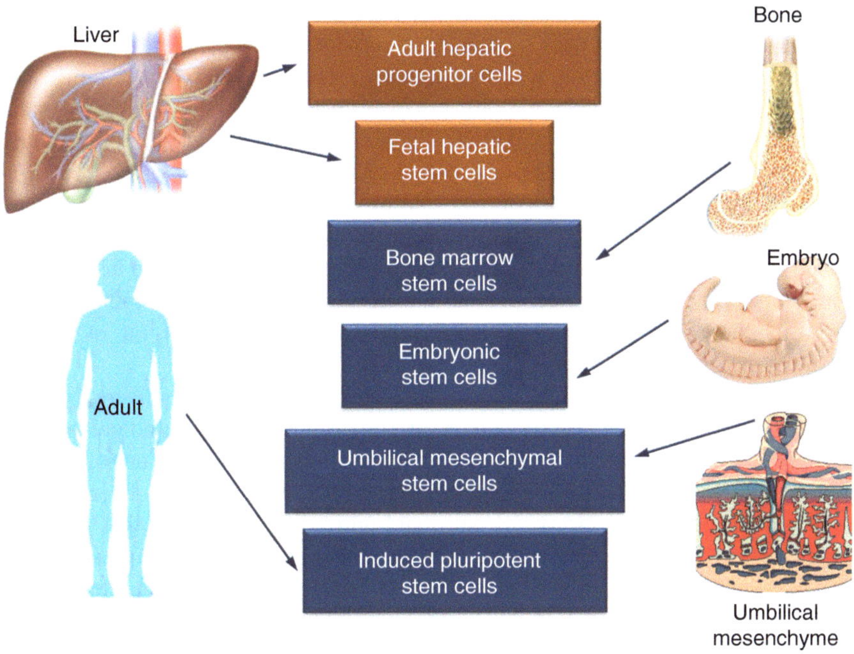

Fig. 18.3 Various origins for the stem cells applicable in liver disease cell therapy. Currently, liver stem cells may come from endogenous hepatic stem cells, including adult liver progenitor cells and fetal hepatic stem cells as indicated in *brown* boxes. In more cases, the cell sources are extrahepatic, such as bone marrow stem cells, embryonic stem cells, induced pluripotent stem cells, and umbilical mesenchymal stem cells as shown in *blue* boxes

18.4.1 Endogenous Hepatic Stem Cells

18.4.1.1 Adult Hepatic Progenitor Cells (Oval Cells)

Massive or chronic liver injuries do harm to the regenerative ability of hepatocytes, even though they can reenter the cell cycle quickly and efficiently after limited injuries have occurred [47]. During severe hepatic injuries, hepatic progenitors are activated. Hepatic progenitors display a bipotent role to differentiate to both hepatocytes and biliary epithelia [48]. However, the identity and in vivo function of this cell population are still controversial. It has been reported that hepatic progenitors may be a potential source for liver carcinoma [49]. In this discovery, a subtype of hepatocellular carcinoma differentially has distinct molecular features such as similar gene expression patterns as fetal hepatoblasts and direct downstream targets of AP-1. These similarities on gene expression patterns suggest that this subtype of hepatocellular carcinoma may arise from adult hepatic stem cells, which have the ability to differentiate to mature hepatocytes or cholangiocytes.

These putative adult liver progenitor cells are usually referred to as the "oval cells." The oval cell responses include four components: activation, proliferation, migration, and differentiation. It is known that various cell types, such as epithelial, hematopoietic, and mesenchymal, are involved in the activation of adult hepatic stem cells [50]. However, it is still unclear which cells play the most important role in transmitting and receiving crucial molecular signals.

The biological characteristics of oval cells provide a clue that isolation and transplantation of these cells may represent a novel source of clinical application to human liver regeneration and repair. It was reported that 90 % of the hepatocytes were replaced by donor oval cells in rat non-fibrotic models of liver injury [51]. In 2009, the data in Wistar rats showed intrahepatic injection of rat oval cells significantly increased the cell survival rate [52].

However, it is extremely difficult to isolate oval cells from human livers, because hepatic progenitors continuously change morphology, phenotype, and cell surface marker expression. Furthermore, most of unveiled surface biomarkers are not strictly specific, which makes the isolation even restricted [53, 54]. Thy-1 has been thought to be a good biomarker. However, more and more data has shown that Thy-1 is expressed by the neighboring stellate cells or progenitors of mesenchymal lineage but not the adult hepatic stem cells themselves [55]. Another important biomarker is the differential expression of AFP, which was thought to be a potential mechanism for distinguishing. But this biomarker is still under controversy.

Besides the isolation of hepatic progenitor cells, there are still some other issues that need to be addressed to achieve "from bench to bedside." The biggest obstacle is the maintenance of adult hepatic stem cells in culture, because it is required to expand small population of cells ex vivo prior to translation. Knowledge in this arena of research is, however, incomplete.

18.4.1.2 Fetal Hepatic Stem Cells

During embryogenesis, hepatic-specified cells begin to proliferate massively once the liver bud is growing [56–61]. These hepatic-specified cells are now referred to as hepatoblasts. Fetal hepatoblasts are able to differentiate into hepatocytes. Compared to adult hepatic stem cells, fetal hepatoblasts have their advantages [62, 63]. First, they have very high proliferation rate. Second, there is less apoptosis after transplantation. Third, fetal hepatoblasts show reduced immunogenic problems. However, the rare availability, sometimes also due to ethical issues, limits the application of this cell type in liver regeneration cell therapy.

For research purpose, some cell lines have been established from mouse embryos [64]. HBC-3, derived from mouse embryonic foregut at ED9.5 [65], grows very well even at low cell density. Met murine hepatocyte (MMH) cell lines have been established from fetal liver at ED14.5 in transgenic mice, which can express human Met receptor [66].

18.4.2 Extrahepatic Stem Cells

18.4.2.1 Bone Marrow Stem Cells

It has been more than 10 years since bone marrow stem cells (BMSCs) were found to have a role in liver repair [28]. It was also reported that some hepatocytes also originate from bone marrow [34, 63]. The mice with fumarylacetoacetate hydrolase enzyme deficiency could survive after transplantation of BM cells derived from wild-type donors. However, the data also showed that it is not easy to achieve stable engraftments of BM-derived hepatocytes [67]. In this study, female HBsAg-tg mice received bone marrow (BM) transplantation from male HBsAg-negative mice. Half of these mice were administered a chemical named retrorsine to block self-proliferation of hepatocytes. Livers were collected 3 and 6 months post-BM transplantation to check the amount of BM-derived hepatocytes. In mice without retrorsine, almost all the new hepatocytes came from hepatocyte replication, but none was derived from BM cells. In mice with retrorsine, 4.8 % of hepatocytes were Y chromosome positive at 3 months. However, this rate dropped to 1.6 % at 6 months. There are no clear reasons given by the investigators. The most possible reason may be less selection pressure during elapsing time

Although the rate of repopulation is not promising in some cases, it is believed that BM cell transplantation can help the regeneration of livers. There are many types of cells in BM. Then, which BM stem cells can repopulate the liver? It remains controversial. In early studies, it seems hematopoietic stem cells (HSCs) were the group involved [30, 68, 69]. In some of these studies [68, 69], coadministration of granulocyte colony-stimulating factor (G-CSF) was reported to be able to enhance HSC engraftments to the liver. In vitro experiments also showed that the medium from injured hepatocytes could induce HSCs in to hepatocytes [38]. However, some studies showed the contribution of mesenchymal stem cells (MSCs) to damaged livers both in vivo and in vitro [37, 70]. In this report, human bone marrow cells were fractionated into MSCs, CD34+ cells, and non-MSCs/CD34- cells. Then, various fractions were directly xenografted into allylalcohol (AA)-treated rat liver. Hepatocyte-like cells were observed only in rat livers with human MSC fractions. This result suggested that MSCs might be another very potent candidate for hepatic differentiation and regeneration therapy.

Many studies showed HSCs potential transdifferentiation ability to hepatocytes, but the reported percentage of hepatocytes derived from HSCs was quite low [71–75]. Although MSCs cannot provide high engraftment ratio either [76, 77], MSCs might be a better candidate for liver regeneration purposes. First, MSCs can be obtained not only from bone marrow but also from many other sources, such as adipose tissue, salivary glands, and pancreatic tissue [78, 79]. Besides the advantage on availability, MSCs can also relieve oxidative stress of damaged liver and facilitate the proliferation of hepatocytes. However, mechanisms underlying such events remain unknown.

18.4.2.2 Embryonic Stem Cells

Embryonic stem cells (ESCs) can be isolated from the inner cell mass of animal embryos. ESCs have the advantages of the most potent differentiation potential and proliferation ability [80]. Although multiple protocols have been studied to transdifferentiate ESCs into functional but immature hepatocytes in vitro [81, 82], the regeneration of injured livers is in very low level with poor hepatocyte function after administered with these ESC-derived precursors [81, 83–85]. In comparison with adult progenitor hepatocytes, ESC-derived hepatic precursors showed less efficiency at regeneration after transplantation to the livers [86]. Some effort has been reported to improve the differentiation of ESCs in vitro with certain signaling factors, such as activin A and wingless-type MMTV integration site family member 3A [87].

18.4.2.3 Umbilical Mesenchymal Stem Cells

Investigations have also been done to study umbilical mesenchymal stem cells, which can be from self- or HLA-matched umbilical cord blood. It was reported that although engrafted umbilical stem cells cannot differentiate into functional hepatocytes, these cells can produce certain cytokines to suppress liver fibrosis [88]. But in Hong et al.'s study, human umbilical mesenchymal stem cells were investigated in vitro to determine whether they are able to differentiate into hepatocyte-like cells [89]. It was found that the cells were morphologically transformed into hepatocyte-like cells. Moreover, about a half of the cells were found to acquire the capability to transport DiI-Ac-LDL, a very important physiological function of hepatocytes. The most critical limitation for this approach is still the availability of appropriate umbilical mesenchymal stem cells.

18.4.2.4 Induced Pluripotent Stem Cells

Same as fetal hepatoblasts, ESCs have limited application due to ethical issues and rare resources. To solve this problem, ESC-like induced pluripotent stem cells (iPSCs) were developed. IPSCs are adult cells reprogrammed by forced expression of transcriptional factors to a pluripotent state. The first iPSCs were developed by Takahashi in 2006 [90], from mouse embryonic fibroblast cultures by introduction of Oct4, Sox2, c-Myc, and Klf4. The developed iPSCs had similar morphology, proliferation, and teratoma as ESCs. In 2007, the first iPSCs from adult human cells were induced by retroviral infection [91, 92]. However, although iPSCs circumvent the ethical issues, they still have other potential problems, such as the risk of teratoma formation and safety of retroviral gene transfer.

18.5 Methods for Stem Cell Delivery

Stem cells can be delivered to the bodies in several different ways, such as systemic infusion, intrahepatic injection, portal vein injection, and some other extracorporeal liver support devices. The principles to choose the delivery methods should be the balance of easy operation, minimal trauma, least side effects, high stem cell survival, and optimal clinical satisfaction. Therefore, each delivery method has its own advantages and disadvantages in special cases.

18.5.1 Systemic Infusion

Systemic infusion is the easiest and the most convenient way to achieve stem cell delivery and does not lead to severe trauma. Many successful cases were reported by this method to treat hepatic diseases through the ability of regeneration of stem cells [93, 94].

In Fang et al.'s work, mice were injected with mesenchymal stem cells via tail vain immediately after exposure to CCl_4 or 1 week after the first dose of CCl_4. After 2 weeks of CCl_4, the authors did not detect cells positive for both ALB and Y chromosome. However, after 5 weeks of CCl_4 plus systematic infusion of mesenchymal stem cells, the engraftment of male Flk1+ mesenchymal stem cells was detected in both groups of mice. Meanwhile, it was found that areas of damaged liver were smaller in the mice injected with mesenchymal stem cells immediately after exposure to CCl_4, compared to those administered with MSCs 1 week later. In another group, similar results were found. Mesenchymal stem cells (MSCs) were systemically administered to CCl_4-induced fibrotic murine model. Both magnetically labeled and unlabeled MSCs showed the potential to differentiate into hepatocyte-like cells.

However, besides the advantages, systemic infusion also has many side effects, such as fever, immune reaction, and donor cell entrapment in the lungs.

18.5.2 Liver Local Administration

18.5.2.1 Intrahepatic Injection

Intrahepatic injection is a very useful local administration method. It is a very efficient and direct way to deposit stem cells to the liver. This strategy has been used to deliver mononuclear cells (MNCs), which were collected from human umbilical cord blood (UCB) in full-term delivery women, to thioacetamide (TAA)-induced hepatic fibrotic rats [95]. In this study, it was reported that 4-week MNCs treatment practically reversed the damaged livers to normal architecture. Although the results are very promising, the biggest shortcoming for intrahepatic injection is potential severe trauma during operation.

18.5.2.2 Portal Vein Injection

Portal vein injection is another important method for stem cell delivery. Although this procedure has the same disadvantages as intrahepatic injection, such as complicated and invasive, portal vein injection has its own advantages. In portal vein injection, infused stem cells can reside in the periportal areas and repopulate faster than intrahepatic injection. Furthermore, for the diseased livers with very bad microenvironment, intrahepatic injection may not be applicable, but portal vein injection can still work. However, portal vein injection also has its own disadvantages compared to intrahepatic injection. It may cause portal hypertension, which can lead to further liver damage. Besides, this procedure may have the same risk as systemic infusion, such as migration of stem cells to systemic veins and embolism of other organs.

There are several successful cases in animal studies for portal vein injection of stem cells [38, 96, 97]. It has been reported that adipose-derived stem cells (ADSCs) were transplanted into liver-injured rats via different routes [97]. It was found that injection via the hepatic portal vein was more efficient than other routes. In another study, CD34+ stem cells from human umbilical cord blood were transduced with a lentiviral vector containing GFP report gene and injected via portal vein into cirrhotic rats. Rats were killed 15 and 60 days post-transplantation, and fluorescence was undetectable in liver sections. These negative results suggest that the stem cells did not engraft in the liver and disappeared from the rats. Although animal data is still controversial, some clinical cases showed promising results on portal vein injection of stem cells [98, 99]. In a recent study in 2010, CD34+ stem cells were isolated, amplified, and then reinjected via patients' portal veins to 48 patients; 36 suffered from chronic end-stage hepatitis C-induced liver disease and 12 harboring end-stage autoimmune liver disease [98]. Treatment was generally well tolerated except for three patients, who developed serious treatment-related complications. There was clinically, biochemically, and statistically significant improvement in a large percentage of patients who received treatment. These results suggest that autologous CD34+ stem cell transplantation via portal vein may be safely administered and have some therapeutic benefit to patients with severe end-stage liver disease.

18.5.2.3 Hepatic Artery Injection

Hepatic artery injection is usually applied to treat liver cancers. Actually, from beginning, this method was developed on the early 1960s to deliver chemotherapy to hepatic colorectal metastasis. There are several reasons to choose hepatic artery for delivery stem cells. The most important reason is, although both of the portal vein and hepatic artery are in charge of blood to the liver, most vessel branches in solid hepatic tumors are derived from hepatic artery [100]. Another reason is the low percentage of blood supply (only 20–30 %) from the hepatic artery (HA) of the total hepatic blood flow. Therefore, the risk of vessel embolization caused by hepatic artery infusion is less than that caused by intraportal vein infusion [100].

There are several successful preclinical and clinical cases of stem cell therapy via hepatic artery in treating liver diseases [101, 102]. In a recent study in 2011 [101], bone marrow mesenchymal stem cells were isolated from autologous bone marrow and infused via hepatic artery to the patients with liver cirrhosis. Liver function and image were monitored for 1 year. Only one case of Takotsubo syndrome happened with early complications. At about 12 months of treatment, two patients developed a cutaneous immunomediated disorder and hepatocellular carcinoma. A reduction in bilirubin was observed at 1 week in almost all the patients. These results suggest applicable effects and reasonable safety of hepatic artery infusion of BM MSCs in cirrhotic patients.

18.5.3 Extrahepatic Administration

18.5.3.1 Intra-Splenic Injection

Intra-splenic injection has been used for a long time for cell delivery to treat liver fibrosis, especially in the cases when the diseased liver is not suitable for cell transplantation. Direct deposit of stem cells into spleen pulp can benefit liver regeneration. Most of the stem cells take effects via translocation from the spleen to the liver via the splenic vein [103]. In this study, animals that received intra-splenic transplantation of syngeneic rat hepatocytes showed transplanted cells migrate from the spleens to the livers via portal veins. The successful cases are in both of preclinical and clinical studies [104, 105]. However, this procedure has the same disadvantages as portal vein injection, such as severe trauma, portal hypertension, and portal vein embolization during stem cell translocation.

18.5.3.2 Intra-Kidney Injection

Kidney capsules are also an alternative place to transplant stem cells for liver disease treatment [106]. However, this approach yields a lower survival because kidney capsules do not have sufficient space for large number of proliferated cells. An optimized place in kidneys for transplantation can be under the bilateral kidney capsule spaces in extracellular matrix [107]. This strategy can increase the survival of transplanted cells. In general, intra-kidney injection will never be a good option to treat liver diseases because of the "long" distance between organs.

18.6 Current Clinical Stem Cell Therapy Studies in Liver Diseases

Currently, the most frequently applied stem cell type under clinical trial is autologous stem cells from bone marrow, not only in hepatic disease treatment but also in other disease therapies. Table 18.2 lists several reported clinical cases on autologous stem cell therapy in liver diseases. This type of cell sources avoids ethical and immunological problems.

Table 18.2 Autologous stem cell therapies for liver diseases with various delivery methods in reported clinical trials

Delivery methods	Type of stem cells	Cell sources	Primary outcome	Therapeutic efficiency	References
Systemic infusion	CD34± enriched mononuclear cells	G-CSF-mobilized peripheral blood	Child-Pugh/MELD	Yes	[115]
	CD34+/CD45/C-kit	Bone marrow from iliac crest	Child-Pugh	Yes	[109]
	MSCs	Bone marrow from iliac crest	MELD	Yes	[116]
Portal vein infusion	CD34+	G-CSF-mobilized peripheral blood	Daily liver growth rate	Yes	[110]
Hepatic artery infusion	CD34+	Bone marrow from iliac crest	MELD	No	[117]
	Mononuclear-enriched cells	Bone marrow from iliac crest	Serum bilirubin, albumin, and INR	Yes	[118]
	CD34+	G-CSF-mobilized peripheral blood	Child-Pugh	Yes	[119]
	CD34+	G-CSF-mobilized peripheral blood	Serum bilirubin and albumin	Yes	[120]
	Mononuclear-enriched cells	Bone marrow from iliac crest	Serum bilirubin, albumin, and MELD	Yes	[121]
Systemic or portal vein infusion	CD13+, CD29+, CD34+, CD44+, CD45+, CD73+, CD105+, and CD166+	Bone marrow from iliac crest	MELD	Yes	[99]
Portal vein or hepatic artery infusion	CD34+	G-CSF-mobilized peripheral blood	Serum bilirubin and albumin	Yes	[73]
	CD34+	G-CSF-mobilized peripheral blood	Serum bilirubin, albumin, and INR	Yes	[98]

There are also cases using the stem cells from siblings of the patients. Northwestern University (633 Clark Street, Evanston, IL 60208), collaborating with Northwest Memorial Hospital, infused stem cells to the patient that have been previously collected from patient's siblings to treat primary biliary cirrhosis (PBC) (ClinicalTrials.gov identifier: NCT00393185). Before infusing stem cells, high dose of cyclophosphamide, fludarabine, and CAMPATH-1H were administered to the patients. Cyclophosphamide [35] and fludarabine [108] are drugs which can reduce the function of patients' immune system. CAMPATH-1H is a protein that is assumed to be able to be against the immune cells that may cause PBC [39]. Although in this case the stem cell type is not autologous, donor still has to be a human leukocyte antigen (HLA)-identical sibling or HLA-matched cord blood donor. HLAs are found on most cells in human body. Immune system components use HLAs as the markers to identify autologous cells. A close HLA match is critical to decrease the possibility that the immune systems will attack the donor's cells after transplantation. Therefore, this clinical trial held by Northwestern University is still pre-restricted by immune problems on donor seeking.

Since autologous BM stem cells are the most applicable source for transplantation in diseased livers, the safety and efficacy after administration in various routes in clinical trials still remains an open-ended question. Terai et al. reported a clinical study about autologous bone marrow cell infusion (ABMI) from the peripheral vein to treat liver cirrhosis [109]. After isolation from autologous bone marrow, mononuclear cells (MNCs) were infused via the peripheral vein. Child-Pugh scores were significantly improved at 4 and 24 weeks after systemic infusion. Meanwhile, there were no apparent adverse effects. From the author's conclusion, ABMI therapy might be a novel therapy to liver cirrhosis. For portal vein delivery, it was reported that CD133+ BM stem cells, infused via portal vein, could enhance hepatic regeneration in patients with malignant liver lesions [110].

However, interpretation of clinical data should be very cautious. First, the patient number enrolled is usually not sufficient to reach adequate statistical significance. One reason may be the invasive procedures required for BM stem cells isolation and administration. Real randomized and blinded trials are necessary ultimately to reveal genuine efficacy. Second, although so many reports have shown that BM stem cells can help liver regeneration, proper administration route has not been determined yet. Most delivery methods are still investigated in animal models. Systemic infusion is least invasive but sometimes with severe immunological issues. The other delivery methods will not only cause big traumas but may encounter problems, like unintentional portal vein embolization. Third, the technology is not precise enough to track transplanted cells in human subjects. Magnetic resonance imaging (MRI) was utilized to track super-magnetic iron oxide-labeled cells, but the makers were diluted after cell division. Technological difficulty to mark various stem cell types was an additional complicating factor [111].

18.7 Concluding Remarks

Although stem cell therapy-aided hepatic regeneration holds great promise and appears immensely provocative for the treatment of liver disease, there are still many problems with this novel therapeutic strategy.

Based on the ongoing clinical trials, to date, autologous stem cell therapy appears to be the most "realistic" cell type for transplantation. The fetal stem cells have many issues related to ethics and sources. Meanwhile, the side effects after stem cell transplantation should not be ignored. Hepatic venoocclusive disease (VOD), lung dysfunction, and graft-versus-host disease (GVHD) are frequent complications after stem cell transplantation [112–114]. Besides side effects, discovery of proper delivery methods has made considerable progress but awaits a major breakthrough. New drugs were discovered and applied to assist stem cell transplantation for liver regeneration in vitro and in vivo, but protocols optimizations remain far from complete.

Future direction of stem cell therapy will be the utility of iPS cells because it can conquer the ethical problems and provide a sustainable source of cells. However, iPS cells may lead to other issues including teratoma formation and gene delivery safety.

In conclusion, stem cell therapy is a very promising strategy to facilitate hepatic regeneration but still has a long way to go because of the realistic issues. The future therapy will overcome these issues by combining the best features of all current protocols.

Acknowledgements We apologize to all of the researchers whose original work was either not cited or was cited in a review article. We thank Dr. Sidhartha Ray for his proofreading and suggestions.

References

1. Guyton A, Hall J (1996) The liver as an organ. In: Textbook of medical physiology. WB Saunders Co., Philadelphia, p 771
2. Fausto N, Campbell JS, Riehle KJ (2006) Liver regeneration. Hepatology 43(S1):S45–S53
3. Fallowfield JA, Kendall TJ, Iredale JP (2006) Reversal of fibrosis: no longer a pipe dream? Clin Liver Dis 10(3):481–497
4. Gallegos-Orozco JF, Vargas HE (2009) Liver transplantation: from child to MELD. Med Clin North Am 93(4):931–950
5. Karp S (2009) Clinical implications of advances in the basic science of liver repair and regeneration. Am J Transplant 9(9):1973–1980
6. Kung JWC, Forbes SJ (2009) Stem cells and liver repair. Curr Opin Biotechnol 20(5):568–574
7. Ogawa S, Miyagawa S (2009) Potentials of regenerative medicine for liver disease. Surg Today 39(12):1019–1025
8. Ehnert S et al (2009) The possible use of stem cells in regenerative medicine: dream or reality? Langenbecks Arch Surg 394(6):985–997
9. Alison M, Islam S, Lim S (2009) Stem cells in liver regeneration, fibrosis and cancer: the good, the bad and the ugly. J Pathol 217(2):282–298

10. Schuppan D, Afdhal NH (2008) Liver cirrhosis. Lancet 371(9615):838–851
11. Lim JK (2010) The liver: biology and pathobiology. J Clin Gastroenterol 44(9):662
12. Starzl TE (1963) Homotransplantation of the liver in humans. Surg Gynecol Obstet 117:659
13. Starzl T et al (1964) Immunosuppression after experimental and clinical homotransplantation of the liver. Ann Surg 160(3):411
14. Moore F et al (1964) Immunosuppression and vascular insufficiency in liver transplantation. Ann N Y Acad Sci 120(1):729–738
15. Starzl TE et al (1967) The use of heterologous antilymphoid agents in canine renal and liver homotransplantation and in human renal homotransplantation. Surg Gynecol Obstet 124(2):301
16. Starzl T et al (2005) History of liver and multivisceral transplantation. In: Transplantation of the liver. Elsevier Saunders, Philadelphia
17. Calne R et al (1978) Cyclosporin A in patients receiving renal allografts from cadaver donors. Lancet 312(8104):1323–1327
18. Starzl TE et al (1981) Liver transplantation with use of cyclosporin A and prednisone. N Engl J Med 305(5):266–269
19. Eghtesad B, Kadry Z, Fung J (2005) Technical considerations in liver transplantation: what a hepatologist needs to know (and every surgeon should practice). Liver Transpl 11(8):861–871
20. Wertheim J et al (2011) Major challenges limiting liver transplantation in the United States. Am J Transplant 11(9):1773–1784
21. Alqahtani SA, Larson AM (2011) Adult liver transplantation in the USA. Curr Opin Gastroenterol 27(3):240
22. Berg C et al (2009) Liver and intestine transplantation in the United States 1998–2007. Am J Transplant 9(4p2):907–931
23. Ghobrial RM et al (2001) A 10-year experience of liver transplantation for hepatitis C: analysis of factors determining outcome in over 500 patients. Ann Surg 234(3):384
24. Berenguer M (2005) What determines the natural history of recurrent hepatitis C after liver transplantation? J Hepatol 42(4):448
25. Dahlke MH et al (2004) Stem cell therapy of the liver—fusion or fiction? Liver Transpl 10(4):471–479
26. Zhao Q et al (2009) Stem/progenitor cells in liver injury repair and regeneration. Biol Cell 101:557–571
27. Li J et al (2011) Therapeutic potential of stem cell in liver regeneration. Front Med 5(1):26–32
28. Petersen B et al (1999) Bone marrow as a potential source of hepatic oval cells. Science 284(5417):1168
29. Oh SH et al (2000) Hepatocyte growth factor induces differentiation of adult rat bone marrow cells into a hepatocyte lineage in vitro. Biochem Biophys Res Commun 279(2):500–504
30. Lagasse E et al (2000) Purified hematopoietic stem cells can differentiate into hepatocytes in vivo. Nat Med 6(11):1229–1234
31. Wang X et al (2003) Cell fusion is the principal source of bone-marrow-derived hepatocytes. Nature 422(6934):897–901
32. Vassilopoulos G, Wang PR, Russell DW (2003) Transplanted bone marrow regenerates liver by cell fusion. Nature 422(6934):901–904
33. Theise ND et al (2000) Liver from bone marrow in humans. Hepatology 32(1):11–16
34. Alison MR et al (2000) Hepatocytes from non-hepatic adult stem cells. Nature 406(6793):257
35. Matthew WD, Sandrock AW Jr (1987) Cyclophosphamide treatment used to manipulate the immune response for the production of monoclonal antibodies. J Immunol Methods 100(1–2):73–82
36. Schwartz RE et al (2002) Multipotent adult progenitor cells from bone marrow differentiate into functional hepatocyte-like cells. J Clin Invest 109(10):1291–1302

37. Sato Y et al (2005) Human mesenchymal stem cells xenografted directly to rat liver are differentiated into human hepatocytes without fusion. Blood 106(2):756–763
38. Jang YY et al (2004) Hematopoietic stem cells convert into liver cells within days without fusion. Nat Cell Biol 6(6):532–539
39. Mottershead M, Neuberger J (2008) Transplantation in autoimmune liver diseases. World J Gastroenterol 14(21):3388
40. Alvarez-Dolado M et al (2003) Fusion of bone-marrow-derived cells with Purkinje neurons, cardiomyocytes and hepatocytes. Nature 425(6961):968–973
41. Camargo FD, Finegold M, Goodell MA (2004) Hematopoietic myelomonocytic cells are the major source of hepatocyte fusion partners. J Clin Invest 113(9):1266–1270
42. Willenbring H et al (2004) Myelomonocytic cells are sufficient for therapeutic cell fusion in liver. Nat Med 10(7):744–748
43. Sgroi A et al (2009) What clinical alternatives to whole liver transplantation? Current status of artificial devices and hepatocyte transplantation. Transplantation 87(4):457
44. Dalgetty DM et al (2009) Progress and future challenges in stem cell-derived liver technologies. Am J Physiol Gastrointest Liver Physiol 297(2):G241–G248
45. Meindl-Beinker NM, Dooley S (2008) Transforming growth factor-β and hepatocyte transdifferentiation in liver fibrogenesis. J Gastroenterol Hepatol 23:S122–S127
46. Anjos-Afonso F, Siapati EK, Bonnet D (2004) In vivo contribution of murine mesenchymal stem cells into multiple cell-types under minimal damage conditions. J Cell Sci 117(23):5655–5664
47. Fausto N (2000) Liver regeneration. J Hepatol 32:19–31
48. Duncan AW, Dorrell C, Grompe M (2009) Stem cells and liver regeneration. Gastroenterology 137(2):466–481
49. Lee JS et al (2006) A novel prognostic subtype of human hepatocellular carcinoma derived from hepatic progenitor cells. Nat Med 12(4):410–416
50. Paku S et al (2001) Origin and structural evolution of the early proliferating oval cells in rat liver. Am J Pathol 158(4):1313
51. Yovchev MI et al (2008) Identification of adult hepatic progenitor cells capable of repopulating injured rat liver. Hepatology 47(2):636–647
52. Wu CX et al (2009) Intrahepatic transplantation of hepatic oval cells for fulminant hepatic failure in rats. World J Gastroenterol 15(12):1506
53. Mancino MG et al (2007) Hepatic "stem" cells: state of the art. Ital J Anat Embryol 112(2):93
54. Roskams T (2006) Different types of liver progenitor cells and their niches. J Hepatol 45(1):1
55. Yovchev MI et al (2007) Novel hepatic progenitor cell surface markers in the adult rat liver. Hepatology 45(1):139–149
56. Zhao R, Duncan SA (2005) Embryonic development of the liver. Hepatology 41(5):956–967
57. Shiojiri N, Lemire JM, Fausto N (1991) Cell lineages and oval cell progenitors in rat liver development. Cancer Res 51(10):2611
58. Meehan RR et al (1984) Pattern of serum protein gene expression in mouse visceral yolk sac and foetal liver. EMBO J 3(8):1881
59. Tee LBG et al (1996) Dual phenotypic expression of hepatocytes and bile ductular markers in developing and preneoplastic rat liver. Carcinogenesis 17(2):251
60. Fausto N (1990) Hepatocyte differentiation and liver progenitor cells. Curr Opin Cell Biol 2(6):1036
61. Brill S, Zvibel I, Reid LM (1995) Maturation-dependent changes in the regulation of liver-specific gene expression in embryonal versus adult primary liver cultures. Differentiation 59(2):95–102
62. Sangan CB, Tosh D (2010) Hepatic progenitor cells. Cell Tissue Res 342(2):131–137.
63. Oertel M et al (2006) Cell competition leads to a high level of normal liver reconstitution by transplanted fetal liver stem/progenitor cells. Gastroenterology 130(2):507–520
64. Rogler LE (1997) Selective bipotential differentiation of mouse embryonic hepatoblasts in vitro. Am J Pathol 150(2):591
65. Rogler CE et al (2007) Clonal, cultured, murine fetal liver hepatoblasts maintain liver specification in chimeric mice. Hepatology 46(6):1971–1978

66. Amicone L et al (1997) Transgenic expression in the liver of truncated met blocks apoptosis and permits immortalization of hepatocytes. EMBO J 16(3):495–503
67. Vig P et al (2006) The sources of parenchymal regeneration after chronic hepatocellular liver injury in mice. Hepatology 43(2):316–324
68. Quintana-Bustamante O et al (2006) Hematopoietic mobilization in mice increases the presence of bone marrow–derived hepatocytes via in vivo cell fusion. Hepatology 43(1):108–116
69. Yannaki E et al (2005) G-CSF-primed hematopoietic stem cells or G-CSF per se accelerate recovery and improve survival after liver injury, predominantly by promoting endogenous repair programs. Exp Hematol 33(1):108–119
70. Lange C et al (2006) Hepatocytic differentiation of mesenchymal stem cells in cocultures with fetal liver cells. World J Gastroenterol 12(15):2394
71. Gilchrist ES, Plevris JN (2010) Bone marrow derived stem cells in liver repair: 10 years down the line. Liver Transpl 16(2):118–129
72. Houlihan DD, Newsome PN (2008) Critical review of clinical trials of bone marrow stem cells in liver disease. Gastroenterology 135(2):438–450
73. Levičar N et al (2008) Long term clinical results of autologous infusion of mobilized adult bone marrow derived CD34+ cells in patients with chronic liver disease. Cell Prolif 41:115–125
74. Tajima F et al (2010) Hepatocyte growth factor mobilizes and recruits hematopoietic progenitor cells into liver through a stem cell factor-mediated mechanism. Hepatol Res 40(7):711–719
75. Yamaguchi K et al (2006) In vivo selection of transduced hematopoietic stem cells and little evidence of their conversion into hepatocytes in vivo. J Hepatol 45(5):681–687
76. Cho KA et al (2009) Mesenchymal stem cells showed the highest potential for the regeneration of injured liver tissue compared with other subpopulations of the bone marrow. Cell Biol Int 33(7):772–777
77. Kuo TK et al (2008) Stem cell therapy for liver disease: parameters governing the success of using bone marrow mesenchymal stem cells. Gastroenterology 134(7):2111–2121, e3
78. Liu ZJ, Zhuge Y, Velazquez OC (2009) Trafficking and differentiation of mesenchymal stem cells. J Cell Biochem 106(6):984–991
79. Aurich H et al (2009) Hepatocyte differentiation of mesenchymal stem cells from human adipose tissue in vitro promotes hepatic integration in vivo. Gut 58(4):570
80. Wang Z et al (2008) Human embryonic stem cells and liver diseases: from basic research to future clinical application. J Dig Dis 9(1):14–19
81. Agarwal S, Holton KL, Lanza R (2008) Efficient differentiation of functional hepatocytes from human embryonic stem cells. Stem Cells 26(5):1117–1127
82. Yamamoto H et al (2003) Differentiation of embryonic stem cells into hepatocytes: biological functions and therapeutic application. Hepatology 37(5):983–993
83. Moriya K et al (2008) Embryonic stem cells reduce liver fibrosis in CCl4-treated mice. Int J Exp Pathol 89(6):401–409
84. Xiao-geng D et al (2008) Effect of intrahepatic transplantation of embryonic stem cells-derived hepatic stem cells on host hepatic function and its safety evaluation. Journal of Clinical Rehabilitative Tissue Engineering Research 12(8):1591
85. Heo J et al (2006) Hepatic precursors derived from murine embryonic stem cells contribute to regeneration of injured liver. Hepatology 44(6):1478–1486
86. Haridass D et al (2009) Repopulation efficiencies of adult hepatocytes, fetal liver progenitor cells, and embryonic stem cell-derived hepatic cells in albumin-promoter-enhancer urokinase-type plasminogen activator mice. Am J Pathol 175(4):1483
87. Wernig M et al (2007) In vitro reprogramming of fibroblasts into a pluripotent ES-cell-like state. Nature 448(7151):318–324
88. Tsai PC et al (2009) The therapeutic potential of human umbilical mesenchymal stem cells from Wharton's jelly in the treatment of rat liver fibrosis. Liver Transpl 15(5):484–495
89. Hong SH et al (2005) In vitro differentiation of human umbilical cord blood-derived mesenchymal stem cells into hepatocyte-like cells. Biochem Biophys Res Commun 330(4): 1153–1161

90. Takahashi K, Yamanaka S (2006) Induction of pluripotent stem cells from mouse embryonic and adult fibroblast cultures by defined factors. Cell 126(4):663–676
91. Okita K, Ichisaka T, Yamanaka S (2007) Generation of germline-competent induced pluripotent stem cells. Nature 448(7151):313–317
92. Yu J et al (2007) Induced pluripotent stem cell lines derived from human somatic cells. Science 318(5858):1917
93. Fang B et al (2004) Systemic infusion of FLK1+ mesenchymal stem cells ameliorate carbon tetrachloride-induced liver fibrosis in mice. Transplantation 78(1):83
94. Ju S et al (2010) In vivo differentiation of magnetically labeled mesenchymal stem cells into hepatocytes for cell therapy to repair damaged liver. Invest Radiol 45(10):625
95. Bassiouny AR et al (2011) Modulation of AP-endonuclease1 levels associated with hepatic cirrhosis in rat model treated with human umbilical cord blood mononuclear stem cells. Int J Clin Exp Pathol 4(7):692
96. Moon Y et al (2009) Multipotent progenitor cells derived from human umbilical cord blood can differentiate into hepatocyte-like cells in a liver injury rat model. Transplant Proc. 41(10):4357–4360
97. Liang L et al (2009) Therapeutic potential and related signal pathway of adipose-derived stem cell transplantation for rat liver injury. Hepatol Res 39(8):822–832
98. Salama H et al (2010) Autologous hematopoietic stem cell transplantation in 48 patients with end-stage chronic liver diseases. Cell Transplant 19(11):1475–1486
99. Kharaziha P et al (2009) Improvement of liver function in liver cirrhosis patients after autologous mesenchymal stem cell injection: a phase I-II clinical trial. Eur J Gastroenterol Hepatol 21(10):1199
100. Ackerman LV (1974) Surgical pathology, 5th edn. Mosby, St. Louis
101. Couto BG et al (2011) Bone marrow mononuclear cell therapy for patients with cirrhosis: a phase 1 study. Liver Int 31(3):391–400
102. Zhu S et al (2011) Effects of intrahepatic bone-derived mesenchymal stem cells autotransplantation on the diabetic beagle dogs. J Surg Res 168(2):213–223
103. Kobayashi N et al (2000) Hepatocyte transplantation in rats with decompensated cirrhosis. Hepatology 31(4):851–857
104. Amer MEM et al (2011) Clinical and laboratory evaluation of patients with end-stage liver cell failure injected with bone marrow-derived hepatocyte-like cells. Eur J Gastroenterol Hepatol 23(10):936
105. Woo DH et al (2012) Direct and indirect contribution of human embryonic stem cell-derived hepatocyte-like cells to liver repair in mice. Gastroenterology 142(3):602–611
106. Ohashi K et al (2000) Sustained survival of human hepatocytes in mice: a model for in vivo infection with human hepatitis B and hepatitis delta viruses. Nat Med 6(3):327–331
107. Ohashi K et al (2005) Stability and repeat regeneration potential of the engineered liver tissues under the kidney capsule in mice. Cell Transplant 14(9):621–627
108. Grabska J, Dasanu CA (2011) Autoimmune phenomena in untreated and treated marginal zone lymphoma. Expert Opin Pharmacother 12(15):2369–79.
109. Terai S et al (2006) Improved liver function in patients with liver cirrhosis after autologous bone marrow cell infusion therapy. Stem Cells 24(10):2292–2298
110. Fürst G et al (2007) Portal vein embolization and autologous CD133+ bone marrow stem cells for liver regeneration: initial experience1. Radiology 243(1):171
111. Schroeder T (2008) Imaging stem-cell-driven regeneration in mammals. Nature 453(7193):345–351
112. Tabbara IA et al (2002) Allogeneic hematopoietic stem cell transplantation: complications and results. Arch Intern Med 162(14):1558
113. Skotnicki A, Krawczyk J (2001) Veno-occlusive disease – an important complication in hematopoietic cells transplantation. Przegl Lek 58(11):995
114. Ferrara JLM et al (2009) Graft-versus-host disease. Lancet 373(9674):1550–1561

115. Yannaki E et al (2006) Lasting amelioration in the clinical course of decompensated alcoholic cirrhosis with boost infusions of mobilized peripheral blood stem cells. Exp Hematol 34(11):1583–1587
116. Newsome PN et al (2003) Human cord blood-derived cells can differentiate into hepatocytes in the mouse liver with no evidence of cellular fusion. Gastroenterology 124(7):1891–1900
117. Mohamadnejad M et al (2007) Phase 1 human trial of autologous bone marrow-hematopoietic stem cell transplantation in patients with decompensated cirrhosis. World J Gastroenterol 13(24):3359
118. Lyra AC et al (2007) Feasibility and safety of autologous bone marrow mononuclear cell transplantation in patients with advanced chronic liver disease. World J Gastroenterol 13(7):1067
119. Pai M et al (2008) Autologous infusion of expanded mobilized adult bone marrow-derived CD34+ cells into patients with alcoholic liver cirrhosis. Am J Gastroenterol 103(8):1952–1958
120. Khan A et al (2008) Safety and efficacy of autologous bone marrow stem cell transplantation through hepatic artery for the treatment of chronic liver failure: a preliminary study. Transplant Proc 40(4):1140–1144
121. Peng L et al (2011) Autologous bone marrow mesenchymal stem cell transplantation in liver failure patients caused by hepatitis B: short-term and long-term outcomes. Hepatology 54(3):820–828

Chapter 19
Application of Microfluidics to Study Stem Cell Dynamics

Huaying Chen and Robert E. Nordon

Abstract Stem cell fate is directed by a complex chemical and mechanical microenvironment composed of secreted factors, extracellular matrix, and direct interactions with other cells. These signals ultimately control stem cell renewal and lineage fate in a developmental context. It may be possible to dissect the role of specific signaling pathways by precise control of microenvironment. However, traditional flask cell culture methods are unable to control microenvironment at microscale. Microfluidic platforms have the potential of mimicking the signals that direct stem cell fate by precise control of the chemical and mechanical milieu of cells at microscale. Furthermore, so called "lab-on-a-chip" technologies can increase research throughput by cost-effect automation of multiple parallel microscale cultures. This chapter will reveal how microfluidics and lab-on-a-chip technologies can be applied to the study of stem cell dynamics.

Keywords Stem cell dynamics • Microfluidics • Computational fluid dynamics • Photolithography • Lab-on-a-chip

19.1 Introduction

Stem cell science offers the promise of delivering a technology that will address the dire shortage of organs and tissues required for regenerative therapies. This great potential resides with the two fundamental properties of stem cells: self-renewal and

H. Chen
Faculty of Engineering, Graduate School of Biomedical Engineering,
The University of New South Wales, Samuels Building, Sydney, NSW 2052, Australia

School of Materials Science and Engineering, Shandong University, Ji'nan, China
e-mail: huaying.chen@uq.edu.au

R.E. Nordon (✉)
Graduate School of Biomedical Engineering, The University of New South Wales,
Samuels Building, Sydney, NSW 2052, Australia
e-mail: r.nordon@unsw.edu.au

M.K. Danquah, R.I. Mahato (eds.), *Emerging Trends in Cell and Gene Therapy*,
DOI 10.1007/978-1-62703-417-3_19,

pluripotency. However, despite dramatic biological advances and several decades of intense research, there are major challenges that lie ahead. A detailed understanding of the dynamics and molecular cues directing stem cell fate is required for stem cell research to become a clinically relevant technology. The stringent and specific microenvironment for stem cell growth has been difficult to establish, characterize, and control using traditional culture and analysis platforms [1]. Flask culture systems are unable to mimic many of the necessary cues required for stem cell development. These include physiological oxygen levels, soluble or immobilized molecular gradients, mechanical stress, and cell to cell interactions in a 3D microenvironment [1]. Transcriptomics and proteomics fail to capture the heterogeneous properties of developing lineage pathways because they measure bulk population properties rather than individual cell properties. Furthermore the dynamical properties of single stem cells is studied most directly by time-lapse video microscopy observation [2]; however, the technology for live-cell tracking is still considered to be low throughput. Therefore, new technologies for single-cell fate mapping are required for this field to progress and are therefore of considerable interest.

Microfluidics or lab-on-a-chip refers to devices for precise manipulation of small volumes of fluid in the microliter to picoliter range. Devices have channels for transferring fluid (gas or liquid) with cross-sectional dimensions in the order of micrometers [3]. Major advantages of microfluidics are the high degree of parallelization enabling high-throughput processing and flexible integration of various functional components allowing high-level automatic analysis [4]. Microfluidic devices have been widely applied for cell-based assay and long-term cell culture, since they have many theoretical advantages over traditional platforms such as well plates, flasks, and Petri dishes. These include low sample consumption, analysis of rapid cell responses to external stimuli (e.g., Ca^{2+} flux), precisely temporal and spatial control over the biophysical and biochemical environment, high-throughput and automatic analyses, and investigation of complex biological processes at the single-cell level [5, 6]. These potential capabilities are required for more quantitative investigation of stem cell properties.

This chapter provides an overview of materials, fabrication, and applications of microfluidics for cellular analysis in stem cell biology. The chapter begins with an introduction to the materials and methods of manufacture. Several important components essential for long-term, on-chip cell culture and manipulation such as pumps, valves, bubble eliminators, and microwells are described. Finally, we describe how these components are adapted for the specific requirements of cell culture and stem cell analysis.

19.2 Materials

The mechanical, optical, and surface properties of microdevice materials are of fundamental importance for developing biocompatible cell-based assays and in particular long-term microfluidic culture. Early on microfluidic devices were manufactured by etching silicon [7] or glass [8, 9] substrates. However, the opacity of silicon in the visible spectrum limits application to live-cell microscopy. Cost and manufacture time using glass substrates are much higher than those utilizing polymer substrates [10]. Following successful replica molding and irreversible bonding of polydimethylsiloxane (PDMS), devices for cellular analysis are most commonly manufactured from PDMS because of its various advantages [11, 12]. These are high-fidelity replication of mold micro-geometry; low curing temperature; reversible and irreversible sealing to glass, silicon, PDMS, and other materials; outstanding optical clarity; and cell biocompatibility [11, 12]. Most importantly, PDMS is an elastomer, with ideal mechanical properties for pneumatic activated on-chip pumps and valves as well as high permeability for O_2 and CO_2 to oxygenate media and buffer pH.

In recent years, there has been an increasing effort to explore alternative materials which may have superior material and manufacturing properties. PDMS is less suitable for high-volume manufacture compared to thermoplastics or other polymers. Microfluidic devices have been manufactured from polystyrene [13–15], poly(methyl methacrylate) (PMMA) [16], polyurethane-methacrylate (PUMA) [17, 18], Ormocomp [19], cyclic olefin polymers (COP), cyclic olefin copolymers (COC) [20–22], and even paper [23, 24]. The fabrication protocol will depend on the physical properties of these materials. Table 19.1 summarizes the relevant manufacturing and bonding methods.

19.3 Fabrication

Various techniques depending on the substrate material have been developed to fabricate lab-on-a-chip devices. Fabrication methods include hot embossing, laser ablation, injection molding, object printing, solid object printing, and micro milling. However, the most commonly applied method is soft lithography using PDMS [27], which consists of mold fabrication using photolithography, PDMS replica molding, and sealing of replicas to glass or PDMS substrates. This section briefly describes standard soft lithography techniques because it is now widely used in many labs for custom manufacture of microfluidic devices.

19.3.1 Photolithography

Photolithography is a high-resolution microprinting process that was originally developed for the microelectronics industry. The methods are simply adapted for manufacture of micron-size fluidic channels. Photoresist is spin coated onto a dehydrated silicon wafer to a desired film thickness. A high-resolution photographic mask specifies the 2D pattern to be printed by exposure of the photoresist to UV light (see Fig. 19.1a). Following exposure, the film is baked again to further

Table 19.1 Materials commonly used for fabricating microfluidics for cellular analyses and relevant manufacture methods

Materials		Manufacture	Bonding methods	References
Silicon or glass wafers		Photolithography and etching by HF-based solution	Adhesive bonding, etc. See details in literature [25]	[7, 8]
Polymers	PDMS	Casting against molds and cured by heat	Oxygen plasma oxidation	[3]
	Polystyrene	Hot embossing or shrink-dink	Thermal diffusion bonding and solvent bonding	[13–15]
	PMMA	Laser ablation, hot embossing, etc.	Thermal bonding	[16]
	PUMA[17, 18]	Casting against molds and cured by exposing to UV light	UV or heat	[17–19]
	Ormocomp[19] (a hybrid ceramic polymer)			
	Cyclic olefin polymers (COP) and copolymers (COC)	Laser ablation Micro milling Injection molding Hot embossing Nano imprinting	Thermal bonding, solvent bonding, etc.	[20–22]
	PMMA/PDMS	Hot embossing and casting	Pressure bonding and UV ozone treatment	[16, 26]

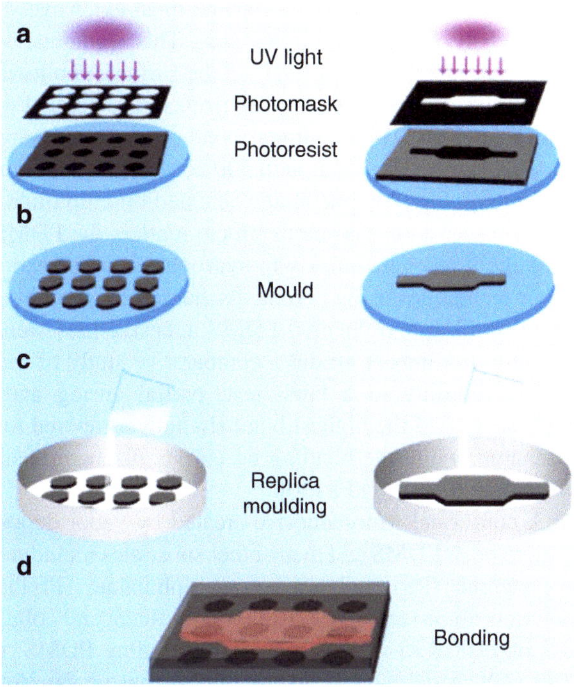

Fig. 19.1 Schematic of the soft lithography process. (**a**) Photolithography, (**b**) mold on the silicon wafer, (**c**) PDMS replica molding, and (**d**) bonding of two PDMS layers

cross-link the resist. For negative or positive photoresists, developer is used to dissolve unexposed or exposed regions, respectively (see Fig. 19.1b). There may be subsequent bakes to fully cure or reflow the remaining photoresist.

19.3.2 Replica Molding and Sealing

The PDMS elastomer base containing silicon hydride groups is thoroughly mixed with the curing agent containing vinyl groups at the weight ratio of 10:1, before the mixture is casted against a photoresist mold fabricated as described above (see Fig. 19.1c). After degassing under vacuum, the PDMS mixture with molds is baked at 65°C for no less than 2 h to cross-link silicon hydride and vinyl groups. Then, the PDMS replica is peeled off from the mold. Sharp punches are used to cut inlet and outlet holes, before bonding to a glass slide or another PDMS layer (see Fig. 19.1d).

PDMS reversibly seals to a planar PDMS or glass surface by van der Waals forces [27]. Reusable PDMS devices were used to patterning surface with cell arrays for biochemical analysis or tissue engineering applications [28]. The reversible bonding of PDMS is water tight but not capable of withstanding pressures larger than 35 kPa [29]. However, for most biological applications especially those integrated with pneumatic pumps and valves, irreversible sealing with much higher bond strength is required. So far the most extensively applied irreversible-sealing approach is oxygen plasma bonding. This technique oxidizes PDMS or glass and introduces silanol groups (OH) on the surfaces. Following the conformal contact of two oxidized substrates, they are irreversibly bonded with covalent siloxane bonds (Si–O–Si) [30]. The maximum bond strength measured when two bonded layers burst (burst pressure) is around 500 kPa [30]. This is much larger than the minimum pressure required by pneumatic actuated on-chip micropumps and valves (40 kPa) [31]. Furthermore, plasma treatment renders the PDMS surface more hydrophilic, although it recovers quickly to hydrophobic when exposed to atmosphere [32–34]. Other techniques using corona discharge [35], partial curing [36, 37], cross-linker variation [31], and uncured PDMS adhesive [38] were also employed for PDMS bonding. Eddings et al. did a comparative study of the above bonding techniques [30]. According to a burst test, partial curing and PDMS adhesive bonding approaches had the highest bond strength compared to oxygen plasma and corona discharge, while the bonding by corona discharge had the lowest strength which was larger than 250 kPa [30].

A conformal nano-adhesive created by vapor deposition was developed by Im et al. to bond PDMS and many other substrates including silicon wafer, glass, quartz, polystyrene (PS), polyethylene terephthalate (PET), polycarbonate (PC), and poly(tetrafluoro ethylene) (PTFE) [39]. Rezai and colleagues review bonding methods of PDMS to various materials including PDMS, glass, PMMA, PS, PC, and PET [40]. A dramatic bond strength increase from 35 to 146 kPa was achieved by microcontact printing Parylene (the trade name of various chemical vapor-deposited poly(p-xylylene) polymers) onto the PDMS surface followed by SF_6, N_2, and O_2 plasma treatment of the assembly [40]. Tsao and DeVoe [41] have reviewed indirect bonding using adhesive and direct bonding by thermal fusion, solvent, and localized welding of a wide range of thermoplastics.

19.4 Microfluidic Components Required for Long-Term Cell Culture

Microfluidics offers the possibility of large-scale integration of hundreds to thousands of micromechanical valves and functional components [42, 43]. This technique enables the automatic implementation of hundreds of cell-based assays in parallel [43]. According to Melin and Quake's review [43], microfluidics with integrated components "has been applied for protein crystallography [44–46], genetic analysis [47], amino acid analysis [48], high-throughput screening [42], bioreactors [49], and single cell analysis [50]." Ng and colleagues reviewed integrated components involving passive chaotic mixers, pneumatically actuated switches and valves, and magnetic filters [51].

Integration of micropumps, valves, gas exchangers, debubblers, cell culture chambers (e.g., microwells), and micro/mini heaters in a single chip allows (a) metered perfusion of media and other reagents which facilitates the regulation of the biochemical and biophysical microenvironments for cell growth, (b) control of cell deposition and retention, (c) scale-down of the culture system from flasks and incubators to microfluidics on a glass slide, and (d) incorporation of time-lapse imaging techniques. Integration of these components in a single device is essential for long-term cell culture and clonal analysis where microenvironment must be controlled precisely. In this section, the application of pumps, valves, bubble traps, and microwells for cell-based assay is reviewed. Gas exchangers for media oxygenation and pH buffering as well as micro/mini heaters for temperature control will be reviewed in Sects. 19.5.4 and 19.5.5, respectively.

19.4.1 Micropumps and Valves

Micropumps integrated within microfluidic devices to supply media, growth factors, and other reagents to individual cell culture chambers guarantee precise temporal and spatial regulation of culture environment [52]. There have been numerous micropump designs reported in the literature. The actuation mechanisms may be categorized into pneumatic pressure [31, 52], electromagnetic actuation [53], electrostatic actuation [54], piezoelectric actuation [55], thermo-penetration [56], recoverable force of a shape memory film [57], evaporation [58], gas permeation [36, 59], a refreshable Braille display [49], and surface tension [60].

Quake's group pioneered the fabrication of pneumatic pressure-driven microvalves and pumps using multilayer soft lithography [31]. This pump consists of a flow channel with a semicircular cross section and three pneumatic channels with rectangular cross section that lie over and are perpendicular to the semicircular flow channel. The semicircular cross section is required for complete occlusion of the fluidic channel when it is deformed by the overlying pneumatic channel. The pneumatic and flow layers are separated by a thin PDMS membrane (usually around 20-μm thick) (see Fig. 19.2). The multilayered device is bonded by both cross-linker variation and O_2 plasma oxidation. Peristaltic flow is driven by activating the three pneumatic channels sequentially. Over the past few years, devices for culture of hESCs have utilized this type of pneumatic pump and valve [61]. This method of flow control has enabled the study of cell culture interactions [62] and tracking of single yeast cells [63].

Grover and co-workers developed latching pneumatic valves (see Fig. 19.3a) which also have a flow layer and pneumatic layer with a thin membrane between two layers [64]. The valves are normally closed (or open) under low pressure which reduces the number of off-chip controllers required for large-scale integration. A four-bit binary demultiplexer was developed to address independent latching valves and to distribute pressure and vacuum pulses to each valve (see Fig. 19.3) [64]. Hulme and colleagues demonstrated a technique for incorporating standardized and prefabricated screw, pneumatic, and solenoid valves into microfluidic structures [65]. Although only a small number of valves were integrated using this method, the prefabricated valves are identical and performed reproducibly in microfluidic devices. Instead of using a posi-

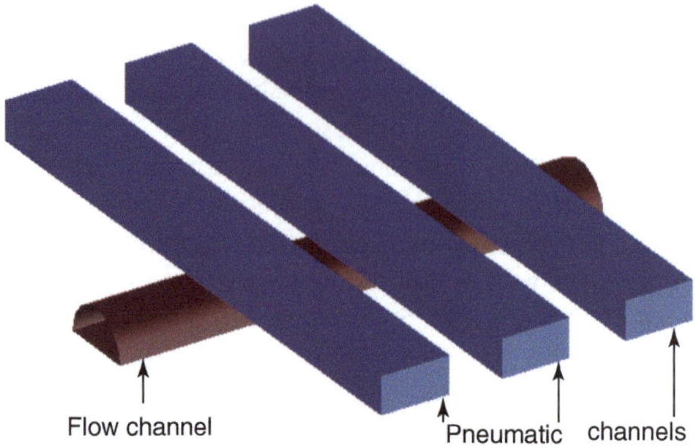

Fig. 19.2 Schematic of a pneumatic, peristaltic micropump. Three rectangular pneumatic channels lie over a semicircular flow channel. After sequential application of pressure in pneumatic channels, the PDMS membrane deforms and displaces volume along the flow channel by peristalsis

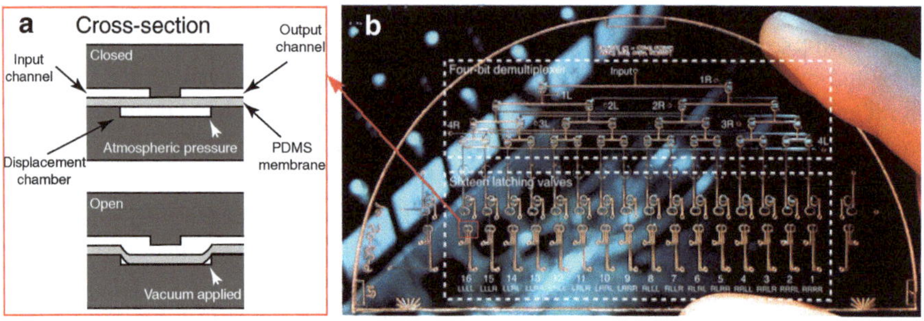

Fig. 19.3 Cross section of a latching valve (**a**) and photo of a demultiplexer with latching valves (**b**). The pressure or vacuum applied from the input port is distributed by demultiplexer (four rows of valves in the *upper dash box*) to each of 16 latching valves [64]. It can address a latching valve every 120 ms (Adapted from [64] with permission)

tive photoresist mold with semicircular cross section to cast the PDMS flow channel, Kwon's group applied a polymer sealing method inside rectangular PDMS channels [66]. In this method, the photo-curable oligomer was injected into rectangular flow channels before gas pressure was applied, and then UV cured in situ over the valve region using a photo mask. After uncured polymer was removed, the channel cross section was perfectly occluded during valve actuation. Based on the latching valves reported by Grover [64], Lai and Folch developed a peristaltic micropump using a single control channel to actuate four microvalves with different sizes [67]. This pump design, similar to that reported by Huang et al. [68], reduces the number of pneumatic channels from three (in Quake's pump) to one, which greatly simplifies the structure and saves the space on an integrated microfluidic chip.

19.4.2 Bubble Traps and Degassing

In microfluidic cell culture systems, air bubbles usually form during system priming, cell suspension loading, and media perfusion processes due to gas permeability of

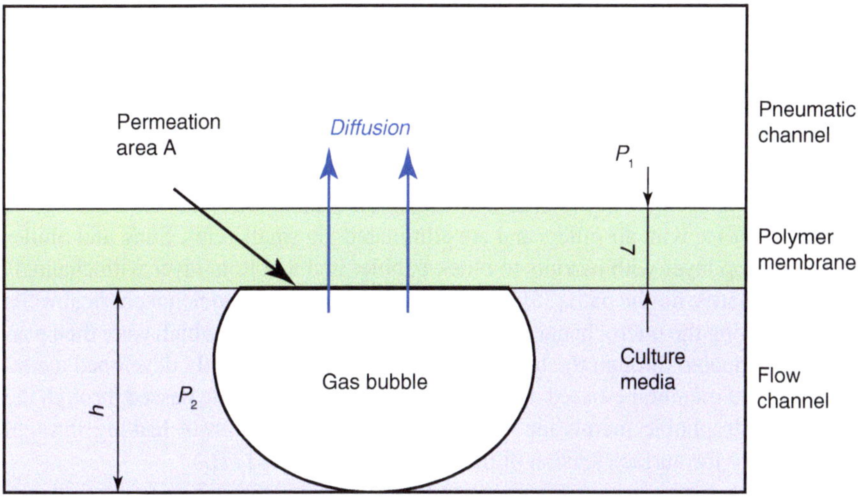

Fig. 19.4 Schematic illustration of the pressure-driven bubble permeation across a polymer membrane. When the pressure (p_2) in the flow channel is larger than the pressure (p_1) in the pneumatic channel, the air bubble permeates through the membrane

tubing and dissolution of gas at pressures below their solubility [6]. Bubbles are detrimental for microscale cell culture leading to blockage of microchannels or shear damage to cells. In this context, various bubble traps or microdevices to remove gas bubbles in microfluidics were developed. These were operated by either the permeation of gas from flow channels into pneumatic channels vented to the atmosphere through a thin membrane [69, 70] or direct flow of gas into a bypass channel connecting the bubble trap [71]. A physical model for pressure-driven bubble elimination in microfluidics was reported by the Park group [72]. As shown in Fig. 19.4, the flux (N) of gas permeating a polymer membrane at steady state is given by

$$N = \frac{P(p_2 - p_1)}{L} \tag{19.1}$$

where P is the permeability of the polymer; p_1 and p_2 are permeate and feed pressure, respectively; and L is the membrane thickness [73]. Ignoring the dissolution of air into culture liquid and bulk polymer, the Park group [72] reported the following exponential decay model to predict the bubble elimination rate with time (Fig. 19.4):

$$A(t) = c_1 \exp\left[\frac{Pk(p_1 - p_2)}{hL}\frac{76T}{273p_{\text{atm}}}t\right] = c_1 e^{ct} + c_2 \tag{19.2}$$

where A is the bubble permeation area; c, c_1, and c_2 are constants; k is correction factor for the effective permeation area; h is the height of the flow channel; t is time; T is absolute room temperature in Kelvin; and p_{atm} is the atmospheric pressure in cmHg [72]. This study provides guidelines for design of membrane-based bubble eliminators in microfluidics.

Several other ingenious designs for bubble elimination have been published. These exploit surface tension properties which are a predominant force at microscale. Cheng and Jiang developed a debubbler to trap and remove bubbles using an array of cylinder-shaped air-liquid interfaces called air pillars [74]. Bubbles entering the device coalesce with air pillars and are eliminated via small vents. Sung and Shuler applied a top layer with barriers to block bubbles and a bottom layer with channels as an alternative fluidic path [75]. Kang and co-workers applied hemispherical wells in series along the microchannel to effectively trap air bubbles which were then passively eliminated through the bulk PDMS [76]. Liu et al. recently developed a nozzle-type and membrane-based debubbler in which the air bubble passed through the porous hydrophobic membrane while the liquid was kept from leaking through pores due to the surface tension of the air-liquid meniscus [77].

19.4.3 Microwells

Cellular analyses based on large cell populations fail to provide information of distinct behaviors of individual cells or rare cells, so single-cell analysis is crucial for studying stem and progenitor cell heterogeneity. In single-cell analysis, individual cells are required to be isolated and located within a desired region for extended periods. Moreover, a large number of cells require analysis before reaching statistical significance because of the stochastic nature of cell division and other cell fate outcomes [78, 79]. Lindström and Andersson-Svahn provided a detailed overview of mechanisms for single-cell isolation, which include mechanical, magnetic, hydrodynamic, optical, dielectrophoretic, and acoustic traps as well as droplets [80].

Microwells have attracted extensive research interest, since they allow robust mechanical isolation, cell aggregation, and cluster formation in a high-throughput manner [81]. Furthermore, large wells (diameter >50 μm) supply enough space for cell spreading and proliferation in long-term cell culture and are compatible with live-cell imaging in situ [82]. This capability enables analyses of hundreds to thousands of clones from a single experiment. Smaller microwells just large enough for a single cell are frequently employed for instant cell analysis [83, 84], single-cell enzyme kinetic analysis [85], deformability of nonadherent cells [86], and short-term time-lapse imaging [87] over a few hours following cell deposition [88]. Charnley et al. have reviewed various applications of microwells for single-cell study [89]. The Lindström group has reviewed the shapes (square, hexagonal, and round), materials (glass, silicon, PDMS, SU-8, etc.), sizes (from a few microns to a few hundred microns), and density of wells in microwell arrays for single-cell analysis [88]. This section reviews experimental studies examining cell deposition into microwells which may increase seeding efficiency and fluid dynamics studies guiding design and optimization of microwell devices.

19.4.3.1 Experimental Study of Cell Deposition

For high-throughput clonal analysis by time-lapse imaging, augmenting the proportion of wells loaded with single cells will increase the throughput of the experiment. Numerous studies have attempted to address this task experimentally. Cells are simply loaded into microwells by sedimentation due to the difference in density of the cell and surrounding fluid under static [85, 90, 91] or flowing conditions [28]. Cells are retained by a low-shear-stress region that develops inside a microwell even with flow [28]. The number of cells within microwells is approximately controlled by cell concentration and flow rate of the cell suspension. Similar to limiting dilution, when the concentration is small enough, microwells are loaded with one cell per well on average. There are various studies looking at how shape, size, and depth of microwells influence single-cell retention statistics. The Takayama group compared single-cell-trapping efficiency of various well geometries including triangle, square, circle, diamond, and cone at the same loading flow rates [82]. Among the tested geometries, triangular wells were the most efficient, 62 % of which were loaded with single cells. Since all parameters except well shapes are constant, this study does not suggest triangular wells are still more efficient when well dimensions and flow rates vary. Our group experimentally studied influence of wall shear stress, plate separation, and well depth on cell capture statistics by imaging and counting nonadherent cells (KG1a) retained in microwells at various media flow rates [92]. For the same device, the mean number of cells per well was inversely related to wall shear stress. At limiting cell-docking frequencies (one cell per well), the number of cells per well was described by a Poisson distribution, therefore indicating that cells were randomly distributed between wells. Mean cell number per well was described by a cubic polynomial model.

Cell seeding studies on their own require a huge amount of systematic experimentation to determine optimal well geometries and dimensions [93]. However, analytical and computational methods to predict cell deposition provide an important design tool for design of microwell devices.

19.4.3.2 Computational Fluid Dynamic Studies

Computational fluid dynamics (CFD) has been extensively applied to study the details of the flow field around a complex microstructure, particularly where there is no analytical solution. CFD analysis and experimental validation are employed to direct the design of microstructures, thereby reducing the labor of experimental testing [94]. Various commercialized CFD software including CFX, CFD-ACE+, Flow 3D, and Fluent has been employed for simulation of microfluidics. Their performances were compared by the Koltay group [95]. Erickson provided an overview of the history and development of CFD study on cellular/particulate transport in microfluidics before 2005 [96].

There have been three main approaches to model the fluid dynamics of cell seeding in a microstructure. Table 19.2 summarizes the simulation methodologies that

Table 19.2 Simulation approaches for cell deposition in microstructures

Approaches	A	B	C
Method	One-way coupled Lagrangian approach to model dynamical cell-fluid interactions	Static microstructure fluid interactions	Static microstructure and cell-fluid interactions
Simulation	Individual cell trajectories	Flow field without cells	Hydrodynamic forces acting upon cells held stationary at various positions to predict cell displacement
Limitations	Computational expensive	No cell-fluid interactions	New geometry and mesh are required as the cell position changes
References	[97, 98, 102]	[82, 93, 99–101]	[103, 104]

were applied to the study of cell deposition in microstructures. Approach A is one-way coupled Lagrangian particle simulation. The injection of cell suspension is simulated where cells are assumed to be rigid spheres and interfere with fluid flow. It was employed to optimize cell trapping in microsieves by the Thorsen and Wang groups [97, 98]. This method is computationally expensive because of dynamical simulation of cell trajectories and their interactions with fluid flow. Additionally, the simulation of rolling and bounce of cells off walls is difficult [93]. Approach B applies a cell-free model where only the fluid flow in the microstructure is investigated. Analytical models that assume simple shear flow are then applied to estimate the viscous forces acting upon cells deposited near the wall. This method has been employed to study cell deposition in grooved microchannels [99] and double grooves [100]. Han and co-workers applied a 2D model to study the flow field in microwells [101]. They demonstrated that for a given parallel plate flow over microwells, recirculation regions become larger in deeper wells. Takayama group modeled the flow fields of triangular, square, circle, diamond, and cone wells (20 μm in depth) with a 280-μm-high channel over wells and found the recirculation region in triangular wells was the largest [82]. Cioffi et al. studied the flow recirculation and wall shear stress in microwells with a depth of either 20 or 80 μm [93]. They experimentally related cell-trapping efficiency to a recirculation region with low wall shear stress. Although approach B is computationally feasible using standard computer hardware, it does not consider the interaction between cells, the fluid, and microstructure. This interaction is significant when the cell volume is on a similar scale to the microstructure.

Approach C employs a cell-containing model where a rigid sphere (representing a cell) is placed at a site of interest in a microstructure. Following simulation, the forces and torques acting on the cell are evaluated. Jang and colleagues described a 2D model of a deep well with a disk (representing a cell) on the well bottom to study the flow field and flow-induced forces on the cell [103]. Our group applied a 3D single-well model with a cell fixed on the downstream bottom of the well to study the torques and forces on the cell (see Fig. 19.5) [92]. Figure 19.5a shows the streamline pattern with flow separation at up- and downstream corners of the microwell. The cell experiences a downward drag force and clockwise torque

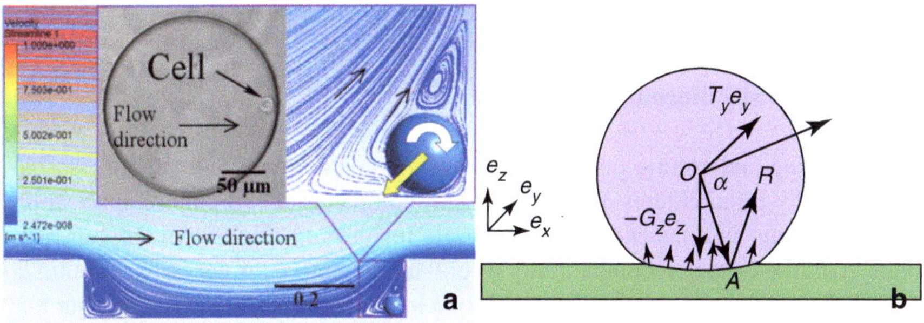

Fig. 19.5 Fluid dynamic simulation of the torque and drag acting upon a sphere fixed at the downstream microwell floor. (**a**) Streamlines for a stable cell deposition position with a phase-contrast micrograph showing deposition of cells in this position (35-μm-deep well and 280-μm plate separation, simulation Re =0.01, cell image Re=0.11 [92]). The *white arrows* and *yellow arrows* in *inset* images indicate the direction of fluid torque and drag, respectively. (**b**) The force and torque on the cell deposited away from the wall of the microwell. *Small arrows* on the cell bottom pointing in different directions are reactive forces distributed over the contacting area due to deformation (Adapted from [92] with permission)

directly proportional to Re. A mechanical equilibrium model was applied to determine the net effect of fluid shear, cell buoyancy, and static wall friction to predict if the cell (sphere) was in a stable position. The fluid dynamic simulation confirmed that flow separation with circulation inside deep microwells hampered cell recovery at high wall shear stress. This simulation method can be used to optimize the design of microwell structures for single-cell deposition.

19.5 Long-Term Cell Culture

The observation of cell fate in culture requires robust, long-term cell cultivation and continuous time-lapse imaging. Long-term, live-cell imaging with traditional culture platforms such as Petri dishes and flasks is challenging and problematic because of (a) difficulty in retaining nonadherent or highly motile cells in the same field of view, (b) disruption of continuous cell tracking when media are replaced, and (c) inaccurate control of microenvironment. Microfluidic platforms have the potential of solving these problems and have been evaluated with various cell types, such as cell lines [98, 105–110], primary rat hepatocytes [111], neural progenitor cells [112], human primary mesenchymal stem cells (hMSCs) [52], and human embryonic stem cells [90], for extended periods of time. We refer the interested reader to other reviews of microfluidic cell culture [113–117]. Paguirigan and Beebe discuss the differences between microfluidic and traditional macroscale cultures [118].

This section discusses the main techniques for long-term, on-chip cell culture reported in the literature. Relevant factors such as biocompatibility, media perfusion, microenvironment, pH buffering, media oxygenation, and temperature regulation are also reviewed.

19.5.1 Biocompatibility

19.5.1.1 PDMS Biocompatibility

Cell counting and sorting only require transient contact with biomaterials, so biocompatibility requirements are less stringent compared to cell culture applications where cells are in contact with biomaterials for days to weeks. A comprehensive understanding of the biocompatibility of polymeric material is necessary for utilization in long-term cell culture. There are two inherent properties of PDMS that may limit its application to cell analysis: (a) the existence of uncured, low molecular weight (LMW) oligomers in bulk PDMS which may be toxic to cells and (b) absorption of small hydrophobic molecules by PDMS [119]. The Whitesides group determined that the weight ratio of uncured oligomers in bulk PDMS was around 5 % [119]. LMW species have been shown to be responsible for the rapid hydrophobic recovery of PDMS surfaces after hydrophilic surface modification [34, 120, 121]. Absorption of small hydrophobic molecules such as steroid hormones and drugs may significantly alter the cell growth milieu [122]. Regehr and colleagues also reported leaching of LMW oligomers into cell culture media and cell membrane [122].

The impact of PDMS material properties on cellular signaling pathways was studied by time-lapse fluorescence imaging [123]. Wlodkowic et al. showed that PDMS was nontoxic compared to its polystyrene counterpart. Our group has also shown that there was no significant difference in growth rates of KG1a cells (an acute myeloid leukemia cell line) cultured on PDMS compared to polystyrene Petri dishes [124]. More extended studies are required to generalize these observations to other cell types such as primary cells and pluripotent stem cell lines.

Solvent extraction [119] and prolonged curing times [120] have been applied to minimize the amount of uncured oligomers in PDMS. However, most organic extraction solvents may cause problems themselves because they are cytotoxic [122]. Furthermore, neither extraction nor prolonged curing can completely remove uncured oligomers [119, 122]. To reduce the absorption of hydrophobic molecules on PDMS surface, investigators have absorbed proteins such as bovine serum albumin and fibronectin [126], silanized surface [126, 127], or grafted polymers [128]. PDMS surfaces coated with Parylenes are believed to prevent leaching and absorption of small molecules [129]. In addition, manufacture of microdevices from denser thermoplastics such as polystyrene does not have significant absorption or leaching problems [14, 130].

19.5.1.2 Tubing Biocompatibility

Tubing connecting the chip to external instruments (i.e., syringe pumps and peristaltic pumps) is necessary for cell-based assays and long-term cell culture. Various types of tubing including Tygon™, Silicone, Teflon™, PEEK, polyethylene, Pharmed™, and Silastic™ are used in microfluidic applications.

Since the flow rate in microfluidics is in the order of microliters/min, depending on tubing diameter, the residence time of perfused reagents within tubing may be several hours. Therefore, tubing that is certified for use in cell culture by manufacturers assumes relatively high flow rates and may be cytotoxic at low flow rates because of extended contact times. Moreover, the large surface area to volume ratio of small caliber tubing can contribute to adsorption and depletion of growth factors and proteins by tubing. Price et al. reported that unwashed Tygon™ tubing reduced the growth of phytoplankton [131]. Park and co-workers demonstrated that peroxide-cured silicone tubing was toxic to the tobacco BY2 cells which are particularly sensitive cell lines [132]. Our group assessed the biocompatibility of Silastic™, Pharmed™, polyethylene, and Tygon™ at microliter/min flow rates (6-h residence time) and showed that polyethylene and Tygon™ tubings were suitable for long-term perfusion culture [124].

19.5.2 Media Perfusion

For long-term on-chip cell culture, media are usually perfused into culture chambers by syringe pumps [110], peristaltic pumps [111, 133], hydraulic pressure [109], or on-chip micropumps [52]. Media perfusion provides controlled supply of growth factors and proteins as well as regulated exposure of fluid-induced forces to cells [6]. In addition, on-chip trypsinization, continuous passage and cell recovery for downstream analyses have been performed by the perfusion system [98, 110]. Generally, there are two perfusion modes: single-pass perfusion (SP) and recirculating perfusion (RP). In SP, media are perfused directly through the cell culture chamber to waste containers, during which the perfused growth factors, cell metabolites, and secreted factors are exposed to cells for short periods of time. So downstream cells are exposed to more metabolites and secreted factors compared to upstream cells. While in RP, media including autocrine factors and cell metabolites are recirculated with retention of these components for longer periods of time.

SP was employed by Tourovskaia et al. to study the differentiation of myoblasts to myotubes [135]. Yu and co-workers developed a microfluidic device to culture NIH 3T3, Hela, and B16 melanoma cell lines with both SP and RP [107]. No significant influence of media perfusion modes on cell growth was found in this study. However, for stem cell types such as HSC and embryonic stem cells, perfusion modes and flow rates appear to have important influences on stem cell fate.

Villa and colleagues reported optimal perfusion conditions for culturing primary mouse embryonic fibroblasts (mEFs) and mouse embryonic stem cells (mESCs) in microfluidic systems [135]. After 1-week culture, both types of cells have similar morphology and viability compared to those cultured in Petri dishes [135]. Cooper-White's group developed a microwell perfusion system to study the expansion of hESC [136]. Moledina et al. more recently showed that autocrine and paracrine factors such as endogenously secreted gp130-activating ligands could generate a gradient of mouse embryonic stem cell fates which depended on the direction of microfluidic flows, position in the flow field, and local cell organization [137]. Levchenko demonstrated experimentally and computationally that mESC viability also depends on secreted factors, which should also be considered when specifying the flow rate for perfusion culture [138].

As discussed above microwell arrays have the advantage of physically isolating stem and progenitor colonies so that progeny shares a common microenvironment. For motile stem cell types such as hematopoietic stem cells, the microwell serves to retain the colony within the same field of view so that individual cell fates can be continuously tracked by live-cell imaging. Furthermore many hundreds of clones can be visually tracked in parallel providing high-content live-cell imaging data for studying the heterogeneity of stem cell fates [5, 139].

Kim and colleagues comprehensively reviewed the microfluidic perfusion culture of mammalian cells from the viewpoints of design, fabrication, operation, and applications in cellular phenotype and function assessment [6]. They provide a practical guide for microfluidic cell culture with media perfusion. The Beebe group defines effective culture volume (ECV) and critical perfusion rate (CPR) as critical parameters for control of microfluidic culture processes; these quantities allow one to set the time interval for media changes and media flow velocity, respectively [140].

The effects of secreted paracrine or autocrine factors need to be carefully distinguished from other physicochemical factors which may influence proliferation and differentiation. We found that perfusion rate has a strong influence on growth rate, and contrary to expectation, over-perfusion may result in cell growth inhibition (see Fig. 19.6). Possible explanations for over-perfusion growth inhibition may be washing out of secreted growth factors or cytotoxicity due to oxidative media degradation or trace amounts of cytotoxic material from the microdevice or tubing.

Microdevices also have application in static cell culture. Shirley et al. bonded a microchannel network to a nano-porous membrane in a Petri dish [141]. The porous membrane allowed free exchange of proteins, nutrients, buffers, and stains between the microchannels and culture media. The Chakraborty group recently reported extended static culture times by functionalizing the inner surfaces of PDMS microchannels to form an air-liquid interface for gas exchange [142]. Although static culture greatly simplifies these experimental systems, there is less control of microenvironment for long-term culture.

19.5.3 Microenvironment

For stem cell culture assay, cells should be maintained in a well-defined microenvironment that mimics an in vivo growth environment. A fully defined stem cell culture microenvironment consists of growth factors, secreted factors, other cells, extracellular matrices, and metabolic substrates. The physical microenvironment includes the viscoelastic properties of the cell substrate which is usually extracellular matrix and fluid shear stress. The ability to temporally and spatially control these in vitro factors allows one to dissect their physiological role or to manipulate stem cell properties for tissue engineering or drug development. A recent review by

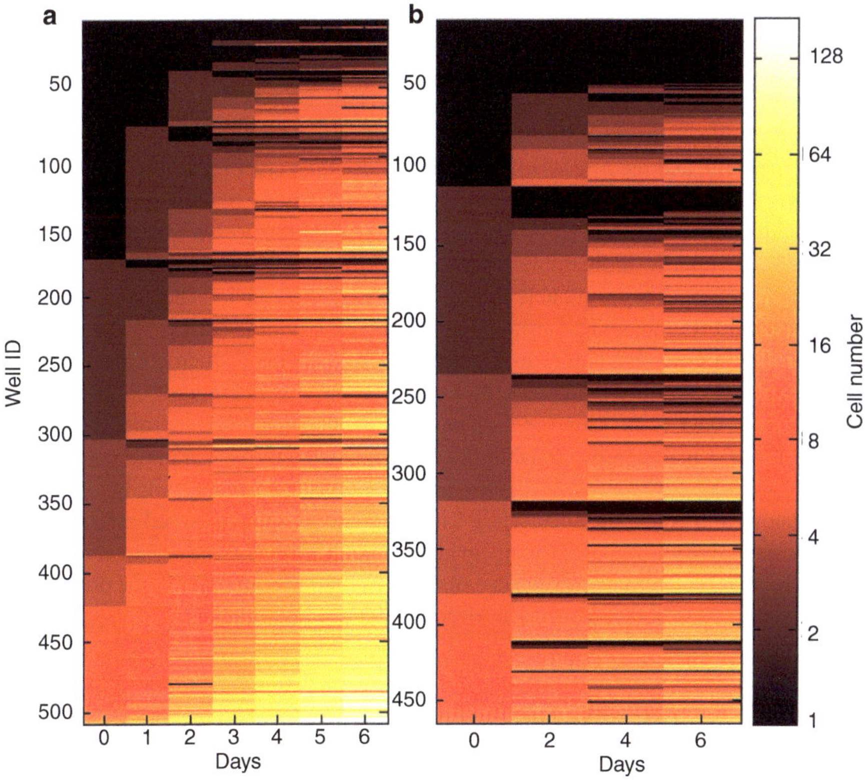

Fig. 19.6 Heat map showing the number of KG1a cells in each well on a microwell array with perfusion rate of (**a**) 0.1 μL/min and (**b**) 0.3 μL/min for 6 days (Reprinted with permission from [124]. Copyright [2011], American Institute of Physics)

the Brzózka group discussed various solutions for mimicking the in vivo microenvironment using microfluidics [143]. These solutions included perfusion cell cultures, extracellular matrix analogues, micro scaffolds, spheroid formation, and cocultures.

19.5.3.1 Biochemical Microenvironment

The cell biochemical milieu is composed of soluble factors such as growth factors, cytokines, hormones, and products of cell metabolism. These molecules can also interact with the insoluble extracellular matrix, which has important roles in guiding tissue architecture [140]. Microfluidic devices have been shown to be valuable in stem cell research because they are able to (a) control delivery and composition of soluble factors, (b) generate steady chemical gradients, and (c) control the partial pressure of dissolved gases [144, 145]. Beebe has also written reviews that examine the regulation of the microenvironment by microfluidics focusing on control of soluble factors, gradients, extracellular matrix (ECM), and gas concentration

[140, 146]. Gupta et al. discusses the control of soluble factors in a review on the study of stem cell biology by microfluidics [147].

Gomez-Storb et al. developed an automated microfluidic system; this was able to create arbitrary culture media formulations in 96 independent culture chambers. This system was able to screen culture conditions such as seeding density, composition of culture medium, and feeding schedules, while individual chambers were observed by time-lapse imaging [52]. A microfluidic system for studying mammalian cells in 3D microenvironments was developed by Lii et al. [148]. This device is capable of real-time and individually addressable control of 3D microenvironments using a parallel array of chambers. They were able to study dynamical interactions between NIH 3T3 fibroblasts and soluble signals from hepatocellular carcinoma cells [62].

Microfluidics also has utility in the control of biochemical microenvironment for tissue explants or micro-organoid culture. This application of microfluidics is aptly called “organ-on-a-chip.” Gunther et al. have devised an ingenious microfluidic design for sealing and maintaining a viable arterial segment on a microfluidic chip [149]. Blake et al. were able to support viability of a rat brain slice by generating laminar flow layers above and below the slice [150]. The Ismagilov group employed microfluidics to investigate mechanisms regulating spatial dynamics of hemostasis (blood clotting) and early patterning of the drosophila embryo [151]. In the future it may be possible to study the 3D assembly of stem cell-derived tissues by microfluidics.

19.5.3.2 Biomechanical Environment

Biophysical properties such as matrix stiffness [152, 153] and fluid shear stress [154, 155] play important roles in differentiation, cell spreading, motility, and haptokinesis. Moraes and colleagues reviewed microfluidic technologies for controlling of mechanical microenvironment [156].

Fluid shear stress plays an important role in determining endothelial cell shape and remodeling vascular architecture. When exposed to fluid shear stress, endothelial cells elongate with their long axis aligning in the flow direction. They also produce vasodilating substances such as nitric oxide (NO), prostacyclin, and C-type natriuretic peptide [155, 157].

For many years the biomechanical microenvironment has been overlooked as a determinant of stem cell behavior. Gilbert and colleagues reported that when muscle stem cells were cultured on a substrate softer than polystyrene, their survival was enhanced and they were able to regenerate and repair damaged muscle [158]. Park et al. reported that the viability and morphology of mouse fibroblasts was higher on stiffer PDMS matrices [159]. Yamamoto and colleagues reported that shear stress induces Flk-1^+ differentiation of mouse embryonic stem cells (mESCs) [155]. Fluid shear also increased expression of the vascular endothelial cell-specific markers Flk-1, Flt-1, vascular endothelial cadherin, and PECAM-1 at both the protein and the mRNA levels [155]. Adamo and colleagues demonstrated that fluid shear stress augmented the expression of *Runx1* in $CD41^+$ c-Kit^+ hematopoietic progenitor cells and increased hematopoietic colony-forming potential using mouse ESCs [160].

A fluid shear gradient was created in a microchannel (see Fig. 19.7a, b) to study the effect of fluid shear stress on cell attachment and migration [154]. For wall shear stress between 4×10^{-4} and 16×10^{-4} Pa, both cell attachment times and migration distances increase (see Fig. 19.7c, d). Such devices will be useful for assessing the role of shear stress in stem cell differentiation and migration.

19.5.3.3 Gradient Generation

Cells direct migration of other cells by secreting chemokines, growth factors, and other factors into their local microenvironment. These biochemical gradients play a central role in directing cell migration during embryonic development or repair/regeneration of damaged tissues by stem cells or their differentiated progeny [161]. Traditional methods for generating chemical gradients such as the Boyden chamber [162], Zigmond chamber [163], Dunn chamber [164], and micropipette-based assays [165, 166] do not offer the level of control provided by microfluidic designs [161, 167].

Microfluidic gradient generators are classified into flow- and diffusion-based designs. Flow-based devices generate gradients by combining fluid streams with different solute concentrations and require convective and diffusive mixing to create gradients. On the other hand, diffusive devices purely rely on the passive flux of molecules between a source (high concentration) and sink (low concentration). According to Fick's law of diffusion at steady state, diffusion-based devices generate

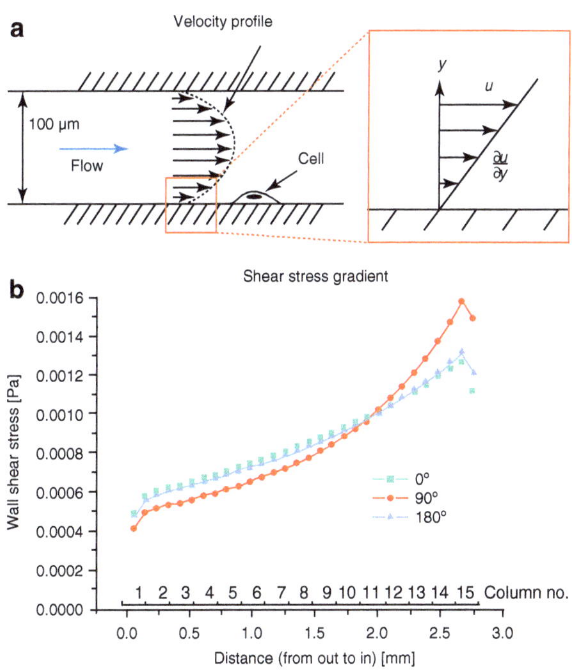

Fig. 19.7 (**a**) Laminar flow in a microchannel generates a parabolic velocity profile with wall shear stress directly proportional to fluid shear rate. (**b**) Shear stress in various microfluidic channels (column 1–15). Relationship of attachment time (**c**) and the migration distance (**d**) to shear stress using L929 cells [154] (Adapted from [154] with permission)

Fig. 19.7 (continued)

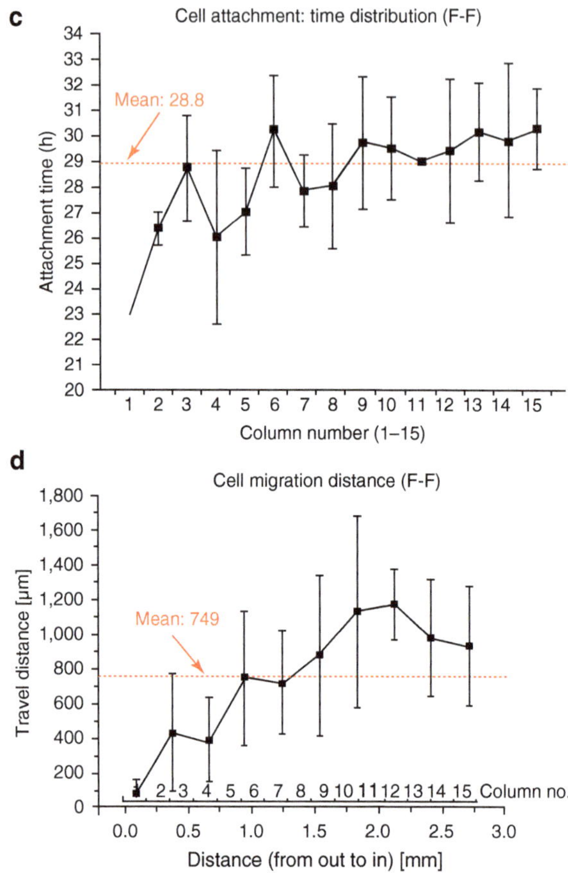

a linear gradient between the source and sink [161]. Gradients can also be generated with solids [168] to control stiffness [169] and gases such as oxygen [170].

There are now many studies that employ microfluidic devices for generating soluble gradients. Sip and co-workers developed a multilayer microfluidic cell culture device enabling mixing and generation of stable chemical gradients [171]. Sahai and colleagues developed a two-layer microfluidic network where pneumatic control channels were bonded to the bottom of a glass Petri dish to generate biochemical gradients by laminar flow [172]. Chung et al. developed a gradient generator that could reversibly bond with a Petri dish to generate gradients of soluble factors [173].

The Jeon group cultured human neural stem cells for a week in epidermal growth factor, fibroblast growth factor 2, and platelet-derived growth factor gradients [174]. They were able to examine the effect of growth factor concentration on neural stem cell proliferation and differentiation by time-lapse imaging. The Lee group utilized a osmosis-driven pump to generate fetal bovine serum gradients for study of human mesenchymal stem cell biology [175]. The same group cultured neural progenitors derived from human embryonic stem cells in continuous cytokine gradients for 8 days [176].

They correlated sonic hedgehog concentrations with neural differentiation, neuronal cell body clusters, and neurite formation [176]. The Beebe group studied human embryonic stem cell colony formation using a microfluidic gradient generator [177].

19.5.4 Control of pH and Media Oxygenation

Metabolic factors play a crucial role in cell development. Mammalian cell culture utilizes $NaHCO_3$ added to media with a 5 % CO_2 in air gas phase [178]. This buffer system effectively maintains the pH in the range of 7.2–7.4. When it is not possible to control the partial pressure of carbon dioxide in microdevices, the buffer 4-(2-hydroxyethyl)-1-piperazineethanesulfonic acid (HEPES) is added to media. The high permeability of PDMS to carbon dioxide allows one to control pH in microfluidics using $NaHCO_3$ buffers. The Groisman group developed two-layer microfluidic devices (see Fig. 19.8) with on-chip gas mixing [170]. The gas and flow layers were separated by a 50-μm PDMS membrane. O_2 and CO_2 supplied to the gas layer permeated into the cell culture network providing precise control of dissolved O_2 and CO_2.

Oxygen plays a central role in modulating transcriptional pathways during development and tissue repair. The oxygen permeability of PDMS has led to many innovative designs for control of oxygen concentration [179–183]. Grist et al. have reviewed optical O_2 sensors for application in microfluidic cell culture [184, 185].

Alder et al. designed a two-layer microfluidic network with a computer-controlled multichannel gas mixer to generate linear, exponential, and non-monotonic O_2 gradients for cell culture [186]. Thomas and colleagues exploited pre-equilibrated aqueous solutions in gas-control channels to regulate the oxygen concentration in static microfluidic chambers [187]. Chen et al. developed a single-layer microfluidic network with a middle channel (1 mm wide) for cell culture and two parallel side channels (100 μm in width) for oxygen generation by chemical reactions. The O_2 released

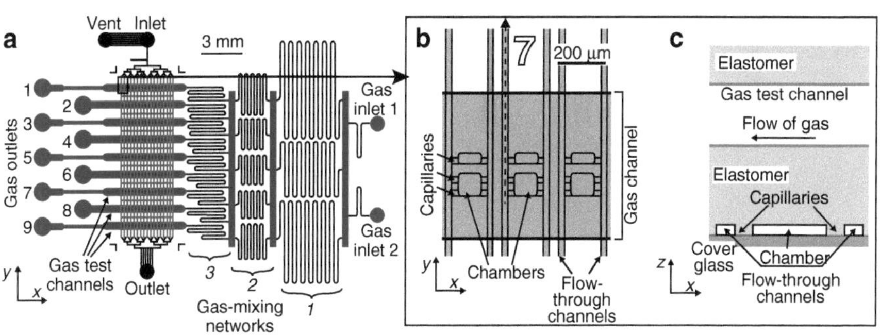

Fig. 19.8 A cell culture chip with an on-chip gas mixing network [170]. (**a**) Top view. Liquid channels for cell culture shown in *black* and gas channels shown in *gray*. An enlarged view of the region inside the *red box* is shown in (**b**) micrograph (top view) and (**c**) schematic cross-sectional view (Adapted from [170] with permission)

during the reaction in pneumatic channels permeated through 50-μm-thick membranes into cell culture channels and could generate O_2 gradients along the channel [188]. This device generates O_2 gradients without connection to an external source.

19.5.5 Temperature Regulation

Maintaining a constant temperature is critical for on-chip cell culture. This is achieved by (a) placing the microfluidic device inside an incubator, (b) enclosing the device within an environmental chamber that controls temperature, or (c) integrating a microheater within the device. Most studies utilize a standard tissue culture incubator without live-cell imaging [98, 189]. Both environmental chambers [190] and embedded microheaters [191, 192] are compatible with live-cell imaging.

Gaitan and Locascio developed an integrated microheating element that was embedded in PDMS microchannels [193]. Glass slide coated with transparent indium tin oxide (ITO) electrical-conducting films generates heat for regulating temperature [185, 191]. As shown in Fig. 19.9, microheaters and gold temperature

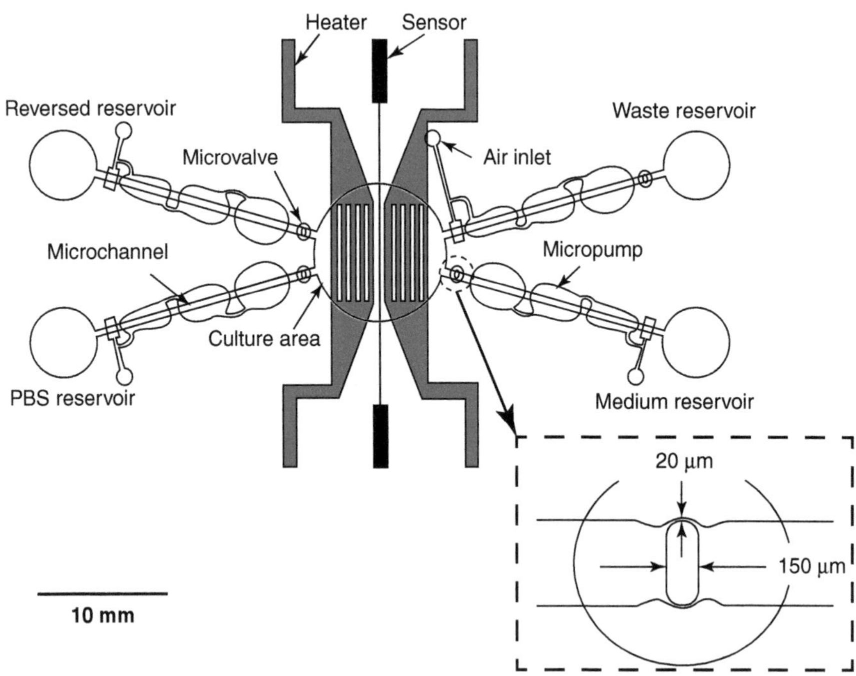

Fig. 19.9 Schematic of an automatic cell culture system [185]. Heaters and sensor are deposited onto the glass slide underneath the PDMS device. The PDMS device has a cell culture area, four micropumps, four micro check valves, microchannels, and reservoirs. *Inset* image shows a micro check valve (Adapted from [185] with permission)

sensors were used for precise temperature regulation (37 ± 0.1 °C) for a system that was mounted on the stage of an inverted microscope [185]. Vigolo and colleagues have generated temperature gradients using a silver-filled epoxy [194]. Other methods for regulating heat production within microdevices include boron-doped polysilicon [195], silver paint [196], and flexible printed circuits [197].

19.6 Time-Lapse Imaging and Clonal Analysis

19.6.1 Time-Lapse Imaging

Compared to discrete observation, continuous imaging of cell fate will enable detailed analysis of stem dynamics [2]. The continuous (or time-lapse) imaging of cell growth refers to the imaging of individual cells at intervals in the order of minutes so that the movement, division, and proliferation of cells are able to be tracked [2]. Schroeder explained the necessity of continuous imaging of cell growth for interpreting observed input and output data [2]. For a discrete observation with one initial cell and four cells at another time point, one does not know the intermediate steps or lineage relations of these cells. Therefore, time-lapse imaging is indispensable for cell lineage analysis [2]. In this section recent developments of microfluidics integrated with time-lapse imaging for cellular analyses will be reviewed.

The monolithic design of lab-on-a-chip facilitates direct observations by time-lapse imaging. Quake's group pioneered this new field by integrating time-lapse imaging with on-chip cell culture. High-content imaging revealed that motility of human MSCs decreased as cells were stimulated with osteogenic medium [52]. King et al. reported the use of fluorescent protein transcriptional reporters, live-cell imaging, and microfluidic perfusion culture to study gene expression programs in living cells [198]. The Lee group studied nanoparticle (drug)-cell interactions using both bright-field and fluorescent time-lapse imaging for 24 h [191]. Lee et al. studied the response of cells to a programmable shear environment by live-cell imaging [199]. Albrecht and co-workers employed a microfluidic system with long-term (120-h) time-lapse imaging of mitotic kinetics and spindle orientation [200]. Gilbert and colleagues cultured MSCs with time-lapse imaging and developed a highly automated algorithm to track lineage relationships [158]. Lecault et al. studied the clonal development of hematopoietic stem cells grown in microwell arrays by time-lapse imaging [5].

In addition, Rieger and Schroeder discuss developments in fluorescent time-lapse imaging and single-cell tracking [201]. Studies of cellular dynamic events at the single-cell level using time-lapse fluorescence microscopy and automated image analysis have been reviewed by Muzzey and Oudenaarden [202]. Vasdekis and co-workers reviewed single-molecule imaging in microfluidics [203]. Chirieleison et al. reviewed live-cell imaging technical issues such as automated microscope stages, environmental control systems, image acquisition systems, and image analysis [204].

19.6.2 Clonal Analysis

Clonal analysis is of fundamental importance for in vitro cell expansion, gene therapy, tissue repair, and study of stem cell biology [205]. For example, hematopoiesis is maintained by a relatively small number of HSCs, and lineage commitment occurs early on in development. So the study of highly purified stem cells at a clonal level is necessary [205–207]. Commonly applied methodologies for clonal tracking involve genetic mosaics, single-cell transplants, retroviral marking, recombination, and single-cell imaging both in vivo and in vitro [208]. In vitro time-lapse imaging has been extensively applied for clonal analysis using conventional culture platforms such as flasks, Petri dishes, and well plates [201, 209].

Recently, the integration of time-lapse imaging with microwell culture has enabled high-throughput clonal analysis. Chin and colleagues studied adult rat neural stem cell proliferation using a microwell culture [112]. Lindström et al. demonstrated the utility of microwell arrays for the clonal analysis of both mESCs and hESCs [210].

To date, there are only a few studies that combine clonal analysis using microfluidic perfusion culture integrated with continuous time-lapse imaging, probably due to the added complexity of combining these technologies within a single platform. Lecault et al. has imaged hematopoietic stem cells cultured in 1,600 microwells with media perfusion every 5 min for 60 h [5]. Chen et al. have also developed microwell perfusion culture. A large number of clones are grown over a relatively small area, so an automatic microscope can scan hundreds of colonies in a few minutes. For example, 500 microwells were scanned at 3-min intervals for 6 days, providing live-cell movies for over 1,000 clones [124]. Continuous live-cell imaging provides a division tree (see Fig. 19.10b), showing lineage relationships for each clone in culture. However, the task of analyzing the large volume of video data (terabytes) is limiting application of high-throughput cell lineage mapping. There are simply not enough human resources to manually track thousands of cells in these movies. So a major effort will be required to automate identification of cell trajectory paths using phase-contrast time-lapse movies [158, 211].

Stem cell fates are stochastic, so appropriate statistical and mathematical methods are required to quantify stem cell differentiation dynamics from hundreds of clonal pedigrees. Nordon and colleagues applied a multi-type branching model to describe the development of human cord blood CD 34^{+} cells and mouse granulocyte-macrophage progenitors at both single- and multiple-cell levels [212]. The multi-type Smith-Martin model is a parametric description of cell differentiation dynamics relating growth factors and microenvironment to the rates of cell mitosis, differentiation, and apoptosis. These *in silico* models describe how microenvironment and developmental history instruct stem cell differentiation dynamics.

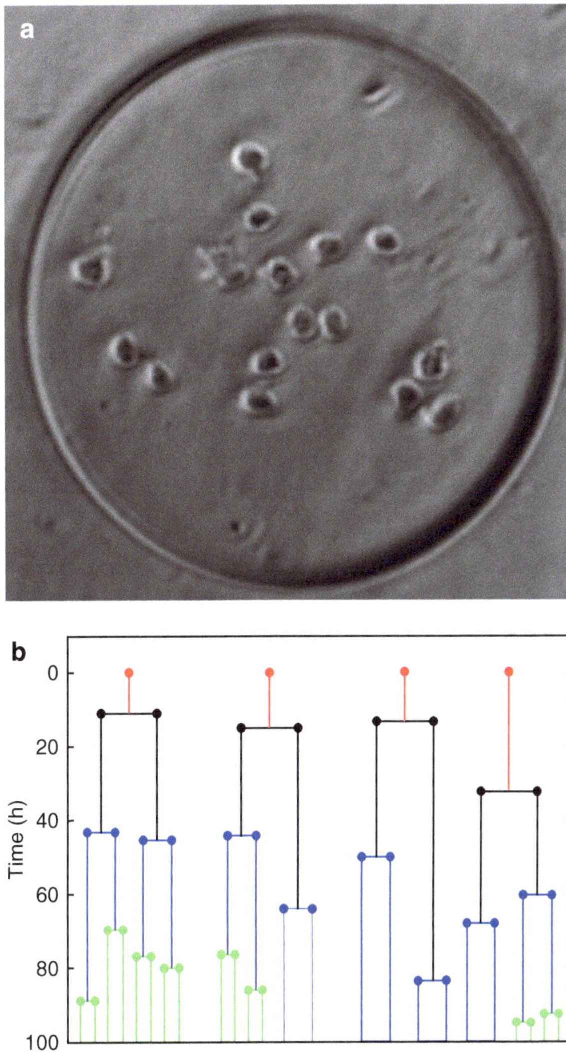

Fig. 19.10 (**a**) Still image from a video showing cell division and movement over 6 days in a microwell (video online) (http://www.ncbi.nlm.nih.gov/pmc/articles/PMC3260560/figure/f5/). The time interval between two frames was 6 min. (**b**) Cell lineages of four individual clones in four wells created by manual processing of image stacks for 100-h cell culture (8,000 images). *Dots* represent cells [124]. (Reprinted with permission from [124]. Copyright [2011], American Institute of Physics)

19.7 Conclusions and Future Research

Microfluidics will provide a valuable tool for stem cell study by enabling precise control of microenvironment with time-lapse imaging of stem cell fates. Lab-on-a-chip has the potential of increasing stem cell research output by parallel integration and automation, with a dramatic reduction in the quantity of reagents required to perform microscale culture experiments. Future research will focus on how to bring this technology to the biologist's lab bench. From this chapter, it is apparent that most of the research effort is still focused on technical questions; however, the number of studies that successfully apply this technology to answer biological questions is growing at a rapid pace. Application of this technology to stem cell research will require training of researchers with interdisciplinary background who will apply this highly quantitative technology to better understand the regulation of stem cell fates.

References

1. Csete M (2010) Q&A: what can microfluidics do for stem-cell research? J Biol 9(1):1
2. Schroeder T (2008) Imaging stem-cell-driven regeneration in mammals. Nature 453(7193):345–351. doi:10.1038/nature07043
3. Sia SK, Whitesides GM (2003) Microfluidic devices fabricated in poly(dimethylsiloxane) for biological studies. Electrophoresis 24(21):3563–3576. doi:10.1002/elps.200305584
4. Lin B, Basuray S (2011) Microfluidics: technologies and applications. Springer, Heidelberg
5. Lecault V, VanInsberghe M, Sekulovic S, Knapp D, Wohrer S, Bowden W, Viel F, McLaughlin T, Jarandehei A, Miller M, Falconnet D, White AK, Kent DG, Copley MR, Taghipour F, Eaves CJ, Humphries RK, Piret JM, Hansen CL (2011) High-throughput analysis of single hematopoietic stem cell proliferation in microfluidic cell culture arrays. Nat Methods 8(7):581–586. doi:10.1038/nmeth.1614
6. Kim L, Toh YC, Voldman J, Yu H (2007) A practical guide to microfluidic perfusion culture of adherent mammalian cells. Lab Chip 7(6):681–694. doi:10.1039/b704602b
7. Harrison DJ, Glavina PG, Manz A (1993) Towards miniaturized electrophoresis and chemical analysis systems on silicon: an alternative to chemical sensors. Sens Actuators B Chem 10(2):107–116. doi:10.1016/0925-4005(93)80033-8
8. Harrison DJ, Manz A, Fan Z, Luedi H, Widmer HM (1992) Capillary electrophoresis and sample injection systems integrated on a planar glass chip. Anal Chem 64(17):1926–1932. doi:10.1021/ac00041a030
9. Harrison DJ, Fluri K, Seiler K, Fan Z, Effenhauser CS, Manz A (1993) Micromachining a miniaturized capillary electrophoresis-based chemical analysis system on a chip. Science 261(5123):895–897
10. Nguyen N-T (2006) Fundamentals and applications of microfluidics. Artech House, Boston
11. Effenhauser CS, Bruin GJM, Paulus A, Ehrat M (1997) Integrated capillary electrophoresis on flexible silicone microdevices: analysis of DNA restriction fragments and detection of single DNA molecules on microchips. Anal Chem 69(17):3451–3457. doi:10.1021/ac9703919
12. Duffy DC, McDonald JC, Schueller OJA, Whitesides GM (1998) Rapid prototyping of microfluidic systems in poly(dimethylsiloxane). Anal Chem 70(23):4974–4984
13. Goral VN, Hsieh YC, Petzold ON, Faris RA, Yuen PK (2011) Hot embossing of plastic microfluidic devices using poly(dimethylsiloxane) molds. J Micromech Microeng 21(1). doi:01700210.1088/0960-1317/21/1/017002
14. Young EWK, Berthier E, Guckenberger DJ, Sackmann E, Lamers C, Meyvantsson I, Huttenocher A, Beebe DJ (2011) Rapid prototyping of arrayed microfluidic systems in polystyrene for cell-based assays. Anal Chem 83(4):1408–1417. doi:10.1021/ac102897h
15. Chen CS, Breslauer DN, Luna JI, Grimes A, Chin WC, Leeb LP, Khine M (2008) Shrinky-dink microfluidics: 3D polystyrene chips. Lab Chip 8(4):622–624. doi:10.1039/b719029h
16. Zhang W, Lin S, Wang C, Hu J, Li C, Zhuang Z, Zhou Y, Mathies RA, Yang CJ (2009) PMMA/PDMS valves and pumps for disposable microfluidics. Lab Chip 9(21):3088–3094
17. Kuo JS, Ng L, Yen GS, Lorenz RM, Schiro PG, Edgar JS, Zhao Y, Lim DSW, Allen PB, Jeffries GDM, Chiu DT (2009) A new USP class VI-compliant substrate for manufacturing disposable microfluidic devices. Lab Chip 9(7):870–876

18. Kuo JS, Zhao Y, Ng L, Yen GS, Lorenz RM, Lim DSW, Chiu DT (2009) Microfabricating high-aspect-ratio structures in polyurethane-methacrylate (PUMA) disposable microfluidic devices. Lab Chip 9(13):1951–1956
19. Sikanen T, Aura S, Heikkilä L, Kotiaho T, Franssila S, Kostiainen R (2010) Hybrid ceramic polymers: new, nonbiofouling, and optically transparent materials for microfluidics. Anal Chem 82(9):3874–3882. doi:10.1021/ac1004053
20. Kuo JS, Chiu DT (2011) Disposable microfluidic substrates: transitioning from the research laboratory into the clinic. Lab Chip 11(16):2656–2665. doi:10.1039/c1lc20125e
21. Park J, Lee D, Kim W, Horiike S, Nishimoto T, Lee SH, Ahn CH (2007) Fully packed capillary electrochromatographic microchip with self-assembly colloidal silica beads. Anal Chem 79(8):3214–3219. doi:10.1021/ac061714g
22. Do J, Ahn CH (2008) A polymer lab-on-a-chip for magnetic immunoassay with on-chip sampling and detection capabilities. Lab Chip 8(4):542–549
23. Martinez AW, Phillips ST, Whitesides GM, Carrilho E (2009) Diagnostics for the developing world: microfluidic paper-based analytical devices. Anal Chem 82(1):3–10. doi:10.1021/ac9013989
24. Robert P (2009) Bioactive paper provides a low-cost platform for diagnostics. Trends Analyt Chem 28(8):925–942. doi:10.1016/j.trac.2009.05.005
25. Niklaus F, Stemme G, Lu JQ, Gutmann RJ (2006) Adhesive wafer bonding. J Appl Phys 99(3). doi:10.1063/1.2168512
26. Ko JS, Yoon HC, Yang HS, Pyo HB, Chung KH, Kim SJ, Kim YT (2003) A polymer-based microfluidic device for immunosensing biochips. Lab Chip 3(2):106–113. doi:10.1039/b301794j
27. McDonald JC, Whitesides GM (2002) Poly(dimethylsiloxane) as a material for fabricating microfluidic devices. Acc Chem Res 35(7):491–499. doi:10.1021/ar010110q
28. Khademhosseini A, Yeh J, Eng G, Karp J, Kaji H, Borenstein J, Farokhzad OC, Langer R (2005) Cell docking inside microwells within reversibly sealed microfluidic channels for fabricating multiphenotype cell arrays. Lab Chip 5(12):1380–1386. doi:10.1039/b508096g
29. McDonald JC, Duffy DC, Anderson JR, Chiu DT, Wu HK, Schueller OJA, Whitesides GM (2000) Fabrication of microfluidic systems in poly(dimethylsiloxane). Electrophoresis 21(1):27–40
30. Eddings MA, Johnson MA, Gale BK (2008) Determining the optimal PDMS-PDMS bonding technique for microfluidic devices. J Micromech Microeng 18(6). doi:10.1088/0960-1317/18/6/067001
31. Unger MA, Chou HP, Thorsen T, Scherer A, Quake SR (2000) Monolithic microfabricated valves and pumps by multilayer soft lithography. Science 288(5463):113–116
32. Hillborg H, Gedde UW (1998) Hydrophobicity recovery of polydimethylsiloxane after exposure to corona discharges. Polymer 39(10):1991–1998. doi:10.1016/s0032-3861(97)00484-9
33. Hillborg H, Ankner JF, Gedde UW, Smith GD, Yasuda HK, Wikstrom K (2000) Crosslinked polydimethylsiloxane exposed to oxygen plasma studied by neutron reflectometry and other surface specific techniques. Polymer 41(18):6851–6863. doi:10.1016/s0032-3861(00)00039-2
34. Bodas D, Khan-Malek C (2007) Hydrophilization and hydrophobic recovery of PDMS by oxygen plasma and chemical treatment – an SEM investigation. Sens Actuator B Chem 123(1):368–373. doi:10.1016/j.snb.2006.08.037
35. Haubert K, Drier T, Beebe D (2006) PDMS bonding by means of a portable, low-cost corona system. Lab Chip 6(12):1548–1549. doi:10.1039/b610567j
36. Eddings MA, Gale BK (2006) A PDMS-based gas permeation pump for on-chip fluid handling in microfluidic devices. J Micromech Microeng 16(11):2396–2402. doi:10.1088/0960-1317/16/11/021
37. Go JS, Shoji S (2004) A disposable, dead volume-free and leak-free in-plane PDMS microvalve. Sens Actuator A Phys 114(2–3):438–444. doi:10.1016/j.sna.2003.12.028
38. Samel B, Chowdhury MK, Stemme G (2007) The fabrication of microfluidic structures by means of full-wafer adhesive bonding using a poly(dimethylsiloxane) catalyst. J Micromech Microeng 17(8):1710–1714. doi:10.1088/0960-1317/17/8/038

39. Im SG, Bong KW, Lee CH, Doyle PS, Gleason KK (2009) A conformal nano-adhesive via initiated chemical vapor deposition for microfluidic devices. Lab Chip 9(3):411–416. doi:10.1039/b812121d
40. Rezai P, Selvaganapathy PR, Rwohl G (2011) Plasma enhanced bonding of polydimethylsiloxane with parylene and its optimization. J Micromech Microeng 21(6). doi:10.1088/0960-1317/21/6/065024
41. Tsao CW, DeVoe DL (2009) Bonding of thermoplastic polymer microfluidics. Microfluid Nanofluid 6(1):1–16. doi:10.1007/s10404-008-0361-x
42. Thorsen T, Maerkl SJ, Quake SR (2002) Microfluidic large-scale integration. Science 298(5593):580–584. doi:10.1126/science.1076996
43. Melin J, Quake SR (2007) Microfluidic large-scale integration: the evolution of design rules for biological automation. Annu Rev Biophys Biomol Struct 36:213–231. doi:10.1146/annurev.biophys.36.040306.132646
44. Hansen CL, Classen S, Berger JM, Quake SR (2006) A microfluidic device for kinetic optimization of protein crystallization and in situ structure determination. J Am Chem Soc 128(10):3142–3143. doi:10.1021/ja0576637
45. Hansen CL, Skordalakes E, Berger JM, Quake SR (2002) A robust and scalable microfluidic metering method that allows protein crystal growth by free interface diffusion. Proc Natl Acad Sci USA 99(26):16531–16536
46. Hansen CL, Sommer MOA, Quake SR (2004) Systematic investigation of protein phase behavior with a microfluidic formulator. Proc Natl Acad Sci USA 101(40):14431–14436. doi:10.1073/pnas.0405847101
47. Liu J, Hansen C, Quake SR (2003) Solving the "world-to-chip" interface problem with a microfluidic matrix. Anal Chem 75(18):4718–4723. doi:10.1021/ac0346407
48. Skelley AM, Scherer JR, Aubrey AD, Grover WH, Ivester RHC, Ehrenfreund P, Grunthaner FJ, Bada JL, Mathies RA (2005) Development and evaluation of a microdevice for amino acid biomarker detection and analysis on Mars. Proc Natl Acad Sci USA 102(4):1041–1046. doi:10.1073/pnas.0406798102
49. Gu W, Zhu XY, Futai N, Cho BS, Takayama S (2004) Computerized microfluidic cell culture using elastomeric channels and Braille displays. Proc Natl Acad Sci USA 101(45):15861–15866. doi:10.1073/pnas.0404353101
50. Marcus JS, Anderson WF, Quake SR (2006) Microfluidic single-cell mRNA isolation and analysis. Anal Chem 78(9):3084–3089. doi:10.1021/ac0519460
51. Ng JMK, Gitlin I, Stroock AD, Whitesides GM (2002) Components for integrated poly(dimethylsiloxane) microfluidic systems. Electrophoresis 23(20):3461–3473
52. Gomez-Sjoberg R, Leyrat AA, Pirone DM, Chen CS, Quake SR (2007) Versatile, fully automated, microfluidic cell culture system. Anal Chem 79(22):8557–8563. doi:10.1021/ac071311w
53. Yamahata C, Lacharme F, Gijs MAM (2005) Glass valveless micropump using electromagnetic actuation. Microelectron Eng 78–79:132–137. doi:10.1016/j.mee.2004.12.018
54. Teymoori MM, Abbaspour-Sani E (2005) Design and simulation of a novel electrostatic peristaltic micromachined pump for drug delivery applications. Sens Actuators A Phys 117(2):222–229. doi:10.1016/j.sna.2004.06.025
55. Kim J-H, Kang CJ, Kim Y-S (2004) A disposable polydimethylsiloxane-based diffuser micropump actuated by piezoelectric-disc. Microelectron Eng 71(2):119–124. doi:10.1016/j.mee.2003.10.005
56. Yoo JC, Moon MC, Choi YJ, Kang CJ, Kim YS (2006) A high performance microfluidic system integrated with the micropump and microvalve on the same substrate. Microelectron Eng 83(4–9):1684–1687. doi:10.1016/j.mee.2006.01.202
57. Xu D, Wang L, Ding G, Zhou Y, Yu A, Cai B (2001) Characteristics and fabrication of NiTi/Si diaphragm micropump. Sens Actuators A Phys 93(1):87–92. doi:10.1016/s0924-4247(01)00628-8
58. Goedecke N, Eijkel J, Manz A (2002) Evaporation driven pumping for chromatography application. Lab Chip 2(4):219–223

59. Hosokawa K, Sato K, Ichikawa N, Maeda M (2004) Power-free poly(dimethylsiloxane) microfluidic devices for gold nanoparticle-based DNA analysis. Lab Chip 4(3):181–185. doi:10.1039/b403930k
60. Zhang T, Cui T (2011) High-performance surface-tension-driven capillary pumping based on layer-by-layer self assembly of TiO2 nanoparticles. International Conference on Solid State Sensors and Actuators - TRANSDUCERS. Beijing. 5–9 June 2011. doi:10.1109/TRANSDUCERS.2011.5969775
61. Kamei KI, Guo SL, Yu ZTF, Takahashi H, Gschweng E, Suh C, Wang XP, Tang JG, McLaughlin J, Witte ON, Lee KB, Tseng HR (2009) An integrated microfluidic culture device for quantitative analysis of human embryonic stem cells. Lab Chip 9(4):555–563. doi:10.1039/b809105f
62. Liu WM, Li L, Wang XM, Ren L, Wang XQ, Wang JC, Tu Q, Huang XW, Wang JY (2010) An integrated microfluidic system for studying cell-microenvironmental interactions versatilely and dynamically. Lab Chip 10(13):1717–1724. doi:10.1039/c001049a
63. Falconnet D, Niemisto A, Taylor RJ, Ricicova M, Galitski T, Shmulevich I, Hansen CL (2011) High-throughput tracking of single yeast cells in a microfluidic imaging matrix. Lab Chip 11(3):466–473. doi:10.1039/c0lc00228c
64. Grover WH, Ivester RHC, Jensen EC, Mathies RA (2006) Development and multiplexed control of latching pneumatic valves using microfluidic logical structures. Lab Chip 6(5):623–631. doi:10.1039/b518362f
65. Hulme SE, Shevkoplyas SS, Whitesides GM (2009) Incorporation of prefabricated screw, pneumatic, and solenoid valves into microfluidic devices. Lab Chip 9(1):79–86. doi:10.1039/b809673b
66. Park W, Han S, Kwon S (2010) Fabrication of membrane-type microvalves in rectangular microfluidic channels via seal photopolymerization. Lab Chip 10(20):2814–2817. doi:10.1039/c005173j
67. Lai HY, Folch A (2011) Design and dynamic characterization of "single-stroke" peristaltic PDMS micropumps. Lab Chip 11(2):336–342. doi:10.1039/c0lc00023j
68. Huang CW, Huang SB, Lee GB (2006) Pneumatic micropumps with serially connected actuation chambers. J Micromech Microeng 16(11):2265–2272. doi:10.1088/0960-1317/16/11/003
69. Skelley AM, Voldman J (2008) An active bubble trap and debubbler for microfluidic systems. Lab Chip 8(10):1733–1737. doi:10.1039/b807037g
70. Johnson M, Liddiard G, Eddings M, Gale B (2009) Bubble inclusion and removal using PDMS membrane-based gas permeation for applications in pumping, valving and mixing in microfluidic devices. J Micromech Microeng 19(9). doi:09501110.1088/0960-1317/19/9/095011
71. Zheng WF, Wang Z, Zhang W, Jiang XY (2010) A simple PDMS-based microfluidic channel design that removes bubbles for long-term on-chip culture of mammalian cells. Lab Chip 10(21):2906–2910. doi:10.1039/c005274d
72. Kang JH, Kim YC, Park JK (2008) Analysis of pressure-driven air bubble elimination in a microfluidic device. Lab Chip 8(1):176–178. doi:10.1039/b712672g
73. Freeman BD, Pinnau I (1999) Polymeric materials for gas separations. In: Polymer membranes for gas and vapor separation, vol 733, ACS symposium series. American Chemical Society, Washington, pp 1–27. doi:10.1021/bk-1999-0733.ch001
74. Cheng DM, Jiang HR (2009) A debubbler for microfluidics utilizing air-liquid interfaces. Appl Phys Lett 95(21). doi:21410310.1063/1.3263944
75. Sung JH, Shuler ML (2009) Prevention of air bubble formation in a microfluidic perfusion cell culture system using a microscale bubble trap. Biomed Microdevices 11(4):731–738. doi:10.1007/s10544-009-9286-8
76. Kang E, Lee DH, Kim CB, Yoo SJ, Lee SH (2010) A hemispherical microfluidic channel for the trapping and passive dissipation of microbubbles. J Micromech Microeng 20(4). doi:045009 10.1088/0960-1317/20/4/045009
77. Liu CC, Thompson JA, Bau HH (2011) A membrane-based, high-efficiency, microfluidic debubbler. Lab Chip 11(9):1688–1693. doi:10.1039/c1lc20089e

78. Raser JM, O'Shea EK (2005) Noise in gene expression: origins, consequences, and control. Science 309(5743):2010–2013. doi:10.1126/science.1105891
79. Rao CV, Wolf DM, Arkin AP (2002) Control, exploitation and tolerance of intracellular noise. Nature 420(6912):231–237. doi:10.1038/nature01258
80. Lindstrom S, Andersson-Svahn H (2010) Overview of single-cell analyses: microdevices and applications. Lab Chip 10(24):3363–3372. doi:10.1039/c0lc00150c
81. Gallego-Perez D, Higuita-Castro N, Sharma S, Reen RK, Palmer AF, Gooch KJ, Lee LJ, Lannutti JJ, Hansford DJ (2010) High throughput assembly of spatially controlled 3D cell clusters on a micro/nanoplatform. Lab Chip 10(6):775–782. doi:10.1039/b919475d
82. Park JY, Morgan M, Sachs AN, Samorezov J, Teller R, Shen Y, Pienta KJ, Takayama S (2010) Single cell trapping in larger microwells capable of supporting cell spreading and proliferation. Microfluid Nanofluid 8(2):263–268. doi:10.1007/s10404-009-0503-9
83. Tokilmitsu Y, Kishi H, Kondo S, Honda R, Tajiri K, Motoki K, Ozawa T, Kadowaki S, Obata T, Fujiki S, Tateno C, Takaishi H, Chayama K, Yoshizato K, Tamiya E, Sugiyama T, Muraguchi A (2007) Single lymphocyte analysis with a microwell array chip. Cytometry A 71A(12):1003–1010. doi:10.1002/cyto.a.20478
84. Deutsch M, Deutsch A, Shirihai O, Hurevich I, Afrimzon E, Shafran Y, Zurgil N (2006) A novel miniature cell retainer for correlative high-content analysis of individual untethered non-adherent cells. Lab Chip 6(8):995–1000. doi:10.1039/b603961h
85. Liu CS, Liu JJ, Gao D, Ding MY, Lin JM (2010) Fabrication of microwell arrays based on two-dimensional ordered polystyrene microspheres for high-throughput single-cell analysis. Anal Chem 82(22):9418–9424. doi:10.1021/ac102094r
86. Rosenbluth MJ, Lam WA, Fletcher DA (2006) Force microscopy of nonadherent cells: a comparison of leukemia cell deformability. Biophys J 90(8):2994–3003. doi:10.1529/biophysj.105.067496
87. Park MC, Hur JY, Cho HS, Park SH, Suh KY (2011) High-throughput single-cell quantification using simple microwell-based cell docking and programmable time-course live-cell imaging. Lab Chip 11(1):79–86. doi:10.1039/c0lc00114g
88. Lindstrom S, Andersson-Svahn H (2011) Miniaturization of biological assays – overview on microwell devices for single-cell analyses. Biochim Biophys Acta 1810(3):308–316. doi:10.1016/j.bbagen.2010.04.009
89. Charnley M, Textor M, Khademhosseini A, Lutolf MP (2009) Integration column: microwell arrays for mammalian cell culture. Integr Biol 1(11–12):625–634. doi:10.1039/b918172p
90. Khademhosseini A, Ferreira L, Blumling J, Yeh J, Karp JM, Fukuda J, Langer R (2006) Co-culture of human embryonic stem cells with murine embryonic fibroblasts on microwell-patterned substrates. Biomaterials 27(36):5968–5977. doi:10.1016/j.biomaterials.2006.06.035
91. Rettig JR, Folch A (2005) Large-scale single-cell trapping and imaging using microwell arrays. Anal Chem 77(17):5628–5634. doi:10.1021/ac0505977
92. Chen H, Rosengarten G, Li M, Nordon RE (2012) Design of microdevices for long-term live cell imaging. J Micromech Microeng (in press)
93. Cioffi M, Moretti M, Manbachi A, Chung BG, Khademhosseini A, Dubini G (2010) A computational and experimental study inside microfluidic systems: the role of shear stress and flow recirculation in cell docking. Biomed Microdevices 12(4):619–626. doi:10.1007/s10544-010-9414-5
94. Çengel YA, Cimbala JM (2010) Fluid mechanics: fundamentals and applications. McGraw-Hill Higher Education, Boston
95. Glatzel T, Litterst C, Cupelli C, Lindemann T, Moosmann C, Niekrawletz R, Streule W, Zengerle R, Koltay P (2008) Computational fluid dynamics (CFD) software tools for microfluidic applications – a case study. Comput Fluids 37(3):218–235. doi:10.1016/j.compfluid.2007.07.014
96. Erickson D (2005) Towards numerical prototyping of labs-on-chip: modeling for integrated microfluidic devices. Microfluid Nanofluid 1(4):301–318. doi:10.1007/s10404-005-0041-z
97. Wang ZH, Kim MC, Marquez M, Thorsen T (2007) High-density microfluidic arrays for cell cytotoxicity analysis. Lab Chip 7(6):740–745. doi:10.1039/b618734j

98. Zhang BY, Kim MC, Thorsen T, Wang ZH (2009) A self-contained microfluidic cell culture system. Biomed Microdevices 11(6):1233–1237. doi:10.1007/s10544-009-9342-4
99. Manbachi A, Shrivastava S, Cioffi M, Chung BG, Moretti M, Demirci U, Yliperttula M, Khademhosseini A (2008) Microcirculation within grooved substrates regulates cell positioning and cell docking inside microfluidic channels. Lab Chip 8(5):747–754. doi:10.1039/b718212k
100. Khabiry M, Chung BG, Hancock MJ, Soundararajan HC, Du YN, Cropek D, Lee WG, Khademhosseini A (2009) Cell docking in double grooves in a microfluidic channel. Small 5(10):1186–1194. doi:10.1002/smll.200801644
101. Han C, Zhang QF, Ma R, Xie L, Qiu TA, Wang L, Mitchelson K, Wang JD, Huang GL, Qiao J, Cheng J (2010) Integration of single oocyte trapping, in vitro fertilization and embryo culture in a microwell-structured microfluidic device. Lab Chip 10(21):2848–2854. doi:10.1039/c005296e
102. Kim MC, Wang ZH, Lam RHW, Thorsen T (2008) Building a better cell trap: applying Lagrangian modeling to the design of microfluidic devices for cell biology. J Appl Phys 103(4). doi:04470110.1063/1.2840059
103. Jang YH, Kwon CH, Kim SB, Selimovic S, Sim WY, Bae H, Khademhosseini A (2011) Deep wells integrated with microfluidic valves for stable docking and storage of cells. Biotechnol J 6(2):156–164. doi:10.1002/biot.201000394
104. Yang J, Li CW, Yang MS (2004) Hydrodynamic simulation of cell docking in microfluidic channels with different dam structures. Lab Chip 4(1):53–59. doi:10.1039/b309940g
105. Hufnagel H, Huebner A, Gulch C, Guse K, Abell C, Hollfelder F (2009) An integrated cell culture lab on a chip: modular microdevices for cultivation of mammalian cells and delivery into microfluidic microdroplets. Lab Chip 9(11):1576–1582. doi:10.1039/b821695a
106. Liu LY, Loutherback K, Liao D, Yeater D, Lambert G, Estevez-Torres A, Sturm JC, Getzenberg RH, Austin RH (2010) A microfluidic device for continuous cancer cell culture and passage with hydrodynamic forces. Lab Chip 10(14):1807–1813. doi:10.1039/c003509b
107. Yu ZTF, Kamei KI, Takahashi H, Shu CJ, Wang XP, He GW, Silverman R, Radu CG, Witte ON, Lee KB, Tseng HR (2009) Integrated microfluidic devices for combinatorial cell-based assays. Biomed Microdevices 11(3):547–555. doi:10.1007/s10544-008-9260-x
108. Leclerc E, Sakai Y, Fujii T (2004) Microfluidic PDMS (polydimethylsiloxane) bioreactor for large-scale culture of hepatocytes. Biotechnol Prog 20(3):750–755. doi:10.1021/bp0300568
119. Wang L, Ni XF, Luo CX, Zhang ZL, Pang DW, Chen Y (2009) Self-loading and cell culture in one layer microfluidic devices. Biomed Microdevices 11(3):679–684. doi:10.1007/s10544-008-9278-0
110. Hung PJ, Lee PJ, Sabounchi P, Lin R, Lee LP (2005) Continuous perfusion microfluidic cell culture array for high-throughput cell-based assays. Biotechnol Bioeng 89(1):1–8. doi:10.1002/bit.20289
111. Powers MJ, Domansky K, Kaazempur-Mofrad MR, Kalezi A, Capitano A, Upadhyaya A, Kurzawski P, Wack KE, Stolz DB, Kamm R, Griffith LG (2002) A microfabricated array bioreactor for perfused 3D liver culture. Biotechnol Bioeng 78(3):257–269. doi:10.1002/bit.10143
112. Chin VI, Taupin P, Sanga S, Scheel J, Gage FH, Bhatia SN (2004) Microfabricated platform for studying stem cell fates. Biotechnol Bioeng 88(3):399–415
113. Wu MH, Huang SB, Cui ZF, Cui Z, Lee GB (2008) A high throughput perfusion-based microbioreactor platform integrated with pneumatic micropumps for three-dimensional cell culture. Biomed Microdevices 10(2):309–319. doi:10.1007/s10544-007-9138-3
114. Shah P, Vedarethinam I, Kwasny D, Andresen L, Dimaki M, Skov S, Svendsen WE (2011) Microfluidic bioreactors for culture of non-adherent cells. Sens Actuator B Chem 156(2):1002–1008. doi:10.1016/j.snb.2011.02.021
115. Rodrigues CAV, Fernandes TG, Diogo MM, da Silva CL, Cabral JMS (2011) Stem cell cultivation in bioreactors. Biotechnol Adv 29(6):815–829. doi:10.1016/j.biotechadv.2011.06.009
116. Wu HW, Lin CC, Lee GB (2011) Stem cells in microfluidics. Biomicrofluidics 5(1). doi:10.1063/1.3528299

117. Zervantonakis IK, Kothapalli CR, Chung S, Sudo R, Kamm RD (2011) Microfluidic devices for studying heterotypic cell-cell interactions and tissue specimen cultures under controlled microenvironments. Biomicrofluidics 5(1). doi:01340610.1063/1.3553237
118. Paguirigan AL, Beebe DJ (2009) From the cellular perspective: exploring differences in the cellular baseline in macroscale and microfluidic cultures. Integr Biol 1(2):182–195. doi:10.1039/b814565b
119. Lee JN, Park C, Whitesides GM (2003) Solvent compatibility of poly(dimethylsiloxane)-based microfluidic devices. Anal Chem 75(23):6544–6554. doi:10.1021/ac0346712
120. Eddington DT, Puccinelli JP, Beebe DJ (2006) Thermal aging and reduced hydrophobic recovery of polydimethylsiloxane. Sens Actuator B Chem 114(1):170–172. doi:10.1016/j.snb.2005.04.037
121. Hillborg H, Gedde UW (1999) Hydrophobicity changes in silicone rubbers. IEEE Trans Dielectr Electr Insul 6(5):703–717. doi:10.1109/94.798127
122. Regehr KJ, Domenech M, Koepsel JT, Carver KC, Ellison-Zelski SJ, Murphy WL, Schuler LA, Alarid ET, Beebe DJ (2009) Biological implications of polydimethylsiloxane-based microfluidic cell culture. Lab Chip 9(15):2132–2139. doi:10.1039/b903043c
123. Wlodkowic D, Faley S, Skommer J, McGuinness D, Cooper JM (2009) Biological implications of polymeric microdevices for live cell assays. Anal Chem 81(23):9828–9833. doi:10.1021/ac902010s
124. Chen H, Li J, Zhang H, Li M, Rosengarten G, Nordon RE (2011) Microwell perfusion array for high-throughput, long-term imaging of clonal growth. Biomicrofluidics 5(4):044117–044113
125. Ostuni E, Chen CS, Ingber DE, Whitesides GM (2001) Selective deposition of proteins and cells in arrays of microwells. Langmuir 17(9):2828–2834. doi:10.1021/la001372o
126. Abate AR, Lee D, Do T, Holtze C, Weitz DA (2008) Glass coating for PDMS microfluidic channels by sol–gel methods. Lab Chip 8(4):516–518
127. Roman GT, Hlaus T, Bass KJ, Seelhammer TG, Culbertson CT (2005) Sol–gel modified poly(dimethylsiloxane) microfluidic devices with high electroosmotic mobilities and hydrophilic channel wall characteristics. Anal Chem 77(5):1414–1422. doi:10.1021/ac048811z
128. Hu S, Ren X, Bachman M, Sims CE, Li GP, Allbritton NL (2004) Surface-directed, graft polymerization within microfluidic channels. Anal Chem 76(7):1865–1870. doi:10.1021/ac049937z
129. Sasaki H, Onoe H, Osaki T, Kawano R, Takeuchi S (2010) Parylene-coating in PDMS microfluidic channels prevents the absorption of fluorescent dyes. Sens Actuators B Chem 150(1):478–482
130. Wang Y, Balowski J, Phillips C, Phillips R, Sims CE, Allbritton NL (2011) Benchtop micromolding of polystyrene by soft lithography. Lab Chip 11(18):3089–3097
131. Price NM, Harrison PJ, Landry MR, Azam F, Hall KJF (1986) Toxic effects of latex and tygon tubing on marine-phytoplankton, zooplankton and bacteria. Mar Ecol Prog Ser 34(1–2):41–49. doi:10.3354/meps034041
132. Park H, Berzin I, De Luis J, Vunjak-Novakovic G (2005) Evaluation of silicone tubing toxicity using tobacco BY2 culture. Vitro Cell Dev Biol Plant 41(4):555–560. doi:10.1079/ivp2005670
133. Chau LT, Rolfe BE, Cooper-White JJ (2011) A microdevice for the creation of patent, three-dimensional endothelial cell-based microcirculatory networks. Biomicrofluidics 5(3). doi:10.1063/1.3609264
134. Tourovskaia A, Figueroa-Masot X, Folch A (2005) Differentiation-on-a-chip: a microfluidic platform for long-term cell culture studies. Lab Chip 5(1):14–19. doi:10.1039/b405719h
135. Villa M, Pope S, Conover J, Fan TH (2010) Growth of primary embryo cells in a microculture system. Biomed Microdevices 12(2):253–261. doi:10.1007/s10544-009-9380-y
136. Titmarsh D, Hidalgo A, Turner J, Wolvetang E, Cooper-White J (2011) Optimization of flowrate for expansion of human embryonic stem cells in perfusion microbioreactors. Biotechnol Bioeng 108(12):2894–2904. doi:10.1002/bit.23260
137. Moledina F, Clarke G, Oskooei A, Onishi K, Gunther A, Zandstra PW (2012) Predictive microfluidic control of regulatory ligand trajectories in individual pluripotent cells. Proc Natl Acad Sci USA 109(9):3264–3269. doi:10.1073/pnas.1111478109

138. Ellison D, Munden A, Levchenko A (2009) Computational model and microfluidic platform for the investigation of paracrine and autocrine signaling in mouse embryonic stem cells. Mol Biosyst 5(9):1004–1012
139. Chen H, Li J, Zhang H, Li M, Rosengarten G, Nordon RE (2011) Microwell perfusion array for high-throughput, long-term imaging of clonal growth. Biomicrofluidics 5. doi:10.1063/1.3669371
140. Young EWK, Beebe DJ (2010) Fundamentals of microfluidic cell culture in controlled microenvironments. Chem Soc Rev 39(3):1036–1048. doi:10.1039/b909900j
141. Masand SN, Mignone L, Zahn JD, Shreiber DI (2011) Nanoporous membrane-sealed microfluidic devices for improved cell viability. Biomed Microdevices 13(6):955–961. doi:10.1007/s10544-011-9565-z
142. Bose N, Das T, Chakraborty D, Maiti TK, Chakraborty S (2012) Enhancement of static incubation time in microfluidic cell culture platforms exploiting extended air-liquid interface. Lab Chip 12(1):69–73. doi:10.1039/c1lc20888h
143. Ziolkowska K, Kwapiszewski R, Brzozka Z (2011) Microfluidic devices as tools for mimicking the in vivo environment. New J Chem 35(5):979–990. doi:10.1039/c0nj00709a
144. Frampton JP, Lai D, Sriram H, Takayama S (2011) Precisely targeted delivery of cells and biomolecules within microchannels using aqueous two-phase systems. Biomed Microdevices 13(6):1043–1051. doi:10.1007/s10544-011-9574-y
145. Chung BG, Lin F, Jeon NL (2006) A microfluidic multi-injector for gradient generation. Lab Chip 6(6):764–768
146. Meyvantsson I, Beebe DJ (2008) Cell culture models in microfluidic systems. Annu Rev Anal Chem 1:423–449. doi:10.1146/annurev.anchem.1.031207.113042
147. Gupta K, Kim DH, Ellison D, Smith C, Kundu A, Tuan J, Suh KY, Levchenko A (2010) Lab-on-a-chip devices as an emerging platform for stem cell biology. Lab Chip 10(16):2019–2031. doi:10.1039/c004689b
148. Lii J, Hsu WJ, Parsa H, Das A, Rouse R, Sia SK (2008) Real-time microfluidic system for studying mammalian cells in 3D microenvironments. Anal Chem 80(10):3640–3647. doi:10.1021/ac8000034
149. Gunther A, Yasotharan S, Vagaon A, Lochovsky C, Pinto S, Yang JL, Lau C, Voigtlaender-Bolz J, Bolz SS (2010) A microfluidic platform for probing small artery structure and function. Lab Chip 10(18):2341–2349. doi:10.1039/C004675b
150. Blake AJ, Pearce TM, Rao NS, Johnson SM, Williams JC (2007) Multilayer PDMS microfluidic chamber for controlling brain slice microenvironment. Lab Chip 7(7):842–849. doi:10.1039/b704754a
151. Kastrup CJ, Runyon MK, Lucchetta EM, Price JM, Ismagilov RF (2008) Using chemistry and microfluidics to understand the spatial dynamics of complex biological networks. Acc Chem Res 41(4):549–558. doi:10.1021/ar700174g
152. Lutolf MP, Gilbert PM, Blau HM (2009) Designing materials to direct stem-cell fate. Nature 462(7272):433–441. doi:10.1038/nature08602
153. Pek YS, Wan ACA, Ying JY (2010) The effect of matrix stiffness on mesenchymal stem cell differentiation in a 3D thixotropic gel. Biomaterials 31(3):385–391. doi:10.1016/j.biomaterials.2009.09.057
154. Park JY, Yoo SJ, Hwang CM, Lee SH (2009) Simultaneous generation of chemical concentration and mechanical shear stress gradients using microfluidic osmotic flow comparable to interstitial flow. Lab Chip 9(15):2194–2202. doi:10.1039/b822006a
155. Yamamoto K, Sokabe T, Watabe T, Miyazono K, Yamashita JK, Obi S, Ohura N, Matsushita A, Kamiya A, Ando J (2005) Fluid shear stress induces differentiation of Flk-1-positive embryonic stem cells into vascular endothelial cells in vitro. Am J Physiol Heart Circ Physiol 288(4):H1915–H1924. doi:10.1152/ajpheart.00956.2004
156. Moraes C, Sun Y, Simmons CA (2011) (Micro)managing the mechanical microenvironment. Integr Biol 3(10):959–971. doi:10.1039/c1ib00056j
157. Davies PF (1995) Flow-mediated endothelial mechanotransduction. Physiol Rev 75(3):519–560

158. Gilbert PM, Havenstrite KL, Magnusson KEG, Sacco A, Leonardi NA, Kraft P, Nguyen NK, Thrun S, Lutolf MP, Blau HM (2010) Substrate elasticity regulates skeletal muscle stem cell self-renewal in culture. Science 329(5995):1078–1081. doi:10.1126/science.1191035
159. Park JY, Yoo SJ, Lee EJ, Lee DH, Kim JY, Lee SH (2010) Increased poly(dimethylsiloxane) stiffness improves viability and morphology of mouse fibroblast cells. BioChip J 4(3):230–236. doi:10.1007/s13206-010-4311-9
160. Adamo L, Naveiras O, Wenzel PL, McKinney-Freeman S, Mack PJ, Gracia-Sancho J, Suchy-Dicey A, Yoshimoto M, Lensch MW, Yoder MC, Garcia-Cardena G, Daley GQ (2009) Biomechanical forces promote embryonic haematopoiesis. Nature 459(7250):1131–1135. doi:10.1038/nature08073
161. Kim S, Kim HJ, Jeon NL (2010) Biological applications of microfluidic gradient devices. Integr Biol 2(11–12):584–603. doi:10.1039/c0ib00055h
162. Boyden S (1962) Chemotactic effect of mixtures of antibody and antigen on polymorphonuclear leucocytes. J Exp Med 115(3):453. doi:10.1084/jem.115.3.453
163. Zigmond SH (1977) Ability of polymorphonuclear leukocytes to orient in gradients of chemotactic factors. J Cell Biol 75(2):606–616. doi:10.1083/jcb.75.2.606
164. Zicha D, Dunn GA, Brown AF (1991) A new direct-viewing chemotaxis chamber. J Cell Sci 99:769–775
165. Gundersen RW, Barrett JN (1979) Neuronal chemotaxis – chick dorsal-root axons turn toward high-concentrations of nerve growth-factor. Science 206(4422):1079–1080. doi:10.1126/science.493992
166. Zheng JQ, Felder M, Connor JA, Poo MM (1994) Turning of nerve growth cones induced by neurotransmitters. Nature 368(6467):140–144. doi:10.1038/368140a0
167. Chung BG, Choo J (2010) Microfluidic gradient platforms for controlling cellular behavior. Electrophoresis 31(18):3014–3027. doi:10.1002/elps.201000137
168. Du Y, Hancock MJ, He J, Villa-Uribe JL, Wang B, Cropek DM, Khademhosseini A (2010) Convection-driven generation of long-range material gradients. Biomaterials 31(9):2686–2694. doi:10.1016/j.biomaterials.2009.12.012
169. Sundararaghavan HG, Monteiro GA, Firestein BL, Shreiber DI (2009) Neurite growth in 3D collagen gels with gradients of mechanical properties. Biotechnol Bioeng 102(2):632–643. doi:10.1002/bit.22074
170. Polinkovsky M, Gutierrez E, Levchenko A, Groisman A (2009) Fine temporal control of the medium gas content and acidity and on-chip generation of series of oxygen concentrations for cell cultures. Lab Chip 9(8):1073–1084
171. Sip CG, Bhattacharjee N, Folch A (2011) A modular cell culture device for generating arrays of gradients using stacked microfluidic flows. Biomicrofluidics 5(2). doi:10.1063/1.3576931
172. Sahai R, Cecchini M, Klingauf M, Ferrari A, Martino C, Castrataro P, Lionetti V, Menciassi A, Beltram F (2011) Microfluidic chip for spatially and temporally controlled biochemical gradient generation in standard cell-culture Petri dishes. Microfluid Nanofluid 11(6):763–771. doi:10.1007/s10404-011-0841-2
173. Chung BG, Park JW, Hu JS, Huang C, Monuki ES, Jeon NL (2007) A hybrid microfluidic-vacuum device for direct interfacing with conventional cell culture methods. BMC Biotechnol 7. doi:10.1186/1472-6750-7-60
174. Chung BG, Flanagan LA, Rhee SW, Schwartz PH, Lee AP, Monuki ES, Jeon NL (2005) Human neural stem cell growth and differentiation in a gradient-generating microfluidic device. Lab Chip 5(4):401–406
175. Park JY, Hwang CM, Lee SH, Lee S-H (2007) Gradient generation by an osmotic pump and the behavior of human mesenchymal stem cells under the fetal bovine serum concentration gradient. Lab Chip 7(12):1673–1680
176. Park JY, Kim S-K, Woo D-H, Lee E-J, Kim J-H, Lee S-H (2009) Differentiation of neural progenitor cells in a microfluidic chip-generated cytokine gradient. Stem Cells 27(11):2646–2654. doi:10.1002/stem.202
177. Abhyankar VV, Beebe DJ (2007) Spatiotemporal micropatterning of cells on arbitrary substrates. Anal Chem 79(11):4066–4073. doi:10.1021/ac062371p

178. Schumacher K, Strehl R, de Vries U, Minuth WW (2002) Advanced technique for long term culture of epithelia in a continuous luminal–basal medium gradient. Biomaterials 23(3):805–815. doi:10.1016/s0142-9612(01)00186-7
179. Zanzotto A, Szita N, Boccazzi P, Lessard P, Sinskey AJ, Jensen KF (2004) Membrane-aerated microbioreactor for high-throughput bioprocessing. Biotechnol Bioeng 87(2):243–254. doi:10.1002/bit.20140
180. de Jong J, Lammertink RGH, Wessling M (2006) Membranes and microfluidics: a review. Lab Chip 6(9):1125–1139. doi:10.1039/b603275c
181. Evenou F, Hamon M, Fujii T, Takeuchi S, Sakai Y (2011) Gas-permeable membranes and co-culture with fibroblasts enable high-density hepatocyte culture as multilayered liver tissues. Biotechnol Prog 27(4):1146–1153. doi:10.1002/btpr.626
182. Evenou F, Fujii T, Sakai Y (2010) Spontaneous formation of stably-attached and 3D-organized hepatocyte aggregates on oxygen-permeable polydimethylsiloxane membranes having 3D microstructures. Biomed Microdevices 12(3):465–475. doi:10.1007/s10544-010-9403-8
183. Lo JF, Sinkala E, Eddington DT (2010) Oxygen gradients for open well cellular cultures via microfluidic substrates. Lab Chip 10(18):2394–2401. doi:10.1039/c004660d
184. Grist SM, Chrostowski L, Cheung KC (2010) Optical oxygen sensors for applications in microfluidic cell culture. Sensors 10(10):9286–9316. doi:10.3390/s101009286
185. Huang CW, Lee GB (2007) A microfluidic system for automatic cell culture. J Micromech Microeng 17(7):1266–1274. doi:10.1088/0960-1317/17/7/008
186. Adler M, Polinkovsky M, Gutierrez E, Groisman A (2010) Generation of oxygen gradients with arbitrary shapes in a microfluidic device. Lab Chip 10(3):388–391. doi:10.1039/b920401f
187. Thomas PC, Raghavan SR, Forry SP (2011) Regulating oxygen levels in a microfluidic device. Anal Chem 83(22):8821–8824. doi:10.1021/ac202300g
188. Chen Y-A, King AD, Shih H-C, Peng C-C, Wu C-Y, Liao W-H, Tung Y-C (2011) Generation of oxygen gradients in microfluidic devices for cell culture using spatially confined chemical reactions. Lab Chip 11(21):3626–3633. doi:10.1039/c1lc20325h
189. Liu K, Pitchimani R, Dang D, Bayer K, Harrington T, Pappas D (2008) Cell culture chip using low-shear mass transport. Langmuir 24(11):5955–5960. doi:10.1021/la8003917
190. Cooksey GA, Elliott JT, Plant AL (2011) Reproducibility and robustness of a real-time microfluidic cell toxicity assay. Anal Chem 83(10):3890–3896. doi:10.1021/ac200273f
191. Hsieh CC, Huang SB, Wu PC, Shieh DB, Lee GB (2009) A microfluidic cell culture platform for real-time cellular imaging. Biomed Microdevices 11(4):903–913. doi:10.1007/s10544-009-9307-7
192. Lin JL, Wang SS, Wu MH, Oh-Yang CC (2011) Development of an integrated microfluidic perfusion cell culture system for real-time microscopic observation of biological cells. Sensors 11(9):8395–8411. doi:10.3390/s110908395
193. Gaitan M, Locascio LE (2004) Embedded microheating elements in polymeric micro channels for temperature control and fluid flow sensing. J Res Nat Inst Stand Technol 109(3):335–344
194. Vigolo D, Rusconi R, Piazza R, Stone HA (2010) A portable device for temperature control along microchannels. Lab Chip 10(6):795–798. doi:10.1039/b919146a
195. Liu C-W, Gau C, Liu C-G, Yang C-S (2005) Design consideration and fabrication of a microchannel system containing a set of heaters and an array of temperature sensors. Sens Actuators A Phys 122(2):177–183. doi:10.1016/j.sna.2005.04.016
196. Wu JB, Cao WB, Wen WJ, Chang DC, Sheng P (2009) Polydimethylsiloxane microfluidic chip with integrated microheater and thermal sensor. Biomicrofluidics 3(1). doi:10.1063/1.3058587
197. Shen K, Chen X, Guo M, Cheng J (2005) A microchip-based PCR device using flexible printed circuit technology. Sens Actuators B Chem 105(2):251–258. doi:10.1016/j.snb.2004.05.069
198. King KR, Wang SH, Irimia D, Jayaraman A, Toner M, Yarmush ML (2007) A high-throughput microfluidic real-time gene expression living cell array. Lab Chip 7(1):77–85. doi:10.1039/b612516f

199. Lee PJ, Gaige TA, Hung PJ (2009) Dynamic cell culture: a microfluidic function generator for live cell microscopy. Lab Chip 9(1):164–166. doi:10.1039/b807682k
200. Albrecht DR, Underhill GH, Resnikoff J, Mendelson A, Bhatia SN, Shah JV (2010) Microfluidics-integrated time-lapse imaging for analysis of cellular dynamics. Integr Biol 2(5–6):278–287. doi:10.1039/b923699f
201. Rieger MA, Schroeder T (2009) Analyzing cell fate control by cytokines through continuous single cell biochemistry. J Cell Biochem 108(2):343–352. doi:10.1002/jcb.22273
202. Muzzey D, van Oudenaarden A (2009) Quantitative time-lapse fluorescence microscopy in single cells. In: Annual review of cell and developmental biology, vol 25. Annual Reviews Inc, Palo Alto, pp 301–327. doi:10.1146/annurev.cellbio.042308.113408
203. Vasdekis AE, Laporte GPJ (2011) Enhancing single molecule imaging in optofluidics and microfluidics. Int J Mol Sci 12(8):5135–5156. doi:10.3390/ijms12085135
204. Chirieleison SM, Bissell TA, Scelfo CC, Anderson JE, Li Y, Koebler DJ, Deasy BM (2011) Automated live cell imaging systems reveal dynamic cell behavior. Biotechnol Prog 27(4):913–924. doi:10.1002/btpr.629
205. Gerrits A, Dykstra B, Kalmykowa OJ, Klauke K, Verovskaya E, Broekhuis MJC, de Haan G, Bystrykh LV (2010) Cellular barcoding tool for clonal analysis in the hematopoietic system. Blood 115(13):2610–2618. doi:10.1182/blood-2009-06-229757
206. Perlingeiro RCR, Kyba M, Daley GQ (2001) Clonal analysis of differentiating embryonic stem cells reveals a hematopoietic progenitor with primitive erythroid and adult lymphoid-myeloid potential. Development 128(22):4597–4604
207. Stachura DL, Svoboda O, Lau RP, Balla KM, Zon LI, Bartunek P, Traver D (2011) Clonal analysis of hematopoietic progenitor cells in the zebrafish. Blood 118(5):1274–1282. doi:10.1182/blood-2011-01-331199
208. Hope K, Bhatia M (2011) Clonal interrogation of stem cells. Nat Methods 8(4):S36–S40
209. Eilken HM, Nishikawa SI, Schroeder T (2009) Continuous single-cell imaging of blood generation from haemogenic endothelium. Nature 457(7231):896–900. doi:10.1038/nature07760
210. Lindstrom S, Eriksson M, Vazin T, Sandberg J, Lundeberg J, Frisen J, Andersson-Svahn H (2009) High-density microwell chip for culture and analysis of stem cells. PLoS One 4(9). doi:e699710.1371/journal.pone.0006997
211. Huth J, Buchholz M, Kraus JM, Schmucker M, von Wichert G, Krndija D, Seufferlein T, Gress TM, Kestler HA (2010) Significantly improved precision of cell migration analysis in time-lapse video microscopy through use of a fully automated tracking system. BMC Cell Biol 11
212. Nordon RE, Ko KH, Odell R, Schroeder T (2011) Multi-type branching models to describe cell differentiation programs. J Theor Biol 277(1):7–18. doi:10.1016/j.jtbi.2011.02.006

Chapter 20
Biomimetic Multiscale Topography for Cell Alignment

Kara E. McCloskey

Abstract Cell-cell communication can be facilitated by manipulation of the topographical surface structure (biomimetics) in which the cells are cultured. This is particularly important for culture of cardiac cells for the enhancing of cell-cell communication. Here, we highlight methods available for generating tissue-like biomimetic cell culture platforms.

Keywords Cell alignment • Topography • Tissue engineering • Cardiomyocytes • Nanofabrication

Abbreviations

CM	Cardiomyocytes
E	Elongation
EC	Endothelial cell
ECM	Extracellular matrix
EF	Elongation factor
ESC	Embryonic stem cells
hESC	Human embryonic stem cells
NNCM	Neonatal cardiomyocytes
P	Perimeter
PDMS	Poly-dimethylsiloxane
PS	Polystyrene

K.E. McCloskey
Graduate Program in Biological Engineering and Small-scale Technologies,
University of California, Merced, CA, USA

School of Engineering, University of California,
5200 North Lake Rd, Merced, CA 95343, USA
e-mail: kmccloskey@ucmerced.edu

M.K. Danquah, R.I. Mahato (eds.), *Emerging Trends in Cell and Gene Therapy*,
DOI 10.1007/978-1-62703-417-3_20,

PVC	Polyvinyl chloride
Ra	Roughness
RIE	Reactive ion etching
SEM	Scanning electron microscopy
SF	Shape factor

20.1 Physiological Significance of Controlling Cell Shape

It is well documented that many cells respond to topographical surface features by changing their proliferation, adhesion or by directed migration and cell orientation, called contact guidance. The physiological significance of controlling cell shape behavior for enhancing cell-tissue function is important in a wide variety of cell types including neurons [1–3], skeletal muscle [4], cardiac muscle [5–15], and even corneal and lens epithelial cells [16, 17]. The alignment of cells in tissue culture systems not only enables a more accurate anatomical state but also provides important cell-cell signaling cues for inducing proper phenotypic and physiologic responses. While most of these studies have addressed microscale topography, recent studies have also demonstrated that topography at the nanoscale provides some critical cues that also play a role in cell alignment [18–21]. Submicron cues have also been implicated in the facilitation of critical cellular functions such as cell morphology, adhesion, gene regulation, and cell-cell communication [18, 21, 22].

20.1.1 Cardiomyocytes

Mature adult cardiomyocytes (CM) exhibit an elongated rod-like shape and contain well-developed contractile apparatus with intercalated discs as electromechanical cell end-to-end couplings. The existence of an organized contractile apparatus in the cardiac cells can be identified by the striated pattern observed after staining for contractile proteins. Gap junctions, located in intercalated discs at the ends of cardiac cells, are responsible for rapid propagation of electrical signals between the cells. However, in standard tissue culture setups, the CM will often undergo dedifferentiation and lose many of these key physiological features, including their elongated shape and defined contractile apparatus. It is thought that the dedifferentiation is due to the lack of appropriate microenvironmental cues in prolonged cultures. It has been shown that the alignment of cardiac cells not only helps limit cell dedifferentiation by increasing the continuity of cell-cell connections but also has a profound effect on impulse initiation and propagation velocities [23]. Moreover, cell culture systems that additionally provide signals to induce anatomical anisotropy of cardiac cells lead to larger cellular coupling resistance in the transverse (along short cell axis) compared with the longitudinal (along long cell axis) direction, resulting in a smaller velocity but larger maximum slope of action potential upstroke [23].

20.1.2 Endothelial Cells

Despite the fact that endothelial cells (EC) cultured on textured surfaces behave more like those on native arteries [24], very few works have shown alignment of endothelial cells on pattern surfaces. Moreover, oriented cells exhibit greater density of focal adhesion contacts and better organization with stronger networks of actin fibers, a highly desirable factor that could aid endothelialization following intravascular stent placement [25]. One study that examined the alignment of EC on rough surfaces (Ra) with peak-to-valley heights up from 0.4 to 1.4 μm did not find very robust EC alignment compared with fibroblasts [26]. Another study found that EC migration increases on grooved surfaces compared with flat controls. However, microgrooved poly-dimethylsiloxane (PDMS) chips generated from silicon molds with channel depths ranging from 200 nm to 5 μm and constant width at 3.5 μm were coated with fibronectin. This work found a maximal EC alignment at 90 % in the channels with a 1 μm depth [27]. A third study observed increased EC migration on grooved gel surfaces, increasing with groove size up to 22 μm [28].

20.1.3 Mechanism of Cell Alignment

Mechanotransduction is the ability of a cell to sense, process, and integrate the mechanical signals found in their microenvironment leading to subsequent changes in morphology, protein synthesis, and/or cell fate. "It is the conversion of mechanical forces into biochemically relevant information" [29] and can include sensing and response to a wide range of physical cues at the interface between the cells and their surrounding matrix [29–31]. This includes the mechanical forces shear stress, strain, and compression, as well as matrix stiffness [32] and surface topography. Although these are clearly different types of physical signals, the sensing mechanisms within the cells are generally the same whether the forces are applied externally or internally. It is the cytoskeletal proteins that are responsible for mediating mechanical effects in nearly all systems [30]. In adherent cell cultures, it is crucial for the cells to maintain their tight association with the diverse connective tissue components that form the extracellular matrix (ECM) because the adhesion receptors that mediate cell-cell (cadherin-dependent linkages) and cell-ECM (integrin-dependent linkages) contacts and the cytoskeletal elements (i.e., F-actin, intermediate filaments, and microtubules) in which the receptors are connected are most strongly implicated in mechanotransduction [33].

Because integrins serve as receptors for ECM binding (Fig. 20.1) and form integrin-anchored focal adhesions through these matrix attachments which connect to cytoskeleton elements in the cell, these focal adhesion complexes are thought to be the main forces at work in the biochemical sensing and response to topographical cues [34]. The integrins first act as conduits for transducing physical forces into chemical cellular responses [35], and then, the forces concentrated at the focal adhesion sites stimulate integrin dimerization and recruitment of focal adhesion proteins paxillin, talin, and vinculin [36], which connect directly to microfilaments and indirectly to microtubules and intermediate filaments [37]. The forces applied to these adhesion

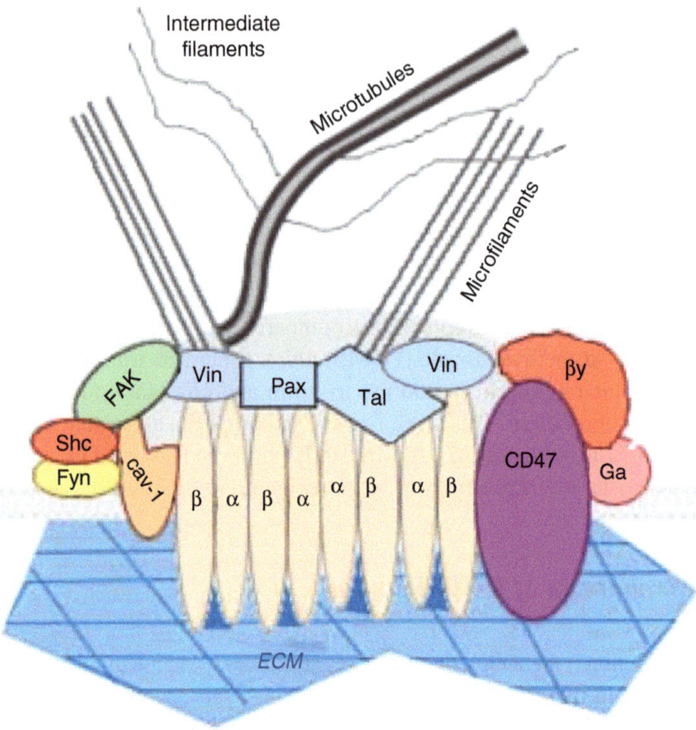

Fig. 20.1 Schematic of proposed integrin-mediated cell signaling. Integrins are receptors for ECM, forming integrin-anchored focal adhesions that connect to cytoskeleton elements. Forces at these sites can then recruit focal adhesion proteins paxillin (*Pax*), talin (*Tal*), and vinculin (*Vin*) that connect to microfilaments, microtubules, and intermediate filaments and activate signaling cascades (Reprinted from [34])

complexes activate integrin-associated signaling cascades that include activation of focal adhesion kinase (FAK), extracellular signal-regulated protein kinase, Shc, Rho, mDIA1, caveolin-1, G proteins, adenylate cyclase, and protein kinase A, reviewed in [37]. It is this activation of G proteins leading to activation or inactivation small G proteins, like Rho, that affects the actin polymerization and actomyosin contractility that enables the profound cytoskeletal rearrangements seen in the cell's response to topographical features [38].

20.1.4 Quantification of Cell Alignment

20.1.4.1 Cell Orientation

Cells are described as aligned, nonaligned, or isometric by measuring the angles between the major axes and the direction of the channels. Here, cells can be defined as "aligned" if their major axes are within ±30° with respect to the wrinkle or channel direction, and the percentage of cells on the channeled surface were quantified. Index of orientation is another method for evaluation of alignment in which cells are defined by the primary orientation with respect to an alignment direction.

20.1.4.2 Cell Elongation

The presentation of the shortest and longest sides of a cell is one way of assessing symmetry. However, the elongation index is a more inclusive parameter for quantifying how cells respond to changes in topography. Elongation factor (EF) can be calculated by the ratio of the maximal diameter (D_{max}), or length, to the minimal diameter (D_{min}), or width of a cell ($EF = D_{max}/D_{min}$) [39]. The shape factor (SF) has also been used to access the geometric shape for endothelial cells [40]. The SF evaluates the perimeter (P) squared with respect to area (A), $SF = P^2/A$. Another length-related index related the perimeter with the elongation (E) of a cell, $E = \pi P/D_{max}$ [41].

20.2 Microfabrication Approaches

The alignment of cells has been studied for the last decade using a variety of microfabrication approaches including microcontact printing, abrasion, photolithography, hot embossing, electrospinning, and laser ablation and nanofabrication approaches including e-beam lithography and nanoimprint lithography.

20.2.1 Abrasion

Polyvinyl chloride (PVC) coverslips are microabraded over the entire surface in a direction parallel to one of the edges using lapping sandpapers with different grit sizes to produce uniformly anisotropic cultures with varying degrees of anisotropy or at two different directions in two adjacent regions to form anisotropic cultures with sharp change in fiber direction. These abraded coverslips can then be cut into circular shapes to fit wells of a standard 12-well tissue culture plates, rinsed in 95 % ethanol, dried using pressurized nitrogen, UV irradiated for 1.5 h to make the PVC surface more hydrophilic, and coated with fibronectin [23].

20.2.2 Microcontact Printing and Photolithography

Briefly, micropatterns are drawn using software in AutoCAD to print high-resolution photomasks. With the use of standard soft lithography techniques, the patterns are then microfabricated in 5 mm layers of photoresist and spin-coated onto a silicon wafer. PDMS molds are cast from the microfabricated pattern at 80 °C for 2 h. These can then be cleaned in ethanol, inked with fibronectin or other matrix protein for 1 h, and dried with N_2 gas. The fibronectin micropattern may then be microcontact-printed from the PDMS stamps onto UV-ozone-treated, PDMS-coated 22 mm glass coverslips for 30 min to allow protein transfer [23]. Photolithography is mostly readily used to fabricate 1–2 μm features but can be used to produce features as small as 400 μm [42].

20.2.3 Hot Embossing

The mold for hot embossing can be fabricated by standard photolithography for the surfaces with 4 mm period grating and deep-UV lithography for the 1 mm period grating, or the trench can be etched by reactive ion etching (RIE) [10]. In RIE, a chemically reactive plasma is used to remove material deposited on wafers. The high-energy ions from the plasma attack the wafer surface, reacting with the material. Hot embossing then creates a polystyrene replica of the silicon mold. In hot embossing, the polystyrene pellets are evenly distributed over an area on top of the mold and covered with another flat wafer. Both the mold and the flat wafer are then treated with an anti-adhesion silane layer (1H,1H,2H,2H-perfluorooctyl-trichlorosilane) to facilitate the separation.

20.2.4 Nanofabrication

Structures with topographical features less then 100 nm are desirable due to the similarity with cellular environments. One of the most common methods to produce nanotopographies is colloidal lithography. Although e-beam lithography can generate much higher resolutions, as small as 4 nm beam diameters, it is only used for generating masters for replication by mechanical transfer to cell culture substrates or biomedical device. Due to the serial writing process to produce the masters, irregular spatial patterns are also difficult. Using colloidal lithography or other self-assembly technologies, regular and irregular sub-100 nm features can be generated on primary patterns [42]. Moreover, because colloids tend to be composed of materials deviating from the base substrate, the heterogeneous chemistries often result in chemical patterning as well as topographical patterning, which can be additionally advantageous.

20.2.5 Alternative Solutions

Most of these approaches are very time consuming, technically challenging and expensive, and, therefore, inaccessible to many biological laboratories. To address this chasm, a tunable, ultrarapid, robust, and inexpensive non-photolithographic fabrication method has been developed for the generation of cell culture substrates with controllable nano- and micro-scale cues. This simple microfabrication method uses prestressed polystyrene (PS) sheets that "shrink" upon heating, commonly known as Shrinky Dinks™. When the polystyrene sheets are coated with a thin layer of metal film (10–100 nm in thickness), the mismatched stiffness between the prestressed polymer sheet and the overlying thin metal film will generate "wrinkles" in the metal during shrinkage-induced heating [43, 44]. The grooves within this metal chip can then be used as a mold for generating PDMS cell culture platforms that retain the topography of the metal mold.

20.3 Case Study

20.3.1 Cell Alignment on Metal-Based "Wrinkles"

The unique self-assembled multiscale topographical substrate can be used to align CM for increased cell-cell signaling and generation of cardiac tissue monolayers. Recently, Luna et al. [11] introduced a tunable, ultrarapid, robust, and inexpensive non-photolithographic fabrication method to create cell culture substrates with controllable nano- and microscale cues. The alignment grooves are created by leveraging the mismatch in stiffness between a prestressed polymer sheet and an overlying thin metal film [43, 44]. When the plastic sheet retracts upon heating, the stiffer metal film buckles in a controllable manner causing "wrinkles."

20.3.1.1 Generation of Wrinkle Molds

Metal wrinkles are fabricated as reported and described [43]. Briefly, gold-palladium is deposited by sputter coating (SEM sputter coater, polaron) at various thicknesses. Heterogeneous wrinkle length scales based on varying thickness of metal coating on prestressed PS sheets (Grafix, Cleveland, OH); coating thicknesses ranged from 15 to 90 nm, with all thicknesses generating "wrinkles" ranging from 20 nm to 10 μm and average wrinkles thicknesses ranging from 800 nm to 1 μm and increasing proportionally with coating thickness [43]. After deposition, PS sheets are induced to thermally shrink by heating to 150–160 °C. Uniaxial wrinkles are generated by constraining two opposite sides with binder clips during heating. The deposited metal layer on top of the PS sheet-generated aligned wrinkles serves as a soft lithography mold for generating PDMS microchips for culturing the cells (Fig. 20.2). The anisotropy and length scale of the wrinkles were determined by performing a fast Fourier transform of the scanning electron micrograph (Fig. 20.2).

20.3.1.2 Generation of PDMS Microchip from Wrinkle Mold

A mixture of 10:1 ratio of PDMS and curing agent (Sylgard 184 Silicone Elastomer Kit, Dow Corning) was poured on the metal mold and set to cure at

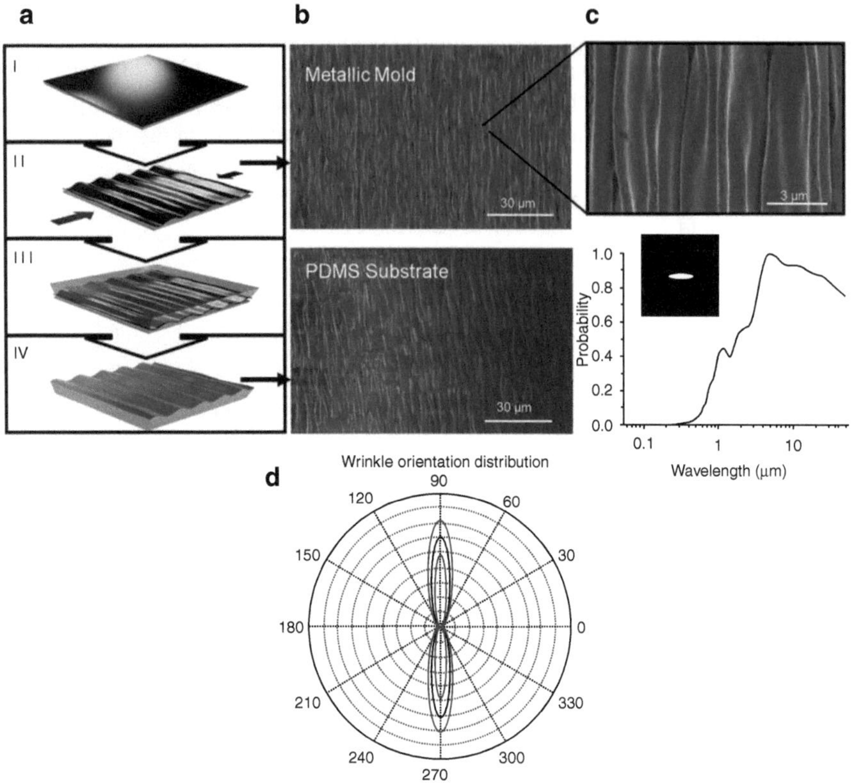

Fig. 20.2 Fabrication and characterization of multiscale wrinkle substrate. (**a**) (*I*) Metallic layer is deposited on prestressed (PS) sheets. (*II*) PS is induced to thermally shrink while constrained from opposite sides to generate aligned anisotropic wrinkles. (*III*) The metal wrinkles are used as a soft lithography mold to generate a PDMS substrate (*IV*) which is used to culture CM. (**b**) Scanning electron micrographs (SEMs) of metal wrinkles (with high-resolution inset) and PDMS substrate. (**c**) The length scale distribution from fast Fourier transform of SEM images. The critical length scale was plotted as a probability function and ranges in critical dimension from the 100 s of nm to several microns. *Inset* shows high degree of anisotropy, as quantified by (**d**) computing a histogram of gradient orientations (*thick lines*) and standard deviation (*thin lines*) (Reprinted from [11])

75 °C. It was then peeled off and cut into to a circle with diameter 15 mm to fit into 24-well plates. Controls with flat PDMS were performed following the same procedure.

20.3.1.3 Cardiac Cell Alignment of Anisotropic Wrinkled PDMS Microchips

Neonatal mouse cardiomyocytes (NMCM) and human embryonic stem cell (hESC)-derived cardiac cells were cultured on laminin- and fibronectin-coated wrinkled PDMS microchips. Then, we stained the NNCM for gap junction protein Cx 43 (Fig. 20.3a, c)

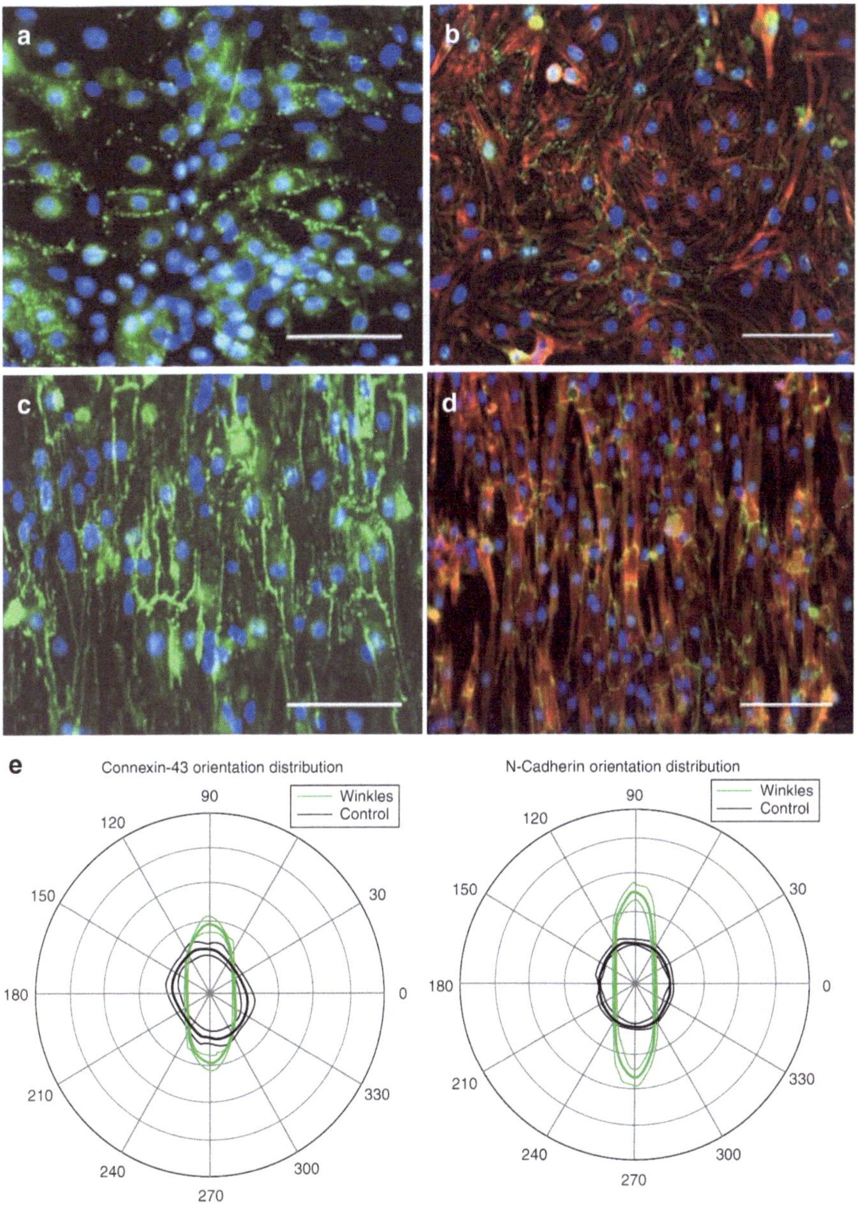

Fig. 20.3 Cardiac-like tissue from aligned CM. Fluorescent micrographs of cardiac cells cultured on (**a**, **b**) control (flat) and (**c**, **d**) wrinkled substrates. (**a**, **c**) Connexin-43 (*green*) and (**b**, **d**) N-cadherin (*green*) and actin (*red*) expression by neonatal mouse cardiomyocytes. *Blue* = nuclear staining DAPI. *Scale bars* = 100 μm. (**e**) Anisotropy analysis of control (*black*) versus *green* (*on wrinkles*) was accomplished by computing a histogram of gradient orientations where the contribution of each pixel was weighted by the gradient magnitude. *Thinner lines* indicate the standard deviation (Reprinted from [11])

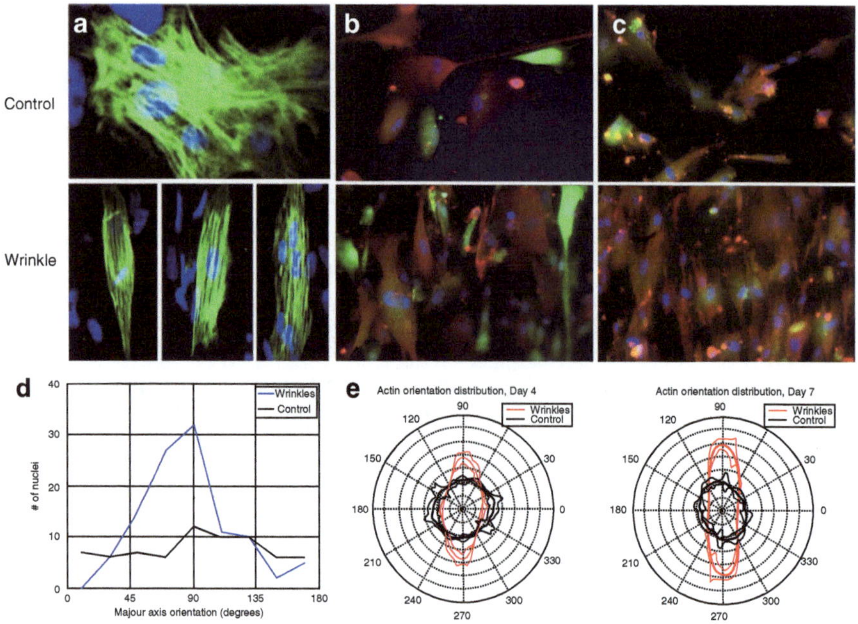

Fig. 20.4 Confocal micrographs hESC-derived CM alignment on wrinkles. (**a**) Human ESC-derived CM were isolated and cultured on flat substrate (*top*) and wrinkled substrates (*bottom*) for 8 days. *Green* = tropomyosin staining, *blue* nuclear staining DAPI. (**b**, **c**) Human ESC-derived CM were also generated using a MLC2v-GFP cell line (*green*) and stained with actin (*red*) and cultured for (**b**) 4 days and (**c**) 7 days on flat (*top*) and wrinkled substrates (*bottom*). (**d**) Image processing was used in order to detect the orientation of the DAPI-labeled nuclei. (**e**) Anisotropy analysis of control (*black*) versus green (on wrinkles) showing that 90° is the direction of wrinkles. The *thinner lines* indicate the standard deviations (Reprinted from [11])

and the fascia adherens protein N-cadherin (Fig. 20.3b, d). Note that the proteins were better localized at the cell-cell junctions on the aligned cells as compared to the peripherally distributed expression on the isotropic cells. Compared with controls (Fig. 20.3a–c, top images), hESC-derived CM (Fig. 20.4a–c, bottom images) also aligned and displayed the typical banding pattern consistent with organized sarcomeric structure patterns. Most importantly, compared with unaligned cells, the alignment of the cardiac cells enables synchronous contraction of the cell culture, mimicking more closely the native heart tissue ([11] #69). This case study demonstrates that both murine neonatal CM (NNCM) and CM derived from human embryonic stem cells (hESC) in vitro align on wrinkled biomimetic surface topography.

20.3.1.4 Endothelial Cell Alignment of Anisotropic Wrinkled PDMS Microchips

Lastly, we explored the alignment of embryonic stem cell (ESC)-derived EC [45, 46] on nanoscale wrinkles. The PDMS chips with 510 nm wrinkled microchips

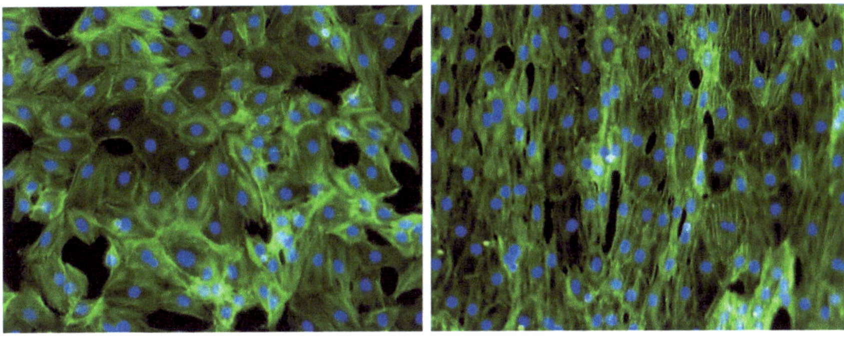

Fig. 20.5 Micrographs of ESC-derived EC alignment on wrinkles. ESC-derived EC were generated, purified, and cultured on flat substrates (*left*) and wrinkled substrates (*right*) for 3 days. *Green* = F-actin staining, *blue* nuclear stain = DAPI

were coated with gelatin and plated with purified ESC-derived EC. After 3 days of culture, 100 % of the ESC-derived EC aligned in the wrinkle direction (Fig. 20.5). The ability to align the EC is a powerful tool for enhancing cell adhesion for coating stents with this antithrombotic surface. Combining EC culture with vascular-like could also provide templates for EC patterning into vascular-like networks for many applications in tissue engineering.

20.4 Summary

The environmental sensing of living cells is a complex biological process displaying many features of "intelligent" systems. The cell can sense a wide range of environmental cues, including both chemical signals and physical forces. The integration of these signals can have effects on cell morphology, contraction, migration, proliferation, and programmed cell death, as regulated by the synthesis of specific proteins that lead to dynamic cell fate responses. The use of biomimetics to mimic and control the cell's micro- and nano-topographical environment is a powerful tool in controlling the cell's microenvironment leading to enhanced physiological responses.

References

1. Recknor JB et al (2004) Oriented astroglial cell growth on micropatterned polystyrene substrates. Biomaterials 25(14):2753–2767
2. Sorensen A et al (2007) Long-term neurite orientation on astrocyte monolayers aligned by microtopography. Biomaterials 28(36):5498–5508
3. Webb A et al (1995) Guidance of oligodendrocytes and their progenitors by substratum topography. J Cell Sci 108(Pt 8):2747–2760

4. Shimizu K, Fujita H, Nagamori E (2009) Alignment of skeletal muscle myoblasts and myotubes using linear micropatterned surfaces ground with abrasives. Biotechnol Bioeng 103(3):631–638
5. Au HT et al (2007) Interactive effects of surface topography and pulsatile electrical field stimulation on orientation and elongation of fibroblasts and cardiomyocytes. Biomaterials 28(29):4277–4293
6. Badie N, Bursac N (2009) Novel micropatterned cardiac cell cultures with realistic ventricular microstructure. Biophys J 96(9):3873–3885
7. Biehl JK et al (2009) Proliferation of mouse embryonic stem cell progeny and the spontaneous contractile activity of cardiomyocytes are affected by microtopography. Dev Dyn 238(8):1964–1973
8. Deutsch J et al (2000) Fabrication of microtextured membranes for cardiac myocyte attachment and orientation. J Biomed Mater Res 53(3):267–275
9. Geisse NA, Sheehy SP, Parker KK (2009) Control of myocyte remodeling in vitro with engineered substrates. In Vitro Cell Dev Biol Anim 45(7):343–350
10. Heidi Au HT et al (2009) Cell culture chips for simultaneous application of topographical and electrical cues enhance phenotype of cardiomyocytes. Lab Chip 9(4):564–575
11. Luna PJ et al (2011) Multi-scale biomimetic topography for the alignment of neonatal and embryonic stem cell-derived heart cells. Tissue Eng Part C Methods 17(5):579–588
12. Motlagh D et al (2003) Microfabricated grooves recapitulate neonatal myocyte connexin43 and N-cadherin expression and localization. J Biomed Mater Res A 67(1):148–157
13. Thomas SP et al (2000) Synthetic strands of neonatal mouse cardiac myocytes: structural and electrophysiological properties. Circ Res 87(6):467–473
14. Yeong WY et al (2010) Multiscale topological guidance for cell alignment via direct laser writing on biodegradable polymer. Tissue Eng Part C Methods 16:1011–1021
15. Zong X et al (2005) Electrospun fine-textured scaffolds for heart tissue constructs. Biomaterials 26(26):5330–5338
16. Rajnicek AM, Foubister LE, McCaig CD (2008) Alignment of corneal and lens epithelial cells by co-operative effects of substratum topography and DC electric fields. Biomaterials 29(13):2082–2095
17. Teixeira AI et al (2006) The effect of environmental factors on the response of human corneal epithelial cells to nanoscale substrate topography. Biomaterials 27(21):3945–3954
18. Dalby MJ et al (2005) Morphological and microarray analysis of human fibroblasts cultured on nanocolumns produced by colloidal lithography. Eur Cell Mater 9:1–8; discussion 8
19. Kim DH et al (2010) Nanoscale cues regulate the structure and function of macroscopic cardiac tissue constructs. Proc Natl Acad Sci USA 107(2):565–570
20. Pot SA et al (2010) Nanoscale topography-induced modulation of fundamental cell behaviors of rabbit corneal keratocytes, fibroblasts, and myofibroblasts. Invest Ophthalmol Vis Sci 51(3): 1373–1381
21. Yim EK et al (2005) Nanopattern-induced changes in morphology and motility of smooth muscle cells. Biomaterials 26(26):5405–5413
22. Koo LY et al (2002) Co-regulation of cell adhesion by nanoscale RGD organization and mechanical stimulus. J Cell Sci 115(Pt 7):1423–1433
23. Bursac N et al (2002) Cardiomyocyte cultures with controlled macroscopic anisotropy: a model for functional electrophysiological studies of cardiac muscle. Circ Res 91(12):e45–e54
24. Goodman SL, Sims PA, Albrecht RM (1996) Three-dimensional extracellular matrix textured biomaterials. Biomaterials 17(21):2087–2095
25. Palmaz JC (1993) Intravascular stents: tissue-stent interactions and design considerations. AJR Am J Roentgenol 160(3):613–618
26. Eisenbarth E et al (2002) Cell orientation and cytoskeleton organisation on ground titanium surfaces. Biomol Eng 19(2–6):233–237
27. Uttayarat P et al (2005) Topographic guidance of endothelial cells on silicone surfaces with micro- to nanogrooves: orientation of actin filaments and focal adhesions. J Biomed Mater Res A 75(3):668–680

28. Palmaz JC, Benson A, Sprague EA (1999) Influence of surface topography on endothelialization of intravascular metallic material. J Vasc Interv Radiol 10(4):439–444
29. Hoffman BD, Grashoff C, Schwartz MA (2011) Dynamic molecular processes mediate cellular mechanotransduction. Nature 475(7356):316–323
30. Chen CS (2008) Mechanotransduction – a field pulling together? J Cell Sci 121(Pt 20):3285–3292
31. Vogel V, Sheetz M (2006) Local force and geometry sensing regulate cell functions. Nat Rev Mol Cell Biol 7(4):265–275
32. Discher DE, Janmey P, Wang YL (2005) Tissue cells feel and respond to the stiffness of their substrate. Science 310(5751):1139–1143
33. Schwartz MA, DeSimone DW (2008) Cell adhesion receptors in mechanotransduction. Curr Opin Cell Biol 20(5):551–556
34. Stolberg S, McCloskey KE (2009) Can shear stress direct stem cell fate? Biotechnol Prog 25(1):10–19
35. Katsumi A et al (2004) Integrins in mechanotransduction. J Biol Chem 279(13):12001–12004
36. Balaban NQ et al (2001) Force and focal adhesion assembly: a close relationship studied using elastic micropatterned substrates. Nat Cell Biol 3(5):466–472
37. Ingber DE, Tensegrity II (2003) How structural networks influence cellular information processing networks. J Cell Sci 116(Pt 8):1397–1408
38. Geiger B, Spatz JP, Bershadsky AD (2009) Environmental sensing through focal adhesions. Nat Rev Mol Cell Biol 10(1):21–33
39. Rao GN et al (1979) Endothelial cell morphology and corneal deturgescence. Ann Ophthalmol 11(6):885–899
40. Collin HB, Grabsch BE (1982) The effect of ophthalmic preservatives on the shape of corneal endothelial cells. Acta Ophthalmol (Copenh) 60(1):93–105
41. Neufeld AH et al (1986) Maintenance of corneal endothelial cell shape by prostaglandin E2: effects of EGF and indomethacin. Invest Ophthalmol Vis Sci 27(10):1437–1442
42. Wood MA (2007) Colloidal lithography and current fabrication techniques producing in-plane nanotopography for biological applications. J R Soc Interface 4(12):1–17
43. Fu C et al (2009) Tunable nanowrinkles on shape memory polymer sheets. Adv Mater 21:4472–4476
44. Nguyen D et al (2009) Tunable shrink-induced honeycomb microwell arrays for uniform embryoid bodies. Lab Chip 9(23):3338–3344
45. Blancas AA, Lauer NE, McCloskey KE (2008) Endothelial differentiation of embryonic stem cells. Curr Protoc Stem Cell Biol Chapter 1:Unit 1F.5
46. Glaser DE et al (2011) Functional characterization of embryonic stem cell-derived endothelial cells. J Vasc Res 48(5):415–428

Chapter 21
Spinal Cord Repair by Means of Tissue Engineered Scaffolds

Mindan Wang, Xiongbiao Chen, and David J. Schreyer

Abstract Spinal cord injury (SCI) leads to devastating and permanent loss of neurological function, affecting all levels below the site of trauma. The injured adult spinal cord has little self-regenerative capacity due to multifactorial reasons. Tissue engineered scaffolds have emerged as a promising approach to promote regeneration of the damaged spinal cord by providing guidance to the regrowing axons. Integration of different therapeutic strategies with scaffolds has achieved substantial reestablishment of functional neural connectivity, with some strategies now being considered for clinical trials. This chapter presents a comprehensive discussion on the development of scaffold-based strategies currently under investigation for spinal cord tissue regeneration. First is a discussion of spinal cord structure, the pathophysiology of spinal cord injury, and various SCI animal models for experimental studies. Second is a detailed literature review and discussion of scaffold

M. Wang, BSc (✉)
Division of Biomedical Engineering,
University of Saskatchewan,
Saskatoon, SK, Canada
e-mail: miw554@mail.usask.ca

X. Chen, PhD
Division of Biomedical Engineering, University of Saskatchewan,
Saskatoon, SK, Canada

Department of Mechanical Engineering, University of Saskatchewan,
Saskatoon, SK, Canada
e-mail: xbc719@mail.usask.ca

D.J. Schreyer, PhD
Division of Biomedical Engineering, University of Saskatchewan,
Saskatoon, SK, Canada

Department of Anatomy and Cell Biology, University of Saskatchewan,
Saskatoon, SK, Canada
e-mail: david.schreyer@usask.ca

M.K. Danquah, R.I. Mahato (eds.), *Emerging Trends in Cell and Gene Therapy*,
DOI 10.1007/978-1-62703-417-3_21,

biomaterials and widely used techniques for scaffold fabrication in the context of SCI repair. This chapter then examines various therapeutic strategies currently used to repair SCI, including cell therapy, extracellular matrix protein/peptide modification, gene therapy, and molecular therapy.

Keywords Cell therapy • Spinal cord injury • Tissue engineered scaffolds • Scaffold biomaterials • Electrospinning

Abbreviation

BDNF	Brain-derived neurotrophic factor
ChABC	Chondroitinase ABC
CNS	Central nervous system
CSPGs	Chondroitin sulfate proteoglycans
DRG	Dorsal root ganglia
ECM	Extracellular matrix
GDNF	Glial cell line-derived neurotrophic factor
HA	Hyaluronic acid or hyaluronan
IKVAV	Ile-Lys-Val-Ala-Val
MAG	Myelin-associated glycoprotein
NT-3	Neurotrophin-3
OMgp	Oligodendrocyte myelin glycoprotein
OPF	Oligo(polyethylene glycol) fumarate
PAN/PVC	Polyacrylonitrile/polyvinylchloride
PCL	Polycaprolactone
PEG	Polyethylene glycol
PHB	Poly-β-hydroxybutyrate
PHEMA	Poly(2-hydroxyethyl methacrylate)
PHEMA-MMA	Poly(2-hydroxyethyl methacrylate-co-methyl methacrylate)
PHPMA	PolyN-(2-hydroxypropyl) methacrylamide
PLA	Poly(D,L-lactic acid)
PLGA	Poly(D,L-lactic-co-glycolic acid)
RAD	Arginine–alanine–aspartate
RGD	Arg-Gly-Asp
SCI	Spinal cord injury
SEM	Scanning electron microscopy
SIS	Small intestinal submucosa
TrkC	Tyrosine receptor kinase C
VEGF	Vascular endothelial growth factor
YIGSR	Tyr-Ile-Gly-Ser-Arg

Objectives

1. To review current knowledge of spinal cord structure, pathophysiology, and spontaneous repair after SCI
2. To survey the different animal models of experimental SCI, with their advantages and limits
3. To learn about various natural and synthetic biomaterials that may be used for the fabrication of tissue scaffolds for SCI repair
4. To understand current approaches to the design and fabrication of scaffolds for SCI repair
5. To consider some of the different therapeutic strategies used with tissue scaffolds for SCI repair

21.1 Introduction

Spinal cord injuries (SCI) have severe clinical consequences due to neuronal cell death at the site of injury as well as the disruption of axonal pathways crossing through the site of injury, leading to the devastating loss of sensory and motor functions essential for daily life. Thousands of people suffer SCI due to motor vehicle accidents, falls, violence, and sports injuries, among other causes, with the highest prevalence in young adults. About 12,000 new cases of SCI occur each year in the United States, with the estimated number of Americans living with a SCI in 2010 at approximately 265,000 [1]. The enormous health care and living expenses for persons with SCI impose a substantial economic burden on global health-care systems. Currently, clinical repair of traumatic spinal cord injury is very limited; statistics show that less than 1 % of patients with SCI undergo complete neurologic recovery by hospital discharge [1]. Acute interventions include anti-inflammatory methylprednisolone therapy and early reduction, decompression, and stabilization [2, 3]. These interventions generally block secondary neuropathological processes, yet the incidence of recovery of neurological function remains highly variable in both acute and chronic SCI, and the return of function is almost never complete.

In the last two decades, research progress has been made in elucidating the fundamental inhibitory mechanisms and pathophysiology that limit regenerative repair after SCI [4–8]. While basic neuroscience continues to be the predominant focus in the field of axon regeneration, emphasis must also be placed on the development of bioengineering strategies by which new basic science advances can realistically and effectively be implemented in patients who sustain SCI. In the past few years, considerable progress has been made in the development of bioengineered tissue scaffolds for the promotion of spinal cord repair. In this chapter, we will take a close look at this progress with emphasis on the pathophysiology of SCI, animal models, biomaterials for the construction of scaffolds, fabrication techniques, and strategies with tissue scaffolds.

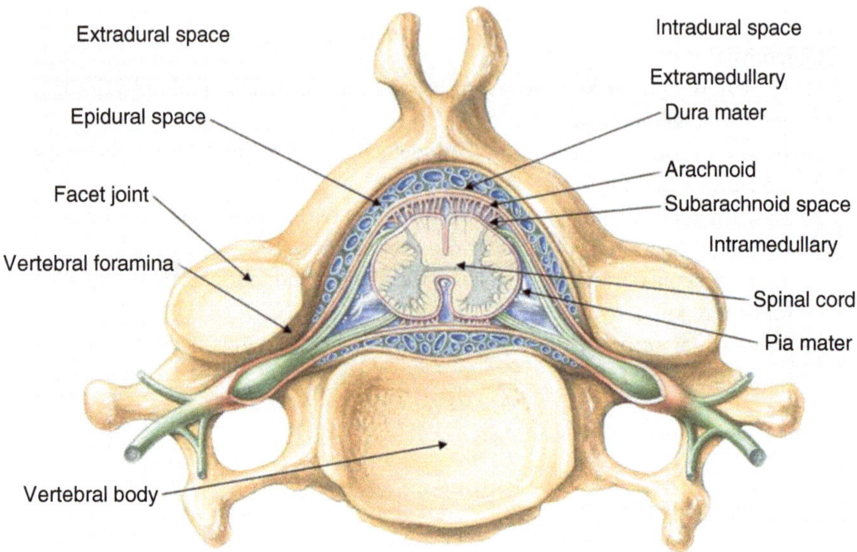

Fig. 21.1 Cross-sectional anatomy of spinal cord [9] (Reprinted with permission from BMJ Publishing Group Ltd.)

21.2 Spinal Cord Injuries

21.2.1 Spinal Cord Structure

The spinal cord extends from the base of the brain through the vertebral canal to the upper lumbar region and is surrounded and protected by the bony vertebral column. Its position is intermediate between the brain and the peripheral nerves, and it conducts sensory information upward and motor information downward to coordinate various body movements and organ functions. The spinal cord itself also contributes to the coordination of certain reflexes and is the site of central locomotor pattern generators that are independent of brain control. The internal structure of the spinal cord (Fig. 21.1) consists of butterfly-shaped gray matter in the center and an external layer of white matter surrounding the gray matter. Within the gray matter are found neurons and the glial cells that support their function. These neurons make synapses with each other, and with other distant neurons, to form the intrinsic neuronal circuitry of the spinal cord. The white matter is composed of bundles of ascending and descending nerve fibers that send and receive information to and from the brain and the peripheral nervous system through the transmission of electrical impulses. The ascending white matter tracts carry sensory information, such as touch, skin temperature, pain, and joint position, from the body upward to the brain, while the descending tracts carry information from the brain downward to initiate movement and control body functions. The white matter axon pathways also contain glial cells, particularly oligodendrocyte cells that provide a covering of insulating myelin to the long nerve fibers. Myelination of axons ensures the efficient

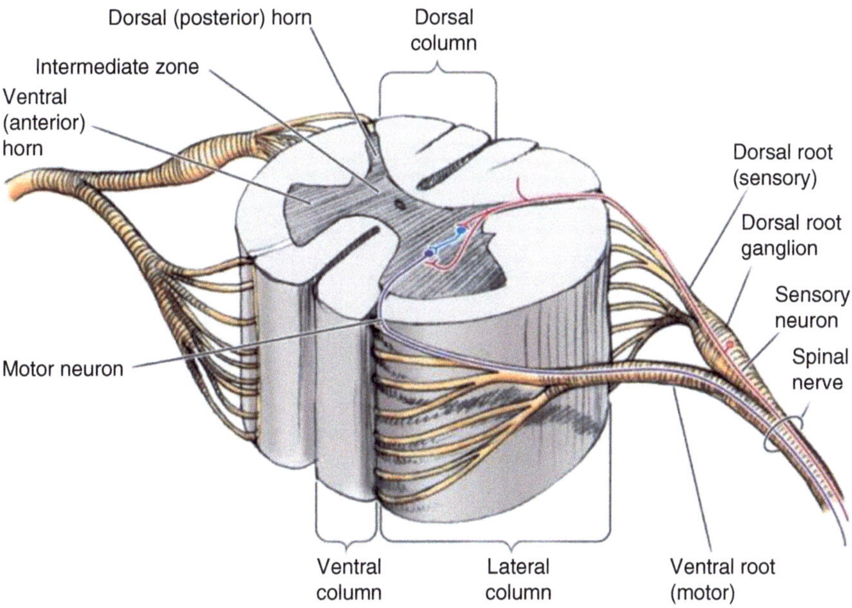

Fig. 21.2 Structure of spinal cord and distribution of dorsal and ventral roots [9] (Reprinted with permission from Sinauer Associates, Inc.)

transmission of electrical impulses along the nerve fibers. Surrounding the entire spinal cord are protective connective tissue membranes called the meninges, including the dura mater, arachnoid mater, and pia mater. The cerebrospinal fluid is found in the central canal and in the space between the arachnoid and pia mater, which is called the subarachnoid space. The cerebrospinal fluid provides a stable chemical environment for the transmission of electrical impulses, and serves as a cushion to protect the delicate nerve tissues against damage inside of the vertebrae.

The human spinal cord has 33 different segments: 8 cervical, 12 thoracic, 5 lumbar, 5 sacral, and 3 coccygeal. At each segment, right and left pairs of spinal nerves emerge to form the major part of the peripheral nervous system. These peripheral nerves carry motor information from the spinal cord to the rest of the body and sensory information from the body back to the spinal cord. Cervical nerves in the neck supply movement and feeling to the arms, neck, and upper trunk and also control breathing. Thoracic nerves in the upper back supply the trunk and abdomen, while lumbar nerves in the lower back supply the legs, bladder, bowel, and sexual organs. The nerve emerging at each spinal cord segment forms as a union of dorsal and ventral nerve roots (Fig. 21.2). Dorsal roots of the peripheral nerves relay sensory information to the posterior horn of the spinal cord gray matter, whereas ventral roots carry motor information from the anterior horn to muscles and visceral organs.

Axon tracts at different positions within the spinal cord white matter convey various types of information (Fig. 21.3). For example, the fasciculus cuneatus and fasciculus gracilis carry ascending touch information, the anterolateral system carries ascending pain and temperature information, and the spinocerebellar tracts provide

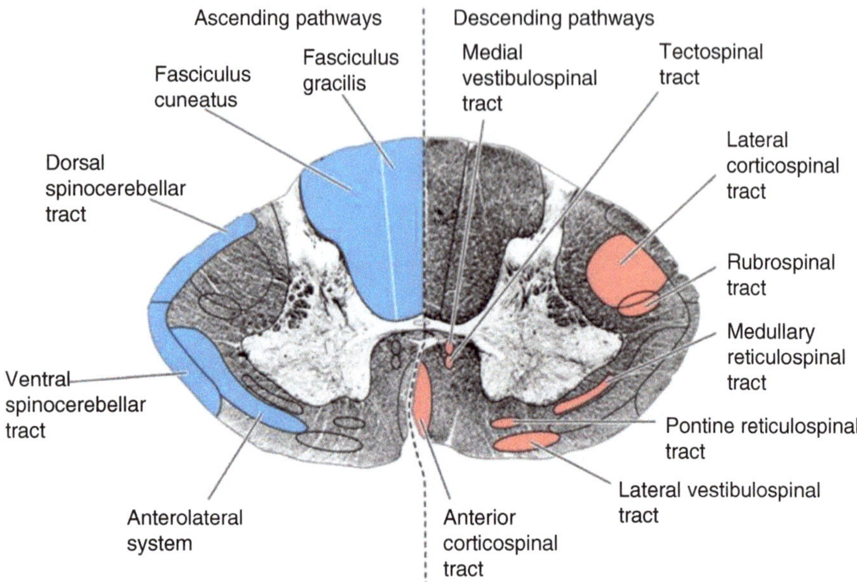

Fig. 21.3 Cross-sectional diagram of the spinal cord, showing distribution of some of the major ascending and descending tracts essential for neurological functions [9] (Reprinted with permission from Oxford University Press)

ascending sensory feedback about body movement. Meanwhile, the corticospinal, rubrospinal, and vestibulospinal tracts carry descending information that ultimately controls muscle movement.

When the spinal cord is injured, damage to gray matter will compromise the local reflex circuitry at that segment. However, the white matter damage is usually of far greater clinical significance. Destruction of a white matter nerve fiber pathway permanently interrupts electrical impulse traffic in both directions. Sensory information from all body regions below the injury cannot reach the brain, and motor information cannot be conveyed to levels below the injury. Communication between different levels of the spinal cord itself is also interrupted. The current emphasis of tissue engineering for SCI repair is to promote regrowth of these essential tracts after injury, which should in turn lead to dramatic restoration of sensory and motor functions in SCI patients.

21.2.2 Pathophysiology of SCI

21.2.2.1 Pathological Processes

The pathological sequelae of SCI can be divided into three major phases: primary, secondary, and chronic. The neurological damage caused at the time of mechanical trauma to the spinal cord is called primary injury. Primary injury can occur in many forms, including contusion, compression, and penetrating trauma. Cases of complete transection of spinal cord are rare, and only about 27 % of injuries are caused by laceration, which usually causes massive tissue loss, cyst formation, and a significant invasion of meningeal cells. The remaining 73 % are caused by compression, with the cord surface remaining intact [10]. Primary injuries cause immediate necrotic death of neurons and transection of nerve fibers (axons) at the lesion site. A subsequent cascade of cellular and biochemical reactions evolves slowly, causing secondary damage to the surrounding tissue that leads to the further loss of tissue and compromise of neurological functions.

Secondary injury of the spinal cord includes microvascular alterations, edema, ischemia, necrosis, free radical formation, lipid peroxidation, excitatory neurotransmitter accumulation, inflammatory responses, and other molecular changes contributing to further neural damage [11–15]. Vascular disruption caused by local ischemia, intravascular thrombosis, vasospasm, hemorrhage, and other possible causes plays a substantial role in secondary damage [16–22]. Lack of blood supply results in deficiencies of oxygen and nutrients in the traumatized tissue and thereby causes a series of injurious events, such as depolarization of the neuronal membrane potential [23, 24]. The depolarized neurons become more electrically active and lead to extensive neurotransmitter release and cyclical excitation of the neighboring neurons in a process called excitotoxicity, which ultimately results in the death of more neurons. Accumulation of free calcium also contributes to this process [24].

After the first few hours following SCI, an inflammatory reaction is triggered by resident microglia and astrocytes and invading macrophages and neutrophils. This inflammation is thought to have both destructive and beneficial roles with respect to spinal cord repair [22, 25]. One negative consequence is that neurons and oligodendrocytes undergo apoptotic cell death, and loss of oligodendrocytes results in the demyelination of surviving axons [26, 27].

Cellular and tissue reactions to SCI can cause the formation of physical and molecular barriers to axonal regeneration. Glial scars are created due to a variety of cellular changes including reactive astrocytes, oligodendrocytes, macrophages, microglia, fibroblasts, and invading Schwann cells from the peripheral nervous system [25, 28–31]. After SCI, astrocytes adopt a reactive hypertrophic phenotype, proliferating and expressing increased levels of glial fibrillary acidic protein and releasing inhibitory extracellular matrix molecules including chondroitin sulfate proteoglycans (CSPGs). Demyelination of surviving axons in the surrounding white matter due to death of oligodendrocytes after SCI leaves myelin debris, which contains several axon growth inhibitory proteins including myelin proteins Nogo-A, myelin-associated glycoprotein (MAG), and oligodendrocyte myelin glycoprotein (OMgp) [32, 33]. Removal of the degenerating myelin debris by phagocytotic cells during the process of Wallerian degeneration is quite slow following traumatic SCI. Other axon growth inhibitory proteins that may be present include ephrin and semaphorin [34, 35]. Removal or neutralization of these inhibitory proteins can result in improved axonal regeneration [36–40].

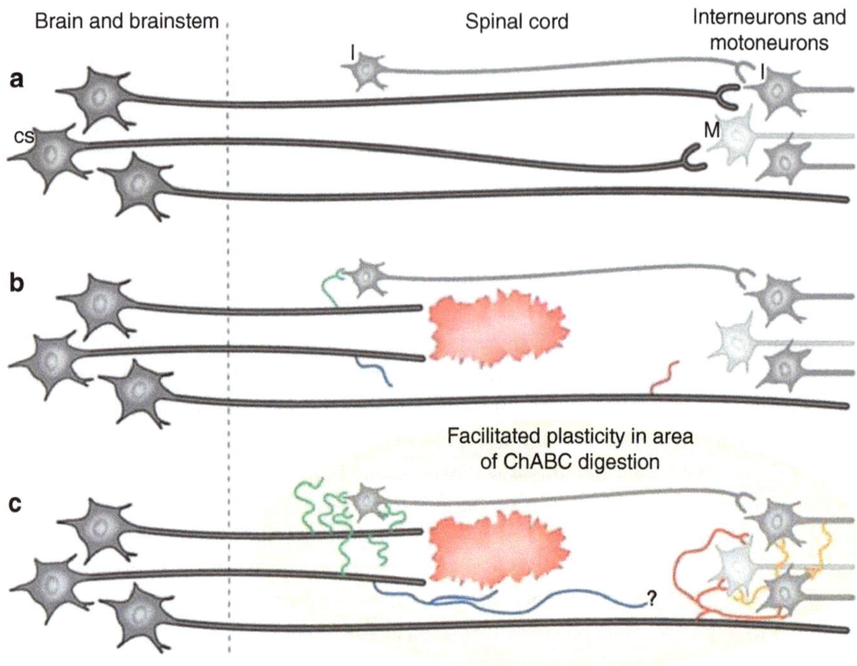

Fig. 21.4 (**a**) Corticospinal neurons project from the cortex and brainstem to networks of interneurons (*I*) and motoneurons (*M*) in the spinal cord. (**b**) A partial injury interrupts the corticospinal axons. Limited axonal plasticity ensues, including some collateral sprouts onto propriospinal interneurons that bypass the lesion, and sprouting of spared fibers below the level of injury. (**c**) Treatment with ChABC greatly enhances localized axonal sprouting, which has thus far been shown for injured corticospinal axons connecting to interneurons (*green*) and growing around the lesion (*blue*). Given the extent of ChABC digestion, one may speculate that collateral sprouting of spared axons past the lesion (*red*) is also enhanced and that sprouting of interneurons distal to the lesion (*brown*) may occur as well. Among all these new connections enabled by ChABC digestion, task-specific grasping training would consolidate those that support grasping, whereas "general" training would consolidate those that support enriched housing-related skills, such as ladder crossing and rope climbing [55] (Reprinted with permission from Elsevier BV)

21.2.2.2 Spontaneous Repair

Despite the negative pathological features that characterize the spinal cord after injury, spontaneous repair processes do occur at lesion sites. Some aspects of the inflammatory response after SCI might be protective, allowing stressed neurons to survive. Furthermore, after injury the reactive astrocytes secrete many cytokines and neurotrophic factors, restore the extracellular ionic environment, and upregulate various cellular surface molecules and extracellular matrix (ECM) molecules such as L1, laminin, and fibronectin [41–46]. Evidence also shows that reactive astrocytes actually protect tissue and preserve function after SCI [47]. Taking advantage of the beneficial features of the cellular response to SCI while neutralizing the obstructive events might improve the outcome of spinal cord regeneration.

There are some cellular responses to injury that suggest the possibility of regenerative repair. Spared axons near the site of injury can undergo "sprouting," and rearrangement of neural pathways within the spinal cord can take place after SCI, though few axons regenerate back to original, distant targets. Corticospinal axons sprout into spared propriospinal tracts to increase connectivity with lumbar motor neurons [48, 49]. The cortical sensorimotor area undergoes functional rearrangement and the rubrospinal system reorganizes to compensate for some of the functional loss due to damaged corticospinal tracts [49, 50].

Neutralization of the inhibitory CSPGs using chondroitinase ABC (ChABC) or addition of certain growth-promoting "neurotrophins" can promote the reconnection of corticospinal neurons with spared spinal cord neurons [51–53]. Combined with specific rehabilitation training, functional recovery can be improved [54]. The axonal plasticity of the corticospinal tract and the improvement of sprouting by ChABC treatment are illustrated in Fig. 21.4. However, the recruitment of sprouted axons does not provide full recovery of spinal cord function.

21.3 Animal Models

Animal models of SCI provide insights into underlying pathophysiological mechanisms as well as the effectiveness of new therapeutic strategies [27, 56]. Many of the features of experimental SCI in rodents, such as cell death, inflammation, axon regeneration, or spinal motor pattern generators, are remarkably similar to humans, and rodents are the most widely used animal models for SCI repair [56]. Rodents also offer the advantages of low cost, few surgical infections, and ease of care and availability. Functional and behavioral analysis techniques have also been well established for rodents.

Experimental tools have been developed to reliably and consistently contuse the spinal cord of rodents to mimic the contusion of human spinal cord. However, the rat contusion model has several shortcomings. In rat and human spinal cord contusion injury there is variable sparing of white matter tracts. This, combined with the occurrence of axonal regeneration and sprouting of distal segments of surviving axons or injured axons in the rat contusion model, complicates interpretation of results from experimental studies. Targeted spinal cord transection models have been considered more appropriate for accurate analysis of axonal regeneration [5, 56]. Furthermore, the rat contusion model appears to primarily affect the dorsal white matter tracts, while both the dorsal and ventral tracts may be affected in humans [57]. Although the pathophysiology of SCI in rodents and humans is quite similar, important differences have been noted, including reduced glial scarring, decreased inflammation and demyelination, elevated Schwann cell infiltration, and prolonged Wallerian degeneration [5]. The functional anatomy of the spinal cord (pathway organization) is also different between rodents and humans, and considering these differences is essential when interpreting experimental data, especially behavioral outcomes [58]. For example, the corticospinal tract is vital for fine motor control in humans and nonhuman primates but less essential in rats [59, 60]. On the other hand, the rubrospinal tract is critical for forelimb movement in rats but is considered to be vestigial in humans [61].

In humans, SCI is most common at the cervical level and it might be more appropriate to carry out functional analysis of recovery from cervical SCI in rodent models of injury [62]. However, low thoracic injury in rodent animal models is more common as this is less traumatic for the animals. In rodents, the close proximity of the lower limbs to the lower thoracic cord also leads to more apparent motor function recovery, which is not the case in humans [63]. One also needs to be aware of the danger of applying experimental results to clinical trials; in one study, patients with incomplete SCI had delayed neurologic worsening after Schwann cell transplantation and a vigorous rehabilitation regime [64].

Other species used as animal models for the study of spinal cord repair include marine species and large mammals, such as cats, dogs, monkeys, and pigs. The disadvantage of marine models is the small working size, precluding surgical maneuvers and device implantation; on the other hand, the ability to manipulate the marine genome and dissect complex molecular events is advantageous. Tests on large mammalian species (e.g., dog, pig, and monkey) are necessary to better evaluate the efficacy and safety of devices, such as bioengineered scaffolds, to regenerate human spinal cord. Ethical issues, extremely high costs, and extensive labor requirements limit the application of these models [5]. The advantage of the cat model for SCI is that locomotor activities, such as walking or standing, can be easily trained; however, the limitation is that locomotion in cats after injury is strongly influenced by the central pattern generator, which is a crucial variance from humans [5].

A number of different animal models of spinal cord injury are in use, including complete transection, hemisection, and contusion/compression injuries. The choice of these models of SCI should be guided by the aim of experiments, with the contusion model matched to the study of physiological responses and pathophysiology of SCI, the hemisection model to the sprouting of intact fibers, and the hemisection/complete transection models to axonal regeneration or implantation of bioengineered scaffolds designed to foster axon growth [5, 56].

21.3.1 Contusion Models

Contusion animal models best mimic the most common human SCI cases. The contusion models are, however, a poor choice for the application of preformed scaffolds as a bridge through the gap due to the irregular geometry of the lesion site. Transplantation of living cells or injectable biomaterials that can gel in situ are perhaps more suitable for contusion models. Moreover, analysis of the axonal regeneration response in contusion models is difficult due to the presence of spared axons and the sprouting of the spared/injured axons.

Several methods have been developed to produce contusions in animal spinal cord. The weight-drop contusion model developed in rats mimics the most prevalent contusion injury in humans and is comparable with human spinal cord injury in terms of functional, electrophysiological, and morphological outcome parameters [57]. A surgical spring-loaded clip or balloon compression is also useful for creating contusions in animal spinal cord models as it can precisely control the length and magnitude of the compressive injury; furthermore, no laminectomy is needed for the balloon compression to injure the spinal cord. Unlike the weight- drop contusion model, however, these two injuries do not mimic true compression injury [57, 65, 66]. Constant displacement of the spinal cord using a modified surgical forceps can also produce reproducible contusion SCI in animals [66]. Due to the high dependency of SCI pathophysiology on the exact nature of the mechanical trauma, more sophisticated methods using computer-controlled impactors to produce contusion in animal spinal cords have been attempted. Such devices are very attractive as a method to create spinal cord contusions due to high reproducibility and the capacity to titrate the degree of injury. However, they have the disadvantages of high costs, variable completeness of the lesion, and suboptimal analytical methods to determine the distinction between the spared and regenerated neural tissues [56].

21.3.2 Hemisection Models

In humans, laceration results in the disruption of dura mater of the spinal cord and can be mimicked by hemisection/partial section animal models using a surgical microknife or microscissors [67, 68]. With hemisection models, either injectable or preformed scaffolds or a section of peripheral nerve can be used to bridge the cavity [69–72]. Animals receiving unilateral hemisection injury maintain bladder and bowel function, which eases the labor-intensive postoperative animal care. Similar to the contusion model, however, hemisection models can also have the problem of differentiating regenerating axons from spared axons as well as collateral sprouting of spared/injured axons.

21.3.3 Full-Transection Models

Full transection of the spinal cord is rare in humans. However, full transection animal models are used to evaluate the efficacy of transplanted devices for promotion of axonal regeneration as they preclude the involvement of spared axons and spontaneous plasticity after SCI. Pathological changes and subsequent neurological outcomes are quite stable after complete spinal cord transection [73]. These features enable distinct differentiation between regenerating axons and spared axons and thereby allow reliable evaluation of therapeutic strategies for spinal cord repair. The disadvantages of transection models include surgical complications, such as instability of the spinal column, as well as greatly increased difficulty in postoperative animal care [5]. Despite this, increasing numbers of experiments are investigating implanted scaffolds for promotion of axonal regeneration and function recovery in complete transection animal models [74–76].

21.4 Scaffold Biomaterials

Implantation of substrates such as cellular grafts can promote regrowth of axons after SCI. However, the growth of axons is highly random and does not extend past the graft site to reenter host tissue [64]. Artificial tissue scaffolds are designed to provide mechanical support for axonal regrowth and to potentially serve as a local delivery system for growth-promoting factors, growth-inhibitor neutralizing agents, or other therapeutic drugs as well as a carrier for supportive cells that might facilitate repair. Biomaterials selected for construction of tissue scaffolds for spinal cord regeneration should meet the following criteria: biocompatibility with the host tissue to avoid adverse immune reactions, an adjustable rate of degradation, and degradation products that are nontoxic [8]. Specific to the needs of spinal cord repair, scaffold biomaterials should ideally also have the ability to reduce astrocytic reaction and fibroblastic gliosis and allow for cell adhesion and axonal regrowth. A variety of biomaterials, both synthetic and natural, have been examined for their suitability to fabricate tissue scaffolds in different tissue engineering strategies.

21.4.1 Natural Biomaterials

21.4.1.1 Collagen

The role that collagen protein might play in spinal cord regeneration is controversial: on one hand, collagen can promote neural cell attachment and growth; on the other hand, it is a component of glial scars, which might impose a physical and chemical barrier for axonal regeneration after SCI [77, 78]. Scars consisting of collagen meshworks have been shown to incorporate factors that are both inhibitory and stimulative for axonal regeneration [79]. Moreover, collagen hydrogels of high concentration can impede axonal regeneration [80]. Despite the unknown cellular and molecular mechanism of collagen in nerve repair, positive outcomes of both in vitro and in vivo studies indicate that its potential application in spinal cord regeneration is quite promising.

Aligned electrospun collagen nanofibers could direct the outgrowth of axons from dorsal root ganglion (DRG) neurons in vitro and a spiral tube of collagen nanofibers supported axon sprouting. Collagen also limited the astrocytic response at the boundary of the lesion site in vivo [81]. Provided that the nanostructure of the collagen scaffold mimics the extracellular matrix of native tissue, 3D scaffolds composed of electrospun collagen nanofibers should show excellent cell compatibility and may be beneficial for the functional recovery of injured spinal cord. A collagen-binding domain can be incorporated into brain-derived neurotrophic factor (BDNF), then the collagen-binding BDNF loaded into collagen scaffolds; when transplanted into the hemisectioned rat spinal cord, such scaffolds resulted in neuroprotection and significant functional recovery [82]. Further modification to the BDNF-loaded collagen scaffolds has also been attempted, through addition of neutralizing antibody to the epidermal growth factor receptor activation (which is downstream of the signaling pathways of such regeneration inhibitors as myelin-associated proteins and CSPGs). The outcome of this triple-functional collagen scaffold was superior compared to the bifunctional one in terms of neural regeneration, recovery of synaptic transmission, and prevention of the formation of glial scars [83]. More corticospinal tract fibers grew into the collagen matrix, and partial functional recovery was achieved despite no corticospinal tracts regrowing into areas caudal to the implant [84]. Transplantation of different chemically cross-linked collagen-based scaffolds into a complete transection model in the rat spinal cord indicated the collagen biomaterial was helpful for SCI repair as it aligned the reparative tissue with the long axis of the spinal cord, reduced scar formation, efficiently delivered living cells containing a transgene for neutrophins, and inhibited the collapse of musculature and connective tissue into the defect [85]. Combined, this evidence demonstrates the potential of collagen as a biomaterial for scaffolds aimed at enhancing spinal cord regeneration.

21.4.1.2 Alginate

Alginate is a naturally occurring, water-soluble, anionic polysaccharide extracted from brown algae. Due to its excellent biocompatibility, low toxicity, bioresorption, and mild gelation by cross-linking with divalent cations, alginate has been widely used for wound dressings and in tissue engineering to encapsulate and implant various growth factors or cell types into lesion sites to enhance the rate of tissue regeneration [86–89]. Chemical modifications such as carbodiimide-mediated amide bond formation of carboxyl groups in alginate with polyethylene glycol diamines, methyl ester L-lysine, or adipic hydrozide enable control over the swelling and mechanical properties of alginate hydrogels [90, 91]. Incorporation of polylactic-co-glycolic acid (PLGA) microspheres loaded with alginate lyase in alginate hydrogels helps to control their degradation rate [92]. Application of alginate-based hydrogels for the regeneration of spinal cord has been widely tested, demonstrating no allergic or inflammatory reactions and providing guidance over axonal regrowth with the formation of functional synapses [93, 94]. Alginate hydrogels with a highly anisotropic capillary structure have been shown to support highly oriented linear axonal regrowth and reinnervation of appropriate target neurons [95]. One concern with respect to the application of alginate hydrogels for spinal cord repair is that the hydrophobic alginate polymer does not promote cell adhesion or neurite outgrowth [96]. Attempts have been made to covalently modify alginate hydrogels with laminin protein, laminin peptides (Tyr-Ile-Gly-Ser-Arg: YIGSR and Ile-Lys-Val-Ala-Val: IKVAV), and fibronectin peptides (Arg-Gly-Asp: RGD) for the construction of three-dimensional culture systems for neural cells and to promote axon outgrowth [97]. Soft alginate hydrogels prepared by cross-linking with substoichiometric concentrations of Ca^{2+}, Ba^{2+}, and Sr^{2+} cations at no greater than 10 % of all potentially available gelation sites were cell adhesive for rat and human neurons, favored neurite outgrowth, and protected neurons against oxidative stress [98]. Microencapsulation of fibroblasts genetically engineered to produce BDNF in alginate hydrogels resulted in outgrowth of axons from DRG neurons in vitro and promoted axonal regrowth and partial recovery of forelimb usage in an injured rat spinal cord [88, 99]. Overall, alginate has considerable potential for application to spinal cord repair, but modification of the physical and chemical properties of alginate hydrogels requires additional preparative steps.

21.4.1.3 Agarose

Agarose is a linear polysaccharide derived from seaweed; specifically, it is an alternating copolymer of 1,4-linked 3,6-anhydro-α-L-galactose and 1,3-linked β-D-galactose. Agarose is thermoresponsive and can gel at temperatures ranging from 17–40 °C, depending on the degree of hydroxyethyl substitution on its side chains. Due to this characteristic, attempts have been made to inject dissolved agarose into SCI lesion sites, where it can be rapidly cooled in situ using liquid nitrogen vapor [71]. The advantage of such injectable scaffolds is that they can conformably fill irregular-shaped lesions and no invasive surgical procedure is required. When loaded with neurotrophic factors, scaffolds tested in a dorsal over-hemisection rat model encouraged axon outgrowth into the scaffolds and reduced inflammatory responses, with decreased reactive astrocytosis and deposition of inhibitory CSPGs [71]. Agarose gels prepared with different concentrations of agarose solution have concentration-dependent stiffness profiles, which affect axon extension by DRG neurons in vitro [100]. Stokolsa and Tuszynski successfully used a freeze-dry process to create agarose nerve guidance scaffolds with uniaxial linear channels (Fig. 21.5), which they tested with and without inclusion of recombinant BDNF in an adult rat model of SCI [101, 102]. Axons grew through the scaffolds in a linear fashion, and the incorporation of BDNF significantly promoted regeneration (Fig. 21.6). Templated agarose scaffolds were also able to orient and provide guidance to local spinal cord axons after SCI [103]. When treated with combination therapies of nerve guidance scaffolds with autologous bone marrow stromal cells expressing neurotrophin-3 (NT-3), lentiviral vectors expressing NT-3 beyond the lesion site, and priming lesions of the sensory neuronal cell body to stimulate the endogenous growth state of the injured neuron, almost all axons entering the scaffolds were able to grow the full length of the lesion cavity, far more than if cell suspension grafts alone were used (Fig. 21.7). Only the group receiving the full combination of treatments had axonal regeneration beyond the lesion site, though the axons were unable to re-penetrate the host spinal cord due to the reactive cell layer between the distal aspect of the scaffold and host tissue [70]. A problem with these scaffolds with uniaxial channels is that the axons cannot penetrate the agarose walls; thus, using implants with anatomical guidance channels positioned for major ascending and descending axon tracts in combination with different therapeutic strategies to promote axonal regeneration would be beneficial.

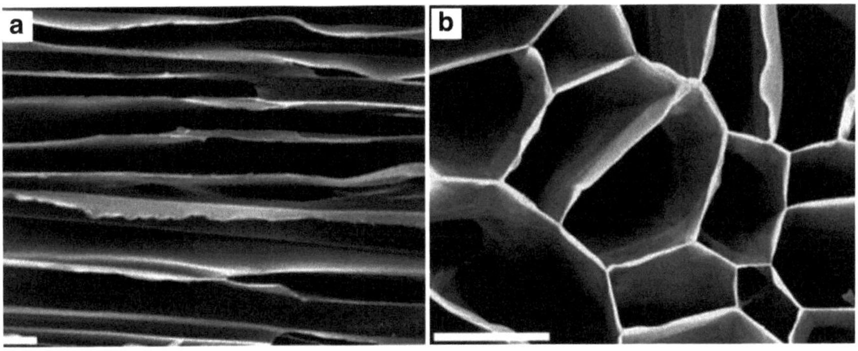

Fig. 21.5 Freeze-dried agarose scaffolds implanted in a complete transection model, promoting extensive linear axonal regeneration. Scanning electron microscopic (SEM) images of agarose scaffolds in (**a**) longitudinal or (**b**) cross-sectional orientation. Scale bar (**a**, **b**) = 100 μm [70] (Reprinted with permission from Elsevier BV)

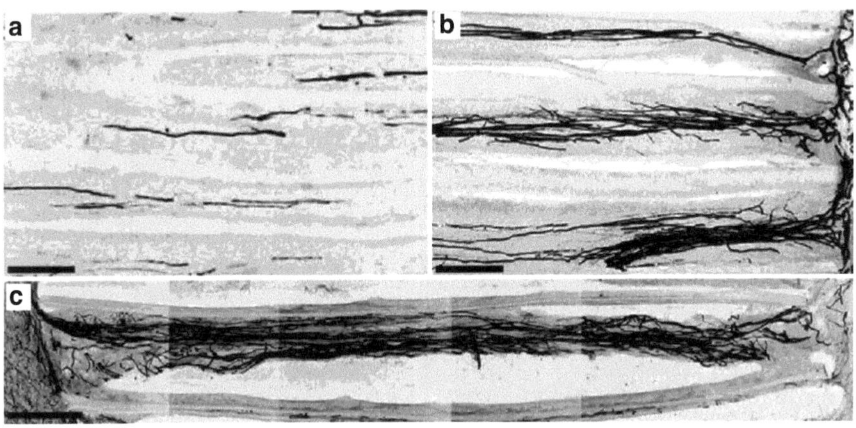

Fig. 21.6 Labeling of axon neurites within the channels of freeze-dried agarose scaffolds. Neurofilament labeling demonstrates penetration and linear growth of axons within channels of freeze-dried agarose scaffolds (**a**) lacking growth factor and (**b**) incorporating 2 μg recombinant human brain-derived neurotrophic factor BDNF into the walls and matrix-filled lumen of individual channels. (**c**) Best example of linear axonal growth through complete length of channel. *Scale bars* = 100 μm [101] (Reprinted with permission from Elsevier BV)

21.4.1.4 Chitosan

Chitosan is a glycosaminoglycan carbohydrate polymer derived from chemical deacetylation of chitin, the major structural polysaccharide found in crustacean and insect exoskeleton. Cells can adhere to chitosan due to its positive charge, which is a function of the degree of alkaline deacetylation [104]. Chitosan has the advantages of being relatively inert, not eliciting a chronic immune response, and maintaining the physical integrity of scaffolds up to 1 year in vivo [105]. Chitosan particles as NT-3 carriers permitted the survival and differentiation of neural stem

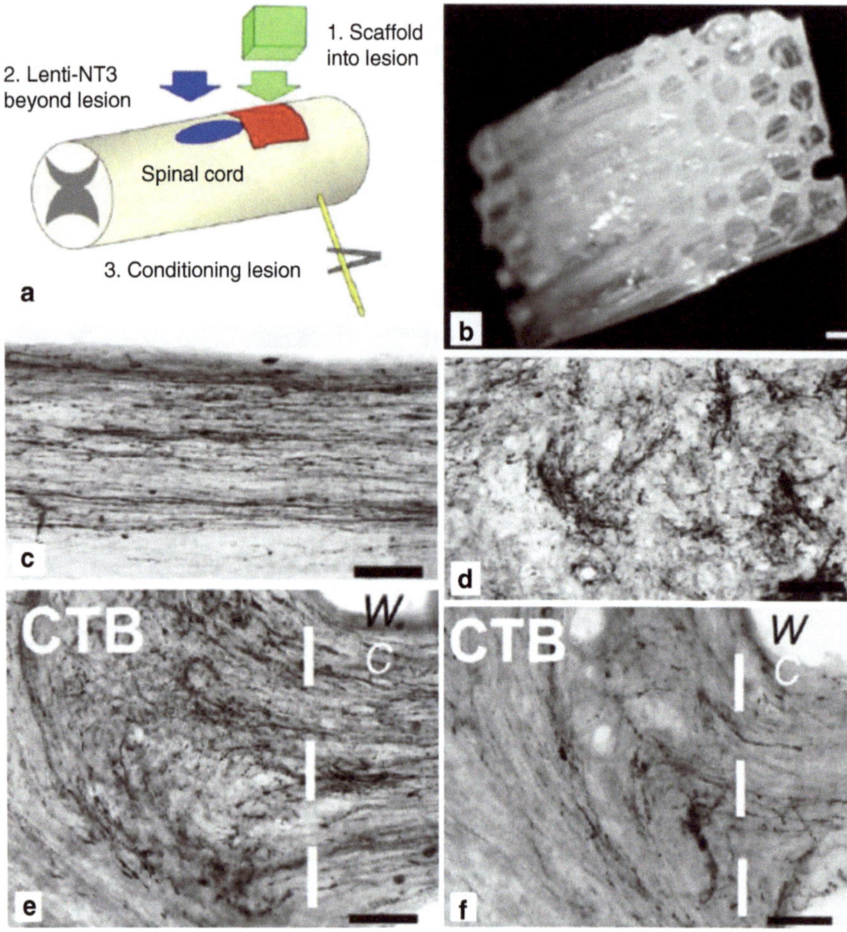

Fig. 21.7 Experimental paradigm and main results of combinatorial treatments. (**a**) Schematic of experimental paradigm illustrates that subjects underwent dorsal column lesions at C4, scaffold implantation (*1*), injection of lentiviral vectors expressing NT-3 rostral to the lesion (*2*), and conditioning lesions (compression) of the sciatic nerve (*3*). Rostral to *left*, caudal to *right*. (**b**) Macroscopic scaffold architecture. (**c**) CTB-labeled sensory axons regenerating into scaffold also exhibit linear orientation corresponding to the rostral–caudal axis of the scaffold implant and spinal cord. (**d**) In contrast, axons labeled by CTB appear randomly oriented in subjects that receive non-organized, cell suspension grafts into lesion cavity without scaffold. (**e**) Numerous CTB-labeled sensory axons exit scaffold in rostral aspect of lesion site when subjected to combinatorial therapy with lentiviral NT-3 injection rostral to the lesion site and conditioning lesions of sciatic nerve. Vertical lines indicate scaffold/lesion interface at rostral aspect of lesion cavity; scaffold to right, axon emergence to left. (**f**) Fewer CTB-labeled sensory axons emerge from scaffold when lenti-NT-3 is not injected rostral to lesion site; this subject did receive a conditioning lesion. *Scale bars* **b**, **c**, **d**, **e**, and **f**, 50 mm [70] (Reprinted with permission from Elsevier BV)

cells into neuronal phenotypes while reducing the consumption of neurotrophic factors in cell transplantation therapy for brain and spinal cord injuries [106]. Recent work by Shoichet and coworkers demonstrated the use of chitosan in

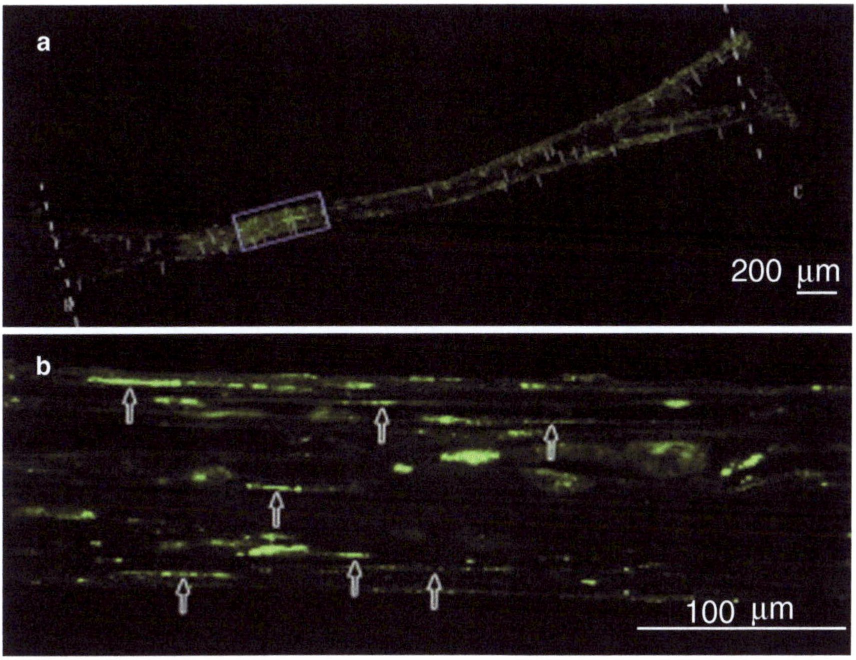

Fig. 21.8 (**a**) Anterograde labeling of the corticospinal tract of a rat implanted with a chitosan tube filled with semifluid type I collagen 12 months after the operation. The labeled regenerated tract entered the lesion area from the rostral end of the tube, traversed the lesioned area, and reentered the host spinal cord from the caudal end of the tube (*arrows*). Tracing is identified by *green* fluorescence. The *white dotted lines* indicate the two ends of the tube. Note that regenerated nerve fibers traverse the entire length of the tube. (**b**) Higher magnification of the area in the box in a [110] (Reprinted with permission from Elsevier BV)

extramedullary and intramedullary conduits or three-dimensional hydrogel scaffolds capable of supporting the survival and differentiation of neural stem/progenitor cells in transected spinal cord [107, 108]. More cells survived in extramedullary chitosan channels seeded with neural stem cells harvested from the brain than in those seeded with stem cells derived from the spinal cord [107]. For spinal cord repair, myelinated and unmyelinated axons as well as blood vessels were present in the extramedullary chitosan channel in a full transection animal model [107]. Radial glial cells, capable of differentiating into astrocytes, oligodendrocytes, neurons, and macrophages in the adult mammalian CNS, migrated from spinal cord stumps into the chitosan channel and acted as cellular scaffolds for axonal guidance [109]. In a recent study, chitosan tubes filled with semifluid type I collagen were shown to provide directional guidance for aligned axonal regrowth and promote nerve regeneration across a gap of 4 mm at the thoracic spinal level (Fig. 21.8). Functional recovery of the essentially paralyzed hind limbs was also promoted, as confirmed by behavioral evaluation [110]. These studies show the promise of chitosan as a biomaterial for scaffold construction to regenerate injured spinal cord.

21.4.1.5 Fibrin and Fibronectin

Fibrin is a protein involved in normal blood clotting. It is produced from fibrinogen by the proteolytic activity of thrombin and forms a cross-linked clot at physiological temperature and pH. Both the gelation and degradation rates can easily be controlled by changing the component concentrations of the mixture [111]. Thus, the fibrin approach can be used as an injectable biomaterial for tissue engineering applications. For example, injection of a commercial fibrin sealant (Tissucol®) into a lesion cavity promoted revascularization and axon growth within the scaffold, resulting in improved locomotor function [112]. Implant of a fibrin scaffold into a subacute dorsal hemisection model of SCI improved axon sprouting and delayed the accumulation of reactive astrocytes surrounding the lesion area [113]. An injectable fibrin/fibronectin composite biomaterial demonstrated more axonal ingrowth than either component alone [114]. Based on beneficial outcomes of plain fibrin scaffolds for spinal cord repair, fibrin has been further developed as a delivery system for trophic factors (NT-3), therapeutic agents such as ChABC, and cells such as bone marrow or neural stem cells, which enhance axon sprouting, reduce the level of inhibitory molecules, increase the number of transplant-derived neurons, and lead to significant behavioral recovery [115–119]. Overall, application of fibrin-based scaffolds for spinal cord repair is quite promising. One main concern in their further development is prolonging their in vivo residence time [120].

Fibronectin is a glycoprotein found in extracellular matrix and plasma. It is involved in many cellular processes, such as tissue repair, blood clotting, cell migration/adhesion, and embryogenesis [121]. A recent study indicates that acute fibronectin treatment of the dorsal spinal white matter after contusion injury may prevent the development of mechanical allodynia but not thermal hyperalgesia. The connecting segment-1 motif of fibronectin was found to contribute to this effect. In addition, fibronectin injection diminished inflammation and blood–spinal cord barrier permeability and blocked the reduction of serotonergic innervation of the superficial dorsal horn, an important descending brainstem system that modulates pain [122]. A neuroprotective effect of fibronectin, resulting in decreased lesion size, apoptosis, and axonal damage, was recently found to be associated with fibronectin peptide PRARIY [72, 123]. When fibronectin mats made from fibrous aggregates of plasma fibronectin were implanted into hemisected rat spinal cords, they integrated well into the host tissue with little or no cavitation and permitted oriented axonal regeneration. Cells such as macrophages, Schwann cells, oligodendrocytes and their precursors, and astrocytes infiltrated the implants and oriented growth of axons occurred into the mats. Implantation of viscous fibronectin gel into the lesioned spinal cord yielded similar results except that a large cavity formed between the implant and the host tissue, probably due to the poor gel-forming characteristics and resultant loss of contact with cavity margins as the protein dissipated [114]. Axonal growth may be associated with laminin deposition and cell infiltration; for example, the interaction of fibronectin with integrin receptors on Schwann cells modulates their infiltration, and these cells may then act as cellular scaffolds for axonal regrowth [123, 124]. However, fibronectin itself can support neurite outgrowth and axonal regeneration of adult brain neurons in vitro [125].

21.4.1.6 Hyaluronic Acid (HA)

Hyaluronic acid (HA) or hyaluronan is a negatively charged, heavily hydrated glycosaminoglycan found in almost all extracellular tissue spaces. It is comprised of repeating disaccharide units of β-1, 4-D-glucuronic acid and β-1, 3-N-acetyl-D-glucosamine, with its carboxylic and hydroxyl functional groups targeted for chemical modification. HA demonstrates excellent biological properties, being biocompatible, biodegradable, and immunoneutral. It has been investigated for its potential in the construction of tissue scaffolds, particularly for SCI repair, as it can inhibit scar formation and is involved in such complex cell-signaling events as cell migration and attachment, angiogenesis, and axon sprouting [126–128]. High molecular weight HA hydrogels implanted in a rat model of spinal dorsal hemisection injury were found to mitigate astrocyte activation and decrease the level of inhibitory CSPG deposition [126]. This effect is associated with the nonadhesive nature of HA hydrogels with respect to inflammatory cell infiltration as well as its interaction with CD44 cell surface receptors on cells such as astrocytes to reduce the inflammatory response [129]. In contrast to high molecular weight HA, low molecular weight HA increased cell proliferation and inflammation [126, 130]. The abundant presence of HA in brain and spinal cord development has inspired researchers to investigate the possibility of controlling neuronal progenitor cell differentiation through fabrication of HA hydrogels with tunable mechanical properties [131–133]. HA-based scaffolds have been tested in animal models for SCI repair [126, 134, 135]. However, as cells do not readily adhere to HA, modifications of HA scaffolds by addition of either ECM components (e.g., laminin, collagen) or cell-adhesive polypeptides (e.g., poly-D-lysine) have been attempted [136–139]. In addition, numerous methods have been developed to chemically modify HA for the controlled release of tethered bioactive agents, such as neutralizing antibody against Nogo receptor-66 to block the activity of myelin-associated axon growth inhibitors [135, 140]; conjugation of Nogo-66 receptor antibody into poly-L-lysine-modified HA resulted in significantly more angiogenesis and axonal regeneration into the implants.

Photo-cross-linked HA scaffolds with varying geometries and controlled degradation rates have been recently fabricated using adapted solid freeform fabrication techniques (Fig. 21.9). Laminin protein has been covalently cross-linked to scaffold surfaces using carbodiimide chemistry, which resulted in the adhesion and survival of Schwann cells onto the scaffold walls [141]. This is quite promising for spinal cord repair, particularly when scaffolds with anatomical guidance channels are combined with controlled distribution of living cells or bioactive agents within the guidance channels. Application of multiphoton lithography has resulted in the successful fabrication of submicron-sized bovine serum albumin protein structures inside HA hydrogels; further modification of the bovine serum albumin protein with laminin-derived polypeptides promoted adhesion and axon outgrowth from cultured DRG neurons [142]. Tests of these HA-based scaffolds in animal models of SCI could shed light on their potential for guiding axonal regeneration and promoting functional recovery.

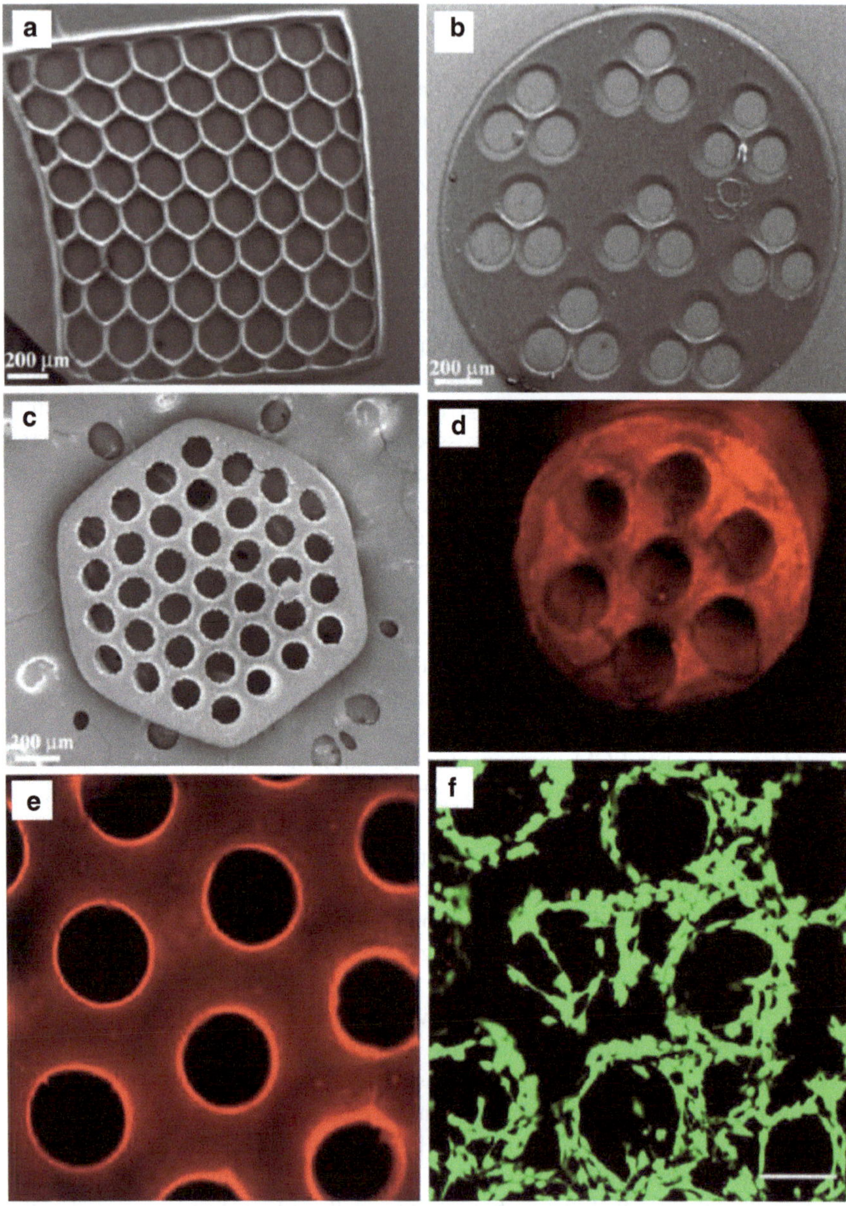

Fig. 21.9 Solid freeform fabrications of hyaluronic acid hydrogels for nerve tissue engineering. SEM micrograph of a single-layer scaffold of glycidyl methacrylate modified hyaluronic acid with (**a**) hexagonal patterns, (**b**) circular patterns with three channels, and (**c**) circular patterns with more than 30 channels. (**d**) Fluorescence micrograph of the cross-section view of the multilumen scaffold. *Red* fluorescent particles were added in the prepolymer solution to better visualize the scaffolds and the internal structure post-fabrication. (**e**) Laminin was covalently conjugated to scaffold surfaces. Fluorescence micrographs show scaffolds immunostained for laminin. (**f**) Schwann cells seeded on laminin-modified HA scaffolds adhered and remained viable (stained *green* with 2 μM of calcein) for 24 h after cell seeding [141] (Reprinted with permission from Springer)

21.4.1.7 Acellular Tissue Grafts

Acellular scaffolds are grafts of native tissue that retain ECM components and organization, but not living cells. Acellular tissue grafts are prepared through various chemical and thermal decellularization methods, such as the freeze–thaw techniques and the cold-preserved method from native tissue. This results in acellular grafts with structural and functional proteins of ECM components and their original three-dimensional distribution, but no immunogenic cellular components [143–145]. Removal of cellular membranes can eliminate the antigens responsible for allograft rejection [146]. Such grafts have been extensively investigated and successfully applied to the repair of skin, bladder, urethra, small bowel, cardiac valve, blood vessel, skeletal muscle, peripheral nerve, and even much more complex organs such as the heart and lung [146–157]. With respect to the CNS, acellular brain scaffolds implanted in vivo onto chick embryo chorioallantoic membrane induced an angiogenic response similar to fibroblast growth factor-2, a well-known angiogenic cytokine [158]. A recent study successfully extracted an acellular scaffold of spinal cord from Sprague–Dawley rats that retained laminin, fibronectin, and type IV collagen in the ECM as confirmed by immunohistochemistry. These scaffolds were judged to be biocompatible in vitro by 3 T3 cell culture and in vivo by immunohistochemical analysis after being implanted into the subcutaneous back skin of rats (Fig. 21.10) [159]. With favorable features including biomimetic microstructure, linear guidance pores, and functional ECM proteins similar to native spinal cord, acellular spinal cord grafts might be useful for spinal cord repair. Indeed, attempts have been made to use extracted acellular peripheral nerve and muscle as scaffolds for regenerating axons of the CNS [160]. Acellular nerve grafts, when implanted into a completely transected rat spinal cord, supported axonal growth to an extent comparable to that of a fresh nerve graft [120]. Chemically extracted acellular muscle shows good integration with the host tissue when implanted into a lateral hemisected adult rat thoracic spinal cord, with sprouting axons growing through the implant in a strikingly parallel and linear fashion (Fig. 21.11), and the appearance of macrophages/microglia and reactive astrocytes was normal [160].

This finding indicates the potential of acellular tissue grafts for spinal cord repair. It remains to be determined which acellular tissue graft provides the best axonal regeneration, and research is needed to confirm advantages of acellular tissue grafts over artificial scaffolds, the latter of which could more easily be modified and functionalized to implement a variety of therapeutic strategies.

21.4.2 Synthetic Biomaterials

In contrast to natural biopolymers, synthetic biomaterials have a larger range of possible chemical modifications that can be tailored for particular applications. Characteristics such as mechanical properties and degradation profile in vivo can

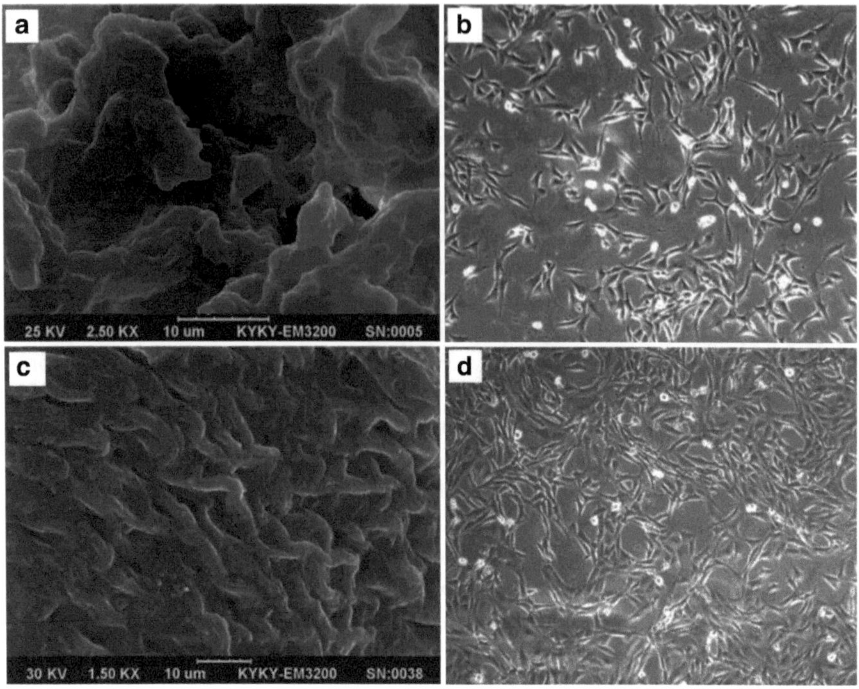

Fig. 21.10 Acellular spinal cord SEM observation and images of coincubation with NIH 3 T3 cells for 72 h. In a cross-section of the scaffold, cells have been removed completely, (**a**) in longitudinal sections (**b**), the ECM and the pore have remained to form three-dimensional network structures. After scaffolds were coincubated with NIH 3 T3 cells for 72 h (**d**), NIH 3 T3 cells showed no signs of cytotoxicity (loss of adherence, nuclear condensation and cell soma contraction) and cells proliferated normally compared with cells in control wells (**c**), expanding from approximately 50–100 % confluency within 72 h [159] (Reprinted with permission from Nature Publishing Group)

more easily be adjusted in synthetic biomaterials. Functionalization of polymers by surface charge modification, topographical modification, and conjugation of trophic factors and adhesive biomolecules has resulted in biomaterials with comparable or even better properties than natural biomaterials. Specifically for spinal cord applications, synthetic polymers such as poly(D,L-lactic acid) (PLA), poly(D,L-lactic-co-glycolic acid) (PLGA), polycaprolactone (PCL), poly(2- hydroxyethyl

Fig. 21.11 Sprouting axons 28 days after surgery, Holmes' silver stain. (**a**), *a–e*: Photomicrographs of a coronal section of the spinal cord, including the chemically extracted acellular muscle implant, 4 weeks after the implantation. The figures show that sprouting axons are present in the rostral, middle, and caudal areas of the scaffold. These axons are distributed in a strikingly parallel and linear manner. The *arrows* indicate the sprouting axons in the graft. (**b**) A photomicrograph of a coronal section in the lesion control group. The figure shows that there are no axons in the lesion cavity. *Scale bars*: (**a**) and (**b**), 625 μm; (*a*) and (*b*), 125 μm; and (*c*), (*d*), and (*e*), 62.5 μm [160] (Reprinted with permission from Wiley Periodicals, Inc)

methacrylate) (PHEMA), polyethylene glycol (PEG), and poly-β-hydroxybutyrate (PHB) have been used. Here, we will mainly focus on those most commonly used: PLA, PLGA, PHEMA, PEG, and polypyrrole.

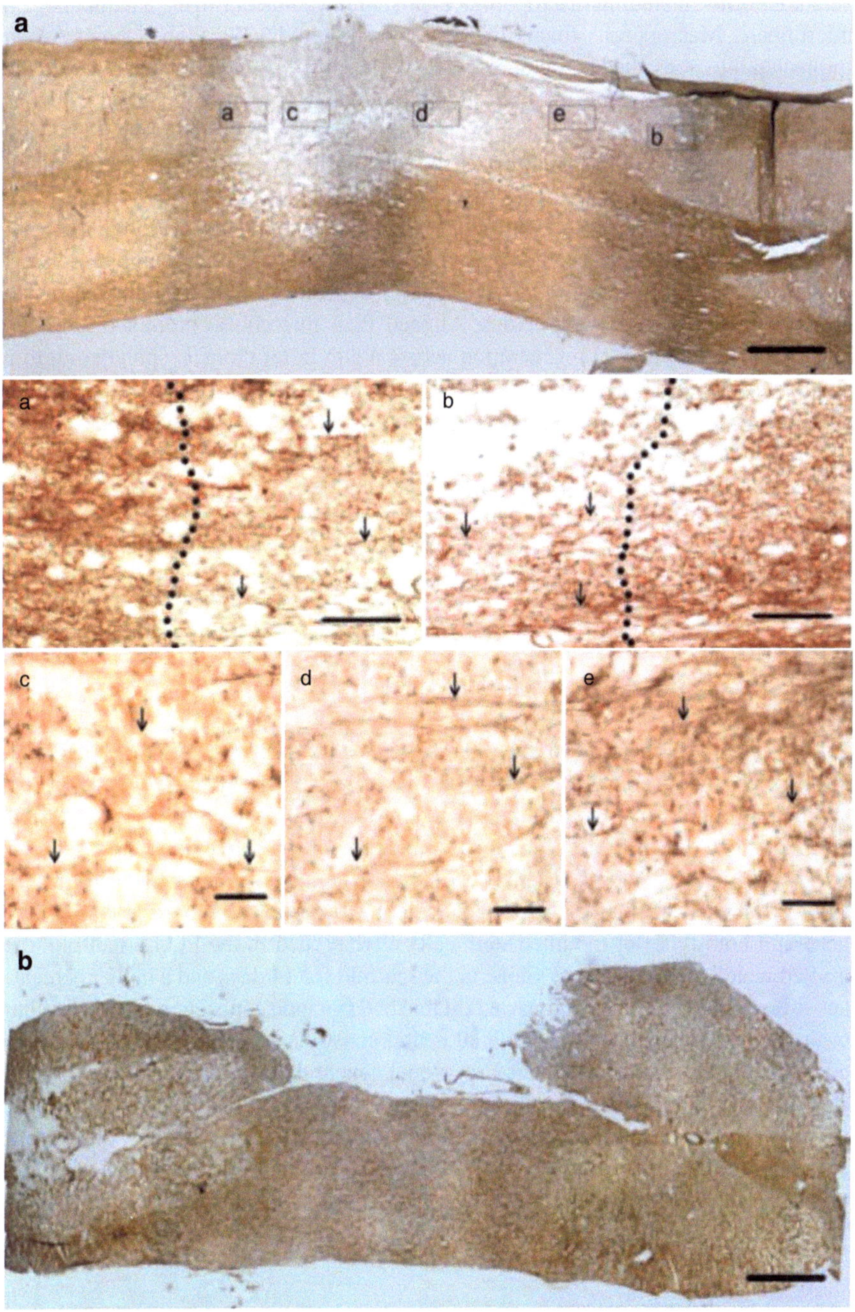

21.4.2.1 Poly(D,L-Lactic Acid) (PLA)

PLA is polyester links of lactic acid that can be hydrolyzed in vivo to release lactide and, thus, is resorbable and biocompatible. It has been used for spinal cord repair in versatile forms, including macroporous foams, multichannel bridges, nanofibers, and microfibers. Macroporous foams of PLA with longitudinally aligned pores, when implanted into transected adult rat thoracic spinal cord, were well tolerated within the injured spinal cord in terms of gliotic and inflammatory response, though more axons were found in the fibrin implant control group [161]. Recently, Tuinstra et al. [162] used multichannel bridges of PLA as the gene delivery system for neurotrophin-encoding lentivirus for spinal cord regeneration after hemisection. In this study, the lentivirus was first complexed with hydroxylapatite nanoparticles and then the complexes deposited into each channel of the bridge using pipet tips. The combined effects of gene delivery and biomaterials were synergistic, leading to increased axon growth and myelination of regrowing axons. Aligned PLA microfibers promoted long-distance rostrocaudal axonal regeneration across a gap in rat thoracic spinal cord, to a significantly greater extent than random fiber and film controls (Fig. 21.12). Without administration of cells, neurotrophins, antibodies, enzymes, or chemical compounds, the PLA microfibers were able to promote robust axonal regeneration, indicating that long-distance axonal regeneration in the CNS is possible without “natural” growth permissive substrates [75]. In the nanoscale range, electrospun aligned PLA nanofibers were found to direct axon outgrowth of DRG neurons and accelerate the processes of axon formation by cultured spinal motor neurons [163, 164].

21.4.2.2 Poly(D,L-Lactic-co-Glycolic Acid) (PLGA)

PLGA is another synthetic polymer that has been widely investigated for its potential in restoring injured spinal cord. It has been used as delivery system for cells, genes, and neurotrophins. In a hemisection rat SCI model, implantation of PLGA scaffolds with neural stem cells led to functional improvement for 1 year and was associated with decreased tissue loss and glial scar [165]. When neural stem cells were genetically transfected with either NT-3 or its receptor tyrosine receptor kinase C (TrkC) gene and both types of modified stem cells were seeded in the PLGA scaffold, the grafted neural stem cells were viable in the scaffold for 14 days and a higher percentage differentiated toward neurons and established synaptic connections [166]. Further evaluation of the effect of this artificial neural construct on axonal regeneration was conducted in a complete transection rat model, where the construct permitted grafted stem cells to differentiate into neuronal phenotype and resulted in synaptogenesis, axonal regeneration, and partial recovery of locomotor function, though axonal regeneration of the corticospinal tract was limited [167]. Composite biomaterials of PLGA and small intestinal submucosa (SIS, derived from the submucosal layer of porcine

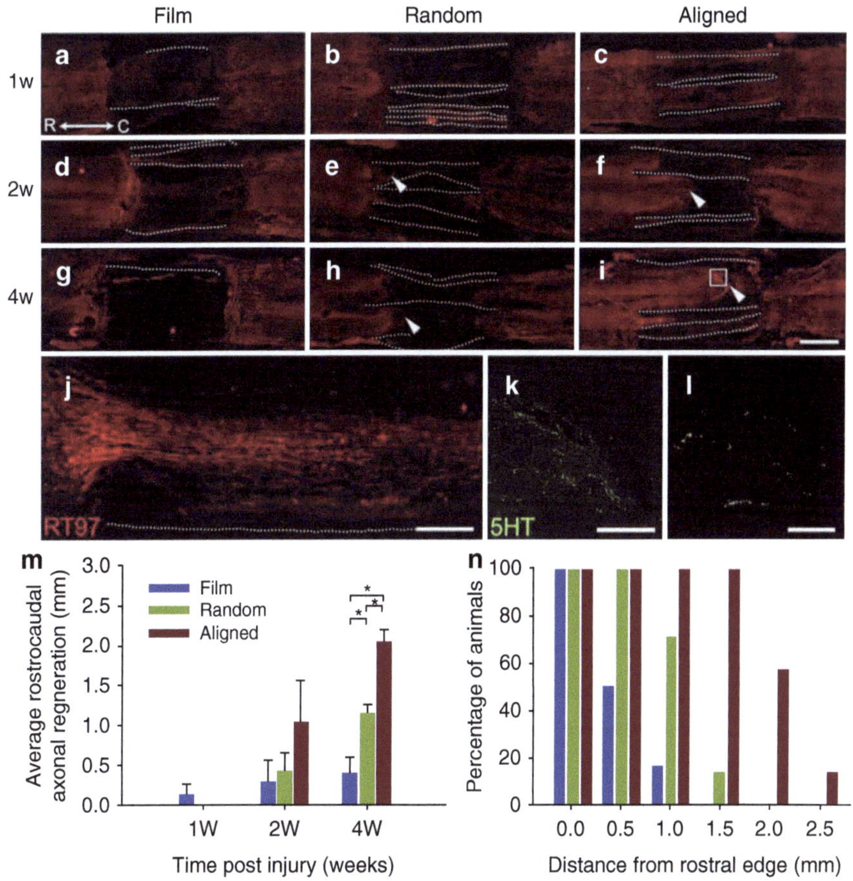

Fig. 21.12 Aligned PLA microfibers promote extensive axonal regeneration. Immunostaining for neurofilament (RT97) was used to visualize axons. Representative horizontal spinal cord sections for film (**a**, **d**, **g**), random (**b**, **e**, **h**), and aligned fiber (**c**, **f**, **i**) conduits. Aligned fibers foster robust, time-dependent rostrocaudal axonal regeneration (**c**, **f**, **i**), whereas the same response is absent in film and random fiber conduits. *Dotted lines* indicate the walls of the conduits. *Arrowheads* (**e**, **f**, **h**, **i**) indicate the regeneration front. (**j**) The axonal regeneration response inside aligned conduits was markedly linear, shown here in a different animal than presented in (**i**). Serotonergic (5HT+) axons were abundant in the robust growth observed inside aligned conduits (**k**, inset from adjacent section of the same animal in **i**). (**l**) Serotonergic axons were present caudal to the graft in 3/21 animals (2 random, 1 aligned fiber). The distance between the rostral edge of the conduit to the "axonal front" was quantified at all time points (**m**). Remarkably, over 4 weeks, aligned fibers promote robust, long-distance regeneration (2,055 ± 150 mm), significantly greater than random fiber (1,162 ± 87 mm) and film (413 ± 199 mm) controls. Notably, at 4 weeks, 100 % (7/7) of the animals from the aligned fiber group had a robust regeneration response present in the middle of the conduit compared to 14.3 % (1/7) and 0 % (0/6) in the random fiber and film groups, respectively (**n**). $^*P<0.05$ by ANOVA. *Scale bars*: (**a**–**i**), 1 μm; (**j**), 500 μm; (**k**), 150 μm; and (**l**), 50 μm [75] (Reprinted with permissions from Elsevier Ltd.)

intestine and consisting of more than 90 % types I and III collagen as well as a wide variety of cytokines) displayed post-implantation inflammation intermediate to that of PLGA- and SIS-based scaffolds [168]. When PLGA/SIS scaffolds seeded with rat bone marrow stem cells were used to bridge completely transected rat spinal cords, it was found that some functional recovery could occur if the gap length was small and that axonal regeneration occurred only in the scaffolds containing stem cells [74]. In addition to cell delivery, PLGA has also been used to support a local, acellular gene delivery system based on lipoplexes. Lipoplex incubation on ECM-coated PLGA resulted in increased gene expression compared to naked plasmid drying methods, and the transgene expression lasted for at least 3 weeks [169]. More recently, PLGA bridges with multiple channels were used as a vehicle for localized delivery of lentivirus encoding the neurotrophins NT-3 or BDNF. Lentivirus was immobilized to nanoparticles and loaded into scaffolds. After transplantation into hemisected rat spinal cord, the multifunctional scaffold resulted in the infiltration of macrophages, Schwann cells, fibroblasts, and astrocytes and promoted significantly more axonal regrowth and myelination of axons [162]. These studies indicate the potential of PLGA scaffolds for gene delivery and synergistic effects toward spinal cord regeneration. Moreover, glial cell line-derived neurotrophic factor (GDNF) loaded within PLGA nanoparticles can be released and maintains its bioactivity, which increased neuronal survival and improved hindlimb locomotor function [170].

21.4.2.3 Poly(2-Hydroxyethyl Methacrylate) (PHEMA)

Nonbiodegradable PHEMA and poly N-(2-hydroxypropyl) methacrylamide (PHPMA) hydrogels have long been used in tissue engineering applications, as they permit the transport of nutrients, gases, and metabolite waste with mechanical properties similar to native soft spinal cord. Particularly for spinal cord repair, implants of PHEMA sponge hydrogels with biomimetic compressive moduli in rats with partial cervical hemisection elicited modest cellular inflammatory responses, minimal scarring, and angiogenesis in the implants. Preloading with BDNF promoted axonal penetration into the gels but not scarring or angiogenesis [171]. Acute and delayed implantation of positively charged PHEMA scaffolds allowed ingrowth of connective tissue elements, blood vessels, neurofilaments, and Schwann cells into the hydrogels with a reduced pseudocyst volume compared to the untreated group [172]. Compared with those with negative or without charge, the positively charged PHEMA scaffolds showed more connective tissue deposition, fewer astrocytic responses, and increased axonal ingrowth into the central parts of the implant [173]. Macroporous scaffolds with internal neurotrophic factor gradients have been produced and found to guide axonal outgrowth in primary neuron cultures [174]. Subsequently, PHEMA hydrogels with longitudinally oriented channels were fabricated using PCL fibers, which were then dissolved completely by sonication in acetone [175]. Copolymerization of PHEMA with 2-aminoethyl methacrylate scaffolds provided the amino groups for covalent incorporation of laminin peptides; the copolymer scaffolds enhanced neural cell adhesion and guided axon outgrowth in vitro [176]. Implantation of a synthetic channel composed of poly(2-hydroxyethyl methacrylate-co-methyl methacrylate) (PHEMA-MMA) into completely transected spinal cord resulted in axonal regeneration from the brainstem [177]. Inclusion of a biomimetic matrix within the channels improved the total number of regenerated axons, but the type of matrix affected their quantity and origin [78].

NeuroGel™, a biocompatible porous PHPMA, has the advantages of porous structure and diffusion properties similar to that of the developing rat brain, low interfacial tension for biological fluids, hemostatic properties, and structural stability [178]. When implanted into cat spinal cord, the hydrogel promoted sustained tissue formation and angiogenesis and supported directional axonal regeneration through the lesion site, accompanied by functional recovery. The regenerated axons were myelinated mainly by Schwann cells. PHPMA hydrogels containing RGD peptides can be used for the delivery of mesenchymal stem cells to treat chronic SCI, with the cells remaining present for up to 5 months. The hydrogels with mesenchymal stem cells prevented tissue atrophy, with infiltration of axons myelinated by Schwann cells, and penetration of blood vessels and astrocytes; the rats also showed some functional improvement with this treatment [179].

21.4.2.4 Polyethylene Glycol (PEG)

PEG is a water-soluble surfactant polymer that has been used as a membrane repair agent. It is biocompatible and has a neuroprotective effect, can repair disrupted plasma membrane, inhibit free radical production, and reduce oxidative stress [180–182]. Recent work by Luo and Shi [183] indicates that PEG also reduces apoptotic cell death after traumatic SCI. PEG was found to improve mitochondrial function and reduce the release of cytochrome *c*, a proapoptotic cell death factor; this might account for its effect on reducing apoptosis. PEG has been shown to recover the anatomical integrity of injured spinal cord and establish conduction of nerve impulses through the lesion with some behavioral recovery [184]. These results indicate that further studies of PEG for applications to spinal cord repair are warranted.

21.4.2.5 Polypyrrole

Electrical stimulation has been widely evaluated for its effect on SCI repair as it can modify cellular activities, including cell migration, cell adhesion, DNA synthesis, and protein secretion [185–187]. Protein adsorption on scaffolds can be enhanced by electrical stimulation, which further promotes cell adhesion and neurite outgrowth [188]. The potential benefits of electrical stimulation inspire the development of electrically conductive scaffolds for tissue engineering applications. Polypyrrole is the most commonly used electrically conductive polymer and has the flexibility to be combined with other polymers, such as electrospun PLGA nanofibers, to form composites [189]. Its capacity to support the proliferation of nerve cells and neurite outgrowth as well as prevent reactive astrocytic proliferation and scar formation has been demonstrated [190, 191]. However, polypyrrole is nondegradable, and studies aimed at fabricating biodegradable polypyrrole scaffolds are needed.

21.5 Fabrication of Scaffolds

21.5.1 Design of Scaffolds

Many interrelated factors need to be taken into consideration when designing scaffolds for spinal cord repair. In general, scaffolds used for spinal cord repair should mimic the structural and biological features of the native ECM of spinal cord [192]. Scaffolds intended for tissue engineering applications should possess the basic requirements of biocompatibility; a biodegradation rate matched to the formation of replacement tissue; nontoxicity; suitable mechanical properties; and high porosity, pore size, interconnectivity, and surface area to volume ratio; and the capacities for cell surface interaction, axon regrowth, and nutrient transport [192]. The macrostructure, microstructure, surface chemistry, surface topography, and elasticity modulus of the scaffolds significantly impact cell behaviors, including cell adhesion, migration, proliferation, and differentiation as well as the orientation and guidance of axonal regeneration and blood vessel formation [103, 164, 193–200]. This is particularly important for scaffolds targeting spinal cord repair, which requires combinatorial strategies such as incorporation of supportive cells, and also promotes the migration of endogenous glial and precursor cells. The various macrostructures of scaffolds widely used for experimental spinal cord regeneration include hydrogels, sponges, single and multichanneled guidance tubes, and nanofiber scaffolds. Solid freeform fabrication of scaffolds with more complex architectures (e.g., branches) might be helpful for complicated nerve injuries [141]. The surface chemistry of the biomaterials also needs to be taken into account as it influences the interaction of the scaffold with macromolecules and cells. Chemical functional groups present on the surface of the biomaterial affect cell behaviors through the interaction of receptors on cell membranes with specific ECM proteins absorbed by different chemical groups [201, 202]. Recent studies of the response of neural stem cells to materials with different chemical groups in a serum-free culture system indicate the chemical groups themselves can impose effects on cell behaviors without the involvement of ECM proteins [203]. Detailed mechanisms of scaffold surface chemistry properties with respect to cell responses are not well understood. Surface topography of scaffolds can influence the orientation of axon growth through contact guidance along surface features such as grooves in substrates at the micro- and nanoscale [204]. Thus, microfabrication techniques, such as photolithography and soft lithography, and nanotechnology, such as electrospinning, are drawing particular interest from researchers with respect to the fabrication of scaffolds with micro-/nanotopographical features. Some excellent review papers provide detailed information about the topographical cues of scaffolds on axonal regeneration [204, 205]. To better mimic the biological properties of native ECM, ECM components are usually included in scaffolds to provide cellular and molecular signaling for better regeneration. Evidence also shows that the mechanical stiffness of scaffolds influences cell behavior, depending on the cell type and the range of moduli presented. Soft substrates promote axon growth but suppress the growth of astrocytes [198]. All of these factors need to be considered and optimized for a given application. Overall, the use of artificial nerve guidance for the repair of SCI is in its infancy, and what constitutes an optimal scaffold remains to be defined [5].

21.5.2 *Fabrication of Scaffolds*

Many novel techniques are currently available for the fabrication of scaffolds with controllable and reproducible macro-architecture, microstructure, and biological properties for spinal cord repair. Conventional techniques such as freeze dry, melt casting, particulate leaching, phase separation, fiber templating, and solvent casting are incapable of producing complicated architecture and usually involve harsh operating conditions that limit the incorporation of bioactive proteins or living cells and may cause toxicity after implantation. Furthermore, conventional scaffold fabrication techniques lack the precise control of pore size, pore geometry, spatial distribution, and interconnectivity of pores. In contrast, novel techniques such as electrospinning, solid freeform fabrication, and self-assembly are gaining more attention for the production of scaffolds that mimic the structural and biological activities of native tissue.

21.5.2.1 Electrospinning

Electrospinning can be used to fabricate scaffolds with diameters in the nano- to micrometer range, which best mimic native ECM and thus provide a suitable environment for cell adhesion, migration, proliferation, and differentiation [206]. In the electrospinning process, polymers are first dissolved by chemical solvents or melting before being spun by a very high voltage to form fibers. Biomaterials used for electrospinning to construct 3D scaffolds for nerve tissue engineering include PLA, PCL, PLGA, poly(3-hydroxybutyrate), poly(3-hydroxybutyrate-co-3-hydroxyvalerate), poly(acrylonitrile-co-methylacrylate), and copolymers of methyl methacrylate and acrylic acid, polydioxanone, polyamide, and chitosan [207]. Natural biomaterials such as gelatin, collagen, and chitosan as well as ECM protein laminin have been combined with synthetic materials for electrospinning, leading to scaffolds with enhanced biocompatibility compared to synthetic materials and improved mechanical properties compared to natural polymers [208–211]. Neural cell proliferation and axon extension can be promoted using electrospun biocomposite materials. The versatility of electrospinning in manipulating the architecture and morphology of scaffolds is achieved through the adjustment of system parameters, such as polymer molecular weight, molecular weight distribution, and solution properties (e.g., viscosity, surface tension, and conductivity) and process parameters, such as flow rate, electric potential, distance between capillary and collector, and motion of collector [212]. Scaffolds with aligned micro- or nanofibers have been fabricated by electrospinning using an insulated sharp needle and a rotating disk collector. The aligned electrospun micro- or nanofibers can enhance extension of axons by sensory and motor neurons in vitro, differentiation of embryonic stem cells, and axonal regeneration of peripheral nerve and spinal cord in vivo [75, 196,

211, 213, 214]. Robust axonal regeneration was observed in a complete thoracic rat spinal cord transection model after grafting of electrospun aligned microfiber-based scaffolds (Fig. 21.13) without incorporation of any bioactive molecules or living cells. The regenerating axons originated from propriospinal neurons of the rostral

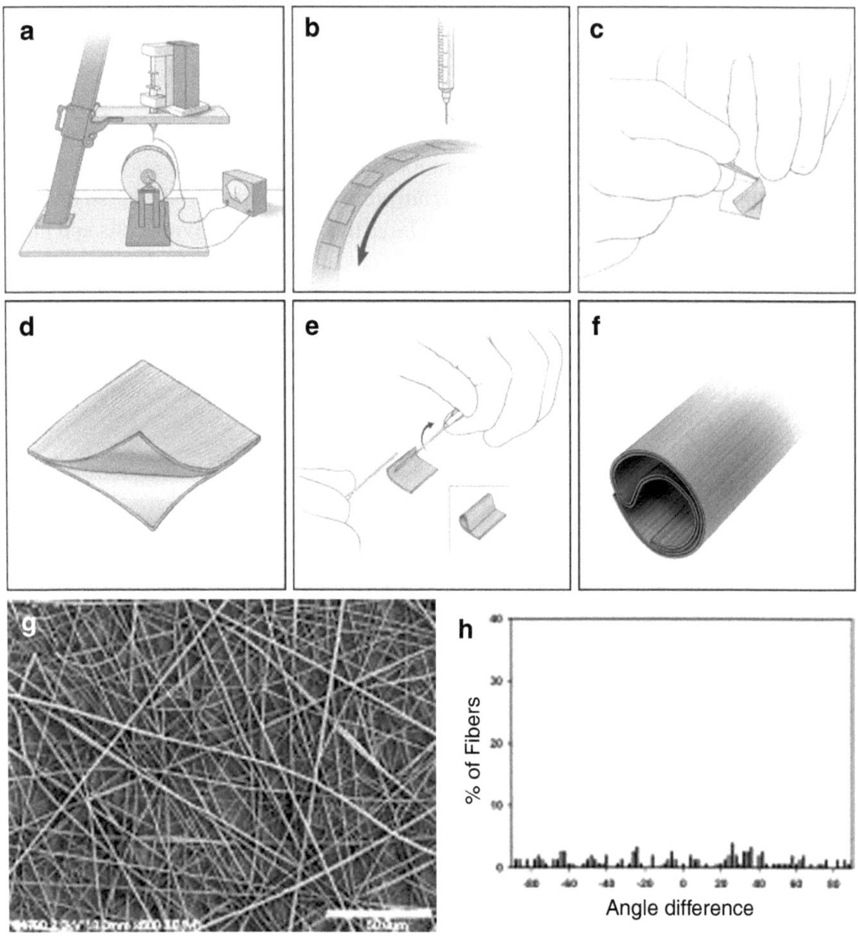

Fig. 21.13 Schematic detailing materials fabrication process of electrospinning and materials characterization. A custom electrospinning apparatus (**a**) was used to generate aligned polymeric fibers. Coverslips were mounted on a grounded target, and a rotation speed of 1,500 rpm was used to align fibers produced by a 15 kV field potential (**b**). Random fibers were generated using a stationary target. For conduit assembly, films with or without electrospun fibers were peeled from coverslips (**c**), placed back to back (**d**), and rolled (**e**) into conduits (**f**). Random (**g**) and aligned (**i**) fibers were visualized by SEM, and alignment was quantified by measuring the angle between a given fiber and the median fiber orientation for 150 fibers per condition (**h** and **j**, respectively). Importantly, fiber alignment was maintained through the process of conduit assembly (**k, l**). (**m**) Macroscopic view of aligned fiber conduit lumen, visualized by mounting a conduit sectioned on the longitudinal axis. (**n**) Coronal view of an aligned fiber conduit, the diameter of all conduits was 2.6 mm. *Scale bars*: 50 μm in (**g**), 100 μm in (**i, k**), and 1 mm in (**m, n**) [75] (Reprinted with permission from Elsevier Ltd.)

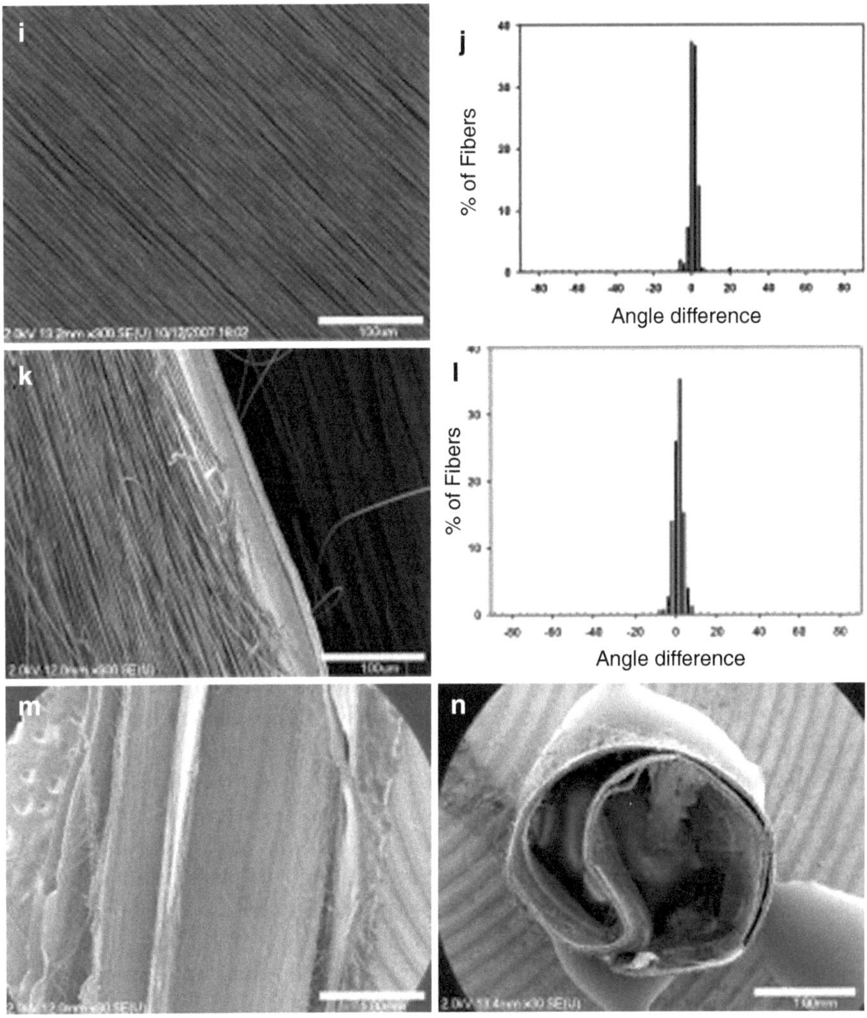

Fig. 21.13 (continued)

spinal cord as well as supraspinal neurons of the reticular formation, red nucleus, raphe, and vestibular nuclei [75]. In another innovative study, muscle cells were electrosprayed while electrospinning poly (es- ter urethane) urea, resulting in scaffolds with a uniform distribution of cells [215]. This indicates the potential to fabricate scaffolds with biological properties using electrospinning techniques. As toxic chemical solvents or high temperatures are required to dissolve most synthetic polymers before electrospinning, the incorporation of cells and/or bioactive molecules when fabricating scaffolds using this technique is an issue that requires further investigation.

21.5.2.2 Self-Assembly

Self-assembly is an alternative fabrication technique to produce nanofiber network scaffolds for CNS regeneration. Advantages of self-assembling nanofibers include diameters within the range of several nanometers to tens of nanometers, high density of incorporation of bioactive peptide sequences, and tissue-like water content. Thus, they better mimic the structure of ECM than electrospun fibers. The self-assembly process is mediated by noncovalent bonds, such as van der Waals forces, hydrogen bonds, and electrostatic forces [216]. Oligopeptides or amphiphilic peptides assemble into nanofibers and form a gelatinous network when exposed to physiological ionic conditions such as tissue culture medium or cerebrospinal fluid. Hydrophilic groups form a sheath, while the hydrophobic backbones form a core [217]. Currently, self-assembling peptides used for nerve tissue engineering include arginine–alanine–aspartate (RAD) 16-I and RAD16-II, IKVAV, and synthesized peptide amphiphile. Primary neurons isolated from the cerebellum and hippocampus of mice, or neuron-like PC12 cells, were able to attach, migrate, proliferate, extend axons, and form synapses when cultured on RAD 16 self- assembling peptides scaffolds [218, 219]. Implants of RAD 16-I scaffolds containing Schwann cells or embryonic neural precursor cells into the dorsal columns of transected rat spinal cord integrated well with the host tissue and led to axonal regeneration. Penetration of blood vessels and migration of host cells into the scaffolds were observed, and some of the transplanted precursor cells differentiated into neurons, astrocytes, and oligodendrocytes [220]. Injectable peptide amphiphile molecules can be applied as a liquid into lesioned spinal cord, where they self-assemble into a nanofiber network to bridge the gap. Neural precursor cells cultured in IKVAV-containing peptide amphiphile nanofibers can differentiate extensively into neurons while astrocytic development is inhibited [221]. An in vivo study of IKVAV peptide amphiphile nanofibers in mouse SCI indicated that the nanofiber network reduced astrogliosis, decreased cell death, and increased the number of oligodendroglia at the lesion site. Axonal regeneration of motor and sensory neurons through the lesion site was observed (Fig. 21.14), resulting in significant behavioral improvement [222]. Despite the beneficial outcomes provided by self-assembling peptide scaffolds, self-assembly requires complicated procedures and techniques, and precise control of the macro-sized pores is difficult [223].

21.5.2.3 Solid Freeform Fabrication

Solid freeform fabrication, also known as rapid prototyping, has gained substantial attention recently for the construction of scaffolds for tissue engineering applications due to its ability to fabricate highly complex, reproducible scaffolds with controllable internal and external architectures and interconnected pore networks. Scaffolds are fabricated in layer-by-layer fashion, controlled by a computer. This method has the advantages of customized design, anisotropic scaffold microstructures, and versatile processing conditions, such as solvent- and/or porogen-free

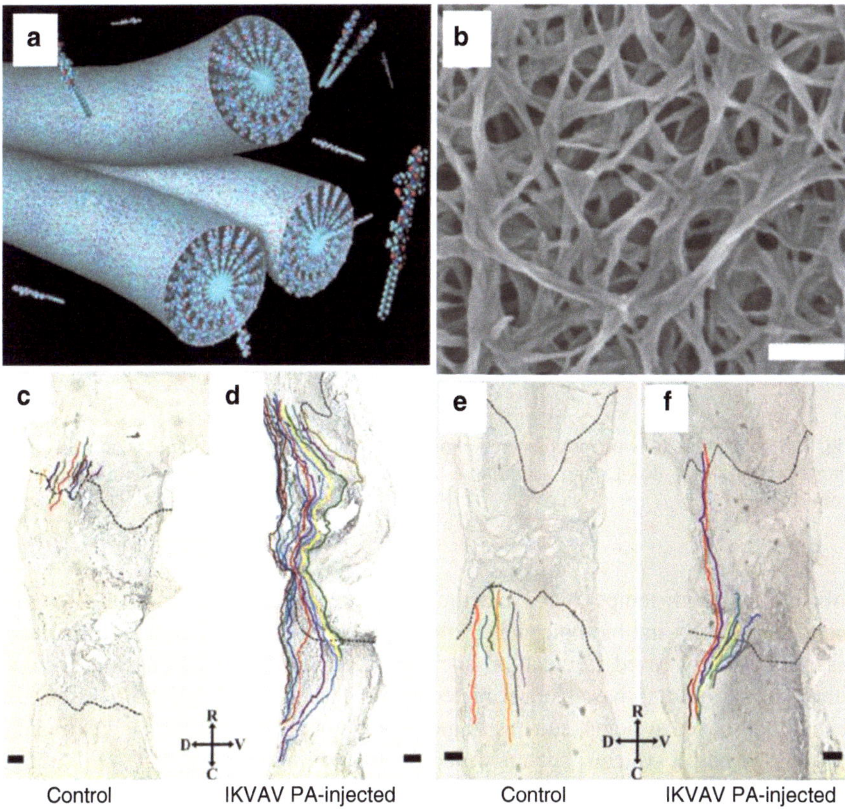

Fig. 21.14 Structure of IKVAV peptide amphiphile scaffold and its promotion of motor and sensory axonal regeneration after SCI. (**a**) Schematic representation showing individual peptide amphiphile molecules assembled into a bundle of nanofibers interwoven to produce the IKVAV peptide amphiphile. (**b**) SEM image shows the network of nanofibers in vitro. Representative Neurolucida tracings of BDA-labeled descending motor fibers within a distance of 500 μm rostral of the lesion in vehicle-injected (**c**) and IKVAV peptide amphiphile-injected (**d**) animals. Representative Neurolucida tracings of BDA-labeled ascending sensory fibers within a distance of 500 μm of the lesion epicenter in vehicle-injected (**e**) and IKVAV peptide amphiphile-injected (**f**) animals. The *dotted lines* demarcate the borders of the lesion. *R* Rostral, *C* caudal, *D* dorsal, *V* ventral. *Scale bars*: (**a**) 200 nm and (**c**–**f**) 100 μm [222] (Reprinted with permission from Society for Neuroscience)

processes and room temperature processing, which are permissive for incorporation of pharmaceutical agents, biological macromolecules, or even living cells during fabrication [224]. Importantly, this method can precisely control the distribution of supportive cells and biomolecules inside the scaffold interior, potentially enabling sophisticated combinatorial strategies to regenerate injured spinal cord. For example, DRG neurites preferentially grow upon and follow discrete laminin-blended chitosan pathways in scaffolds produced using the method [225]. The handling of natural biopolymers usually does not involve harsh conditions such as organic

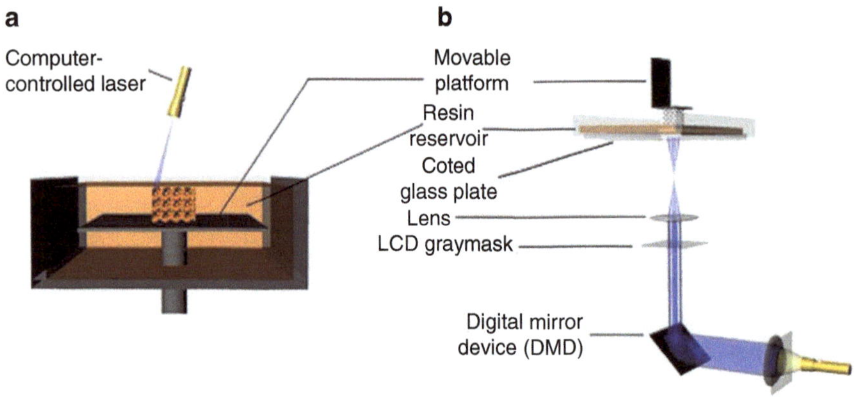

Fig. 21.15 Schemes of two types of stereolithography setups. (**a**) A bottom-up system with scanning laser. (**b**) A top-down setup with digital light projection [230] (Reprinted with permission from Elsevier Ltd.)

solvents and high temperature, and thus they are particularly promising for the inclusion of cells and biomolecules. Solid freeform fabrication techniques include laser technology-based fabrication, such as selective laser sintering and stereolithography; systems based on print technology, such as 3D printing; assembly technology-based systems, such as shape deposition manufacturing; and extrusion technology-based systems, such as fused deposition modeling, direct 3D plotting, multiphase jet solidification, and precise extrusion manufacturing. Solid freeform fabrication techniques can also be used to create a negative mould within which a scaffold can be cast using any desired polymeric and/or ceramic biomaterials. Detailed descriptions of the advantages and limitations of each freeform fabrication technique have been discussed by others [224, 226–228]. Here, we focus mainly on stereolithography and 3D plotting techniques, as these allow for the incorporation of cells and biomolecules for soft tissue applications such as spinal cord repair.

Stereolithography uses radical initiated polymerization of biomaterials to construct 3D scaffolds. Compared to other fabrication techniques, these methods have high resolution and accuracy and are capable of fabricating micro-sized structures with submicron resolution [229]. Two types of stereolithography setups are used depending on the build orientation and method of illumination (Fig. 21.15). One uses a computer-controlled laser beam to illuminate the liquid resin from above and the scaffold is built from the bottom-up on a support platform. The other is a top-down approach, where light is projected on a transparent, nonadhering plate from underneath and the support platform is dipped into the resin from above [230]. In such setups, a digital mirror device, an array of up to several millions of mirrors that can rotate independently to an on or off state, can be used for simultaneous photopolymerization of partial and entire layers of a scaffold via projection. Precise and complex internal architectures, including pore size and shape, can be controlled by the digital mirror device [231]. Multilayer scaffolds have been fabricated using digital mirror-based systems from photo-cross-linkable poly(ethylene glycol) diacrylates.

Murine bone marrow-derived cells incorporated into the scaffolds during fabrication or seeded onto the fibronectin-functionalized scaffolds post-fabrication survived and retained osteogenic differentiation ability, as indicated by efficient matrix mineralization [231]. In another study, gelatin methacrylate was used to fabricate scaffolds by projection stereolithography and the porous scaffolds allowed for uniform distribution of seeded human umbilical vein endothelial cells. These cells were able to proliferate to high density and maintained their endothelial phenotype [232]. HA, after being modified by photopolymerizable methacrylate groups, can be used in stereolithography to produce architecturally complex structures, such as scaffolds with branched tubes. A gradient of fluorescent microparticles in the conduit was also achieved, indicating the potential to create defined gradients of bioactive molecules within scaffolds using this technique [141], possibly to direct cell movement. Further studies of solid freeform fabricated scaffolds are required to explore the sophisticated strategies for directing axonal regeneration in SCI animal models.

3D plotting is another fabrication method that shows promise for tissue engineering applications due to its ability to fabricate hydrogel scaffolds and integrate supporting cells and bioactive molecules during the fabrication process without the involvement of toxic chemicals or high temperatures. Biomaterials are dispensed in a liquid medium, such as melts, solutions, pastes, thermosets, filled polymers, or reactive oligomers [233]. Hydrogel scaffolds have been successfully fabricated with limited geometry using 3D plotting systems to shape agarose, fibrin, chitosan, and alginate [233–236]. Incorporation of living cells in the dispensing process of 3D plotting is not yet common, and few in vivo studies have been carried out to evaluate the ability of these scaffolds to promote regeneration in nervous system tissues. In one study, 3D tubular structures were fabricated with 3D bioplotting from a starch and PCL composite material, then a polysaccharide-based gellan gum hydrogel injected into the central area of the structures. An in vitro cytotoxicity assay using a culture of oligodendrocyte-like cells and in vivo assessment in a hemisection rat SCI model indicated that the hybrid scaffolds were noncytotoxic [237]. Future work could focus on scaffolds fabricated by 3D plotting with distributions of cells and biomolecules designed for axonal regeneration. Precise control of the scaffold architecture can be guided by established models for flow rate, scaffold porosity, pore size, mechanical properties, and cell damage to avoid a trial and error process and enable the optimization of geometry, process parameters for mechanical control, biological properties, and minimization of cell damage [238–241]. Biofabrication of scaffolds using 3D plotting still has problems with respect to insufficient resolution and accuracy, maintenance of sterility and avoiding cell damage during the fabrication process, and limited biomaterial choices.

21.6 Strategies with Scaffolds

The consensus of recent literature is that synergistic strategies will be required to restore the complex structural and functional properties of injured spinal cord. Combinatorial strategies include the management of scaffold properties, cell delivery, molecular adjustment, gene therapy, hostile environment engineering, and augmentation of the intrinsic growth state of the neuron. Interactions between the scaffolds and the host tissue are important as they impact cell behaviors such as cell adhesion, survival, proliferation, differentiation, axon growth, tissue infiltration, cell death, accumulation of reactive astrocytes, and glial scar formation [222]. The current focus of tissue engineered scaffolds for SCI repair is to best mimic the physical, chemical, biological, and micro- or nanostructural properties of the ECM, which could most favorably encourage axonal regeneration and functional recovery after injury.

Many studies have demonstrated the advantages of synergistic effects of scaffolds with other therapeutic strategies [242–244]. Combinatorial therapies of enhancing the intrinsic growth capacity of sensory neurons, engraftment of stem or precursor cells, and delayed neurotrophic factor (NT-3) delivery beyond the lesion site, when initiated from 6 weeks to as long as 15 months after SCI, were able to support bridging axonal regeneration, whereas individual components of the full combination failed to elicit bridging [245]. However, the number of axons reentering the distal host tissue was quite limited and regrowing axons within the cellular matrix were not linear. Application of scaffolds with anisotropic channels or linear nanofibers might be helpful to promote more and better oriented axonal regeneration through the defect site. In this section, we will mainly focus on the cell therapy, biomodification, gene therapy, and molecular therapy with scaffolds for SCI repair.

21.6.1 Cell Therapy with Scaffolds

Critical gap length, defined as the length of the gap between uninjured segments of neural tissue at which the frequency of axonal reconnection is 50 %, can be lengthened by incorporation of supportive cells that secrete various trophic factors, ECM proteins, and anti-inflammatory cytokines. Cellular grafts alone result in low cell survival rates and random axon regrowth reentering host tissue [246]. Scaffolds can serve as a cell delivery system to maintain cell behavior and provide guidance for axonal regeneration provided by modification of surface chemistry, topography, and ECM proteins/peptides. Cells can be loaded onto the scaffolds by direct cell incorporation during the scaffold fabrication process, suspension of cells within an incorporated matrix (e.g., fibrin or Matrigel), or cell seeding on porous/multichannel scaffolds. Various cells of different types and sources loaded onto scaffolds have been investigated to regenerate injured spinal cord. Here, we focus on those cell types that have been evaluated in human clinical trials, including stem or progenitor cells and non-stem cells such as olfactory ensheathing cells and Schwann cells.

21.6.1.1 Neural Stem/Progenitor Cells

Neural stem cells are multipotent cells found in the CNS of mammals that have the ability to self-renew and multidifferentiate into both glial cells and neurons.

One issue with neural stem cell delivery by tissue engineered scaffolds for SCI is ensuring their long-term survival, proliferation, and differentiation. Direct cellular injection results in differentiation of stem cells into glial lineages rather than neurons [247]; topical application of neurotrophic factors, immunosuppression, cell adhesion ligands, and adjusting the characteristics of the artificial scaffolds can overcome this problem [131, 248–250]. For example, implantation of neural stem cells genetically modified to produce NT-3 or co-implanting them with Schwann cells overexpressing NT-3 resulted in improved anatomical repair and/or functional recovery in rat SCI [251, 252]. Additional studies have further demonstrated the beneficial effects offered by cografting neural stem cells and Schwann cells overexpressing TrkC and NT-3, respectively, in complete rat thoracic spinal cord transection models [253]. Mechanical properties of the substrates also affect stem cell differentiation. Generally speaking, 3D scaffolds with Young's moduli similar to that of native CNS tissue result in a preferable differentiation of stem cells into a neural phonotype [131, 254, 255]. The origin of the neural stem cells also needs to be taken into consideration, as the number of surviving cells originating from the subependymal region of the lateral ventricles can be approximately five times that of cells harvested from the spinal cord, and less astrocytic differentiation and more oligodrocytic differentiation were observed in the former group [107]. The underlying reasons for this finding are not known. In addition, no neurons were generated by the differentiation of the neural stem cells in vivo despite the fact that neurogenesis occurs in vitro [108]. Collagen, chitosan, fibrin, and PLA scaffolds have been used as delivery systems for neural stem cells, with improved recovery of structure and function of neural tissue compared to control groups [107, 119, 167, 256]. However, few studies have been done to promote the survival and maturity of newly generated neurons as well as the recruitment of these neurons for functional circuitry. Suppressing expression of the low-affinity neurotrophin receptor p75NTR may reduce the apoptosis in neurons derived from stem cells [257]. No clinical trials of neural stem cells in human SCI have been undertaken, probably due to the ethical concerns regarding their origins, practical issues of isolation and directed differentiation, and the unclear mechanisms for functional benefit (immunomodulation and angiogenesis) that they may provide [258].

21.6.1.2 Mesenchymal Stromal/Stem Cells

Mesenchymal stem cells can be isolated from the stromal compartment of bone marrow or Wharton's jelly of the umbilical cord. They hold promise for SCI repair as they have antiapoptotic, anti-inflammatory, and immunomodulatory effects to suppress neural apoptosis in vitro, mesodermal differentiation potential into multiple cellular phenotypes in vivo, as well as the ability to secrete several neurotrophic factors [258–261]. The neuroprotective effect of mesenchymal stem cells may contribute to their beneficial effect in experimental SCI models, with most studies using intraspinal, intrathecal, and systemic delivery [262–265]. Addition of neurotrophic factors such as NT-3 further increased the therapeutic effects of mesenchymal stem cells [266]. Combinatorial strategies of mesenchymal stem cell transplantation with an NT-3 gradient and peripheral nerve conditioning lesions to enhance central axon regenerative capacity facilitated axonal regeneration 1 year after SCI [245]. Collectively, these findings in preclinical studies shed light on the potential for clinical treatment of patients with SCI using human bone marrow cells. Clinical studies of autologous transplantation of mesenchymal stem cells in acute and chronic SCI are underway [258, 267]. Unfortunately, human mesenchymal stem cells derived from bone marrow when implanted into the lesion site of SCI models have low survival rates over long periods [268, 269]. Delivery via scaffolds or increased p75NTR expression may increase the survival of mesenchymal stem cells in vivo [257]. Chitosan, templated agarose, collagen, Matrigel, PLGA, and gelatin sponge scaffolds have been combined with delivery of mesenchymal stem cells for SCI repair, leading to reduced cavity formation, attenuated inflammatory response, and improved functional recovery [70, 76, 270–272]. Human mesenchymal stem cells seeded on multichannel PLGA scaffolds can survive for as long as 8 weeks after transplantation and are able to differentiate into neural cells [76]. In a complete rat spinal cord transection model, the combinatorial strategy of mesenchymal stem cells within PLGA/SIS scaffolds resulted in significantly more hindlimb locomotion recovery compared with plain scaffold implants; axonal regeneration only occurred in rats implanted with human stem cell-seeded scaffolds [74]. In another study, RGD-modified HPMA hydrogels seeded with mesenchymal stem cells extensively improved motor and sensory functional outcomes in chronic SCI [179].

21.6.1.3 Olfactory Ensheathing Cells

Olfactory ensheathing cells are an unusual mature glial cell type derived from the nerve fiber layer of the olfactory bulb and nasal olfactory mucosa. They have been extensively investigated with respect to SCI repair due to their ability to facilitate the normal, lifelong, repeated regeneration of replacement olfactory axons from the peripheral nasal olfactory mucosa to the CNS environment of the olfactory bulb [273]. Olfactory ensheathing cells can remyelinate regrowing axons and secrete trophic factors such as NGF, BDNF, and vascular endothelial growth factor (VEGF) [274–276]. Evidence for promotion of axonal regeneration and functional recovery by olfactory ensheathing cells has been presented in several studies but cannot be independently confirmed, with some studies finding olfactory ensheathing cells do not have axonal growth-promoting properties after SCI [277–280]. Three clinical trials have been conducted to treat human SCI using olfactory ensheathing cell transplantation [281–283]. This therapeutic strategy was safe, with no deleterious motor changes, posttraumatic syringomyelia, or tumor formation for 3 years following transplantation; however, functional recover y was quite limited. For example, in the most recent of such studies, only one of six SCI human subjects treated with OEC transplantation had demonstrated sensory gain, but no functional recovery; the others showed no detectable change in clinical and functional tests [283]. The results appear disappointing in consideration of the promising animal experiments. There have been few studies regarding the combinatorial effect of olfactory ensheathing cells within tissue engineered scaffolds for SCI repair. PLGA conduits have been used for olfactory ensheathing cell delivery for 10 mm-defect sciatic nerve of rats. The conduit was found to be compatible with olfactory ensheathing cells, but there was no functional recovery [284]. Starch/PCL scaffolds, electrospun silk fibroin scaffolds, and collagen–heparan sulfate biological scaffolds have been studied with olfactory ensheathing cell culture in vitro with the aim to apply the olfactory ensheathing cell-loaded scaffolds to SCI repair [285–287].

21.6.1.4 Schwann Cells

Schwann cells, the major glial cell type of the peripheral nervous system, have been investigated for restoring injured spinal cord since 1981, when transplantation of purified Schwann cells was first carried out [288]. Schwann cells show particular promise as they can be harvested from patients for an autologous transplant. Not only do they provide remyelination of regrowing axons when transplanted into injured spinal cord, but the infiltration and remyelination by endogenous Schwann cells from the peripheral nerves into the spinal cord lesion site has been observed, suggesting that host Schwann cells might contribute to the recovery of injured spinal cord [289]. Transplantation of Schwann cells can reduce the size of spinal cysts as well as secrete various trophic factors and cell adhesion molecules, leading to a more permissive environment for axonal growth and neural survival [290]. Despite the beneficial outcome provided by Schwann cell transplantation, decreasing cell numbers after transplantation pose a problem and some studies show insufficient axonal regeneration of brainstem spinal axons using Schwann cells alone; axons moreover do not leave the Schwann cell graft to reenter the distal host spinal cord. Thus, strategies combining scaffolds with bioactive molecules or other supportive cells are usually used. Schwann cells seeded in Matrigel within polyacrylonitrile/polyvinylchloride (PAN/PVC) nerve guidance channels in a spinal cord transection injury have been shown to survive and increase axonal remyelination and axonal regeneration of propriospinal but not supraspinal tracts [291]. Further administration of methylprednisolone or delivery of BDNF and NT-3 by this approach has also been evaluated [292–294]. Other scaffolds used for Schwann cell delivery include PHB tubular conduits, multichannel tubular PLA and PLGA scaffolds, self-assembling peptide nanofiber scaffolds, poly(ε-caprolactone fumarate) hydrogel, oligo(polyethylene glycol) fumarate (OPF) hydrogel, or positively charged OPF hydrogel [295–299]. Despite the increased axon remyelination and axonal regeneration often seen using Schwann cell therapy, regrowing axons were unable to reenter the distal host tissue and thus functional recovery was either not observed or quite limited. Physical properties of the scaffolds, particularly inappropriate mechanical properties, may be blamed. Another possibility is the low integration and migration of transplanted Schwann cells from the implantation site, especially due to the inhibitory effects of reactive astrocytes

[300]. By engineering the transplanted Schwann cells to overexpress the polysialylated neural cell adhesion molecule or combining with delivery of trophic factors, ChABC, or olfactory ensheathing cells, integration and migration of Schwann cells into the spinal cord can be improved, resulting in better axonal regeneration, remyelination, and functional restoration [293, 301–305]. However, Schwann cell therapy appears to provoke an extensive astrocytic response, and source limitations and the necessity to amplify the cell number in vitro may further restrict the clinical application of Schwann cells for SCI repair.

21.6.2 ECM Protein/Peptide Modification

ECM proteins can influence such cellular activities as cell migration, axonal guidance, synaptogenesis, cell survival, differentiation, and myelination [306]. Incorporation of ECM proteins or peptides derived from them, either by coating, covalent cross-linking, or blending, may enhance cell adhesion and differentiation and cell migration and axon outgrowth. Collagen, fibronectin, and laminin have been the most widely investigated ECM proteins for biomodification of scaffolds to achieve better axonal regeneration [211, 307–309]. They act through integrin receptors on cell membranes [72, 139, 309–311]. Limitations of modification with whole ECM molecules include batch variation and the risk of disease transmission [312]. Thus, synthetically prepared amino acid sequences from the domains of ECM proteins responsible for cell–substrate interactions are used as alternatives. RGD, YIGSR, and IKVAV are the three most commonly used cell adhesion peptides in tissue engineering to promote the biocompatibility and interaction of scaffolds with cells [313, 314]. However, which functionalization technique and protein/peptide species and concentration are most efficient for each different scaffold application is still not known. Blended electrospinning of laminin and PLA is a facile and efficient method to modify nanofibers compared with covalent immobilization and physical adsorption, though all functionalization techniques led to increased axon extension [211]. However, coating of alginate hydrogels with laminin showed very little differentiation and axon outgrowth of NB2a neuroblastoma cells, whereas both numbers and lengths of axon outgrowth increased with increasing peptide density on YIGSR covalently conjugated alginate substrates [315]. In another study, covalent modification of a 3D alginate hydrogel with laminin or its peptide IKVAV resulted in the survival, attachment, and neurite outgrowth of neurons after 1 month culture within 3D hydrogel constructs; neurons on the hydrogel modified with RGD did not support long-term neuronal survival and attachment [97]. Self-assembly of IKVAV-containing peptide amphiphile induced very rapid differentiation of neural stem cells into neurons while discouraging the development of astrocytes, which was considered associated with the amplification of bioactive epitope presentation to cells by the self-assembled nanofibers [221]. Despite the potential of peptide modification for scaffold biomodification, another problem with ECM protein/peptide modification is that inclusion of a single ECM protein or peptide can only mediate one or a few biological interactions; a more comprehensive biomimetic implant might be capable of providing numerous biological interactions. Based on this, fabricating scaffolds from composite biomaterials comprised of various ECM molecules might enable several effects to be integrated toward the goal of improving axonal regeneration. Certain fabrication techniques are amenable to the differential placement of such molecules in precise micropatterns. However, combination of different ECM proteins/peptides should be made upon careful consideration, as the combinatorial effects might not turn out to be synergistic.

21.6.3 Gene Therapy with Scaffolds

Gene therapy has emerged as a promising analytical strategy for SCI repair as it can be used as a method to articulate the underlying mechanisms for the lack of self-regenerative capacity by identifying gene candidates responsible for promoting growth or overcoming growth inhibition. Researchers have identified the three myelin inhibitory molecules (Nogo-A, MAG, and OMgp) as well as their in vivo interaction and relative potencies either using mutants or by knockout of the correspondent genes [316, 317]. Screening candidate genes for regeneration is also useful for spinal cord repair in that overexpression of axon growth-promoting genes in the spinal cord could enhance regeneration. Identification of these growth-promoting genes can be achieved through the overexpression of candidate genes in vivo after viral transduction of a targeted neuronal population, on the condition that the axons of transduced neurons can be identified and their enhanced growth can be detected in the spinal cord. A dual promoter lentiviral vector has been developed for simultaneous CNS neuronal expression of a potential axon growth-promoting gene candidate and a fluorescent protein-based axonal tracer [318], which could help enhance the efficiency and sensitivity of in vivo screens for regeneration-promoting genes in the injured CNS.

Gene therapy has also been used to achieve long-term local delivery of therapeutic molecules into the lesion site. Therapeutic molecules delivered successfully by this technique for SCI include growth factors, such as neurotrophins and neuropoietic cytokines, as well as factors to neutralize the inhibitors of axonal regeneration. Two main methods are used to deliver therapeutic biomolecules by gene therapy. One is through direct (in vivo) delivery of genes to the spinal cord by recombinant viral vectors, such as adeno-associated virus and lentivirus; the other approach is to transplant supportive cells that have been genetically manipulated in vitro to express various growth factors before being implanted into the injured spinal cord. The latter approach is referred as ex vivo delivery of genes. A variety of cells including Schwann cells, fibroblasts, olfactory ensheathing cells, and various stem cells have been used for the ex vivo delivery of genes into injured spinal cord. The in vivo and ex vivo delivery of various biomolecules has been summarized by Bo et al. [319]. For example, NT-3 genetically modified human umbilical stem cells, when transplanted into contusion spinal injury rat models, significantly improved locomotor functional recovery in comparison with a group treated with nonmodified stem cells. Increased intensity of serotonergic axons, increased volume of spared myelination, and decreased area of cystic cavity were also observed in the NT-3 genetically modified group [266]. Cograft of neural stem cells and Schwann cells overexpressing TrkC and NT-3, respectively, significantly improved relay of the cortical motor- and somatosensory-evoked potentials as well as ameliorated hindlimb deficits when compared to controls [253]. In another study, provision of an NT-3 gradient by injection of a lentivirus expressing NT-3, combined with a conditional lesion and transplantation of bone marrow stem cells in a chronic SCI model, resulted in the regrowth of axons that reached the distal host tissue [245]. Controlled and specific gene expression is required, as inappropriately high or prolonged expression of exogenous proteins could cause adverse effects. For example, unregulated expression of VEGF by the delivery of VEGF gene may induce pathological angiogenesis or promote tumor growth, diabetic proliferative retinopathy, and rupture of atherosclerotic plaques. Strategies to more effectively control gene expression for SCI repair include hypoxia-inducible gene expression systems using the erythropoietin enhancer and RTP promoter [320–323]. Combination of direct and ex vitro gene delivery with nerve guidance scaffolds may be more beneficial due to the combined advantages of precise placement of gene expression at the site required for nerve guidance and tissue formation. Templated agarose scaffolds have been used as a delivery system for NT-3 genetically modified mesenchymal stem cells to bridge the gap caused by transection of ascending spinal cord dorsal column sensory axons [70]. Combined with the NT-3 gradient created beyond the lesion site by lentiviral vectors and priming lesions of the sensory neuronal cell body to stimulate the endogenous growth state of the injured neuron, the bridging scaffolds resulted in the linear, highly organized regeneration of long-tract sensory axons, with most axons entering the scaffolds and continuing to grow the full length of the lesion cavity.

Scaffolds have also been used for the localized delivery of lentivirus for gene delivery [162, 169]. For example, multichannel scaffolds of PLGA have been used to deliver neurotrophin encoding lentivirus within the channels, with transgene expression lasting for at least 4 weeks and leading to increased axon growth and myelination [162]. Spatial control of gene expression within a scaffold can be achieved by localized inducer release from the scaffold [324]. Adsorption of DNA to specific regions of a tissue engineered scaffold is being investigated to induce cells to express different genes depending on their location within the scaffold; the result may be scaffolds that can stimulate specific axonal populations with varying signaling requirements [325–327].

21.6.4 Molecular Therapy with Scaffolds

In addition to delivering supportive cells and gene, scaffolds have also been widely used as drug delivery systems for trophic factors, the anti-inflammatory drug methylprednisolone, and antibodies, aimed at providing a more permissive environment for axonal regeneration. Delivery of several molecules by scaffolds is usually required for better outcomes with respect to spinal cord regeneration due to highly complicated structural and functional features of the spinal cord.

Incorporation of neurotrophic factors is beneficial due to their neuroprotective effect for neural cell survival, their ability to influence glial development and enhance axonal regeneration, and their promotion of plasticity of spared axons after SCI and remyelination of axons [328–331]. Trophic factors investigated for SCI include NGF, BDNF, NT-3, NT-4/5, GDNF, and epidermal growth factor. The de- livery of neurotropic factors by scaffolds for SCI repair has been reviewed in detail by others [332, 333]. Prior research has elucidated the sensitivity of certain spinal tracts to specific neurotrophic factors, with NGF promoting the growth of nociceptive axons; BDNF modifying the growth of motor neurons; NT-3 affecting the corticospinal tracts and dorsal sensory axons; NT-4/5, the proprioceptive and motor modifying inputs; and GDNF the proprioceptive, dorsal sensory, and nociceptive neurons [334]. Chen et al. [335] provide a comprehensive review of the various systems used for the delivery of single or multiple trophic factors in tissue engineering (Table 21.1). Future work with respect to trophic factors for tissue engineering will require delivery of multiple trophic factors with controlled spatiotemporal patterns. The use of platelet-rich plasma or gene therapy could also be used for controlled release of multiple trophic factors. In addition, generation of gradients of multiple trophic factors in vivo is necessary for axonal regeneration and should be further investigated.

Nerve guidance scaffolds can be combined with delivery of antibodies to neutralize the hostile extracellular environment after SCI, which is a major factor accounting for poor regeneration of the spinal cord. Current strategies are mainly targeting the CSPGs and myelin-associated inhibitors including Nogo, OMpg, and MAG. After SCI, CSPGs are upregulated and many studies indicate that they act as a barrier to regenerating axons by inducing growth cone collapse [336]. Application of ChABC could liberate the chondroitin sulfate glycosaminoglycans chains from the core protein and thereby reduce the inhibition by CSPGs for axonal regeneration. ChABC could also promote sprouting/plasticity of uninjured systems and neuroprotection of injured projection neurons. Mechanisms underlying the beneficial effects of ChABC have been discussed by Bradbury and Carter [337]. Combinatorial strategies of CSPG digestion with scaffolds and other therapeutic strategies have been shown to promote the regeneration of injured axons into distal targets, with some accompanied functional recovery [304, 338]. However, the intrathecal application of ChABC using osmotic minipumps results in a rapid deterioration of enzymatic activity. Future studies using scaffolds as the delivery system may contribute to longer-term enzymatic activity of ChABC.

Scaffolds have also been used as the delivery system for antibodies to myelin-associated inhibitors. HA hydrogels have been covalently modified with poly-L-lysine and Nogo-66 receptor antibodies, leading to attachment, survival, and neurite extension of neural cells in vitro [339]. In an in vivo study of rat lateral spinal cord hemisection, the composite hydrogels led to significantly more neurofilament positive axons within the hydrogels compared to the controls; more cells and myelinated

Table 21.1 Schematic diagrams of various delivery strategies for neurotrophic factors and their representative release profiles [335]

Strategies	Images	Release profiles	Delivery models (examples)
Incorporation (bolus injection or growth factors directly adsorbed in vehicles)		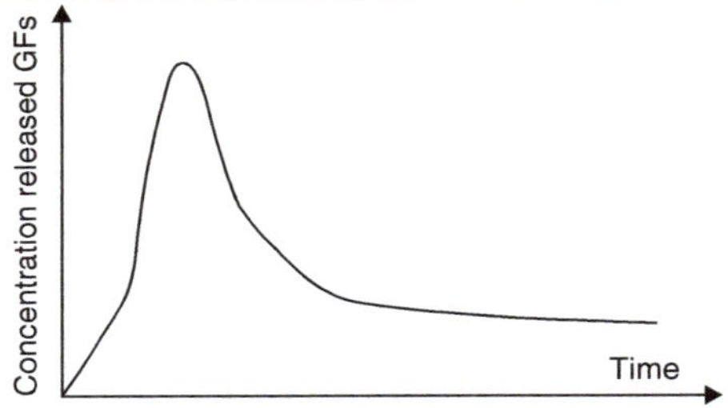	Burst release (frequently investigated porous scaffolds, hydrogels, etc.)
Non-covalent immobilisation (growth factors encapsulated into low crosslinked microspheres)	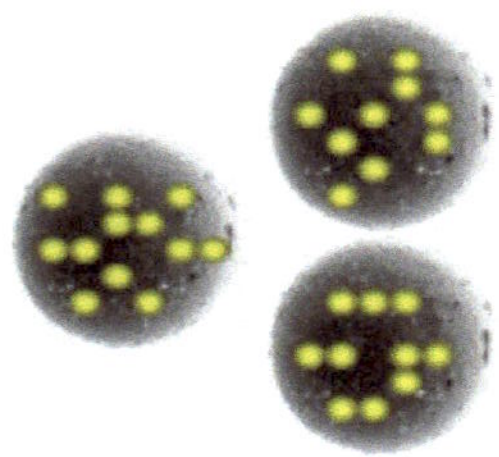	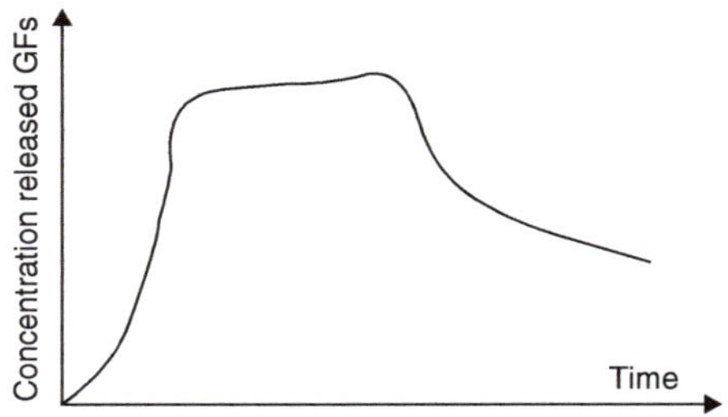	Sustained release (frequently investigated microspheres, nanoparticles, etc.)
Non-covalent immobilisation (growth factors encapsulated into high crosslinked microspheres)		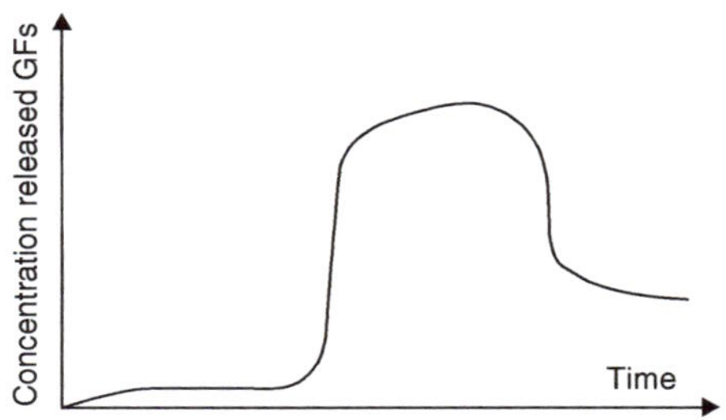	Delayed release (well-designed microspheres, core-shell vehicles, etc.)

Non-covalent immobilisation (growth factors delivered by both scaffolds and microspheres)	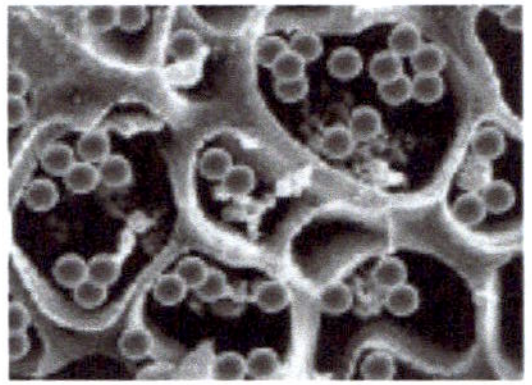	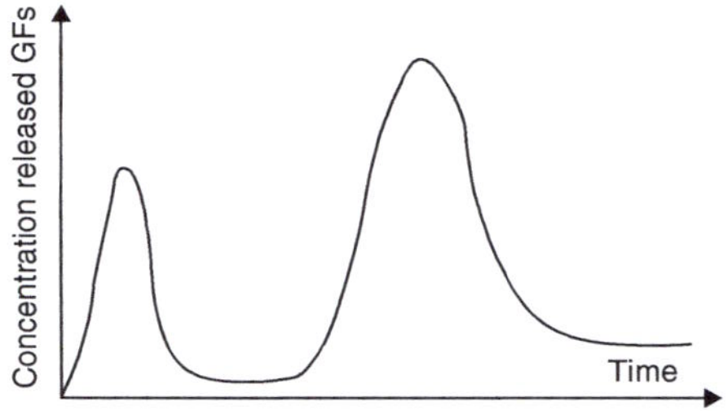	Pulse-like release (well-designed composites, multifunctional vehicles, etc.)
Covalent immobilisation (growth factors chemically crosslinked via difunctional linkers)	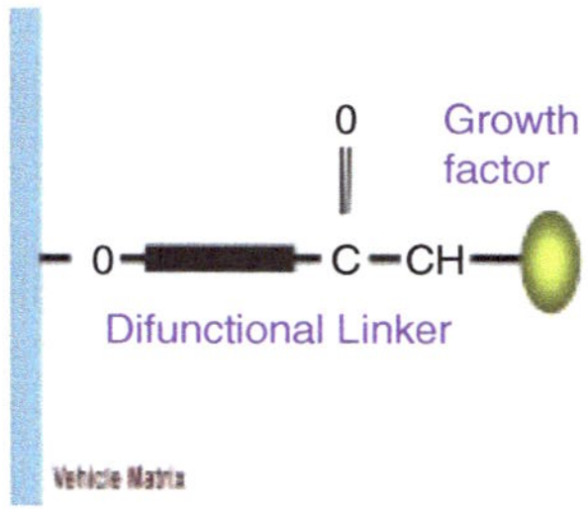	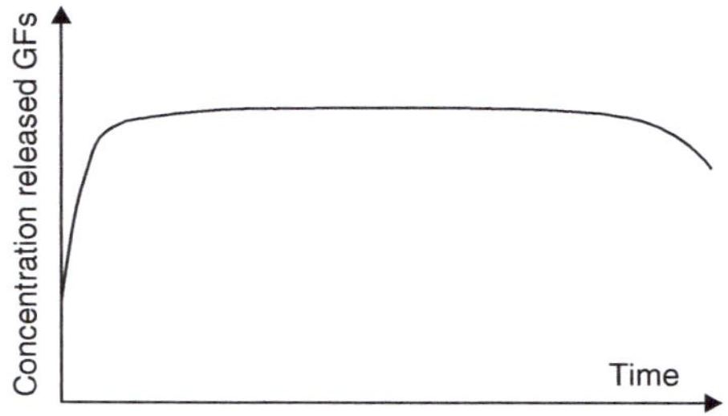	Continuous release (fibrillar collagen, alginate gel, etc.)
Covalent immobilisation or chemical modification (environment-sensitive materials)	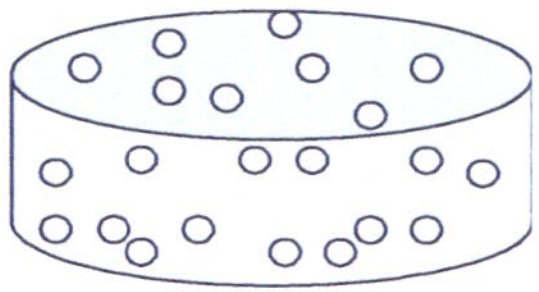	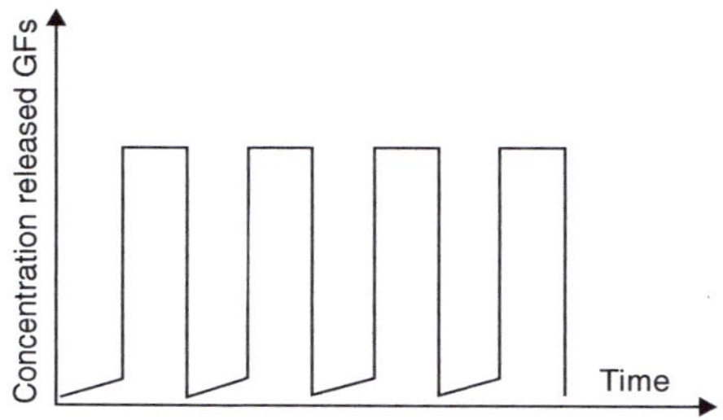	Pulsatile release (intelligent hydrogels, on-off delivery systems, etc.)

axons were also present in the implant [135]. Combination of antibody delivery with other therapeutic strategies is required for better regeneration of injured spinal cord. However, not all combinations of individual strategies that have proven effective for promoting regeneration will lead to complementary or synergistic effects. The mechanisms underlying different therapies must be taken into account in the design of combined therapies so as to achieve positive interactions [340].

21.7 Summary

1. Spinal cord is essential for establishing functional interaction between the brain and the periphery for the execution of various body movement and organ functions.
2. Different axon tracts in various locations of spinal cord white matter convey different types of information.
3. The severely limited self-regenerative capacity of spinal cord is primarily due to the physical, molecular, and cellular non-permissive environment after SCI. This poor environment occurs as a result of both primary and secondary injuries.
4. Contusion injury models reflect the most frequent type of human SCI, while hemisection/complete transection animal models are more suitable for evaluating the efficacy and safety of axon regeneration therapeutic strategies.
5. Scaffold materials for SCI should be biocompatible, biodegradable, and non-toxic. They should also tend to reduce astrocytic reaction and fibroblastic gliosis, and allow for cell adhesion and axonal regrowth.
6. Different natural and synthetic biomaterials have been applied to construct scaffolds for SCI, and modifications of materials may be needed.
7. The macrostructure, microstructure, surface chemistry, surface topography, and mechanical properties of scaffolds have substantial effects on axonal regeneration and functional recovery of SCI.
8. Electrospinning and self-assembly techniques can fabricate scaffolds at micro- and nanoscales to mimic native ECM, leading to enhanced axonal regeneration and functional recovery of injured spinal cord.
9. Stereolithography and 3D plotting are two solid freeform fabrication techniques that allow for the incorporation of living cells and bioactive proteins during the fabrication process.
10. A range of cell types, genes, ECM proteins/peptides, and molecules have been delivered by scaffolds, and application of multifactorial strategies is commonly used in attempts to better promote spinal cord regeneration.

References

1. National Spinal Cord Injury Statistical Center (NSCIS) (2011) Spinal cord injury facts and figures at a glance, February 2011. Available at https://www.nscisc.uab.edu/public_content/pdf/Facts%202011%20Feb%20Final.pdf. Accessed 18 May 2012

2. Fehlings MG, Sekhon LH, Tator C (2001) The role and timing of decompression in acute spinal cord injury: what do we know? What should we do? Spine (Phila Pa 1976) 26:S101–S110
3. Fehlings MG, Tator CH (1999) An evidence-based review of decompressive surgery in acute spinal cord injury: rationale, indications, and timing based on experimental and clinical studies. J Neurosurg 91:1–11
4. Schnell L, Schwab ME (1990) Axonal regeneration in the rat spinal cord produced by an antibody against myelin-associated neurite growth inhibitors. Nature 343:269–272
5. van Blitterswijk C, Hubbell J, Cancedda R, de Bruijn JD, Lindahl A, Sohier J, Williams D (2008) Tissue engineering, 1st edn. Academic, London, p 760
6. Neumann S, Bradke F, Tessier-Lavigne M, Basbaum AI (2002) Regeneration of sensory axons within the injured spinal cord induced by intraganglionic cAMP elevation. Neuron 34:885–893
7. Neumann S, Woolf CJ (1999) Regeneration of dorsal column fibers into and beyond the lesion site following adult spinal cord injury. Neuron 23:83–91
8. Imaizumi T, Lankford KL, Waxman SG, Greer CA, Kocsis JD (1998) Transplanted olfactory ensheathing cells remyelinate and enhance axonal conduction in the demyelinated dorsal columns of the rat spinal cord. J Neurosci 18:6176–6185
9. Blumenfeld H (2002) Neuroanatomy through clinical cases, 1st edn. Sinauer Associates, Sunderland
10. Oudega M, Moon LD, de Almeida Leme RJ (2005) Schwann cells for spinal cord repair. Braz J Med Biol Res 38:825–835
11. Samadikuchaksaraei A (2007) An overview of tissue engineering approaches for management of spinal cord injuries. J Neuroeng Rehabil 4:15
12. Winkler T, Sharma HS, Gordh T, Badgaiyan RD, Stalberg E, Westman J (2002) Topical application of dynorphin A (1–17) antiserum attenuates trauma induced alterations in spinal cord evoked potentials, microvascular permeability disturbances, edema formation and cell injury: an experimental study in the rat using electrophysiological and morphological approaches. Amino Acids 23:273–281
13. Bao F, John SM, Chen Y, Mathison RD, Weaver LC (2006) The tripeptide phenylalanine-(D) glutamate-(D) glycine modulates leukocyte infiltration and oxidative damage in rat injured spinal cord. Neuroscience 140:1011–1022
14. Park E, Velumian AA, Fehlings MG (2004) The role of excitotoxicity in secondary mechanisms of spinal cord injury: a review with an emphasis on the implications for white matter degeneration. J Neurotrauma 21:754–774
15. Conti A, Cardali S, Genovese T, Di Paola R, La Rosa G (2003) Role of inflammation in the secondary injury following experimental spinal cord trauma. J Neurosurg Sci 47:89–94
16. Dohrmann GJ, Wagner FC Jr, Bucy PC (1972) Transitory traumatic paraplegia: electron microscopy of early alterations in myelinated nerve fibers. J Neurosurg 36:407–415
17. Rivlin AS, Tator CH (1978) Regional spinal cord blood flow in rats after severe cord trauma. J Neurosurg 49:844–853
18. Koyanagi I, Tator CH, Lea PJ (1993) Three-dimensional analysis of the vascular system in the rat spinal cord with scanning electron microscopy of vascular corrosion casts. Part 2: acute spinal cord injury. Neurosurgery 33:285–291; discussion 292
19. Anthes DL, Theriault E, Tator CH (1996) Ultrastructural evidence for arteriolar vasospasm after spinal cord trauma. Neurosurgery 39:804–814
20. Noble LJ, Wrathall JR (1989) Correlative analyses of lesion development and functional status after graded spinal cord contusive injuries in the rat. Exp Neurol 103:34–40
21. Tator CH, Koyanagi I (1997) Vascular mechanisms in the pathophysiology of human spinal cord injury. J Neurosurg 86:483–492
22. Wang MD, Zhai P, Chen XB, Schreyer DJ, Sun XD, Cui FZ (2011) Bioengineered scaffolds for spinal cord repair. Tissue Eng Part B Rev 17:177–194
23. Holtz A, Nystrom B, Gerdin B, Olsson Y (1990) Neuropathological changes and neurological function after spinal cord compression in the rat. J Neurotrauma 7:155–167
24. Szydlowska K, Tymianski M (2010) Calcium, ischemia and excitotoxicity. Cell Calcium 47:122–129
25. Jones TB, McDaniel EE, Popovich PG (2005) Inflammatory-mediated injury and repair in the traumatically injured spinal cord. Curr Pharm Des 11:1223–1236

26. Shuman SL, Bresnahan JC, Beattie MS (1997) Apoptosis of microglia and oligodendrocytes after spinal cord contusion in rats. J Neurosci Res 50:798–808
27. Schwab ME, Bartholdi D (1996) Degeneration and regeneration of axons in the lesioned spinal cord. Physiol Rev 76:319–370
28. Guest JD, Hiester ED, Bunge RP (2005) Demyelination and Schwann cell responses adjacent to injury epicenter cavities following chronic human spinal cord injury. Exp Neurol 192:384–393
29. Jones LL, Yamaguchi Y, Stallcup WB, Tuszynski MH (2002) NG2 is a major chondroitin sulfate proteoglycan produced after spinal cord injury and is expressed by macrophages and oligodendrocyte progenitors. J Neurosci 22:2792–2803
30. Jones LL, Margolis RU, Tuszynski MH (2003) The chondroitin sulfate proteoglycans neurocan, brevican, phosphacan, and versican are differentially regulated following spinal cord injury. Exp Neurol 182:399–411
31. Bruce JH, Norenberg MD, Kraydieh S, Puckett W, Marcillo A, Dietrich D (2000) Schwannosis: role of gliosis and proteoglycan in human spinal cord injury. J Neurotrauma 17:781–788
32. McKerracher L, Winton MJ (2002) Nogo on the go. Neuron 36:345–348
33. Filbin MT (2003) Myelin-associated inhibitors of axonal regeneration in the adult mammalian CNS. Nat Rev Neurosci 4:703–713
34. De Winter F, Oudega M, Lankhorst AJ, Hamers FP, Blits B, Ruitenberg MJ, Pasterkamp RJ, Gispen WH, Verhaagen J (2002) Injury-induced class 3 semaphorin expression in the rat spinal cord. Exp Neurol 175:61–75
35. Bundesen LQ, Scheel TA, Bregman BS, Kromer LF (2003) Ephrin-B2 and EphB2 regulation of astrocyte-meningeal fibroblast interactions in response to spinal cord lesions in adult rats. J Neurosci 23:7789–7800
36. Wang GY, He XJ, Yuan PW, Li HP, Chang R (2011) Semaphorin 3A expression in spinal cord injured rats after olfactory ensheathing cell transplantation. Neural Regen Res 6:756–761
37. Bregman BS, Kunkelbagden E, Schnell L, Dai HN, Gao D, Schwab ME (1995) Recovery from spinal cord injury mediated by antibodies to neurite growth inhibitors. Nature 378:498–501
38. Merkler D, Metz GAS, Raineteau O, Dietz V, Schwab ME, Fouad K (2001) Locomotor recovery in spinal cord-injured rats treated with an antibody neutralizing the myelin- associated neurite growth inhibitor Nogo-A. J Neurosci 21:3665–3673
39. Bradbury EJ, Moon LDF, Popat RJ, King VR, Bennett GS, Patel PN, Fawcett JW, SB MM (2002) Chondroitinase ABC promotes functional recovery after spinal cord injury. Nature 416:636–640
40. Omoto S, Ueno M, Mochio S, Yamashita T (2011) Corticospinal tract fibers cross the ephrin-B3-negative part of the midline of the spinal cord after brain injury. Neurosci Res 69:187–195
41. Aubert I, Ridet JL, Gage FH (1995) Regeneration in the adult mammalian CNS – guided by development. Curr Opin Neurobiol 5:625–635
42. Levison SW, Ducceschi MH, Young GM, Wood TL (1996) Acute exposure to CNTF in vivo induces multiple components of reactive gliosis. Exp Neurol 141:256–268
43. Sykova E, Svoboda J, Simonova Z, Jendelova P (1992) Role of astrocytes in ionic and volume homeostasis in spinal-cord during development and injury. Prog Brain Res 94:47–56
44. Alonso G, Privat A (1993) Reactive astrocytes involved in the formation of lesional scars differ in the mediobasal hypothalamus and in other forebrain regions. J Neurosci Res 34:523–538
45. Frisen J, Arvidsson U, Lindholm T, Fried K, Verge VMK, Cullheim S, Hokfelt T, Risling M (1993) trkC expression in the injured rat spinal cord. Neuroreport 5:349–352
46. Lasalle GL, Rougon G, Valin A (1992) The embryonic form of neural cell -surface molecule (E-NCAM) in the rat hippocampus and its reexpression on glial-cells following kainic acid-induced status epilepticus. J Neurosci 12:872–882
47. Faulkner JR, Herrmann JE, Woo MJ, Tansey KE, Doan NB, Sofroniew MV (2004) Reactive astrocytes protect tissue and preserve function after spinal cord injury. J Neurosci 24:2143–2155
48. Raineteau O, Fouad K, Bareyre FM, Schwab ME (2002) Reorganization of descending motor tracts in the rat spinal cord. Eur J Neurosci 16:1761–1771

49. Bareyre FM, Kerschensteiner M, Raineteau O, Mettenleiter TC, Weinmann O, Schwab ME (2004) The injured spinal cord spontaneously forms a new intraspinal circuit in adult rats. Nat Neurosci 7:269–277
50. Raineteau O, Schwab ME (2001) Plasticity of motor systems after incomplete spinal cord injury. Nat Rev Neurosci 2:263–273
51. Kwok JCF, Afshari F, Garcia-Alias G, Fawcett J (2008) Proteoglycans in the central nervous system: plasticity, regeneration and their stimulation with chondroitinase ABC. Restor Neurol Neurosci 26:131–145
52. Massey JM, Hubscher CH, Wagoner MR, Decker JA, Amps J, Silver J, Onifer SM (2006) Chondroitinase ABC digestion of the perineuronal net promotes functional collateral sprouting in the cuneate nucleus after cervical spinal cord injury. J Neurosci 26:4406–4414
53. Vavrek R, Girgis J, Tetzlaff W, Hebert GW, Fouad K (2006) BDNF promotes connections of corticospinal neurons onto spared descending interneurons in spinal cord injured rats. Brain 129:1534–1545
54. Garcia-Alias G, Barkhuysen S, Buckle M, Fawcett JW (2009) Chondroitinase ABC treatment opens a window of opportunity for task-specific rehabilitation. Nat Neuosci 12:1145–1151
55. Tetzlaff W, Fouad K, Kwon B (2009) Be careful what you train for. Nat Neuosci 12:1145–1151
56. Talac R, Friedman JA, Moore MJ, Lu L, Jabbari E, Windebank AJ, Currier BL, Yaszemski MJ (2004) Animal models of spinal cord injury for evaluation of tissue engineering treatment strategies. Biomaterials 25:1505–1510
57. Metz GAS, Curt A, van de Meent H, Klusman I, Schwab ME, Dietz V (2000) Validation of the weight-drop contusion model in rats: a comparative study of human spinal cord injury. J Neurotrauma 17:1–17
58. Darian-Smith C (2009) Synaptic plasticity, neurogenesis, and functional recovery after spinal cord injury. Neuroscientist 15:149–165
59. Lawrence DG, Kuypers HG (1968) The functional organization of the motor system in the monkey. I. The effects of bilateral pyramidal lesions. Brain 91:1–14
60. Lemon RN, Griffiths J (2005) Comparing the function of the corticospinal system in different species: organizational differences for motor specialization? Muscle Nerve 32:261–279
61. Nathan PW, Smith MC (1982) The rubrospinal and central tegmental tracts in man. Brain 105:223–269
62. Dietz V (2008) Ready for human spinal cord repair? Brain 131:2240–2242
63. Rahimi-Movaghar V, Yazdi A, Mohammadi M (2008) Usefulness of the tail -flick reflex in the prognosis of functional recovery in paraplegic rats. Surg Neurol 70:323–325; discussion 325
64. Saberi H, Moshayedi P, Aghayan HR, Arjmand B, Hosseini SK, SH E-R, Rahimi-Movaghar V, Raza M, Firouzi M (2008) Treatment of chronic thoracic spinal cord injury patients with autologous Schwann cell transplantation: an interim report on safety considerations and possible outcomes. Neurosci Lett 443:46–50
65. Borgens RB, Shi R (2000) Immediate recovery from spinal cord injury through molecular repair of nerve membranes with polyethylene glycol. FASEB J 14:27–35
66. Borgens RB (2001) Cellular engineering: molecular repair of membranes to rescue cells of the damaged nervous system. Neurosurgery 49:370–378; discussion 378–379
67. Bunge RP, Puckett WR, Becerra JL, Marcillo A, Quencer RM (1993) Observations on the pathology of human spinal cord injury. A review and classification of 22 new cases with details from a case of chronic cord compression with extensive focal demyelination. Adv Neurol 59:75–89
68. Kakulas BA (1999) A review of the neuropathology of human spinal cord injury with emphasis on special features. J Spinal Cord Med 22:119–124
69. Oudega M, Hagg T (1996) Nerve growth factor promotes regeneration of sensory axons into adult rat spinal cord. Exp Neurol 140:218–229
70. Gros T, Sakamoto JS, Blesch A, Havton LA, Tuszynski MH (2010) Regeneration of long – tract axons through sites of spinal cord injury using templated agarose scaffolds. Biomaterials 31:6719–6729

71. Jain A, Kim YT, McKeon RJ, Bellamkonda RV (2006) In situ gelling hydrogels for conformal repair of spinal cord defects, and local delivery of BDNF after spinal cord injury. Biomaterials 27:497–504
72. King VR, Hewazy D, Alovskaya A, Phillips JB, Brown RA, Priestley JV (2010) The neuroprotective effects of fibronectin mats and fibronectin peptides following spinal cord injury in the rat. Neuroscience 168:523–530
73. Stokes BT, Jakeman LB (2002) Experimental modelling of human spinal cord injury: a model that crosses the species barrier and mimics the spectrum of human cytopathology. Spinal Cord 40:101–109
74. Kang KN, Lee JY, da Kim Y, Lee BN, Ahn HH, Lee B, Khang G, Park SR, Min BH, Kim JH, Lee HB, Kim MS (2011) Regeneration of completely transected spinal cord using scaffold of poly(D, L-lactide-co-glycolide)/small intestinal submucosa seeded with rat bone marrow stem cells. Tissue Eng Part A 17:2143–2152
75. Hurtado A, Cregg JM, Wang HB, Wendell DF, Oudega M, Gilbert RJ, McDonald JW (2011) Robust CNS regeneration after complete spinal cord transection using aligned poly - L-lactic acid microfibers. Biomaterials 32:6068–6079
76. Kang KN, da Kim Y, Yoon SM, Lee JY, Lee BN, Kwon JS, Seo HW, Lee IW, Shin HC, Kim YM, Kim HS, Kim JH, Min BH, Lee HB, Kim MS (2012) Tissue engineered regeneration of completely transected spinal cord using human mesenchymal stem cells. Biomaterials 33:4828–4835
77. Harley BA, Spilker MH, Wu JW, Asano K, Hsu HP, Spector M, Yannas IV (2004) Optimal degradation rate for collagen chambers used for regeneration of peripheral nerves over long gaps. Cells Tissues Organs 176:153–165
78. Tsai EC, Dalton PD, Shoichet MS, Tator CH (2006) Matrix inclusion within synthetic hydrogel guidance channels improves specific supraspinal and local axonal regeneration after complete spinal cord transection. Biomaterials 27:519–533
79. Klapka N, Muller HW (2006) Collagen matrix in spinal cord injury. J Neurotrauma 23:422–435
80. Labrador RO, Buti M, Navarro X (1998) Influence of collagen and laminin gels concentration on nerve regeneration after resection and tube repair. Exp Neurol 149:243–252
81. Liu T, Houle JD, Xu J, Chan BP, Chew SY (2012) Nanofibrous collagen nerve conduits for spinal cord repair. Tissue Eng Part A 18:1057–1066
82. Han QQ, Sun WJ, Lin H, Zhao WX, Gao Y, Zhao YN, Chen B, Xiao ZF, Hu W, Li Y, Yang B, Dai JW (2009) Linear ordered collagen scaffolds loaded with collagen -binding brain-derived neurotrophic factor improve the recovery of spinal cord injury in rats. Tissue Eng Part A 15:2927–2935
83. Han QQ, Jin W, Xiao ZF, Ni HB, Wang JH, Kong J, Wu J, Liang WB, Chen L, Zhao YN, Chen B, Dai JW (2010) The promotion of neural regeneration in an extreme rat spinal cord injury model using a collagen scaffold containing a collagen binding neuroprotective protein and an EGFR neutralizing antibody. Biomaterials 31:9212–9220
84. Houweling DA, Lankhorst AJ, Gispen WH, Bar PR, Joosten EA (1998) Collagen containing neurotrophin-3 (NT-3) attracts regrowing injured corticospinal axons in the adult rat spinal cord and promotes partial functional recovery. Exp Neurol 153:49–59
85. Cholas RH, Hsu HP, Spector M (2012) The reparative response to cross-linked collagen- based scaffolds in a rat spinal cord gap model. Biomaterials 33:2050–2059
86. Pielesz A, Katarzyna B, Klimczak M (2008) Physicochemical properties of commercial active alginate dressings. Polim Med 38:3–17
87. Perets A, Baruch Y, Weisbuch F, Shoshany G, Neufeld G, Cohen S (2003) Enhancing the vascularization of three-dimensional porous alginate scaffolds by incorporating controlled release basic fibroblast growth factor microspheres. J Biomed Mater Res A 65: 489–497
88. Tobias CA, Han SS, Shumsky JS, Kim D, Tumolo M, Dhoot NO, Wheatley MA, Fischer I, Tessler A, Murray M (2005) Alginate encapsulated BDNF-producing fibroblast grafts permit recovery of function after spinal cord injury in the absence of immune suppression. J Neurotrauma 22:138–156

89. Kim JO, Choi JY, Park JK, Kim JH, Jin SG, Chang SW, Li DX, Hwang MR, Woo JS, Kim JA, Lyoo WS, Yong CS, Choi HG (2008) Development of clindamycin -loaded wound dressing with polyvinyl alcohol and sodium alginate. Biol Pharm Bull 31:2277–2282
90. Eiselt P, Lee KY, Mooney DJ (1999) Rigidity of two-component hydrogels prepared from alginate and poly(ethylene glycol)-diamines. Macromolecules 32:5561–5566
91. Lee KY, Rowley JA, Eiselt P, Moy EM, Bouhadir KH, Mooney DJ (2000) Controlling mechanical and swelling properties of alginate hydrogels independently by cross-linker type and cross-linking density. Macromolecules 33:4291–4294
92. Ashton RS, Banerjee A, Punyani S, Schaffer DV, Kane RS (2007) Scaffolds based on degradable alginate hydrogels and poly(lactide-co-glycolide) microspheres for stem cell culture. Biomaterials 28:5518–5525
93. Suzuki K, Suzuki Y, Ohnishi K, Endo K, Tanihara M, Nishimura Y (1999) Regeneration of transected spinal cord in young adult rats using freeze-dried alginate gel. Neuroreport 10:2891–2894
94. Suzuki Y, Kitaura M, Wu SF, Kataoka K, Suzuki K, Endo K, Nishimura Y, Ide C (2002) Electrophysiological and horseradish peroxidase-tracing studies of nerve regeneration through alginate-filled gap in adult rat spinal cord. Neurosci Lett 318:121–124
95. Prang P, Muller R, Eljaouhari A, Heckmann K, Kunz W, Weber T, Faber C, Vroemen M, Bogdahn U, Weidner N (2006) The promotion of oriented axonal regrowth in the injured spinal cord by alginate-based anisotropic capillary hydrogels. Biomaterials 27:3560–3569
96. Novikova LN, Mosahebi A, Wiberg M, Terenghi G, Kellerth JO, Novikov LN (2006) Alginate hydrogel and matrigel as potential cell carriers for neurotransplantation. J Biomed Mater Res A 77:242–252
97. Frampton JP, Hynd MR, Shuler ML, Shain W (2011) Fabrication and optimization of alginate hydrogel constructs for use in 3D neural cell culture. Biomed Mater 6:015002–015019
98. Matyash M, Despang F, Mandal R, Fiore D, Gelinsky M, Ikonomidou C (2012) Novel soft alginate hydrogel strongly supports neurite growth and protects neurons against oxidative stress. Tissue Eng Part A 18:55–66
99. Francis NL, Shanbhag MS, Fischer I, Wheatley MA (2011) Influence of alginate cross-linking method on neurite response to microencapsulated neurotrophin-producing fibroblasts. J Microencapsul 28:353–362
100. Balgude AP, Yu X, Szymanski A, Bellamkonda RV (2001) Agarose gel stiffness determines rate of DRG neurite extension in 3D cultures. Biomaterials 22:1077–1084
101. Stokols S, Tuszynski MH (2006) Freeze-dried agarose scaffolds with uniaxial channels stimulate and guide linear axonal growth following spinal cord injury. Biomaterials 27:443–451
102. Stokols S, Tuszynski MH (2004) The fabrication and characterization of linearly oriented nerve guidance scaffolds for spinal cord injury. Biomaterials 25:5839–5846
103. Stokols S, Sakamoto J, Breckon C, Holt T, Weiss J, Tuszynski MH (2006) Templated agarose scaffolds support linear axonal regeneration. Tissue Eng 12:2777–2787
104. Nisbet DR, Crompton KE, Horne MK, Finkelstein DI, Forsythe JS (2008) Neural tissue engineering of the CNS using hydrogels. J Biomed Mater Res B Appl Biomater 87:251–263
105. Kim H, Tator CH, Shoichet MS (2011) Chitosan implants in the rat spinal cord: biocompatibility and biodegradation. J Biomed Mater Res A 97:395–404
106. Yang Z, Duan H, Mo L, Qiao H, Li X (2010) The effect of the dosage of NT-3/chitosan carriers on the proliferation and differentiation of neural stem cells. Biomaterials 31:4846–4854
107. Nomura H, Zahir T, Kim H, Katayama Y, Kulbatski I, Morshead CM, Shoichet MS, Tator CH (2008) Extramedullary chitosan channels promote survival of transplanted neural stem and progenitor cells and create a tissue bridge after complete spinal cord transection. Tissue Eng Part A 14:649–665
108. Zahir T, Nomura H, Guo XD, Kim H, Tator C, Morshead C, Shoichet M (2008) Bioengineering neural stem/progenitor cell-coated tubes for spinal cord injury repair. Cell Transplant 17:245–254
109. Nomura H, Kim H, Mothe A, Zahir T, Kulbatski I, Morshead CM, Shoichet MS, Tator CH (2010) Endogenous radial glial cells support regenerating axons after spinal cord transection. Neuroreport 21:871–876

110. Li X, Yang Z, Zhang A, Wang T, Chen W (2009) Repair of thoracic spinal cord injury by chitosan tube implantation in adult rats. Biomaterials 30:1121–1132
111. Lord ST (2007) Fibrinogen and fibrin: scaffold proteins in hemostasis. Curr Opin Hematol 14:236–241
112. Petter-Puchner AH, Froetscher W, Krametter-Froetscher R, Lorinson D, Redl H, van Griensven M (2007) The long-term neurocompatibility of human fibrin sealant and equine collagen as biomatrices in experimental spinal cord injury. Exp Toxicol Pathol 58:237–245
113. Johnson PJ, Parker SR, Sakiyama-Elbert SE (2010) Fibrin-based tissue engineering scaffolds enhance neural fiber sprouting and delay the accumulation of reactive astrocytes at the lesion in a subacute model of spinal cord injury. J Biomed Mater Res A 92:152–163
114. King VR, Alovskaya A, Wei DY, Brown RA, Priestley JV (2010) The use of injectable forms of fibrin and fibronectin to support axonal ingrowth after spinal cord injury. Biomaterials 31:4447–4456
115. Hyatt AJ, Wang D, Kwok JC, Fawcett JW, Martin KR (2010) Controlled release of chondroitinase ABC from fibrin gel reduces the level of inhibitory glycosaminoglycan chains in lesioned spinal cord. J Control Release 147:24–29
116. Itosaka H, Kuroda S, Shichinohe H, Yasuda H, Yano S, Kamei S, Kawamura R, Hida K, Iwasaki Y (2009) Fibrin matrix provides a suitable scaffold for bone marrow stromal cells transplanted into injured spinal cord: a novel material for CNS tissue engineering. Neuropathology 29:248–257
117. Johnson PJ, Parker SR, Sakiyama-Elbert SE (2009) Controlled release of neurotrophin-3 from fibrin-based tissue engineering scaffolds enhances neural fiber sprouting following subacute spinal cord injury. Biotechnol Bioeng 104:1207–1214
118. Taylor SJ, Rosenzweig ES, McDonald JW 3rd, Sakiyama-Elbert SE (2006) Delivery of neurotrophin-3 from fibrin enhances neuronal fiber sprouting after spinal cord injury. J Control Release 113:226–235
119. Johnson PJ, Tatara A, McCreedy DA, Shiu A, Sakiyama-Elbert SE (2010) Tissue-engineered fibrin scaffolds containing neural progenitors enhance functional recovery in a subacute model of SCI. Soft Matter 6:5127–5137
120. Khaing ZZ, Schmidt CE (2012) Advances in natural biomaterials for nerve tissue repair. Neurosci Lett 519:103–114
121. Yamada KM, Olden K (1978) Fibronectins – adhesive glycoproteins of cell surface and blood. Nature 275:179–184
122. Lin CY, Lee YS, Lin VW, Silver J (2011) Fibronectin inhibits chronic pain development after spinal cord injury. J Neurotrauma 29:589–599
123. King VR, Henseler M, Brown RA, Priestley JV (2003) Mats made from fibronectin support oriented growth of axons in the damaged spinal cord of the adult rat. Exp Neurol 182:383–398
124. King VR, Phillips JB, Hunt-Grubbe H, Brown R, Priestley JV (2006) Characterization of non-neuronal elements within fibronectin mats implanted into the damaged adult rat spinal cord. Biomaterials 27:485–496
125. Tonge DA, de Burgh HT, Docherty R, Humphries MJ, Craig SE, Pizzey J (2012) Fibronectin supports neurite outgrowth and axonal regeneration of adult brain neurons in vitro. Brain Res 1453:8–16
126. Khaing ZZ, Milman BD, Vanscoy JE, Seidlits SK, Grill RJ, Schmidt CE (2011) High molecular weight hyaluronic acid limits astrocyte activation and scar formation after spinal cord injury. J Neural Eng 8:046033
127. Peattie RA, Nayate AP, Firpo MA, Shelby J, Fisher RJ, Prestwich GD (2004) Stimulation of in vivo angiogenesis by cytokine-loaded hyaluronic acid hydrogel implants. Biomaterials 25:2789–2798
128. Hashizume M, Mihara M (2010) High molecular weight hyaluronic acid inhibits IL-6-induced MMP production from human chondrocytes by up-regulating the ERK inhibitor, MKP-1. Biochem Biophys Res Commun 403:184–189

129. Schimizzi AL, Massie JB, Murphy M, Perry A, Kim CW, Garfin SR, Akeson WH (2006) High-molecular-weight hyaluronan inhibits macrophage proliferation and cytokine release in the early wound of a preclinical postlaminectomy rat model. Spine J 6:550–556
130. Campo GM, Avenoso A, Campo S, D'Ascola A, Nastasi G, Calatroni A (2010) Molecular size hyaluronan differently modulates toll-like receptor-4 in LPS-induced inflammation in mouse chondrocytes. Biochimie 92:204–215
131. Seidlits SK, Khaing ZZ, Petersen RR, Nickels JD, Vanscoy JE, Shear JB, Schmidt CE (2010) The effects of hyaluronic acid hydrogels with tunable mechanical properties on neural progenitor cell differentiation. Biomaterials 31:3930–3940
132. Oohira A, Matsui F, Matsuda M, Shoji R (1986) Developmental change in the glycosaminoglycan composition of the rat brain. J Neurochem 47:588–593
133. Meszar Z, Felszeghy S, Veress G, Matesz K, Szekely G, Modis L (2008) Hyaluronan accumulates around differentiating neurons in spinal cord of chicken embryos. Brain Res Bull 75:414–418
134. Horn EM, Beaumont M, Shu XZ, Harvey A, Prestwich GD, Horn KM, Gibson AR, Preul MC, Panitch A (2007) Influence of cross-linked hyaluronic acid hydrogels on neurite outgrowth and recovery from spinal cord injury. J Neurosurg Spine 6:133–140
135. Wei YT, He Y, Xu CL, Wang Y, Liu BF, Wang XM, Sun XD, Cui FZ, Xu QY (2010) Hyaluronic acid hydrogel modified with nogo-66 receptor antibody and poly-L-lysine to promote axon regrowth after spinal cord injury. J Biomed Mater Res B Appl Biomater 95:110–117
136. Brannvall K, Bergman K, Wallenquist U, Svahn S, Bowden T, Hilborn J, Forsberg-Nilsson K (2007) Enhanced neuronal differentiation in a three-dimensional collagen-hyaluronan matrix. J Neurosci Res 85:2138–2146
137. Tian WM, Hou SP, Ma J, Zhang CL, Xu QY, Lee IS, Li HD, Spector M, Cui FZ (2005) Hyaluronic acid-poly-D-lysine-based three-dimensional hydrogel for traumatic brain injury. Tissue Eng 11:513–525
138. Zhang H, Wei YT, Tsang KS, Sun CR, Li J, Huang H, Cui FZ, An YH (2008) Implantation of neural stem cells embedded in hyaluronic acid and collagen composite conduit promotes regeneration in a rabbit facial nerve injury model. J Transl Med 6:67
139. Hou S, Xu Q, Tian W, Cui F, Cai Q, Ma J, Lee IS (2005) The repair of brain lesion by implantation of hyaluronic acid hydrogels modified with laminin. J Neurosci Methods 148:60–70
140. Tian WM, Zhang CL, Hou SP, Yu X, Cui FZ, Xu QY, Sheng SL, Cui H, Li HD (2005) Hyaluronic acid hydrogel as Nogo-66 receptor antibody delivery system for the repairing of injured rat brain: in vitro. J Control Release 102:13–22
141. Suri S, Han LH, Zhang W, Singh A, Chen S, Schmidt CE (2011) Solid freeform fabrication of designer scaffolds of hyaluronic acid for nerve tissue engineering. Biomed Microdevices 13:983–993
142. Seidlits SK, Schmidt CE, Shear JB (2009) High-resolution patterning of hydrogels in three dimensions using direct-write photofabrication for cell guidance. Adv Funct Mater 19:3543–3551
143. Dubovy P, Svizenska I, Klusakova I, Zitkova A, Houst'Ava L, Haninec P (2001) Laminin molecules in freeze-treated nerve segments are associated with migrating Schwann cells that display the corresponding alpha6beta1 integrin receptor. Glia 33:36–44
144. Evans PJ, Mackinnon SE, Levi AD, Wade JA, Hunter DA, Nakao Y, Midha R (1998) Cold preserved nerve allografts: changes in basement membrane, viability, immunogenicity, and regeneration. Muscle Nerve 21:1507–1522
145. Ide C, Tohyama K, Tajima K, Endoh K, Sano K, Tamura M, Mizoguchi A, Kitada M, Morihara T, Shirasu M (1998) Long acellular nerve transplants for allogeneic grafting and the effects of basic fibroblast growth factor on the growth of regenerating axons in dogs: a preliminary report. Exp Neurol 154:99–112
146. Hudson TW, Zawko S, Deister C, Lundy S, Hu CY, Lee K, Schmidt CE (2004) Optimized acellular nerve graft is immunologically tolerated and supports regeneration. Tissue Eng 10:1641–1651

147. Nagao RJ, Lundy S, Khaing ZZ, Schmidt CE (2011) Functional characterization of optimized acellular peripheral nerve graft in a rat sciatic nerve injury model. Neurol Res 33:600–608
148. Krekoski CA, Neubauer D, Zuo J, Muir D (2001) Axonal regeneration into acellular nerve grafts is enhanced by degradation of chondroitin sulfate proteoglycan. J Neurosci 21:6206–6213
149. Takami Y, Matsuda T, Yoshitake M, Hanumadass M, Walter RJ (1996) Dispase/detergent treated dermal matrix as a dermal substitute. Burns 22:182–190
150. Sutherland RS, Baskin LS, Hayward SW, Cunha GR (1996) Regeneration of bladder urothelium, smooth muscle, blood vessels and nerves into an acellular tissue matrix. J Urol 156:571–577
151. Bolland F, Korossis S, Wilshaw SP, Ingham E, Fisher J, Kearney JN, Southgate J (2007) Development and characterisation of a full-thickness acellular porcine bladder matrix for tissue engineering. Biomaterials 28:1061–1070
152. Parnigotto PP, Gamba PG, Conconi MT, Midrio P (2000) Experimental defect in rabbit urethra repaired with acellular aortic matrix. Urol Res 28:46–51
153. Pahari MP, Raman A, Bloomenthal A, Costa MA, Bradley SP, Banner B, Rastellini C, Cicalese L (2006) A novel approach for intestinal elongation using acellular dermal matrix: an experimental study in rats. Transplant Proc 38:1849–1850
154. Knight RL, Wilcox HE, Korossis SA, Fisher J, Ingham E (2008) The use of acellular matrices for the tissue engineering of cardiac valves. Proc Inst Mech Eng H 222:129–143
155. Conconi MT, Nico B, Mangieri D, Tommasini M, di Liddo R, Parnigotto PP, Nussdorfer GG, Ribatti D (2004) Angiogenic response induced by acellular aortic matrix in vivo. Anat Rec A Discov Mol Cell Evol Biol 281:1303–1307
156. Marzaro M, Conconi MT, Perin L, Giuliani S, Gamba P, De Coppi P, Perrino GP, Parnigotto PP, Nussdorfer GG (2002) Autologous satellite cell seeding improves in vivo bio-compatibility of homologous muscle acellular matrix implants. Int J Mol Med 10:177–182
157. Zhong H, Chen B, Lu S, Zhao M, Guo Y, Hou S (2007) Nerve regeneration and functional recovery after a sciatic nerve gap is repaired by an acellular nerve allograft made through chemical extraction in canines. J Reconstr Microsurg 23:479–487
158. Ribatti D, Conconi MT, Nico B, Baiguera S, Corsi P, Parnigotto PP, Nussdorfer GG (2003) Angiogenic response induced by acellular brain scaffolds grafted onto the chick embryo chorioallantoic membrane. Brain Res 989:9–15
159. Guo SZ, Ren XJ, Wu B, Jiang T (2010) Preparation of the acellular scaffold of the spinal cord and the study of biocompatibility. Spinal Cord 48:576–581
160. Zhang XY, Xue H, Liu JM, Chen D (2011) Chemically extracted acellular muscle: a new potential scaffold for spinal cord injury repair. J Biomed Mater Res A 100:578–587
161. Patist CM, Mulder MB, Gautier SE, Maquet V, Jerome R, Oudega M (2004) Freeze -dried poly(D, L-lactic acid) macroporous guidance scaffolds impregnated with brain-derived neurotrophic factor in the transected adult rat thoracic spinal cord. Biomaterials 25:1569–1582
162. Tuinstra HM, Aviles MO, Shin S, Holland SJ, Zelivyanskaya ML, Fast AG, Ko SY, Margul DJ, Bartels AK, Boehler RM, Cummings BJ, Anderson AJ, Shea LD (2011) Multifunctional, multichannel bridges that deliver neurotrophin encoding lentivirus for regeneration following spinal cord injury. Biomaterials 33:1618–1626
163. Gertz CC, Leach MK, Birrell LK, Martin DC, Feldman EL, Corey JM (2010) Accelerated neuritogenesis and maturation of primary spinal motor neurons in response to nanofibers. Dev Neurobiol 70:589–603
164. Corey JM, Lin DY, Mycek KB, Chen Q, Samuel S, Feldman EL, Martin DC (2007) Aligned electrospun nanofibers specify the direction of dorsal root ganglia neurite growth. J Biomed Mater Res A 83:636–645
165. Teng YD, Lavik EB, Qu X, Park KI, Ourednik J, Zurakowski D, Langer R, Snyder EY (2002) Functional recovery following traumatic spinal cord injury mediated by a unique polymer scaffold seeded with neural stem cells. Proc Natl Acad Sci USA 99:3024–3029

166. Xiong Y, Zeng YS, Zeng CG, Du BL, He LM, Quan DP, Zhang W, Wang JM, Wu JL, Li Y, Li J (2009) Synaptic transmission of neural stem cells seeded in 3 -dimensional PLGA scaffolds. Biomaterials 30:3711–3722
167. Du BL, Xiong Y, Zeng CG, He LM, Zhang W, Quan DP, Wu JL, Li Y, Zeng YS (2011) Transplantation of artificial neural construct partly improved spinal tissue repair and functional recovery in rats with spinal cord transection. Brain Res 1400:87–98
168. Kim MS, Ahn HH, Shin YN, Cho MH, Khang G, Lee HB (2007) An in vivo study of the host tissue response to subcutaneous implantation of PLGA- and/or porcine small intestinal submucosa-based scaffolds. Biomaterials 28:5137–5143
169. De Laporte L, Yan AL, Shea LD (2009) Local gene delivery from ECM-coated poly(lactide-co-glycolide) multiple channel bridges after spinal cord injury. Biomaterials 30:2361–2368
170. Wang YC, Wu YT, Huang HY, Lin HI, Lo LW, Tzeng SF, Yang CS (2008) Sustained intraspinal delivery of neurotrophic factor encapsulated in biodegradable nanoparticles following contusive spinal cord injury. Biomaterials 29:4546–4553
171. Bakshi A, Fisher O, Dagci T, Himes BT, Fischer I, Lowman A (2004) Mechanically engineered hydrogel scaffolds for axonal growth and angiogenesis after transplantation in spinal cord injury. J Neurosurg Spine 1:322–329
172. Hejcl A, Urdzikova L, Sedy J, Lesny P, Pradny M, Michalek J, Burian M, Hajek M, Zamecnik J, Jendelova P, Sykova E (2008) Acute and delayed implantation of positively charged 2-hydroxyethyl methacrylate scaffolds in spinal cord injury in the rat. J Neurosurg Spine 8:67–73
173. Hejcl A, Lesny P, Pradny M, Sedy J, Zamecnik J, Jendelova P, Michalek J, Sykova E (2009) Macroporous hydrogels based on 2-hydroxyethyl methacrylate. Part 6: 3D hydrogels with positive and negative surface charges and polyelectrolyte complexes in spinal cord in- jury repair. J Mater Sci Mater Med 20:1571–1577
174. Moore K, MacSween M, Shoichet M (2006) Immobilized concentration gradients of neurotrophic factors guide neurite outgrowth of primary neurons in macroporous scaffolds. Tissue Eng 12:267–278
175. Flynn L, Dalton PD, Shoichet MS (2003) Fiber templating of poly(2 -hydroxyethyl methacrylate) for neural tissue engineering. Biomaterials 24:4265–4272
176. Yu TT, Shoichet MS (2005) Guided cell adhesion and outgrowth in peptide-modified channels for neural tissue engineering. Biomaterials 26:1507–1514
177. Tsai EC, Dalton PD, Shoichet MS, Tator CH (2004) Synthetic hydrogel guidance channels facilitate regeneration of adult rat brainstem motor axons after complete spinal cord transection. J Neurotrauma 21:789–804
178. Woerly S, Doan VD, Sosa N, de Vellis J, Espinosa A (2001) Reconstruction of the transected cat spinal cord following NeuroGel implantation: axonal tracing, immunohistochemical and ultrastructural studies. Int J Dev Neurosci 19:63–83
179. Hejcl A, Sedy J, Kapcalova M, Toro DA, Amemori T, Lesny P, Likavcanova-Masinova K, Krumbholcova E, Pradny M, Michalek J, Burian M, Hajek M, Jendelova P, Sykova E (2010) HPMA-RGD hydrogels seeded with mesenchymal stem cells improve functional outcome in chronic spinal cord injury. Stem Cells Dev 19:1535–1546
180. Luo J, Borgens R, Shi R (2002) Polyethylene glycol immediately repairs neuronal membranes and inhibits free radical production after acute spinal cord injury. J Neurochem 83:471–480
181. Luo J, Shi R (2004) Diffusive oxidative stress following acute spinal cord injury in guinea pigs and its inhibition by polyethylene glycol. Neurosci Lett 359:167–170
182. Shi R, Borgens RB (2000) Anatomical repair of nerve membranes in crushed mammalian spinal cord with polyethylene glycol. J Neurocytol 29:633–643
183. Luo J, Shi R (2007) Polyethylene glycol inhibits apoptotic cell death following traumatic spinal cord injury. Brain Res 1155:10–16
184. Borgens RB, Shi R, Bohnert D (2002) Behavioral recovery from spinal cord injury following delayed application of polyethylene glycol. J Exp Biol 205:1–12
185. Brown MJ, Loew LM (1994) Electric Field-directed fibroblast locomotion involves cell-surface molecular reorganization and is calcium-independent. J Cell Biol 127:117–128

186. Li XF, Kolega J (2002) Effects of direct current electric fields on cell migration and actin filament distribution in bovine vascular endothelial cells. J Vasc Res 39:391–404
187. Ozawa H, Abe E, Shibasaki Y, Fukuhara T, Suda T (1989) Electric fields stimulate DNA synthesis of mouse osteoblast-like cells (Mc3t3-E1) by a mechanism involving calcium ions. J Cell Physiol 138:477–483
188. Kotwal A, Schmidt CE (2001) Electrical stimulation alters protein adsorption and nerve cell interactions with electrically conducting biomaterials. Biomaterials 22:1055–1064
189. Lee JY, Bashur CA, Goldstein AS, Schmidt CE (2009) Polypyrrole-coated electrospun PLGA nanofibers for neural tissue applications. Biomaterials 30:4325–4335
190. Shi GX, Rouabhia M, Wang ZX, Dao LH, Zhang Z (2004) A novel electrically conductive and biodegradable composite made of polypyrrole nanoparticles and polylactide. Biomaterials 25:2477–2488
191. Stauffer WR, Cui XT (2006) Polypyrrole doped with 2 peptide sequences from laminin. Biomaterials 27:2405–2413
192. Tabesh H, Amoabediny G, Nik NS, Heydari M, Yosefifard M, Siadat SO, Mottaghy K (2009) The role of biodegradable engineered scaffolds seeded with Schwann cells for spinal cord regeneration. Neurochem Int 54:73–83
193. Scott JB, Afshari M, Kotek R, Saul JM (2011) The promotion of axon extension in vitro using polymer-templated fibrin scaffolds. Biomaterials 32:4830–4839
194. Wong DY, Leveque JC, Brumblay H, Krebsbach PH, Hollister SJ, Lamarca F (2008) Macroarchitectures in spinal cord scaffold implants influence regeneration. J Neurotrauma 25:1027–1037
195. Cho YI, Choi JS, Jeong SY, Yoo HS (2010) Nerve growth factor (NGF)-conjugated electrospun nanostructures with topographical cues for neuronal differentiation of mesenchymal stem cells. Acta Biomater 6:4725–4733
196. Xie J, Willerth SM, Li X, Macewan MR, Rader A, Sakiyama-Elbert SE, Xia Y (2009) The differentiation of embryonic stem cells seeded on electrospun nanofibers into neural lineages. Biomaterials 30:354–362
197. Wang Y, Yao M, Zhou J, Zheng W, Zhou C, Dong D, Liu Y, Teng Z, Jiang Y, Wei G, Cui X (2011) The promotion of neural progenitor cells proliferation by aligned and randomly oriented collagen nanofibers through beta1 integrin/MAPK signaling pathway. Biomaterials 32:6737–6744
198. Georges PC, Miller WJ, Meaney DF, Sawyer ES, Janmey PA (2006) Matrices with compliance comparable to that of brain tissue select neuronal over glial growth in mixed cortical cultures. Biophys J 90:3012–3018
199. Engler AJ, Sen S, Sweeney HL, Discher DE (2006) Matrix elasticity directs stem cell lineage specification. Cell 126:677–689
200. Leipzig ND, Shoichet MS (2009) The effect of substrate stiffness on adult neural stem cell behavior. Biomaterials 30:6867–6878
201. Keselowsky BG, Collard DM, Garcia AJ (2005) Integrin binding specificity regulates biomaterial surface chemistry effects on cell differentiation. Proc Natl Acad Sci USA 102:5953–5957
202. Keselowsky BG, Collard DM, Garcia AJ (2004) Surface chemistry modulates focal adhesion composition and signaling through changes in integrin binding. Biomaterials 2(5):5947–5954
203. Ren YJ, Zhang H, Huang H, Wang XM, Zhou ZY, Cui FZ, An YH (2009) In vitro behavior of neural stem cells in response to different chemical functional groups. Biomaterials 30:1036–1044
204. Li GN, Hoffman-Kim D (2008) Tissue-engineered platforms of axon guidance. Tissue Eng Part B Rev 14:33–51
205. Yu LMY, Leipzig ND, Shoichet MS (2008) Promoting neuron adhesion and growth. Materials Today 11:36–43
206. Martins A, Araujo JV, Reis RL, Neves NM (2007) Electrospun nanostructure d scaffolds for tissue engineering applications. Nanomedicine (Lond) 2:929–942
207. Cao H, Liu T, Chew SY (2009) The application of nanofibrous scaffolds in neural tissue engineering. Adv Drug Deliv Rev 61:1055–1064

208. Prabhakaran MP, Venugopal JR, Chyan TT, Hai LB, Chan CK, Lim AY, Ramakrishna S (2008) Electrospun biocomposite nanofibrous scaffolds for neural tissue engineering. Tissue Eng Part A 14:1787–1797
209. Ghasemi-Mobarakeh L, Prabhakaran MP, Morshed M, MH N-E, Ramakrishna S (2008) Electrospun poly(epsilon-caprolactone)/gelatin nanofibrous scaffolds for nerve tissue engineering. Biomaterials 29:4532–4539
210. Schnell E, Klinkhammer K, Balzer S, Brook G, Klee D, Dalton P, Mey J (2007) Guidance of glial cell migration and axonal growth on electrospun nanofibers of poly-epsilon-caprolactone and a collagen/poly-epsilon-caprolactone blend. Biomaterials 28:3012–3025
211. Koh HS, Yong T, Chan CK, Ramakrishna S (2008) Enhancement of neurite outgrowth using nano-structured scaffolds coupled with laminin. Biomaterials 29:3574–3582
212. Yang F, Murugan R, Wang S, Ramakrishna S (2005) Electrospinning of nano/micro scale poly(L-lactic acid) aligned fibers and their potential in neural tissue engineering. Biomaterials 26:2603–2610
213. Wang HB, Mullins ME, Cregg JM, McCarthy CW, Gilbert RJ (2010) Varying the diameter of aligned electrospun fibers alters neurite outgrowth and Schwann cell migration. Acta Biomater 6:2970–2978
214. Kim YT, Haftel VK, Kumar S, Bellamkonda RV (2008) The role of aligned polymer fiber-based constructs in the bridging of long peripheral nerve gaps. Biomaterials 29:3117–3127
215. Stankus JJ, Guan J, Fujimoto K, Wagner WR (2006) Microintegrating smooth muscle cells into a biodegradable, elastomeric fiber matrix. Biomaterials 27:735–744
216. Zhang S (2003) Fabrication of novel biomaterials through molecular self-assembly. Nat Biotechnol 21:1171–1178
217. Zhao XB, Pan F, Xu H, Yaseen M, Shan HH, Hauser CAE, Zhang SG, Lu JR (2010) Molecular self-assembly and applications of designer peptide amphiphiles. Chem Soc Rev 39:3480–3498
218. Holmes TC, de Lacalle S, Su X, Liu G, Rich A, Zhang S (2000) Extensive neurite outgrowth and active synapse formation on self-assembling peptide scaffolds. Proc Natl Acad Sci USA 97:6728–6733
219. Semino CE, Kasahara J, Hayashi Y, Zhang SG (2004) Entrapment of migrating hippocampal neural cells in three-dimensional peptide nanofiber scaffold. Tissue Eng 10:643–655
220. Gelain F, Bottai D, Vescovi A, Zhang SG (2006) Designer self-assembling peptide nanofiber scaffolds for adult mouse neural stem cell 3-dimensional cultures. PLoS One 1(1):e119.
221. Silva GA, Czeisler C, Niece KL, Beniash E, Harrington DA, Kessler JA, Stupp SI (204) Selective differentiation of neural progenitor cells by high-epitope density nanofibers. Science 303:1352–1355
222. Tysseling-Mattiace VM, Sahni V, Niece KL, Birch D, Czeisler C, Fehlings MG, Stupp SI, Kessler JA (2008) Self-assembling nanofibers inhibit glial scar formation and promote axon elongation after spinal cord injury. J Neurosci 28:3814–3823
223. Shin H, Jo S, Mikos AG (2003) Biomimetic materials for tissue engineering. Biomaterials 24:4353–4364
224. Leong KF, Cheah CM, Chua CK (2003) Solid freeform fabrication of three -dimensional scaffolds for engineering replacement tissues and organs. Biomaterials 24:2363–2378
225. Zhu N, Li MG, Guan YJ, Schreyer DJ, Chen XB (2010) Effects of laminin blended with chitosan on axon guidance on patterned substrates. Biofabrication 2:045002–045009
226. Hutmacher DW, Sittinger M, Risbud MV (2004) Scaffold-based tissue engineering: rationale for computer-aided design and solid free-form fabrication systems. Trends Biotechnol 22:354–362
227. Sachlos E, Czernuszka JT (2003) Making tissue engineering scaffolds work. Review: the application of solid freeform fabrication technology to the production of tissue engineering scaffolds. Eur Cell Mater 5:29–39; discussion 39–40
228. Yeong WY, Chua CK, Leong KF, Chandrasekaran M (2004) Rapid prototyping in tissue engineering: challenges and potential. Trends Biotechnol 22:643–652
229. Maruo S, Ikuta K (2002) Submicron stereolithography for the production of freely movable mechanisms by using single-photon polymerization. Sens Actuators A Phys 100:70–76

230. Melchels FPW, Feijen J, Grijpma DW (2010) A review on stereolithography and its applications in biomedical engineering. Biomaterials 31:6121–6130
231. Lu Y, Mapili G, Suhali G, Chen SC, Roy K (2006) A digital micro -mirror device-based system for the microfabrication of complex, spatially patterned tissue engineering scaffolds. J Biomed Mater Res A 77A:396–405
232. Gauvin R, Chen YC, Lee JW, Soman P, Zorlutuna P, Nichol JW, Bae H, Chen S, Khademhosseini A (2012) Microfabrication of complex porous tissue engineering scaffolds using 3D projection stereolithography. Biomaterials 33:3824–3834
233. Landers R, Hubner U, Schmelzeisen R, Mulhaupt R (2002) Rapid prototyping of scaffolds derived from thermoreversible hydrogels and tailored for applications in tissue engineering. Biomaterials 23:4437–4447
234. Khalil S, Nam J, Sun W (2005) Multi-nozzle deposition for construction of 3D biopolymer tissue scaffolds. Rapid Prototyping J 11:9–17
235. Geng L, Feng W, Hutmacher DW, Wong YS, Loh HT, Fuh YH (2005) Direct writing of chitosan scaffolds using a robotic system. Rapid Prototyping J 11:90–97
236. Cohen DL, Malone E, Lipson H, Bonassar LJ (2006) Direct freeform fabrication of seeded hydrogels in arbitrary geometries. Tissue Eng 12:1325–1335
237. Silva NA, Salgado AJ, Sousa RA, Oliveira JT, Pedro AJ, Leite-Almeida H, Cerqueira R, Almeida A, Mastronardi F, Mano JF, Neves NM, Sousa N, Reis RL (2010) Development and characterization of a novel hybrid tissue engineering-based scaffold for spinal cord injury repair. Tissue Eng Part A 16:45–54
238. Chen XB, Li MG, Ke H (2008) Modeling of the flow rate in the dispensing-based process for fabricating tissue scaffolds. J Manuf Sci Eng 130:021003–021009
239. Li MG, Tian XY, Chen XB (2009) Modeling of flow rate, pore size, and porosity for the dispensing-based tissue scaffolds fabrication. J Manuf Sci Eng 131:034501–034505
240. Woodfield TB, Malda J, de Wijn J, Peters F, Riesle J, van Blitterswijk CA (2004) Design of porous scaffolds for cartilage tissue engineering using a three-dimensional fiber-deposition technique. Biomaterials 25:4149–4161
241. Chang R, Nam J, Sun W (2008) Effects of dispensing pressure and nozzle diameter on cell survival from solid freeform fabrication-based direct cell writing. Tissue Eng Part A 14: 41–48
242. Lu P, Yang H, Jones LL, Filbin MT, Tuszynski MH (2004) Combinatorial therapy with neurotrophins and cAMP promotes axonal regeneration beyond sites of spinal cord injury. J Neurosci 24:6402–6409
243. Pearse DD, Pereira FC, Marcillo AE, Bates ML, Berrocal YA, Filbin MT, Bunge MB (2004) cAMP and Schwann cells promote axonal growth and functional recovery after spinal cord injury. Nat Med 10:610–616
244. Houle JD, Tom VJ, Mayes D, Wagoner G, Phillips N, Silver J (2006) Combining an autologous peripheral nervous system "bridge" and matrix modification by chondroitinase allows robust, functional regeneration beyond a hemisection lesion of the adult rat spinal cord. J Neurosci 26:7405–7415
245. Kadoya K, Tsukada S, Lu P, Coppola G, Geschwind D, Filbin MT, Blesch A, Tuszynski MH (2009) Combined intrinsic and extrinsic neuronal mechanisms facilitate bridging axonal regeneration one year after spinal cord injury. Neuron 64:165–172
246. Blesch A, Lu P, Tuszynski MH (2002) Neurotrophic factors, gene therapy, and neural stem cells for spinal cord repair. Brain Res Bull 57:833–838
247. Cao QL, Zhang YP, Howard RM, Walters WM, Tsoulfas P, Whittemore SR (2001) Pluripotent stem cells engrafted into the normal or lesioned adult rat spinal cord are restricted to a glial lineage. Exp Neurol 167:48–58
248. Martino G, Pluchino S (2006) The therapeutic potential of neural stem cells. Nat Rev Neurosci 7:395–406
249. Kelly CM, Precious SV, Scherf C, Penketh R, Amso NN, Battersby A, Allen ND, Dunnett SB, Rosser AE (2009) Neonatal desensitization allows long-term survival of neural xenotransplants without immunosuppression. Nat Methods 6:271–273

250. Yan J, Xu L, Welsh AM, Chen D, Hazel T, Johe K, Koliatsos VE (2006) Combined immunosuppressive agents or CD4 antibodies prolong survival of human neural stem cell grafts and improve disease outcomes in amyotrophic lateral sclerosis transgenic mice. Stem Cells 24:1976–1985
251. Lu P, Jones LL, Snyder EY, Tuszynski MH (2003) Neural stem cells constitutively secrete neurotrophic factors and promote extensive host axonal growth after spinal cord injury. Exp Neurol 181:115–129
252. Guo JS, Zeng YS, Li HB, Huang WL, Liu RY, Li XB, Ding Y, Wu LZ, Cai DZ (2007) Cotransplant of neural stem cells and NT −3 gene modified Schwann cells promote the recovery of transected spinal cord injury. Spinal Cord 45:15–24
253. Wang JM, Zeng YS, Wu JL, Li Y, Teng YD (2011) Cograft of neural stem cells and Schwann cells overexpressing TrkC and neurotrophin-3 respectively after rat spinal cord transection. Biomaterials 32:7454–7468
254. Banerjee A, Arha M, Choudhary S, Ashton RS, Bhatia SR, Schaffer DV, Kane RS (2009) The influence of hydrogel modulus on the proliferation and differentiation of encapsulated neural stem cells. Biomaterials 30:4695–4699
255. Teixeira AI, Ilkhanizadeh S, Wigenius JA, Duckworth JK, Inganas O, Hermanson O (2009) The promotion of neuronal maturation on soft substrates. Biomaterials 30:4567–4572
256. Yu H, Cao B, Feng M, Zhou Q, Sun X, Wu S, Jin S, Liu H, Lianhong J (2010) Combinated transplantation of neural stem cells and collagen type I promote functional recovery after cerebral ischemia in rats. Anat Rec (Hoboken) 293:911–917
257. Edalat H, Hajebrahimi Z, Movahedin M, Tavallaei M, Amiri S, Mowla SJ (2011) p75NTR suppression in rat bone marrow stromal stem cells significantly reduced their rate of apoptosis during neural differentiation. Neurosci Lett 498:15–19
258. Fehlings MG, Vawda R (2011) Cellular treatments for spinal cord injury: the time is right for clinical trials. Neurotherapeutics 8:704–720
259. Scuteri A, Cassetti A, Tredici G (2006) Adult mesenchymal stem cells rescue dorsal root ganglia neurons from dying. Brain Res 1116:75–81
260. Crigler L, Robey RC, Asawachaicharn A, Gaupp D, Phinney DG (2006) Human mesenchymal stem cell subpopulations express a variety of neuro-regulatory molecules and promote neuronal cell survival and neuritogenesis. Exp Neurol 198:54–64
261. Wilkins A, Kemp K, Ginty M, Hares K, Mallam E, Scolding N (2009) Human bone marrow-derived mesenchymal stem cells secrete brain-derived neurotrophic factor which promotes neuronal survival in vitro. Stem Cell Res 3(1):63–70.
262. Akiyama Y, Radtke C, Honmou O, Kocsis JD (2002) Remyelination of the spinal cord following intravenous delivery of bone marrow cells. Glia 39:229–236
263. Uccelli A, Benvenuto F, Laroni A, Giunti D (2011) Neuroprotective features of mesenchymal stem cells. Best Pract Res Clin Haematol 24:59–64
264. Zurita M, Vaquero J, Bonilla C, Santos M, De Haro J, Oya S, Aguayo C (2008) Functional recovery of chronic paraplegic pigs after autologous transplantation of bone marrow stromal cells. Transplantation 86:845–853
265. Ohta M, Suzuki Y, Noda T, Ejiri Y, Dezawa M, Kataoka K, Chou H, Ishikawa N, Matsu-moto N, Iwashita Y, Mizuta E, Kuno S, Ide C (2004) Bone marrow stromal cells infused into the cerebrospinal fluid promote functional recovery of the injured rat spinal cord with reduced cavity formation. Exp Neurol 187:266–278
266. Shang AJ, Hong SQ, Xu Q, Wang HY, Yang Y, Wang ZF, Xu BN, Jiang XD, Xu RX (2011) NT-3-secreting human umbilical cord mesenchymal stromal cell transplantation for the treatment of acute spinal cord injury in rats. Brain Res 1391:102–113
267. Sykova E, Homola A, Mazanec R, Lachmann H, Konradova SL, Kobylka P, Padr R, Neuwirth J, Komrska V, Vavra V, Stulik J, Bojar M (2006) Autologous bone marrow transplantation in patients with subacute and chronic spinal cord injury. Cell Transplant 15:675–687
268. Cizkova D, Rosocha J, Vanicky I, Jergova S, Cizek M (2006) Transplants of human mesenchymal stem cells improve functional recovery after spinal cord injury in the rat. Cell Mol Neurobiol 26:1167–1180

269. Parr AM, Tator CH, Keating A (2007) Bone marrow-derived mesenchymal stromal cells for the repair of central nervous system injury. Bone Marrow Transplant 40:609–619
270. Chen X, Yang Y, Yao J, Lin W, Li Y, Chen Y, Gao Y, Gu X, Wang X (2011) Bone marrow stromal cells-loaded chitosan conduits promote repair of complete transection injury in rat spinal cord. J Mater Sci Mater Med 22:2347–2356
271. Lu D, Mahmood A, Qu C, Hong X, Kaplan D, Chopp M (2007) Collagen scaffolds populated with human marrow stromal cells reduce lesion volume and improve functional outcome after traumatic brain injury. Neurosurgery 61:596–602; discussion 602–593
272. Koda M, Kamada T, Hashimoto M, Murakami M, Shirasawa H, Sakao S, Ino H, Yoshinaga K, Koshizuka S, Moriya H, Yamazaki M (2007) Adenovirus vector-mediated ex vivo gene transfer of brain-derived neurotrophic factor to bone marrow stromal cells promotes axonal regeneration after transplantation in completely transected adult rat spinal cord. Eur Spine J 16:2206–2214
273. Doucette R (1991) PNS-CNS transitional zone of the first cranial nerve. J Comp Neurol 312:451–466
274. Boruch AV, Conners JJ, Pipitone M, Deadwyler G, Storer PD, Devries GH, Jones KJ (2001) Neurotrophic and migratory properties of an olfactory ensheathing cell line. Glia 33:225–229
275. Au E, Roskams AJ (2003) Olfactory ensheathing cells of the lamina propria in vivo and in vitro. Glia 41:224–236
276. Kocsis JD, Lankford KL, Sasaki M, Radtke C (2009) Unique in vivo properties of olfactory ensheathing cells that may contribute to neural repair and protection following spinal cord injury. Neurosci Lett 456:137–142
277. Steward O, Sharp K, Selvan G, Hadden A, Hofstadter M, Au E, Roskams J (2006) A re-assessment of the consequences of delayed transplantation of olfactory lamina propria following complete spinal cord transection in rats. Exp Neurol 198:483–499
278. Deumens R, Koopmans GC, Honig WM, Maquet V, Jerome R, Steinbusch HW, Joosten EA (2006) Chronically injured corticospinal axons do not cross large spinal lesion gaps after a multifactorial transplantation strategy using olfactory ensheathing cell/olfactory nerve fibroblast-biomatrix bridges. J Neurosci Res 83:811–820
279. Lu P, Yang H, Culbertson M, Graham L, Roskams AJ, Tuszynski MH (2006) Olfactory ensheathing cells do not exhibit unique migratory or axonal growth-promoting properties after spinal cord injury. J Neurosci 26:11120–11130
280. Tetzlaff W, Okon E, Karimi-Abdolrezaee S, Hill C, Sparling J, Plemel J, Plunet W, Tsai E, Baptiste D, Smithson L, Kawaja M, Fehlings M, Kwon B (2011) A systematic review of cellular transplantation therapies for spinal cord injury. J Neurotrauma 28:1611–1682
281. Huang H, Chen L, Wang H, Xiu B, Li B, Wang R, Zhang J, Zhang F, Gu Z, Li Y, Song Y, Hao W, Pang S, Sun J (2003) Influence of patients' age on functional recovery after trans – plantation of olfactory ensheathing cells into injured spinal cord injury. Chin Med J (Engl) 116:1488–1491
282. Lima C, Pratas-Vital J, Escada P, Hasse-Ferreira A, Capucho C, Peduzzi JD (2006) Olfactory mucosa autografts in human spinal cord injury: a pilot clinical study. J Spinal Cord Med 29:191–203; discussion 204–196
283. Mackay-Sim A, Feron F, Cochrane J, Bassingthwaighte L, Bayliss C, Davies W, Fronek P, Gray C, Kerr G, Licina P, Nowitzke A, Perry C, Silburn PA, Urquhart S, Geraghty T (2008) Autologous olfactory ensheathing cell transplantation in human paraplegia: a 3 -year clinical trial. Brain 131:2376–2386
284. Li BC, Jiao SS, Xu C, You H, Chen JM (2010) PLGA conduit seeded with olfactory ensheathing cells for bridging sciatic nerve defect of rats. J Biomed Mater Res A 94:769–780
285. Tang ZP, Liu N, Li ZW, Xie XW, Chen Y, Shi YH, Zeng WG, Wang SX, Chen J, Yang J, Pan DJ (2010) In vitro evaluation of the compatibility of a novel collagen-heparan sulfate biological scaffold with olfactory ensheathing cells. Chin Med J (Engl) 123:1299–1304
286. Shen Y, Qian Y, Zhang H, Zuo B, Lu Z, Fan Z, Zhang P, Zhang F, Zhou C (2010) Guidance of olfactory ensheathing cell growth and migration on electrospun silk fibroin scaffolds. Cell Transplant 19:147–157

287. Silva NA, Sousa RA, Pires AO, Sousa N, Salgado AJ, Reis RL (2011) Interactions between Schwann and olfactory ensheathing cells with a starch/polycaprolactone scaffold aimed at spinal cord injury repair. J Biomed Mater Res A 100A:470–476
288. Duncan ID, Aguayo AJ, Bunge RP, Wood PM (1981) Transplantation of rat Schwann cells grown in tissue culture into the mouse spinal cord. J Neurol Sci 49:241–252
289. Hill CE, Moon LD, Wood PM, Bunge MB (2006) Labeled Schwann cell transplantation: cell loss, host Schwann cell replacement, and strategies to enhance survival. Glia 53:338–343
290. Oudega M, Xu XM (2006) Schwann cell transplantation for repair of the adult spinal cord. J Neurotrauma 23:453–467
291. Xu XM, Guenard V, Kleitman N, Bunge MB (1995) Axonal regeneration into Schwann cell-seeded guidance channels grafted into transected adult rat spinal cord. J Comp Neurol 351:145–160
292. Xu XM, Guenard V, Kleitman N, Aebischer P, Bunge MB (1995) A combination of BDNF and NT-3 promotes supraspinal axonal regeneration into Schwann cell grafts in adult rat thoracic spinal cord. Exp Neurol 134:261–272
293. Bamber NI, Li H, Lu X, Oudega M, Aebischer P, Xu XM (2001) Neurotrophins BDNF and NT-3 promote axonal re-entry into the distal host spinal cord through Schwann cell-seeded mini-channels. Eur J Neurosci 13:257–268
294. Chen A, Xu XM, Kleitman N, Bunge MB (1996) Methylprednisolone administration improves axonal regeneration into Schwann cell grafts in transected adult rat thoracic spinal cord. Exp Neurol 138:261–276
295. Olson HE, Rooney GE, Gross L, Nesbitt JJ, Galvin KE, Knight A, Chen B, Yaszemski MJ, Windebank AJ (2009) Neural stem cell- and Schwann cell-loaded biodegradable polymer scaffolds support axonal regeneration in the transected spinal cord. Tissue Eng Part A 15:1797–1805
296. Hurtado A, Moon LD, Maquet V, Blits B, Jerome R, Oudega M (2006) Poly (D, L -lactic ac- id) macroporous guidance scaffolds seeded with Schwann cells genetically modified to secrete a bifunctional neurotrophin implanted in the completely transected adult rat thoracic spinal cord. Biomaterials 27:430–442
297. Novikova LN, Pettersson J, Brohlin M, Wiberg M, Novikov LN (2008) Biodegradable poly-beta-hydroxybutyrate scaffold seeded with Schwann cells to promote spinal cord repair. Biomaterials 29:1198–1206
298. He J, Zhou W, Zhou X, Zhong X, Zhang X, Wan P, Zhu B, Chen W (2008) The anatase phase of nanotopography titania plays an important role on osteoblast cell morphology and proliferation. J Mater Sci Mater Med 19:3465–3472
299. Guo J, Su H, Zeng Y, Liang YX, Wong WM, Ellis-Behnke RG, So KF, Wu W (2007) Reknitting the injured spinal cord by self-assembling peptide nanofiber scaffold. Nanomedicine 3:311–321
300. Afshari FT, Kwok JC, Fawcett JW (2010) Astrocyte-produced ephrins inhibit schwann cell migration via VAV2 signaling. J Neurosci 30:4246–4255
301. Bachelin C, Zujovic V, Buchet D, Mallet J, Baron-Van Evercooren A (2010) Ectopic expression of polysialylated neural cell adhesion molecule in adult macaque Schwann cells promotes their migration and remyelination potential in the central nervous system. Brain 133:406–420
302. Papastefanaki F, Chen J, Lavdas AA, Thomaidou D, Schachner M, Matsas R (2007) Grafts of Schwann cells engineered to express PSA-NCAM promote functional recovery after spinal cord injury. Brain 130:2159–2174
303. Iannotti C, Li H, Yan P, Lu X, Wirthlin L, Xu XM (2003) Glial cell line-derived neurotrophic factor-enriched bridging transplants promote propriospinal axonal regeneration and enhance myelination after spinal cord injury. Exp Neurol 183:379–393
304. Fouad K, Schnell L, Bunge MB, Schwab ME, Liebscher T, Pearse DD (2005) Combining Schwann cell bridges and olfactory-ensheathing glia grafts with chondroitinase promotes locomotor recovery after complete transection of the spinal cord. J Neurosci 25:1169–1178
305. Chau CH, Shum DK, Li H, Pei J, Lui YY, Wirthlin L, Chan YS, Xu XM (2004) Chondroitinase ABC enhances axonal regrowth through Schwann cell-seeded guidance channels after spinal cord injury. FASEB J 18:194–196

306. Reichardt LF, Tomaselli KJ (1991) Extracellular matrix molecules and their receptors: functions in neural development. Annu Rev Neurosci 14:531–570
307. Hashemi SM, Soudi S, Shabani I, Naderi M, Soleimani M (2011) The promotion of nanofibers immobilized with collagen for neural stem cells culture. J Mater Sci Mater Med 19:847–854
308. Li WS, Guo Y, Wang H, Shi DJ, Liang CF, Ye ZP, Qing F, Gong J (2008) Electrospun nanofibers immobilized with collagen for neural stem cells culture. Journal of Materials Science Materials in Medicine 19:847–854
309. Prabhakaran MP, Venugopal JR, Ramakrishna S (2009) Mesenchymal stem cell differentiation to neuronal cells on electrospun nanofibrous substrates for nerve tissue engineering. Biomaterials 30:4996–5003
310. Yuan Y, Dopheide SM, Ivanidis C, Salem HH, Jackson SP (1997) Calpain regulation of cytoskeletal signaling complexes in von Willebrand factor - stimulated platelets. Distinct roles for glycoprotein Ib-V-IX and glycoprotein IIb-IIIa (integrin alphaIIbbeta3) in von Willebrand factor-induced signal transduction. J Biol Chem 272:21847–21854
311. Yu X, Dillon GP, Bellamkonda RB (1999) A laminin and nerve growth factor-laden three-dimensional scaffold for enhanced neurite extension. Tissue Eng 5:291–304
312. He J, Wang XM, Spector M, Cui FZ (2012) Scaffolds for central nervous system tissue engineering. Frontiers of Materials Science 6:1–25
313. Suzuki M, Itoh S, Yamaguchi I, Takakuda K, Kobayashi H, Shinomiya K, Tanaka J (2003) Tendon chitosan tubes covalently coupled with synthesized laminin peptides facilitate nerve regeneration in vivo. J Neurosci Res 72:646–659
314. Cui FZ, Tian WM, Hou SP, Xu QY, Lee IS (2006) Hyaluronic acid hydrogel immobilized with RGD peptides for brain tissue engineering. J Mater Sci Mater Med 17:1393–1401
315. Dhoot NO, Tobias CA, Fischer I, Wheatley MA (2004) Peptide-modified alginate surfaces as a growth permissive substrate for neurite outgrowth. J Biomed Mater Res A 71:191–200
316. Cafferty WBJ, Duffy P, Huebner E, Strittmatter SM (2010) MAG and OMgp synergize with Nogo-A to restrict axonal growth and neurological recovery after spinal cord trauma. J Neurosci 30:6825–6837
317. Lee JK, Zheng B (2011) Role of myelin-associated inhibitors in axonal repair after spinal cord injury. Exp Neurol 235:33–42
318. Low K, Blesch A, Herrmann J, Tuszynski MH (2010) A dual promoter lentiviral vector for the in vivo evaluation of gene therapeutic approaches to axon regeneration after spinal cord injury. Gene Ther 17:577–591
319. Bo XN, Wu DS, Yeh J, Zhang Y (2011) Gene therapy approaches for neuroprotection and axonal regeneration after spinal cord and spinal root injury. Curr Gene Ther 11:101–115
320. Liu ML, Oh JS, An SS, Pennant WA, Kim HJ, Gwak SJ, Yoon do H, Kim KN, Lee M, Ha Y (2010) Controlled nonviral gene delivery and expression using stable neural stem cell line transfected with a hypoxia-inducible gene expression system. J Gene Med 12:990–1001
321. Choi BH, Ha Y, Ahn CH, Huang X, Kim JM, Park SR, Park H, Park HC, Kim SW, Lee M (2007) A hypoxia-inducible gene expression system using erythropoietin 3′ untranslated region for the gene therapy of rat spinal cord injury. Neurosci Lett 412:118–122
322. Lee M, Choi D, Choi MJ, Jeong JH, Kim WJ, Oh S, Kim YH, Bull DA, Kim SW (2006) Hypoxia-inducible gene expression system using the erythropoietin enhancer and 3′-untranslated region for the VEGF gene therapy. J Control Release 115:113–119
323. Ryu JK, Lee M, Choi MJ, Kim HA, Jin HR, Kim WJ, Yin GN, Song KM, Kwon MH, Suh JK (2012) Gene therapy with an erythropoietin enhancer-mediated hypoxia-inducible gene expression system in the corpus cavernosum of mice with high-cholesterol diet-induced erectile dysfunction. J Androl 33:845–853
324. Baraniak PR, Nelson DM, Leeson CE, Katakam AK, Friz JL, Cress DE, Hong Y, Guan J, Wagner WR (2011) Spatial control of gene expression within a scaffold by localized inducer release. Biomaterials 32:3062–3071

325. De Laporte L, Huang A, Ducommun MM, Zelivyanska ML, Aviles MO, Adler AF, Shea LD (2010) Patterned transgene expression in multiple-channel bridges after spinal cord injury. Acta Biomater 6:2889–2897
326. Houchin-Ray T, Swift LA, Jang JH, Shea LD (2007) Patterned PLG substrates for localized DNA delivery and directed neurite extension. Biomaterials 28:2603–2611
327. De Laporte L, Yang Y, Zelivyanskaya ML, Cummings BJ, Anderson AJ, Shea LD (2009) Plasmid releasing multiple channel bridges for transgene expression after spinal cord injury. Mol Ther 17:318–326
328. Aguayo AJ, Clarke DB, Jelsma TN, Kittlerova P, Friedman HC, Bray GM (1996) Effects of neurotrophins on the survival and regrowth of injured retinal neurons. Ciba Found Symp 196:135–144; discussion 144–138
329. Saito A, Tominaga T, Chan PH (2005) Neuroprotective role of neurotrophins: relationship between nerve growth factor and apoptotic cell survival pathway after cerebral ischemia. Curr Atheroscler Rep 7:268–273
330. Rosenberg SS, Ng BK, Chan JR (2006) The quest for remyelination: a new role for neurotrophins and their receptors. Brain Pathol 16:288–294
331. Zhou L, Baumgartner BJ, Hill-Felberg SJ, McGowen LR, Shine HD (2003) Neurotrophin-3 expressed in situ induces axonal plasticity in the adult injured spinal cord. J Neurosci 23:1424–1431
332. Madigan NN, McMahon S, O'Brien T, Yaszemski MJ, Windebank AJ (2009) Current tissue engineering and novel therapeutic approaches to axonal regeneration following spinal cord injury using polymer scaffolds. Respir Physiol Neurobiol 169:183–199
333. McCreedy DA, Sakiyama-Elbert SE (2012) Combination therapies in the CNS: engineering the environment. Neurosci Lett 519:115–121
334. Lu P, Tuszynski MH (2008) Growth factors and combinatorial therapies for CNS regeneration. Exp Neurol 209:313–320
335. Chen FM, Zhang M, Wu ZF (2010) Toward delivery of multiple growth factors in tissue engineering. Biomaterials 31:6279–6308
336. Benowitz LI, Yin Y (2007) Combinatorial treatments for promoting axon regeneration in the CNS: strategies for overcoming inhibitory signals and activating neurons' intrinsic growth state. Dev Neurobiol 67:1148–1165
337. Bradbury EJ, Carter LM (2011) Manipulating the glial scar: chondroitinase ABC as a therapy for spinal cord injury. Brain Res Bull 84:306–316
338. Hwang DH, Kim HM, Kang YM, Joo IS, Cho CS, Yoon BW, Kim SU, Kim BG (2011) Combination of multifaceted strategies to maximize the therapeutic benefits of neural stem cell transplantation for spinal cord repair. Cell Transplant 20:1361–1379
339. Hou S, Tian W, Xu Q, Cui F, Zhang J, Lu Q, Zhao C (2006) The enhancement of cell adherence and inducement of neurite outgrowth of dorsal root ganglia co-cultured with hyaluronic acid hydrogels modified with Nogo-66 receptor antagonist in vitro. Neuroscience 137: 519-529
340. Maier IC, Ichiyama RM, Courtine G, Schnell L, Lavrov I, Edgerton VR, Schwab ME (2009) Differential effects of anti-Nogo-A antibody treatment and treadmill training in rats with incomplete spinal cord injury. Brain 132:1426–1440

Chapter 22
Stem Cells in Ligament Tissue Engineering

Mukai Chimutengwende-Gordon and Wasim S. Khan

Abstract Injured ligaments have a poor capacity for healing due to their relative avascularity. Ligament reconstruction is well established for injuries such as anterior cruciate ligament rupture. However, the use of autografts and allografts for ligament reconstruction may be associated with a number of complications, and outcomes are variable. Ligament tissue engineering using stem cells is a novel technique that has the potential to provide an unlimited source of tissue. The process of tissue engineering involves the use of stem cells, growth factors, mechanical loading, a bioreactor, a biomimetic scaffold and gene therapy. In vitro and in vivo studies on ligament tissue engineering have shown some promising results; however, clinical research in this field is needed.

Keywords Ligament injury • Ligament reconstruction • Anterior cruciate ligament rupture • Ligament tissue engineering • Cell therapy

Abbreviations

ACL	Anterior cruciate ligament
BMPs	Bone morphogenic proteins
EGF	Epidermal growth factor

M. Chimutengwende-Gordon
Division of Surgery, University College of London,
London, UK

W.S. Khan, MBChB, MSc, MRCS, PhD (✉)
Division of Surgery, University College of London,
London, UK

Division of Surgery, Institute of Orthopaedic and Musculoskeletal Sciences,
Royal National Orthopaedic Hospital, University College London,
Stanmore, Middlesex HA7 4LP, UK
e-mail: wasimkhan@doctors.org.uk

M.K. Danquah, R.I. Mahato (eds.), *Emerging Trends in Cell and Gene Therapy*,
DOI 10.1007/978-1-62703-417-3_22,

FGFβ	Basic fibroblast growth factor
GDF	Growth and differentiation growth factor
IGF	Insulin-like growth factor
MCL	Medial collateral ligaments
MSCs	Mesenchymal stem cells
PCL	Polycaprolactone
PDGF	Platelet-derived growth factor
PGA	Polyglycolic acid
PLA	Polylactic acid
PLLA	Poly-L-lactic acid
TGFα	Transforming growth factor alpha
TGFβ	Transforming growth factor beta
VEGF	Vascular endothelial growth factor

22.1 Introduction

Ligament injuries account for a significant proportion of musculoskeletal injuries and result in disability and morbidity to patients worldwide [1]. Ligament injuries are commonly associated with sporting or overuse injuries [2]. For example, a tear or rupture of the anterior cruciate ligament (ACL) is one of the commonest sports injuries (particularly in football) [3]. Seventy percent of ACL tears occur as a result of repeatedly performed noncontact mechanisms such as sudden deceleration, landing and pivoting manoeuvres [3]. More than 200,000 ACL reconstructions are performed yearly in the United States, and the number being performed is increasing in frequency [4–6]. The cost of treating injuries to the cruciate ligaments is relatively high and has previously been estimated to be almost US $3,000 per patient [7]. The total expenditure on ACL reconstructions in a year has been estimated as exceeding $5 billion [8, 9].

Current treatment regimens for ligament injuries depend on the degree of injury and the patient's activity level, symptoms and effect on quality of life. There are three stages of ligament injury. Grade I injuries are mild sprains that are not associated with ligament laxity. Grade II injuries show moderately increased joint laxity. Grade III injuries are severe and associated with complete ligament disruption and significant laxity [10]. Treatment may consist of nonoperative management with pain relief and rehabilitation. However, operative management with autografts, allografts and synthetic grafts is often undertaken [11]. Ligaments are poorly vascularized and have a limited capacity for healing. When healing does occur the composition of the healed tissue is different to normal tissue and the biomechanical properties of the healed tissue are usually inferior [2]. Despite appropriate treatment, the ligament may not necessarily achieve its pre-injury characteristics or function and outcomes are variable. Additionally, the reconstructive surgery itself may be associated with disadvantages. Autografts may be associated with donor site morbidity. Allografts carry the risk of immunological reactions and infection. Synthetic grafts may be complicated by foreign body reactions [2].

Tissue engineering has a potentially very useful role in the specialty of orthopaedic surgery in general, as musculoskeletal tissues are often injured or lost in trauma and disease and may demonstrate limited healing potential [12]. Tissue engineering could be used to repair and regenerate tissue such as bone, cartilage, tendon as well as ligament. In vivo injection of appropriate cells into the injured ligament in conjunction with the use of biomimetic scaffolds and bioreactors is a strategy that could potentially accelerate the process of tissue repair [12].

This chapter discusses the characteristics of ligamentous tissue and approaches that are being developed to repair and regenerate ligament such as stem cell therapy, use of growth factors, gene therapy and mechanical stimulation.

22.2 Ligament Function Structure and Healing

Ligaments span a joint and connect one bone to another. Ligaments passively stabilize joints and help in guiding joints through their normal range of motion when a tensile load is applied. Ligaments also play a role in joint proprioception. When ligaments are strained they invoke neurological feedback signals that activate muscular contraction, and this appears to play a role in proprioception. Ligaments consist of dense bands of collagenous tissue. The surface of a ligament is often covered by an outer layer known as the epiligament. The epiligament merges into the periosteum of the bone around the attachment site of the ligament. Beneath the epiligament the ligament is organized into bundles of parallel fibres. The epiligament is more vascular and more cellular with more sensory and proprioceptive nerves than the underlying ligament [13].

Microscopically the ligament is composed of cells and an extracellular matrix. The cells are fibroblasts and account for approximately 20 % of the tissue. The extracellular matrix accounts for approximately 80 % of the tissue. The fibroblasts are responsible for synthesis of the matrix which consists of approximately 70 % water and 30 % collagen, ground substance and elastin. Type I collagen accounts for 85 % of the collagen in ligaments. Type I collagen has an enormous tensile strength enabling fibrils to be stretched without being broken. Less than 10 % of the collagen in ligaments is type III. This is more often found in healing tissues before most of it is converted to type I collagen. Very small amounts of collagen types VI, V, XI and XIV are present. The collagen accounts for 75 % of the dry weight. The remaining 25 % consists of proteoglycans, elastin and other proteins and glycoproteins such as actin, laminin and integrin [2, 13].

The collagen bundles are aligned along the long axis of the ligament and have a periodic change in direction along the length known as the crimp pattern. Crimp is thought to play a biomechanical role. It is likely that with increased loading, some areas of the ligament 'uncrimp' which allows the ligament to elongate without sustaining damage [13, 14].

As mentioned earlier, regeneration and healing of ligaments after injury is often poor due to their relatively avascular nature. Healing of ligaments can be divided into four stages. Firstly, there is a haemorrhagic stage in which the ligament ends retract and a blood clot forms and fills the gap. Cytokines are released within the clot and a heavily cellular infiltrate of polymorphonuclear leucocytes and lymphocytes appear within several hours [2, 10, 12, 13].

The second stage is the inflammatory stage in which macrophages appear by 24–48 h. By 72 h the wound also contains platelets and multipotential mesenchymal cells. Macrophages phagocytose necrotic tissues as well as secreting growth factors such as basic fibroblast growth factor (FGFβ), transforming growth factor alpha and beta (TGFα and TGFβ) and platelet-derived growth factor (PDGF). Platelets release PDGF, TGFβ and epidermal growth factor (EGF). These growth factors are chemotactic for fibroblasts and other cells, stimulate fibroblast proliferation and synthesize types I, III and V collagen and non-collagenous proteins. The growth factors also induce neovascularization and formation of granulation tissue [10, 13].

During the proliferative stage (stage 3), fibroblasts produce dense, cellular, collagenous connective tissue binding the torn ligament ends. This 'scar tissue' is initially disorganized. Capillary buds begin to form. After a few weeks, the collagen becomes quite well aligned with the long axis of the ligament. However, this tissue contains more type III collagen in relation to type I and more type V collagen. The collagen fibrils also have smaller diameters [10, 13].

The fourth stage consists of remodelling and maturation of the tissue. There is a gradual decrease in the cellularity of the tissue. Defects in the scar become filled in and the matrix becomes more dense and longitudinally orientated. The matrix begins to become more like normal ligament and continues to mature for at least a year. However, this tissue never achieves the morphological or mechanical characteristics of normal pre-injury ligament. There is a persistently decreased collagen fibril diameter and failure of collagen cross-links to mature as well as altered proteoglycan profiles (increased biglycan and decreased decorin protein and mRNA levels). There are also differences in the collagen types, altered cell connections, increased vascularity, abnormal innervation and increased cellularity and vascularity [10, 13].

During the remodelling stage, the viscolelastic properties recover to up to 20 % of normal. The tissue also has inferior creep properties (i.e. deformation properties under constant or cyclic loading). A rabbit model looking at healing of the medial collateral ligament demonstrated that ligament scars creep tissue as much as normal medial collateral ligaments (MCL) during cyclic and static loads that are only a fraction of the loads. Extensive creep could result in joint laxity. The resultant tissue has half the normal failure load and absorbs less energy before failing [10, 13].

22.3 Cell Sources for Ligament Tissue Engineering

Reparative cells could be recruited from host tissue through the specific attachment of tissue-engineered scaffolds. However, seeding cells could further improve the functionality of tissue-engineered constructs [15]. Cellular interaction between local tissue host cells and donor cells while extracellular matrix is being excreted may result in accelerated ligament healing. The seeded cells are involved in attracting reparative and or progenitor cells through chemotaxis signals. They also lay down extracellular matrix which results in initiation of further recruitment of reparative and/or progenitor cells. Additionally, they incorporate and release endogenous growth factors to elicit an immune response [15].

It is important to select the appropriate cell type for the specific application in order for the tissue-engineered product to have the best outcome. However, little is known about the optimal cell source for ligament tissue engineering. The cell type selected must show enhanced proliferation and production of an appropriate extracellular matrix and must be able to survive in an intraarticular environment in the patient's knee [16]. Mesenchymal stem cells (MSCs) have the ability to proliferate and differentiate into a variety of mesenchymal cell phenotypes including osteoblasts, chondroblasts, myoblasts and fibroblasts [12]. Culture conditions can be designed to direct MSC differentiation into the desired mesenchymal phenotype [9]. The potential use of mesenchymal stem cells to regenerate ligament tissue will be discussed in Sect. 22.4.

Primary fibroblasts derived from ligaments such as the ACL or MCL are another option. ACL fibroblasts can be harvested in diagnostic arthroscopic procedures after ACL rupture. As the MCL is extraarticular, it could be easily harvested partially without impairing its function in the long term [12, 15].

A study by Cooper et al. investigated the cellular response of primary rabbit connective tissue fibroblasts from four sources (Achilles tendon, patellar tendon, medial collateral ligament and anterior cruciate ligament) to a novel three-dimensional poly-L-lactic acid (PLLA)-braided scaffold for ACL tissue engineering. The fibroblasts from all four sources had similar morphological appearances in culture on tissue culture polystyrene. However, the cellular growth is different according to the cell source. They concluded that ACL fibroblasts were the most suited for ACL tissue engineering [17].

Bellincampi et al. investigated skin fibroblasts as a potential source for ligament tissue engineering as skin fibroblasts are known to have a greater healing potential and may be easily retrieved in a clinical setting. ACL and skin fibroblasts were harvested, cultured, labelled, seeded on collagen fibre scaffolds in vitro and implanted into the autogenous knee joint in a rabbit model. The cells remained viable for at least 4–6 weeks after implantation. They concluded that both skin and ACL fibroblasts survived in an intraarticular environment, but the potential of ACL fibroblasts to improve neoligament formation may be limited by a poor intrinsic healing capacity [18]. Tremblay et al. implanted a bioengineered ACL graft seeded with autologous living dermal fibroblasts into goat knee joints for 6 months. Histological and ultrastructural analysis demonstrated a highly organized ligamentous structure with vascularization, innervation and organized Sharpey's fibres and collagen at the osseous insertion sites of the grafts [19]. Morbidity associated with harvesting of the skin is a potential limitation of using skin fibroblasts as a source for ligament tissue engineering. Additionally, the performance of skin fibroblasts for ligament tissue engineering may be affected as the physiological environment of skin fibroblasts is different to that of ligaments [12, 15].

22.4 Mesenchymal Stem Cell Therapy

Although the use of primary fibroblasts for ligament tissue engineering is a logical approach, the use of stem cells may be more efficient. It has been shown in a rabbit model that MSCs have a significantly higher proliferation rate and collagen production than ACL and MCL fibroblasts and that MSCs could survive for at least 6 weeks in the knee joint [15]. Eijk et al. seeded bone marrow stromal cells, skin fibroblasts and ACL fibroblasts at different seeding densities onto braided poly (L-lactide/glycolide) scaffolds. The cells were cultured for up to 12 days. All cell types readily attached to the scaffold. On day 12, the scaffolds seeded MSCs showed the highest DNA content and collagen production. Scaffolds seeded with ACL fibroblasts showed the lowest DNA content and collagen production [16].

MSCs may differentiate into ligament fibroblasts after 2 weeks [12]. MSCs may be isolated from a variety of adult tissues including the bone marrow (obtained from aspiration of the iliac crest). Other potential sources of MSCs include adipose tissue (see Fig. 22.1), cord blood and possibly synovial fluid in ligament regeneration [21]. An alternative approach is the use of embryonic stem cells which are derived from

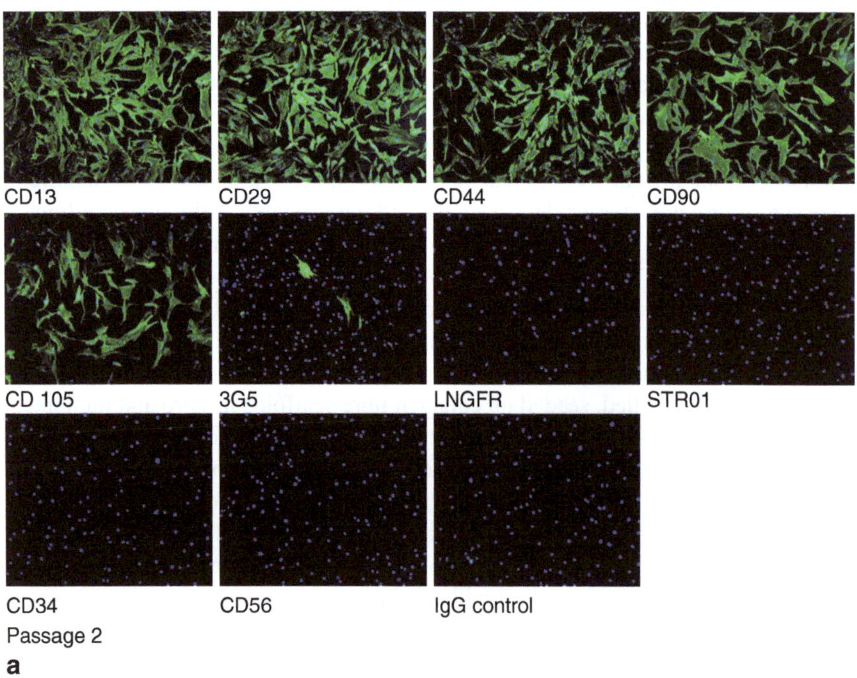

Fig. 22.1 Cell surface epitope characterization of passage 2 (**a**), passage 10 (**b**) and passage 18 (**c**) fat pad-derived MSCs using a panel of antibodies. Cell surface staining using FITC-conjugated secondary antibody (*green*) and DAPI (*blue*) shows that the cells stained strongly for CD13, CD29, CD44, CD90 and CD105 and poorly for LNGFR, STRO1, CD34 and CD56. Occasional cells stained positively for 3G5. No staining was observed for the IgG control [20]

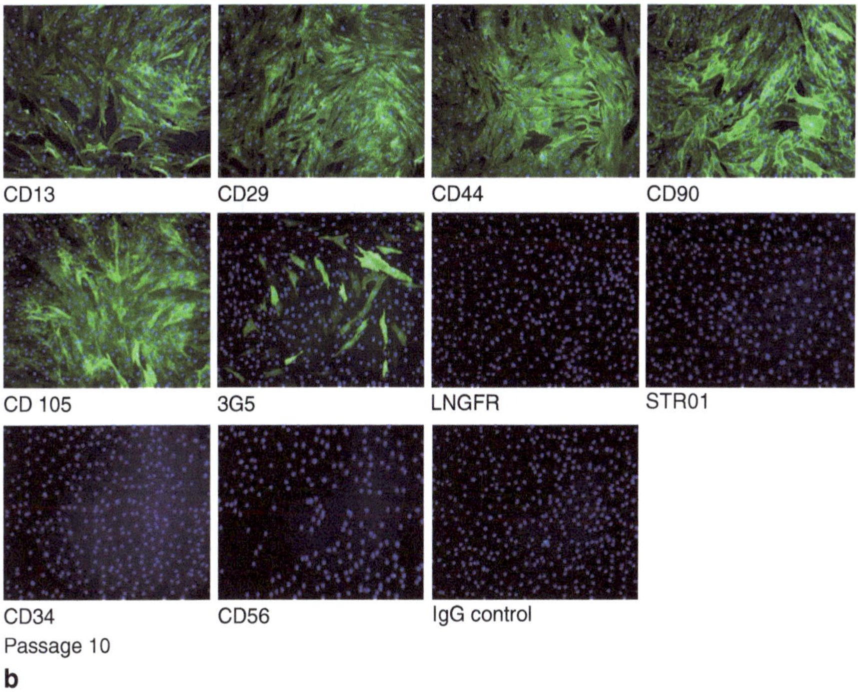

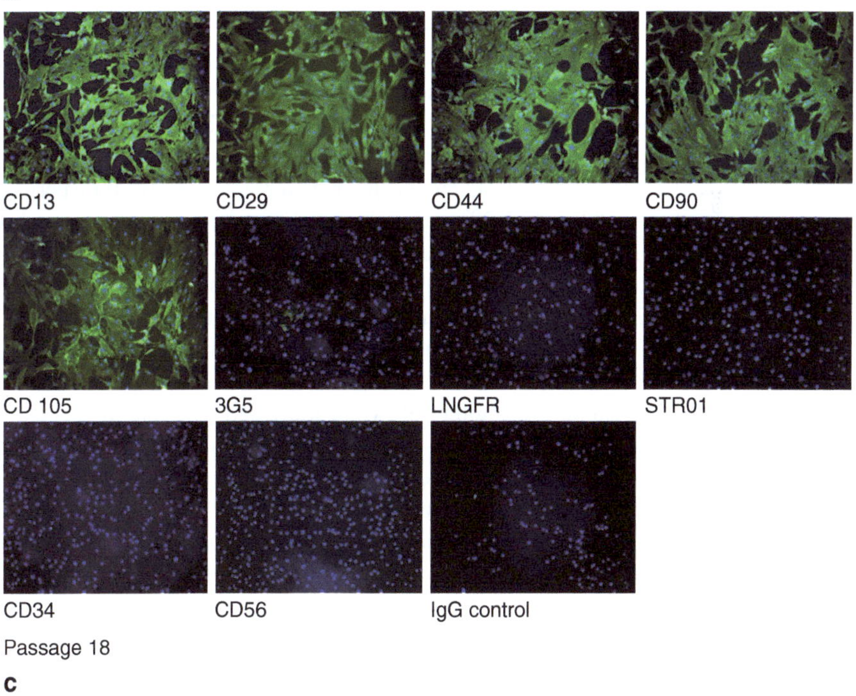

Fig. 22.1 (continued)

the inner cell mass of the blastocyst and are capable of unlimited undifferentiated proliferation and have been shown to differentiate into all types of somatic cells. However, the use of embryonic stem cells is associated with several disadvantages including technical difficulties, immunogenicity, tumour formation in vivo, uncertainty regarding the long-term outcome and ethical considerations [12, 22].

Adult mesenchymal stem cells have the advantage of possessing immunomodulatory properties. Although these immunomodulatory properties have not been fully explained, they make MSCs potential candidates for cellular therapy in an allogeneic setting. Transplantation of MSCs into an allogeneic host may not require immunosuppressive therapy. Adult MSCs express intermediate levels of class I major histocompatibility complex proteins but do not express human leucocyte antigen (class II) antigens on the cell surface [12, 23–25]. MSCs have been shown to have an indirect inhibitory effect on T-cells which is mediated by regulatory antigen-presenting cells with T-cell suppressive properties [24].

22.5 Bioreactor Systems

The differentiation of MSCs into fibroblasts may be accelerated by the use of a bioreactor which provides a controlled biomimetic optimum environment for cell functions. Bioreactors are a key component of tissue engineering [26]. They use various combinations of chemical, mechanical, electrical or magnetic stimulation to guide differentiation, proliferation and tissue development. In the case of ligament tissue engineering, a bioreactor may be used to accelerate the process of differentiation of MSCs into the fibroblastic lineage [12]. The body may be used as a bioreactor when a cell-scaffold composite is implanted directly into the injured site. Another approach is to culture the cell-scaffold composite in a bioreactor ex vivo for a period of time before transplantation [27, 28].

In order for a bioreactor to function successfully, there are several basic design principles that need to be fulfilled. Firstly, a bioreactor should maintain precise control of the physiological environment of the tissue culture, including control of variables such as temperature, oxygen concentrations, pH, nutrients, media flow rate, metabolite concentrations and specific tissue markers within close limits. Bioreactors should also be able to support the culture of two or more cell types simultaneously particularly when engineering complex tissues. It is also essential that the bioreactor is designed to operate under strict aseptic conditions in order to prevent any contamination of the tissues by influx of microorganisms [29].

Chemical stimulation techniques are employed by using chemicals such as growth factors. Growth factors are polypeptides that support various terminal phenotypes and regulate stem cell differentiation and proliferation. Examples of growth factors include TGFβ, bone morphogenic proteins (BMPs), fibroblast growth factors (FGFs), EGF, vascular endothelial growth factor (VEGF), PDGF, growth and differentiation growth factor (GDF) and insulin-like growth factor (IGF) [12, 27, 29].

Mechanical stimulation techniques involve subjecting a scaffold to mechanical stresses resembling the in vivo environment. It is used to induce differentiation of

MSCs into the fibroblast lineage. Intracellular signalling cascades are activated by triggering the cell surface stretch receptors leading to synthesis of the necessary extracellular matrix proteins [12]. The effects of mechanical stimulation are dependent on the magnitude, duration and frequency of mechanical stress [30]. Additionally, mechanical stimulation has been shown to affect extracellular matrix synthesis and remodelling. Enzyme activity and growth factor expression, collagen type I, collagen type III, elastin and tenascin-C expression in MSCs have been shown to be increased with the application of mechanical loads [28].

Coculture may also be used to induce differentiation of MSCs because of its ability to promote cell communications [12]. Direct coculture of MSCs with fibroblasts induces MSCs to differentiate into fibroblast-like cells [31]. Cell-to-cell interactions in the microenvironment play a key role in regulating the differentiation of MSCs in the healing process. Additionally, specific regulatory signals released from fibroblasts have been shown to support the selective differentiation of MSCs towards ligament fibroblasts in a two-dimensional transwell insert coculture system [30]. Fan et al. demonstrated that specific regulatory signals released from fibroblasts in a three-dimensional coculture can also enhance the differentiation of MSCs for ligament tissue engineering [32].

Electromagnetic stimulation has been shown to have positive results. For example, Fung et al. showed that low-energy laser therapy can enhance the mechanical strength of healing MCL in rats and increase collagen fibril size [33].

Although various commercial bioreactor systems are available, some may not be applicable to ligament tissue engineering as the design lacks the specificity to meet the requirements for engineering of ligament tissue [9]. Altman et al. designed a bioreactor to permit the controlled application of ligament-like multidimensional mechanical strains to undifferentiated cells embedded in a collagen gel. They used mechanical stimulation in vitro to induce the differentiation of mesenchymal progenitor cells from bone marrow into a ligament cell lineage in preference to bone or cartilage cell lineages [26, 34]. Kahn et al. designed a bioreactor for tissue engineering of ligament tissue that imposed mechanical conditions close to the physiological movement of the ACL. The bioreactor consisted of a mechanical part allowing movement to be applied on scaffolds, two culture chambers, a perfusion flow system to renew nutrients in the culture medium, a heating enclosure as well as an electronic component to manage movement and to regulate heating [35].

22.6 Scaffolds

Biomaterial scaffolds provide a structural and logistic template in which new tissue formation and remodelling can occur [9]. Scaffolds are designed to support cell attachment, survival, migration and differentiation as well as to control transport of nutrients, metabolites and regulatory molecules to and from the cells [22]. A scaffold should be made of a biocompatible, biodegradable material and should be able to bridge any complex three-dimensional anatomical defect. This may be achieved using surgical experience or through sophisticated computer mapping systems [12].

The scaffold should ideally possess adequate strength post implantation to be effective as a load-bearing construct and degrade at a rate matching the rate of new tissue deposition. The scaffold should also have sufficient void volume for cell infiltration and extracellular matrix to promote gradual load transfer from the scaffold to the neotissue [36]. Porous scaffolds enhance tissue regeneration by delivering biofactors. However, pores that are too large would compromise the mechanical properties of the scaffold [12]. Currently all materials used in ligament tissue engineering are polymers [37]. Polymers may be naturally derived, e.g. gelatin, small intestine submucosal extracellular matrix or silk. Synthetic polymers include polyesters such as polyglycolic acid.

Collagen used in laboratories is usually derived from the bovine submucosa and intestine from rats tails in small quantities. The derived collagen requires processing to remove foreign antigens, to improve its mechanical strength and sometimes to slow down the degradation rate by cross-linking. The predominant chemical cross-linking agents used in research are glutaraldehyde, formaldehyde, polyepoxy compounds, acyl azide, carbodiimides and hexamethylene diisocyanate. Potential toxic residues are a disadvantage. Physical methods include drying, heating or exposure to ultraviolet or gamma radiation [37]. Fibroblasts have been shown to attach, proliferate and secrete new collagen when seeded on collagen fibre scaffolds [38]. In vivo, it has been demonstrated that fibroblast-seeded collagen scaffolds may remain viable after implantation into the knee joint for prolonged periods [18]. Examples of commercially available biological collagen-based scaffolds include Restore (derived from porcine small intestine), GraftJacket (from human cadaver dermis), Permacol (from porcine dermis) and Bio-Blanket (from bovine dermis) [39]. Advantages of collagen include the ability to alter resorption rate and mechanical properties of scaffolds through cross-linking and low antigenicity. The scaffolds experience an early decrease in mechanical strength followed by tissue remodelling between by 20 weeks resulting in a strength gain similar to autografts [14].

Silk has the advantage of possessing good biocompatibility, slow biodegradability and excellent tensile strength and toughness [9, 28]. Silk fibroin is a protein excreted by silkworms and isolated from sericin [28]. Silk fibroin has similar mechanical properties to functional ACL when organized into an appropriate wire-rope geometry. Silk scaffolds also support cell attachment and spreading by providing an appropriate three-dimensional culture environment. Silk fibres lose the majority of their tensile strength within 1 year in vivo and fail to be recognized in 2 years [37]. Silk-fibre matrices have been shown to support adult stem cell differentiation towards ligament lineages [40]. A composite scaffold fabricated from silk and collagen tested in a rabbit MCL defect model was shown to improve structural and functional ligament repair by regulating ligament matrix gene expression and collagen fibril assembly [41].

Synthetic polymers that have been investigated for ligament repair include polyglycolic acid (PGA), polylactic acid (PLA), their copolymers and polycaprolactone (PCL). PLA is a commonly used synthetic scaffold which easily degrades within the human body by forming lactic acid. PCL and PGA degrade in a similar way to PLA but exhibit different rates of degradation. An advantage of using a synthetic polymer is that there is no limit to the supply of grafts and no risk of disease. These polymers are designed to degrade over time. Their mechanical properties may be controlled by altering the degree of polymer crystallinity, changing the polymer molecular weight or changing the ratio of each polymer in the copolymer [12, 14].

22.7 Gene Transfer Technology

Gene transfer technology may be used to sustain sufficient quantities of growth factor within the local tissue [12]. Gene transfer is a method to deliver genetic material and information to cells to alter their synthesis or function. Genes can be introduced into cells using retroviral and adenoviral vectors as carriers, liposomes or with a gene gun. The genes can be placed in the cell outside ex vivo or in vivo. The target cells can be made to produce or increase expression of growth factors or suppress the synthesis of endogenous proteins [10]. Wei et al. surgically implanted bone marrow-derived MSCs transfected with adenovirus vector encoding TGF-β1, VEGF or TGF-β1/VEGF into experimental ACL grafts in rabbits. They found that this significantly promoted angiogenesis compared to non-transfected control cells. The best mechanical properties were achieved at 24 weeks [42]. Hildebrand et al. used a retroviral ex vivo and an adenoviral in vivo technique to introduce and express the LacZ marker gene in the MCL and ACL of rabbits. LacZ gene expression was detected and shown to last between 10 days and 3 weeks in the MCL and ACL with the use of the retrovirus and between 3 and 6 weeks in the MCL and at least 6 weeks in the ACL with the adenoviruses [43]. Menetrey et al. showed the feasibility of gene transfer to a normal ACL using direct, fibroblast-mediated and myoblast-mediated approaches. Adenoviral particles were directly injected into the ACL of rabbits. Rabbit myoblasts and ACL fibroblasts were transduced with recombinant adenoviral particles carrying the LacZ reporter gene, and these were also injected into the ACL of rabbits. The persistence of gene expression lasting up to 6 weeks was observed for the direct and myoblast-mediated gene transfers. Fibroblast-mediated gene transfer showed low efficiency with gene expression persisting for 1 week in the ligament and 2 weeks in the synovial tissue surrounding the ligament. Only a few cells located in the synovium were positive for the marker gene at 3 weeks post injection [44]. A number of other studies have indicated that using gene therapy to improve ligament healing is a promising approach [28, 45–47].

22.8 Conclusion

Ligament injuries may be challenging to treat. Results of ligament reconstruction with grafts are variable. Considerable progress has been made in generating tissue-engineered ligaments. Important areas for future development include improving the biomechanical properties of tissue-engineered ligaments, improving the characteristics of scaffold materials and increasing the strength of ligament-bone junctions of implanted engineered ligament. Studies on the generation of tissue-engineered ligaments have generally been in vitro preliminary studies or trials in animal models. In the future, large clinical trials, in particular randomized controlled trials, assessing tissue-engineered ligaments should be performed. The use of tissue-engineered ligaments would potentially have significant health-care implications. In view of the ageing population, the number of patients who will benefit from the use of tissue-engineered ligaments is likely to increase with time.

References

1. Al-Rashid M, Khan WS (2011) Stem cells and ligament repair. In: Berhardt L (ed) Advances in medicine and biology. Nova Science Publishers, Inc, Hauppauge, pp 343–347
2. Pastides P, Khan W (2011) Tendon and ligament injuries: the evolving role of stem cells and tissue engineering. Br J Med Med Res 1(4):569–580
3. Papoutsidakis A (2011) Predisposing factors for anterior cruciate ligament injury. Br J Sports Med 45(2):e2
4. Spindler KP, Wright RW (2008) Clinical practice. Anterior cruciate ligament tear. N Engl J Med 359(20):2135–2142
5. Lyman S, Koulouvaris P, Sherman S, Do H, Mandl LA, Marx RG (2009) Epidemiology of anterior cruciate ligament reconstruction: trends, readmissions, and subsequent knee surgery. J Bone Joint Surg Am 91(10):2321–2328
6. Shelton WR, Fagan BC (2011) Autografts commonly used in anterior cruciate ligament reconstruction. J Am Acad Orthop Surg 19(5):259–264
7. de Loes M, Dahlstedt LJ, Thomee R (2000) A 7-year study on risks and costs of knee injuries in male and female youth participants in 12 sports. Scand J Med Sci Sports 10(2):90–97
8. Pennisi E (2002) Tending tender tendons. Science 295(5557):1011
9. Vunjak-Novakovic G, Altman G, Horan R, Kaplan DL (2004) Tissue engineering of ligaments. Annu Rev Biomed Eng 6:131–156
10. Woo SL, Hildebrand K, Watanabe N, Fenwick JA, Papageorgiou CD, Wang JH (1999) Tissue engineering of ligament and tendon healing. Clin Orthop Relat Res 367(Suppl):S312–S323
11. Hoffman A, Gross G (2006) Tendon and ligament engineering: from cell biology to in vivo application. Regen Med 1(4):563–574
12. Yates EW, Rupani A, Foley GT, Khan WS, Cartmell S, Anand SJ (2012) Ligament tissue engineering and its potential role in anterior cruciate ligament reconstruction. Stem Cells Int 2012:438125
13. Frank C (2004) Ligament structure, physiology and function. J Musculoskelet Neuronal Interact 4(2):199–201
14. Freeman J, Kwansa A (2008) Recent advancements in ligament tissue engineering: the use of various techniques and materials for ACL repair. Recent Patents Biomed Eng 1:18–23
15. Ge Z, Goh JC, Lee EH (2005) Selection of cell source for ligament tissue engineering. Cell Transplant 14(8):573–583
16. Van Eijk F, Saris DB, Riesle J, Willems WJ, Van Blitterswijk CA, Verbout AJ, Dhert WJ (2004) Tissue engineering of ligaments: a comparison of bone marrow stromal cells, anterior cruciate ligament, and skin fibroblasts as cell source. Tissue Eng 10(5–6):893–903
17. Cooper JA Jr, Bailey LO, Carter JN, Castiglioni CE, Kofron MD, Ko FK, Laurencin CT (2006) Evaluation of the anterior cruciate ligament, medial collateral ligament, Achilles tendon and patellar tendon as cell sources for tissue-engineered ligament. Biomaterials 27(13):2747–2754
18. Bellincampi LD, Closkey RF, Prasad R, Zawadsky JP, Dunn MG (1998) Viability of fibroblast-seeded ligament analogs after autogenous implantation. J Orthop Res 16(4):414–420

19. Tremblay P et al (2011) Potential of skin fibroblasts for application to anterior cruciate ligament tissue engineering. Cell Transplant 20(4):535–542
20. Khan WS, Adesida AB, Tew SR, Longo UG, Hardingham TE (2012) Fat pad-derived mesenchymal stem cells as a potential source for cell-based adipose tissue repair strategies. Cell Prolif 45(2):111–120
21. McGonagle D, Jones E (2008) A potential role for synovial fluid mesenchymal stem cells in ligament regeneration. Rheumatology (Oxford) 47(8):1114–1116
22. Chimutengwende-Gordon M, Khan WS (2012) Advances in the use of stem cells and tissue engineering applications in bone repair. Curr Stem Cell Res Ther 7(2):122–126
23. Le Blanc K, Ringden O (2005) Immunobiology of human mesenchymal stem cells and future use in hematopoietic stem cell transplantation. Biol Blood Marrow Transplant 11(5):321–334
24. Beyth S, Borovsky Z, Mevorach D, Liebergall M, Gazit Z, Aslan H, Galun E, Rachmilewitz J (2005) Human mesenchymal stem cells alter antigen-presenting cell maturation and induce T-cell unresponsiveness. Blood 105(5):2214–2219
25. Chamberlain G, Fox J, Ashton B, Middleton J (2007) Concise review: mesenchymal stem cells: their phenotype, differentiation capacity, immunological features, and potential for homing. Stem Cells 25(11):2739–2749
26. Altman G, Lu H, Horan R, Calabro T, Ryder D, Kaplan D (2002) Advanced bioreactor with controlled application of multi-dimensional strain for tissue engineering. J Biomech Eng 124:742–749
27. Goh JC, Ouyang HW, Teoh SH, Chan CK, Lee EH (2003) Tissue-engineering approach to the repair and regeneration of tendons and ligaments. Tissue Eng 9(Suppl 1):S31–S44
28. Yilgor C, Yilgor HP, Huri G (2012) Tissue engineering strategies in ligament regeneration. Stem Cells Int 2012:374676
29. Oragui E, Nannaparaju M, Khan WS (2011) The role of bioreactors in tissue engineering for musculoskeletal applications. Open Orthop J 5(Suppl 2):267–270
30. Lee IC, Wang JH, Lee YT, Young TH (2007) The differentiation of mesenchymal stem cells by mechanical stress or/and co-culture system. Biochem Biophys Res Commun 352(1):147–152
31. Ball SG, Shuttleworth AC, Kielty CM (2004) Direct cell contact influences bone marrow mesenchymal stem cell fate. Int J Biochem Cell Biol 36(4):714–727
32. Fan H, Liu H, Toh SL, Goh JC (2008) Enhanced differentiation of mesenchymal stem cells co-cultured with ligament fibroblasts on gelatin/silk fibroin hybrid scaffold. Biomaterials 29(8):1017–1027
33. Fung DT, Ng GY, Leung MC, Tay DK (2003) Effects of a therapeutic laser on the ultrastructural morphology of repairing medial collateral ligament in a rat model. Lasers Surg Med 32(4):286–293
34. Altman GH, Horan RL, Martin I, Farhadi J, Stark PR, Volloch V, Richmond JC, Vunjak-Novakovic G, Kaplan DL (2002) Cell differentiation by mechanical stress. FASEB J 16(2):270–272
35. Kahn CJ, Vaquette C, Rahouadj R, Wang X (2008) A novel bioreactor for ligament tissue engineering. Biomed Mater Eng 18(4–5):283–287
36. Teh TK, Toh SL, Goh JC (2011) Aligned hybrid silk scaffold for enhanced differentiation of mesenchymal stem cells into ligament fibroblasts. Tissue Eng Part C Methods 17(6):687–703
37. Ge Z, Yang F, Goh JC, Ramakrishna S, Lee EH (2006) Biomaterials and scaffolds for ligament tissue engineering. J Biomed Mater Res A 77(3):639–652
38. Dunn MG, Liesch JB, Tiku ML, Zawadsky JP (1995) Development of fibroblast-seeded ligament analogs for ACL reconstruction. J Biomed Mater Res 29(11):1363–1371
39. Chen J, Xu J, Wang A, Zheng M (2009) Scaffolds for tendon and ligament repair: review of the efficacy of commercial products. Expert Rev Med Devices 6(1):61–73
40. Altman GH, Horan RL, Lu HH, Moreau J, Martin I, Richmond JC, Kaplan DL (2002) Silk matrix for tissue engineered anterior cruciate ligaments. Biomaterials 23(20):4131–4141
41. Chen X, Qi YY, Wang LL, Yin Z, Yin GL, Zou XH, Ouyang HW (2008) Ligament regeneration using a knitted silk scaffold combined with collagen matrix. Biomaterials 29(27):3683–3692

42. Wei X, Mao Z, Hou Y, Lin L, Xue T, Chen L, Wang H, Yu C (2011) Local administration of TGFbeta-1/VEGF165 gene-transduced bone mesenchymal stem cells for Achilles allograft replacement of the anterior cruciate ligament in rabbits. Biochem Biophys Res Commun 406(2):204–210
43. Hildebrand KA, Deie M, Allen CR, Smith DW, Georgescu HI, Evans CH, Robbins PD, Woo SL (1999) Early expression of marker genes in the rabbit medial collateral and anterior cruciate ligaments: the use of different viral vectors and the effects of injury. J Orthop Res 17(1):37–42
44. Menetrey J, Kasemkijwattana C, Day CS, Bosch P, Fu FH, Moreland MS, Huard J (1999) Direct-, fibroblast- and myoblast-mediated gene transfer to the anterior cruciate ligament. Tissue Eng 5(5):435–442
45. Lou J, Tu Y, Burns M, Silva MJ, Manske P (2001) BMP-12 gene transfer augmentation of lacerated tendon repair. J Orthop Res 19(6):1199–1202
46. Pascher A et al (2004) Enhanced repair of the anterior cruciate ligament by in situ gene transfer: evaluation in an in vitro model. Mol Ther 10(2):327–336
47. Steinert AF, Weber M, Kunz M, Palmer GD, Noth U, Evans CH, Murray MM (2008) In situ IGF-1 gene delivery to cells emerging from the injured anterior cruciate ligament. Biomaterials 29(7):904–916

Chapter 23
Biomaterials for Bone Tissue Engineering

Qizhi Chen

Abstract Biomaterials play a critical role in bone engineering, working as an artificial extracellular matrix to support regeneration. From the materials science point of view, natural bone is ceramic-polymer composite. It is not surprising that huge efforts have been invested into the development of bioceramics and composites that mimic that of native bone. This chapter provides a comprehensive review on the biomaterials used in bone tissue engineering, including bioceramics, polymers and composites. The rational of bone tissue engineering is briefly introduced first. This is followed by systematic review of bioceramic (e.g. calcium phosphates, hydroxyapatite and bioactive glasses), biomedical polymers (e.g. polylactic acid, polyglycolic acid and their copolymers) and polymer-based ceramic-filled composites. Each section includes discussions of the material's biocompatibility and biodegradability and two essential features of biomaterials in most tissue engineering applications, followed by a detailed description of its mechanical properties. Finally, the major achievements and remaining challenges for biomaterials used in bone tissue engineering are summarised.

Keywords Biomaterials • Bone tissue engineering • Bioceramics • Biomedical polymers

Q. Chen
Department of Materials Engineering,
Monash University,
Clayton Campus, Room 151, Building 19,
Clayton, VIC 3800, Australia
e-mail: qizhi.chen@monash.edu

M.K. Danquah, R.I. Mahato (eds.), *Emerging Trends in Cell and Gene Therapy*,
DOI 10.1007/978-1-62703-417-3_23, © Springer Science+Business Media New York 2013

23.1 Bone Tissue Engineering

Tissue engineering has emerged as a distinct scientific field from the historical evolution of medicine. In this revolutionary development, the fundamental of health care in the conventional practice of medicine remains with us in tissue engineering, that is, the body heals itself [1]. This principle is rooted in the truth that the organs of the body have an ability to regenerate and recover when they are diseased or injured. The regeneration, however, can only occur within limits such that recovery is almost impossible in many cases without medical manipulation. In conventional medical treatment, surgeons support a patient's vital functions by optimising the environment most conductive to healing, and physicians attempt to neutralise hostile factors and at the same time enhance the supply of oxygen and nutrients that the body needs for the healing process. Surgeons eliminate hostile factors through excising the necrotic or malign tissue that is the source of unfavourable chemical agents, reconstruct tissue through the suture of the remaining tissue, auto-/allo-/xenotransplantation or implantation of prosthesis and manipulate the local environment to help the body heal itself by, for example, medication and blood supply [2].

In tissue engineering, organs being transplanted or prosthesis being implanted, living cells are harvested and expanded in vitro, and a designed scaffold is used to dictate the regeneration of the shape and function of the desired tissue by providing structural cues. Then the scaffold which is cultured with sufficient cells is implanted, and the tissue engineers and surgeons manipulate the local environment. Under ideal conditions, this will then enable the body to heal itself. It is when the attention of medical treatment focused on the regeneration of living tissues for the body in the laboratory (i.e. ex vivo) that the reconstructive surgery came to be called tissue engineering [2]. In summary, tissue engineering induces the regeneration ability of the host body through a designed scaffold that is populated with cells and signalling molecules, aiming at regenerating functional tissue as an alternative to conventional organ transplantation and tissue reconstruction.

The above definition of tissue engineering is a specific concept. Tissue engineering has been generally defined as the application of principles and methods of engineering and life sciences to obtain a fundamental understanding of structure-function relationships in normal and pathological mammalian tissue and the development of biological substitutes to restore, maintain or improve tissue function [3]. Other similar definitions exist. In 1993, Langer and Vacanti [4] defined tissue engineering as an interdisciplinary field that applies the principles of engineering and life sciences toward the development of biological substitutes that restore, maintain or improve tissue function. In 1995, Galletti et al. [5] defined tissue engineering as the basic science and development of biological substitutes for implantation into the body or the fostering of tissue remodelling for the purpose of replacing, repeating, regenerating, reconstructing or enhancing biological function.

There are several clinical reasons to develop bone tissue engineering. Although transplantation and implantation are standard methods in the conventional clinical treatment, shortcomings are encountered with their usage [6]. Firstly, the application of bone grafts is limited by the size of the defect and the viability of the host body.

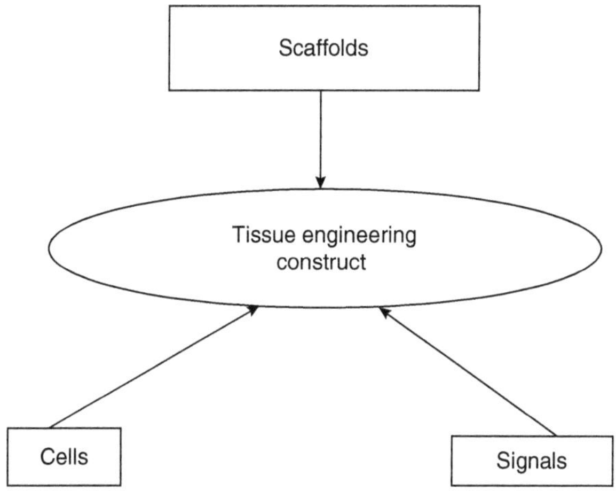

Fig. 23.1 The tissue engineering triad [2]

There can be significant donor site morbidity in autografting and donor tissue scarcity for allografting. Allografting also introduces the risk of disease and infection. Secondly, the revolution of implantation, which has led to a remarkable increase in the quality of life for millions of patients in the last 30 years, has run the course. The orthopaedic prostheses have an excellent 15 years survivability of 75–85 %. However, there is a requirement of longer than 30 years survivability by the aging population [7–9].

Tissue engineering will ultimately have a more profound impact on our life than we can now appreciate. Its technical significance lies in that this treatment will address the transplantation crisis caused by donor scarcity, immune rejection and pathogen transfer [10]. This revolution will also reach a goal of more than 30 years implant survivability.

23.2 Tissue Engineering Approaches

The approaches of tissue engineering are established on the fact that living bodies have the potential of regeneration and on the supposition that the employment of natural biology (e.g. cells and biomolecules) of the living body will maximise the capacity for regeneration and allow for greater success in developing therapeutic strategies aimed at the replacement, repair, maintenance and enhancement of tissue function [11].

In essence, tissue engineering is a technique of imitating nature. Natural tissues consist of three components: cells, extracellular matrix (ECM) and signalling systems. The ECM is made up of a complex of cell secretions immobilised in spaces, thus forming a scaffold for its cells. Hence, it is natural that the engineered tissue construct is a triad [2], the three constitutes of which correspond to the abovementioned three basic components of natural tissues. Figure 23.1 illustrates the triad, that is, living cells, scaffolds and signal molecules.

New functional living tissue is generated by living cells in the triad system, like in a natural biological system. But the regeneration in the engineered system is

Table 23.1 Two approaches of tissue engineering [6, 12]

1. *Acellular approach*
This approach relies on guided regeneration of tissue materials that serve as templates for ingrowth of host cells and tissue in vivo
2. *Cellular approach*
This approach relies on cells that have been cultured with scaffold in vitro and then implanted as part of an engineered device. The success of such a cell-based approach for tissue engineering of bone repair is critically dependent on the developments of an ECM-like scaffold for cell delivery

achieved with the guide of a scaffold. Such scaffolds can be natural, man-made or a composite of both. The use of signalling molecules has a potential to markedly increase scaffold effectiveness. Living cells can migrate into the implant after implantation (acellular approach) or can be associated with the matrix in cell culture before implantation (cellular approach). Such cells can be isolated as fully differentiated cells of the tissue they hoped to recreate, or they can be manipulated to produce the desired function when isolated from other tissues or stem cell sources. These two types of approaches in tissue engineering are summarised in Table 23.1 [6, 12]. In both approaches, the tissue-like matrix (also called scaffolds or templates) to which specific cell types are attached either in vivo or in vitro is one of the most important components in the engineering of new functional tissues.

23.3 Challenges in Tissue Engineering

Tissue engineering involves many disciplines, including microanatomy; cell, molecular and developmental biology; immunology; materials science; and branches of engineering. Hence, the advancement of tissue engineering depends on the progresses of science and technology gained in these fields. Being a very much fledgling discipline, tissue engineering encounters a variety of challenges, which can be grouped into three categories associated with the science and technology of cells, materials and interaction between them, as summarised in Table 23.2 [11]. The challenges that the material scientists encounter are linked with the required properties of ideal scaffolds. An ideal scaffold should be a mimic ECM of the tissue that is to be engineered. For bone regeneration, the biggest challenge is a scaffold suitable to replace large cortical bone defects and capable of load transmission. The specific criteria for an ideal scaffold for bone regeneration are listed in Table 23.3.

23.4 Biomaterials for Bone Tissue Engineering

The first step in achieving a successful scaffold is to design and produce a bone-matrix-like biomaterial. Natural bone matrix is a composite composed of biological ceramic (a natural apatite) and biological polymer. Table 23.4 gives a brief description

Table 23.2 Three categories of challenges in tissue engineering [11]

1. *Challenges associated with cells*
The understanding of cells and cell technology, including cell sourcing, the manipulation of cell function and the future use of stem cell technology. The discovery that embryonic stem cells can be recovered from human foetal tissue and propagated for long period without losing their toti- or pluripotency has a huge impact on tissue engineering. Stem cells, together with signalling molecules, play an important role in tissue and organ development. How to direct their differentiation is a subject of high current interest
2. *Challenges associated with biomaterials and scaffolds*
The design and fabrication of tissue-like materials to provide a scaffold or template. It has been documented that there are certain biocompatible materials that enable cells to be seeded onto a synthetic scaffold and that some types of cells are capable of undergoing subsequent differentiation to generate new functional tissue after being cultivated in vitro and implanted with scaffold into living bodies. Bone cells are in this category. One of the challenges in bone tissue engineering is to develop ECM-like scaffolds that can deliver cells, provide proper mechanical stability and be degradable at the desired rate until replaced by newly formed bone
3. *Challenges associated with interaction between cells and scaffolds*
Integration into living systems. The interface between the cells and the scaffold must be clearly understood so that the interface can be optimised. Their design characteristics are major challenges for the field of bone tissue engineering and should be considered at a molecular chemical level

Table 23.3 Criteria of an ideal scaffold for bone engineering [13–15]

1. *Ability to deliver cells*
The material should not only be biocompatible (i.e. harmless) but also foster cell attachment, differentiation and proliferation
2. *Osteoconductivity*
It would be best if the material encourages osteoconduction with host bone. Osteoconductivity not only eliminates the formation of encapsulating tissue but also brings about a strong bond between the scaffold and host bone
3. *Biodegradability*
The composition of the material, combined with the porous structure of the scaffold, should lead to biodegradation in vivo at rates appropriate to tissue regeneration
4. *Mechanical properties*
The mechanical strength of the scaffold, which is determined by both the properties of the biomaterial and the porous structure, should be sufficient to provide mechanical stability to constructs in load-bearing sites prior to synthesis of new extracellular matrix by cells
5. *Porous structure*
The scaffold should have an interconnected porous structure with porosity >90 % and diameters between 300 and 500 μm for cell penetration, tissue ingrowth and vascularisation and nutrient delivery
6. *Fabrication*
The material should possess desired fabrication capability, for example, being readily produced into irregular shapes of scaffolds that match the defects in bone of individual patients
7. *Commercialisation*
The synthesis of the material and fabrication of the scaffold should be suitable for commercialisation

Table 23.4 Composition of natural bone matrix

1. *Biological ceramic*
Carbonated hydroxyapatite $Ca_{10}(PO_4)_6(OH)_2$ accounts for nearly 2/3 of the weight of bone. The inorganic component provides compressive strength to bone
2. *Biological polymer*
Roughly 1/3 of the weight of bone is from collagen fibres. Collagen fibres are tough and flexible and thus tolerate stretching, twisting and bending

of bone-matrix composition. It is not surprising that polymers, ceramics or their composites have been chosen for bone repair [1, 2]. They can be either synthetic or naturally occurring. Table 23.5 lists synthetic and natural scaffold biomaterials that have been most widely investigated for bone regeneration, some of which are well established and clinically applicable. In Table 23.5, the naturally occurring polymers are grouped into (1) carbohydrates (polysaccharides) and (2) proteins, which are two of four types of biological molecules [the other two types are (3) nucleic acids and (4) lipids] [16, 17]. In the following sections, the biocompatibility, biodegradability and mechanical properties of these scaffold materials are reviewed. They are the most important factors to be considered in a design of a bone regeneration scaffold.

23.5 Bioceramics: Calcium Phosphates

23.5.1 Biocompatibility

Since almost 2/3 of the weight of bone is hydroxyapatite $Ca_{10}(PO_4)_6(OH)_2$, it seems logical to use this ceramic as the major component of scaffold materials for bone tissue engineering. Actually, hydroxyapatite and related calcium phosphates (CaP) (e.g. β-tricalcium phosphate) have been intensively investigated [18, 19, 22]. As expected, calcium phosphates have an excellent biocompatibility due to their close chemical and crystal resemblance to bone mineral [20, 21]. Although they have not shown osteoinductive ability, they certainly possess osteoconductive properties as well as a remarkable ability to bind directly to bone [33–36]. Huge amounts of in vivo and in vitro assessments have reported that calcium phosphates, no matter which forms (bulk, coating, powder or porous) and which phases (crystalline or amorphous) they are in, always support the attachment, differentiation and proliferation of cells (such as osteoblasts and mesenchymal cells), with hydroxyapatite being the best one among them [37]. While the excellent biological performance of hydroxyapatite and related calcium phosphates has been well documented, the slow biodegradation of their crystalline phases and the weak mechanical strength of their amorphous states limit their application in engineering of new bone tissue, especially at load-bearing sites.

Table 23.5 List of promising scaffold biomaterials for bone regeneration

Biomaterial	Abbreviation	Application
1. *Ceramic* [18, 19]		
Calcium phosphate [20–22]	CaP	
Hydroxyapatite	HA	Dental
Tricalcium phosphate	TCP	Drug delivery
Biphasic calcium phosphate: HA and TCP	BCP	Scaffolds
Bioactive glasses [23–26]		Dental
Bioglass®		Drug delivery
Phosphate glass		Scaffolds
Bioactive glass-ceramic [27, 28]		Dental
Apatite-wollastonite	A/W	Drug delivery
Ceravital®		Scaffolds
2. *Polymer* [29–32]		
Synthetic degradable polymer		
Bulk biodegradable polymer		Sutures
Aliphatic polyester		Dental
Poly(lactic acid)	PLA	Orthopaedic
Poly($_{D-}$lactic acid)	PDLA	Drug delivery
Poly($_{L-}$lactic acid)	PLLA	Scaffolds
Poly($_{D,L-}$lactic acid)	PDLLA	
Poly(glycolic acid)	PGA	
Poly(lactic-co-glycolic acid)	PLGA	
Poly(ε-caprolactone)	PCL	
Poly(hydroxyalkanoate)	PHA	
Poly(3- or 4-hydroxybutyrate)	PHB	
Poly(3-hydroxyoctanoate)	PHO	
Poly(3-hydroxyvalerate)	PHV	
Poly (*p*-dioxanone)	PPD or PDS	
Poly(propylene fumarate)	PPF	
Poly(1,3-trimethylene carbonate)	PTMC	
Surface bioerodible polymer		Drug delivery
Poly(ortho ester)	POE	
Poly(anhydride)	PA	
Poly(phosphazene)	PPHOS	
Natural degradable polymer		
Polysaccharide		
Hyaluronan	HyA	
Alginate		
Gelatine		
Chitosan		
Protein		
Collagen		
Fibrin		
3. *Composite* [6]		
Selected combinations of above ceramics and polymers, as discussed in the relevant context		

Table 23.6 Comparison of mechanical properties of calcium phosphates and human bone

Ceramics	Compressive strength/MPa	Tensile strength/MPa	Elastic modulus/GPa	Fracture toughness/ $MPa\sqrt{m}$	References
Calcium phosphates	20–900	30–200	30–103	<1.0	[40, 43]
Hydroxyapatite	>400	~40	~100	~1.0	[40, 43]
Cortical bone	130–180	50–151	12–18	6–8	[44–47]

23.5.2 Degradability

Typically, crystalline calcium phosphates have long degradation time in vivo, often on the order of months, even years [38]. The dissolution rates of synthetic hydroxyapatite depend on the type and concentration of the buffered or unbuffered solutions, pH of the solution, degree of the saturation of the solution, solid/solution ratio, the length of suspension in the solution and the composition and crystallinity of the hydroxyapatite. In the case of crystalline hydroxyapatite, the degree of micro- and macroporosities, defect structure and amount and type of other phases present also have significant influence [2]. Crystalline hydroxyapatite exhibits the slowest degradation rate, compared with other calcium phosphates. The dissolution rate decreases in the following order [39]:

Other amorphous CaP > amorphous HA > other crystalline CaP > crystalline HA

23.5.3 Mechanical Properties

In the body, the mechanical properties of natural bone change with their biological location because the crystallinity, porosity and composition of bone adjust according to their biological environment. The properties of synthetic calcium phosphates vary significantly with their crystallinity, grain size, porosity and composition (e.g. calcium deficiency) as well. In general, the mechanical properties of synthetic calcium phosphates decrease significantly with increasing content of amorphous phase, microporosity and grain size. High crystallinity, low porosity and small grain size tend to give higher stiffness, higher compressive and tensile strength and greater fracture toughness [40, 41]. It has been reported that the flexural strength and fracture toughness of dense hydroxyapatite are much lower in a dry condition than in a wet condition [42].

If we compare the properties of hydroxyapatite and related calcium phosphates with those of bone (Table 23.6), we find that bone has a reasonably good compressive strength though it is lower than that of hydroxyapatite and better tensile strength and significantly better fracture toughness than hydroxyapatite. The apatite crystals in bone tissue make it strong enough to tolerate compressive loading. The high tensile strength and fracture toughness of bone are attributed to the tough and flexible collagen fibres. Hence, calcium phosphates alone cannot be used for load-bearing scaffolds in spite of its good biocompatibility and osteoconductivity.

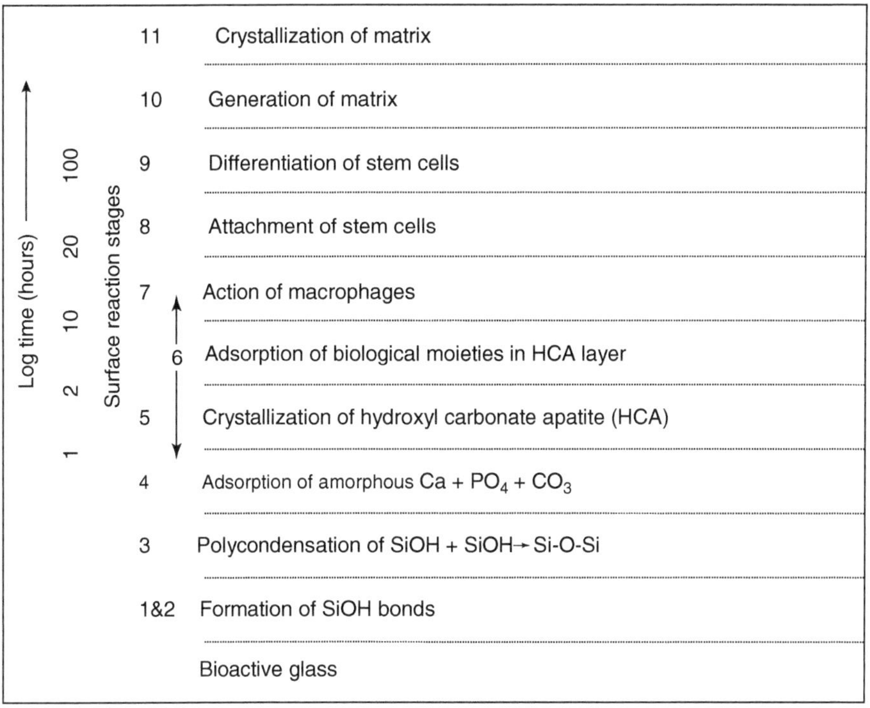

Fig. 23.2 Sequence of interfacial reactions involved in forming a bond between bone and bioactive ceramics and glasses [52, 70]

23.6 Bioceramics: Bioactive Silicate Glasses

23.6.1 Biocompatibility

As early as in 1969, Hench and colleagues discovered that certain silicate glass compositions had excellent biocompatibility as well as the ability of bone bonding [24–26]. Through interfacial and cell-mediated reactions, bioactive glass develops a calcium-deficient, carbonated calcium phosphate surface layer that allows it to chemically bond to host bone. This bone-bonding behaviour is referred to as bioactivity and has been associated with the formation of a carbonated hydroxyapatite layer on the glass surface when implanted or in contact with biological fluids [48–51]. The stages that are involved in forming the bone bond of bioactive glasses and bioactive glass-ceramics were summarised by Hench as shown in Fig. 23.2. Although many details remain unknown at present, it is clearly recognised that for a bond with bone tissue to occur, a layer of biologically active carbonated hydroxyapatite (CHA) must form (stages 4 and 5). This conclusion is based on the finding that CHA is the only common characteristic of all the known bioactive implant materials [52]. Bioactivity is not an exclusive property of bioactive silicate glasses. Hydroxyapatite and related calcium phosphates also show an excellent ability to bond to bone, as discussed above. The capability of an implant to form a biological interface with surrounding tissue is critical in elimination of scaffold loosening. Bioactive glasses have also been found to support enzyme activity [53–56] and vascularisation [57, 58]; foster osteoblast adhesion, growth and differentiation; and induce the differentiation of mesenchymal cells into osteoblasts [59–61] and osteoconductivity [62].

A significant finding for the development of bone engineering is that the dissolution products from bioactive glasses exert a genetic control over osteoblast cycle and rapid expression of genes that regulate osteogenesis and the production of growth factors [63]. Silicon has been found to play a key role in the bone mineralisation and gene activation, which has led to the substitution of silicon for calcium into synthetic hydroxyapatite. Investigations in vivo have shown that bone ingrowth into silicon-substituted HA granules was remarkably greater than that into pure HA [64].

The above-mentioned advantages make the well-known 45S5 Bioglass® successfully applied in clinic as treatment of periodontal disease (PerioGlas®) and as a bone filler material (NovaBone®) [50]. Bioglass® implants have also been used to replace damaged middle ear bones, restoring hearing to patients [18, 50]. Recently, bioactive glasses have gained attention as promising scaffold materials [14, 65–69]. But this application has encountered a hurdle caused by an apparent conflict between the properties of biodegradability and mechanical reliability, which will be discussed in Sects. 23.6.2 and 23.6.3, respectively.

23.6.2 Composition and Biodegradability

The basic constituents of the most bioactive glasses are SiO_2, Na_2O, CaO and P_2O_5. The well-known 45S5 Bioglass® (first bioactive composition) contains 45 % SiO_2, 24.5 % Na_2O, 24.4 % CaO and 6 % P_2O_5, in weight percent. The bioreactivity of the material is composition dependent. Hench and co-workers [52] have systematically studied a series of glasses in the four-component systems with a constant 6 wt% P_2O_5 content. This work is summarised in the ternary SiO_2-Na_2O-CaO diagram shown in Fig. 23.3. In region A, the glasses are bioactive and bond to bone. In region B, glasses are nearly inert when implanted. Compositions in region C are resorbed within 10–30 days in tissue. In region D, the compositions are not technically practical.

The key advantage that makes bioactive glasses promising scaffold materials is the possibility of controlling a range of chemical properties and thus the rate of bioresorption. The structure and chemistry of glasses, in particular sol–gel-derived glasses [48, 49], can be tailored at a molecular level by varying either composition or thermal or environmental processing history. It is possible to design glasses with degradation properties specific to a particular application of bone tissue engineering.

However, it was reported that crystallisation of bioactive glasses decreased the level of bioactivity [71] and even turned a bioactive glass into an inert material [72].

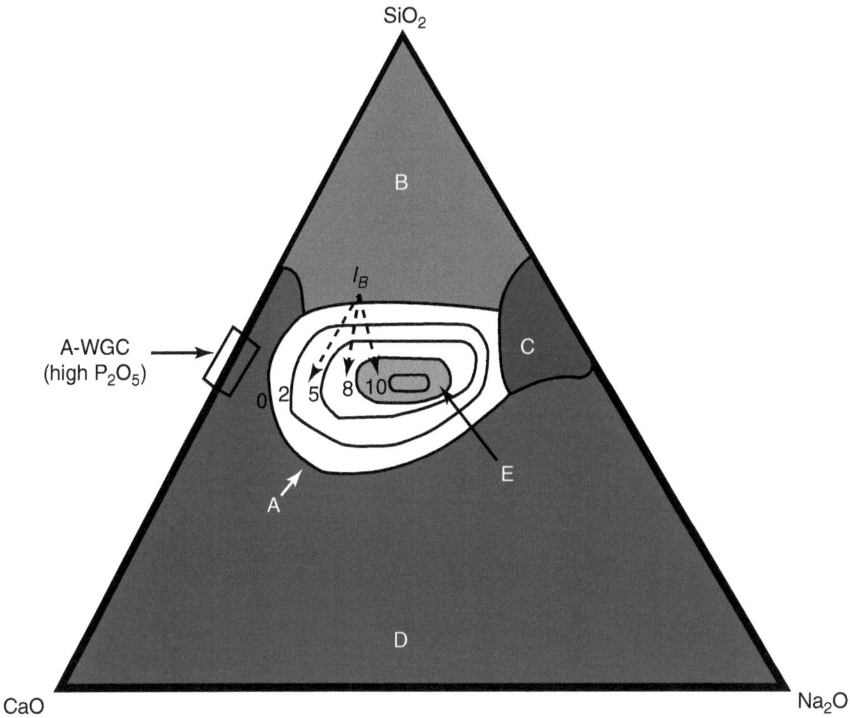

Fig. 23.3 Compositional dependence (in weight %) of bone bonding and soft tissue bonding of bioactive glasses and glass-ceramics. Bioactivity index I_B is defined as $I_B = 100 / t_{0.5}$,where $t_{0.5}$ is the time taken for 50 % of the interface to bond to bone. All compositions have a constant 6 wt% of P_2O_5. In region *A*, the glasses are bioactive and bond to bone. In region *B*, glasses are nearly inert when implanted. Compositions in region *C* are resorbed within 10–30 days in tissue. In region *D*, the compositions are not technically practical. In the region where $I_B > 8$ (called region *E*), soft tissue bonding occurs. Apatite-wollastonite glass-ceramic (A-WGC) has higher P_2O_5 content [52] (Color figure online)

This is one of disadvantages that limit the application of bioactive glasses as scaffold materials, as full crystallisation happens prior to significant densification upon heat treatment (i.e. sintering) [73]. Extensive sintering is necessary to densify the struts of a scaffold, which would otherwise be made up of loosely packed particles and thus too fragile to handle.

23.6.3 Mechanical Properties

The primary disadvantage of bioactive glasses is their mechanical weakness and low fracture toughness (Table 23.7) due to their amorphous structure. Hence, bioactive glasses alone have limited application in load-bearing situations owing to poor

Table 23.7 Mechanical properties of hydroxyapatite, 45S5 Bioglass®, glass-ceramics and human cortical bone

Ceramics	Compression strength/MPa	Tensile strength/MPa	Elastic modulus/GPa	Fracture toughness/ $MPa\sqrt{m}$	References
Hydroxyapatite	>400	~40	~100	~1.0	[40, 43]
45S5 Bioglass®	~500	42	35	0.5–1	[43, 74]
A-W	1,080	215(bend)	118	2.0	[27]
Parent glass of A-W	NA	72 (bend)	NA	0.8	[27]
Bioverit® I	500	140–180 (bend)	70–90	1.2–2.1	[75]
Cortical bone	130–180	50–151	12–18	6–8	[44–47]

mechanical strength that mismatches with surrounding bone. However, these materials can be used in combination with polymers to form composite materials having bone repair potential [61].

23.7 Bioceramics: Glass-Ceramics

Glass-ceramics are fine-grained polycrystalline materials formed when glasses of suitable compositions are heat treated and thus undergo controlled crystallisation to the lower-energy, crystalline state. Only specific glass compositions are suitable precursors for glass-ceramics, such as some bioactive glasses. Some glasses are too stable and difficult to crystallise (e.g. window glass), whereas others crystallise too readily in an uncontrollable manner resulting in undesirable microstructures [76].

Usually, a glass-ceramic is not fully crystalline; typically, the microstructure is 50–95 vol.% crystalline with the remainder being residual glass. The mechanical properties of glass-ceramics are superior to those of the parent glass [76]. Almost all bioactive glasses can be strengthened by the formation of crystalline particles upon heat treatment into a glass-crystal region of its phase diagram [77]. The resultant glass-ceramics can exhibit better mechanical properties than both the parent glass and sintered crystalline ceramics (Table 23.7). There are many biomedical glass-ceramics available for the repair of damaged bone. Among them, apatite-wollastonite (A-W), Ceravital® and Bioverit® glass-ceramics have been intensively investigated [18, 19, 27–29, 75–81].

23.7.1 A-W Glass-Ceramic

In A-W glass-ceramic, the glass matrix is reinforced by β-wollastonite ($CaSiO_3$) and a small amount of apatite phase, which precipitate successively at 870 and 900 °C, respectively [27]. Some mechanical properties of this glass-ceramic have been listed in Table 23.7. The high bending strength (215 MPa) of A-W glass-ceramic is due to the precipitation of the wollastonite as well as apatite. These two precipitates also give the glass-ceramic a higher fracture toughness than both the glass and ceramic phases. It is believed that the wollastonite effectively prevents straight propagation of cracks, causing them to turn or branch out [27, 78, 79].

A-W glass-ceramic is capable of binding tightly to living bone in a few weeks after implantation, and the implants do not deteriorate in vivo [80]. The excellent bone-bonding ability of A-W glass-ceramic is attributed to the glass matrix and apatite precipitates, whereas the in vivo stability as a whole plant is due to the inertness of β-wollastonite. Although the long-term integrity in vivo is desirable in the application of non-resorbable prosthesis, it does not match the goal of tissue engineering which demands biodegradable scaffolds.

23.7.2 Ceravital® Glass-Ceramics [81]

"Ceravital" was coined to mean a number of different compositions of glasses and glass-ceramics and not only one product. Their basic network components include SiO_2, $Ca(PO_2)_2$, CaO, Na_2O, MgO and K_2O, with ceramic additions being Al_2O_3, Ta_2O_5, TiO_2, B_2O_3, $Al(PO_3)_3$, SrO, La_2O_3 or Gd_2O_3. This material system was developed as solid fillers in the load-bearing conditions for the replacement of bone and teeth. It turned out, however, that their mechanical properties do not serve the purpose.

The surface bioreactivity of Ceravital® products is such that the long-term stability of the materials is eventually endangered by the process. However, this degradability is a favourite property in tissue engineering application.

23.7.3 Bioverit® Glass-Ceramics [75]

Bioverit® products are mica-apatite glass-ceramics. Mica crystals (aluminium silicate minerals) give the materials good machinability, and apatite crystals ensure the bioactivity of the implants. The mechanical properties of Bioverit® materials (Table 23.7) allow them to be used as fillers in dental application. As regards bioreactivity, Bioverit® implants show a hydrolytic stability in vivo.

23.8 Naturally Occurring Biopolymers

Much research effect has been focused on naturally occurring polymers such as demineralised bone extracellular matrix (ECM) [6, 82], purified collagen [83, 84] and chitosan [85] for tissue engineering applications. Theoretically, naturally occurring polymers should not cause foreign material response when implanted in

Table 23.8 List of naturally occurring polymers and their main application fields [87]

Polymer	Source	Main application fields
Collagen	Tendons and ligament	Multi-applications, including bone tissue engineering
Collagen-GAG (alginate) copolymers		Artificial skin grafts for skin replacement
Albumin	In blood	Transporting protein, used as coating to form a thromboresistant surface
Hyaluronic acid	In the ECM of all higher animals	An important starting material for preparation of new biocompatible and biodegradable polymers that have applications in drug delivery, tissue engineering and viscosupplementation
Fibrinogen-fibrin	Purified from plasma in blood	Multi-applications, including bone tissue engineering
Chitosan	Shells of shrimp and crabs	Multi-applications, including bone tissue engineering
Polyhydroxyalkanoates	Fermentation	Cardiovascular and bone tissue engineering

humans. They provide a natural substrate for cellular attachment, proliferation and differentiation in its native state. For the above-mentioned reasons, naturally occurring polymers could be a favourite substrate for tissue engineering [29]. Table 23.8 presents some of the naturally occurring polymers, their sources and applications. Among them, collagen and chitosan are most widely investigated for bone engineering and are briefly introduced here.

23.8.1 Collagen and ECM-Based Materials

The most commonly used naturally occurring polymers have been the structural protein collagen. Biomaterials derived from ECM include collagen and other naturally occurring structural and functional proteins. Natural polymers must be modified and sterilised before clinic use. All methods of stabilisation and sterilisation can moderately or severely alter the rate of in vivo degradation and change the mechanical and physical properties of the native polymers. Each method has certain advantages and disadvantages and thus should be selectively utilised for scaffolds of specifically sited bone tissue engineering [86].

23.8.2 Chitosan

The use of chitosan for bone tissue engineering has been widely investigated [85, 88]. This is in part due to the apparent osteoconductive properties of chitosan. Mesenchymal stem cells cultured in the presence of chitosan have demonstrated an increased differentiation to osteoblasts compared to cell cultured in the absence of chitosan [89]. It is also speculated that chitosan may enhance osteoconduction in vivo by entrapping growth factors at the wound site [90, 91].

23.9 Synthetic Polymers

Although naturally occurring polymers possess the above-mentioned advantages, their poor mechanical properties and variable physical properties with different sources of the protein matrices have hampered progress with these approaches. Concerns have also arisen regarding immunogenic problems associated with the introduction of foreign collagen [38].

Following the developmental efforts using naturally occurring polymers as scaffolds, much attention has been paid to synthetic polymers. Synthetic polymers are thought to have a future in tissue engineering due to not only their excellent processing characteristics, which can ensure the off-the-shelf availability, but also their advantage of being biocompatible and biodegradable [38, 92]. Synthetic polymers have predictable and reproducible mechanical and physical properties (e.g. tensile strength, elastic modulus and degradation rate) and can be manufactured with great precision. Although they are unfamiliar to cells and many suffer some shortcomings, such as eliciting persistent inflammatory reactions, being eroded, incompliant or unable to integrate with host tissues, they may be replaced in vivo in a timely fashion by native tissue. It has become widely realised that an ideal tissue-engineered bone substitute should be a synthetic scaffold, which is biocompatible and provides for cell attachment, proliferation and maturation; has mechanical properties to match those of the tissues at the site of implantation; and degrades at rates to match tissue replacement. Table 23.9 lists selected properties of synthetic, biocompatible polymers that have been intensively investigated as scaffold materials for tissue engineering, type I collagen fibres being included for comparison.

23.9.1 Bulk-Degradable Polymers

23.9.1.1 Saturated Poly-α-hydroxy Esters (PLA, PGA and PCL)

The biodegradable synthetic polymers most often utilised for three-dimensional scaffold in tissue engineering are the poly(α-hydroxy acids), including poly(lactic acid) (PLA) and poly(glycolic acid) (PGA), as well as poly(lactic-co-glycolide) (PLGA) copolymers [4, 93]. PLA exists in three forms: $_{\text{L-}}$PLA (PLLA), $_{\text{D-}}$PLA (PDLA) and racemic mixture of $_{\text{D,L-}}$PLA (PDLLA).

These polymers remain popular for a variety of reasons, among which biocompatibility and biodegradability are the first. These materials have chemical properties that allow hydrolytic degradation through de-esterification. Once degraded, the monomeric components of each polymer are removed by natural pathways: PGA

Table 23.9 Physical properties of synthetic, biocompatible and biodegradable polymers investigated as scaffold materials

Polymers	Melting point T_m/°C	Glass transition point T_g/°C	Degradation time/months	Tensile or compressive* strength/MPa	Modulus/GPa	Ultimate elongation (%)	References
1. *Bulk-degradable polymers*							
PDLLA	Amorphous	55–60	12–16	Pellet: 35–150* Film or disk: 29–35	Film or disk: 1.9–2.4	Pellet: 0.5–8.0 Film or disk: 5.0–6.0	[92, 108, 109]
PLLA	173–178	60–65	>24	Pellet: 40–120 Film or disk: 28–50 Fibre: 870–2,300	Film or disk: 1.2–3.0 Fibre: 10–16	Pellet: 2.0–10.0 Film or disk: 2.0–6.0 Fibre: 12–26	[92, 108]
PGA	225–230	35–40	6–12	Fibre: 340–920	Fibre: 7–14	Fibre: 15–25	[92, 110, 111]
PLGA	Amorphous	45–55	Adjustable	41.4–55.2	1.4–2.8	3–10	[29]
PPF			Bulk	2–30*			[29, 31]
PCL	58	−72	Bulk	10–15	0.15–0.33	400–1,200	[112–114]
PHB	177	4	Very slow	25–45	1.5–1.8	25	[115, 116]
2. *Surface-eroding polymers*							
Poly(anhydrides)	150–200		Surface	25–27 30–40*	0.14–1.4	NA	[29, 31, 117]
Poly(ortho esters)	30–100		Surface	4–16*	2.5–4.4	NA	[29, 118]
Polyphosphazene	−66–50	242	Surface	NA	NA	NA	[119, 120]
3. *Type I collagen*			Bulk	Uncross-linked fibre: 0.91–7.2 Cross-linked fibre: 46.8–68.8	Uncross-linked fibre: $1.8–46 \times 10^{-3}$ Cross-linked fibre: 0.383–0.766	Uncross-linked fibre: 24.1–68.0 Cross-linked fibre: 11.6–15.6	[121]

can be converted to other metabolites or eliminated by other mechanisms, and PLA can be cleared through tricarboxylic acid cycle. The body already contains highly regulated mechanisms for completely removing monomeric components of lactic and glycolic acids. Due to these properties, PLA and PGA have been used in products such as degradable sutures and have been proved by US food and drug administration (FDA) [29]. Other properties of special interest are that they have a very good processability and that a wide range of degradation rates, physical, mechanical and other properties can be achieved by PLA and PGA of various molecular weights and its copolymers. However, these polymers undergo a bulk erosion process in contact with body fluids such that they can cause scaffolds to fail prematurely. In addition, abrupt release of these acidic degradation products can cause a strong inflammatory response [94–99].

In general, PGA degrades faster than PLA, as listed in Table 23.9. Their degradation rates decrease in the following order:

$$\xrightarrow[\text{Degradation rates decrease}]{\text{PGA>PDLLA>PLLA}}$$

Table 23.9 also lists the mechanical properties of type I collagen, which is the major organic component of extracellular matrix in bone. The strength and ductility (e.g. ultimate elongation) of PLA and PGA are comparable to those of type I collagen fibres.

PDLLA has been extensively investigated as a biomedical coating material because of its excellent features with respect to implant coating [29, 100]. In addition to its high mechanical stability [101], PDLLA also shows excellent biocompatibility in vivo and good osteoinductive potential [102]. PDLLA of low molecular weight can be combined with drugs like growth factors [102], antibiotics [100] or thrombin inhibitor [103] to establish a locally acting drug-delivery system. It is because of these desirable features that much more attention has recently been paid to PDLLA for applying it as a scaffold material for tissue engineering.

Highly porous 3D scaffolds made of Bioglass®-filled PDLLA and PLGA were first fabricated by Boccaccini et al. [66]. Since then, an increasing number of publications have emerged on this subject. Porous PDLLA foams and Bioglass®-filled PDLLA composite foams have both been fabricated, using thermally induced phase-separation (TIPS) technique [104, 105]. Bioglass®-filled PDLLA composite foams exhibit high bioactivity, assessed by the formation of HA on the strut surfaces upon immersion in SBF [106]. It has also been shown that the foams support the migration, adhesion, spreading and viability of MG-63 cells (osteosarcoma cell line) [107].

Poly(ε-caprolactone) (PCL) is also an important member of the aliphatic polyester family. It has been used to effectively entrap antibiotic drugs, and thus a construct made with PCL can be considered as a drug-delivery system, being used to enhance bone ingrowth and regeneration in the treatment of bone defects [122, 123]. The degradation of PCL and its copolymers involves similar mechanisms to PLA, proceeding in two stages: random hydrolytic ester cleavage and weight loss through the diffusion of oligomeric species from the bulk. It has been found that the degradation of PCL system with a high molecular weight ($\bar{M}_n$ of 50,000) is remarkably slow, requiring 3 years for complete removal from the host body [124].

23.9.1.2 Poly(hydroxyalkanoates) (PHB, PHBV, P4HB, PHBHHx and PHO)

Recently, another type of polyesters, poly(hydroxyalkanoates) (PHAs), has been suggested for tissue engineering due to their controllable biodegradation and high biocompatibility [116]. They are aliphatic polyesters as well but produced by microorganisms under unbalanced growth conditions [125, 126]. They are generally biodegradable (via hydrolysis) and thermo-processable, making them attractive as biomaterials for applications in medical devices and tissue engineering. Over the past years, PHA, particularly poly(3-hydroxybutyrate) (PHB), copolymers of 3-hydroxybutyrate and 3-hydroxyvalerate (PHBV), poly(4-hydroxybutyrate) (P4HB), copolymers of 3-hydroxybutyrate and 3-hydroxyhexanoate (PHBHHx) and poly(3-hydroxyoctanoate) (PHO) were demonstrated to be suitable for tissue engineering and are reviewed in detail in ref [116, 127].

Dependent on the property requirement by different applications, PHA polymers can be either blended, surface modified or composed with other polymers, enzymes or inorganic materials to further adjust their mechanical properties or biocompatibility. The blending among the several PHA themselves can change dramatically the material properties and biocompatibility [116, 125].

PHB is of particular interest for bone tissue application as it was demonstrated to produce a consistent favourable bone tissue adaptation response with no evidence of an undesirable chronic inflammatory response after implantation periods up to 12 months. Bone is formed close to the material and subsequently becomes highly organised, with up to 80 % of the implant surface lying in direct apposition to new bone. The materials showed no evidence of extensive structural breakdown in vivo during the implantation period of the study [128].

However, a drawback of some PHA polymers is their limited availability and the time-consuming extraction procedure from bacteria cultures that is required for obtaining sufficient processing amounts as described in the literature [116, 129]. Therefore, the extraction process might be a challenge to a cost-effective industrial upscale production for large amounts of some PHA polymers.

23.9.1.3 Poly(propylene fumarate) (PPF)

Poly(propylene fumarate) (PPF) is an unsaturated linear polyester. Similarly to PLA and PGA, the degradation products of PPF via hydrolysis (i.e. propylene glycol and fumaric acid) are biocompatible and readily removed from the body. The double bond along the backbone of the polymer permits cross-linking in situ, which causes a mouldable composite to harden within 10–15 min. Mechanical properties and degradation time of the composite may be controlled by varying the PPF molecular weight. Therefore, preservation of the double bonds and control of molecular weight during PPF synthesis are critical issues [130]. PPF has been suggested for use as scaffold for guided tissue regeneration, often as part of an injectable bone replacement composite [131], and has been used as a substrate for osteoblast culture [132].

23.9.2 Surface Bioeroding Polymers

There is a family of hydrophobic polymers that undergo a heterogeneous hydrolysis process that is predominantly confined to the polymer-water interface. This property is referred to as surface eroding as opposed to bulk-degrading behaviour. These surface bioeroding polymers have been intensively investigated as drug-delivery vehicles. The surface-eroding characteristics offer three key advantages over bulk degradation when used as scaffold materials: (1) retention of mechanical integrity over the degradative lifetime of the device, owing to the maintenance of mass-to-volume ratio; (2) minimal toxic effects (i.e. local acidity), owing to lower solubility and concentration of degradation products; and (3) significantly enhanced bone ingrowth into the porous scaffolds, owing to the increment in pore size as the erosion proceeds [133].

23.9.2.1 Poly(anhydrides)

Poly(1,3-bis-p-carboxyphenoxypropane anhydride) [134] and poly(erucic acid dimer anhydride) [135] are biodegradable polymers for controlled drug delivery in a form of implant or injectable microspheres. Studies in rabbits have shown that the osteocompatibility of poly(anhydrides) that undergo photocuring is comparable to PLA and that the implants of poly(anhydrides) show enhanced integration with surrounding bone in comparison to PLA controls [136].

23.9.2.2 Poly(ortho esters) (POE)

POE scaffolds were coated with cross-linked acidic gelatine to improve surface properties for cell attachment. Preliminary in vitro and in vivo results revealed that POE showed no inflammation and had little or no effect on bone formation, while PLA provoked a chronic inflammatory response and inhibited bone formation [137, 138].

23.9.2.3 Polyphosphazenes

These polymers have shown promise as bioerodible materials capable of controlled degradation and sustained drug delivery for therapeutic [119, 139] and bone regeneration [140, 141]. Their tailored side groups enable a wide variety of hydrolytic properties to be designed into selected polymers for applications in biological environments without the release of harmful degradation products at physiological concentration.

23.10 Biocomposites

From a biological perspective, it is a natural strategy to combine polymers and ceramics to fabricate scaffolds for bone tissue engineering because native bone is the combination of a naturally occurring polymer and biological apatite. From the materials science point of view, a single material type does not always provide the necessary mechanical and/or chemical properties desired for a particular application. In these instances, composite materials designed to combine the advantages of both materials may be most appropriate. Polymers and ceramics that degrade in vivo should be chosen for designing biocomposites for tissue engineering scaffolds. While massive release of acidic degradation from polymers causes inflammatory reactions [94, 142, 143], the basic degradation of calcium phosphate or bioactive glasses would buffer the acidic by-products of polymers and may thereby help to avoid the formation of an unfavourable environment for cells due to a decreased pH. Mechanically, bioceramics are much stronger than polymers and play a critical role in providing mechanical stability to constructs prior to synthesis of new bone matrix by cells. However, as mentioned above, ceramics and glasses are very fragile due to their intrinsic brittleness and flaw sensitivity. To capitalise on their advantages and minimise their shortcomings, ceramic and glass materials can be combined with various polymers to form composite biomaterials for osseous regeneration. Table 23.10 lists selected ceramic/glass-polymer composites, which were designed as biomedical devices or scaffold materials for bone tissue engineering, and their mechanical properties.

In general, all these synthetic composites have good biocompatibility. Kikuchi et al., for instance, combined TCP with PLA to form a polymer-ceramic composite, which was found to possess the osteoconductivity of β-TCP and the degradability of PLA [144].

The research team led by Laurencin synthesised porous scaffolds containing PLGA and HA, which was reported to combine the degradability of PLGA with the bioactivity of HA, fostering cell proliferation and differentiation as well as mineral formation [145–147]. The composites of bioactive glass-PLA were observed to form calcium phosphate layers on their surfaces and support rapid and abundant growth of human osteoblasts and osteoblast-like cells when culture in vitro [66, 104–107, 148–154].

Table 23.10 Biocomposites designed for bone tissue engineering

Biocomposite									
Ceramic	Polymer	Percentage of ceramic (%)	Porosity (%)	Pore size/μm	Compressive (C), tensile (T), flexural (F), bending (B) strength/MPa	Modulus/MPa	Ultimate strain (%)	Toughness/ kJ/m^2	References
1. *Dense composites*									
HA fibre	PDLLA	2–10.5 (vol.)	Not applicable		45 (F)	1.75–2.47×10^3			[156]
	PLLA	10–70 (wt.)			50–60 (F)	6.4–12.8×10^3	0.7–2.3		[155]
HA	PLGA	40–85 (vol.)			22 (F)	1.1×10^3		5.29	[157–159]
	Chitosan	40–85 (vol.)			12 (F)	2.15×10^3		0.092	[158]
	Chitosan + PLGA	40–85 (vol.)			43 (F)	2.6×10^3		9.77	[158]
	PPhos	85–95 (wt.)							[160]
	Collagen	50–72 (wt.)							[161]
β-TCP	PLLA-co-PEH	75 (wt.)			51 (F)	5.18×10^3			[144]
	PPF	25 (wt.)			7.5–7.7 (C)	191–134			[162]
A/W	PE	10–50 (vol.)			18–28 (B)	0.9–5.7×10^3			[163–165]
$Ca_3(CO_3)_2$	PLLA	30 (wt.)	50	3.5–6×10^3			[166]	Human cortical bone	70(wt.)
50–150(T)	12–18×10^3			[44–47]	130–180 (C)				

(continued)

Table 23.10 (continued)

Biocomposite									
Ceramic	Polymer	Percentage of ceramic (%)	Porosity (%)	Pore size/μm	Compressive (C), tensile (T), flexural (F), bending (B) strength/ MPa	Modulus/MPa	Ultimate strain (%)	Toughness/ kJ/m²	References
2. *Porous composites*									
Amorphous CaP	PLGA	28–75 (wt.)	75	>100		65			[167, 168]
β-TCP	Chitosan-gelatin	10–70 (wt.)		322–355	0.32–0.88 (C)	3.94–10.88			[169]
HA	PLLA	50 (wt.)	85–96	100×300	0.39 (C)	10–14			[170]
	PLGA	60–75 (wt.)	81–91	800–1,800	0.07–0.22 (C)	2–7.5			[171]
	PLGA		30–40	110–150		337–1,459			[147]
Bioglass®	PLGA	75 (wt.)	43	89	0.42 (C)	51			[65, 148, 149]
	PLLA	20–50 (wt.)	77–80	~100(macro) ~10(micro)	1.5–3.9 (T)	137–260	1.1–13.7		[150]
	PLGA	0.1–1 (wt.)		50–300					[151]
	PDLLA	5–29 (wt.)	94	~100(macro) 10–50(micro)	0.07–0.08	0.65–1.2	7.21–13.3		[106, 107, 152]
Phosphate glass	PLA-PDLLA	40 (wt.)	93–97	98–154	0.017–0.020 (C)	0.075–0.12			[153]
A/W	PDLLA	20–40 (wt.)	85.5–95.2						[154]
Human cancellous bone					4–12 (C)	100–500	1.65–2.11		[172–174]

A comparison between the dense composites and cortical bone indicates that the most promising synthetic composite seems to be HA fibre-reinforced PLA composites [155], which however exhibit mechanical property values close to the lower values of the cortical bone. Up to now, the best composite scaffolds reported in literature seem to be those from Bioglass® and PLLA or PDLLA [104, 105, 150, 151]. They have a well-defined porous structure, and at the same time, their mechanical properties are close to (but lower than) those of cancellous bone.

23.11 Summary

While the ideal tissue-engineered bone substitute should be a material, which is bioresorbable and biocompatible and supports cell attachment, proliferation and maturation and which is ultimately resorbed once new bone has formed, allowing this bone to undergo remodelling, this goal has not been achieved so far. Material scientists must continue to strive to design and fabricate a synthetic material so as to make the dream of a "tissue-engineered bone substitute" a reality. To design a composite scaffold, it is necessary to weight up the "pros and cons" of the potential precursor materials, which are summarised in Table 23.11. Among the bioactive ceramics and glasses listed in Table 23.11, bioactive (silicate) glasses offer remarkable advantages. The ability to enhance vascularisation, the role of silicon in rapid gene expression that regulates osteogenesis and the tailorable degradation rate make bioactive glasses promising scaffold materials over others, and thus they are the material of choice as the inorganic component of composite scaffolds in this study. Although bioactive glasses are brittle with low fracture toughness (Table 23.7), these materials can be used in combination with polymers to form composite materials.

It can be argued that the controllable biodegradability of bioactive glasses makes them advantageous over HA and related crystalline calcium phosphates. Nano-sized carbonated HA is a stable component of natural bone, though it metabolises like all tissues. Hence, it would be fundamentally wrong if one expected HA to degrade fast in a physiological environment. In fact, it has been well documented in the literature that HA degrades very slowly, nearly inert [40]. This should make HA less favoured as a scaffold material for use in tissue engineering. The degradation rates of amorphous HA and TCP are high, but they are too fragile to build a 3D porous network.

Between the two types of polymers, the bulk-degradable type is more promising than the surface-erosive group, considering that being replaced by new bone tissue is one of the important criteria of an ideal scaffold material (Table 23.3). Among these bulk-degradable polymers, amorphous PDLLA is one of the most interesting materials as a polymer component of scaffolds because it can be combined with biomolecules, such as growth factors [102] and antibiotics [100], to establish a locally acting drug-delivery system. It is expected that the local drug-delivery system will promote bone regeneration and eliminate inflammatory responses upon scaffold degradation.

Table 23.11 Advantages and disadvantages of synthetic scaffold biomaterials in bone tissue engineering

Biomaterial	Positive	Negative
Calcium phosphates (e.g. HA, TCP and biphase CaP)	1. Excellent biocompatibility 2. Supporting cell activity 3. Good osteoconductivity	1. Brittle 2. They biodegrade too slowly in the crystalline state and are mechanically too weak in the amorphous state
Bioactive silicate glasses	1. Excellent biocompatibility 2. Supporting cell activity 3. Good osteoconductivity 4. Vascularisation 5. Rapid gene expression 6. Tailorable degradation rate	1. Mechanically brittle and weak
Bioactive glass-ceramics (e.g. A-W)	1. Excellent biocompatibility 2. Supporting cell activity 3. Good osteoconductivity	1. Brittle 2. Too slow degradation rate
Bulk biodegradable polymers	1. Good biocompatibility	1. Inflammatory caused by acid degradation products
Poly(lactic acid)	2. Biodegradable (with a wide range of degradation rates)	2. Accelerated degradation rates cause collapse of scaffolds
Poly(glycolic acid)	3. Bioresorbable	
Poly(lactic-co-glycolic acid)	4. Good processability	
Poly(propylene fumarate)	5. Good ductility	
Surface bioerodible polymers	1. Good biocompatibility	1. They cannot be completely replaced by new bone tissue
Poly(ortho esters)	2. Retention of mechanical integrity over the degradative lifetime of the device	
Poly(anhydrides) Poly(phosphazene)	3. Significantly enhanced bone ingrowth into the porous scaffolds, owing to the increment in pore size	
Composites (containing bioactive phases)	1. Excellent biocompatibility	1. Still not as good as natural bone matrix
	2. Supporting cell activity 3. Good osteoconductivity 4. Tailorable degradation rate 5. Improved mechanical properties	2. Fabrication techniques can be complex

References

1. Vacanti CA (2000) Foreword. In: Lanza RP, Langer R, Vacanti JP (eds) Principles of tissue engineering, 2nd edn. Academic, California, p xxix
2. Bell E (2000) Tissue engineering in perspective. In: Lanza RP, Langer R, Vacanti JP (eds) Principles of tissue engineering, 2nd edn. Academic, California, pp xxxv–xli

3. Skalak R, Fox CF (eds) Tissue engineering: proceedings a workshop, Granlibackken, Latke Tahoe, California, New York. Alan Liss, 26 Feb 1987
4. Langer R, Vacanti JP (1993) Tissue engineering. Science 260(5110):920–926
5. Galletti PM, Hellman KB, Nerem RM (1995) Tissue engineering: from basic science to products. Tissue Eng 1:147–149
6. Burg KJL, Porter S, Kellam JF (2000) Biomaterial developments for bone tissue engineering. Biomaterials 21(23):2347–2359
7. Hench LL (1998) Biomaterials: a forecast for the future. Biomaterials 19(16):1419–1423
8. Jones JR, Hench LL (2001) Materials perspective – biomedical materials for new millennium: perspective on the future. Mater Sci Technol 17(8):891–900
9. Hench LL, Polak JM (2002) Third-generation biomedical materials. Science 295(5557):1014–1017
10. Naughton GK (1995) Emerging developments in tissue engineering and cell technology. Tissue Eng 1:211–219
11. Nerem RM (2000) The challenge of imitating nature. In: Lanza RP, Langer R, Vacanti JP (eds) Principles of tissue engineering, 2nd edn. Academic, California, pp 9–15
12. Vacanti JP, Vacanti CA (2000) The history and scope of tissue engineering. In: Lanza RP, Langer R, Vacanti JP (eds) Principles of tissue engineering, 2nd edn. Academic, California, pp 3–7
13. Bruder SP, Caplan AI (2000) Bone regeneration through cellular engineering. In: Lanza RP, Langer R, Vacanti JP (eds) Principles of tissue engineering, 2nd edn. Academic, California, pp 683–696
14. Jones JR, Hench LL (2003) Regeneration of trabecular bone using porous ceramics. Curr Opin Solid State Mater Sci 7(4–5):301–307
15. Jones JR, Boccaccini AR (2005) Cellular ceramics in biomedical applications: tissue engineering. In: Scheffler M, Colombo P (eds) Cellular ceramics: structure, manufacturing, processing and applications. Wiley-VCH Verlag GmbH & Co. KGaA, Weinheim, pp 550–573
16. Solomons TWG, Fryhle CB (1997) Organic chemistry, 6th edn. Wiley, New York
17. Berg JM, Tymoczko JL, Stryer L (2002) Biochemistry, 5th edn. W.H. Freeman and Company, New York
18. Hench LL, Wilson J (1999) An introduction to bioceramics, 2nd edn. Word Scientific, London
19. Yamamuro T, Hench LL, Wilson J (1990) Handbook of bioactive ceramics. CRC Press, Boca Raton
20. Jarcho M, Kay JF, Gumaer KI, Doremus RH, Drobeck HP (1977) Tissue, cellular and subcellular events at a bone-ceramic hydroxyapatite interface. J Bioeng 1(2):79–92
21. Jarcho M (1981) Calcium-phosphate ceramics as hard tissue prosthetics. Clin Orthop Relat Res 157:259–278
22. LeGeros RZ, LeGeros JP Calcium phosphate ceramics: past, present and future. In: Bioceramics 15: proceedings of the 15th international symposium on ceramics in medicine, Sydney, 4–8 Dec 2002, pp 3–10
23. Hench LL, Andersson O (1999) Bioactive glasses. In: Hench LL, Wilson J (eds) An introduction to bioceramics, 2nd edn. Word Scientific, London, pp 41–62
24. Hench LL, Splinter RJ, Allen WC (1971) Bonding mechanisms at the interface of ceramic prosthetic materials. J Biomed Mater Res Symp 2(part 1):117–141
25. Hench LL, Paschall HA (1973) Direct chemical bond of bioactive glass-ceramic materials to bone and muscle. J Biomed Mater Res Symp 4:25–42
26. Hench LL, Paschall HA (1974) Histochemical response at a biomaterial's interface. J Biomed Mater Res Symp 5(part 1):49–64
27. Kokubo T (1999) A/W glass-ceramic: processing and properties. In: Hench LL, Wilson J (eds) An introduction to bioceramics, 2nd edn. Word Scientific, London, pp 75–88
28. Kokubo T, Ito S, Sakka S, Yamamuro T (1986) Formation of a high-strength bioactive glass ceramic in the system Mgo-Cao-Sio2-P2o5. J Mater Sci 21(2):536–540
29. Seal BL, Otero TC, Panitch A (2001) Polymeric biomaterials for tissue and organ regeneration. Mater Sci Eng R-Rep 34(4–5):147–230

30. Atala A, Lanza RP (eds) (2002) Methods of tissue engineering. Academic, California, pp 505–574
31. Gunatillake PA, Adhikari R (2003) Biodegradable synthetic polymers for tissue engineering. Eur Cell Mater 5:1–16
32. Garlotta D (2001) A literature review of poly(lactic acid). J Polymers Environ 9(2):63–84
33. Driskell TD, Hassler CR, McCoy LR (1973) The significance of resorbable bioceramics in the repair of bone defects. In: proceeding of the 26th annual conference medicine and biology, pp 199–206
34. Denissen HW, Degroot K, Makkes PC, Vandenhooff A, Klopper PJ (1980) Tissue-response to dense apatite implants in rats. J Biomed Mater Res 14(6):713–721
35. Hollinger JO, Battistone GC (1986) Biodegradable bone repair materials – synthetic-polymers and ceramics. Clin Orthop Relat Res 207:290–305
36. Hammerle CHF, Olah AJ, Schmid J, Fluckiger L, Gogolewski S, Winkler JR, Lang NP (1997) The biological effect of natural bone mineral on bone neoformation on the rabbit skull. Clin Oral Implants Res 8(3):198–207
37. Brown S, Clarke I, Williams P (2001) Bioceramics 14: proceedings of the 14th international symposium on ceramics in medicine, international symposium on ceramics in medicine, Palm Springs, CA, Nov 2001, pp 213–269
38. Vacanti CA, Bonassar LJ, Vacanti JP (2000) Structure tissue engineering. In: Lanza RP, Langer R, Vacanti JP (eds) Principles of tissue engineering, 2nd edn. Academic, California, pp 671–682
39. Oonishi H, Kushitani S, Iwaki H (1995) Comparative bone formation in several kinds of bioceramic granules. In: Wilson J, Hench LL, Greenspan D (eds) Bioceramics 8: proceedings of the 8th international symposium on ceramics in medicine, Tokyo, Japan, 1995. Elsevier Science Ltd, pp 137–144
40. LeGeros RZ, LeGeros JP (1999) Dense hydroxyapatite. In: Hench LL, Wilson J (eds) An introduction to bioceramics, 2nd edn. Word Scientific, London, pp 139–180
41. Kokubo T (1999) Novel biomedical materials based on glasses. In: Shackelford JF (ed) Bioceramics -applications of ceramic and glass materials in medicine, vol 293, Materials science forum. Trans Tech Publications Ltd, Switzerland, pp 65–82
42. de Groot K, Lein CPAT, Wolke JGC, de Bliek-Hogervost JMA (1990) Chemistry of calcium phosphate bioceramics. In: Yamamuro T, Hench LL, Wilson J (eds) Handbook of bioactive ceramics, vol 11. CRC press, Boca Raton, pp 3–16
43. Hench LL (1999) Bioactive glasses and glasses-ceramics. In: Shackelford JF (ed) Bioceramics – applications of ceramic and glass materials in medicine. Trans Tech Publication, Switzerland, pp 37–64
44. Nalla RK, Kinney JH, Ritchie RO (2003) Mechanistic fracture criteria for the failure of human cortical bone. Nat Mater 2(3):164–168
45. Zioupos P, Currey JD (1998) Changes in the stiffness, strength, and toughness of human cortical bone with age. Bone 22(1):57–66
46. Keaveny TM, Hayes WC (1993) Mechanical properties of cortical and trabecular bone. In: Hall BK (ed) Bone. A treatise, vol 7, Bone growth. CRC Press, Boca Raton, pp 285–344
47. Moore WR, Graves SE, Bain GI (2001) Synthetic bone graft substitutes. Aust N Z J Surg 71(6):354–361
48. Pereira MM, Clark AE, Hench LL (1994) Calcium-phosphate formation on sol–gel-derived bioactive glasses in-vitro. J Biomed Mater Res 28(6):693–698
49. Hench LL (1997) Sol–gel materials for bioceramic applications. Curr Opin Solid State Mater Sci 2(5):604–610
50. Hench LL (1998) Bioceramics. J Am Ceram Soc 81(7):1705–1728
51. Wilson J, Pigott GH, Schoen FJ, Hench LL (1981) Toxicology and biocompatibility of bioglasses. J Biomed Mater Res 15(6):805–817
52. Hench LL, Wilson J (1984) Surface-active biomaterials. Science 226(4675):630–636
53. Lobel KD, Hench LL (1996) In-vitro protein interactions with a bioactive gel-glass. J Sol–Gel Sci Technol 7(1–2):69–76

54. Lobel KD, Hench LL (1998) In vitro adsorption and activity of enzymes on reaction layers of bioactive glass substrates. J Biomed Mater Res 39(4):575–579
55. Ohgushi H, Dohi Y, Yoshikawa T, Tamai S, Tabata S, Okunaga K, Shibuya T (1996) Osteogenic differentiation of cultured marrow stromal stem cells on the surface of bioactive glass ceramics. J Biomed Mater Res 32(3):341–348
56. Aksay IA, Weiner S (1998) Biomaterials – is this really a field of research? Curr Opin Solid State Mater Sci 3(3):219–220
57. Day RM, Boccaccini AR, Shurey S, Roether JA, Forbes A, Hench LL, Gabe SM (2004) Assessment of polyglycolic acid mesh and bioactive glass for soft-tissue engineering scaffolds. Biomaterials 25(27):5857–5866
58. Keshaw H, Forbes A, Day RM (2005) Release of angiogenic growth factors from cells encapsulated in alginate beads with bioactive glass. Biomaterials 26(19):4171–4179
59. Schepers E, Declercq M, Ducheyne P, Kempeneers R (1991) Bioactive glass particulate material as a filler for bone-lesions. J Oral Rehabil 18(5):439–452
60. Gatti AM, Valdre G, Andersson OH (1994) Analysis of the in-vivo reactions of a bioactive glass in soft and hard-tissue. Biomaterials 15(3):208–212
61. Roether JA, Gough JE, Boccaccini AR, Hench LL, Maquet V, Jerome R (2002) Novel bioresorbable and bioactive composites based on bioactive glass and polylactide foams for bone tissue engineering. J Mater Sci Mater Med 13(12):1207–1214
62. Oonishi H, Kushitani S, Yasukawa E, Iwaki H, Hench LL, Wilson J, Tsuji EI, Sugihara T (1997) Particulate bioglass compared with hydroxyapatite as a bone graft substitute. Clin Orthop Relat Res 334:316–325
63. Xynos ID, Edgar AJ, Buttery LDK, Hench LL, Polak JM (2000) Ionic products of bioactive glass dissolution increase proliferation of human osteoblasts and induce insulin-like growth factor II mRNA expression and protein synthesis. Biochem Biophys Res Commun 276(2):461–465
64. Xynos ID, Hukkanen MVJ, Batten JJ, Buttery LD, Hench LL, Polak JM (2000) Bioglass$_{®}$ 45S5 stimulates osteoblast turnover and enhances bone formation in vitro: implications and applications for bone tissue engineering. Calcif Tissue Int 67(4):321–329
65. Laurencin CT, Lu HH, Khan Y (2002) Processing of polymer scaffolds: polymer-ceramic composite foams. In: Atala A, Lanza RP (eds) Methods of tissue engineering. Academic, California, pp 705–714
66. Boccaccini AR, Notingher I, Maquet V, Jerome R (2003) Bioresorbable and bioactive composite materials based on polylactide foams filled with and coated by Bioglass® particles for tissue engineering applications. J Mater Sci Mater Med 14(5):443–450
67. Livingston T, Ducheyne P, Garino J (2002) In vivo evaluation of a bioactive scaffold for bone tissue engineering. J Biomed Mater Res 62(1):1–13
68. Kaufmann E, Ducheyne P, Shapiro IM (2000) Evaluation of osteoblast response to porous bioactive glass (45S5) substrates by RT-PCR analysis. Tissue Eng 6(1):19–28
69. Yuan HP, de Bruijn JD, Zhang XD, van Blitterswijk CA, de Groot K (2001) Bone induction by porous glass ceramic made from Bioglass$_{®}$ (45S5). J Biomed Mater Res 58(3):270–276
70. Hench LL (1997) Theory of bioactivity: the potential for skeletal regeneration. Anales De Quimica 93(1):S44–S48
71. Peitl O, LaTorre GP, Hench LL (1996) Effect of crystallization on apatite-layer formation of bioactive glass 45S5. J Biomed Mater Res 30(4):509–514
72. Li P, Yang Q, Zhang F, Kokubo T (1992) The effect of residual glassy phase in a bioactive glass-ceramic on the formation of its surface apatite layer in vitro. J Mater Sci Mater Med 3(6):452–456
73. Clupper DC, Hench LL (2003) Crystallization kinetics of tape cast bioactive glass 45S5. J Non-Cryst Solids 318(1–2):43–48
74. Hench LL, Kokubo T (1998) Properties of bioactive glasses and glass-ceramics. In: Black J, Hastings G (eds) Handbook of biomaterial properties. Chapman & Hall, London, pp 355–363
75. Holand W, Vogel W (1993) Machinable and phosphate glass-ceramics. In: Hench LL, Wilson J (eds) An introduction to bioceramics, vol Singapore. World Scientific, Singapore, pp 125–137

76. Rawlings RD, Wu JP, Boccaccini AR (2006) Glass-ceramics: their production from wastes – a review. J Mater Sci 41:733–761
77. Kingery WD, Bowen HK, Uhlmann DR (1976) Introduction to ceramics, 2nd edn. Wiley, New York
78. Calver A, Hill RG, Stamboulis A (2004) Influence of fluorine content on the crystallization behavior of apatite-wollastonite glass-ceramics. J Mater Sci 39(7):2601–2603
79. Rafferty A, Clifford A, Hill R, Wood D, Samuneva B, Dimitrova-Lukacs M (2000) Influence of fluorine content in apatite-mullite glass-ceramics. J Am Ceram Soc 83(11): 2833–2838
80. Yamamuro T (1993) A/W glass-ceramic: clinical applications. In: Hench LL, Wilson J (eds) An introduction to bioceramics. World Scientific, Singapore, pp 89–103
81. Grodd UM, Muller-Mai C, Voigt C (1993) Ceravital® bioactive glass-ceramics. In: Hench LL, Wilson J (eds) An introduction to bioceramics. World Scientific, Singapore, pp 105–123
82. Russell JL, Block JE (1999) Clinical utility of demineralized bone matrix for osseous defects, arthrodesis, and reconstruction: impact of processing techniques and study methodology. Orthopedics 22(5):524–531
83. Yaylaoglu MB, Yildiz C, Korkusuz F, Hasirci V (1999) A novel osteochondral implant. Biomaterials 20(16):1513–1520
84. Du C, Cui FZ, Zhu XD, de Groot K (1999) Three-dimensional nano-HAp/collagen matrix loading with osteogenic cells in organ culture. J Biomed Mater Res 44(4):407–415
85. Brown CD, Hoffman AS (2002) Modification of natural polymer: chitosan. In: Atala A, Lanza RP (eds) Methods of tissue engineering. Academic, California, pp 565–574
86. Badylak SE (2002) Modification of natural polymers: collagen. In: Atala A, Lanza RP (eds) Methods of tissue engineering. Academic, California, pp 505–514
87. Atala A, Lanza RP (eds) (2002) Methods of tissue engineering. Academic, California
88. Wan ACA, Khor E, Hastings GW (1998) Preparation of a chitin-apatite composite by in situ precipitation onto porous chitin scaffolds. J Biomed Mater Res 41(4):541–548
89. Klokkevold PR, Vandemark L, Kenney EB, Bernard GW (1996) Osteogenesis enhanced by chitosan (poly-N-acetyl glycosaminoglycan) in vitro. J Periodontol 67(11):1170–1175
90. Muzzarelli RAA, Zucchini C, Ilari P, Pugnaloni A, Belmonte MM, Biagini G, Castaldini C (1993) Osteoconductive properties of methylpyrrolidinone chitosan in an animal-model. Biomaterials 14(12):925–929
91. Muzzarelli RAA, Muzzarelli C (2005) Chitosan chemistry: relevance to the biomedical sciences. In: Polysaccharides 1: structure, characterization and use, vol 186, Advances in polymer science. Springer, Berlin, pp 151–209
92. Middleton JC, Tipton AJ (2000) Synthetic biodegradable polymers as orthopedic devices. Biomaterials 21(23):2335–2346
93. Langer R, Vacanti JP, Vacanti CA, Atala A, Freed LE, Vunjak-Novakovic G (1995) Tissue engineering biomedical applications. Tissue Eng 1:151–161
94. Bergsma EJ, Rozema FR, Bos RRM, Debruijn WC (1993) Foreign-body reactions to resorbable poly(L-lactide) bone plates and screws used for the fixation of unstable zygomatic fractures. J Oral Maxillofac Surg 51(6):666–670
95. Tams J, Rozema FR, Bos RRM, Roodenburg JLN, Nikkels PGJ, Vermey A (1996) Poly(L-lactide) bone plates and screws for internal fixation of mandibular swing osteotomies. Int J Oral Maxillofac Surg 25(1):20–24
96. Martin C, Winet H, Bao JY (1996) Acidity near eroding polylactide-polyglycolide in vitro and in vivo in rabbit tibial bone chambers. Biomaterials 17(24):2373–2380
97. Suuronen R, Pohjonen T, Hietanen J, Lindquist C (1998) A 5-year in vitro and in vivo study of the biodegradation of polylactide plates. J Oral Maxillofac Surg 56(5):604–614
98. Tatakis DN, Trombelli L (1999) Adverse effects associated with a bioabsorbable guided tissue regeneration device in the treatment of human gingival recession defects. A clinicopathologic case report. J Periodontol 70(5):542–547
99. Bostman OM, Pihlajamaki HK (2000) Adverse tissue reactions to bioabsorbable fixation devices. Clin Orthop Relat Res 371:216–227

100. Gollwitzer H, Ibrahim K, Meyer H, Mittelmeier W, Busch R, Stemberger A (2003) Antibacterial poly(D, L-lactic acid) coating of medical implants using a biodegradable drug delivery technology. J Antimicrob Chemother 51(3):585–591
101. Schmidmaier G, Wildemann B, Stemberger A, Haas NP, Raschke M (2001) Biodegradable poly(D, L-lactide) coating of implants for continuous release of growth factors. J Biomed Mater Res 58(4):449–455
102. Schmidmaier G, Wildemann B, Bail H, Lucke M, Fuchs T, Stemberger A, Flyvbjerg A, Haas NP, Raschke M (2001) Local application of growth factors (insulin-like growth factor-1 and transforming growth factor-beta 1) from a biodegradable poly(D, L-lactide) coating of osteosynthetic implants accelerates fracture healing in rats. Bone 28(4):341–350
103. Herrmann R, Schmidmaier G, Markl B, Resch A, Hahnel I, Stemberger A, Alt E (1999) Antithrombogenic coating of stents using a biodegradable drug delivery technology. Thromb Haemost 82(1):51–57
104. Maquet V, Boccaccini AR, Pravata L, Notingher I, Jerome R (2003) Preparation, characterization, and in vitro degradation of bioresorbable and bioactive composites based on Bioglass®-filled polylactide foams. J Biomed Mater Res A 66A(2):335–346
105. Maquet V, Boccaccini AR, Pravata L, Notingher I, Jerome R (2004) Porous poly(alpha-hydroxyacid)/Bioglass® composite scaffolds for bone tissue engineering. I: preparation and in vitro characterisation. Biomaterials 25(18):4185–4194
106. Blaker JJ, Gough JE, Maquet V, Notingher I, Boccaccini AR (2003) In vitro evaluation of novel bioactive composites based on Bioglass$_{®}$-filled polylactide foams for bone tissue engineering scaffolds. J Biomed Mater Res A 67A(4):1401–1411
107. Verrier S, Blaker JJ, Maquet V, Hench LL, Boccaccini AR (2004) PDLLA/Bioglass$_{®}$ composites for soft-tissue and hard-tissue engineering: an in vitro cell biology assessment. Biomaterials 25(15):3013–3021
108. Lu LC, Mikos AG (1999) Poly(lactic acid). In: Mark JE (ed) Polymer data handbook. Oxford Press, Oxford, pp 527–533
109. Yang SF, Leong KF, Du ZH, Chua CK (2001) The design of scaffolds for use in tissue engineering. Part 1. Traditional factors. Tissue Eng 7(6):679–689
110. Lu LC, Mikos AG (1999) Poly(glycolic acid). In: Mark JE (ed) Polymer data handbook. Oxford Press, Oxford, pp 566–569
111. Ramakrishna S, Huang ZM, Kumar GV, Batchelor AW, Mayer J (2004) An introduction to biocomposites. Imperial College Press, London
112. Iroh JO (1999) Poly(epsilon-caprolactone). In: Mark JE (ed) Polymer data handbook. Oxford Press, Oxford, pp 361–362
113. Baji A, Wong SC, Srivatsan TS, Njus GO, Mathur G (2006) Processing methodologies for polycaprolactone-hydroxyapatite composites: a review. Mater Manufacturing Processes 21(2):211–218
114. Calandrelli L, Immirzi B, Malinconico M, Luessenheide S, Passaro I, Di Pasquale R, Oliva A (2004) Natural and synthetic hydroxyapatite filled PCL: mechanical properties and biocompatibility analysis. J Bioactive Compatible Polymers 19(4):301–313
115. Ramsay BA, Langlade V, Carreau PJ, Ramsay JA (1993) Biodegradability and mechanical-properties of poly-(beta-hydroxybutyrate-Co-beta-hydroxyvalerate) starch blends. Appl Environ Microbiol 59(4):1242–1246
116. Chen GQ, Wu Q (2005) The application of polyhydroxyalkanoates as tissue engineering materials. Biomaterials 26(33):6565–6578
117. Yoda N (2003) Synthesis of polyanhydrides II: new aromatic polyanhydrides with high melting points and fibre-forming properties. Die Markromolekulare Chemie 32:1–12
118. Kellomaki M, Heller J, Tormala P (2000) Processing and properties of two different poly (ortho esters). J Mater Sci Mater Med 11(6):345–355
119. Magill JH (1999) Poly(phosphazenes), bioerodible. In: Mark JE (ed) Polymer data handbook. Oxford Press, Oxford, pp 746–749
120. Laurencin CT, Lakshmi S Polyphosphazene nanofibers for biomedical applications: preliminary studies. In: proceeding of the NATOASI conference, Antalya, Turkey, 1–12 Sep 2003, pp 281–300

121. Kumudine C, Premachandra JK (1999) Collagen. In: Mark JE (ed) Polymer data handbook. Oxford Press, Oxford, pp 70–77
122. Pitt CG, Chasalow FI, Hibionada YM, Klimas DM, Schindler A (1981) Aliphatic polyesters.1. The degradation of poly(epsilon-caprolactone) in vivo. J Appl Polymer Sci 26(11):3779–3787
123. Pitt CG, Gratzl MM, Kimmel GL, Surles J, Schindler A (1981) Aliphatic polyesters.2. The degradation of poly(Dl-lactide), poly(epsilon-caprolactone), and their copolymers in vivo. Biomaterials 2(4):215–220
124. Gabelnick HL (1983) Biodegradable implants: alternative approaches. In: Mishell DR (ed) Advanced in human fertility and reproductive endocrinology, vol 2, Long acting steroid contraception. Raven, New York, pp 149–173
125. Doi Y, Kitamura S, Abe H (1995) Microbial synthesis and characterization of poly(3-hydroxybutyrate-Co-3-hydroxyhexanoate). Macromolecules 28(14):4822–4828
126. Li HY, Du RL, Chang J (2005) Fabrication, characterization, and in vitro degradation of composite scaffolds based on PHBV and bioactive glass. J Biomater Appl 20(2):137–155
127. Misra SK, Valappil SP, Roy I, Boccaccini AR (2006) Polyhydroxyalkanoate (PHA)/inorganic phase composites for tissue engineering applications. Biomacromolecules 7(8):2249–2258
128. Doyle C, Tanner ET, Bonfield W (1991) In vitro and in vivo evaluation of polyhydroxybutyrate and of polyhydroxybutyrate reinforced with hydroxyapatite. Biomaterials 12(9): 841–847
129. Verma S, Bhatia Y, Valappil SP, Roy I (2002) A possible role of poly-3-hydroxybutyric acid in antibiotic production in Streptomyces. Arch Microbiol 179(1):66–69
130. Payne RG, Mikos AG (2002) Synthesis of synthetic polymers: poly(propylene fumarate). In: Atala A, Lanza RP (eds) Methods of tissue engineering. Academic, California, pp 649–652
131. Yaszenski MJ, Payne RG, Hayes WC, Langer R, Aufdemorte TB, Mikos AG (1995) The ingrowth of new bone tissue and initial mechanical properties of a degrading polymeric composite scaffold. Tissue Eng 1:41–52
132. Peter SJ, Lu L, Kim DJ, Stamatas GN, Miller MJ, Yaszemski MJ, Mikos AG (2000) Effects of transforming growth factor beta 1 released from biodegradable polymer microparticles on marrow stromal osteoblasts cultured on poly(propylene fumarate) substrates. J Biomed Mater Res 50(3):452–462
133. Shastri VP, Zelikin A, Hildgen P (2002) Synthesis of synthetic polymers: poly(anhydrides). In: Atala A, Lanza RP (eds) Methods of tissue engineering. Academic, California, pp 609–617
134. Domb AJ, Langer R (1999) Poly(1,3-bis-p-carboxyphenoxypropane anhydride). In: Mark JE (ed) Polymer data handbook. Oxford Press, Oxford, pp 303–305
135. Domb AJ, Langer R (1999) Poly(erucic acid dimmer anhydride). In: Mark JE (ed) Polymer data handbook. Oxford Press, Oxford, pp 457–459
136. Shastri V, Marini P, Padera R, Kirchain S, Tarcha P, Langer R (1999) Osteocompatibility of photopolymerizable anhydride networks. Mater Res Soc Symp Proc 530:93–98
137. Solheim E, Sudmann B, Bang G, Sudmann E (2000) Biocompatibility and effect on osteogenesis of poly(ortho ester) compared to poly(DL-lactic acid). J Biomed Mater Res 49(2): 257–263
138. Andriano KP, Gurny R, Heller J (2002) Synthesis of synthetic polymers: Poly(ortho esters). In: Atala A, Lanza RP (eds) Methods of tissue engineering. Academic, California, pp 619–627
139. Allcock HR (2002) Syntheses of synthetic polymers: polyphosphazenes. In: Atala A, Lanza RP (eds) Methods of tissue engineering. Academic, California, pp 597–608
140. Laurencin CT, Norman ME, Elgendy HM, Elamin SF, Allcock HR, Pucher SR, Ambrosio AA (1993) Use of polyphosphazenes for skeletal tissue regeneration. J Biomed Mater Res 27(7):963–973
141. Laurencin CT, ElAmin SF, Ibim SE, Willoughby DA, Attawia M, Allcock HR, Ambrosio AA (1996) A highly porous 3-dimensional polyphosphazene polymer matrix for skeletal tissue regeneration. J Biomed Mater Res 30(2):133–138
142. Bergsma JE, Debruijn WC, Rozema FR, Bos RRM, Boering G (1995) Late degradation tissue-response to poly(L-lactide) bone plates and screws. Biomaterials 16(1):25–31

143. Temenoff JS, Lu L, Mikos AG (2000) Bone tissue engineering using synthetic biodegradable polymer scaffolds. In: Davies JE (ed) Bone engineering. EM Squared, Toronto, pp 455–462
144. Kikuchi M, Tanaka J, Koyama Y, Takakuda K (1999) Cell culture test of TCP/CPLA composite. J Biomed Mater Res 48(2):108–110
145. Attawia MA, Herbert KM, Laurencin CT (1995) Osteoblast-like cell adherence and migration through 3-dimensional porous polymer matrices. Biochem Biophys Res Commun 213(2):639–644
146. Laurencin CT, Attawia MA, Elgendy HE, Herbert KM (1996) Tissue engineered bone-regeneration using degradable polymers: the formation of mineralized matrices. Bone 19(1):S93–S99
147. Devin JE, Attawia MA, Laurencin CT (1996) Three-dimensional degradable porous polymer-ceramic matrices for use in bone repair. J Biomater Sci Polym Ed 7(8):661–669
148. Stamboulis AG, Boccaccini AR, Hench LL (2002) Novel biodegradable polymer/bioactive glass composites for tissue engineering applications. Adv Eng Mater 4(3):105–109
149. Lu HH, El-Amin SF, Scott KD, Laurencin CT (2003) Three-dimensional, bioactive, biodegradable, polymer-bioactive glass composite scaffolds with improved mechanical properties support collagen synthesis and mineralization of human osteoblast-like cells in vitro. J Biomed Mater Res A 64A(3):465–474
150. Zhang K, Wang YB, Hillmyer MA, Francis LF (2004) Processing and properties of porous poly(L-lactide)/bioactive glass composites. Biomaterials 25(13):2489–2500
151. Blaker JJ, Day RM, Maquet V, Boccaccini AR (2004) Novel bioresorbable poly(lactide-co-glycolide) (PLGA) and PLGA/Bioglass® composite tubular foam scaffolds for tissue engineering applications. In: Advanced materials forum Ii, vol 455–456, Materials science forum. Trans Tech Publications Ltd, Zurich-Uetikon, pp 415–419
152. Blaker JJ, Maquet V, Jerome R, Boccaccini AR, Nazhat SN (2005) Mechanical properties of highly porous PDLLA/Bioglass® composite foams as scaffolds for bone tissue engineering. Acta Biomater 1(6):643–652
153. Navarro M, Ginebra MP, Planell JA, Zeppetelli S, Ambrosio L (2004) Development and cell response of a new biodegradable composite scaffold for guided bone regeneration. J Mater Sci Mater Med 15(4):419–422
154. Li HY, Chang J (2004) Preparation and characterization of bioactive and biodegradable wollastonite/poly(D, L-lactic acid) composite scaffolds. J Mater Sci Mater Med 15(10): 1089–1095
155. Kasuga T, Ota Y, Nogami M, Abe Y (2001) Preparation and mechanical properties of polylactic acid composites containing hydroxyapatite fibers. Biomaterials 22(1):19–23
156. Deng XM, Hao JY, Wang CS (2001) Preparation and mechanical properties of nanocomposites of poly(D, L-lactide) with Ca-deficient hydroxyapatite nanocrystals. Biomaterials 22(21):2867–2873
157. Xu HHK, Simon CG (2004) Self-hardening calcium phosphate cement-mesh composite: reinforcement, macropores, and cell response. J Biomed Mater Res A 69A(2):267–278
158. Xu HHK, Quinn JB, Takagi S, Chow LC (2004) Synergistic reinforcement of in situ hardening calcium phosphate composite scaffold for bone tissue engineering. Biomaterials 25(6):1029–1037
159. Xu HHK, Simon CG Jr (2004) Self-hardening calcium phosphate composite scaffold for bone tissue engineering. J Orthop Res 22(3):535–543
160. Greish YE, Bender JD, Lakshmi S, Brown PW, Allcock HR, Laurencin CT (2005) Low temperature formation of hydroxyapatite-poly(alkyl oxybenzoate)phosphazene composites for biomedical applications. Biomaterials 26(1):1–9
161. Rodrigues CVM, Serricella P, Linhares ABR, Guerdes RM, Borojevic R, Rossi MA, Duarte MEL, Farina M (2003) Characterization of a bovine collagen-hydroxyapatite composite scaffold for bone tissue engineering. Biomaterials 24(27):4987–4997
162. Peter SJ, Miller ST, Zhu GM, Yasko AW, Mikos AG (1998) In vivo degradation of a poly(propylene fumarate) beta-tricalcium phosphate injectable composite scaffold. J Biomed Mater Res 41(1):1–7

163. Juhasz JA, Best SM, Kawashita M, Miyata N, Kokubo T, Nakamura T, Bonfield W (2003) Mechanical properties of glass-ceramic A-W – polyethylene composites: effect of filler content. In: Bioceramics 15, vol 240–2, Key engineering materials. Trans Tech Publications Ltd, Zurich-Uetikon, pp 947–950
164. Juhasz JA, Best SM, Bonfield W, Kawashita M, Miyata N, Kokubo T, Nakamura T (2003) Apatite-forming ability of glass-ceramic apatite-wollastonite – polyethylene composites: effect of filler content. J Mater Sci Mater Med 14(6):489–495
165. Juhasz JA, Best SM, Brooks R, Kawashita M, Miyata N, Kokubo T, Nakamura T, Bonfield W (2004) Mechanical properties of glass-ceramic A-W-polyethylene composites: effect of filler content and particle size. Biomaterials 25(6):949–955
166. Kasuga T, Maeda H, Kato K, Nogami M, Hata K, Ueda M (2003) Preparation of poly(lactic acid) composites containing calcium carbonate (vaterite). Biomaterials 24(19):3247–3253
167. Ambrosio AMA, Sahota JS, Khan Y, Laurencin CT (2001) A novel amorphous calcium phosphate polymer ceramic for bone repair: 1. Synthesis and characterization. J Biomed Mater Res 58(3):295–301
168. Khan YM, Katti DS, Laurencin CT (2004) Novel polymer-synthesized ceramic composite-based system for bone repair: an in vitro evaluation. J Biomed Mater Res A 69A(4): 728–737
169. Yin YJ, Ye F, Cui JF, Zhang FJ, Li XL, Yao KD (2003) Preparation and characterization of macroporous chitosan-gelatin beta-tricalcium phosphate composite scaffolds for bone tissue engineering. J Biomed Mater Res A 67A(3):844–855
170. Zhang RY, Ma PX (1999) Poly(alpha-hydroxyl acids) hydroxyapatite porous composites for bone-tissue engineering. I. Preparation and morphology. J Biomed Mater Res 44(4): 446–455
171. Guan LM, Davies JE (2004) Preparation and characterization of a highly macroporous biodegradable composite tissue engineering scaffold. J Biomed Mater Res A 71A(3):480–487
172. Giesen EBW, Ding M, Dalstra M, van Eijden T (2001) Mechanical properties of cancellous bone in the human mandibular condyle are anisotropic. J Biomech 34(6):799–803
173. Yeni YN, Fyhrie DP (2001) Finite element calculated uniaxial apparent stiffness is a consistent predictor of uniaxial apparent strength in human vertebral cancellous bone tested with different boundary conditions. J Biomech 34(12):1649–1654
174. Yeni YN, Hou FJ, Vashishth D, Fyhrie DP (2001) Trabecular shear stress in human vertebral cancellous bone: intra- and inter-individual variations. J Biomech 34(10):1341–1346

Chapter 24
Gene Therapy for the Inner Ear: Progress and Prospects

Hong Sun, Aji Huang, Hong Liu, and Shousong Cao

Abstract Inner ear diseases increasingly affected the quality of life of patients. The current diagnostic protocol used in auditory area has been in place for over half a century and is insensitive to detect alterations in hearing status at early stage. Furthermore, no effective treatment is available for inner ear diseases although some patients may benefit from hearing aid with a hearing amplifier or cochlear implantation. Understanding the molecular mechanisms and investigating the fundamental process of hearing disorder could help us to avoid hearing defects or even to cure the disease by gene-based or molecular therapy. Currently, gene therapy for inner ear disease has become an emerging field of study. The medicine for treatment of inner ear diseases is undergoing a revolutionary change since the completion of the human genome sequencing. Various new discoveries and advanced technologies have been made in inner ear gene therapy such as viral-based or nonviral-based gene vectors, delivery strategies, and therapeutic genes and targets as well as the animal models for study of gene therapy for inner ear. Gene therapy may become an effective treatment in clinic for inner ear diseases in the future. This chapter is to summarize current significant advances and technological challenges for inner ear gene therapy

H. Sun (✉) • H. Liu
Department of Otolaryngology, Head & Neck Surgery,
Xiangya Hospital, Central South University,
87 Xiang Ya Road, Changsha, Hunan 410008, China
e-mail: shjhaj@vip.163.com

A. Huang
Department of Breast Surgery,
Affiliated Cancer Hospital, Medical
Center of Fudan University, Shanghai 200032, China

S. Cao (✉)
Department of Medicine, Roswell Park Cancer Institute,
Elm and Carlton Streets, Buffalo, NY 14263, USA

Central South University,
Changsha, Hunan, China
e-mail: shousong.cao@roswellpark.org

M.K. Danquah, R.I. Mahato (eds.), *Emerging Trends in Cell and Gene Therapy*, DOI 10.1007/978-1-62703-417-3_24,

and to discuss the previous works in detail by examining and analyzing the publications of inner ear gene therapy from the literature and patent documents and identify the promising methods, novel techniques, and vital research projects. Additionally, we also discuss the progress and prospects of inner ear gene therapy, the advances and shortages with possible solutions within this field of research.

Keywords Inner ear • Hearing loss • Gene therapy • Gene transfer • Gene vector • Cochlea

24.1 Introduction

24.1.1 Background of Inner Ear and Hearing Disorders

Inner ear, the innermost part of the vertebrate ear, is a highly differentiated peripheral auditory organ. It contains bony labyrinth, membranous labyrinth, a hollow cavity with a system composed of two basic parts, and lies in the temporal bones of both mammalian animals and human being. It consists of cochlea comprising organ of Corti, which plays the role of detecting the mechanical vibrates of sound waves, and vestibular organ containing the utricle, saccule, and ampullae, which plays the role of detecting linear motion (utricle and saccule) and curvilinear motion (ampullae). The sensory epitheliums of the cochlea and vestibule can transform mechanical stimuli to bioelectrical impulses, and the impulses could be then recognized by the brain. In detail, the afferent neurons contacting the bottoms of the hair cells in these sensory epitheliums transfer the impulses from hair cells to the auditory or vestibular nerve center located in the temporal lobe of the cerebrum via auditory brain stem. Consequently, the inner ear is extremely important for detecting hearing and balance stimuli which is crucial for an individual's responses to the environments, especially in special situations.

Hearing disorder is differentiated based on the age of onset, site of trauma, and diversity severity [4]. Many people suffer hearing disorders such as hearing loss, tinnitus, presbycusis, and vertigo. There are three main types of organic hearing impairment: sensorineural hearing loss (SNHL), conductive hearing loss, and mixed hearing loss (combination of sensorineural and conductive loss) [59]. SNHL is the most common form of deafness and accounts for about 90 % of all hearing loss. SNHL mostly affects the sensory epithelia in the organ of Corti of the inner ear, and it may be further divided into two types, i.e., cochlear deafness and retrocochlear deafness (central deafness). About 60 % of hearing defects, which have a genetic basis and an important part of gene-based hearing loss, did not present typical syndrome. Instead, most of them are inherited in an autosomal recessive mode [52]. Currently, disabling hearing impairment (deafness) affected 250 million of people in the world in 2001, and it was much higher than that of previously estimated. About two thirds of the 250 million affected people were from the developing countries. There are 110 million of deaf persons from Southeast Asia of a population of 1.5 billion, with 7.3 % overall prevalence of deafness. A population-based survey of ear and hearing disorders for 6,626 persons in Guizhou province of China demonstrated that the prevalence of hearing impairment was 17.1 % (the standardized rate was 17.6 % for the whole country) and hearing disability was 6.1 % (the standardized rate was 6.5 % for the whole country), the rate slightly lower than the global population of 7.3 % [90]. About 35 million Americans had detectable hearing loss, affecting approximately 1.7 % of the people under the age of 18, 31.4 % over the age of 65, and 40–50 % over the age of 75 and older [53]. Two to three out of every 1,000 children were born deaf or hard of hearing in the United States. The main cause of hearing disability is SNHL, usually resulting from hypoxia-ischemia of the cochlea, ototoxicity, noise injury, virus infection, and heredity or gene mutation. SNHL could be inherited as an autosomal dominant or recessive pattern, with 90 % as autosomal recessive. Unfortunately, no substantial therapy could reverse SNHL so far. Hearing loss affects more people than any other disease and negatively impacts the quality of life of numerous patients and their families although it is not a life-threatening disease. Therefore, it is in urgent need to find effective treatment for this disease. To develop pharmacological treatment or molecular therapeutics for hearing and balance disorders, we should firstly accurately diagnose the given defect(s) and advance easily diagnosable disease process because similar hearing impairments may be caused by various pathological mechanisms or genetic deficits. For instance, either loss of sensorial hair cells, spiral ganglion neurons, or lesions of the cochlear lateral wall may lead to a hearing loss.

24.1.2 Initiation and the Critical Challenges of Inner Ear Gene Therapy

Since there are no specially efficient and effective pharmaceutical treatment for inner ear disease, inner ear gene therapy is considered as a possible and prospective method to correct various hearing disorders. We are able to better understand the genetic defects, which cause various kinds of hereditary deafness, and to identify the molecular basis of degenerative diseases due to the advance of biochemistry and genetics and the development of molecular biological techniques. In the auditory system, the hair cells and spiral ganglion neurons are the common pathological sites of inner ear diseases and also the major targets for inner ear gene therapy [8].

In the early 1970s, the concept of gene therapy was initially raised by Osterman and colleagues [58]; afterward, enormous development and progress have been made in this field with the therapeutic genes, the vector for gene transfection, the target cells/organs, the routes of gene delivery, animal models for gene therapy, prevention and management of the complications, and the related ethical problems caused by gene therapy [58, 63]. Gene therapy may be regarded as an approach to treat diseases by introducing a desired foreign gene or gene regulatory elements such as RNA interference into the target cells to replace or fix the cell's defective gene or make the transfected cell to express the products of the desired therapeutic gene to reverse or cure a disease by the transfected therapeutic gene [49]. In other words, it could be defined as regulated expression of an introduced gene that achieves a biological effect and/or produces a therapeutic effect in the desired cell or tissue by a genetically engineered vector. The general process is probably described as several significant steps. Initially, the gene therapy should be based on desired biological effects. Secondly, the process should be easily accessible, and the therapeutic gene could enter the target cell or tissue at adequate level. Thirdly, the therapeutic gene ought to target considerable and enough cell population. Finally, the regulation of the transgene should ensure a consistent expression of the gene and the safety of the host [88]. Considering the risk of degradation of the foreign nucleic acid by the nuclease in the body, the foreign gene needs to be delivered to the appropriate place. Therefore, the gene must be carried by a vehicle to help it entering the target cells and protect it from the endosomes, disconnect each other in the cytoplasm, and finally enter the nucleus. Such vehicle is currently defined as gene vector [98]. Gene vector is a crucial part among the four major elements of gene therapy, the gene vector, the route of gene administration, the therapeutic gene, and the target cells.

About 25 years after the initiation of gene therapy research, the study of gene therapy for the inner ear was started in the 1990s. In 1994, Fujiyoshi and colleagues firstly reported the study of therapy for hearing disorders [19]. They developed the myelin basic protein (MBP) transgenic mice by microinjecting an MBP cosmid clone into the pronucleus of fertilized eggs of shiverer mice and then replaced the autosomal recessive mutation (deletion) gene by the transgene for MBP. Interestingly, they found that the MBP-transgenic mice recovered MBP up to 25 % of normal levels, and significantly higher myelinated axons were present in the transgenic mice compared to control mice. Additionally, the inter-peak latencies of auditory brainstem response were shortened in the transgenic mice compared to the control mice. In the following years, other research groups also reported the successful transfection of foreign gene into inner ear using replication-deficient viral vectors in vitro and in vivo [43, 64]. The inner ear, especially the cochlea, is a highly differentiated and extremely precise and sensitive electrophysiological organ with finespun anatomical structure; any improper intervention during the process of gene transfection may result in an undesired morphological and/or functional damage to the inner ear. As a result, it is a technical problem to transfect exogenous gene into the inner ear without affecting the morphological and physiological of inner ear, besides facing the same problems of gene transduction with other organs or tissues, such as the safety and transfection efficiency of the vectors. For instance, the blood-labyrinth barrier prevents macromolecules moving from peripheral blood into the inner ear, and the invasive approach of gene transfection may cause the morphological and/or functional damage to the inner ear which is a disaster to the patient. Up to now, a lot of

Table 24.1 Summary of the critical challenges, progress, and prospects for inner ear gene therapy

Summary		
Critical challenges	Progress	Prospects
Delivering the therapeutic gene to the target cells or tissues	Hair cell regeneration in mammal cochlea with Math1 gene transfection	Breakthrough of transfection efficiency for nonviral vectors or multiplex gene vectors within 5 years
Delivering the therapeutic gene safely and effectively	Bactofection with bacteria as a vector	Intact RWM approach to replace labyrinth drilling or RWM injection for inner ear gene delivery
Controlling exactly the degree and efficacy of gene expression	Multiplex gene vectors	Intrauterine gene therapy for the treatment of hereditary or congenital deafness with improved endoscopic technique and related instruments
Targeting gene expression to the desired place and maintaining a appropriate duration	Inorganic nanoparticle vectors	Regeneration of functional hair cells in the mammal cochlea via cochlear cell's transdifferentiation
Overcoming the toxicity of vectors for gene delivery	Delivering therapeutic genes via intrauterine approach or along with an implanted cochlear electrode	The first case of successful inner ear gene therapy may be reported within 5–8 years
Keeping gene expression persistently after gene therapy	Delivering therapeutic genes into inner ear via intact round window membrane	Gene therapy for inner ear may become a common practice for the treatment of inner ear diseases within 15–20 years
Translating the basic research into clinical practice	Developing new therapeutic genes such as otospiralin, connexins, and XIAP	

exciting advances and great progress have already been made in the field of research for inner ear gene therapy. The study of inner ear gene therapy has become an emerging field, which may bring a glimmer of hope for successful treatment of hearing disability. However, there still have been many obstacles and challenges which limit the rapid development of gene therapy for inner ear into the common clinical practice. We need to address several critical questions before we could move it into the clinic: (1) how to deliver the therapeutic gene to the target cells or tissues, (2) how to exactly handle the degree and efficacy of gene expression, (3) how to control the gene expression to a desired degree and place, (4) how to overcome the toxicity of vectors for gene delivery, and (5) how to keep persistent gene expression after gene therapy [24, 57, 67, 68, 71, 85]. Furthermore, the advances and progress of inner ear gene therapy are still at the experimental stage, and it has a long way to go for translating the success of laboratory research into clinical practice. The critical challenges, progress, and prospects for inner ear gene therapy are summarized in Table 24.1.

24.2 Holistic Progress and Therapeutic Approaches for Inner Ear Gene Therapy

24.2.1 The Vectors for Inner Ear Gene Therapy

In general, the choice of vector is the crucial element in gene therapy process. Based on the existence of a blood-labyrinth barrier, which is similar to the blood–brain barrier, therapeutic gene linked to vector is hard to enter the cochlea, which makes systemic delivery of pharmaceuticals less efficient and effect. This is also one of the main obstacles which affected translating gene therapy technique from basic research into clinical practice. Moreover, many molecules with therapeutic effect are hard to access to the inner ear owing to tight junctions between the cells, substantial barriers among tissues of cochlea. Consequently, it is common to deliver drug to inner ear locally rather than systemically [83]. Additionally, there is no ideal gene delivery system for in vivo gene therapy so far [89].

An ideal gene vector for delivery of therapeutic gene into the inner ear ought to be concentrated in a considerable volume and exclusively expressed in the targeted cells or tissues in order to avoid hydraulic trauma and minimized the damage to the host when delivering to the perilymph or endolymph [77]. Briefly, an ideal gene delivery should possess the following features: (1) reaching the target tissues/organs in vivo, preferably with recognition of the specific target cells; (2) crossing the membranous barriers of the cell and deliver its cargo intracellularly; (3) easily controlling the intensity and duration of foreign gene expression with the precondition of high efficiency of expression; (4) able to be biodegraded; (5) a low incidence of immune responses from the host; and (6) being manufactured on a commercial scale and easy for clinical use. Specifically, the gene delivery vectors could be divided into two major types on the basis of their original nature: the viral and nonviral vectors. Nowadays, scientific researchers are increasingly to pay more attention to nonviral vectors. Nonviral vectors could be made by relatively simple process without immunogenicity or carcinogenicity. Furthermore, these vectors have virtually unlimited loading capacity and could be produced on a large scale. Virus is a natural invader of cells and has very intensive capacity of infecting cells obtaining through a long process of evolution [40]. So, not surprisingly, virus has been used as the gene vehicle in the earliest studies of gene therapy [58, 63]. To overcome the shortcomings, the virus was modified before gene transfection by deleting partial sequences related to its replication for preventing from harming the host. The novel replication-deficient virus was used as a gene vector without the capacity of replication. Although the transfection efficiency of the viral vector is high, its clinical application is still limited owing to various significant shortcomings including immunogenicity and carcinogenicity, difficulty in production, lack of selectivity for specific target cells, and the possibility of causing diseases due to reversion of the engineered replication-deficient virus to a wild-type virus [6, 44]. The common gene vectors used for inner ear gene therapy are summarized in Table 24.2.

Table 24.2 Summary of the commonly available vectors for gene transfer

Nonviral vectors	Viral vectors	Multiplex gene vectors
Type:	*Type*:	*Type*:
Cationic polymer	Adenovirus	Combining with viral and nonviral elements
Cationic liposome	Adeno-associated virus	
Inorganic nanoparticles	Retrovirus	
	Hemagglutinating virus	
Advantage:	*Advantage*:	*Advantage*:
Ability to form a complex with a plasmid carrying various desired genes	High transfection efficiency	Having the advantages of both viral and nonviral vectors
No immunogenicity, or carcinogenicity, and low or even no toxicity to the host	Stable expression in the host cells	High transfection efficiency
Easy to produce and store with prospective economic benefit	Easy and inexpensive to produce	

24.2.1.1 Viral Vectors

Various types of replication-deficient viruses have been modified and developed as viral vector system so far. The commonly used viruses contain adenovirus, adeno-associated virus (AAV), retrovirus including lentivirus, herpes simplex virus (HSV), and hemagglutinating virus of Japan (HVJ, Sendai virus)-a member of the paramyxovirus family and so on. For instance, Bermingham and colleagues effectively activated the regeneration of cochlear hair cells in the matured ear of the mammals by using the adenovirus-mediated atonal gene [6]. Additionally, Lalwani and colleagues transfected adeno-associated virus (AAV) into the cochlea of guinea pigs to evaluate the effect of introducing foreign genetic material into the inner ear [43]. They found that AAV presented much deeper staining reactivity in the spiral limbus, cochlear lateral wall, spiral ganglion neurons, and the basic membrane in the treated cochlea than the control ear. This positive result has brought significant interest for a series of similar studies about gene therapy for the peripheral auditory system. However, there is no evidence of regeneration of the mammalian auditory sensory epithelium in vivo except a low degree of regeneration for the vestibular epithelium in the rodents just after birth until now. Consequently, it was indicated that the highly differentiated hair cells and neurons of the mammalian inner ear cannot be replaced through the cellular regeneration if they are damaged after birth. Previous studies had shown full regeneration of avian stato-acoustic epithelia after inner ear damage, so the attempts have been made for many years to activate potential capacity of regenerating functional sensory epithelium of the inner ear after impairment in the mammal. Liu and colleagues found that adult utricular sensory epithelia of the mouse displayed the characteristic features as stem cells, the capacity of self-renewal and expressing marker genes for the development of inner ear and the nervous system [44]. This experiment provided positive evidence for the possibility of sensory epithelium regeneration in the mammalian inner ear.

The formation of the regenerated hair-cell-like cells strongly predicts that the inner ear stem cells are pluripotent and the damaged hair cells may be replaced by the regenerated cells via proper artificial interventions. In 1999, Bermingham and colleagues found that mouse atonal homolog 1 (Math1, also known as Atoh1) was crucial for generating cochlear hair cells [6]. Math1 is a basic helix-loop-helix (bHLH) transcription factor homolog of the Drosophila atonal gene. These homologues of bHLH with similar structure and function were called as atonal-related factors or bHLH transcription factors including Math1, Cath1 (chicken atonal homolog 1), Xath1 (Xenopus atonal homolog 1), and Hath1 (human atonal homolog 1). Be concise, Math1 is a kind of positive regulator of hair-cell differentiation during cochlear development and merely expressed in the developing stage of the hair cells. Kawamoto and colleagues successfully developed endolymphatic perfusion of adenoviral vector loaded with Math1 gene into mature guinea pig cochlea and found that Math1 was expressed in the supporting cells of the organ of Corti and its adjacent nonsensory epithelial cells [37]. Interestingly, the immature hair cells appeared in the organ of Corti, and new hair cells were present in the regions of interdental cells, inner sulcus, and Hensen cell. In addition, the axons of ganglion neurons were attached to some of the newly developed hair cells. As a result, this finding could be regarded as a potentially prospective strategy for restoring hearing capacity, which is to induce a phenotypic transdifferentiation of nonsensory cells retaining the competence of response to Math1 or other atonal-related factors in the damaged inner ear to regenerate new hair cells or sensory epithelia with normal morphological and functional properties. Besides, the bHLH-related inhibitors of differentiation and DNA-binding (Id) proteins, originally isolated from Drosophila as a proneural gene for chordotonal organs, are known to negatively regulate many bHLH transcription factors, including Math1, in a number of different systems. The bHLH transcription factors modulate the development of several systems of both vertebrate and invertebrate and also play an important role in the differentiation of inner ear hair cells [33, 37]. For instance, Izumikawa and colleagues reported an experiment that they observed numerous newly generated cochlear hair cells in the out hair cell region of the cochlea of guinea pigs after delivery of atonal gene by the adenoviral vectors into the damage cochlea with the destroyed hair cells by coadministration of kanamycin and ethacrynic acid prior to the experiment [30]. More importantly, the average threshold of auditory brainstem response (ABR) of the group of treated animals was significantly lower than that of the control group, which indicated that transfection of atonal gene into the inner ear could help hearing recovery. This is the first report of successful regeneration of hair cells with hearing improvement in the experimentally profoundly deafened mature mammalian cochlea. However, the success of atonal gene transfection is currently debated and discussed with caution due to the difficult reproducibility of the experiment. This study would strongly imply the feasibility of hair cell regeneration in mammalian cochlea if the result could be confirmed by other separate laboratories. There were reports of the regeneration of vestibular hair cells or ectopic vestibular hair cell-like cells in the rodent by delivery of adenovirus vector-mediated Math1 gene [28, 76]. However, no other successful experiment has been reported on

Atoh1 induced cochlear hair cells regeneration based on the similar condition from other laboratories so far. Therefore, it could not translate the laboratory success into the clinical setting for generation of cochlear hair cells with natural morphology and function immediately. The newly developed or regenerated cochlear hair cells induced by foreign gene are not the same natural and functional hair cells in terms of spatial location, histological and ultrastructural morphology, as well as the physiological and biochemical properties. It has not been clarified for all the cellular signal pathways and regulating elements necessary for natural hair cell regeneration which developed during the long process of evolution just as the observed hair cell regeneration in the avian inner ear. The sequences of Math1 and Hath1 genes are publicly available in the GenBank. A novel adenoviral gene delivery vector containing several elements has been reported [34]: first, a genome including adenoviral 5′ ITR, 3′ ITR, and encapsidation signal; second, a DNA sequence (therapeutic gene) encoding a heterologous protein or polypeptide; and third, a promoter for controlling the expression of DNA sequence. In addition, at least a portion of the DNA sequences of E1, E4, E2a and/or E2 has been deleted to eliminate the harmful function of the proteins. Therefore, the vector retains at least partial functions of the adenoviral protein as well as the advantages of viral vector. The vector has the advantages for minimizing the host's immunological response to the vector, prolonging the duration of vector's existence, and increasing gene expression. Wadell and colleagues invented a new viral vector by using the adenovirus type 11p (Ad 11p) and type 4p (Ad 4p) as the backbone of the vector system [89]. This vector was able to deliver transfected gene into the cells of neural origin, especially for human neuro-origin cells. The adenovirus type 11p is suitable for gene vector because of the relatively low prevalence in population, the capacity of high affinity, and excessive infection. Kaneda and colleagues invented an injectable pharmaceutical preparation of gene therapy for hearing impairment [35]. It comprised a virus envelope vector (using HVJ as the 12 backbone) and a plasmid DNA inserted with a hepatocyte growth factor (HGF) gene as the therapeutic gene. Overexpression of HGF was observed in the cochlea after injection of the vector loading with plasmid inserted with HGF gene into the subarachnoid space of the deaf Sprague–Dawley rats induced by kanamycin. It was confirmed by the immunohistochemical and auditory examinations that the HGF gene transfection via cerebrospinal fluid (CSF) had protective effect on the cochlea from kanamycin ototoxicity. We hope to see more reports with the same method of administration to replicate and confirm the result.

24.2.1.2 Nonviral Vectors

The fundamental nonviral vector system containing the vector's backbone and a plasmid DNA which can be inserted with desired nucleic acid sequences such as a therapeutic and/or reporter gene. In general, nonviral vectors have some special advantages for gene delivery: (1) ability to form a complex with a plasmid carrying various desired genes; (2) possibly no immmunogenicity, nor carcinogenicity, and

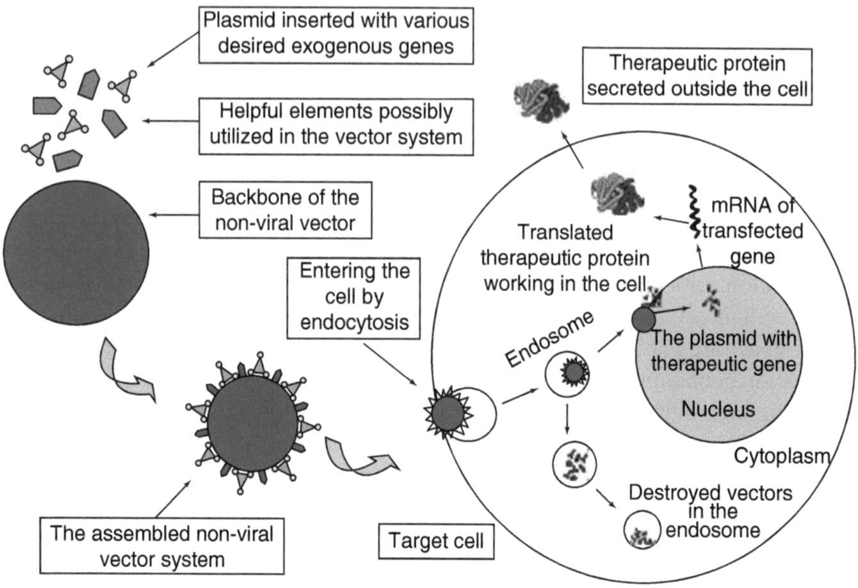

Fig. 24.1 Schematic of gene therapy with a nonviral vector system (Adapted from Sun et al. [81])

low or even no toxicity to the host; (3) could become a targeting vector when binding with cell-specific ligand [11]. The commonly used nonviral vectors consist of cationic polymer, cationic liposome, and other inorganic nanoparticles. The vector combines with its cargo (the foreign nucleic acid) with negative charges on the surface via electrostatic effect and compresses it to a smaller size for better protection and transportation. However, the low efficiency of transfection and transient expression in the host are the major disadvantages for nonviral vectors. A scheme for transfection process of nonviral gene vectors adopted from our previous publication is illustrated in Fig. 24.1 [81].

As novel nonviral vectors with wide application prospects, hydroxyapatite nanoparticles (nHAs) have been used for gene therapy in our group [32, 79, 91]. The nHAs have many advantages for gene therapy, including excellent biocompatibility, low cytotoxic effect, non-immunogenicity, non-oncogenicity, and unlimited loading capacity. Additionally, owing to its ability to enable the construction of various vectors with multiple functions, nHAs are suitable for connecting different molecular groups if necessary [79]. Our laboratory has used nHAs as nonviral vectors for transfecting NT-3 gene into the primarily cultured cochlear spiral ganglion neurons of neonatal mice in vitro and the cochlear spiral ganglion neurons of adult guinea pigs in vivo. To our knowledge, this was the first report of a therapeutic gene being successfully transfected into the mammalian inner ear with an inorganic nanoparticle vector. The nHAs exhibited excellent biocompatibility, possibly due to its natural mineral form of calcium apatite to mimic the mammalian bone mineral compartment [25]. However, the utmost disadvantages for nHAs, as for all nonviral vectors, are the limited transfection efficiency and short transient expression in the host cells.

One widely used nonviral vector for gene therapy is cationic polymers, especially polyethyleneimine (PEI) and polyamidoamine dendrimer (PAMAM-D). Bangham and colleagues used polymer-chitosan, a natural form of cationic, as transfection vectors [5]. It is a polycationic polysaccharide extracted from aquatic products, which has excellent biocompatibility. Owing to its poor water solubility due to the strong hydrogen bonds between neighboring chitosan molecules, the directly usage of natural chitosan for gene delivery is limited to some extent. Some researchers have investigated to treat materials with organic or inorganic acid to dissolve chitosan in water [50]. Mori and colleagues invented a modified cationic polymer, which has the insoluble and biodegradable advantages [48]. Therefore, the nucleic acid, connected with cationic polymer, would be easily released from the complex in the body when administered in vivo. This water-insoluble biodegradable polymer vector is a gelatin with hydrogel insolubilized in water via cross-linking effect. The sustained and relatively controlled DNA release increases transfection efficiency and the duration of the therapeutic gene in the target cells. The vector was used to transfect green fluorescent protein (GFP) into dendritic cells separated from human peripheral blood. The transfection efficiency was reported to be 77 %, which is relatively higher compared to the commonly used nonviral vectors. A complex compound with polymer and various ligands may be labeled as a multifunctional gene vector. It has distinguished capacity, such as targeting, dependent upon the ligand(s) conjugated with the polymer. Polymeric amino acid vectors for gene delivery were synthesized using traditional chemical synthetic methods in the past. So the sequences and molecular weight of the polymers were randomly varied, making them difficult to attach functional motifs such as targeting ligands at the precise locations. Zaki and colleagues invented a novel nonviral vector used genetic engineering technique [98]. Such vector was genetically engineered polymer transcribed from a single gene with nucleic acid-binding protein. The main structure of nucleic acid-binding protein contained at least one tandem repeat of a cationic amino acid-containing monomer (CAACM), in which the cationic amino acids (usually lysine and/or arginine) were positively charged and bound to negatively charged nucleic acids at pH 7.4. This vector is called nucleic acid-binding protein-based polymer (NABP) or amino acid-based polymer. It can be enhanced by linking to specific elements such as a target ligand, an endosome disrupting moiety, or a nuclear localization sequence. However, the transfection efficiency of the polymer was significantly lower than the commercial liposomes in three tested cultured cell lines.

Another kind of nonviral vector widely used for gene therapy is liposome. In 1965, Bangham and colleagues initially described the structure and basic properties of liposome [5]. From then on, liposome has become an increasingly important vector for drug and gene delivery, especially considering its advantage in protection of biological molecules from degradation with the increase of cellular uptake. Cationic liposome is one of the most commonly used liposome formulation for delivering anionic molecules such as DNA [86].

If we linked a nonviral vector to an appropriate tracer, the process of expression and distribution of a foreign gene in the host could be monitored in vivo. The positive effect of such vector is to allow monitoring the expression and distribution of foreign gene in vivo almost at the real time. For instance, the liposome-based vectors linked with radioactive isotope as the tracer agent were used to monitor gene expression by positron emission tomography, gamma camera, or single-photon emission computed tomography. Sen and colleagues invented a vector system encoding with fusion protein comprising a tracer molecule to monitor the blood level of therapeutic protein at high sensitivity by labeling the glucagon-originated peptide region with a fusion protein expressed by the host cells [26, 73]. It could provide an effective approach for gene therapy in the clinic in a negative feedback fashion if the gene delivery system is able to monitor the process of foreign gene expression in vivo.

Inorganic nanoparticles were increasingly used as delivery vectors for gene therapy recently. Inorganic nanoparticles could be utilized both in vitro and in vivo for gene delivery. The wide application of inorganic nanoparticles as nonviral vectors for gene therapy may partially be limit due to the lower transfection efficiency compared to liposome or polymer to some extent [48, 74, 79]. The inorganic nanoparticles, however, have the advantages of easy preparation, relative convenience for storage, and low cost, and more importantly, it could be manufactured on a commercial scale. It would become more popular or even an ideal gene vector if its transfection efficiency could be significantly increased. Hence, inorganic nanoparticles are worthy of further investigation and exploration. To enhance the transfection efficiency of inorganic nanoparticles system, our group performed a conjunct compound hydroxyapatite nanoparticles (nHAs) with polyethylenimine (PEI) [96]. In our recent study, PEI-nHAs was loaded with the recombinant plasmid pEGFPC2-NT3 and then directly administrating into the intact round window membranes (RWMs) in chinchillas. We utilized EGFP as reporting signal to analyze the effect and efficiency of transfection in inner ear, and the samples were detected and evaluated under the confocal microscopy post experiment. Surprisingly, the results indicated that abundant and condensed EGFP green fluorescence was present in the transitional zone and the region of dark cells on both sides of the crista ampullaris and around the macula of the utricle. Salem and colleagues constructed a nonviral gene delivery system based on multisegment bimetallic nanorods [72]. It could simultaneously bind with compacted DNA plasmids and targeting ligands to deliver therapeutic molecules into the inner ear of rat. This approach helps to control precisely over the composition, size, and multifunctionality of the gene delivery system. Kopke and colleague invented a vector with superparamagnetic iron oxide nanoparticles (SNP) composed of magnetite (Fe_3O_4) for therapeutic molecule delivery into the inner ear of rat [42]. The embedded SNP coated with either dextran, silica, or poly (D, L,-Lactide-co-glycolide) were placed in the round window membrane (RWM) niche of the rat. The RWM of the experimental ear of the rat was positioned horizontally upward and placed the head on the surface of the center of a 4-in. cube NdFeB48 magnet, as the magnet pole faced on the opposite side of the rat's head. The experiment demonstrated that the forces generated by permanent magnetic fields could sufficiently pull SNP into and across RWM entering the perilymphatic space of the cochlea. No significant toxicity was observed in the tested rat.

Consideration of the expression level of therapeutic gene affecting the therapeutic status, it is important to enhance the ability of delivery system to deliver sufficient level of therapeutic gene into the target cells or tissues. Roy and colleague used nerve growth factor-derived peptide (hNgf-EE)-mediated nanoparticles (NPs) to target cells of the inner ear [69]. They found that the NPs specifically targeted the spiral ganglion neurons with higher binding affinity and without any adverse toxicity to the host. The results indicate a selective cell target by multifunctional nanoparticles and demonstrate the superiority of ligands mediated vectors. Based on published reports and our own experience, it could improve the delivery system for the inner ear gene therapy by using target-specific promoters and tissue- or cell-specific ligands [4]. Consequently, our group plan to modulate the skeleton of non-viral vector with polyethylene glycol and polyethylenimine and then link the vector with selected ligands which have specific affinity to the target cell or organ in inner ear currently. This method may significantly enhance the therapeutic effects and transfection efficiency of inner ear gene therapy.

"Multiplex gene vectors," a kind of novel complex vectors consisting of both viral and nonviral elements, were developed by several laboratories. These new gene vectors may improve the outcomes of gene therapy in both laboratories and clinical settings. Rozenberg and colleagues invented a new gene vector called as "targeted artificial gene delivery (TAGD)" [70]. It contained a multifunctional artificial surface moiety surrounding with a recombinant viral particle with a therapeutic gene, namely, the recombinant core. This novel functional artificial surface moiety comprised at least one of the elements, immunoprotective, targeting, or cell-entry element. Consequently, the vector system was capable of specifically binding to the target cell and delivering the core into the cell. The immunoprotective element is a synthetic polymer moiety comprising a poly (ethylene glycol) and a copolymer of glutamic acid with leucine. The targeting element binds to a receptor highly expressed on the surface of the targeted cells than the normal cells. The targeting moiety is a peptide or peptidomimetic ligand for a cell surface receptor. Finally, the cell-entry element is a membrane-destabilizing moiety comprising an amphiphilic α-helix derived from the C-terminal domain of a viral envelope protein. The membrane-destabilizing moiety also comprises a copolymer of glutamic acid and leucine. Yu and Matsumoto constructed a multiplex gene vector system comprised a cationic polymer with desired nucleic acid (e.g., therapeutic gene) and a lipid-based vesicle encapsulating a membrane active agent, such as viral envelope proteins or membrane active peptides, to enhance efficiency of foreign gene transfection into eukaryotic cells [97]. The viral envelope protein encapsulated by lipid-based vesicle was vesicular stomatitis virus G (VSVG) envelope protein. This protein is a transmembrane glycoprotein and induces membrane fusion at acidic pH in the absence of other viral components so it could increase the quantity of the vector particles entering the target cells. The cationic polymer with VSVG vesicle increased transfection efficiencies more than 100 times compared to that with cationic polymer alone. The multiplex gene vectors possess the advantages of both viral vectors and nonviral vectors. Therefore, it may become a prospective approach for gene therapy in the future. A novel technique has been developed to deliver cargo-carrying nanoparticles into target cells with intracellular bacteria [2]. The process of transferring plasmid DNA into the target cells using bacteria as a nonviral carrier is called "bactofection." The nanoparticles containing plasmid DNA (inserted with GFP gene) were linked to the surface of the Listeria monocytogenes through the specific combination of biotin and avidin, which can penetrate into mammalian cells in a non-phagocytic process. Listeria monocytogenes served as both the carrier and cell-entry element of the vector system to bring the cargo-carrying nanoparticles into the target cells and consequently expressing the desired gene. The cargo-carrying bacteria which the authors named as "microbots" had been successfully transfected and expressed the reporter gene in various cultured cell lines and in the living mice in vivo. The "microbots" may become a new promising approach to deliver different types of cargo (genes, drugs, and other biological active molecules) into a variety of cultured cells and live animals.

24.2.2 Delivery Routes/Strategies for Inner Ear Gene Therapy

The routes of delivery therapeutic gene into the inner ear are also very important for inner ear gene therapy. Considering the factors that vestibular end organs and the cochlea are isolate, the fluid spaces of the ear are made up of separately cochlear endolymph and perilymph, and the cochlear hair cells and spiral ganglions are extremely sensitive to trauma, it is extremely difficult for vector introduction [29]. Basically, an ideal route for inner ear gene delivery should possess several features as below. First, the cargo-carrying gene vectors can effectively access the inner ear without harm to the peripheral vestibular or auditory organs. Second, any impairment of the inner ear induced by the process should be minimal and acceptable compared to the benefit from gene therapy. Third, the administration to the inner ear should be convenient, easy for operation and control. The delivery routes/strategies for the inner ear gene therapy are summarized in Table 24.3.

24.2.2.1 Round Window Membrane Permeation

Round window is the merely membranously sealed window on the bony labyrinth, and the round window membrane (RWM) is consists of three layers. The inner layer is a continuation of the epithelial layer of the perilymphatic space; the middle layer contains a large amount of collagen and elastic fibers as well as fibrocytes; and the outer epithelial layer is non-ciliated but often contains microvilli [47]. Goycoolea and colleagues firstly reported about RWM permeability to macromolecules with

Table 24.3 Summary of delivery routes/strategies for the inner ear gene therapy

Routes	Advantage	Disadvantage
Labyrinthine drilling or RWM injection	Maximum drug/gene entering	Higher risks of inducing damage and infection of the inner ear
	Minimum systemic interference	Difficulty of operation
Intrathecal injection	Relatively convenient	Diffusing effect may harm to other part of the central nervous system
	No harm to inner ear	
Systemic delivery	Convenient	Difficult to pass the blood labyrinthine barrier
	No harm to inner ear	
Intratympanic approach or intact RWM permeation	Relatively high drug/gene entering	Difficult for operation
	Minimum systemic interference	Risk of the infection of the middle ear
	Minimal harm to the inner ear	May slight harm to RWM and inner ear
Intrauterine approach	Starting treatment in embryonic stage	May harm to the embryo and/or the mother
		Difficulty for operation
		Needing expensive equipments
Cochlear implant (electrode)	Convenience	–

Modified from the Table 24.2 in Sun et al. [81]

tritiated normal human serum albumin in cat model of otitis media [22]. The study demonstrated the feasibility of diffusion of smaller molecules (toxins and enzymes) through an intact RWM. A large number of studies confirmed that the corresponding exogenous substances can be detected in the perilymph after placing albumin, steroid, antibiotics, anesthetics, and toxins on the RWM, respectively [2, 47, 61, 62]. Additionally, some experiments have shown that the transfection efficiency with this approach could be further increased with the help of facilitating agents. Wang and colleagues investigated a method to increase the permeability of RWM to adeno-associated viral (rAAV) vector by partial digestion of RWM with collagenase solution in guinea pigs [93]. The study showed that elevated delivery of rAAV across the partially digested RWM with increased transfection efficacy to a satisfactory level comparing to directly delivering rAAV via intact RWM. The evaluation of auditory function showed that this enzymatic manipulation did not cause permanent hearing loss if applied appropriately. Morphological observations also showed that the damage to RWM caused by partial digestion could be spontaneously healed within 4 weeks. These experiments imply that RWM is a potential approach for delivery of the biologically active molecules (drug or toxin) into the inner ear. Higher perilymphatic drug concentrations have been detected after delivering the drugs intratympanically through the intact RWM compared to other routes such as peritoneum and bloodstream in many laboratories. The molecules used for the studies of RWM delivery include antibiotics, poly lactic/glycolic acid (PLGA), and methylprednisolone [3, 7, 61, 84]. The intact RWM routine (intratympanic pathway) may provide higher perilymphatic concentration of the exogenous molecules with low drug dosage due to bypassing the blood-labyrinth barrier. Meanwhile, it induces slight but acceptable structural impairment of the inner ear with minimal systemic interference although it is relatively inconvenient for clinical use compared with the oral or intravenous route. However, the molecular weight of various complexes of exogenous gene, the backbone of the vector, and the helper element is much higher than that of the tested drugs mentioned above; it is more difficult to deliver sufficient quantity of these complexes into the inner ear through intact RWM. Jero and colleagues placed a gelfoam cube absorbed with liposome or adenoviral vector on the RWM of the mouse; transgenes were successfully expressed in a variety of cochlear cells or tissues [31]. Recently, several independent studies from different groups have confirmed the feasibility of inner ear gene transfer mediated by viral or nonviral vectors via intact RWM [82, 100]. The study of ultrastructure of RWM revealed that the paracellular pathway is the major route for the gene vector to penetrate through RWM [100]. The efficiency of foreign gene transfection will be significantly increased if RWM is pretreated or simultaneously treated with a facilitating agent, such as histamine, local anesthetic phenol, or other chemicals [10, 82]. The possible mechanism for the effects of facilitating agents on RWM may be due to damage of the RWM epithelium by the facilitator to enlarge the mini-space among cells. RWM offers an atraumatic route of administration to the inner ear. The intact RWM route may become an ideal approach for inner ear gene transfection compared to invasive gene delivery methods such as labyrinth drilling and RWM injection.

24.2.2.2 Perilymphatic or Endolymphatic Perfusion

A conventional route for therapeutic gene delivery is perilymphatic or endolymphatic perfusion of gene vectors through a tiny hole drilled on the bony wall of labyrinth. Alternatively, some researchers administrated drugs via RWM injection with or without the help of an osmotic micro-pump. This approach may allow maximum drug/gene entering the inner ear with minimum systemic interference. Numerous reports have confirmed successful delivery of various foreign genes [2, 36, 44, 56, 78, 94] or genetic engineered therapeutic cells [55] into inner ears mediated by various vectors via this routine. The osmotic micro-pump may provide continuous drug/gene supply for several days to several months depending on the requirements of the therapy and the capacity of the micro-pump. Therefore, inner ear perfusion with osmotic micro-pump could be a considerable option for continuing and steady delivery of foreign gene, especially for perilymphatic infusion [45]. Actually, the invasive approach could directly deliver the foreign genes into inner ear with relatively higher efficiency and easier control to compensate the disadvantage of transient gene expression mediated by nonviral gene vectors. However, this method could also increase the risks of damage and infection of the inner ear, which has limited its broad use in the clinic.

24.2.2.3 Intrathecal Injection

Stover and colleagues have transferred adenovirus-mediated lacZ reporter genes into unilateral cochlea of guinea pigs with intrathecal injection and observed the whole process of transgene expressions in the contralateral (untransfected) cochlea and cerebrospinal fluid (CSF) [78]. Interestingly, successful transduction of both cochleae was achieved by direct injection of adenovirus-mediated lacZ reporter gene into the CSF intrathecally, while no transduction was observed when injecting the reporter gene into the bloodstream. Additionally, it has been confirmed that the systemic route of foreign gene administration was unfeasible owing to the existence of the blood-labyrinth barrier. As a result, CSF could be regarded as a considerable delivery routine for the inner ear gene transfer, and the cochlear aqueduct may be the most likely route for virus into the contralateral cochlea. Another novel strategy for inner ear gene therapy was invented by Oshima and colleagues. They combined human hepatocyte growth factor (HGF) gene with HVJ envelope (HVJ-E) vector and delivered the complex into the inner ear of rats by intrathecal injection of the viral vectors into CSF via cisterna magna [56]. Transgene expression was detected in the spiral ganglion cells (SGCs) of the deaf rats induced by kanamycin, and hearing impairment was protected or significantly recovered by HGF gene transfer before or 2 weeks after kanamycin treatment. Nevertheless, the quantity of vectors entering the inner ear is obviously limited via the CSF route, and the foreign gene transfected into the central nervous system (CNS) may result in unexpected side effects, even at the risk of harm to the CNS. Consequently, this new approach should be used with caution.

24.2.2.4 Applied with Cochlear Implant Electrode

Another novel approach for delivering foreign genes into the inner ear is cochlear implant electrodes. Cochlear implant electrodes were coated with guinea pig fibroblasts transfected by an adenoviral vector with a brain-derived neurotrophic factor (BDNF) gene, and the BDNF-secreting cells were coated by agarose gel [65]. The study demonstrated that the BDNF expressing electrodes were able to preserve more spiral ganglion neurons of cochlea than the control after 48 days of implantation. This result indicated the feasibility of combining cochlear implant therapy with ex vivo gene transfer to enhance the survival of the spiral ganglion neurons. Additionally, Gubbels and colleagues invented another potential method for delivering foreign gene into inner ear [23]. The researches have successfully transfected foreign genes into the cochlea at the otocyst of a developing embryonic mouse via the utero. In detail, a plasmid containing genes encoding atonal homolog 1 and enhanced green fluorescent protein (GFP) was microinjected through the mouse uterus into the fluid-filled cavity of the embryonic on day 11.5 (E11.5) otic vesicle, and then the ventral progenitor cells of the organ of Corti in the plasmid-filled otocyst was electroporatingly transfected for the foreign gene with a directional square-wave pulse train. Expression of hair cell marker myosin 7a (Myo7a) by Atoh1/GFP$^+$

cells was detected in the otocyst 24 h after the electroporation. These Atoh1/GFP⁺/Myo7a⁺ cells were presented in the base, mid-base, and apex of the transfected cochleae. The cochlear stereotyped pattern of the hair cells in the one inner and three outer rows was altered by the overexpression of the Atoh1/GFP⁺/Myo7a⁺ cells named as supernumerary cells. There were phalloidin-positive epithelial protrusions on the apical surfaces of the cells at E18.5, that resembled immature stereociliary bundles, lasted for 1 month after the birth. Furthermore, the experiment confirmed that the cochlear morphology and hearing ability of the transfected mouse were not affected by the intrauterine gene transfer. Therefore, this may be a promising new approach for the inner ear gene delivery.

24.2.3 The Therapeutic Genes for Inner Ear

Based on the mechanisms or processes of different inner ear disease, the therapeutic genes for the inner ear gene therapy may be described as several types: (1) cell or tissue protectors, (2) modulators of gene expression, (3) inhibitors of adverse factors, and (4) activators of cell transdifferentiation. These therapeutic molecules have rapidly increased in number as the research in this field has been dramatically advanced in recent years. Table 24.4 summarizes the therapeutic genes used for inner ear gene therapy currently.

24.2.3.1 Neurotrophic Factors

Cochlear hair cells play an important role to convert sound waves into electrical signals in spiral ganglion neurons (SGNs) for transmission to the brain. Additionally, cochlear hair cells also influence the status of SGNs by supporting neurotrophic factors. Thus, loss of cochlear hair cells may lead to the gradual degeneration of SGNs. Consequently, it is significant to focus on the maintenance of neurotrophic factors for inner ear therapy [66]. Neurotrophic factors are a large group of biologically active peptides; most of them are capable of protecting epithelial cells and spiral ganglion neurons of inner ear from the damage caused by various pathogenic factors and promote the recovery from cochlear injure. Briefly, the neurotrophic factors play an important role in cellular differentiation, proliferation, development, neuronal plasticity, and the cellular survival, not only in embryonic stage but also throughout the entire lifetime. Additionally, neurotrophic factors in the inner ear are mainly produced by hair cells to maintain the normal function and survival of cochlear hair cells and neurons. Direct infusion of foreign neurotrophic factors into the inner ear could protect SGNs and cochlear hair cells from ototoxic drugs, noised-induced trauma, or other damage in different animals [18, 20, 21, 87, 95]. Currently, more than 20 neurotrophic factors have been revealed with protective effects on inner ear cells, which belong to one of the following groups: (1) neurotrophins (NTs) family including nerve growth factor (NGF), BDNF, and neurotrophin 3–7 (NT-3, NT-4/5, NT-7);

Table 24.4 Summary of the therapeutic genes used for the inner ear gene therapy

Gene	Main target	Protective effect
Neurotrophic factors: 1. Neurotrophin family 2. Glial cell line-derived neurotrophic factor family 3. Ciliary neurotrophic factor family 4. Fibroblast growth factor family 5. Other neurotrophically acting factors: (a) Epidermal growth factor (EGF) (b) Transforming growth factor (TGF) (c) Platelet-derived growth factor (PDGF) (d) Insulin-like growth factor (IGF) (e) Hepatocyte growth factor (HGF)	Hair cells, neurons	Comprehensive protection, acting as anti-apoptosis, antioxidation, and modulation of neuronal physiological and biochemical activities
Otospiralin	Nonsensory epithelial cells of the inner ear	Maintaining normal structure and function of the fibrocytes and other nonsensory epithelial cells in inner ear
Anti-apoptotic agent: XIAP, IAPs, and Bcl-2 family	Hair cells, neurons, other inner ear cells	Anti-apoptosis
Connexins: Such as CX26 (GJB2), CX30 (GJB6), CX31 (GJB3), and CX43 (GJA1)	Gap junctions in inner ear	Maintaining normal structure and function of the gap junction of the inner ear
Atonal-related factors or bHLH transcription factors: Such as Math1, Cath1, Xath1, and Hath1	Supporting cells of the organ of Corti and its adjacent nonsensory epithelial cells	Inducing a phenotypic transdifferentiation from nonsensory cells to the hair cells

Modified from the Table 24.1 in Sun et al. [81]

(2) glial cell line-derived neurotrophic factor (GDNF) family; (3) ciliary neurotrophic factor (CNTF) family; (4) fibroblast growth factor (FGF) family including at least 17 members, with acid fibroblast growth factor (aFGF or FGF-1) and basic fibroblast growth factor (bFGF or FGF-2) as the most extensively studied factors; and (5) other neurotrophically acting factors (miscellany) including but not limited to epidermal growth factor (EGF), transforming growth factor (TGF), platelet-derived growth factor (PDGF), insulin-like growth factor (IGF), and HGF. It was reported that either NT-3 or NT-3^{+} BDNF could be infused into the scala tympani of the ototoxically

damaged guinea pig cochleae resulting in more than 90 % survival of auditory neurons compared with only 14–24 % neuronal survival in the untreated control [56, 75]. Additionally, neurotrophic factors especially NT-3 and BDNF were demonstrated having the capability of protecting the inner ear hair cells and neurons via anti-apoptosis, antioxidation, and modulation of neuronal physiological and biochemical activities. A large number of documents have demonstrated the successful transfections and expressions of neurotrophic factors into the inner ear mediated by both viral and nonviral vectors in vitro and in vivo [32, 54, 75]. It is well established that either BDNF or NT-3 may have protective effects on inner ear hair cells and neurons even given 1 month after the deafness existence [46]. Consequently, the neurotrophic factor genes were regarded as the preferred therapeutic genes for inner ear gene therapy so far. Moreover, other neurotrophic factors such as TGF, GDNF, FGF, CNTF, and HGF also possessed similar protective effects as BDNF and NT-3 on inner ear hair cells and neurons [38, 51, 56, 60].

24.2.3.2 Atonal Homolog 1

As mentioned previously, atonal homolog 1 (encoded by Atoh1), a basic helix-loop-helix transcription factor required for the development of cochlear hair cells, was originally isolated from Drosophila chordotonal organs [12, 33]. It plays an important role in the differentiation of hair cells of the developing inner ear, and it is essential for the generate hair cells [6, 30, 37]. The homologues of homolog 1 were named as atonal-related factors or bHLH transcription factors which include Math1, Cath1, Xath1, and Hath1, with Math1 being the most extensively studied factor. Math1 is a positive regulator of hair cells differentiation during cochlear development and is expressed only in the developing stage of the hair cells. Gubbels and colleagues transferred therapeutic gene Atoh1 by intrauterine approach and successfully produced functional supernumerary hair cells in the mouse cochlea [23]. The experiment showed that newborn hair cells present stereociliary bundles, which attracted neuronal processes and accompanying synapse. Moreover, the cochlear hair cells induced by Athoh1 exhibited the similar range of current amplitudes, sensitivity, and adaptation as normal hair cells. This positive result showed that functional cochlear hair cells could be induced by manipulation of cells for overexpression of transcription factors in the postnatal mammalian cochlea. Kawamoto and colleagues investigated the phenomenon of overexpression of Math1 in cochlear nonsensory cells of mature guinea pig [37]. They transfected adenovirus with the Math1 gene by administrating into the endolymph of the adult guinea pig cochlea in vivo caused Math1 overexpression in nonsensory cochlear cells, based on the condition that Math1 protein was present in the region of supporting cells of the organ of Corti and in nearby nonsensory epithelial cells. After vector-mediated overexpression of Math1, newborn cochlear hair cells were presented in the region of organ of Corti and other supporting cells regions. Furthermore, the axons of ganglion neuron attached to the newly formed hair cells, indicating that newborn cochlear hair cells were ectopically positioned and have the ability to attract auditory nerve fibers. The experiments indicated that there is a possibility that transfection of the genes encoding atonal-related factors for the inner ear gene therapy may lead to an exciting breakthrough in the regeneration of inner ear hair cells in the mammalians.

24.2.3.3 XIAP

The X-linked inhibitor of apoptosis protein (XIAP), a member of the inhibitor of apoptosis protein family, is the most convincing anti-apoptotic agent that may be used for inner ear gene therapy so far. XIAP is an extremely potent suppressor of apoptosis, and it selectively binds and inhibits caspase-3, caspase-7, and caspase-9 [17]. It was recently discovered that XIAP inhibited apoptosis of the cochlear cells in various conditions such as age-related hearing loss [92] and drug-induced ototoxicity [14]. It has been demonstrated that XIAP could significantly reversed the severity of hearing loss induced by cisplatin [14]. In detail, the rats were treated with cisplatin after delivery of XIAP gene into the inner ear of rats by an adeno-associated viral (AAV) vector via RWM injection for at least 2 months, and the auditory-evoked brainstem response (ABR) threshold and out hair cells were investigated 72 h after cisplatin treatment. The data showed that transfection of the AAV encoding with XIAP into inner ear could significantly protect the hosts from cisplatin-induced ototoxicity by the anti-apoptosis effects of XIAP, and the ABR-threshold shift and hair-cell loss were attenuated by as much as 78 and 45 %, respectively, comparing with contralateral (untreated) cochleae. The results indicated that XIAP may provide a potent, specific, and long-term protection for the cochlear cells, and XIAP could become a promising gene for inner ear gene therapy.

24.2.3.4 Otospiralin

Another novel therapeutic gene is otospiralin, a newly developed ear-specific protein produced by fibrocytes from the nonsensory epithelial regions of the inner ear, specifically by spiral ligament and spiral limbus in the cochlea, and the maculae and semicircular canals of the vestibule. These mesenchymal nonsensory epithelial tissues surrounding the neuroepithelium in the inner ear play an important role for ionic balance to maintain the normal structure and function of the inner ear. Otospiralin is a novel 6.4 kDa protein with unknown function, and it shares the protein motif with the gag p30 core shell nucleocapsid protein of type C retroviruses [15]. Delprat and colleagues demonstrated a rapid decrease of the compound action potentials and irreversible deafness by downregulation of otospiralin with cochlear perfusion of antisense oligonucleotides of otospiralin in the guinea pigs. Some researchers indicated that hair-cell loss and degeneration of the organ of Corti might be the possible mechanism for the deafness. Consequently, otospiralin is essential for the survival of the cochlear sensory epithelium. The same research group further found that knocking out the Otos encoding otospiralin leading to degeneration of type II and IV fibrocytes and moderate hearing loss in the mice [16]. These findings suggested that loss of otospiralin would induce fibrocyte damage and lead to both structural and functional impairment of the inner ear. The loss of hair cells also indicated the importance of supporting cells for hair cells survival. It is not just for "supporting" the hair cells in the right spatial place. Obvious decrease of cisplatin-induced apoptosis was observed in the cultured spiral ligament fibrocytes by upregulating the otospiralin gene expression through adenoviral vector-mediated gene transfection [99]. In summary, otospiralin may be a prospective protective molecule for inner ear, and it is possible to employ the specific antibody to otospiralin as the targeting element of gene vector for inner ear gene therapy based on the specificity of otospiralin to the inner ear.

24.2.3.5 Connexins

Connexins have been demonstrated to be crucial for maintenance of hearing capacity and could be a potential new target for inner ear gene therapy. Connexins are the major proteins of gap junctions, A lot of hearing impairments were caused by mutations of genes encoding connexins, such as non-syndromic hereditary deafness which counts for 70 % of the inherited hearing impairment [13]. The most frequently mutations of connexins are from Cx26 gene encoding connexin 26, which accounts for almost 49 % of nonsymdromic deafness. More than 100 causative mutations in CX26 (GJB2) have been detected and counted for a majority of prelingual deafness [27]. Other relatively common mutations were detected in CX30 (GJB6), CX31 (GJB3), and CX43 (GJA1) [13]. A susceptibility gene of non-syndromic sensorineural autosomal deafness was identified to link to chromosome 13q11-12 (DFNB1), where the Cx26 gene is localized [39]. Therefore, connexin 26 is regarded as an important component in the cochlea. Connexin 26 exists in gap junctions and connects many types of cells in the cochlea including the epithelial cells and the connective tissues [41]. Thus, it will certainly affect the intracellular and/or intercellular internal environments or signaling pathways if the cochlear gap junctions were damaged by the mutations of connexin genes. Sun and colleagues have investigated the lesion pattern and time course of cellular degeneration in the cochlea of conditional Cx26 (cCx26) null and Cx30 null mice and observed that cellular degeneration in the cochlea of cCx26 null mice was dramatically more rapid and widespread than that in Cx30 null mice [80]. The result indicates that different deafness mechanisms may exist in spite of co-assembly of Cx26 and Cx30 in formation of the gap junctions in the cochlea. Additionally, one of the functions of the gap junction systems is to recirculate K^+ ions from hair cells to the strial marginal cells. It would disrupt the ionic balance in the cochlea and result in hearing loss if the recirculation of K^+ ions was interrupted [41]. Furthermore, connexins play an important role in maintaining the morphology and the function of the inner ear. Therefore, connexins may be a curable treatment for a large number of hereditary deafness in the embryonic or even early postnatal stage by replacing the mutated connexin gene of the patients with a specific therapeutic gene. Unfortunately, we are still looking forward to the first successful case to be reported.

24.3 Summary and Future Directions for Inner Ear Gene Therapy

Several interesting clinical trials have been reported using quite promising techniques in the cases of ex vivo gene therapy, in which cells were removed from the patient, treated and then returned to the patient. These clinical trials have shown some promise [1, 9, 71]. These positive results profoundly encouraged the scientific community and clinicians to work together to pursue further advanced technologies and pay more attention on the field of gene therapy for human disease including the inner ear. Actually, during the last decade, numerous advances and progress have been made in gene therapy for the inner ear, especially for the development of gene vector system. As more attention has been paid to nonviral vectors in the last couples of years, dramatic achievements and great progress related to nonviral gene delivery system have been made. Moreover, the inorganic nanoparticle gene vectors have been developed rapidly in the recent years owing to its specific advantages, compare to other conventional vectors [5, 79]. Three types of vectors may have greater prospects. The first one is the bacterial vector which mediates bactofection. The vector has no carcinogenesis because the host's genome will not be integrated by the bacteria. The second one is the "multiplex gene vector" which has the advantages of both viral and nonviral gene delivery systems and is constructed by biochemical and genetic engineering techniques. The capabilities of the so-called multiplex gene vectors could be tremendously increased compared with a pure nonviral vector. The major advantage of the "multiplex gene vector" is their almost infinite structural variations, which could affect physicochemical and biological properties of the vectors. Such vectors may become an ideal tool for gene therapy in the future. The third one is the labeled gene vector with obvious commercial values. It could be utilized as an ex vivo and in vivo gene vector with the function of real-time monitoring. On the basis of recent progress in gene vector development, it is estimated that breakthrough for transfection efficiency of nonviral vectors or multiplex gene vectors could take place within 5 years. It will take about 15 years for the "multiplex gene vector" to become a routine tool for the physicians to carry out gene therapy for the patient.

The intact RWM is regarded as the most promising route for delivering therapeutic gene to the inner ear, due to its utmost safe, effective advantages, and easy operation in clinical practice. It is reasonable to believe that intact RWM approach may finally and permanently replace the commonly used transfection approach of labyrinth drilling or RWM injection for inner ear gene delivery. Some practices of the inner ear gene therapy may need to be carried out at the embryonic stage for the treatment of hereditary or congenital deafness in the future. In these circumstances, the intrauterine approach may be the best choice, especially when the endoscopic technique and the employed instruments are improved with more accurate and better operation. Considering the large population of the hereditary and congenital deafness, the in utero approach for inner ear gene delivery is

Table 24.5 Common methods for evaluation of auditory status post inner ear gene therapy

Electric response audiometry	
Vestibular function	Vestibular evoked myogenic potential
Auditory pathway	Auditory brainstem response, compound action potential
Cochlear lymph	Endocochlear potential
Cochlear hair cell	Cochlear microphonics, summating potentials
Spiral ganglion neurons	Single-unit recording
Outer hair cell	Cochlear otoacoustic emissions
Morphology and pathology analysis for inner ear	
Structure	Light microscopy, immunocytochemistry confocal microscopy
Ultrastructure	Transmission electron microscopy, scanning electron microscopy

worthy of further study. However, this strategically important project has not been paid enough attention until now.

Neurotrophic factors, especially NT-3 and BDNF, are well-known comprehensive protectors for inner ear hair cells and neurons. In addition to the well-known major protective factors, otospiralin and XIAP are newly discovered as possible protectors for the inner ear cells. The dream to regenerate functional hair cells in the mammalian cochlea may someday become true via the cochlear cell's transdifferentiation. Neurotrophic factors, inhibitors of apoptosis, antioxidants, otospiralin, and the atonal-related factors may become the mainstream of therapeutic molecules for inner ear gene therapy. Those therapeutic genes should be selected depending on the base of different types of disease. As the advance and progress of science and technology, new therapeutic targets (such as connexins) and novel therapeutic molecules will be rapidly discovered and developed. More and more patients with hearing loss will seek the treatment of gene therapy. All the patients should be regularly followed up and evaluated after gene therapy for inner ear. Table 24.5 summarized the methods for evaluation of auditory status post inner ear gene therapy. Based on our knowledge to the current status of inner ear gene therapy, we predict that the first successful case of inner ear gene therapy may be reported within 5–8 years. Inner ear gene therapy may become one of the common choices by the otologists for the treatment of inner ear diseases in about 15–20 years.

The expanding development of the molecular biology, improvements of technical tools, and advances in gene transfer technology ought to impulse in the progress of gene therapy. In the near future, gene therapy for inner ear will be combined with stem cell therapy, conventional drug treatment, as well as surgical operation, synergistically to prevent, restore, and even cure human from hearing disorder.

Acknowledgments This paper was partially supported by the National Natural Science Foundation of China (30772402) and the Science and Technology Commission of Hunan Province (Grant number: 2007SK2001).

Disclosure Statement The authors have no conflicts of interest to declare.

References

1. Aiuti A, Slavin S, Aker M, Ficara F, Deola S, Mortellaro A, Morecki S, Andolfi G, Tabucchi A, Carlucci F, Marinello E, Cattaneo F, Vai S, Servida P, Miniero R, Roncarolo MG, Bordignon C (2002) Correction of ADA-SCID by stem cell gene therapy combined with nonmyeloablative conditioning. Science 296:2410–2413
2. Akin D, Sturgis J, Ragheb K, Sherman D, Burkholder K, Robinson JP, Bhunia AK, Mohammed S, Bashir R (2007) Bacteria-mediated delivery of nanoparticles and cargo into cells. Nat Nanotechnol 2:441–449
3. Anniko M, Hellstrom S, Schmidt SH, Spandow O (1989) Toxic effects on inner ear of noxious agents passing through the round window membrane. Acta Otolaryngol Suppl 457:49–56
4. Atar O, Avraham KB (2005) Therapeutics of hearing loss: expectations vs reality. Drug Discov Today 10:1323–1330
5. Bangham AD, Standish MM, Watkins JC (1965) Diffusion of univalent ions across the lamellae of swollen phospholipids. J Mol Biol 13:238–252
6. Bermingham NA, Hassan BA, Price SD, Vollrath MA, Ben-Arie N, Eatock RA, Bellen HJ, Lysakowski A, Zoghbi HY (1999) Math1: an essential gene for the generation of inner ear hair cells. Science 284:1837–1841
7. Bird PA, Begg EJ, Zhang M, Keast AT, Murray DP, Balkany TJ (2007) Intratympanic versus intravenous delivery of methylprednisolone to cochlear perilymph. Otol Neurotol 28: 1124–1130
8. Brough DE (2007) Methods of gene therapy for treating disorders of the ear by administering a vector encoding an antonal-associated factor. US2007141029. http://www.investorvillage.com/mbthread.asp?mb=1195&tid=2586647&showall=1. Accessed 12 March 2013
9. Cavazzana-Calvo M, Hacein-Bey S, de Saint Basile G, Gross F, Yvon E, Nusbaum P, Selz F, Hue C, Certain S, Casanova JL, Bousso P, Deist FL, Fischer A (2000) Gene therapy of human severe combined immunodeficiency (SCID)-X1 disease. Science 288:669–672
10. Chandrasekhar SS, Rubinstein RY, Kwartler JA, Gatz M, Connelly PE, Huang E, Baredes S (2000) Dexamethasone pharmacokinetics in the inner ear: comparison of route of administration and use of facilitating agents. Otolaryngol Head Neck Surg 122:521–528
11. Chattopadhyay D, Mitra S, Maitra A (2005) A method using inorganic nanoparticles as non-viral vectors for gene therapy. WO2005123142
12. Chen P, Johnson JE, Zoghbi HY, Segil N (2002) The role of Math1 in inner ear development: uncoupling the establishment of the sensory primordium from hair cell fate determination. Development 129:2495–2505
13. Cohen-Salmon M, del Castillo FJ, Petit C (2005) Connexins responsible for hereditary deafness – the tale unfolds. In: Winterhager E (ed) Gap junctions in development and disease. Springer, Berlin/Heidelberg, pp 111–134
14. Cooper LB, Chan DK, Roediger FC, Shaffer BR, Fraser JF, Musatov S, Selesnick SH, Kaplitt MG (2006) AAV-mediated delivery of the caspase inhibitor XIAP protects against cisplatin ototoxicity. Otol Neurotol 27:484–490
15. Delprat B, Boulanger A, Wang J, Beaudoin V, Guitton MJ, Venteo S, Dechesne CJ, Pujol R, Lavigne-Rebillard M, Puel JL, Hamel CP (2002) Downregulation of otospiralin, a novel inner ear protein, causes hair cell degeneration and deafness. J Neurosci 22:1718–1725
16. Delprat B, Ruel J, Guitton MJ, Hamard G, Lenoir M, Pujol R, Puel JL, Brabet P, Hamel CP (2005) Deafness and cochlear fibrocyte alterations in mice deficient for the inner ear protein otospiralin. Mol Cell Biol 25:847–853
17. Deveraux QL, Takahashi R, Salvesen GS, Reed JC (1997) X-linked IAP is a direct inhibitor of cell-death proteases. Nature 388:300–304

18. Ernfors P, Duan ML, ElShamy WM, Canlon B (1996) Protection of auditory neurons from aminoglycoside toxicity by neurotrophin-3. Nat Med 2:463–467
19. Fujiyoshi T, Hood L, Yoo TJ (1994) Restoration of brain stem auditory-evoked potentials by gene transfer in shiverer mice. Ann Otol Rhinol Laryngol 103:449–456
20. Gillespie LN (2003) Regulation of axonal growth and guidance by the neurotrophin family of neurotrophic factors. Clin Exp Pharmacol Physiol 30:724–733
21. Gillespie LN, Clark GM, Bartlett PF, Marzella PL (2003) BDNF-induced survival of auditory neurons in vivo: cessation of treatment leads to accelerated loss of survival effects. J Neurosci Res 71:785–790
22. Goycoolea MV, Paparella MM, Goldberg B, Carpenter AM (1980) Permeability of the round window membrane in otitis media. Arch Otolaryngol 106:430–433
23. Gubbels SP, Woessner DW, Mitchell JC, Ricci AJ, Brigande JV (2008) Functional auditory hair cells produced in the mammalian cochlea by in utero gene transfer. Nature 455:537–541
24. Hallauer PL, Hastings KE (2000) Human cytomegalovirus IE1 promoter/enhancer drives variable gene expression in all fiber types in transgenic mouse skeletal muscle. BMC Genet 1:1
25. Han Y, Li S, Wang X, Bauer I, Yin M (2007) Sonochemical preparation of hydroxyapatite nanoparticles stabilized by glycosaminoglycans. Ultrason Sonochem 14:286–290
26. Hanawa H (2006) Vector for gene therapy and method of quantifying target protein in mammal or cultured cells with the administration of the vector for gene therapy. US2006223767. http://www.techmanage.net/pdfs/October06PatentApps.pdf. Accessed 12 March 2013
27. Hoang Dinh E, Ahmad S, Chang Q, Tang W, Stong B, Lin X (2009) Diverse deafness mechanisms of connexin mutations revealed by studies using in vitro approaches and mouse models. Brain Res 1277:52–69
28. Huang Y, Chi F, Han Z, Yang J, Gao W, Li Y (2009) New ectopic vestibular hair cell-like cells induced by Math1 gene transfer in postnatal rats. Brain Res 1276:31–38
29. Husseman J, Raphael Y (2009) Gene therapy in the inner ear using adenovirus vectors. Adv Otorhinolaryngol 66:37–51
30. Izumikawa M, Minoda R, Kawamoto K, Abrashkin KA, Swiderski DL, Dolan DF, Brough DE, Raphael Y (2005) Auditory hair cell replacement and hearing improvement by Atoh1 gene therapy in deaf mammals. Nat Med 11:271–276
31. Jero J, Mhatre AN, Tseng CJ, Stern RE, Coling DE, Goldstein JA, Hong K, Zheng WW, Hoque AT, Lalwani AK (2001) Cochlear gene delivery through an intact round window membrane in mouse. Hum Gene Ther 12:539–548
32. Jiang M, Zhang YQ, He GX, Sun H (2007) Protective effect of NT-3 gene mediated by hydroxyapatite nanoparticle on the cochlea of guinea pigs injured by excitotoxicity. Zhong Nan Da Xue Xue Bao Yi Xue Ban 32:563–567
33. Jones JM, Montcouquiol M, Dabdoub A, Woods C, Kelley MW (2006) Inhibitors of differentiation and DNA binding (Ids) regulate Math1 and hair cell formation during the development of the organ of Corti. J Neurosci 26:550–558
34. Kadan M, Gaziglia M, Trapnell B (1996) Improved adenoviral vectors and produced cells. WO 96/18418, PCT/US95/15947
35. Kaneda Y, Oshima K, Morishita R, Kubo T (2006) Drug for auditory dysfunction. US7390482. http://www.freepatentsonline.com/7390482.html. Accessed 12 March 2013
36. Kanzaki S, Shiotani A, Inoue M, Hasegawa M, Ogawa K (2007) Sendai virus vector-mediated transgene expression in the cochlea in vivo. Audiol Neurootol 12:119–126
37. Kawamoto K, Ishimoto S, Minoda R, Brough DE, Raphael Y (2003) Math1 gene transfer generates new cochlear hair cells in mature guinea pigs in vivo. J Neurosci 23:4395–4400
38. Kawamoto K, Yagi M, Stover T, Kanzaki S, Raphael Y (2003) Hearing and hair cells are protected by adenoviral gene therapy with TGF-beta1 and GDNF. Mol Ther 7:484–492
39. Kelsell DP, Dunlop J, Stevens HP, Lench NJ, Liang JN, Parry G, Mueller RF, Leigh IM (1997) Connexin 26 mutations in hereditary non-syndromic sensorineural deafness. Nature 387: 80–83
40. Kesser BW, Hashisaki GT, Holt JR (2008) Gene transfer in human vestibular epithelia and the prospects for inner ear gene therapy. Laryngoscope 118:821–831

41. Kikuchi T, Adams JC, Miyabe Y, So E, Kobayashi T (2000) Potassium ion recycling pathway via gap junction systems in the mammalian cochlea and its interruption in hereditary nonsyndromic deafness. Med Electron Microsc 33:51–56
42. Kopke RD, Wassel RA, Mondalek F, Grady B, Chen K, Liu J, Gibson D, Dormer KJ (2006) Magnetic nanoparticles: inner ear targeted molecule delivery and middle ear implant. Audiol Neurootol 11:123–133
43. Lalwani AK, Walsh BJ, Reilly PG, Muzyczka N, Mhatre AN (1996) Development of in vivo gene therapy for hearing disorders: introduction of adeno-associated virus into the cochlea of the guinea pig. Gene Ther 3:588–592
44. Liu YH, Ke XM, Qin Y, Gu ZP, Xiao SF (2007) Adeno-associated virus-mediated Bcl-xL prevents aminoglycoside-induced hearing loss in mice. Chin Med J (Engl) 120:1236–1240
45. Luebke AE, Steiger JD, Hodges BL, Amalfitano A (2001) A modified adenovirus can transfect cochlear hair cells in vivo without compromising cochlear function. Gene Ther 8:789–794
46. Miller JM, Le Prell CG, Prieskorn DM, Wys NL, Altschuler RA (2007) Delayed neurotrophin treatment following deafness rescues spiral ganglion cells from death and promotes regrowth of auditory nerve peripheral processes: effects of brain-derived neurotrophic factor and fibroblast growth factor. J Neurosci Res 85:1959–1969
47. Miriszlai E, Benedeczky I, Csapo S, Bodanszky H (1978) The ultrastructure of the round window membrane of the cat. ORL J Otorhinolaryngol Relat Spec 40:111–119
48. Mori H, Tabata Y, Ando K (2006) Nucleic acid-containing complex. TW267380B
49. Mulligan RC (1993) The basic science of gene therapy. Science 260:926–932
50. Nah JW, Jung TR, Jang MK, Jeong J-II (2011) Water soluble chitosan nanoparticle for delivering an anticancer agent and preparing method. US7883723 B2. http://www.google.com/patents/US7883723. Accessed 12 March 2013.
51. Nakaizumi T, Kawamoto K, Minoda R, Raphael Y (2004) Adenovirus-mediated expression of brain-derived neurotrophic factor protects spiral ganglion neurons from ototoxic damage. Audiol Neurootol 9:135–143
52. Nance WE (2003) The genetics of deafness. Ment Retard Dev Disabil Res Rev 9:109–119
53. National Institute on Deafness and Other Communication Disorders NIH (2007) Statistics about hearing disorders, ear infections and deafness. Available at: www.nidcd.nih.gov/health/statistics/hearing.asp. Accessed 12 March 2013
54. Noushi F, Richardson RT, Hardman J, Clark G, O'Leary S (2005) Delivery of neurotrophin-3 to the cochlea using alginate beads. Otol Neurotol 26:528–533
55. Okano T, Nakagawa T, Kita T, Endo T, Ito J (2006) Cell-gene delivery of brain-derived neurotrophic factor to the mouse inner ear. Mol Ther 14:866–871
56. Oshima K, Shimamura M, Mizuno S, Tamai K, Doi K, Morishita R, Nakamura T, Kubo T, Kaneda Y (2004) Intrathecal injection of HVJ-E containing HGF gene to cerebrospinal fluid can prevent and ameliorate hearing impairment in rats. FASEB J 18:212–214
57. Osten P, Grinevich V, Cetin A (2007) Viral vectors: a wide range of choices and high levels of service. Handb Exp Pharmacol 178:177–202
58. Osterman JV, Waddell A, Aposhian HV (1970) DNA and gene therapy: uncoating of polyoma pseudovirus in mouse embryo cells. Proc Natl Acad Sci U S A 67:37–40
59. Petersen MB, Willems PJ (2006) Non-syndromic, autosomal-recessive deafness. Clin Genet 69:371–392
60. Pickles JO, Harter C, Rebillard G (1998) Fibroblast growth factor receptor expression in outer hair cells of rat cochlea. Neuroreport 9:4093–4095
61. Plontke SK, Mynatt R, Gill RM, Borgmann S, Salt AN (2007) Concentration gradient along the scala tympani after local application of gentamicin to the round window membrane. Laryngoscope 117:1191–1198
62. Plontke SK, Biegner T, Kammerer B, Delabar U, Salt AN (2008) Dexamethasone concentration gradients along scala tympani after application to the round window membrane. Otol Neurotol 29:401–406
63. Qasba PK, Aposhian HV (1971) DNA and gene therapy: transfer of mouse DNA to human and mouse embryonic cells by polyoma pseudovirions. Proc Natl Acad Sci U S A 68:2345–2349

64. Raphael Y, Frisancho JC, Roessler BJ (1996) Adenoviral-mediated gene transfer into guinea pig cochlear cells in vivo. Neurosci Lett 207:137–141
65. Rejali D, Lee VA, Abrashkin KA, Humayun N, Swiderski DL, Raphael Y (2007) Cochlear implants and ex vivo BDNF gene therapy protect spiral ganglion neurons. Hear Res 228:180–187
66. Richardson RT, Noushi F, O'Leary S (2006) Inner ear therapy for neural preservation. Audiol Neurootol 11:343–356
67. Riordan JR, Rommens JM, Kerem B, Alon N, Rozmahel R, Grzelczak Z, Zielenski J, Lok S, Plavsic N, Chou JL et al (1989) Identification of the cystic fibrosis gene: cloning and characterization of complementary DNA. Science 245:1066–1073
68. Rosenecker J, Huth S, Rudolph C (2006) Gene therapy for cystic fibrosis lung disease: current status and future perspectives. Curr Opin Mol Ther 8:439–445
69. Roy S, Johnston AH, Newman TA, Glueckert R, Dudas J, Bitsche M, Corbacella E, Rieger G, Martini A, Schrott-Fischer A (2010) Cell-specific targeting in the mouse inner ear using nanoparticles conjugated with a neurotrophin-derived peptide ligand: potential tool for drug delivery. Int J Pharm 390:214–224
70. Rozenberg Y, Medvedkin V, Anderson WF (2001) Targeted artificial gene delivery. WO 01/12235, PCT/US00/22619
71. Ryan AF, Dazert S (2009) Gene therapy for the inner ear: challenges and promises. Adv Otorhinolaryngol 66:1–12
72. Salem AK, Searson PC, Leong KW (2003) Multifunctional nanorods for gene delivery. Nat Mater 2:668–671
73. Sen L, Gambhir SS (2007) A method for noninvasively and quantitatively monitoring therapeutic and diagnostic transgene expression induced by ex vivo and in vivo gene targeting in organs, tissues and cells. WO2007109335
74. Son SJ, Bai X, Lee SB (2007) Inorganic hollow nanoparticles and nanotubes in nanomedicine part 1. Drug/gene delivery applications. Drug Discov Today 12:650–656
75. Staecker H, Kopke R, Malgrange B, Lefebvre P, Van de Water TR (1996) NT-3 and/or BDNF therapy prevents loss of auditory neurons following loss of hair cells. Neuroreport 7:889–894
76. Staecker H, Praetorius M, Baker K, Brough DE (2007) Vestibular hair cell regeneration and restoration of balance function induced by math1 gene transfer. Otol Neurotol 28:223–231
77. Staecker H, Praetorius M, Brough DE (2011) Development of gene therapy for inner ear disease: using bilateral vestibular hypofunction as a vehicle for translational research. Hear Res 276:44–51
78. Stover T, Yagi M, Raphael Y (2000) Transduction of the contralateral ear after adenovirus-mediated cochlear gene transfer. Gene Ther 7:377–383
79. Sun H, Jiang M, Zhu SH (2008) In vitro and in vivo studies on hydroxyapatite nanoparticles as a novel vector for inner ear gene therapy. Zhonghua Er Bi Yan Hou Tou Jing Wai Ke Za Zhi 43:51–57
80. Sun Y, Tang W, Chang Q, Wang Y, Kong W, Lin X (2009) Connexin30 null and conditional connexin26 null mice display distinct pattern and time course of cellular degeneration in the cochlea. J Comp Neurol 516:569–579
81. Sun H, Huang A, Cao S (2011) Current status and prospects of gene therapy for the inner ear. Hum Gene Ther 22:1311–1322
82. Suzuki M, Yamasoba T, Suzukawa K, Kaga K (2003) Adenoviral vector gene delivery via the round window membrane in guinea pigs. Neuroreport 14:1951–1955
83. Swan EE, Mescher MJ, Sewell WF, Tao SL, Borenstein JT (2008) Inner ear drug delivery for auditory applications. Adv Drug Deliv Rev 60:1583–1599
84. Tamura T, Kita T, Nakagawa T, Endo T, Kim TS, Ishihara T, Mizushima Y, Higaki M, Ito J (2005) Drug delivery to the cochlea using PLGA nanoparticles. Laryngoscope 115: 2000–2005
85. Toniatti C, Bujard H, Cortese R, Ciliberto G (2004) Gene therapy progress and prospects: transcription regulatory systems. Gene Ther 11:649–657

86. Vagle K, Wang W, Vargeese C, Zhang Y, Chen T (2008) Lipid nanoparticle based compositions and methods for the delivery of biologically active molecules. US2008/0020058 A1. http://ip.com/patapp/US20080020058. Accessed 12 March 2013
87. Van de Water TR, Staecker H, Ernfors P, Moonen G, Lefebvre PP (1996) Neurotrophic factors as pharmacological agents for the treatment of injured auditory neurons. Ciba Found Symp 196:149–162
88. Van de Water TR, Staecker H, Halterman MW, Federoff HJ (1999) Gene therapy in the inner ear. Mechanisms and clinical implications. Ann NY Acad Sci 884:345–360
89. Wadell G, Mei YF, Segerman A, Skog J, Kristina L (2002) Viral vector for gene therapy. WO02053759
90. Wang Y, Yang C, Xu S, Xue X, Li F, Liu Y, Ye Q (2007) Report on the study of who ear and hearing disorders survey protocol in Guizhou province. Lin Chung Er Bi Yan Hou Tou Jing Wai Ke Za Zhi 21:731–734
91. Wang G, Zhu S, Tan G, Zhou K, Huang S, Zhao Y, Li Z, Huang B (2008) Study on biocompatibility of hydroxyapatite/high density polyethylene (HA/HDPE) nano-composites artificial ossicle. Sheng Wu Yi Xue Gong Cheng Xue Za Zhi 25:607–610
92. Wang J, Menchenton T, Yin S, Yu Z, Bance M, Morris DP, Moore CS, Korneluk RG, Robertson GS (2010) Over-expression of X-linked inhibitor of apoptosis protein slows presbycusis in C57BL/6J mice. Neurobiol Aging 31:1238–1249
93. Wang H, Murphy R, Taaffe D, Yin S, Xia L, Hauswirth WW, Bance M, Robertson GS, Wang J (2012) Efficient cochlear gene transfection in guinea-pigs with adeno-associated viral vectors by partial digestion of round window membrane. Gene Ther 19:255–263
94. Wenzel GI, Xia A, Funk E, Evans MB, Palmer DJ, Ng P, Pereira FA, Oghalai JS (2007) Helper-dependent adenovirus-mediated gene transfer into the adult mouse cochlea. Otol Neurotol 28:1100–1108
95. Wise AK, Richardson R, Hardman J, Clark G, O'Leary S (2005) Resprouting and survival of guinea pig cochlear neurons in response to the administration of the neurotrophins brain-derived neurotrophic factor and neurotrophin-3. J Comp Neurol 487:147–165
96. Wu XH, Sun H, Xing XW, Huang LH, Huang SP (2010) Surface modification and DNA-binding assessment of nano-hydroxyapatite. Nan Fang Yi Ke Da Xue Xue Bao 30:2233–2236
97. Yu L, Matsumoto K (2006) Vector for transfection of eukaryotic cells. US2006/0074045 A1. http://ip.com/pat/US20060074045. Accessed 12 March 2013
98. Zaki M, Arash H, Hamidreza G (2007) Recombinant protein polymer vectors for systemic gene delivery. US2007/0098702 A1. www.freepatentsonline.com/y2007/0098702.html. Accessed 12 March 2013
99. Zhuo XL, Wang Y, Zhuo WL, Zhang YS, Wei YJ, Zhang XY (2008) Adenoviral-mediated up-regulation of Otos, a novel specific cochlear gene, decreases cisplatin-induced apoptosis of cultured spiral ligament fibrocytes via MAPK/mitochondrial pathway. Toxicology 248:33–38
100. Zou J, Saulnier P, Perrier T, Zhang Y, Manninen T, Toppila E, Pyykko I (2008) Distribution of lipid nanocapsules in different cochlear cell populations after round window membrane permeation. J Biomed Mater Res B Appl Biomater 87:10–18

Chapter 25
Preclinical Studies on Growth Plate Cartilage Regeneration Using Chondrocytes or Mesenchymal Stem Cells

Rosa Chung and Cory J. Xian

Abstract The undesirable bony tissue repair that occur following growth plate cartilage injuries can cause serious orthopaedic problems such as limb length discrepancy and bone angulation deformities which rely on extremely invasive surgical procedures for correction. Currently, no biological therapy exists to prevent the faulty repair and to induce regeneration of the injured growth plate cartilage. In the search for a biological alternative, earlier studies have tried direct transplants of chondrocytes or cartilage tissues into the injured growth plate showing some limited success and no clinical application. In recent years, more interest has been shown towards utilising multipotent mesenchymal stem cells (MSCs) for growth plate injury repair. Using different types of growth plate injury repair models, a number of studies have investigated efficacy of promoting regeneration by directly transferring MSCs or using cells embedded in scaffolds and chondrogenic growth factors such as transforming growth factor-β3 (TGF-β3) and insulin-like growth factor-I (IGF-I). Although studies in rabbit models have shown some promise of inducing growth plate regeneration using MSCs, further studies using large animal models and with clinical trials are required to develop a practical MSC-based therapy for inducing growth plate regeneration and preventing bone growth defects.

Keywords Growth plate cartilage injury • Growth plate cartilage repair • Chondrocytes • Mesenchymal stem cells • Biomaterial scaffolds

---

R. Chung • C.J. Xian, PhD (✉)
Sansom Institute for Health Research, School of Pharmacy and Medical Sciences,
University of South Australia, City East Campus,
GPO 2471, Adelaide, SA 5001, Australia
e-mail: cory.xian@unisa.edu.au

M.K. Danquah, R.I. Mahato (eds.), *Emerging Trends in Cell and Gene Therapy*,
DOI 10.1007/978-1-62703-417-3_25,

25.1 The Growth Plate, Growth Injuries and Associated Bone Growth Defects

The growth plate is situated at the ends of long bones and is solely responsible for longitudinal growth. Bone lengthening is achieved via the process of endochondral ossification. Unlike direct bone formation (intramembranous ossification), endochondral ossification is a two-step process that involves a cartilaginous intermediate formed by chondrocytes in the growth plate. However, being a cartilaginous tissue, the growth plate is the weakest area of the long bone and often prone to injury. Previous clinical studies have reported that growth plate injuries are involved in 20 % of all fractures in young children and adolescents [1]. Unfortunately about 30 % of growth plate-related cases are not healed properly but by an unwanted faulty bony repair tissue [2, 3]. This bony repair at the injured growth plate is what is causing orthopaedic problems such as limb lengthening discrepancy and bone angulation deformities [3, 4]. For clinicians, the Salter-Harris classification system is often used to distinguish the different types of growth plate injuries and hence predict their associated prognoses. Generally speaking, types I–II growth plate injuries do not require any corrective surgery, as they do not often result in bony tissue formation. However, for types III–V, a faulty bony repair is usually the outcome [5].

In recent years, a number of studies in a rat growth plate injury model have been conducted attempting to understand the cellular and molecular mechanisms underlying the undesirable bony repair at growth plate injured site. These studies have established distinct injury repair phases leading to bony repair, namely, the inflammatory, fibrogenic, osteogenic, bone bridge formation and remodelling responses occurring on days 1–3, 3–7 and 7–14 and 10–25, respectively [6–14]. Similarly, these distinct phases of injury repair responses were also observed in other growth plate fracture models including murine, rabbits and pigs [15–17]. Although further studies are needed to elucidate mechanisms for the bony repair of injured growth plate, the above recent studies in rat models have demonstrated involvement of both intramembranous (direct) ossification and endochondral ossification (indirect) bone formation during the bony repair.

25.2 Current Surgical Interventions/Corrections

Dependent on their types and severity, some bone growth defects following growth plate injuries may require clinical/surgical interventions [18]. For injuries which result in very slight limb length discrepancies, a shoe lift device is often enough to compensate. For other more serious injuries, the resulting angulation and length discrepancies will require corrective surgeries. In most cases the patient must refrain from using the injured leg in order to minimise the angulation deformity [19]. One established method for correcting severe angulation involves a technique known as a wedge osteotomy [20]. For limb length discrepancy, the most common means of correction would involve a surgical lengthening procedure often involving a surgically created fracture at the diaphysis and then the gradual lengthening of the affected bone to match the patient's other leg [19, 21]. This procedure is most often achieved with a large external frame called the Ilizarov frame [21, 22]. Although being somewhat successful, this procedure is extremely invasive, riddled with associated problems such as infections at the pin sites, further fractures, compartment syndrome, time period involved (up to 6 months or more), pain associated with the surgery and the need for repetitive treatments in growing patients until they reach skeletal maturity [23]. More recently, a new internal nail bone lengthening device called "Fit Bone" has been developed, which is said to be less invasive and requires less time [24].

In addition, bone bridge resection followed by the insertion of interpositional material, called Langenskiold procedure, is also one method used to correct growth plate injury-induced bone defects. Some of the interpositional materials examined or used clinically include fat, silastic and bone wax [20, 25, 26]. This technique originates from initial work done by Langenskiold in 1967, and with this procedure it was reported that 82 % of 38 cases resulted in some beneficial effects from using autogenous fat as interpositional material [20]. Similarly, Bright et al. also reported a 70 % good to excellent success rate with another interpositional material, silastic, in treating limb length discrepancies and angulations [25]. However, even with further studies using a myriad of different materials and showing some success, limitations such as effects on bone growth as well as overall safety issues or long-term effects still need to be addressed. Currently no biological therapy is available for preventing the bony repair and for promoting regeneration of the injured growth plate.

25.3 Early Attempts with Chondrocyte or Tissue Implantation

As an attempt to develop biological treatments, chondrocytes have been tried as another biological interpositional material for growth plate cartilage regeneration. An early study by Bentley and Greer reported some success following the delivery of allogeneic chondrocytes into the growth plate injury sites of New Zealand white rabbits [27]. Taken from the growth plate, these implanted chondrocytes were able to fill the void and form a columnar-type structure as in normal growth plate. In addition, these cells did not cause any immunological reactions and began to show signs of endochondral ossification [27]. In a study by Lee et al. , allogeneic chondrocytes were transplanted embedded in agarose gel in a rabbit growth plate injury model [28]. Although the study found some success in halting growth arrest, it was unable to provide significant correction of deformation as well as the restoration of growth of the bone [28]. Unfortunately, this success was not reciprocated in a larger ovine model as Hansen et al. found no effect on bony bridge formation following the direct transplantation of chondrocytes to the growth plate injury site in young sheep [29]. Another earlier sheep study by Foster et al. showed some success of this chondrocyte implantation approach in preventing bony tissue formation following a growth plate injury [30]. However, analysis of the chondrocytes (transplanted embedded in a collagen substrate) showed poor survival with the longest survival period being approximately 4 weeks.

Overall, although the chondrocyte transplantation approach has shown some success in rabbit models, it has limited success in large animal models and has no clinical application. In addition, limitations of using chondrocytes include the supply, difficulty and morbidity involved in chondrocyte harvest (usually collected from articular surfaces). Furthermore, chondrocytes are known to dedifferentiate following prolonged culture during in vitro expansion. Alternatively, Yoo et al. tried perichondrium-derived chondrocytes [31] as perichondrium cells have been shown to have the ability to differentiate into chondrocytes. Thus, Yoo et al. differentiated perichondrium cells into chondrocytes ex vivo before implanting them (embedded in fibrin beads) into the injury site in a rabbit growth plate injury model [31]. However, with this approach, Yoo et al. was able to show limited success with only a slight effect on preventing angular deformity and limb length discrepancy [31]. Thus, with limited success and disadvantages of this chondrocyte transplantation approach, including those associated with harvest and expansion, in recent years, more studies have turned to investigating efficacy of using multipotent MSCs for regeneration of injured growth plate cartilage.

25.4 MSC Cell-Based Therapies and Transplantation

25.4.1 Mesenchymal Stem Cells

There are a myriad of previous studies that have highlighted the multipotent abilities of MSCs differentiating into a number of cell types including those of bone, cartilage and fat. In addition, MSCs make an ideal cell source as they are readily abundant and have been successfully isolated from various sources including adipose tissues [32–34], skeletal muscle [35, 36], periosteum [37, 38] and bone marrow [39, 40]. However, bone marrow-derived MSCs in particular hold a great interest for the regeneration of articular and growth plate cartilage and have been shown in vivo to be more likely to form hyaline cartilage in comparison to MSCs derived from other sources such as adipose tissues [41]. Many previous studies have demonstrated the capability of bone marrow-derived MSCs to undergo successful chondrogenic differentiation in vitro by controlling the culture conditions [39, 42–44]. Furthermore, although making up only a fraction of the population of total cells present in the bone marrow, these cells have been demonstrated to be easily isolated and expanded in vitro [39, 40].

Chen et al. demonstrated the successful treatment of growth plate defects with transplanted MSCs isolated from the periosteum [45]. Chen et al. suggested that due to their high proliferation rate, they were an ideal source for donor cells [45]. Using a rabbit model, Hui et al. compared various sources of MSCs embedded in fibrin glue and found that MSCs sourced from the bone marrow as well as the periosteum showed greater potential for growth plate cartilage regeneration in comparison to those derived from fat [46]. In addition, Planka et al. compared whether autogenous or allogeneic MSC transplantation has any effect on their ability to form cartilage within the growth plate injury site and found that both autologous and allogeneic MSCs resulted in the formation of hyaline chondrocytes within the growth plate injury site in a rabbit model [47]. No overall differences were noted for correcting tibia length and angulation defects. Similar results were also observed in a guinea pig growth plate injury model [48]. Recently, in a larger animal growth plate injury model, autologous bone marrow-derived MSCs embedded in gelfoam were implanted into the growth plate defect in young sheep [49]. However, unlike the rabbit models, the bone marrow MSCs failed to form cartilage and instead caused an increase in the fibrous tissue formation within the growth plate injury site of sheep [49]. Overall, although the potential of transplanted MSCs have been demonstrated in rabbit growth plate injury models, these positive results have failed to be replicated in large animal model. Therefore, further investigations into the feasibility and efficacy of MSC transplantation for growth plate regeneration are needed in large animal models.

25.4.2 Growth Factors

In addition to finding the appropriate source of stem cells, successful cartilage tissue engineering also requires the correct signalling molecules to ensure chondrogenesis [50]. Growth factors such as platelet-derived growth factor (PDGF), fibrogenic growth factor (FGF-2), transforming growth factors (TGF-β1 and TGF-β3), bone morphogenic protein-2 and bone morphogenic protein-7 (BMP-2, BMP-7), epidermal growth factor (EGF) and insulin-like growth factor (IGF-I) have all been shown to be important during the migration, proliferation and/or chondrogenic differentiation of MSCs (Table 25.1). In particular, TGF-β1 and TGF-β3 as well as FGF-2 and IGF-I have been shown to be the more potent chondrogenic growth factors. A myriad of studies have shown that both TGF-β1 and TGF-β3 are important during chondrogenesis [68–70]. McCarty et al. found that the addition of TGF-β1 as well as the combination of TGF-β1 and BMP-7 significantly increased collagen-2 and aggrecan expression in in vitro chondrogenic pellet culture [39]. TGF-β3 has also been shown to stimulate extracellular matrix synthesis in a rabbit model of acute articular cartilage injury [66], and one study in young rabbits with growth plate defects found that the addition of MSCs and TGF-β3 into the defects caused a significant decrease in angular deformity [71]. However, this result was not replicated when a similar experiment was conducted in the ovine tibial growth plate injury model [49].

Insulin-like growth factor (IGF-I) is important in normal growth plate physiology particularly in the differentiation and maturation of growth plate chondrocytes.

Table 25.1 Chondrogenic growth factors

Growth factor	Effect on MSC and chondrogenesis	References
Platelet-derived growth factor (PDGF)	Induces the proliferation and migration of MCS	[8, 39, 51, 52]
Insulin-like growth factor (IGF-I)	Increases MSC proliferation Enhance chondrogenesis (more effective in combination with TGF-β1)	[53–55]
Fibroblast growth factor (FGF-2)	Enhances MSC differentiation Enhances chondrogenesis Enhances cell proliferation and proteoglycan synthesis	[56–58]
Bone morphogenic protein-2, Bone morphogenic protein-7 (BMP-2, BMP-7)	BMP-2 increases proliferation in vitro at high concentrations Stimulates cartilage repair in vivo Enhances chondrogenesis of synovial MSCs in combination with TGF-β1	[39, 59–62]
Epidermal growth factor (EGF)	Enhances MSC proliferation	[39, 63]
Transforming growth factor-β1 (TGF-β1)	Increases chondrogenesis in combination with dexamethasone Induces chondrogenic differentiation	[39, 64, 65]
Transforming growth factor-β3 (TGF-β3)	Increases migration of MSCs Stimulates articular cartilage repair Involved in chondrocyte differentiation	[66, 67]

IGF-I has also been shown to have a significant influence on the induction of chondrogenic differentiation from MSCs [55, 72]. The chondrogenic properties of IGF-I have been shown to significantly amplify when used in combination with TGF-β1 [54, 73] as well as BMP-7 [74]. Interestingly, this synergistic chondrogenic effect was achieved only when MSCs were exposed to TGF-β1 prior to IGF-I [53]. This study suggests that the efficiency and success of any engineered cartilage can be strongly influenced by the controlled sequential treatment of growth factors [44].

In addition to TGF-β1, TGF-β3 and IGF-I, FGF-2 has also been reported as possessing chondrogenic properties. FGF-2 has demonstrated its ability to enhance mesenchymal differentiation of MSCs [58] as well as improving the cartilage healing process of an osteochondral lesion in rabbits [52]. In addition, in vitro work has also found that the treatment of bone marrow MSCs with FGF-2 resulted in an increase in proliferation and proteoglycan synthesis [57]. Further studies are needed to explore other potential potent chondrogenic growth factors for successful cartilage regeneration. However, what is important is the knowledge that supplementation with the appropriate growth factor or growth factor combination is vital for the success of using a MSC-based therapy for growth plate cartilage regeneration.

25.4.3 Biomaterial Scaffolds

Athanasiou et al. outlined three main basic components for successful tissue engineering, namely, the cells, the signalling molecules and the scaffold [50]. In order to stimulate chondrogenesis of MSCs, the material by which the scaffold is derived from could play a large role in determining success of cartilage tissue engineering. Other than stimulating chondrogenesis, the scaffold enables stable support for the MSCs as an absence of a scaffold decreased viability of MSCs over time [75]. Di Martino et al. outlined several key points that are important when designing an appropriate scaffold [76], such as its biocompatibility, appropriate pore size (to allow movement of cells) and bio-absorbability or bio-degradability. In addition, the type of scaffolds can also determine the seeding as well as the degree of proliferation and migration to encourage appropriate cellular organisation as well as matrix production [77, 78]. Interestingly, Wise et al. observed that a smaller fibre diameter seemed to enhance the chondrogenic potential of MSCs [79].

An important aspect to consider when designing a scaffold is its main material. A myriad of studies have utilised different scaffolding materials for cartilage tissue repair, ranging from those synthetically created to naturally available. Some of synthetically constructed materials previously used in cartilage repair studies include polyethylene (PEG) [80] as well as poly(glycolic acid-co-lactic acid) (PGLA) [81]. However, while most of these scaffolds have been used and shown some success in articular cartilage regeneration studies [82, 83], it is unclear whether they can be potentially used for growth plate regeneration. In addition, although synthetically created scaffolds offer the freedom for controlling aspects such as pore size, some of their limitations include relatively weak cell adhesion properties as well as concerns for poorer biocompatibility [84, 85]. On the other hand, many studies have also utilised scaffolds consisting of naturally present biomaterials. These include core ingredients such as chitosan, agarose, alginate, fibrin and hyaluronan [59, 86–88]. These protein or carbohydrate-based natural materials offer a more biocompatible, biodegradable as well as a more natural microenvironment for MSCs [89].

Chitosan is a natural polysaccharide derived from the shells of crustaceans [90]. Chitosan has been shown to be highly biocompatible and similar to glycosaminoglycans (GAGs) normally present within the extracellular matrix of cartilage making it ideal for all types of cartilage repair [91]. Using an MSC and scaffold construct consisting of chitosan and collagen, Planka et al. saw some success in minimising growth arrest and angulation deformity in a miniature pig growth plate injury model [48]. A similar construct combining chitosan and MSC also resulted in restoration of a large growth plate defect in a rabbit model [92]. Chen et al. combined MSC and agarose for their rabbit growth plate injury study and observed that treatment decreased growth arrest and angular deformity normally associated with growth plate injury [45]. Although not yet tested in a growth plate injury model, another promising naturally existing scaffold material is hyaluronic acid. Recently, hyaluronic acid enhanced chondrogenesis of MSCs [68] and could promote formation of neocartilage in vitro [93].

Interestingly, in a recent study, [26] utilised a scaffold-free tissue-engineered construct on partial growth arrest in a rabbit growth plate injury model. Using MSCs derived from the synovial of the same species, [26] found the implanted MSCs were able to differentiate into proliferative and pre-hypertrophic-like chondrocytes in vivo [26]. Furthermore, in comparison to MSCs embedded in scaffolds, a non-scaffold construct resulted in decreased bony repair tissue formation as well as allowing longitudinal bone growth of limbs [26]. Therefore, although some of the above studies produced promising results with some scaffold materials, the most ideal scaffold/cell/growth factor combination for optimally inducing growth plate cartilage regeneration is still yet to be uncovered.

25.5 Conclusion

Growth plate injuries are common and the associated bone growth defects still remain great experimental and clinical challenges, for which currently no biological therapy exists. In search for a biological solution, previous attempts involving direct chondrocyte transplantation found limited success experimentally and no application clinically in promoting growth plate regeneration. Although several studies in rabbit models have demonstrated success of inducing growth plate regeneration using MSCs, further studies using large animal models and with clinical trials are required to investigate whether a practical MSC-based therapy can be defined and be useful for inducing growth plate regeneration and preventing bone growth defects in children.

Acknowledgements The authors wish to thank Bone Growth Foundation, National Health and Medical Research Council of Australia and Channel-7 Children's Research Foundation (South Australia) for funding some of our own research work discussed in this chapter. The authors thank Dr Bruce K Foster (Women's and Children's Hospital, Adelaide, Australia) for the useful discussion.

References

1. Mizuta T et al (1987) Statistical analysis of the incidence of physeal injuries. J Pediatr Orthop 7(5):518–523
2. Barmada A, Gaynor T, Mubarak SJ (2003) Premature physeal closure following distal tibia physeal fractures: a new radiographic predictor. J Pediatr Orthop 23(6):733–739
3. Ogden JA (1984) Growth slowdown and arrest lines. J Pediatr Orthop 4(4):409–415
4. Wattenbarger JM, Gruber HE, Phieffer LS (2002) Physeal fractures, part I: histologic features of bone, cartilage, and bar formation in a small animal model. J Pediatr Orthop 22(6): 703–709
5. Basener CJ, Mehlman CT, DiPasquale TG (2009) Growth disturbance after distal femoral growth plate fractures in children: a meta-analysis. J Orthop Trauma 23(9):663–667

6. Arasapam G et al (2006) Roles of COX-2 and iNOS in the bony repair of the injured growth plate cartilage. J Cell Biochem 99(2):450–461
7. Chung R, Cool JC, Scherer MA, Foster BK, Xian CJ (2006) Roles of neutrophil-mediated inflammatory response in the bony repair of injured growth plate cartilage in young rats. J Leukoc Biol 80(6):1272–1280
8. Chung R et al (2009) Potential roles of growth factor PDGF-BB in the bony repair of injured growth plate. Bone 44(5):878–885
9. Macsai CE, Foster BK, Xian CJ (2008) Roles of Wnt signalling in bone growth, remodelling, skeletal disorders and fracture repair. J Cell Physiol 215(3):578–587
10. Macsai CE et al (2012) Microarray expression analysis of genes and pathways involved in growth plate cartilage injury responses and bony repair. Bone 50(5):1081–1091
11. Ngo TQ et al (2006) Expression of bone morphogenic proteins and receptors at the injured growth plate cartilage in young rats. J Histochem Cytochem 54(8):945–954
12. Xian CJ et al (2004) Intramembranous ossification mechanism for bone bridge formation at the growth plate cartilage injury site. J Orthop Res 22(2):417–426
13. Zhou FH, Foster BK, Sander G, Xian CJ (2004) Expression of proinflammatory cytokines and growth factors at the injured growth plate cartilage in young rats. Bone 35:1307–1315
14. Zhou FH et al (2006) TNF-alpha mediates p38 MAP kinase activation and negatively regulates bone formation at the injured growth plate in rats. J Bone Miner Res 21(7):1075–1088
15. Jaramillo D et al (1990) Posttraumatic growth-plate abnormalities: MR imaging of bony-bridge formation in rabbits. Radiology 175(3):767–773
16. Lee MA, Nissen TP, Otsuka NY (2000) Utilization of a murine model to investigate the molecular process of transphyseal bone formation. J Pediatr Orthop 20(6):802–806
17. Wirth T et al (1994) The implantation of cartilaginous and periosteal tissue into growth plate defects. Int Orthop 18(4):220–228
18. Bostock SH, Peach BG (1996) Spontaneous resolution of an osseous bridge affecting the distal tibial epiphysis. J Bone Joint Surg Br 78(4):662–663
19. Foster BK, Johnstone EW (2000) Management of growth plate injuries. In: Benson M, Fixsen J, MacNicol M, Parsch K (eds) Paediatric orthopaedics and fractures. Harcourt Publishers, London
20. Langenskiold A (1981) Surgical treatment of partial closure of the growth plate. J Pediatr Orthop 1(1):3–11
21. Peterson HA (1984) Partial growth plate arrest and its treatment. J Pediatr Orthop 4(2):246–258
22. Paley D et al (1997) Femoral lengthening over an intramedullary nail. A matched-case comparison with Ilizarov femoral lengthening. J Bone Joint Surg Am 79(10):1464–1480
23. Dahl MT, Gulli B, Berg T (1994) Complications of limb lengthening. A learning curve. Clin Orthop Relat Res (301):10–18
24. Baumgart R (2009) The reverse planning method for lengthening of the lower limb using a straight intramedullary nail with or without deformity correction. A new method. Oper Orthop Traumatol 21(2):221–233
25. Bright RW (1974) Operative correction of partial epiphyseal plate closure by osseous-bridge resection and silicone-rubber implant. An experimental study in dogs. J Bone Joint Surg Am 56(4):655–664
26. Yoshida K et al (2012) Treatment of partial growth arrest using an in vitro-generated scaffold-free tissue-engineered construct derived from rabbit synovial mesenchymal stem cells. J Pediatr Orthop 32(3):314–321
27. Bentley G, Greer RB 3rd (1971) Homotransplantation of isolated epiphyseal and articular cartilage chondrocytes into joint surfaces of rabbits. Nature 230(5293):385–388
28. Lee EH et al (1998) Treatment of growth arrest by transfer of cultured chondrocytes into physeal defects. J Pediatr Orthop 18(2):155–160
29. Hansen AL et al (1990) Growth-plate chondrocyte cultures for reimplantation into growth-plate defects in sheep. Characterization of cultures. Clin Orthop Relat Res (256):286–298

30. Foster BK et al (1990) Reimplantation of growth plate chondrocytes into growth plate defects in sheep. J Orthop Res 8(4):555–564
31. Yoo WJ et al (2005) Implantation of perichondrium-derived chondrocytes in physeal defects of rabbit tibiae. Acta Orthop 76(5):628–636
32. Awad HA et al (2004) Chondrogenic differentiation of adipose-derived adult stem cells in agarose, alginate, and gelatin scaffolds. Biomaterials 25(16):3211–3222
33. Dragoo JL et al (2003) Tissue-engineered cartilage and bone using stem cells from human infrapatellar fat pads. J Bone Joint Surg Br 85(5):740–747
34. Peterson B et al (2005) Healing of critically sized femoral defects, using genetically modified mesenchymal stem cells from human adipose tissue. Tissue Eng 11(1–2):120–129
35. Deans RJ, Moseley AB (2000) Mesenchymal stem cells: biology and potential clinical uses. Exp Hematol 28(8):875–884
36. Jankowski RJ, Deasy BM, Huard J (2002) Muscle-derived stem cells. Gene Ther 9(10):642–647
37. Fukumoto T et al (2003) Combined effects of insulin-like growth factor-1 and transforming growth factor-beta1 on periosteal mesenchymal cells during chondrogenesis in vitro. Osteoarthritis Cartilage 11(1):55–64
38. Nakamura K et al (1999) p38 mitogen-activated protein kinase functionally contributes to chondrogenesis induced by growth/differentiation factor-5 in ATDC5 cells. Exp Cell Res 250(2):351–363
39. McCarty RC et al (2009) Characterisation and developmental potential of ovine bone marrow derived mesenchymal stem cells. J Cell Physiol 219(2):324–333
40. Vinatier C et al (2009) Cartilage tissue engineering: towards a biomaterial-assisted mesenchymal stem cell therapy. Curr Stem Cell Res Ther 4(4):318–329
41. Park J et al (2006) Transgene-activated mesenchymal cells for articular cartilage repair: a comparison of primary bone marrow-, perichondrium/periosteum- and fat-derived cells. J Gene Med 8(1):112–125
42. Caplan AI (1994) The mesengenic process. Clin Plast Surg 21(3):429–435
43. Prockop DJ (1997) Marrow stromal cells as stem cells for nonhematopoietic tissues. Science 276(5309):71–74
44. Xian CJ, Foster BK (2006) Repair of injured articular and growth plate cartilage using mesenchymal stem cells and chondrogenic gene therapy. Curr Stem Cell Res Ther 1(2):213–229
45. Chen F et al (2003) Cultured mesenchymal stem cell transfers in the treatment of partial growth arrest. J Pediatr Orthop 23(4):425–429
46. Hui JH et al (2005) Comparative study of the ability of mesenchymal stem cells derived from bone marrow, periosteum, and adipose tissue in treatment of partial growth arrest in rabbit. Tissue Eng 11(5–6):904–912
47. Planka L et al (2008) Allogeneic and autogenous transplantations of MSCs in treatment of the physeal bone bridge in rabbits. BMC Biotechnol 8:70
48. Planka L et al (2008) Use of allogenic stem cells for the prevention of bone bridge formation in miniature pigs. Physiol Res 58:885–893
49. McCarty RC et al (2010) Application of autologous bone marrow derived mesenchymal stem cells to an ovine model of growth plate cartilage injury. Open Orthop J 4:204–210
50. Athanasiou KA et al (2001) Basic science of articular cartilage repair. Clin Sports Med 20(2):223–247
51. Hayashi M et al (2008) Weekly intra-articular injections of bone morphogenetic protein-7 inhibits osteoarthritis progression. Arthritis Res Ther 10(5):R118
52. Maehara H et al (2010) Repair of large osteochondral defects in rabbits using porous hydroxyapatite/collagen (HAp/Col) and fibroblast growth factor-2 (FGF-2). J Orthop Res 28(5):677–686
53. Indrawattana N et al (2004) Growth factor combination for chondrogenic induction from human mesenchymal stem cell. Biochem Biophys Res Commun 320(3):914–919
54. Longobardi L et al (2006) Effect of IGF-I in the chondrogenesis of bone marrow mesenchymal stem cells in the presence or absence of TGF-beta signaling. J Bone Miner Res 21(4):626–636

55. Neidel J, Schulze M, Sova L (1994) Insulin-like growth factor I accelerates recovery of articular cartilage proteoglycan synthesis in culture after inhibition by interleukin 1. Arch Orthop Trauma Surg 114(1):43–48
56. Gouttenoire J et al (2004) Modulation of collagen synthesis in normal and osteoarthritic cartilage. Biorheology 41(3–4):535–542
57. Stewart AA et al (2007) Effect of fibroblast growth factor-2 on equine mesenchymal stem cell monolayer expansion and chondrogenesis. Am J Vet Res 68(9):941–945
58. Tsutsumi S et al (2001) Retention of multilineage differentiation potential of mesenchymal cells during proliferation in response to FGF. Biochem Biophys Res Commun 288(2):413–419
59. Kuo CK et al (2006) Cartilage tissue engineering: its potential and uses. Curr Opin Rheumatol 18(1):64–73
60. Kumar S, Nagy TR, Ponnazhagan S (2010) Therapeutic potential of genetically modified adult stem cells for osteopenia. Gene Ther 17(1):105–116
61. Miyamoto C et al (2007) Osteogenic protein-1 with transforming growth factor-beta1: potent inducer of chondrogenesis of synovial mesenchymal stem cells in vitro. J Orthop Sci 12(6):555–561
62. Pountos I et al (2010) The effect of bone morphogenetic protein-2, bone morphogenetic protein-7, parathyroid hormone, and platelet-derived growth factor on the proliferation and osteogenic differentiation of mesenchymal stem cells derived from osteoporotic bone. J Orthop Trauma 24(9):552–556
63. Tamama K et al (2006) Epidermal growth factor as a candidate for ex vivo expansion of bone marrow-derived mesenchymal stem cells. Stem Cells 24(3):686–695
64. Johnston B, Hering TM, Caplan AI, Goldberg VM, Yoo JU (1998) In vitro chondrogenesis of bone marrow-derived mesenchymal progenitor cells. Exp Cell Res 238:256–272
65. Iwasaki M et al (1993) Transforming growth factor-beta 1 stimulates chondrogenesis and inhibits osteogenesis in high density culture of periosteum-derived cells. Endocrinology 132(4):1603–1608
66. Lee CH et al (2010) Regeneration of the articular surface of the rabbit synovial joint by cell homing: a proof of concept study. Lancet 376(9739):440–448
67. Goessler UR et al (2005) In-vitro analysis of the expression of TGFbeta -superfamily-members during chondrogenic differentiation of mesenchymal stem cells and chondrocytes during dedifferentiation in cell culture. Cell Mol Biol Lett 10(2):345–362
68. Bian L et al (2011) Enhanced MSC chondrogenesis following delivery of TGF-beta3 from alginate microspheres within hyaluronic acid hydrogels in vitro and in vivo. Biomaterials 32(27):6425–6434
69. Dickhut A et al (2010) Chondrogenesis of human mesenchymal stem cells by local transforming growth factor-beta delivery in a biphasic resorbable carrier. Tissue Eng Part A 16(2):453–464
70. Tuli R et al (2003) Transforming growth factor-beta-mediated chondrogenesis of human mesenchymal progenitor cells involves N-cadherin and mitogen-activated protein kinase and Wnt signaling cross-talk. J Biol Chem 278(42):41227–41236
71. Ahn JI et al (2004) Stem cell repair of physeal cartilage. J Orthop Res 22(6):1215–1221
72. Li J et al (2012) Dynamic compression of rabbit adipose-derived stem cells transfected with insulin-like growth factor 1 in chitosan/gelatin scaffolds induces chondrogenesis and matrix biosynthesis. J Cell Physiol 227(5):2003–2012
73. Worster AA et al (2001) Chondrocytic differentiation of mesenchymal stem cells sequentially exposed to transforming growth factor-beta1 in monolayer and insulin-like growth factor-I in a three-dimensional matrix. J Orthop Res 19(4):738–749
74. Loeser RF, Pacione CA, Chubinskaya S (2003) The combination of insulin-like growth factor 1 and osteogenic protein 1 promotes increased survival of and matrix synthesis by normal and osteoarthritic human articular chondrocytes. Arthritis Rheum 48(8):2188–2196
75. Jeong JH et al (2009) Human mesenchymal stem cells implantation into the degenerated coccygeal disc of the rat. Cytotechnology 59:55–64

76. Di Martino A, Sittinger M, Risbud MV (2005) Chitosan: a versatile biopolymer for orthopaedic tissue-engineering. Biomaterials 26(30):5983–5990
77. Moroni L et al (2006) Fiber diameter and texture of electrospun PEOT/PBT scaffolds influence human mesenchymal stem cell proliferation and morphology, and the release of incorporated compounds. Biomaterials 27(28):4911–4922
78. Ragetly GR et al (2010) Cartilage tissue engineering on fibrous chitosan scaffolds produced by a replica molding technique. J Biomed Mater Res A 93(1):46–55
79. Wise JK et al (2009) Chondrogenic differentiation of human mesenchymal stem cells on oriented nanofibrous scaffolds: engineering the superficial zone of articular cartilage. Tissue Eng Part A 15(4):913–921
80. Liu SQ et al (2010) Biomimetic hydrogels for chondrogenic differentiation of human mesenchymal stem cells to neocartilage. Biomaterials 31(28):7298–7307
81. Fan H et al (2010) TGF-beta3 immobilized PLGA-gelatin/chondroitin sulfate/hyaluronic acid hybrid scaffold for cartilage regeneration. J Biomed Mater Res A 95(4):982–992
82. Haisch A et al (2002) A tissue-engineering model for the manufacture of auricular-shaped cartilage implants. Eur Arch Otorhinolaryngol 259(6):316–321
83. Uematsu K et al (2005) Cartilage regeneration using mesenchymal stem cells and a three-dimensional poly-lactic-glycolic acid (PLGA) scaffold. Biomaterials 26(20):4273–4279
84. Engler AJ et al (2006) Matrix elasticity directs stem cell lineage specification. Cell 126(4):677–689
85. Spain TL, Agrawal CM, Athanasiou KA (1998) New technique to extend the useful life of a biodegradable cartilage implant. Tissue Eng 4(4):343–352
86. Lisignoli G et al (2005) Cellular and molecular events during chondrogenesis of human mesenchymal stromal cells grown in a three-dimensional hyaluronan based scaffold. Biomaterials 26(28):5677–5686
87. Kessler MW, Grande DA (2008) Tissue engineering and cartilage. Organogenesis 4(1):28–32
88. Park SH et al (2005) Tissue-engineered cartilage using fibrin/hyaluronan composite gel and its in vivo implantation. Artif Organs 29(10):838–845
89. Noth U, Steinert AF, Tuan RS (2008) Technology insight: adult mesenchymal stem cells for osteoarthritis therapy. Nat Clin Pract Rheumatol 4(7):371–380
90. Ragetly GR et al (2010) Effect of chitosan scaffold microstructure on mesenchymal stem cell chondrogenesis. Acta Biomater 6(4):1430–1436
91. VandeVord PJ et al (2002) Evaluation of the biocompatibility of a chitosan scaffold in mice. J Biomed Mater Res 59(3):585–590
92. Li L et al (2004) Chitin as a scaffold for mesenchymal stem cells transfers in the treatment of partial growth arrest. J Pediatr Orthop 24(2):205–210
93. Chung C et al (2009) The influence of degradation characteristics of hyaluronic acid hydrogels on in vitro neocartilage formation by mesenchymal stem cells. Biomaterials 30(26):4287–4296

Chapter 26
Cell and Gene Transfer Strategies for Vascularization During Skin Wound Healing

Benoit Hendrickx, Margot Den Hondt, Kristoff Verdonck, Jan J. Vranckx, and Aernout Luttun

Abstract Adequate vascularization is pivotal to skin wound healing. Therefore, designing efficient revascularization strategies based on the mechanisms behind electromechanical stimulation of wound vascularization would be beneficial to the growing number of patients in need of improved wound healing. Recent attention has centered on applying gene/protein transfer and cell differentiation/transplantation approaches to stimulate and mimic the molecular events occurring during wound revascularization. Although both gene/protein transfer and cell differentiation/transplantation are faced with important challenges, researchers have made tremendous advances and shown both strategies to be a promising approach.

B. Hendrickx, MD, PhD
Department of Cardiovascular Sciences,
Molecular and Vascular Biology Research Unit, KU Leuven,
Leuven, Belgium

Department of Development and Regeneration, Cell and Gene Therapy
Applications Research Unit, Katholieke Universiteit Leuven, University Hospitals,
Herestraat 49, Leuven, B-3000, Belgium

M. Den Hondt, MD • J.J. Vranckx, MD, PhD (✉)
Department of Development and Regeneration, Cell and Gene Therapy Applications
Research Unit, Katholieke Universiteit Leuven, University Hospitals,
Herestraat 49, Leuven, B-3000, Belgium
e-mail: jan.vranckx@uzleuven.be

K. Verdonck, MD
Department of Cardiovascular Sciences, Molecular and Vascular Biology Research Unit,
KU Leuven, Leuven, Belgium

A. Luttun, PhD (✉)
Department of Cardiovascular Sciences, Molecular and Vascular Biology
Research Unit, Endothelial Cell Biology Unit, Katholieke Universiteit Leuven,
Campus Gasthuisberg, Onderwijs & Navorsing 1, Herestraat 49, B-3000, Leuven, Belgium
e-mail: aernout.luttun@med.kuleuven.be

M.K. Danquah, R.I. Mahato (eds.), *Emerging Trends in Cell and Gene Therapy*,
DOI 10.1007/978-1-62703-417-3_26,

In this chapter, we give an overview of the myriad of molecular players involved in neovascularization. We also discuss the molecular mechanisms of neovascularization during wound healing and provide an in-depth review on neovascular strategies and techniques for wound healing and tissue-engineered skin equivalents.

Keywords Skin wound healing • Vasculogenesis • Angiogenesis • Endothelial cells • Vascular endothelial growth factor • Gene transfer • Stem/progenitor cells • Skin tissue engineering

26.1 Introduction

Skin covers our entire body surface and therefore is considered as our largest organ spanning 1.6–1.8 m^2 in an adult human being [1]. It constitutes a protective barrier against physical, chemical, or bacterial threats, maintains our body temperature, and also serves as a sensory end organ that reacts to cold, heat, pressure, or injury. A large breach or wound in this protective barrier, especially one that encompasses all layers of the skin (a so-called full-thickness wound), represents a dangerous – if not mortal – threat if not healed appropriately. Hence, strategies to accelerate or improve – repair of wounded skin are of utmost importance and can be lifesaving. In order to design such strategies, it is obligatory to understand the anatomy of the skin as well as the natural course of the repair process.

Three skin layers can be distinguished: the outer layer called "epidermis," the middle layer or "dermis," and the supporting "hypodermis." The dermis can further be divided in an upper papillary layer and a lower reticular layer. Like most organs, the skin is invested with an elaborate blood vascular network that feeds skin cells with oxygen and nutrients and removes waste products [2]. Indeed, 1 cm^2 of skin contains up to 0.7 m of blood vessels. While the epidermis is relying on diffusion from the dermal plexus for its oxygen supply (and hence is avascular), both dermis and hypodermis contain a blood vessel network for active oxygen transport. This network is built as two horizontal plexi, one more superficial just underneath the dermo-epidermal junction and another deeper at the dermal-subcutaneous junction. Both plexi are connected by paired ascending arterioles and descending venules. From the arterial arm of the upper plexus, a papillary capillary network sprouts. The density of this network significantly decreases with age [3]. The capillaries servicing the sweat glands and the hair follicles, on the other hand, originate from the lower plexus. To ensure a proper interstitial fluid balance, the dermis also contains a network of blind-ending lymphatic capillaries that connect to larger collectors in the muscle layer underneath the hypodermis which recirculate tissue fluid or "lymph" and extravasated immune cells to the blood vascular system [4].

When the integrity of skin is compromised, such as during trauma or burns, a repair process is initiated within minutes. This complex wound healing response can be somewhat artificially divided into four different partially overlapping phases during which distinct cell types, growth factors, and matrix components interact with the intention to restore the barrier [5–7]. During the first phase, the immediate response, or hemostatic phase, excessive blood loss is prevented by activation of the clotting cascade in which platelets and fibrin form a vascular plug. Next, an inflammatory phase ensues during which neutrophils, macrophages, and lymphocytes infiltrate the wound area to remove cell debris and bacteria. Thirdly, the proliferative or granulation tissue formation phase is characterized by deposition of a temporary matrix of collagen (the main structural component of the dermis) and other proteins by (myo)fibroblasts, ingrowth of new blood vessels, re-epithelialization (involving keratinocyte migration and proliferation), and wound contraction (by myofibroblasts). Finally, during the maturation or resolution phase, new blood vessels mature, and the provisional collagen network is remodeled to better resemble the structure of native dermis and regain tensile strength and elasticity. Unfortunately, unlike wounds during fetal development, even under the best healing conditions, large postnatal wounds will leave a scar which maintains only 70 % of the tensile strength of healthy skin [8]. Furthermore, this stepwise process can be perturbed at different stages leading to aberrant wound healing. For instance, when inflammation persists, wounds become chronic, such as those in ischemic limbs. Another example is when blood vessels and/or collagenous fibrosis persists in the maturation phase resulting in hypertrophic scars or keloids [9, 10].

26.2 Neovascularization During Wound Healing

26.2.1 Mechanisms of Neovascularization

Ingrowth of new blood vessels into the wound area is one of the central events that nurture the wound healing process. Blood vessels are built of mainly two different cell types. The endothelial cells (EC) line the inside of the tube, while the peri-endothelial cells (called smooth muscle cells or SMC in larger vessels and pericytes in capillaries) surround the endothelium and confer contractile properties. There are two ways according to which new blood vessels can grow in a wound: (sprouting) angiogenesis and vasculogenesis [11]. The former refers to a sequence of cellular and molecular events by which new vessels sprout from preexisting vessels and subsequently mature by acquiring a SMC/pericyte coat. The latter encompasses the "de novo" assembly of endothelial tubes from circulating endothelial progenitor cells (EPC). Recently, a third mechanism was proposed that could explain the early appearance of patent blood vessels in the wound bed. Kilarski and colleagues proposed that rapid vascularization of wounds can be accounted for by contraction-driven mechanical translocation of existing intact vascular loops, which was subsequently termed "looping angiogenesis" by Benest and Augustin [12, 13].

26.2.2 Molecular Players in Neovascularization

Making blood vessels is not an individual sport but involves many different team players, such as growth factors/receptors, downstream signaling pathways, junctional molecules, and extracellular matrix (ECM) components. Identifying them will offer potential targets for gene/protein therapy to stimulate wound vascularization and hence wound healing. In the following chapter, we give an overview of the myriad of factors involved in angiogenic and vasculogenic blood vessel formation. Thus far, only a limited number of molecular players in blood vessel formation have been tested for their efficacy in wound revascularization (Table 26.1). For some factors, their involvement in wound healing in general and revascularization in particular has been studied in mice genetically deficient for or overexpressing these factors (e.g., basic fibroblast growth factor or bFGF [71], placental growth factor or PlGF [72], hypoxia inducible factor-1α or HIF-1α [73, 74], α3β1 integrin [75], α2β1 integrin [76], platelet-derived growth factor-D or PDGF-D/vascular endothelial growth factor-E or VEGF-E [77], endothelial nitric oxide synthase or eNOS [78], inducible NOS or iNOS [79–81], monocyte chemoattractant protein-1 or MCP-1 [82], CXCR2 [83], granulocyte-macrophage colony-stimulating factor or GM-CSF [84–86], thrombospondin-1 [87], thrombospondin-2 [88]) or by blocking antibodies (e.g., neuropilin-1 [89], αvβ3 integrin [90], VEGF-A [91]), but not all of these have been therapeutically tested in preclinical wound models. Furthermore, only a small number of these growth factors have been clinically tested (e.g., PDGF-B, epidermal growth factor or EGF, and bFGF) [92], and only one growth factor, i.e., PDGF-B, has been FDA approved for treatment of diabetic foot ulcers (becaplermin or Regranex®), although this was mainly for its beneficial effects on re-epithelialization rather than its effect on revascularization [93].

26.2.3 Sprouting Angiogenesis

The formation of a new blood vessel by angiogenic sprouting involves different steps, each of them driven by certain categories of growth factors. For an elaborate review of the angiogenic process, we refer the reader to [94–98]. Below, we only highlight the protagonists that play a part in this complex process.

The wound bed is severely hypoxic, with oxygen levels lower than 1.5 % in the wound center [99, 100]. This lack of oxygen triggers expression of HIF-1α in the exposed cells which in turn boosts the expression of many HIF-responsive angiogenic factors, including VEGF-A (mostly the $VEGF_{165}$ isoform) which will orchestrate the ensuing process of endothelial sprouting [101]. A first requirement for sprouting is the breakdown of the basement membrane and ECM surrounding EC and SMC, mainly mediated by matrix metalloproteinases (MMP), such as MT1-MMP [97, 98]. Angiopoietin-2, a ligand for Tie-2 stored in EC, is important for SMC detachment (reviewed in [102]). Next, certain EC (the so-called tip cells) are selected for taking the lead position in the nascent sprout, while others

Table 26.1 (Lymph)angiogenic and vasculogenic factors tested in wound healing

Factor	Transfer method	Model	Effect	Reference(s)
A. *Monotherapy*				
HIF-1α	Electroporation	Db/db mice	Increased levels of angiogenic factors (VEGF-A, PlGF, PDGF-B, angiopoietin-2); increased circulating angiogenic cells	[14]
PlGF	Adenovirus	Streptozotocin-induced diabetic mice	Increased vascularization; increased levels of PDGF-B, VEGF-A, and bFGF; increased fibroblast migration in vitro	[15]
PHD2	siRNA inhibition	Db/db mice	HIF-1α stabilization; increased levels of VEGF-A and bFGF; increased vessel density	[16]
HIF-1α	Plasmid encoding constitutively active HIF-1α	Db/db mice	Increased vascularization; increased expression of VEGF-A, heme oxygenase-1, iNOS	[17]
IGF-1	Recombinant protein	Diabetic mice	Increased HIF-1α protein synthesis	[18]
Notch1/Dll4	Function blocking antibodies	C57Bl/6 mice	Decreased lymphangiogenesis	[19]
PDGF-C	Recombinant protein	Diabetic mice	Mesenchymal cell proliferation	[20]
SDF-1α	Protein on alginate hydrogel scaffold	Mice	Increased vascularization	[21]
S1P	Subcutaneous protein injection	Normal and db/db mice; rats	Increased vascularization	[22]
PDGF-BB	Protein in microspheres seeded on nanofibrous scaffolds	Rats	Increased blood vessel number; increased expression of CXCL1, 2, 5	[23]
PDGF-BB	Protein on collagen membranes	Rabbits	Increased formation of capillaries	[24]
PDGF-BB	Protein in gel	Streptozotocin-induced diabetic rats	Increased angiogenesis	[25]

(continued)

Table 26.1 (continued)

Factor	Transfer method	Model	Effect	Reference(s)
PDGF-BB	Recombinant protein	Patients with diabetic foot ulcers	Reduced time to healing; more complete healing	[26]
PDGF-B	Lentivirus	Db/db mice	Increased vascularization; improved collagen organization	[27]
KGF-1 (FGF-7)	Liposomes	Rats	Improved neovascularization; increased FGF, IGF-1	[28]
bFGF	Recombinant protein in a chitosan gel	Db/db mice	Increased number of vessels	[29, 30]
FGF-1	Recombinant protein on collagen scaffold	Rabbits	Increased angiogenesis	[31]
FGF-1	Recombinant protein on fibrin scaffold	Rabbits	Increased angiogenesis	[32]
bFGF	Recombinant protein	Db/db mice	Increased capillary density	[33]
PDGF-BB	Recombinant protein	Db/db mice	Increased capillary density	[33]
SDF-1α	Lentivirus	Db/db mice	Increased granulation tissue formation; increased epithelialization	[34]
NO	Nanoparticles	BalbC mice	Increased angiogenesis (by increased release of TNFα)	[35]
SHH	Topical application of recombinant protein	Normal and diabetic mice	Increased eNOS expression	[36]
NO	Topical NO-gel	CF-1 mice	Increased angiogenesis; increased re-epithelialization; increased collagen production; increased follicular regeneration	[37]
Angiopoietin-1	AAV	Db/db mice	Increased vessel number; increased expression of eNOS (but not VEGF-A)	[38]
Angiopoietin-1	Adenovirus encoding cartilage oligomeric protein (COMP)-Ang1 or protein	Db/db mice	Enhanced angiogenesis; enhanced lymphangiogenesis	[39]

EPO	Recombinant protein	Mice	Increased angiogenesis; increased epithelial proliferation; increased maturation of ECM; increased levels of VEGF-A, iNOS and eNOS	[40]
eNOS	Adenoviral gene transfer	Diabetic mice	Increased eNOS protein levels and NOS activity	[41]
MnSOD	Adenoviral gene transfer	Diabetic mice	Increased eNOS protein levels and NOS activity	[41]
VEGF-A(165)	AAV	Mice	Increased angiogenesis; increased epithelial proliferation; increased ECM maturation; increased eNOS and iNOS levels	[42]
EPO	Topical application of recombinant protein	Diabetic rats	Increased microvascular density, VEGF-A, collagen content	[43]
EPO	Repetitive or single high- or low-dose recombinant protein	Nude mice	Single high-dose induced maturation of microvascular networks and accelerated epithelialization; repetitive high-dose caused rheological malfunction and lack of maturation of microvascular networks	[44]
EPO	Topical application of recombinant protein	Mice	Increased VEGF-A expression; increased collagen content	[45]
EPO	Topical administration of recombinant protein	Db/db mice	Increased VEGF-A mRNA and protein and increased CD31 mRNA and protein	[46]
EPO	Recombinant protein	Rats	Increased VEGF-A protein levels and increased microvascular density	[47]
GM-CSF	Intradermal injection of recombinant protein	Patients with non-healing venous leg ulcers	Increased blood vessel density; increased VEGF-A (but not PlGF levels)	[48]
IGF-1	Liposomal delivery of cDNA	Rats	Increased epidermal regeneration; increased angiogenesis (increased VEGF-A concentration)	[49]
KGF-1 (FGF-7)	Liposomal delivery of cDNA	Rats	Improved epidermal regeneration; increased collagen deposition; increased neovascularization and concomitant VEGF-A concentrations	[50]

(continued)

Table 26.1 (continued)

Factor	Transfer method	Model	Effect	Reference(s)
TGFβ1	Recombinant protein (high dose)	Pigs	Increased angiogenesis but abnormal epithelial differentiation	[51]
VEGF-A(165)	AAV	Rats	Increased vessel formation (capillaries and larger vessels)	[52]
TGFβ1	cDNA + electrical field	Db/db mice	Increased epithelialization; increased collagen content; increased angiogenesis	[53]
VEGF-A(165)	Ex vivo, tetracycline inducible, plasmid + liposomes	Pigs	Increased vascularization; increased re-epithelialization	[54]
SDF-1α	Local injection of recombinant protein in wound base	Streptozotocin-induced diabetic mice	Increased homing of EPC	[55]
cAMP	Topical application	Db/db mice	Increased EPC recruitment; increased wound vascularization; increased levels of VEGF-A and SDF-1α	[56]
VEGF-A	Subcutaneous injection of minicircle plasmid combined with cationic dendrimer	Diabetic mice	Well-ordered dermal structure	[57]
VEGF-A(165)	Minicircle DNA plasmid transferred by microbubble destruction method	Streptozotocin-induced diabetic mice	Increased blood flow	[58, 59]
VEGF-C	Adenovirus	Db/db mice	Increased angiogenesis; increased lymphangiogenesis	[60]
HIF-2α	Adenovirus	Mice	Increased expression of VEGF-A, Flt1, Flk1, Tie2; promotion of vascular maturation	[61]
VEGF-A(165)	Topical application of recombinant protein	Db/db mice	Increased levels of PDGF-B and bFGF; increased granulation tissue; induction of leaky vasculature; increased mobilization of EPC	[62]
VEGF-A(165)	Adenovirus	Diabetic mice	Promotion of angiogenesis	[63]

VEGF-A	ex vivo retroviral gene transfer into keratinocytes	Nude mice	Increased vascularization	[64]
B. *Combinatorial approaches*				
VEGF-A + FGF-4	AAV	Db/db mice	Increased vascularity; increased MMP-9 and VEGFR1 expression in fibroblasts	[65]
IGF-1 + KGF	Liposomes	Rats	Increased vascularization; increased VEGF-A expression	[66, 67]
VEGF-E/PlGF-1 chimera	Conditioned media from adenovirally infected mouse embryonic fibroblasts	Normal and diabetic mice	Increased angiogenesis; decreased lymphangiogenesis	[68]
EPO + fibronectin	Protein in cream	Normal and diabetic mice	Increased angiogenesis; increased eNOS	[69]
VEGF-A + bFGF + PDGF-B	Subcutaneous injection of proteins (before wound induction; "priming")	Mice	Increased vessel density	[70]

Many of these (growth) factors were tested in more (pre)clinical studies; however, only those that reported an effect on vessel formation are listed here.

(the so-called "stalk" cells) will follow and elongate the sprout. This selection process is achieved by lateral inhibition through Notch. Notch activation in collaboration with bone morphogenetic protein (BMP) signaling in neighboring cells confers a stalk cell phenotype to EC; hence, inhibition of Notch leads to uncoordinated "hypersprouting" and to dysfunctional blood vessels [103]. To form a correct endothelial network, an EC sprout needs to follow a certain path in order to meet another sprouting EC for fusion. To this end, tip cells use sensors (called "filopodia") to scan their environment for attractive or repulsive cues. Such guidance cues are provided by at least four ligand/receptor pairs: Slits/roundabout4 [104], ephrinB2/EphB4 (reviewed in [96]), semaphorin3E/PlexinD1 [105], and netrins/Unc5B [106].

Meanwhile, stalk cells trailing behind the tip cells proliferate in response to VEGF-A and thereby elongate the sprout, for instance, by pushing the sprout forward through secretion of matrix components, such as Egfl7 [107]. VEGF-A signaling in stalk cells is dampened by Notch signaling [108–110], while Notch signaling itself is controlled by SIRT1, an NAD^+-dependent acetylase [111]. Wnt signaling confers stability to the stalk by inducing tight intercellular junctions [112]. Another task of stalk cells is lumen formation. How this process exactly takes place remains largely unknown; however, it is assumed that lumen formation during angiogenic sprouting resembles that occurring during coalescence of endothelial precursors during vasculogenesis. The latter involves at least three coordinated activities (reviewed in [113, 114]), i.e., the increased cellular contact (mediated by junctional molecules, e.g., vascular tubulogenesis is severely affected in VE-cadherin knockout mice [115]), the establishment of apical-basal polarity (involving translocation of junctions to the lateral side, exocytosis to the apical surface of negatively charged anti-adhesive proteins, such as CD34-sialomucins that form a cell-cell repulsive glycocalyx and rearrangement of the cytoskeleton), and interaction with the surrounding matrix (e.g., fibronectin [116, 117]) through integrins (e.g., β_1 integrins [117, 118]).

When two sprouting tip cells meet, they fuse and connect to form a continuous lumen. This connection is stabilized by VE-cadherin [119]. Interestingly, for proper fusion, EC are assisted by another cell type, macrophages, which function as "bridging" cells [120]. Recently, it was shown that another VEGF family member, VEGF-C, secreted by these macrophages mediates fusion stabilization and phenotypic conversion of the fusing tip cells to stalk-like cells [121].

Next, the newly formed EC branch is stabilized by deposition of ECM and recruitment of SMC/pericytes. EC-secreted PDGF-BB attracts these mural cells to their abluminal surface. Pericyte/EC interaction additionally involves PlGF (a VEGF family member that is specifically important for non-physiological blood vessel formation [72, 122]), FGF9 [123], angiopoietin-1/Tie receptor signaling, S1P/sphingosine-1 phosphate receptor (S1PR) signaling, and Notch3 signaling (reviewed in [124, 125]). Upon vessel stabilization, flow is reestablished in the new branch and the VEGF-A gradient, and the hypoxia that triggered it is eliminated. The EC are then subjected to another phenotypic transition: they become quiescent again, and their straight and firm alignment resembles the phalanx formation of ancient Greek soldiers and are hence called "phalanx cells" [126]. Some of the molecular signals that mediate this phenotypic conversion have been revealed, including prolyl-hydroxylase domain-2 (PHD2; an oxygen sensor that stabilizes HIF [126]), BMP-9/Alk1, thrombospondin, FGF, HoxC9 [127], microRNA (miR)-132 [128], angiopoietin-2/FOXO1 [129], and angiopoietin-1/Tie2 (reviewed in [97, 98, 102]).

26.2.4 Vasculogenesis

Compared to sprouting angiogenesis, the distinct steps in vasculogenesis, especially after birth, have been less elaborately studied. Most research efforts have been spent on identifying signals that recruit EPC to and retain/arrest them at the target site of neovascularization (reviewed in [130–132]). As we will explain in more detail below (see Sect. 26.2.4.2.2), since their initial description in 1997 by Asahara et al. [133], the literature concerning EPC has been confusing as their true nature, their origin, and surface marker expression have been subject for discussion [134–141]. Therefore, some of the recruitment signals mentioned below may be common to different EPC types, while others may only apply for a subset of them. Moreover, similar recruitment signals may exist for other stem/progenitor cells, such as those of the hematopoietic lineage.

Originally, it was thought that the principal – if not the only – source of EPC is the bone marrow (BM), where upon mobilization, they have to relocate from a quiescent to a proliferative niche, a process encompassing PlGF, MMP-9, and membrane-bound c-kit ligand [142, 143]. Mobilization to the circulation additionally involves nitric oxide (NO) production through eNOS expressed by osteoblasts and EC in the BM niche microenvironment [132, 144] and proteinases (e.g., MMP [145], cathepsins [146], elastase) released from neutrophils which cleave adhesion or retention molecules that prevent mobilization, e.g., stromal cell-derived factor-1α (SDF-1α; also known as CXCL12) [147, 148]. Once in the circulation, VEGF or PlGF produced at the hypoxic site are important recruitment signals [149]. Other described natural agents with EPC-mobilizing capacity are GM-CSF, granulocyte colony-stimulating factor (GCSF), bFGF, PDGF-C [150]. SDF-1α, Dickkopf (Dkk)-1 (a Wnt inhibitor), growth hormone, estrogens, insulin growth factor-1 (IGF-1), erythropoietin, MCP-1, CCL2, CXCL7, growth-regulated oncogene-α (Gro-α; also known as CXCL1), dibutyryl cAMP [56], CCL5/CCR5 [151], and angiopoieitin-1 (reviewed in [130–132, 141, 152, 153]). Also biophysical stimulation of EPC mobilization by exposure to hyperbaric oxygen has been described, a procedure that was approved by the FDA as adjuvant therapy for wound healing in diabetic patients [55, 154–156].

Once the cells arrive at the neovascularization site, they need to incorporate into newly forming vessels. However, as we will discuss below, not all EPC incorporate into the vessel wall. Some of them are strategically localized around the growing vessel after extravasation. Nevertheless, independent of their incorporation into the vessel wall, EPC first need to be arrested at the luminal side of vessels at the site of neovascularization. This involves interaction with adhesion molecules (i.e., selectins and β1- and β2-integrins; reviewed in [130]). Interestingly, the expression of integrins by EPC can be exploited in tissue engineering constructs in which integrin-binding motifs can be incorporated to better retain EPC. For progenitors that do not incorporate into the vessel wall but extravasate, it was shown that SDF-1α serves as a retention signal to guarantee persistent perivascular positioning of these cells from where they can release growth factors for communication with resident vascular cells [157, 158]. In diabetic wounds, SDF-1α concentrations in the wound were shown to be reduced, resulting in defective homing of BM-derived cells to the wound bed (reviewed in [154]).

Recent studies have questioned the impact of BM-derived contribution of EPC to wound vascularization [159] or to postnatal neovascularization in general (reviewed in [134, 140, 141]). One of the potential explanations for the limited contribution of BM-derived EPC is the existence of non-BM sources. Indeed, using a mouse parabiosis model, Aicher et al. demonstrated that peripheral organs, such as liver and intestines, also contribute to the pool of circulating EPC [160]. Furthermore, in the last 6 years, a strong case has been made for the existence of vascular wall-resident stem/progenitor cells that represent a local and immediately available source of EPC and other stem/progenitor cells for vascular repair (reviewed in [141, 161–167]). For the latter, long-distance chemoattractants are no longer required for bringing the cells to the site of neovascularization.

26.2.5 MicroRNAs: New Kids on the Block in Regulating EC and Their Progenitors

Epigenetic regulation of gene expression adds another level of complexity to gene and protein expression. Recently, it has been argued that posttranscriptional regulation through RNA interference by microRNA (miRNA) is an impactful determinant of gene/protein expression in the cardiovascular system (reviewed in [168–171]). MiRNAs are small noncoding RNA that base-pair with (partially) complementary sequences in the 3′ untranslated regions (UTR) of target genes to regulate gene expression posttranscriptionally. MiRNA often show tissue-specific distribution, and each miRNA may target up to several hundred mRNA. Some of the miRNA (e.g., miR126 and miR92a) have been preferentially described in EC and may thus play a role during neovascularization in vivo [172, 173]. Furthermore, abnormal regulation of miRNA has been associated with endothelial precursor dysfunction and, more broadly, with (vascular) differentiation [174], pluripotency (i.e., the ability to differentiate into all cell types of the body), survival [156], and self-renewal of stem cells (reviewed in [175–177]). In addition to being involved in neovascularization, miRNA have been implicated in other aspects of the wound healing process, such as inflammation, fibroblast proliferation and senescence, keratinocyte differentiation and proliferation, and hair follicle regeneration [178] (reviewed in [179–181]). Furthermore, in addition to miRNA, other epigenetic mechanisms, such as DNA (de)methylation, have been shown to determine endothelial differentiation of stem cells [182].

26.3 Dermal and Epidermal Healing Through Neovascularization

Early wound healing studies and our recent studies have demonstrated that the wound bed initially is severely hypoxic [100, 183]. This acute hypoxia is necessary to initiate many events in the healing process. However, since this lack of oxygen is in disproportion to the high metabolic needs of cells that take part in the wound repair process, oxygen levels need to be restored after this initial hypoxic phase. We recently showed that revascularization of the wound bed with exogenous addition of blood outgrowth endothelial cells (BOEC), a subset of EPC (see Sect. 26.2.4.2.2), efficiently reduces the amount of hypoxic cells in the wound bed [100].

In order to better understand how prolonged hypoxia affects dermal wound healing, we performed culture experiments with keratinocytes (the protagonist cells in epidermal repair) and dermal fibroblasts (the main cell type involved in dermal healing) under hypoxic or normoxic conditions [100]. Interestingly, while others had shown that acute hypoxia induces keratinocyte motility [184], we found that prolonged hypoxia negatively affects keratinocyte migration and proliferation. Sustained hypoxia also significantly impaired collagen organization in dermal fibroblast sheets, perhaps through modulation of MMP-1 expression [100, 185]. Restoring oxygen levels to normoxia significantly improved keratinocyte migration/proliferation and collagen organization, suggesting that revascularization boosts the healing process by providing oxygen to the wound repairing cells.

Another mechanism according to which neovascularization may improve wound healing is that the newly delivered EC secrete growth factors that communicate with the wound repair cells. Many growth factors have been reported to affect keratinocyte (e.g., keratinocyte growth factor or KGF, interleukin-6, hepatocyte growth factor or HGF, GM-CSF), fibroblast (e.g., MMP-9, MMP-14, MMP-1), and endothelial or inflammatory cell (e.g., VEGF-A, PlGF, angiopoietin-2, MCP-1, bFGF) behavior. We found that many of these were produced by BOEC in vitro [100]. It remains to be determined which of these effectively contribute to improved wound healing mediated by BOEC in vivo. When depleting PlGF from BOEC, they supported neovascularization and re-epithelialization to a significantly lesser extent (Verdonck K, et al. 2010), suggesting that the beneficial effects of BOEC on wound healing were at least in part owing to their secretion of PlGF.

Finally, contraction of the wounds is one of the mechanisms to ensure rapid wound closure. While myofibroblast-mediated wound contraction was suggested to contribute to fast neovascularization of the wound bed (see Sect. 26.2.1) [13], to our knowledge, there have been no systematic studies addressing the reverse question whether neovascularization directly affects wound contraction, for instance, by influencing myofibroblast behavior. In our studies, we found that increased vascular ingrowth by BOEC transplantation did not increase the wound contraction rate [100]. Accordingly, the deficient maturation of blood vessels by the natural angiogenesis inhibitor endostatin did not delay wound contraction. One study reported that the angiogenic chemokine chicken chemotactic and angiogenic factor (cCAF) stimulated the differentiation of fibroblasts into myofibroblasts in vitro and increased the number of myofibroblasts and thereby accelerated wound contraction in vivo. However, it is not clear whether the pro-angiogenic effect and the effect on myofibroblasts were directly linked in vivo [186]. In vitro studies have suggested that under long-term exposure to inflammatory cytokines, dermal microvascular EC may irreversibly transdifferentiate into myofibroblasts; however, it remains to be determined if such cell transitions also occur in vivo [187].

26.4 Neovascularization Strategies for Wound Healing

In the previous chapter, we demonstrated that neovascularization favorably affects epidermal and dermal wound healing. Therefore, many studies have been dedicated to designing optimal strategies to improve wound vascularization. These can be roughly divided in three categories: gene (protein) transfer, cellular approaches (involving differentiated or stem/progenitor cells that actively participate in forming new blood vessels), and electromechanical stimulation.

26.4.1 Gene/Protein-Based Neovascularization Strategies

The first studies aimed at mimicking the molecular events occurring during wound revascularization were related to supplying recombinant proteins to the wound bed (Table 26.1). However, topical administration of high doses of recombinant growth factors as proteins has major shortcomings: potential systemic side effects; short shelf life; low bioavailability; enzymatic inactivation, denaturation, and oxidation; and inefficient delivery to target cells. In order to achieve a therapeutic effect, mostly high and repetitive doses are required, which makes this a very costly treatment modality. Some of these problems have been overcome by the development of controlled release systems, e.g., by chemical immobilization of the factors to the matrix or by physical encapsulation in the delivery system (e.g., nanoparticles; reviewed in [188–190]). Furthermore, these release systems can be modified so that release can be triggered upon demand (e.g., by changing the temperature, pH, exposure to light, or electrical fields) [189].

Gene transfer offers an attractive way for direct delivery to the healing wound and is intended to introduce genetic material encoding growth factors directly into the target cells which then results in protein synthesis. Hence, gene transfer can offer targeted local and persistent delivery of de novo synthesized growth factor to the wound environment over many days. Skin is a good candidate tissue for gene transfer not only because of its obvious accessibility but also for its large capacity for regeneration, including vascular regeneration. Gene transfer was originally designed to treat congenital defects (e.g., cystic fibrosis, hemophilia, and severe combined immunodeficiency or SCID) but has meanwhile found other applications, one of them being revascularization therapy, including during wound healing. One bottleneck that is determinant for the efficacy of gene transfer is the delivery system. Development of such delivery systems still is an intensive field of investigation since no single gene transfer strategy is optimal for all medical applications: all have their specific attributes each with advantages and disadvantages depending on the target tissues (Tables 26.2 and 26.3). The different kinds of gene transfer can be categorized according to the type of delivery system used. In case of "gene therapy," the gene is permanently incorporated into the host cell DNA, leading to a lasting expression (unless gene silencing occurs or genetically modified cells die); in other cases (called "gene medicine"), the gene is only transiently present, and hence its expression is likewise temporary. In the case of wound healing, the end goal is rather to temporarily boost neovascularization, after which this response has to be dampened so that the novel blood vessel network can mature and stabilize. Indeed, as mentioned above, a prolonged and continuous "hypervascularization" response may lead to the formation of hypertrophic scars or keloids. Gene transfer can be achieved in vivo by applying the genetic material onto the wound bed or by subcutaneous injection. Alternatively, gene transfer for revascularization can be done "ex vivo" in non-vascular cells, followed by transplantation of these genetically modified cells into the wound (Table 26.1). As the main task of these non-vascular cells is not to actively participate in blood vessel formation, we do not categorize this approach under "cellular neovascularization strategies." In case the ex vivo manipulated cells are of vascular origin, then we categorize this under "combined gene and cellular approaches" (see Sect. 26.4.3).

26.4.1.1 Viral Gene Transfer

Viral gene transfer is the original and therefore most established technology for gene delivery. As part of their replicative cycle, viruses use the cellular machinery for expression and replication of their own genome. Viral transduction strategies are based on this natural ability of viruses to infect cells. However, in order to avoid the production of infective viral particles, specific sequences of the viral genome are deleted in viral vectors used for gene transfer. Different virus types exist, and a functionally important layer of classification is based on integration into host DNA (Table 26.2). Two virus types have the ability to do so, i.e., retroviruses and lentiviruses, the former targeting only dividing cells and the latter also transducing non-proliferating cells. The permanent nature of the resulting gene expression may not be ideal for wound neovascularization. Furthermore, the integration in the genome is random, which may lead to insertional mutagenesis or silencing of the transgene. Indeed, Fischer et al. reported the development of T cell leukemia after ex vivo retroviral gene transfer and subsequent transplantation in X-linked SCID patients [193]. Hence, other virus types can be considered, including adenoviruses, adeno-associated viruses (AAV), or herpes simplex viruses. While AAV can be stably integrated in the host genome, recombinant AAV vectors developed for gene transfer remain episomal because of the deletion

Table 26.2 Viral gene transfer methods

Viral vector	Genome	Packing ability (kb)	Expression mode in host cell	Target cell and viral titers	Disadvantages
Retrovirus	Single-stranded RNA	8	Stable (chromosomal integration)	Actively dividing cells[a] 10^6–10^7 appm	Risk for insertional mutagenesis Unstable viral particles Intermediate titers
Lentivirus	Single-stranded RNA	7	Stable (chromosomal integration)	Dividing and non-dividing cells 10^4–10^5 appm	Risk for insertional mutagenesis Low titers
Adenovirus	Double-stranded DNA	7.5–35	Transient (episomal)	Dividing and non-dividing cells 10^{10}–10^{12} appm	Potential strong host immune response
Adeno-associated virus	Single-stranded DNA[b]	4.6	Transient (episomal)	Dividing and non-dividing cells 10^4–10^5 appm	Small packing ability Requires helper virus for replication that must be eliminated from the final stock
Herpes simplex virus 1	Double-stranded DNA	30–50	Transient (episomal)	Mainly neurotropic cells 10^{11}–10^{12} appm	Potential wild-type virus breakthrough New HSV-1 amplicon vectors have low antigenicity but also low titers Restricted target cell tropism

appm active particles per ml in the produced purified viral stocks

[a]The inability of retroviruses to infect non-dividing cells is perhaps a lesser concern in the context of wound revascularization since EC and SMC do proliferate during the formation of new vessels.

[b]Since the requirement of conversion from single- to double-stranded DNA for successful expression in target cells was an important barrier to efficient transduction with AAV vectors, McCarthy et al. developed double-stranded AAV vectors that exploit a hairpin intermediate of the AAV replication cycle, thereby mediating 10- to 100-fold higher levels of transgenic expression in vitro and in vivo [191].

Table 26.3 Non-viral gene transfer methods

Method	Advantage(s)	Disadvantage(s)	Gene transfer efficiency
Naked DNA/plasmid injection	Simple	Nonspecific, unstable	Low
Gene bombardment	Simple	Risk for mechanical damage	Low
	Large amounts of DNA can be delivered	Limited penetration depth[a]	
Microseeding	Large amounts of DNA can be delivered	Risk for inflammatory reaction by punctures	Low (naked DNA or plasmid)
		Limited penetration depth	High (viral vectors)
Electroporation/ nucleoporation/ sonoporation	Nontoxic	Complex equipment	Low-intermediate
	Large amounts of DNA can be delivered	Limited cell viability	
		Feasibility in vivo (nucleoporation)?	
Cationic liposomes	Low immunogenicity	Potentially cytotoxic	Variable (5–90 %)
	Simple, local		
	Large amounts of DNA can be delivered		
Human artificial chromosomes	Large gene inserts can be incorporated	Endogenous genes need to be removed from the HAC vector as this may affect physiological gene expression	High
	Stable episomal maintenance		

[a]Dileo et al. developed a modified gene gun with higher discharge speed leading to better penetration until the dermal layer [192].

of the *rep* and *cap* genes from the viral genome. AAV has been applied to overexpress VEGF-A in wounds [42, 52] (Table 26.1). Interestingly, the efficiency with which vascular cells can be targeted by AAV is largely dependent on the serotype (8 of which have been described), type 2 being relatively inefficient and type 1 and 5 being very suitable [194]. Adenoviral PDGF-B transfer was one of the first viral wound healing strategies, although improved vascular stabilization through recruitment of SMC/pericytes was not an end point in these studies [195]. While inflammation may often accompany adenoviral gene transfer because of the capsid proteins, AAV transfer has been associated with lesser immunogenicity (Table 26.2). A disadvantage of AAV vectors is their limited (4.6 kb) capacity to harbor foreign DNA (Table 26.2). This has been overcome by novel techniques such as viral DNA dimerization. For the latter, the gene to be inserted is split into two separate parts and separately packaged into two vectors. After being co-infected into target cells, these two AAV vectors then form head-to-tail heterodimers through identical sequence homology of the inverted terminal repeats, thus rejoining the split gene into one continuous DNA molecule [196]. Finally, herpes simplex virus (HSV)-1 vectors, on the other hand, can incorporate large transgenes (up to 50 kb; Table 26.2).

26.4.1.2 Non-viral Gene Transfer

All non-viral gene transfer strategies (Table 26.3) lead to temporary gene expression, and transfer efficiency is in general lower than for viral methods. Perhaps the most "simple" way to deliver a gene to a target tissue is to inject "naked" DNA into skin or to topically apply it to the wound; however, the fragility, size, and electrical charge of the DNA represent significant barriers to efficient transfer into cells. Therefore, technically sophisticated modifications of this technique have been devised, such as "microseeding," "gene bombardment," gene electroporation, or gene sonoporation. During microseeding, a technique designed by Eriksson et al., the gene of interest is injected in the target cells by using oscillating solid microneedles mounted on a modified tattooing machine whereby the penetration depth of the needles can be varied depending on the needle design [197]. In addition to using it for DNA transfer, the microseeding technique can also be used for delivery of (adeno)viral vectors [198]. Gene bombardment (also known as particle-mediated or ballistic gene transfer), on the other hand, is accomplished by using a "gene gun" during which the DNA is coated onto micrometer scale gold or tungsten particles which are propelled into the target tissue resulting in expression for several days in and around the targeted tissue [199]. Disadvantages however are limited transfection depth and rate. Using a modified gene gun with higher discharge speed, Dileo et al. achieved a higher level of gene expression in both epidermis and dermis [192]. Application of an electrical field to wounds (for electroporation) in combination with DNA greatly enhanced *TGF-β1* gene transfer with increased angiogenesis in wounds of diabetic mice [53]. A more recent improved variant of the electroporation technique is "nucleoporation" although this procedure is likely not applicable in vivo but will be of interest for ex vivo applications. The technique involves a combination of optimized electrical parameters and media specific for primary cell types. Since the DNA is delivered straight into the nucleus, expression can occur without delay. In a recent study comparing different modes of non-viral gene transfer, we showed that nucleoporation was the most efficient gene transfer method for dermal fibroblasts [200]. While electroporation and nucleoporation use electrical pulses to modify the permeability of the cell membrane, sonoporation or cellular sonication encompasses the use of ultrasound waves to facilitate uptake of DNA into the target cell. Sonoporation employs the acoustic cavitation of microbubbles to enhance DNA delivery [201]. The technique has been tested in preclinical studies to deliver angiogenic genes to wounds [58, 59] and is under active study for targeted gene transfer in vivo in patients using an ultrasonic transducer [202].

In most cases, DNA transfer is achieved by using plasmids that still contain bacterial elements (e.g., the origin of replication) that may be perceived as foreign and hence destroyed by the mammalian target cell leading to silencing of expression.

Minicircles are small circular plasmid derivatives in which all prokaryotic vector sequences have been deleted. This resulted in a significantly increased expression level of the inserted gene compared to standard plasmid DNA [203]. Their small size (4 kb) also facilitates their delivery into cells. Minicircle DNA encoding VEGF-A has been tested for its efficiency to induce blood vessel formation in animal wound models [57–59].

The use of cationic liposomes ("lipofection") is another efficient means of delivering genes into target cells [204]. Because of their positive charge, they can form non-covalent complexes with negatively charged residues in DNA, an association that protects the DNA from degradation. The excess positively charged groups of the complex can then interact with the negatively charged cell membrane leading to facilitated uptake by endocytosis. Such an approach efficiently targeted EC with IGF-1 cDNA in vivo in injured rat skin, resulting in their increased proliferation [49]. Recently, these liposome-based gene transfer methods have also been used for viral vector transfer. Such combinations resulted in increased transfection efficiency and lower immunogenicity due to the "shielding" of adenoviral vectors with the polycationic lipids and therefore represent promising future possibilities [205, 206].

The latest technique in non-viral gene transfer is the use of human artificial chromosomes (HAC), which have been heralded as the most promising non-viral vectors of the future. They represent "mini-chromosomes" containing specific DNA fragments that will enter the cell and permanently reside there as a stable episome. HAC transfer the inserted DNA into daughter cells during cell division. The possibility to incorporate very large gene inserts is an important asset of this method [207, 208].

Importantly, like any other cells, vascular cells, such as EC or their precursors, have different propensities for efficient gene transfer by any of these non-viral gene transfer methods. Indeed, in the comparative study mentioned above, we directly compared different non-viral gene transfer methods in human BOEC, showing that Effectene, a type of liposome transfer, was most efficient [200].

26.4.1.3 Constitutive Versus Regulable Expression

Since the formation of blood vessels requires the action of several growth factors at certain specific steps during the process, creating the opportunity to regulate the expression of the delivered growth factor in time would allow to better mimic the natural course of molecular events during blood vessel formation. Therefore, instead of putting the expression of the gene under the control of a constitutive promoter, several inducible promoter systems have been designed that can be regulated by exogenously supplied drugs. The same systems have also been used to create conditional knockout mice or mice with inducible overexpression.

To date, mainly five major drug-responsive systems exist for inducing the expression of a gene. The first four systems all employ a minimal mammalian cell promoter (which by itself exhibits little basal activity and is fused to a cis-corresponding DNA-binding element) and a hybrid transactivator whose transactivating activity is regulated by a drug. Tetracycline (or doxycycline)-dependent systems are perhaps the most commonly known. In its original format, the tTA transactivator is a hybrid between the transactivating domain of HSV-1 VP16 and a tetracycline repressor (tetR), and the minimal promoter is based on the human cytomegalovirus (hCMV) promoter [209]. The system is available in two main configurations, "Tet-off" and "Tet-on," in which the tTA transactivator can only bind to the target minimal promoter in the absence or presence of doxycycline, respectively. A few years later, Yao et al. developed a new system not relying on the hCMV promoter and tTA but on the wild-type hCMV major immediate-early promoter and tetR itself (a system called "T-REx") [210]. Later a T-REx-encoding replication-defective HSV-1 recombinant vector was developed which allowed for an up to 1,000-fold tetracycline-regulated gene expression [211]. Another system is based on Ecdysone, an insect steroid hormone [212]. Upon administration of an Ecdysone-analog such as muristerone A, the Ecdysone receptor/VP16 fusion protein is activated. Subsequent heterodimerization with the retinoid X receptor then causes binding of the complex to the Ecdysone response element leading to transcription of the target transgene. The RU486/antiprogestin-mifepristone system is based on a fusion protein of the ligand-binding domain of the human progesterone receptor, the yeast transcriptional activator GAL4 DNA-binding domain, and the HSV protein VP16 transactivation domain [213]. This fusion protein activates target genes containing GAL4 binding sites in response to progesterone antagonists such as R486/mifepristone. Estrogen inducible systems are very similar to the progesterone-inducible systems, the former using the estrogen receptor in the fusion protein [214]. Finally, the drug rapamycin (also known as sirolimus, an inhibitor of mammalian target of rapamycin, mTOR) has the ability to induce dimerization of two cellular proteins, FKBP12 and FRAP. FKBP12/FRAP heterodimers can bind corresponding DNA response elements and activate transcription of a downstream target gene (reviewed in [215]). However, rapamycin has been shown to have anti-angiogenic (as well as anti-lymphangiogenic) effects, one of the reasons why it is now being tested in cancer clinical trials [216]. Therefore, this system is not suitable for regulable gene transfer for stimulation of blood vessel growth in wounds.

The Cre-Lox system based on a single enzyme, Cre recombinase, that recombines a pair of short target sequences called *Lox* sequences is a conditional expression modality that does not rely on a minimal promoter [217]. Cre-mediated recombination can be made inducible by using a tamoxifen-responsive variant of the Cre recombinase, i.e., CreERT2. The Cre enzyme and the original *Lox* site called the *LoxP* sequence are derived from a bacteriophage P1 [217]. While the Cre-Lox system is most routinely used to shut down the expression of a certain gene, the system can also be adapted to induce expression, i.e., by introducing a "floxed" stop codon cassette in front of the gene of interest. Cre-induced excision of the stop codon cassette will then initiate the expression of the gene. The disadvantage of this system is that the induction of gene expression is irreversible, in contrast to the other 4 systems where expression is no longer induced in the absence of the drug.

26.4.1.4 Monotherapy Versus Combination Therapy

As evident from our description of the molecular players in (wound) vascularization, the application of a single growth factor does not mimic the multifactorial nature of the process. Indeed, transfer of multiple genes, preferably at strategic time points, is the ultimate goal [190]. Proof of principle of such multimodal approaches has been delivered in studies testing the efficacy of subcutaneous implantation of different formulation of growth factors. For instance, a combination of VEGF-A and KGF or VEGF-A and angiopoietin-1 in a hydrogel increased the angiogenic response [218, 219]. Combining VEGF-A with bFGF in porous collagen-based scaffolds resulted in a higher blood vessel density than either factor alone [220, 221]. Some of the controlled release systems mentioned above can be designed to secrete growth factors sequentially rather than simultaneously [189, 190]. For instance, sequential release of VEGF-A and PDGF-B increased vascular density upon subcutaneous injection in rats [222] and favorably affected blood vessel numbers, size, and maturity upon implantation in ischemic muscles in mice [223]. Also several preclinical wound studies have tested combinations of factors with different activities in the blood vessel formation process resulting in improved vascularization (Table 26.1).

One interesting possibility is the use of products containing "natural" combinations of endogenous growth factors, such as platelet-rich plasma (PRP) or platelet-derived wound healing factors (PDWHF) [224]. Platelets play important roles during the natural course of wound healing, one of them being the secretion of many growth factors, adhesive molecules, and lipids that regulate the migration, proliferation, and function of keratinocytes, fibroblasts, and vascular cells. Some of the up to 30 (growth) factors stored by platelets in their α-granules are PDGF-B, TGFβ, VEGF-A, bFGF, EGF, IGF-1, IGF-2, IL-8, and GM-CSF, many of which have been shown to have a pro-angiogenic or pro-vasculogenic effect [225–227]. A recent study showed that PRP also contains pro-angiogenic short peptides such as UN3 [228]. PRP is blood plasma that has been enriched with high concentrations of platelets that release these growth factors from their storage pools upon activation with thrombin and/or calcium chloride added to the plasma. Several techniques and devices have been developed to concentrate platelets in PRP with different degrees of enrichment [226]. We and others have tested PRP combined with cell therapy in preclinical models of wound healing [229, 230]. PRP has been clinically tested for a number of reparative applications, including healing of chronic ulcers and burns, with variable efficiency [190, 226, 231, 232]. This variable outcome may be related to the different concentrations of platelets and derived growth factors in PRP preparations, which calls for a better characterization of these products [231]. Furthermore, the precise effects of PRP on the wound healing process remain incompletely understood. Definitive proof for clinical efficacy will require additional placebo-controlled and double-blinded clinical trials [190, 226]. Another endogenous growth factor combination product is amnion-derived cellular cytokine solution (ACCS) which is a cytokine-rich solution secreted from amnion-derived multipotent progenitor cells containing multiple naturally appearing factors at physiological concentrations (e.g., PDGF, VEGF-A,

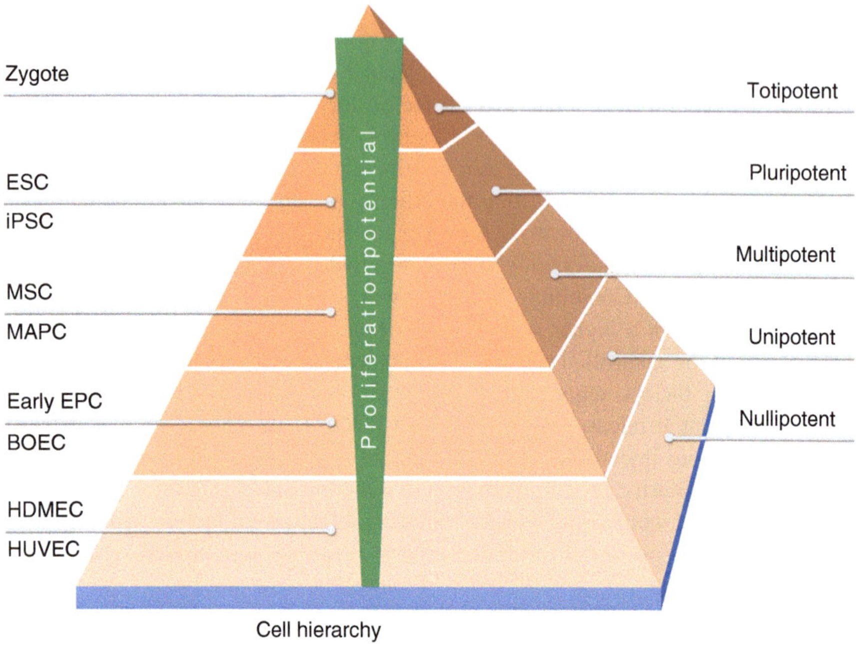

Fig. 26.1 Hierarchy of cell sources for vascularization of skin wounds and tissue-engineered skin equivalents. Candidate cell populations for vascularization can be categorized according to their differentiation potential which inversely correlates with proliferation capacity. Cells with the broadest differentiation potential are at the upper part of the pyramid, whereas fully differentiated, mature endothelial cells (EC) are at the base of the pyramid. The cell on top of the pyramid is the zygote, which is called "totipotent" as it gives rise to all cells of the embryo in addition to all cell types of the extraembryonic tissues. *ESC* embryonic stem cell, *iPSC* induced pluripotent stem cell, *MSC* mesenchymal stem cell, *MAPC* multipotent adult progenitor cell, *EPC* endothelial progenitor cell, *BOEC* blood outgrowth endothelial cell, *HDMEC* human dermal microvascular endothelial cell, *HUVEC* human umbilical cord vein endothelial cell

TGFβ2, angiogenin, tissue inhibitor of metalloproteinase (TIMP)-1, and TIMP-2) [233]. ACCS has shown promising effects on healing burns, the primary benefit being accelerated re-epithelialization [234, 235]. Effects on blood vessel formation were however not evaluated in these studies.

26.4.2 Cellular Neovascularization Strategies

Many cell types have been implicated in the formation of blood vessels, ranging from mature EC to pluripotent embryonic stem cells (ESC). In the next chapter, we give an overview of the different cell types that have been used to promote wound vascularization, using their differentiation status and potential as a means of classification (Fig. 26.1). Mature EC and unipotent EPC are committed to the endothelial lineage and will in vivo behave like EC, whereas multipotent and pluripotent stem/progenitor cells have the intrinsic capacity to differentiate into multiple cell lineages and will therefore need to be stimulated into becoming an EC, either in vitro, before application to the wound, or in situ after application, through communication with other wound cells and/or the ECM. Furthermore, these cells may – upon proper stimulation – also contribute to pericytes or vascular SMC that can stabilize the inner endothelial layer of blood vessels. Like for endothelium, unipotent circulating progenitors with the capacity to form SMC or pericytes have also been described, but will not be discussed here since there have been no reports on their use in wound vascularization (for a general review on these cells, we refer to [236–238]).

The prototypical example of a multipotent stem cell is the hematopoietic stem cell (HSC), which generates all lineages of the blood. Although these cells or their differentiated progeny have an adjuvant paracrine function in wound neovascularization [239] – a process called "hemangiogenesis" (reviewed in [240–242]) – or the formation of bioengineered vascular networks [243], we will not further discuss them here. Furthermore, during development, the extra- and intraembryonic formation of new blood vessels is anatomically and chronically closely linked to hematopoiesis in the form of a bipotential (EC-hematopoietic) precursor (the "hemangioblast") or intimal Runx1^{+} EC with blood-forming capacity ("hemogenic endothelium" in the floor of the dorsal aorta; reviewed in [244]). According to certain in vitro models, the hemangioblast would also have the potential to give rise to SMC (reviewed in [245, 246]). However, it is currently not clear whether an equivalent bi- or tripotential cell persists during adulthood and if so, whether it would be feasible to isolate and use it therapeutically, for instance, in wound revascularization.

Importantly, when considering the use of a cell for therapeutic revascularization, the choice is not only dependent on its inherent potential to physically contribute to this process by vascular differentiation and incorporation. Indeed, as described above for HSC, cells may also have an important adjuvant role in blood vessel formation by secreting growth factors that communicate with endogenous vascular cells. As such, these cells serve as a "natural" delivery device for these factors. Finally, in addition to these mechanistic criteria, several issues related to clinical applicability may co-determine the choice of the optimally suited cell type [11]. Such issues are ease of harvest, possibility for expansion to clinically used amounts, genetic stability, non-immunogenicity (in which case, autologous cells or allogeneic cells with immunomodulatory effects will be advantageous), and the possibility to prepare the cells in formulations free of xenobiotic culture media components.

26.4.2.1 Human Dermal Microvascular EC and HUVEC

Human Dermal Microvascular EC

Human dermal microvascular EC (HDMEC) are at the lowest step of the differentiation potency ladder since they are terminally differentiated cells ("nullipotent"; Fig. 26.1). Since skin is highly vascular (see Sect. 26.1), a common and obvious source for harvesting mature EC is the dermis itself. Their isolation was initially described from neonatal foreskin [247], but HDMEC can be obtained from adult skin as well [248]. Their harvest is however complicated by low numbers, short life span, and fibroblast contamination [249]. Several techniques have been proposed to purify HDMEC populations, including the use of Percoll gradients [250], anti-E-selectin monoclonal antibody-coated magnetic beads [249], or EN-4 panning [251]. Furthermore, there have been efforts to design serum-free culture techniques for HDMEC to offer a safer cell product for use in humans [251]. Despite these technical improvements, the use of HDMEC has been limited to preclinical and in vitro studies, such as seeding of human dermal fibroblast sheets [252] or subcutaneous implantation in immunodeficient mice [248, 253].

HUVEC

Using the human umbilical vein wall as a source, EC (HUVEC) are easy to be harvested from there, can be expanded to large numbers – most likely because of their fetal origin and hence more juvenile character – and can be easily cryopreserved until further use [254]. Therefore, these cells have served as a reference EC line in many vascularization studies, including in the context of wound healing and skin tissue engineering [100, 255]. Given their fetal origin, the cells would have to be used in an allogeneic setting. Besides contributing to vasculogenesis, they produce a combination of trophic factors that interact with wound repairing cells, including vascular cells [100, 256]. We recently tested HUVEC in a full-thickness wound model in nude mice but could not demonstrate an increase in wound bed revascularization, despite their ability to incorporate in newly forming vessels. Furthermore, despite their production of several factors for communication with keratinocytes and fibroblasts, no trophic effect on dermal or epidermal healing was apparent [100]. Interestingly, it was shown that culturing HUVEC in a three-dimensional spheroid system as opposed to the classical two-dimensional culture improved their survival in a hypoxic environment (in this case, ischemic limb muscle) and enhanced their production of (angiogenic) growth factors [257]. Similarly, HUVEC spheroid implantation in Matrigel also supported HUVEC survival and vascular tube formation in immunocompromised mice [258]. Testing this spheroid system in wound healing is therefore very appealing. Furthermore, in another approach, survival of HUVEC was also prolonged by overexpression of caspase-resistant Bcl-2, which allowed for the HUVEC-based vessels to become stabilized by SMC [259].

26.4.2.2 Endothelial Progenitors

It has long been thought that the "de novo" formation of blood vessels through recruitment and incorporation of endothelial progenitors – called "angioblasts" – was something that only occurred during development. This thinking changed in 1997 when Asahara et al. first described that new blood vessels in the adult can also form by recruitment of BM-derived EPC – the postnatal equivalents of the embryonic "angioblasts" – that incorporate into newly forming vessels and differentiate

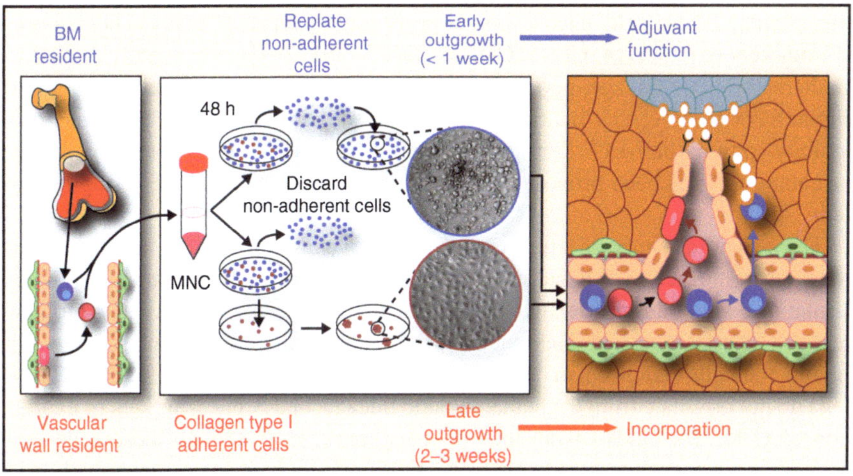

Fig. 26.2 Defining EPC based on differential adherence and outgrowth kinetics in culture. As a means to reach a consensus concerning the definition of endothelial progenitor cells (EPC) and as an alternative to the use of surface markers for this definition, differential adherence and outgrowth dynamics have been introduced as novel criteria. According to this new way of categorizing EPC, two distinct populations can be distinguished: those emerging early in the culture dish after replating the initially non-adherent portion (*blue cells*) of peripheral blood mononuclear cells (*MNC*) which can be mobilized from the bone marrow (*BM*) through diverse recruitment signals and those growing out late (after 2–3 weeks; *red cells*) after plating the collagen I-adherent fraction of peripheral blood MNC which are currently thought to be residing in the intima of blood vessels. For the latter, it is unknown which signals and events are involved in their mobilization into the peripheral blood. Both fractions have different functional behaviors (late outgrowth cells incorporate into growing vessels, early outgrowth cells do not but have an adjuvant role) and expression characteristics, the most typical difference being the expression of hematopoietic markers in early EPC, which is lacking in late outgrowth EC

into EC [133]. These unipotent endothelial progenitors represent a higher step of the differentiation potential ladder (Fig. 26.1). The groundbreaking findings by Asahara et al. set the stage for an exponentially growing number of studies that investigated the involvement of BM as a reservoir for circulating endothelial progenitors that participate in neovascularization. Unfortunately, as often happens in a booming field, the initial enthusiasm related to EPC meanwhile has been tempered, and the wealth of rapidly emerging studies has caused confusion about the true nature of these progenitors, their origin, their precise characteristics (e.g., surface marker expression), and the degree to which they directly contribute to neovascularization [134–141].

There have been efforts to resolve these confusing issues about EPC by shifting their definition from one based on the expression of surface markers to one based on differential adherence and outgrowth kinetics in culture and their capacity to functionally behave like EC, i.e., to become part of the endothelial lining of a new vessel [137, 139, 260]. Based on this new definition, mainly two types of EPC can be distinguished (Fig. 26.2). The first category originates from the initially non-adherent mononuclear fraction of circulating blood and emerges early (in less than a week) after plating. These cells have a limited proliferation capacity, express hematopoietic markers (e.g., CD14, CD45, CD34, CD133 [261–264]), are clonally related to the hematopoietic lineage, poorly form vascular tubes, do not actively incorporate into nascent blood vessels but rather trophically support them, and are referred to as "colony-forming unit-EC" [260]. Given their rather trophic role on the preexisting vasculature, it has been proposed to rename these EPC as "circulating angiogenic cells" [135].

The second category of "true" endothelial progenitors is present in the immediately adherent fraction of blood mononuclear cells, grows out later (after 2–4 weeks of plating), is highly proliferative, forms vascular tubes, and incorporates into new vessels. They are called "late outgrowth endothelial cells," "blood outgrowth endothelial cells" (BOEC), or "endothelial colony-forming cells" [136, 260]. These cells do not originate from the hematopoietic lineage, and hence their lack of expression of CD45 (or CD14) is an important hallmark. Even though these cells lack expression of progenitor markers, such as AC133, they are also functionally and molecularly different from mature EC (e.g., HUVEC) [100]. One potential caveat about the long-term cultured late outgrowth EC is that these cells may change upon culture and hence be quite different from their in vivo ancestor [135]. Nevertheless, we showed that, except for a decrease in CD34 upon passaging, the expression pattern of these cells is relatively stable throughout the culture period [100]. Although initial reports doubted their paracrine angiogenic effect [265], we and others have clearly demonstrated their trophic effect on angiogenesis [100, 266]. Indeed, we have examined the growth factor secretion profile of BOEC in more detail and showed that they express significant amounts of VEGF-A, PlGF, PDGF-BB, angiopoietin-2, MCP-1, bFGF, KGF, IL-6, HGF, and GM-CSF [100]. It is not clear what is the exact origin of BOEC; however, the current hypothesis is that while early EPC may be recruited from the BM, BOEC may actually reside in the intima of established large or small blood vessels and hence form a local pool of highly proliferative EC (progenitors) readily available for vascular maintenance and regeneration [138, 141, 267] (Fig. 26.2). Under certain conditions, these cells detach from the vessel wall and hence can be harvested from circulating blood. Both cord blood and adult blood have been demonstrated to contain BOEC, the former being a more abundant source. The proliferative disadvantage of peripheral blood-derived BOEC can however be overcome by VEGF-A stimulation [268]. Unlike for adult BOEC, karyotypic aberrations have been reported for cord blood-derived BOEC [269]. On the other hand, cord blood-derived BOEC formed more durable and stable vessels than their peripheral blood-derived counterparts [270].

Both topical [271] and systemic [272] applications of early EPC have proven to be efficient in increasing wound bed vascularity and wound healing in mouse models, and their beneficial effect was augmented in hyperbaric oxygen conditions [273]. Also in pathologic situations in which there is a decreased revascularization capacity in the host (e.g., diabetes), early EPC have proven their ability to boost angiogenesis and re-epithelialization rates, in part through activation of the Wnt signaling pathway [274]. Topical application of peripheral blood-derived BOEC in a nude mouse wound healing model resulted in significant participation in hybrid vessel formation, enhanced vascular expansion, increased SMC coating of host vessels, and improved re-epithelialization and dermal collagen organization [100]. Recently, Reinisch et al. have described a large-scale, animal protein-free expansion strategy for BOEC that supports their proliferation and preserves their functional characteristics, even after cryopreservation, which makes their use in humans more feasible [275]. Given their complementary mode of action, combined application of early EPC and BOEC seems highly attractive [276]. Indeed, co-transplantation of both EPC types had a synergistic effect on revascularization of ischemic limbs [277], a scenario that could also be tested in wound healing studies.

26.4.2.3 Multipotent or Pluripotent Stem Cells

Stem cells or progenitors in the adult can have the ability to differentiate into multiple cell types, and this differentiation potential may not be limited to their tissue of origin. Among those more-than-unipotent progenitors, we distinguish two main subtypes (Fig. 26.1): multipotent stem/progenitor cells, i.e., those that make many but not all cell types of the body, and pluripotent stem/progenitors, i.e., those that can give rise to all ~220 differentiated cell types that make up a living organism. As an accompanying note, we want to explain our use of the term "stem/progenitor cells" throughout this chapter. One of the requirements to call a cell a "stem cell" is the demonstration of self-renewal – which in principle can only be rigorously proven in vivo [278]. "Progenitor" is used when self-renewal is no longer a feature of the cell. Since for many multipotent cells the issue of self-renewal has not been appropriately addressed, we use the term stem/progenitor cells to designate these cells.

Mesenchymal Stem Cells

Mesenchymal stem cells (MSC), a term originally coined by Caplan in 1991 [279], represent a heterogeneous group of multipotent stem/progenitor cells that can be harvested from several tissues, including BM, skeletal muscle, brain, skin [280], kidney [281], dental pulp [282], adipose tissue, umbilical cord (blood), several compartments of the umbilical cord (i.e., the amniotic membrane [283], the cord vessel wall [284, 285], and Wharton's jelly or the mucous connective tissue of the cord all contain MSC; reviewed in [286]), amniotic fluid [287], endometrium [288], and additional fetal and adult tissues [289–291]. Recently, cells with MSC characteristics were even found in debrided skin of burns patients [292]. Because of the multitude of studies on MSC and the heterogeneity among these cell types, the Mesenchymal and Tissue Stem Cell Committee of the International Society for Cellular Therapy has proposed a number of minimal criteria that need to be met in order to name a cell "MSC" [293]: plastic adherence, defined surface marker expression, and trilineage differentiation potential (osteoblasts, adipocytes, and chondroblasts). When considering the ease of harvest, there are however mainly three sources that are appealing for therapeutic use: BM, adipose tissue, and umbilical cord (blood). The first two can be used in an autologous setting while umbilical cord (blood) will be allogeneic (unless the blood is taken into a blood bank). The procurement of BM carries some degree of non-trivial morbidity, making the use of adipose tissue or umbilical cord (blood) more appealing from that perspective. In general, MSC act on wound healing through transdifferentiation or cell fusion to wound healing cells [294], production of cytokines and growth factors [295], maintenance of the ECM, modulation of the immune system, and wound contraction [289, 296] (recently reviewed in [297–299]). There is evidence that adipose-derived MSC and BM-MSC kept in culture for a long time undergo spontaneous transformation and form tumors in vivo, which might jeopardize clinical application [300, 301]. A recent report showed that the acquisition of aneuploidy not necessarily leads to spontaneous transformation [302]. Nevertheless, it is commendable to only use lower passages of these cells. On the other hand, MSC display immunosuppressive characteristics through interactions with natural killer (NK) cells, B and T lymphocytes, and monocytes, making them compatible with allogeneic transplantation [303]. Standard MSC culture protocols include the use of fetal calf serum, possibly leading to zoonoses in the host. Platelet lysate has been advocated as a surrogate for fetal calf serum in isolation of BM-MSC, but it is not clear yet whether this isolation procedure gives rise to cells with identical properties as BM-MSC obtained through the standard isolation protocol [304, 305].

As for the effect of MSC on blood vessel formation, at least three possibilities have been described in the literature. Direct contribution to EC was reported in vitro and/or in vivo [280, 282, 306–316]. In this context, it is relevant to consider that adipose tissue is a rich source of EC (progenitors) [317] and consequently that the stromal vascular fraction from which adipose-derived MSC (also known as "processed lipo-aspirate (PLA) cells," "adipose-derived stromal cells," or "adipose-derived stem cells") are derived in fact is a heterogeneous population that contains, in addition to MSC, also vascular EC, the latter which may account for "direct EC differentiation" [318, 319]. Recent studies, including our own [320–323] (reviewed in [324]), have rather favored alternative pro-angiogenic mechanisms for MSC. First, there is increasing evidence that they mainly have a trophic function on blood vessel growth by secreting numerous growth factors [295, 297, 323, 325–328]. Interestingly, this trophic mechanism may differ depending on the source of MSC. For instance, while adipose-derived MSC highly depend on plasmin-mediated and only limitedly on MMP-mediated ECM breakdown, BM-MSC completely depend on membrane-type MMP for inducing capillary expansion [329, 330]. Despite these different trophic mechanisms, the overall capacity of MSC to modulate blood vessel formation in vivo seems similar for different tissue-specific MSC [331]. Secondly, recent studies have reported that MSC in many tissues in fact represent (a subset of) pericytes, a claim mostly based on co-expression of mesenchymal (e.g., CD29, CD44, CD73, CD90, CD105) and pericyte markers (e.g., NG2, SMC α-actin, PDGFRβ, caldesmon, calponin) in vitro and/or in vivo [332–335]. Hence, a direct contribution to the supporting pericyte/SMC layer rather than the endothelial layer of newly formed blood vessels seems plausible. This supportive mode of contribution to blood vessel formation has been termed "angiopoiesis" as opposed to "vasculogenesis" that implies a physical contribution to endothelium [278].

Several clinical and preclinical studies have tested the in vivo potential of MSC in revascularization, including in wound healing (recently reviewed in [297, 299, 336, 337]), alone or in combination with other cells. Here, we only highlight a number of these studies. BM-MSC were shown to augment vascularization in mice as well as in patients with acute wounds when sprayed in a fibrin glue [338]. When adipose-derived MSC were locally injected into wounds of diabetic mice, vessels were seen earlier and in higher amounts in the granulation tissue, and VEGF levels were also higher [339]. As mentioned above, BM-derived or adipose-derived MSC can also differentiate towards perivascular cells, thereby increasing vessel stability [309, 332, 335, 340]. Lipo-aspirate-derived MSC are currently used in a phase I/II clinical trial in the treatment of perianal fistula [341, 342] and in phase IV trials for restoration of lumpectomy defects in the breast. These trials so far demonstrate the safety and the feasibility of the clinical use of these cells [341, 342]. MSC derived from perivascular tissue of umbilical cord were applied together with a fibrin sealant onto wounds in immunodeficient BalbC mice, thereby accelerating re-epithelialization and improving dermal collagen organization. An effect on blood vessel formation was not evaluated in this study [285]. Combining human cord blood-derived BOEC and BM- or cord blood-derived MSC in a subcutaneously implanted Matrigel plug in mice resulted in a functional human vessel network that was stabilized by MSC-derived perivascular cells, suggestive for the benefit of a combined BOEC-MSC approach [331, 343]. MSC derived from the amniotic membrane of umbilical cord are currently being tested in clinical trials of patients with burns or diabetic patients with foot ulcers [283].

Non-MSC Multipotent Adult Progenitor Cells

As BM is a rich source for stem/progenitor cells, many other non-MSC cell types (e.g., multipotent adult progenitor cells or MAPC, marrow-isolated adult multilineage inducible or MIAMI cells, BM-derived multipotent stem cells or BMSC) with multipotent differentiation capacity could also be derived from BM [344–346] (reviewed in [347]). In addition, other tissues, such as umbilical cord blood, also host such multipotent stem/progenitor cells, called unrestricted somatic stem cells (USSC) [348]. MAPC are BM-derived multipotent cells, displaying a vast expansion capacity combined with a broad differentiation potential that encompasses cells from the three embryonic germ layers [345]. These MAPC are morphologically, molecularly, and functionally different from BM-MSC and can in vitro differentiate into EC that contribute to vasculogenesis in vivo [349–352]. Recently, it was shown that the endothelial differentiation process of MAPC is subject to epigenetic regulation since inhibition of DNA methyl transferases and histone deacetylases induced EC differentiation [353]. While MAPC represent an almost inexhaustible source of cells that can be used in wound vascularization, the long-term procedure required to derive MAPC makes an autologous approach in an acute setting not feasible. Nevertheless, similar to MSC, MAPC also seem to have immunosuppressive capacity, which may be favorable in an allogeneic setting [354, 355]. Despite their large expansion capacity, no karyotypic instabilities have been reported for human MAPC. Our recent data reveal that MAPC transplantation improves wound vascularization as well as re-epithelialization in a full-thickness wound model in immunodeficient mice (Hendrickx et al. 2009).

Embryonic Stem Cells

Embryonic stem cells (ESC) are derived from the inner cell mass of the blastocyst-stage embryo, first successfully isolated from mice [356] and much later from humans [357]. These cells are pluripotent (Fig. 26.1), meaning that they can give rise to all (~220) differentiated somatic cell types, including all skin cell types. The fact that derivation of ESC requires the destruction of an embryo results in ethical restraints on their (clinical) use. ESC have an unlimited self-renewal capacity and therefore represent an inexhaustible source of vascular cells for wound revascularization. One of the most stringent tests to demonstrate pluripotency – and a testimony to their unrestricted proliferation potential – is the generation of a benign tumor ("teratoma") upon implantation in an immunodeficient host. In fact, the very same test is routinely used to assess tumorigenicity prior to clinical testing. Thus, teratoma formation represents another hazard of using ESC for clinical purposes [358]. It was recently shown that BM-derived angiogenic macrophages may contribute to ESC-derived teratoma formation by secreting migration inhibitory factor (MIF) [359]. To circumvent this problem, robust differentiation of ESC and purification of EC progeny prior to transplantation is an absolute requirement. Alternatively, physical encapsulation of EC-differentiated ESC is a way to avoid teratoma formation; however, under those conditions, direct contribution to neovascularization by vascular incorporation is also prevented [360]. Given their origin, ESC can only be used in an allogeneic setting. While somatic cell nuclear transfer using the nucleus of a somatic cell of the patient has been proposed as an alternative to offer the possibility to use ESC in an autologous way, the further development of this technology has met with significant opposition due to technical limitations and ethical concerns, the latter related to the possibility for reproductive cloning [361]. A perhaps more feasible strategy to overcome immune rejection of transplanted ESC (derivatives) is co-transplantation with immunomodulatory MSC, although this did not seem very effective [362].

Endothelial differentiation of ESC has been achieved by using different protocols. Furthermore, ESC can also be coaxed into a pericyte or SMC-like cell, perhaps through a bipotential EC-SMC intermediate [363–366]. As vascular differentiation procedures from ESC have been extensively reviewed just recently [315, 366–368], here we only summarize the most important issues. In general, three types of protocols have been described for vascular cell differentiation, i.e., those based on an embryoid body intermediate (three-dimensional method; usually with low differentiation efficiency), those using two-dimensional coculture with feeder cells or mature EC, and those using two-dimensional monocultures in defined chemical conditions followed by culture manipulations. Exposure to mechanical in addition to chemical signals has been explored to improve vascular differentiation [369]. Many variations exist on these main principles, such that no standardized protocol has been developed. While recent adaptations and refinement of the methods have yielded high (up to 80 % [370]) purity of vascular differentiated cells, this still requires an additional purification step as the contamination with as little as 10,000 remaining undifferentiated ESC can be sufficient to give rise to teratomas [371]. This purification step can either be positive selection based on markers expressed on the wanted differentiated progeny or negative selection based on the expression of unique markers on undifferentiated cells. Such a negative selection marker (SSEA-5) was recently identified [372]. Several studies have pursued the goal of developing differentiation protocols devoid of xenobiotic substances since this is another pre-requisite for clinical use [370]. Finally, from comparative studies between mature EC lineages and endothelial progenitors at different stages of differentiation, it has been shown that the degree of EC maturation is also critical for therapeutic efficacy, the fully differentiated state, or the very immature state being not efficient [364, 373, 374]. Therefore, achieving an intermediate differentiation state is the desired end goal of vascular differentiation from ESC.

EC or SMC derived from ESC have been tested for their contribution in revascularization in multiple preclinical models of ischemia (including peripheral vascular disease, myocardial infarction, and stroke; reviewed in [366, 367]), mostly showing that a combination of ESC-EC and ESC-SMC delivers a more stable and durable vascular network than either cell population alone. While other cutaneous cell types (e.g., keratinocytes [375]) differentiated from ESC have been tested for their efficacy for improving wound healing, we have no record in the current literature of the use of ESC-derived vascular differentiated cells in wound healing. Recently, $AC133^{+}KDR^{+}$ endothelial precursors were derived from ESC, and their conditioned medium (containing a number of pro-angiogenic factors, e.g., VEGF-A and bFGF) was shown to promote wound healing [376].

Induced Pluripotent Stem Cells

As first described by Takahashi et al. in mice, induced pluripotent stem cells (iPSC) are ESC-like cells obtained through reprogramming of skin fibroblasts (or other somatic cells) by transduction with a combination of only four genes (*Oct4*, *Sox2*, *c-Myc*, and *Klf4*) [377]. In 2007, iPSC were also successfully derived from human fibroblasts by using a combination of different genes (*Oct3/4*, *Lin28*, *nanog*, and *Sox2*) [378]. As this avoids the destruction of embryos, the use of iPSC is not linked to major ethical issues. Moreover, unlike ESC, they can be derived from the patient, making an autologous approach feasible. While iPSC are not entirely similar to ESC, they also carry the risk of teratoma formation and genetic instability [379]. Moreover, the need for reprogramming by genetic manipulation through viral integration renders an additional risk for transformation. Therefore, based on the observation that sustained expression of the introduced pluripotency factors is not required for stable reprogramming, alternative transient methods have been designed to induce reprogramming, such as non-viral (e.g., protein, miRNA, minicircle [380], or small molecule) transfer or even transgene removal whereby the transgenes are excised using Cre recombinase (reviewed in [368, 381]).

Given the inherent risk for teratoma formation, also for iPSC, in vitro vascular differentiation protocols have been developed [382–385]. These vascular (EC) derivatives were successfully tested in immunodeficient murine models of peripheral arterial disease [385, 386], but not yet in wound healing models.

26.4.3 Combined Gene/Protein and Cellular Approaches

Recently, the combination of cellular and gene/protein transfer approaches has emerged as a promising avenue to improve vascularization of chronic and acute skin wounds or ischemic tissues. Such a combinatorial approach can be applied for at least three reasons, which we illustrate here with some examples (Fig. 26.3). First, the ex vivo transfected/transduced cells can be used as a delivery device for angiogenic/vasculogenic factors to overcome the limited success of protein delivery or to have a more directed way of gene delivery. Indeed, if the manipulated cells are vascular, then they will occupy a strategic position – i.e., in or around growing endothelial sprouts – to deliver the factors in close vicinity of the endogenous vascular target cells, which may lead to a more efficient effect on blood vessel formation. Furthermore, gene transfer of angiogenic factors in vascular cells may improve their incorporation in nascent blood vessels [387]. For instance, overexpression of VEGF-A in EPC increased their vascular incorporation. Secondly, the combination of gene transfer and cell therapy can serve as a way to improve survival/engraftment/proliferation or to overcome dysfunctionality of the transplanted cells, which is often a problem in an autologous setting. Alternatively, preconditioning the cells by exposure to recombinant proteins may also achieve the same effect. Indeed, many studies have pointed out that using autologous cells could be problematic since the patient's own progenitors may be dysfunctional and present in lower numbers (reviewed in [132, 135, 137]) requiring additional measures to increase their functionality [135, 388]. For instance, aged MSC were shown to have a decreased capacity to induce wound revascularization [389]. Di Rocco et al. transplanted adipose-derived MSC following ex vivo SDF-1α gene transfer in wounds of diabetic mice, showing more persistent cell survival/engraftment [390]. Similarly, preconditioning of MSC with recombinant SDF-1α protein had a beneficial effect on cell survival in infarcted myocardium [391]. Overexpression of manganese superoxide dismutase in EPC from diabetic mice restored their ability to stimulate wound revascularization [392]. Overexpression of a variant of v-myc and AKT in adipose-derived MSC resulted in improved proliferation and increased secretion of VEGF-A in wounds [393]. Gene transfer of peroxisome proliferator-activated receptor-γ coactivator-1α (PGC-1α) enhanced engraftment and the pro-angiogenic effects of MSC in ischemic mouse limbs [394]. Third, in case of multi- or pluripotent stem/progenitor cells, gene transfer (e.g., of transcription factors; reviewed in [395]) can mediate transdifferentiation of the stem/progenitor cells to the endothelial or SMC/pericyte lineage. For instance, Duffy et al. demonstrated that overexpression of ephrinB2 in MSC induces an early endothelial phenotype and increases their potential

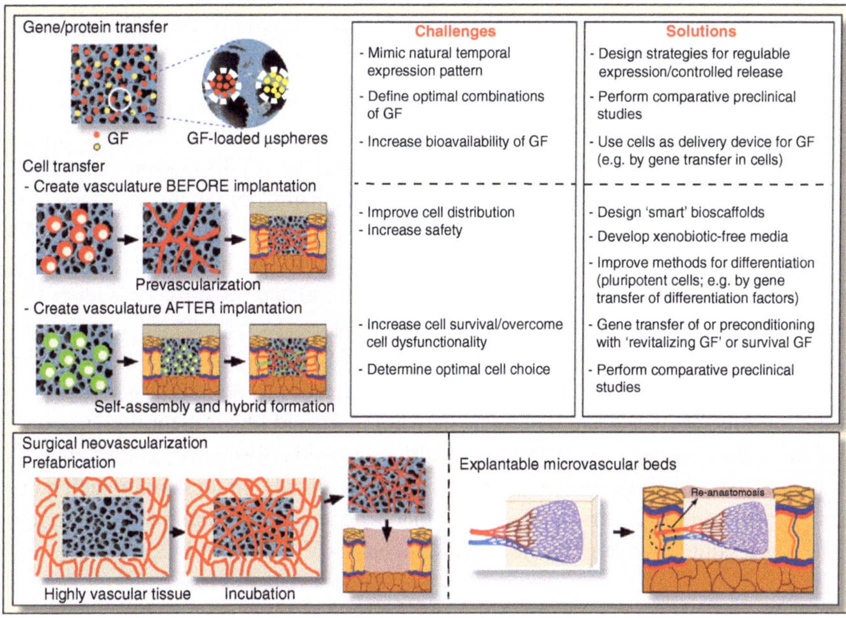

Fig. 26.3 Approaches for vascularization of skin equivalents and their challenges. Largely three types of approaches can be distinguished to vascularize skin equivalents: gene/protein transfer, cell transfer, or surgical techniques. When using cells, two strategies can be followed, one being the formation of a vascular network before implantation of the scaffold (called "prevascularization") and another being the seeding of vascular cells onto the scaffold just before or after its placement into the wound (in which case, vessels are assembled in situ, usually as hybrids between host cells (in *red*) and transplanted cells (in *green*)). Both gene/protein and cell-based techniques are faced with several challenges (*central box*) for which a number of solutions have been developed (*right box*). In some cases, the solution consists of combining gene/protein transfer with cell transplantation (e.g., the use of cells as "natural" delivery devices for growth factors or GF, the use of gene transfer to improve cell differentiation which is important for pluripotent cells, or the use of gene/protein transfer to improve survival or restore function of the transplanted cells). Surgical revascularization ("prefabrication") can be achieved by implanting the scaffold in a highly vascularized region followed by an incubation period before transfer of the vascularized scaffold into the wound. Another surgical technique is microvascular bed explantation followed by re-anastomosis to the host vasculature in the wound edges. *μspheres* microspheres (Adapted from Romano Di Peppe et al. [63])

to form vessel-like structures [396]. Adenoviral overexpression of soluble Frizzled1 in MSC increased their perivascular location and expression of α-SMC-actin, resulting in vessel maturation in a Matrigel implantation model [397]. As described above, for pluripotent stem cells, robust predifferentiation is an absolute requirement to avoid teratoma formation.

In addition to combining gene/protein transfer with cellular therapy by using the cells as a delivery device for gene/protein expression, the two therapeutic modalities can also be combined in another way. For instance, we recently showed that incorporation of a natural (angiogenic) growth factor combination (i.e., PRP) together with endothelial progenitor cells in a gel was more efficient in revascularizing porcine skin wounds than using PRP alone [230].

26.4.4 Electromechanical Stimulation of Blood Vessel Formation

Vacuum-assisted therapy (VAC) or microdeformational therapy is an upcoming promising technology in wound healing. Multiple (pre)clinical studies have shown that one of the effects of VAC is increased blood vessel formation in wounds (reviewed in [398]), although the underlying mechanisms remain unclear. One possibility is that applying topical negative pressure to the wound area stimulates the production of endogenous angiogenic growth factors, such as VEGF-A [399]. Furthermore, also growth of lymph vessels was stimulated most likely leading to a more efficient removal of excess wound fluid [400].

Extreme low-frequency electromagnetic fields (ELF-EMF) also have been considered for their beneficial effect on the wound healing process (reviewed in [401]). One of the reasons why such manipulations may aid in wound repair is through a neoangiogenic effect, i.e., by stimulating EC proliferation, tube formation, and production of angiogenic growth factors (e.g., bFGF [402]). In a study on temporal punch biopsies from human volunteers, it was shown that degenerate electrical waveform upregulates expression of angiogenic molecules, including VEGF-A and CD31 [403].

26.5 Neovascularization Strategies for Tissue-Engineered Skin Equivalents

For full-thickness skin defects larger than 4 cm in diameter, boosting blood vessel growth alone may not be sufficient and may not lead to the desired healing result [404]. The latter is mainly due to excessive wound contraction, which may be particularly harmful if wounds are located at joint regions, where contraction may limit mobility, or in the face, where contraction will lead to a poor esthetic outcome. Therefore, in such cases, a skin graft is needed to ensure sufficient coverage and to limit wound contraction. While split-thickness skin grafts – consisting of epidermis and part of the dermis and currently still the "gold standard" for extensive wounds – provide immediate epidermal coverage, they do not prevent excessive contraction. Furthermore, in case of extensive wound surfaces, the remaining intact donor sites will not be sufficient to provide enough graft material for complete coverage. Tissue-engineered skin equivalents have been proposed to overcome donor graft shortage (reviewed in [404]). Importantly, to prevent contraction, epidermal substitutes are not effective. Instead, natural (human- or animal-derived) or synthetic dermal substitutes (e.g., AlloDerm®, Integra®) or combined epidermal/dermal substitutes (e.g., Apligraf®, OrCel®) of sufficient thickness are needed for that purpose. These skin equivalents can be completely acellular or can be seeded with autologous or allogeneic skin cells, mostly keratinocytes and/or fibroblasts. Yet another approach is the design of a completely cellular dermal construct that acquires sufficient thickness through multilayering of fibroblasts that secrete their own extracellular matrix [100]. Several of these substitutes have been developed and approved for clinical use in burns patients or patients with chronic ulcers (for a recent overview, we refer to [405]).

However, while the increased thickness of these constructs offers sufficient mechanical resistance to limit wound contraction, as they are thicker than the in vivo diffusion limit of oxygen, survival and integration of these skin equivalents is critically dependent on fast and functional vascularization. The critical thickness of engineered tissue constructs that invokes the need for vascularization was determined to be 2 mm [406]. Several strategies to provide blood vessels in tissue-engineered skin equivalents can be envisioned. Again, they can be divided in non-cellular and cellular techniques (reviewed in [407]), and the two modalities can be also combined for reasons as mentioned above (see Sect. 26.4.3; Fig. 26.3). A specific strategy to provide vessels in tissue-engineered constructs, including skin, is "surgical neovascularization" or prefabrication, in which the tissue scaffold is implanted in a certain location that more extensively supports vascular ingrowth (e.g., because of implantation around a vascular pedicle) than the defect location (Fig. 26.3). After a certain "incubation" time, the vascularized scaffold is transferred to the target tissue. Another technology is the use of explantable microvascular beds (EMB; Fig. 26.3), e.g., the omentum, that can be microsurgically removed, ex vivo manipulated (e.g., by seeding additional cells onto them), and transferred to the target location, followed by microsurgical re-anastomosis to the vasculature of the target tissue. We will not further discuss this here and refer the reader to recent reviews on these surgical techniques [407, 408].

26.5.1 Non-cellular Neovascularization Strategies: Growth Factors

Incorporation of angiogenic growth factors is the most common way to stimulate vascular ingrowth into acellular, synthetic skin equivalents. The success of such an approach is mostly dependent on a prolonged bioactivity of the growth factors, which can be achieved by a slow release system (e.g., incorporation in microspheres (Fig. 26.3) or binding to fibrin-based or polyethylene glycol-based materials) [409]. Alternatively, the application of the dermal scaffold can be combined with repeated topical administration of the growth factor, as recently performed with SDF-1α. This combination improved scaffold neovascularization and reduced wound contraction [410]. The increased dermal vascularization may not only lead to reduced contraction but also may improve the re-epithelialization capacity of subsequently applied epidermal cells [411]. As mentioned above, combination of several angiogenic growth factors may also here be more effective than single growth factors, and sequential release of these factors may mimic more reliably the temporal sequence of events that occurs during neovascularization. In another embodiment, growth factors can be released or overexpressed by non-vascular skin cells incorporated into the tissue-engineered skin construct. Supp et al. demonstrated that cultured skin substitutes containing keratinocytes overexpressing VEGF-A caused less wound contraction than non-overexpressing substitutes [412]. Similarly, overexpression of FGF-7 (KGF) in keratinocytes seeded onto an acellular human dermis significantly improved neovascularization of this bioengineered construct upon implantation in wounds, most likely by stimulating the secretion of VEGF-A by the keratinocytes [413]. None of the currently available clinically approved skin equivalents features growth factor overexpression.

26.5.2 *Cellular Neovascularization Strategies*

When using cells for neovascularization of skin equivalents, largely two scenarios are possible [409]. The first one is to create endothelial cell-lined vascular structures into the scaffold before its application to the wound, the so-called prevascularization technique (Fig. 26.3). While this results in faster inosculation to the host vascular network [414], this approach sometimes raises the problem of unequal cell distribution in the skin substitute leading to incomplete vascular networks. A better distribution may be obtained by low-pressure centrifugation [415, 416]. Improved vascular distribution and vascular cell colonization can also be achieved by optimizing the structural/biological composition and the biophysical/biomechanical properties of the scaffolds (reviewed in [417]) and/or by applying several patterning technologies, such as microcontact printing, micromachining, laser-guided writing, and photolithography (reviewed in [407]). These scaffolds with an improved interactive relationship with implanted (and also surrounding) cells have been called "smart" bioscaffolds. Another challenge for prevascularization may be the stabilization of the preformed endothelial networks with SMC [407, 418]. Interestingly, a recent study using a crossover wild-type/GFP skin transplantation model underscored the importance of prevascularization and revealed that up to ~70 % of the preexisting vessels in the graft (mostly those in the periphery) become replaced by ingrowing vessels from the surrounding host muscular wound bed but that there is also a temporary angiogenic response originating from the central preexisting vessels in the graft [419].

Alternatively, the vascular (progenitor) cells can be homogeneously applied onto the skin construct immediately before or after it has been placed in the wound. The latter approach relies on self-assembly of the applied cells into lumenized structures and their subsequent stabilization (Fig. 26.3). This type of vessel growth may encompass the formation of hybrid vascular structures composed of a mixture of seeded cells and ingrowing host cells. In the following chapters, rather than giving an extensive overview of the exponentially growing number of studies related to cell-based strategies for vascularization of tissue-engineered skin equivalents, we highlight a number of exemplary and/or landmark papers.

26.5.2.1 Mature EC: HDMEC or HUVEC

Seeding of HDMEC onto a porous poly-L-lactic acid sponge leads to the formation of a functional human blood vessel network within 10 days, stabilized by mouse perivascular cells within 21 days after subcutaneous transplantation in SCID mice [253]. When used with fibroblasts and keratinocytes in a cultured skin substitute,

Supp et al. found that HDMEC first form multicellular aggregates in vitro, followed by organization into linear and circular vascular-like structures upon implantation of the skin substitute in an athymic mouse wound model. However, the authors did not document a functional connection between the host blood vessel network and the HDMEC vessel network [248]. As mentioned above (see Sect. 26.4.2.1.1), the main drawback for use of these cells to vascularize skin substitutes remains however their low yields, and therefore it is unlikely that HDMEC will widely be used for vascularization of tissue-engineered skin substitutes.

When combined in vitro with collagen, fibroblasts, and keratinocytes, HUVEC formed vascular tubular networks [420, 421], even more so upon stimulation with angiogenic growth factors like VEGF-A or bFGF [422]. In vivo, HUVEC transduced with Bcl-2 (to protect them against apoptosis) efficiently formed blood vessel networks in collagen/fibronectin gel plugs implanted in subcutaneous pockets of immunodeficient mice [259]. Together with the in vitro experience, this has lead to the development of vascularized skin equivalents, built of HUVEC and keratinocytes seeded in a decellularized dermis [423] or in a fibroblast-containing collagen sponge [414] that showed quick inosculation with the host vascular network upon transplantation in mice models. Interestingly, when aiming at reconstruction of the deepest skin layer, the hypodermis, HUVEC may play an important role as well. This layer consists mainly of fat, and the main challenge of reconstruction of a significant volume of fat tissue is quick and adequate perfusion. Even though in vitro studies suggested a synergistic effect on angiogenesis between adipose-derived MSC and HUVEC [424], some authors were unable to demonstrate this synergy in vivo in fibrin plugs [425]. This may however be related to the relative small size of the plugs used in this study since co-transplantation of HUVEC with adipose-derived MSC in bigger plugs showed more volume maintenance and less necrosis than with adipose-derived MSC alone [426]. As mentioned above, given their origin, in the absence of cell banking, HUVEC can only be used in an allogeneic setting when considering clinical applications.

26.5.2.2 Endothelial Progenitors

In vitro studies showed that BOEC are capable of forming vascular networks in different matrices in a similar or superior way compared to HUVEC [427–429]. Progenitor-derived EC were more efficient for seeding tissue-engineered vascular conduits than HUVEC or human saphenous vein EC [430]. Peripheral blood-derived BOEC seeding of decellularized human dermal matrices covered with keratinocytes in a mouse wound model resulted in the formation of a functional hybrid vascular network [266, 431]. Cord blood-derived BOEC seemed more potent in their vasculogenic abilities than their adult blood-derived counterparts [266, 270, 429]. When compared to HDMEC, peripheral blood- or cord blood-derived BOEC induced a higher vascular density when co-implanted with SMC in Matrigel [429]. In many of the studies mentioned above, the formation of a durable vasculature by BOEC incorporated in a matrix scaffold was indeed dependent on co-implantation with mural cells (e.g., saphenous vein SMC or 10 T1/2 cells) [270, 429]. We have combined adult blood-derived BOEC with a dermal substitute made out of multilayered dermal fibroblast sheets and found BOEC to be effective in both angiogenesis and vasculogenesis to stimulate re-epithelialization and to improve dermal matrix organization [100]. Recently, animal-free culture protocols have been described for BOEC [275], keratinocytes, and fibroblasts [432], which opens possibilities for development of a completely autologous vascularized skin substitute.

26.5.2.3 Multipotent or Pluripotent Stem/Progenitor Cells

Among the different types of multipotent or pluripotent stem/progenitor cells, MSC from various sources have been most extensively tested for their ability to improve vascularization of tissue-engineered skin constructs. Markowicz et al. noted increased vascularization as well as enhanced collagen production when BM-derived MSC were seeded in a collagen sponge in vivo [433]. Liu et al. seeded BM-MSC on collagen-glycosaminoglycan scaffolds and applied them to burn wounds in pigs. Vascular content was significantly increased resulting in a better re-epithelialization and a reduced wound contraction [434]. The same lab later compared different scaffold types (small intestinal mucosa, acellular dermal matrix, and collagen-chondroitin sulfate-hyaluronic acid) for their efficiency to support vascularization, revealing that all three tested scaffolds supported blood vessel growth, however, to a different extent. Furthermore, the scaffolds with the best vascularization capacity (i.e., small intestinal mucosa and acellular dermal matrix) also had a more pronounced stimulating effect on VEGF secretion by the adipose-derived MSC seeded in these scaffolds in vitro, which correlated with a better vascularization in vivo in a murine skin wound model [435]. Autologous adipose-derived MSC, injected in full-thickness skin grafts in rats, improved skin graft survival by increasing graft vascularization through endothelial differentiation and angiogenic growth factor production [436]. When seeded into an acellular dermal matrix (Alloderm®) or a silk fibroin-chitosan scaffold, adipose-derived MSC survived for at least 2 weeks and directly contributed to the blood vessel formation by differentiating into EC (in addition to fibroblastic and keratinocytic differentiation), thereby significantly improving wound healing in mice [437, 438]. Recently, when BM-MSC were seeded in a collagen-based biomimetic hydrogel, they were found to differentiate into vascular cells (both EC and pericytes) in addition to fibroblasts, but not keratinocytes, and levels of angiogenic growth factors were increased within the wound bed [439]. BM-MSC were also effective in boosting angiogenesis – most likely by secreting VEGF-A – in a diabetic wound model in rats, when impregnated in a collagen-based artificial dermal substitute [440]. Similarly, dermal vascularization – and hence healing – of a non-healing ulcer of a diabetic patient was significantly improved when applying BM-MSC on a biodegradable collagen membrane seeded with autologous skin fibroblasts [441]. We recently found MAPC to (mainly trophically) support vascularization and accelerate re-epithelialization of self-assembled human dermal fibroblasts sheets (Hendrickx et al. 2009).

In contrast to MSC, only few studies have tested pluripotent stem cells for their ability to support vascularization in tissue-engineered constructs, and none of them have been used in skin substitutes in particular. EC differentiated from human ESC enhanced patent vascular formation in PLLA (poly-(L-lactic acid))/PLGA (poly-lactic-glycolic acid) scaffolds when implanted in vivo in immunodeficient mice [442]. Co-seeding with embryonic fibroblasts before implantation in SCID mice improved stabilization of the endothelial networks [443]. EC derived from murine ESC were successful in forming an intimal layer of cells onto a tissue-engineered vascular media consisting of SMC and collagen [444]. In another approach, Huang et al. seeded murine ESC-derived cells containing 30 % of Flk-1+ cells onto a compliant polyurethane tube and exposed them to mechanical stress loading (a combination of wall shear stress and circumferential strain) resulting in simultaneous differentiation of EC and SMC, which seemed to organize themselves in discrete layers [445]. Human ESC-derived EC formed lumenized vessels when seeded in a porous scaffold and implanted in nude mice. Furthermore, implantation of a collagen gel containing these cells into infracted hearts of nude rats supported the generation of a vascular network that functionally connected with the heart vasculature [446]. Similarly, human ESC-derived EC directly participated in neovascularization upon transplantation in a bioactive hydrogel in infarcted rat hearts, thereby significantly limiting infarct size and improving heart function [447].

26.6 Conclusions and Future Challenges

Like the majority of tissues in our body, skin is a highly vascularized tissue. Therefore, adequate vascularization is a cornerstone of skin wound healing and skin tissue engineering. Insufficient oxygenation of wounds, such as in diabetic patients, leads to chronic non-healing ulcers. Lack of blood vessels in tissue-engineered skin for patients with burns, ischemic wounds, or surgical wounds hampers its survival and prevents its successful integration into the skin wound. Knowing the cellular and molecular mechanisms behind blood vessel formation is of utmost importance as it will offer the possibility to design efficient revascularization strategies for the growing number of patients in need of improved wound healing. These strategies can be mainly categorized in gene/protein transfer and cell differentiation/transplantation approaches. The mechanisms behind electromechanical stimulation of wound vascularization and their overall benefit for wound healing, on the other hand, remain ill-defined.

As for gene/protein or cell-mediated approaches, both strategies are faced with important challenges (Fig. 26.3). The main challenge of protein/gene therapy is to mimic the natural course of the complex molecular events that occur during blood vessel formation. This implies the application of combinations of growth factors rather than monotherapy and the possibility to regulate gene expression or protein delivery in time in order to have the growth factor present within the appropriate phase of the neovascularization process. With regard to cellular approaches, the biggest challenge will be to make the appropriate cell choice, which will require additional comparative preclinical studies. Furthermore, the choice will not only depend on the biological properties of the cells themselves but also – and perhaps to a larger extent – of the type of clinical emergency. For instance, acute interventions in patients with trauma or extensive burns will require the instant availability of sufficient amounts of cells, which will preclude most of the autologous cell sources (which usually require a significant time for their derivation and ex vivo expansion). In other clinical cases, such as chronic wounds, skin reconstruction and revascularization can be delayed, which leaves sufficient time for isolation and expansion of autologous cells. Another challenge with cell therapy is to design safe cell products that are neither immunogenic nor tumorigenic, the latter particularly relevant for pluripotent stem cells. Finally, another success-limiting factor for cell therapy is the poor survival of cells when confronted with the hostile wound environment. Smart tissue engineering or combination of cell therapy with gene/protein transfer may offer solutions for this. Future efforts in the field of skin tissue engineering constructs will also have to include the design of standardized methods for in vitro and in vivo evaluation of their functionality, as this will accelerate their transition from the bench to the bedside [448]. Successfully facing these many challenges will require a multidisciplinary approach and will only be possible upon dynamic interaction between clinicians, bioengineers, and biologists, much like wound revascularization itself is a process involving different actors.

References

1. Bender AE, Bender DA (1995) A dictionary of food and nutrition. Oxford University Press, New York
2. Braverman IM (2000) The cutaneous microcirculation. J Investig Dermatol Symp Proc 5:3–9
3. Li L et al (2006) Age-related changes of the cutaneous microcirculation in vivo. Gerontology 52:142–153
4. Alitalo K, Tammela T, Petrova TV (2005) Lymphangiogenesis in development and human disease. Nature 438:946–953
5. Gurtner GC, Werner S, Barrandon Y, Longaker MT (2008) Wound repair and regeneration. Nature 453:314–321
6. Shaw TJ, Martin P (2009) Wound repair at a glance. J Cell Sci 122:3209–3213
7. Barrientos S, Stojadinovic O, Golinko MS, Brem H, Tomic-Canic M (2008) Growth factors and cytokines in wound healing. Wound Repair Regen 16:585–601
8. Rhett JM et al (2008) Novel therapies for scar reduction and regenerative healing of skin wounds. Trends Biotechnol 26:173–180
9. Kose O, Waseem A (2008) Keloids and hypertrophic scars: are they two different sides of the same coin? Dermatol Surg 34:336–346
10. van der Veer WM et al (2011) Time course of the angiogenic response during normotrophic and hypertrophic scar formation in humans. Wound Repair Regen 19:292–301
11. Hendrickx B, Vranckx JJ, Luttun A (2011) Cell-based vascularization strategies for skin tissue engineering. Tissue Eng Part B Rev 17:13–24
12. Benest AV, Augustin HG (2009) Tension in the vasculature. Nat Med 15:608–610
13. Kilarski WW, Samolov B, Petersson L, Kvanta A, Gerwins P (2009) Biomechanical regulation of blood vessel growth during tissue vascularization. Nat Med 15:657–664

14. Liu L et al (2008) Age-dependent impairment of HIF-1alpha expression in diabetic mice: correction with electroporation-facilitated gene therapy increases wound healing, angiogenesis, and circulating angiogenic cells. J Cell Physiol 217:319–327
15. Cianfarani F et al (2006) Placenta growth factor in diabetic wound healing: altered expression and therapeutic potential. Am J Pathol 169:1167–1182
16. Wetterau M et al (2011) Topical prolyl hydroxylase domain-2 silencing improves diabetic murine wound closure. Wound Repair Regen 19:481–486
17. Mace KA, Yu DH, Paydar KZ, Boudreau N, Young DM (2007) Sustained expression of Hif-1alpha in the diabetic environment promotes angiogenesis and cutaneous wound repair. Wound Repair Regen 15:636–645
18. Yu DH, Mace KA, Hansen SL, Boudreau N, Young DM (2007) Effects of decreased insulin-like growth factor-1 stimulation on hypoxia inducible factor 1-alpha protein synthesis and function during cutaneous repair in diabetic mice. Wound Repair Regen 15:628–635
19. Niessen K et al (2011) The Notch1-Dll4 signaling pathway regulates mouse postnatal lymphatic development. Blood 118:1989–1997
20. Gilbertson DG et al (2001) Platelet-derived growth factor C (PDGF-C), a novel growth factor that binds to PDGF alpha and beta receptor. J Biol Chem 276:27406–27414
21. Henderson PW et al (2011) Stromal-derived factor-1 delivered via hydrogel drug-delivery vehicle accelerates wound healing in vivo. Wound Repair Regen 19:420–425
22. Kawanabe T, Kawakami T, Yatomi Y, Shimada S, Soma Y (2007) Sphingosine 1-phosphate accelerates wound healing in diabetic mice. J Dermatol Sci 48:53–60
23. Jin Q et al (2008) Nanofibrous scaffolds incorporating PDGF-BB microspheres induce chemokine expression and tissue neogenesis in vivo. PLoS One 3:e1729
24. Sun W et al (2007) Collagen membranes loaded with collagen-binding human PDGF-BB accelerate wound healing in a rabbit dermal ischemic ulcer model. Growth Factors 25:309–318
25. Li H et al (2008) Research of PDGF-BB gel on the wound healing of diabetic rats and its pharmacodynamics. J Surg Res 145:41–48
26. Steed DL (2006) Clinical evaluation of recombinant human platelet-derived growth factor for the treatment of lower extremity ulcers. Plast Reconstr Surg 117:143S–149S; discussion 150S-151S
27. Lee JA et al (2005) Lentiviral transfection with the PDGF-B gene improves diabetic wound healing. Plast Reconstr Surg 116:532–538
28. Pereira CT, Herndon DN, Rocker R, Jeschke MG (2007) Liposomal gene transfer of keratinocyte growth factor improves wound healing by altering growth factor and collagen expression. J Surg Res 139:222–228
29. Obara K et al (2005) Acceleration of wound healing in healing-impaired db/db mice with a photocrosslinkable chitosan hydrogel containing fibroblast growth factor-2. Wound Repair Regen 13:390–397
30. Obara K et al (2003) Photocrosslinkable chitosan hydrogel containing fibroblast growth factor-2 stimulates wound healing in healing-impaired db/db mice. Biomaterials 24:3437–3444
31. Pandit A, Ashar R, Feldman D, Thompson A (1998) Investigation of acidic fibroblast growth factor delivered through a collagen scaffold for the treatment of full-thickness skin defects in a rabbit model. Plast Reconstr Surg 101:766–775
32. Pandit AS, Feldman DS, Caulfield J, Thompson A (1998) Stimulation of angiogenesis by FGF-1 delivered through a modified fibrin scaffold. Growth Factors 15:113–123
33. Greenhalgh DG, Sprugel KH, Murray MJ, Ross R (1990) PDGF and FGF stimulate wound healing in the genetically diabetic mouse. Am J Pathol 136:1235–1246
34. Badillo AT, Chung S, Zhang L, Zoltick P, Liechty KW (2007) Lentiviral gene transfer of SDF-1alpha to wounds improves diabetic wound healing. J Surg Res 143:35–42
35. Han G et al (2012) Nitric oxide-releasing nanoparticles accelerate wound healing by promoting fibroblast migration and collagen deposition. Am J Pathol 180:1465–1473
36. Luo JD et al (2009) Sonic hedgehog improves delayed wound healing via enhancing cutaneous nitric oxide function in diabetes. Am J Physiol Endocrinol Metab 297:E525–E531

37. Zhu H, Wei X, Bian K, Murad F (2008) Effects of nitric oxide on skin burn wound healing. J Burn Care Res 29:804–814
38. Bitto A et al (2008) Angiopoietin-1 gene transfer improves impaired wound healing in genetically diabetic mice without increasing VEGF expression. Clin Sci (Lond) 114(707–718)
39. Cho CH et al (2006) COMP-angiopoietin-1 promotes wound healing through enhanced angiogenesis, lymphangiogenesis, and blood flow in a diabetic mouse model. Proc Natl Acad Sci USA 103:4946–4951
40. Galeano M et al (2006) Recombinant human erythropoietin improves angiogenesis and wound healing in experimental burn wounds. Crit Care Med 34:1139–1146
41. Luo JD, Wang YY, Fu WL, Wu J, Chen AF (2004) Gene therapy of endothelial nitric oxide synthase and manganese superoxide dismutase restores delayed wound healing in type 1 diabetic mice. Circulation 110:2484–2493
42. Galeano M et al (2003) Effect of recombinant adeno-associated virus vector-mediated vascular endothelial growth factor gene transfer on wound healing after burn injury. Crit Care Med 31:1017–1025
43. Hamed S et al (2010) Topical erythropoietin promotes wound repair in diabetic rats. J Invest Dermatol 130:287–294
44. Sorg H et al (2009) Effects of erythropoietin in skin wound healing are dose related. FASEB J 23:3049–3058
45. Sayan H, Ozacmak VH, Guven A, Aktas RG, Ozacmak ID (2006) Erythropoietin stimulates wound healing and angiogenesis in mice. J Invest Surg 19:163–173
46. Galeano M et al (2004) Recombinant human erythropoietin stimulates angiogenesis and wound healing in the genetically diabetic mouse. Diabetes 53:2509–2517
47. Buemi M et al (2004) Recombinant human erythropoietin stimulates angiogenesis and healing of ischemic skin wounds. Shock 22:169–173
48. Cianfarani F et al (2006) Granulocyte/macrophage colony-stimulating factor treatment of human chronic ulcers promotes angiogenesis associated with de novo vascular endothelial growth factor transcription in the ulcer bed. Br J Dermatol 154:34–41
49. Jeschke MG, Schubert T, Klein D (2004) Exogenous liposomal IGF-I cDNA gene transfer leads to endogenous cellular and physiological responses in an acute wound. Am J Physiol Regul Integr Comp Physiol 286:R958–R966
50. Jeschke MG et al (2002) Non-viral liposomal keratinocyte growth factor (KGF) cDNA gene transfer improves dermal and epidermal regeneration through stimulation of epithelial and mesenchymal factors. Gene Ther 9:1065–1074
51. Lynch SE, Colvin RB, Antoniades HN (1989) Growth factors in wound healing. Single and synergistic effects on partial thickness porcine skin wounds. J Clin Invest 84:640–646
52. Deodato B et al (2002) Recombinant AAV vector encoding human VEGF165 enhances wound healing. Gene Ther 9:777–785
53. Lee PY, Chesnoy S, Huang L (2004) Electroporatic delivery of TGF-beta1 gene works synergistically with electric therapy to enhance diabetic wound healing in db/db mice. J Invest Dermatol 123:791–798
54. Dickens S, Vermeulen P, Hendrickx B, Van den Berge S, Vranckx JJ (2008) Regulable vascular endothelial growth factor165 overexpression by ex vivo expanded keratinocyte cultures promotes matrix formation, angiogenesis, and healing in porcine full-thickness wounds. Tissue Eng Part A 14:19–27
55. Gallagher KA et al (2007) Diabetic impairments in NO-mediated endothelial progenitor cell mobilization and homing are reversed by hyperoxia and SDF-1 alpha. J Clin Invest 117:1249–1259
56. Asai J, Takenaka H, Katoh N, Kishimoto S (2006) Dibutyryl cAMP influences endothelial progenitor cell recruitment during wound neovascularization. J Invest Dermatol 126:1159–1167
57. Kwon MJ et al (2012) Effective healing of diabetic skin wounds by using nonviral gene therapy based on minicircle vascular endothelial growth factor DNA and a cationic dendrimer. J Gene Med 14:272–278

58. Ko J et al (2011) Comparison of EGF with VEGF non-viral gene therapy for cutaneous wound healing of streptozotocin diabetic mice. Diabetes Metab J 35:226–235
59. Yoon CS et al (2009) Sonoporation of the minicircle-VEGF(165) for wound healing of diabetic mice. Pharm Res 26:794–801
60. Saaristo A et al (2006) Vascular endothelial growth factor-C accelerates diabetic wound healing. Am J Pathol 169:1080–1087
61. Takeda N et al (2004) Endothelial PAS domain protein 1 gene promotes angiogenesis through the transactivation of both vascular endothelial growth factor and its receptor, Flt-1. Circ Res 95:146–153
62. Galiano RD et al (2004) Topical vascular endothelial growth factor accelerates diabetic wound healing through increased angiogenesis and by mobilizing and recruiting bone marrow-derived cells. Am J Pathol 164:1935–1947
63. Romano Di Peppe S et al (2002) Adenovirus-mediated VEGF(165) gene transfer enhances wound healing by promoting angiogenesis in CD1 diabetic mice. Gene Ther 9:1271–1277
64. Rio MD et al (1999) Nonviral transfer of genes to pig primary keratinocytes. Induction of angiogenesis by composite grafts of modified keratinocytes overexpressing VEGF driven by a keratin promoter. Gene Ther 6:1734–1741
65. Jazwa A et al (2010) Combined vascular endothelial growth factor-A and fibroblast growth factor 4 gene transfer improves wound healing in diabetic mice. Genet Vaccines Ther 8:6
66. Jeschke MG, Herndon DN (2007) The combination of IGF-I and KGF cDNA improves dermal and epidermal regeneration by increased VEGF expression and neovascularization. Gene Ther 14:1235–1242
67. Jeschke MG, Klein D (2004) Liposomal gene transfer of multiple genes is more effective than gene transfer of a single gene. Gene Ther 11:847–855
68. Zheng Y et al (2007) Chimeric VEGF-ENZ7/PlGF specifically binding to VEGFR-2 accelerates skin wound healing via enhancement of neovascularization. Arterioscler Thromb Vasc Biol 27:503–511
69. Hamed S et al (2011) Fibronectin potentiates topical erythropoietin-induced wound repair in diabetic mice. J Invest Dermatol 131:1365–1374
70. Ackermann M et al (2011) Priming with a combination of proangiogenic growth factors improves wound healing in normoglycemic mice. Int J Mol Med 27:647–653
71. Ortega S, Ittmann M, Tsang SH, Ehrlich M, Basilico C (1998) Neuronal defects and delayed wound healing in mice lacking fibroblast growth factor 2. Proc Natl Acad Sci USA 95:5672–5677
72. Carmeliet P et al (2001) Synergism between vascular endothelial growth factor and placental growth factor contributes to angiogenesis and plasma extravasation in pathological conditions. Nat Med 7:575–583
73. Sarkar K et al (2012) Tie2-dependent knockout of HIF-1 impairs burn wound vascularization and homing of bone marrow-derived angiogenic cells. Cardiovasc Res 93:162–169
74. Zhang X et al (2010) Impaired angiogenesis and mobilization of circulating angiogenic cells in HIF-1alpha heterozygous-null mice after burn wounding. Wound Repair Regen 18:193–201
75. Mitchell K et al (2009) Alpha3beta1 integrin in epidermis promotes wound angiogenesis and keratinocyte-to-endothelial-cell crosstalk through the induction of MRP3. J Cell Sci 122:1778–1787
76. Zweers MC et al (2007) Integrin alpha2beta1 is required for regulation of murine wound angiogenesis but is dispensable for reepithelialization. J Invest Dermatol 127:467–478
77. Uutela M et al (2004) PDGF-D induces macrophage recruitment, increased interstitial pressure, and blood vessel maturation during angiogenesis. Blood 104:3198–3204
78. Lee PC et al (1999) Impaired wound healing and angiogenesis in eNOS-deficient mice. Am J Physiol 277:H1600–H1608
79. Chin LC et al (2011) The influence of nitric oxide synthase 2 on cutaneous wound angiogenesis. Br J Dermatol 165:1223–1235

80. Yamasaki K et al (1998) Reversal of impaired wound repair in iNOS-deficient mice by topical adenoviral-mediated iNOS gene transfer. J Clin Invest 101:967–971
81. Most D, Efron DT, Shi HP, Tantry US, Barbul A (2002) Characterization of incisional wound healing in inducible nitric oxide synthase knockout mice. Surgery 132:866–876
82. Low QE et al (2001) Wound healing in MIP-1alpha(−/−) and MCP-1(−/−) mice. Am J Pathol 159:457–463
83. Devalaraja RM et al (2000) Delayed wound healing in CXCR2 knockout mice. J Invest Dermatol 115:234–244
84. Fang Y, Gong SJ, Wang Y, Xu YH, Bao SS (2007) Effect of GMCSF-absence on neovascularization during wound healing. Zhonghua Zheng Xing Wai Ke Za Zhi 23:233–235
85. Fang Y, Gong SJ, Xu YH, Hambly BD, Bao S (2007) Impaired cutaneous wound healing in granulocyte/macrophage colony-stimulating factor knockout mice. Br J Dermatol 157:458–465
86. Mann A, Breuhahn K, Schirmacher P, Blessing M (2001) Keratinocyte-derived granulocyte-macrophage colony stimulating factor accelerates wound healing: stimulation of keratinocyte proliferation, granulation tissue formation, and vascularization. J Invest Dermatol 117:1382–1390
87. Streit M et al (2000) Thrombospondin-1 suppresses wound healing and granulation tissue formation in the skin of transgenic mice. EMBO J 19:3272–3282
88. Kyriakides TR, Tam JW, Bornstein P (1999) Accelerated wound healing in mice with a disruption of the thrombospondin 2 gene. J Invest Dermatol 113:782–787
89. Matthies AM, Low QE, Lingen MW, DiPietro LA (2002) Neuropilin-1 participates in wound angiogenesis. Am J Pathol 160:289–296
90. Brooks PC, Clark RA, Cheresh DA (1994) Requirement of vascular integrin alpha v beta 3 for angiogenesis. Science 264:569–571
91. Howdieshell TR et al (2001) Antibody neutralization of vascular endothelial growth factor inhibits wound granulation tissue formation. J Surg Res 96:173–182
92. Fu X, Li X, Cheng B, Chen W, Sheng Z (2005) Engineered growth factors and cutaneous wound healing: success and possible questions in the past 10 years. Wound Repair Regen 13:122–130
93. Riedel K et al (2006) Current status of genetic modulation of growth factors in wound repair. Int J Mol Med 17:183–193
94. Adams RH, Alitalo K (2007) Molecular regulation of angiogenesis and lymphangiogenesis. Nat Rev Mol Cell Biol 8:464–478
95. Carmeliet P, De Smet F, Loges S, Mazzone M (2009) Branching morphogenesis and antiangiogenesis candidates: tip cells lead the way. Nat Rev Clin Oncol 6:315–326
96. Eilken HM, Adams RH (2010) Dynamics of endothelial cell behavior in sprouting angiogenesis. Curr Opin Cell Biol 22:617–625
97. Potente M, Gerhardt H, Carmeliet P (2011) Basic and therapeutic aspects of angiogenesis. Cell 146:873–887
98. De Smet F, Segura I, De Bock K, Hohensinner PJ, Carmeliet P (2009) Mechanisms of vessel branching: filopodia on endothelial tip cells lead the way. Arterioscler Thromb Vasc Biol 29:639–649
99. LaVan FB, Hunt TK (1990) Oxygen and wound healing. Clin Plast Surg 17:463–472
100. Hendrickx B et al (2010) Integration of blood outgrowth endothelial cells in dermal fibroblast sheets promotes full thickness wound healing. Stem Cells 28:1165–1177
101. Andrikopoulou E et al (2011) Current Insights into the role of HIF-1 in cutaneous wound healing. Curr Mol Med 11:218–235
102. Augustin HG, Koh GY, Thurston G, Alitalo K (2009) Control of vascular morphogenesis and homeostasis through the angiopoietin-Tie system. Nat Rev Mol Cell Biol 10:165–177
103. Moya IM et al (2012) Stalk cell phenotype depends on integration of Notch and Smad1/5 signaling cascades. Dev Cell 22:501–514
104. Jones CA et al (2008) Robo4 stabilizes the vascular network by inhibiting pathologic angiogenesis and endothelial hyperpermeability. Nat Med 14:448–453

105. Gu C et al (2005) Semaphorin 3E and plexin-D1 control vascular pattern independently of neuropilins. Science 307:265–268
106. Lu X et al (2004) The netrin receptor UNC5B mediates guidance events controlling morphogenesis of the vascular system. Nature 432:179–186
107. Schmidt M et al (2007) EGFL7 regulates the collective migration of endothelial cells by restricting their spatial distribution. Development 134:2913–2923
108. Jakobsson L et al (2010) Endothelial cells dynamically compete for the tip cell position during angiogenic sprouting. Nat Cell Biol 12:943–953
109. Harrington LS et al (2008) Regulation of multiple angiogenic pathways by Dll4 and Notch in human umbilical vein endothelial cells. Microvasc Res 75:144–154
110. Boulton ME, Cai J, Grant MB (2008) Gamma-Secretase: a multifaceted regulator of angiogenesis. J Cell Mol Med 12:781–795
111. Guarani V et al (2011) Acetylation-dependent regulation of endothelial Notch signalling by the SIRT1 deacetylase. Nature 473:234–238
112. Phng LK et al (2009) Nrarp coordinates endothelial Notch and Wnt signaling to control vessel density in angiogenesis. Dev Cell 16:70–82
113. Zeeb M, Strilic B, Lammert E (2010) Resolving cell-cell junctions: lumen formation in blood vessels. Curr Opin Cell Biol 22:626–632
114. Iruela-Arispe ML, Davis GE (2009) Cellular and molecular mechanisms of vascular lumen formation. Dev Cell 16:222–231
115. Carmeliet P et al (1999) Targeted deficiency or cytosolic truncation of the VE-cadherin gene in mice impairs VEGF-mediated endothelial survival and angiogenesis. Cell 98:147–157
116. Jin SW, Beis D, Mitchell T, Chen JN, Stainier DY (2005) Cellular and molecular analyses of vascular tube and lumen formation in zebrafish. Development 132:5199–5209
117. Francis SE et al (2002) Central roles of alpha5beta1 integrin and fibronectin in vascular development in mouse embryos and embryoid bodies. Arterioscler Thromb Vasc Biol 22:927–933
118. Zovein AC et al (2010) Beta1 integrin establishes endothelial cell polarity and arteriolar lumen formation via a Par3-dependent mechanism. Dev Cell 18:39–51
119. Almagro S et al (2010) The motor protein myosin-X transports VE-cadherin along filopodia to allow the formation of early endothelial cell-cell contacts. Mol Cell Biol 30:1703–1717
120. Fantin A et al (2010) Tissue macrophages act as cellular chaperones for vascular anastomosis downstream of VEGF-mediated endothelial tip cell induction. Blood 116:829–840
121. Tammela T et al (2011) VEGFR-3 controls tip to stalk conversion at vessel fusion sites by reinforcing Notch signalling. Nat Cell Biol 13:1202–1213
122. Luttun A et al (2002) Revascularization of ischemic tissues by PlGF treatment, and inhibition of tumor angiogenesis, arthritis and atherosclerosis by anti-Flt1. Nat Med 8:831–840
123. Frontini MJ et al (2011) Fibroblast growth factor 9 delivery during angiogenesis produces durable, vasoresponsive microvessels wrapped by smooth muscle cells. Nat Biotechnol 29:421–427
124. Bergers G, Song S (2005) The role of pericytes in blood-vessel formation and maintenance. Neuro Oncol 7:452–464
125. Gaengel K, Genove G, Armulik A, Betsholtz C (2009) Endothelial-mural cell signaling in vascular development and angiogenesis. Arterioscler Thromb Vasc Biol 29:630–638
126. Mazzone M et al (2009) Heterozygous deficiency of PHD2 restores tumor oxygenation and inhibits metastasis via endothelial normalization. Cell 136:839–851
127. Stoll SJ, Bartsch S, Augustin HG, Kroll J (2011) The transcription factor HOXC9 regulates endothelial cell quiescence and vascular morphogenesis in zebrafish via inhibition of interleukin 8. Circ Res 108:1367–1377
128. Anand S et al (2010) MicroRNA-132-mediated loss of p120RasGAP activates the endothelium to facilitate pathological angiogenesis. Nat Med 16:909–914
129. Goettsch W et al (2008) Flow-dependent regulation of angiopoietin-2. J Cell Physiol 214:491–503
130. Hristov M, Zernecke A, Liehn EA, Weber C (2007) Regulation of endothelial progenitor cell homing after arterial injury. Thromb Haemost 98:274–277

131. Aicher A, Zeiher AM, Dimmeler S (2005) Mobilizing endothelial progenitor cells. Hypertension 45:321–325
132. Leone AM et al (2009) From bone marrow to the arterial wall: the ongoing tale of endothelial progenitor cells. Eur Heart J 30:890–899
133. Asahara T et al (1997) Isolation of putative progenitor endothelial cells for angiogenesis. Science 275:964–967
134. Ahn GO, Brown JM (2009) Role of endothelial progenitors and other bone marrow-derived cells in the development of the tumor vasculature. Angiogenesis 12:159–164
135. Fadini GP, Losordo D, Dimmeler S (2012) Critical reevaluation of endothelial progenitor cell phenotypes for therapeutic and diagnostic use. Circ Res 110:624–637
136. Ingram DA, Caplice NM, Yoder MC (2005) Unresolved questions, changing definitions, and novel paradigms for defining endothelial progenitor cells. Blood 106:1525–1531
137. Pearson JD (2009) Endothelial progenitor cells – hype or hope? J Thromb Haemost 7:255–262
138. Richardson MR, Yoder MC (2011) Endothelial progenitor cells: quo vadis? J Mol Cell Cardiol 50:266–272
139. Luttun A, Verfaillie CM (2007) Will the real EPC please stand up? Blood 109:1795–1796
140. Timmermans F et al (2009) Endothelial progenitor cells: identity defined? J Cell Mol Med 13:87–102
141. Watt SM, Athanassopoulos A, Harris AL, Tsaknakis G (2010) Human endothelial stem/progenitor cells, angiogenic factors and vascular repair. J R Soc Interface 7(Suppl 6):S731–S751
142. Hattori K et al (2002) Placental growth factor reconstitutes hematopoiesis by recruiting VEGFR1(+) stem cells from bone-marrow microenvironment. Nat Med 8:841–849
143. Heissig B et al (2002) Recruitment of stem and progenitor cells from the bone marrow niche requires MMP-9 mediated release of kit-ligand. Cell 109:625–637
144. Aicher A et al (2003) Essential role of endothelial nitric oxide synthase for mobilization of stem and progenitor cells. Nat Med 9:1370–1376
145. Cheng XW et al (2007) Mechanisms underlying the impairment of ischemia-induced neovascularization in matrix metalloproteinase 2-deficient mice. Circ Res 100:904–913
146. Urbich C et al (2005) Cathepsin L is required for endothelial progenitor cell-induced neovascularization. Nat Med 11:206–213
147. Levesque JP, Takamatsu Y, Nilsson SK, Haylock DN, Simmons PJ (2001) Vascular cell adhesion molecule-1 (CD106) is cleaved by neutrophil proteases in the bone marrow following hematopoietic progenitor cell mobilization by granulocyte colony-stimulating factor. Blood 98:1289–1297
148. Levesque JP, Hendy J, Takamatsu Y, Simmons PJ, Bendall LJ (2003) Disruption of the CXCR4/CXCL12 chemotactic interaction during hematopoietic stem cell mobilization induced by GCSF or cyclophosphamide. J Clin Invest 111:187–196
149. Li B et al (2006) VEGF and PlGF promote adult vasculogenesis by enhancing EPC recruitment and vessel formation at the site of tumor neovascularization. FASEB J 20:1495–1497
150. Li X et al (2005) Revascularization of ischemic tissues by PDGF-CC via effects on endothelial cells and their progenitors. J Clin Invest 115:118–127
151. Ishida Y et al (2012) Pivotal role of the CCL5/CCR5 interaction for recruitment of endothelial progenitor cells in mouse wound healing. J Clin Invest 122:711–721
152. Dimmeler S (2010) Regulation of bone marrow-derived vascular progenitor cell mobilization and maintenance. Arterioscler Thromb Vasc Biol 30:1088–1093
153. Zampetaki A, Kirton JP, Xu Q (2008) Vascular repair by endothelial progenitor cells. Cardiovasc Res 78:413–421
154. Velazquez OC (2007) Angiogenesis and vasculogenesis: inducing the growth of new blood vessels and wound healing by stimulation of bone marrow-derived progenitor cell mobilization and homing. J Vasc Surg 45(Suppl A):A39–A47
155. Thom SR et al (2011) Vasculogenic stem cell mobilization and wound recruitment in diabetic patients: increased cell number and intracellular regulatory protein content associated with hyperbaric oxygen therapy. Wound Repair Regen 19:149–161

156. Yu X, Cohen DM, Chen CS (2012) miR-125b is an adhesion-regulated microRNA that protects mesenchymal stem cells from anoikis. Stem Cells 30:956–964
157. Ruiz de Almodovar C, Luttun A, Carmeliet P (2006) An SDF-1 trap for myeloid cells stimulates angiogenesis. Cell 124:18–21
158. Grunewald M et al (2006) VEGF-induced adult neovascularization: recruitment, retention, and role of accessory cells. Cell 124:175–189
159. Bluff JE, Ferguson MW, O'Kane S, Ireland G (2007) Bone marrow-derived endothelial progenitor cells do not contribute significantly to new vessels during incisional wound healing. Exp Hematol 35:500–506
160. Aicher A et al (2007) Nonbone marrow-derived circulating progenitor cells contribute to postnatal neovascularization following tissue ischemia. Circ Res 100:581–589
161. Ergun S, Tilki D, Klein D (2011) Vascular wall as a reservoir for different types of stem and progenitor cells. Antioxid Redox Signal 15:981–995
162. Majesky MW, Dong XR, Hoglund V, Daum G, Mahoney WM Jr (2012) The adventitia: a progenitor cell niche for the vessel wall. Cells Tissues Organs 195:73–81
163. Majesky MW, Dong XR, Hoglund V, Mahoney WM Jr, Daum G (2011) The adventitia: a dynamic interface containing resident progenitor cells. Arterioscler Thromb Vasc Biol 31:1530–1539
164. Pacilli A, Pasquinelli G (2009) Vascular wall resident progenitor cells: a review. Exp Cell Res 315:901–914
165. Psaltis PJ, Harbuzariu A, Delacroix S, Holroyd EW, Simari RD (2011) Resident vascular progenitor cells – diverse origins, phenotype, and function. J Cardiovasc Transl Res 4:161–176
166. Torsney E, Xu Q (2011) Resident vascular progenitor cells. J Mol Cell Cardiol 50:304–311
167. Chen CW, Corselli M, Peault B, Huard J (2012) Human blood-vessel-derived stem cells for tissue repair and regeneration. J Biomed Biotechnol 2012:597439
168. Cordes KR, Srivastava D (2009) MicroRNA regulation of cardiovascular development. Circ Res 104:724–732
169. Caporali A, Emanueli C (2011) MicroRNA regulation in angiogenesis. Vascul Pharmacol 55:79–86
170. Fichtlscherer S, Zeiher AM, Dimmeler S (2011) Circulating microRNAs: biomarkers or mediators of cardiovascular diseases? Arterioscler Thromb Vasc Biol 31:2383–2390
171. Ohtani K, Dimmeler S (2011) Control of cardiovascular differentiation by microRNAs. Basic Res Cardiol 106:5–11
172. Wang S et al (2008) The endothelial-specific microRNA miR-126 governs vascular integrity and angiogenesis. Dev Cell 15:261–271
173. Bonauer A et al (2009) MicroRNA-92a controls angiogenesis and functional recovery of ischemic tissues in mice. Science 324:1710–1713
174. Kane NM et al (2012) Role of microRNAs 99b, 181a, and 181b in the differentiation of human embryonic stem cells to vascular endothelial cells. Stem Cells 30:643–654
175. Sen CK (2011) MicroRNAs as new maestro conducting the expanding symphony orchestra of regenerative and reparative medicine. Physiol Genomics 43:517–520
176. Howard L, Kane NM, Milligan G, Baker AH (2011) MicroRNAs regulating cell pluripotency and vascular differentiation. Vascul Pharmacol 55:69–78
177. Guo L, Zhao RC, Wu Y (2011) The role of microRNAs in self-renewal and differentiation of mesenchymal stem cells. Exp Hematol 39:608–616
178. Teta M et al (2012) Inducible deletion of epidermal Dicer and Drosha reveals multiple functions for miRNAs in postnatal skin. Development 139:1405–1416
179. Bavan L, Midwood K, Nanchahal J (2011) MicroRNA epigenetics: a new avenue for wound healing research. BioDrugs 25:27–41
180. Shilo S, Roy S, Khanna S, Sen CK (2007) MicroRNA in cutaneous wound healing: a new paradigm. DNA Cell Biol 26:227–237
181. Zou Z et al (2010) More insight into mesenchymal stem cells and their effects inside the body. Expert Opin Biol Ther 10:215–230

182. Banerjee S, Bacanamwo M (2010) DNA methyltransferase inhibition induces mouse embryonic stem cell differentiation into endothelial cells. Exp Cell Res 316:172–180
183. Hunt TK, Twomey P, Zederfeldt B, Dunphy JE (1967) Respiratory gas tensions and pH in healing wounds. Am J Surg 114:302–307
184. O'Toole EA et al (1997) Hypoxia increases human keratinocyte motility on connective tissue. J Clin Invest 100:2881–2891
185. Kan C, Abe M, Yamanaka M, Ishikawa O (2003) Hypoxia-induced increase of matrix metalloproteinase-1 synthesis is not restored by reoxygenation in a three-dimensional culture of human dermal fibroblasts. J Dermatol Sci 32:75–82
186. Feugate JE, Li Q, Wong L, Martins-Green M (2002) The cxc chemokine cCAF stimulates differentiation of fibroblasts into myofibroblasts and accelerates wound closure. J Cell Biol 156:161–172
187. Chaudhuri V, Zhou L, Karasek M (2007) Inflammatory cytokines induce the transformation of human dermal microvascular endothelial cells into myofibroblasts: a potential role in skin fibrogenesis. J Cutan Pathol 34:146–153
188. Zhang S, Uludag H (2009) Nanoparticulate systems for growth factor delivery. Pharm Res 26:1561–1580
189. Lee K, Silva EA, Mooney DJ (2011) Growth factor delivery-based tissue engineering: general approaches and a review of recent developments. J R Soc Interface 8:153–170
190. Chen FM, Zhang M, Wu ZF (2010) Toward delivery of multiple growth factors in tissue engineering. Biomaterials 31:6279–6308
191. McCarty DM, Monahan PE, Samulski RJ (2001) Self-complementary recombinant adeno-associated virus (scAAV) vectors promote efficient transduction independently of DNA synthesis. Gene Ther 8:1248–1254
192. Dileo J, Miller TE Jr, Chesnoy S, Huang L (2003) Gene transfer to subdermal tissues via a new gene gun design. Hum Gene Ther 14:79–87
193. Hacein-Bey-Abina S et al (2003) A serious adverse event after successful gene therapy for X-linked severe combined immunodeficiency. N Engl J Med 348:255–256
194. Chen S et al (2005) Efficient transduction of vascular endothelial cells with recombinant adeno-associated virus serotype 1 and 5 vectors. Hum Gene Ther 16:235–247
195. Liechty KW et al (1999) Adenoviral-mediated overexpression of platelet-derived growth factor-B corrects ischemic impaired wound healing. J Invest Dermatol 113:375–383
196. Sun L, Li J, Xiao X (2000) Overcoming adeno-associated virus vector size limitation through viral DNA heterodimerization. Nat Med 6:599–602
197. Eriksson E et al (1998) In vivo gene transfer to skin and wound by microseeding. J Surg Res 78:85–91
198. Vranckx JJ et al (2005) In vivo gene delivery of Ad-VEGF121 to full-thickness wounds in aged pigs results in high levels of VEGF expression but not in accelerated healing. Wound Repair Regen 13:51–60
199. Eming SA et al (1999) Particle-mediated gene transfer of PDGF isoforms promotes wound repair. J Invest Dermatol 112:297–302
200. Dickens S et al (2010) Nonviral transfection strategies for keratinocytes, fibroblasts, and endothelial progenitor cells for ex vivo gene transfer to skin wounds. Tissue Eng Part C Methods 16:1601–1608
201. Dijkmans PA et al (2004) Microbubbles and ultrasound: from diagnosis to therapy. Eur J Echocardiogr 5:245–256
202. Nyborg WL et al (2006) Emerging therapeutic ultrasound. World Scientific, Hackensack
203. Chen ZY, He CY, Ehrhardt A, Kay MA (2003) Minicircle DNA vectors devoid of bacterial DNA result in persistent and high-level transgene expression in vivo. Mol Ther 8:495–500
204. Felgner PL, Ringold GM (1989) Cationic liposome-mediated transfection. Nature 337:387–388
205. Jacobsen F et al (2006) Polybrene improves transfection efficacy of recombinant replication-deficient adenovirus in cutaneous cells and burned skin. J Gene Med 8:138–146

206. Yotnda P et al (2002) Bilamellar cationic liposomes protect adenovectors from preexisting humoral immune responses. Mol Ther 5:233–241
207. Kazuki Y et al (2011) Refined human artificial chromosome vectors for gene therapy and animal transgenesis. Gene Ther 18:384–393
208. Oshimura M, Katoh M (2008) Transfer of human artificial chromosome vectors into stem cells. Reprod Biomed Online 16:57–69
209. Gossen M et al (1995) Transcriptional activation by tetracyclines in mammalian cells. Science 268:1766–1769
210. Yao F et al (1998) Tetracycline repressor, tetR, rather than the tetR-mammalian cell transcription factor fusion derivatives, regulates inducible gene expression in mammalian cells. Hum Gene Ther 9:1939–1950
211. Yao F, Theopold C, Hoeller D, Bleiziffer O, Lu Z (2006) Highly efficient regulation of gene expression by tetracycline in a replication-defective herpes simplex viral vector. Mol Ther 13:1133–1141
212. No D, Yao TP, Evans RM (1996) Ecdysone-inducible gene expression in mammalian cells and transgenic mice. Proc Natl Acad Sci USA 93:3346–3351
213. Wang Y, O'Malley BW Jr, Tsai SY, O'Malley BW (1994) A regulatory system for use in gene transfer. Proc Natl Acad Sci USA 91:8180–8184
214. Braselmann S, Graninger P, Busslinger M (1993) A selective transcriptional induction system for mammalian cells based on Gal4-estrogen receptor fusion proteins. Proc Natl Acad Sci USA 90:1657–1661
215. Mills AA (2001) Changing colors in mice: an inducible system that delivers. Genes Dev 15:1461–1467
216. Fasolo A, Sessa C (2012) Targeting mTOR pathways in human malignancies. Curr Pharm Des 18:2766–2777
217. Sauer B, Henderson N (1988) Site-specific DNA recombination in mammalian cells by the Cre recombinase of bacteriophage P1. Proc Natl Acad Sci USA 85:5166–5170
218. Peattie RA et al (2006) Dual growth factor-induced angiogenesis in vivo using hyaluronan hydrogel implants. Biomaterials 27:1868–1875
219. Riley CM et al (2006) Stimulation of in vivo angiogenesis using dual growth factor-loaded crosslinked glycosaminoglycan hydrogels. Biomaterials 27:5935–5943
220. Dogan AK, Gumusderelioglu M, Aksoz E (2005) Controlled release of EGF and bFGF from dextran hydrogels in vitro and in vivo. J Biomed Mater Res B Appl Biomater 74: 504–510
221. Nillesen ST et al (2007) Increased angiogenesis and blood vessel maturation in acellular collagen-heparin scaffolds containing both FGF2 and VEGF. Biomaterials 28:1123–1131
222. Richardson TP, Peters MC, Ennett AB, Mooney DJ (2001) Polymeric system for dual growth factor delivery. Nat Biotechnol 19:1029–1034
223. Chen RR, Silva EA, Yuen WW, Mooney DJ (2007) Spatio-temporal VEGF and PDGF delivery patterns blood vessel formation and maturation. Pharm Res 24:258–264
224. Knighton DR, Ciresi KF, Fiegel VD, Austin LL, Butler EL (1986) Classification and treatment of chronic nonhealing wounds. Successful treatment with autologous platelet-derived wound healing factors (PDWHF). Ann Surg 204:322–330
225. Borrione P, Gianfrancesco AD, Pereira MT, Pigozzi F (2010) Platelet-rich plasma in muscle healing. Am J Phys Med Rehabil 89:854–861
226. Pallua N, Wolter T, Markowicz M (2010) Platelet-rich plasma in burns. Burns 36:4–8
227. Yu W, Wang J, Yin J (2011) Platelet-rich plasma: a promising product for treatment of peripheral nerve regeneration after nerve injury. Int J Neurosci 121:176–180
228. Demidova-Rice TN, Wolf L, Deckenback J, Hamblin MR, Herman IM (2012) Human platelet-rich plasma- and extracellular matrix-derived peptides promote impaired cutaneous wound healing in vivo. PLoS One 7:e32146
229. Blanton MW et al (2009) Adipose stromal cells and platelet-rich plasma therapies synergistically increase revascularization during wound healing. Plast Reconstr Surg 123:56S–64S

230. Vermeulen P et al (2009) A plasma-based biomatrix mixed with endothelial progenitor cells and keratinocytes promotes matrix formation, angiogenesis, and reepithelialization in full-thickness wounds. Tissue Eng Part A 15:1533–1542
231. Anitua E et al (2008) Effectiveness of autologous preparation rich in growth factors for the treatment of chronic cutaneous ulcers. J Biomed Mater Res B Appl Biomater 84:415–421
232. Foster TE, Puskas BL, Mandelbaum BR, Gerhardt MB, Rodeo SA (2009) Platelet-rich plasma: from basic science to clinical applications. Am J Sports Med 37:2259–2272
233. Steed DL et al (2008) Amnion-derived cellular cytokine solution: a physiological combination of cytokines for wound healing. Eplasty 8:e18
234. Bergmann J et al (2009) The effect of amnion-derived cellular cytokine solution on the epithelialization of partial-thickness donor site wounds in normal and streptozotocin-induced diabetic swine. Eplasty 9:e49
235. Payne WG et al (2010) Effect of amnion-derived cellular cytokine solution on healing of experimental partial-thickness burns. World J Surg 34:1663–1668
236. Daniel JM, Sedding DG (2011) Circulating smooth muscle progenitor cells in arterial remodeling. J Mol Cell Cardiol 50:273–279
237. Orlandi A, Bennett M (2010) Progenitor cell-derived smooth muscle cells in vascular disease. Biochem Pharmacol 79:1706–1713
238. Sirker AA, Astroulakis ZM, Hill JM (2009) Vascular progenitor cells and translational research: the role of endothelial and smooth muscle progenitor cells in endogenous arterial remodelling in the adult. Clin Sci (Lond) 116(283–299)
239. Templin C et al (2009) Ex vivo expanded haematopoietic progenitor cells improve dermal wound healing by paracrine mechanisms. Exp Dermatol 18:445–453
240. Carmeliet P, Luttun A (2001) The emerging role of the bone marrow-derived stem cells in (therapeutic) angiogenesis. Thromb Haemost 86:289–297
241. Lyden D et al (2001) Impaired recruitment of bone-marrow-derived endothelial and hematopoietic precursor cells blocks tumor angiogenesis and growth. Nat Med 7:1194–1201
242. Takakura N (2006) Role of hematopoietic lineage cells as accessory components in blood vessel formation. Cancer Sci 97:568–574
243. Melero-Martin JM et al (2010) Host myeloid cells are necessary for creating bioengineered human vascular networks in vivo. Tissue Eng Part A 16:2457–2466
244. Huber TL (2010) Dissecting hematopoietic differentiation using the embryonic stem cell differentiation model. Int J Dev Biol 54:991–1002
245. Huber TL, Kouskoff V, Fehling HJ, Palis J, Keller G (2004) Haemangioblast commitment is initiated in the primitive streak of the mouse embryo. Nature 432:625–630
246. Keller G (2005) Embryonic stem cell differentiation: emergence of a new era in biology and medicine. Genes Dev 19:1129–1155
247. Davison PM, Bensch K, Karasek MA (1980) Isolation and growth of endothelial cells from the microvessels of the newborn human foreskin in cell culture. J Invest Dermatol 75: 316–321
248. Supp DM, Wilson-Landy K, Boyce ST (2002) Human dermal microvascular endothelial cells form vascular analogs in cultured skin substitutes after grafting to athymic mice. FASEB J 16:797–804
249. Richard L, Velasco P, Detmar M (1998) A simple immunomagnetic protocol for the selective isolation and long-term culture of human dermal microvascular endothelial cells. Exp Cell Res 240:1–6
250. Kubota Y, Kleinman HK, Martin GR, Lawley TJ (1988) Role of laminin and basement membrane in the morphological differentiation of human endothelial cells into capillary-like structures. J Cell Biol 107:1589–1598
251. Gupta K, Ramakrishnan S, Browne PV, Solovey A, Hebbel RP (1997) A novel technique for culture of human dermal microvascular endothelial cells under either serum-free or serum-supplemented conditions: isolation by panning and stimulation with vascular endothelial growth factor. Exp Cell Res 230:244–251

252. Sorrell JM, Baber MA, Caplan AI (2007) A self-assembled fibroblast-endothelial cell co-culture system that supports in vitro vasculogenesis by both human umbilical vein endothelial cells and human dermal microvascular endothelial cells. Cells Tissues Organs 186: 157–168
253. Nor JE et al (2001) Engineering and characterization of functional human microvessels in immunodeficient mice. Lab Invest 81:453–463
254. Polchow B et al (2012) Cryopreservation of human vascular umbilical cord cells under good manufacturing practice conditions for future cell banks. J Transl Med 10:98
255. Unger RE, Krump-Konvalinkova V, Peters K, Kirkpatrick CJ (2002) In vitro expression of the endothelial phenotype: comparative study of primary isolated cells and cell lines, including the novel cell line HPMEC-ST1.6R. Microvasc Res 64:384–397
256. Hannan RL et al (1988) Endothelial cells synthesize basic fibroblast growth factor and transforming growth factor beta. Growth Factors 1:7–17
257. Bhang SH et al (2012) Three-dimensional cell grafting enhances the angiogenic efficacy of human umbilical vein endothelial cells. Tissue Eng Part A 18:310–319
258. Alajati A et al (2008) Spheroid-based engineering of a human vasculature in mice. Nat Methods 5:439–445
259. Schechner JS et al (2000) In vivo formation of complex microvessels lined by human endothelial cells in an immunodeficient mouse. Proc Natl Acad Sci USA 97:9191–9196
260. Yoder MC et al (2007) Redefining endothelial progenitor cells via clonal analysis and hematopoietic stem/progenitor cell principals. Blood 109:1801–1809
261. Awad O et al (2006) Differential healing activities of CD34+ and CD14+ endothelial cell progenitors. Arterioscler Thromb Vasc Biol 26:758–764
262. Case J et al (2007) Human CD34+ AC133+ VEGFR-2+ cells are not endothelial progenitor cells but distinct, primitive hematopoietic progenitors. Exp Hematol 35:1109–1118
263. Kim SY et al (2005) Differentiation of endothelial cells from human umbilical cord blood AC133-CD14+ cells. Ann Hematol 84:417–422
264. Rehman J, Li J, Orschell CM, March KL (2003) Peripheral blood "endothelial progenitor cells" are derived from monocyte/macrophages and secrete angiogenic growth factors. Circulation 107:1164–1169
265. Sieveking DP, Buckle A, Celermajer DS, Ng MK (2008) Strikingly different angiogenic properties of endothelial progenitor cell subpopulations: insights from a novel human angiogenesis assay. J Am Coll Cardiol 51:660–668
266. Shepherd BR et al (2006) Vascularization and engraftment of a human skin substitute using circulating progenitor cell-derived endothelial cells. FASEB J 20:1739–1741
267. Yoder MC (2010) Is endothelium the origin of endothelial progenitor cells? Arterioscler Thromb Vasc Biol 30:1094–1103
268. van Beem RT et al (2009) Blood outgrowth endothelial cells from cord blood and peripheral blood: angiogenesis-related characteristics in vitro. J Thromb Haemost 7:217–226
269. Corselli M et al (2008) Clinical scale ex vivo expansion of cord blood-derived outgrowth endothelial progenitor cells is associated with high incidence of karyotype aberrations. Exp Hematol 36:340–349
270. Au P et al (2008) Differential in vivo potential of endothelial progenitor cells from human umbilical cord blood and adult peripheral blood to form functional long-lasting vessels. Blood 111:1302–1305
271. Suh W et al (2005) Transplantation of endothelial progenitor cells accelerates dermal wound healing with increased recruitment of monocytes/macrophages and neovascularization. Stem Cells 23:1571–1578
272. Sander AL et al (2011) Systemic transplantation of progenitor cells accelerates wound epithelialization and neovascularization in the hairless mouse ear wound model. J Surg Res 165:165–170
273. Goldstein LJ et al (2006) Endothelial progenitor cell release into circulation is triggered by hyperoxia-induced increases in bone marrow nitric oxide. Stem Cells 24:2309–2318

274. Barcelos LS et al (2009) Human CD133+ progenitor cells promote the healing of diabetic ischemic ulcers by paracrine stimulation of angiogenesis and activation of Wnt signaling. Circ Res 104:1095–1102
275. Reinisch A, Strunk D (2009) Isolation and animal serum free expansion of human umbilical cord derived mesenchymal stromal cells (MSCs) and endothelial colony forming progenitor cells (ECFCs). J Vis Exp 8(32):pii.1525
276. Krenning G, van Luyn MJ, Harmsen MC (2009) Endothelial progenitor cell-based neovascularization: implications for therapy. Trends Mol Med 15:180–189
277. Yoon CH et al (2005) Synergistic neovascularization by mixed transplantation of early endothelial progenitor cells and late outgrowth endothelial cells: the role of angiogenic cytokines and matrix metalloproteinases. Circulation 112:1618–1627
278. Bianco P (2011) Back to the future: moving beyond "mesenchymal stem cells". J Cell Biochem 112:1713–1721
279. Caplan AI (1991) Mesenchymal stem cells. J Orthop Res 9:641–650
280. Vishnubalaji R et al (2012) In vitro differentiation of human skin-derived multipotent stromal cells into putative endothelial-like cells. BMC Dev Biol 12:7
281. Chen J et al (2008) Kidney-derived mesenchymal stem cells contribute to vasculogenesis, angiogenesis and endothelial repair. Kidney Int 74:879–889
282. Marchionni C et al (2009) Angiogenic potential of human dental pulp stromal (stem) cells. Int J Immunopathol Pharmacol 22:699–706
283. Branski LK, Gauglitz GG, Herndon DN, Jeschke MG (2009) A review of gene and stem cell therapy in cutaneous wound healing. Burns 35:171–180
284. Romanov YA, Svintsitskaya VA, Smirnov VN (2003) Searching for alternative sources of postnatal human mesenchymal stem cells: candidate MSC-like cells from umbilical cord. Stem Cells 21:105–110
285. Zebardast N, Lickorish D, Davies JE (2010) Human umbilical cord perivascular cells (HUCPVC): a mesenchymal cell source for dermal wound healing. Organogenesis 6:197–203
286. Yang S, Huang S, Feng C, Fu X (2012) Umbilical cord-derived mesenchymal stem cells: strategies, challenges, and potential for cutaneous regeneration. Front Med 6:41–47
287. Anker PS I't et al (2004) Isolation of mesenchymal stem cells of fetal or maternal origin from human placenta. Stem Cells 22:1338–1345
288. Spitzer TL et al (2012) Perivascular human endometrial mesenchymal stem cells express pathways relevant to self-renewal, lineage specification, and functional phenotype. Biol Reprod 86:58
289. Herdrich BJ, Lind RC, Liechty KW (2008) Multipotent adult progenitor cells: their role in wound healing and the treatment of dermal wounds. Cytotherapy 10:543–550
290. da Silva Meirelles L, Chagastelles PC, Nardi NB (2006) Mesenchymal stem cells reside in virtually all post-natal organs and tissues. J Cell Sci 119:2204–2213
291. Otto WR, Wright NA (2011) Mesenchymal stem cells: from experiment to clinic. Fibrogenesis Tissue Repair 4:20
292. Natesan S, Wrice NL, Baer DG, Christy RJ (2011) Debrided skin as a source of autologous stem cells for wound repair. Stem Cells 29:1219–1230
293. Dominici M et al (2006) Minimal criteria for defining multipotent mesenchymal stromal cells. The International Society for Cellular Therapy position statement. Cytotherapy 8:315–317
294. Fu X, Li H (2009) Mesenchymal stem cells and skin wound repair and regeneration: possibilities and questions. Cell Tissue Res 335:317–321
295. Chen L, Tredget EE, Wu PY, Wu Y (2008) Paracrine factors of mesenchymal stem cells recruit macrophages and endothelial lineage cells and enhance wound healing. PLoS One 3:e1886
296. Hanson SE, Bentz ML, Hematti P (2010) Mesenchymal stem cell therapy for nonhealing cutaneous wounds. Plast Reconstr Surg 125:510–516
297. Hocking AM, Gibran NS (2010) Mesenchymal stem cells: paracrine signaling and differentiation during cutaneous wound repair. Exp Cell Res 316:2213–2219

298. Li H, Fu X (2012) Mechanisms of action of mesenchymal stem cells in cutaneous wound repair and regeneration. Cell Tissue Res 348:371–377
299. Sorrell JM, Caplan AI (2010) Topical delivery of mesenchymal stem cells and their function in wounds. Stem Cell Res Ther 1:30
300. Miura M et al (2006) Accumulated chromosomal instability in murine bone marrow mesenchymal stem cells leads to malignant transformation. Stem Cells 24:1095–1103
301. Rubio D et al (2005) Spontaneous human adult stem cell transformation. Cancer Res 65:3035–3039
302. Tarte K et al (2010) Clinical-grade production of human mesenchymal stromal cells: occurrence of aneuploidy without transformation. Blood 115:1549–1553
303. Koc ON et al (2002) Allogeneic mesenchymal stem cell infusion for treatment of metachromatic leukodystrophy (MLD) and Hurler syndrome (MPS-IH). Bone Marrow Transplant 30:215–222
304. Bernardo ME, Locatelli F, Fibbe WE (2009) Mesenchymal stromal cells. Ann N Y Acad Sci 1176:101–117
305. Schallmoser K et al (2007) Human platelet lysate can replace fetal bovine serum for clinical-scale expansion of functional mesenchymal stromal cells. Transfusion 47:1436–1446
306. Cao Y et al (2005) Human adipose tissue-derived stem cells differentiate into endothelial cells in vitro and improve postnatal neovascularization in vivo. Biochem Biophys Res Commun 332:370–379
307. Miranville A et al (2004) Improvement of postnatal neovascularization by human adipose tissue-derived stem cells. Circulation 110:349–355
308. Planat-Benard V et al (2004) Plasticity of human adipose lineage cells toward endothelial cells: physiological and therapeutic perspectives. Circulation 109:656–663
309. Sasaki M et al (2008) Mesenchymal stem cells are recruited into wounded skin and contribute to wound repair by transdifferentiation into multiple skin cell type. J Immunol 180:2581–2587
310. Wu KH et al (2007) In vitro and in vivo differentiation of human umbilical cord derived stem cells into endothelial cells. J Cell Biochem 100:608–616
311. Chen MY, Lie PC, Li ZL, Wei X (2009) Endothelial differentiation of Wharton's jelly-derived mesenchymal stem cells in comparison with bone marrow-derived mesenchymal stem cells. Exp Hematol 37:629–640
312. Gang EJ et al (2006) In vitro endothelial potential of human UC blood-derived mesenchymal stem cells. Cytotherapy 8:215–227
313. Miao Z et al (2006) Isolation of mesenchymal stem cells from human placenta: comparison with human bone marrow mesenchymal stem cells. Cell Biol Int 30:681–687
314. Oswald J et al (2004) Mesenchymal stem cells can be differentiated into endothelial cells in vitro. Stem Cells 22:377–384
315. Cerqueira MT, Marques AP, Reis RL (2012) Using stem cells in skin regeneration: possibilities and reality. Stem Cells Dev 21:1201–1214
316. Colazzo F, Chester AH, Taylor PM, Yacoub MH (2010) Induction of mesenchymal to endothelial transformation of adipose-derived stem cells. J Heart Valve Dis 19:736–744
317. Szoke K, Beckstrom KJ, Brinchmann JE (2012) Human adipose tissue as a source of cells with angiogenic potential. Cell Transplant 21:235–250
318. Zimmerlin L et al (2010) Stromal vascular progenitors in adult human adipose tissue. Cytometry A 77:22–30
319. Locke M, Feisst V, Dunbar PR (2011) Concise review: human adipose-derived stem cells: separating promise from clinical need. Stem Cells 29:404–411
320. Nakagami H et al (2005) Novel autologous cell therapy in ischemic limb disease through growth factor secretion by cultured adipose tissue-derived stromal cells. Arterioscler Thromb Vasc Biol 25:2542–2547
321. Moon MH et al (2006) Human adipose tissue-derived mesenchymal stem cells improve postnatal neovascularization in a mouse model of hindlimb ischemia. Cell Physiol Biochem 17:279–290

322. Schlosser S et al (2012) Paracrine effects of mesenchymal stem cells enhance vascular regeneration in ischemic murine skin. Microvasc Res 83:267–275
323. Roobrouck VD et al (2011) Differentiation potential of human postnatal mesenchymal stem cells, mesoangioblasts, and multipotent adult progenitor cells reflected in their transcriptome and partially influenced by the culture conditions. Stem Cells 29:871–882
324. Chamberlain G, Fox J, Ashton B, Middleton J (2007) Concise review: mesenchymal stem cells: their phenotype, differentiation capacity, immunological features, and potential for homing. Stem Cells 25:2739–2749
325. Kinnaird T et al (2004) Marrow-derived stromal cells express genes encoding a broad spectrum of arteriogenic cytokines and promote in vitro and in vivo arteriogenesis through paracrine mechanisms. Circ Res 94:678–685
326. Wu Y, Zhao RC, Tredget EE (2010) Concise review: bone marrow-derived stem/progenitor cells in cutaneous repair and regeneration. Stem Cells 28:905–915
327. Yew TL et al (2011) Enhancement of wound healing by human multipotent stromal cell conditioned medium: the paracrine factors and p38 MAPK activation. Cell Transplant 20:693–706
328. Boomsma RA, Geenen DL (2012) Mesenchymal stem cells secrete multiple cytokines that promote angiogenesis and have contrasting effects on chemotaxis and apoptosis. PLoS One 7:e35685
329. Ghajar CM et al (2010) Mesenchymal cells stimulate capillary morphogenesis via distinct proteolytic mechanisms. Exp Cell Res 316:813–825
330. Kachgal S, Putnam AJ (2011) Mesenchymal stem cells from adipose and bone marrow promote angiogenesis via distinct cytokine and protease expression mechanisms. Angiogenesis 14:47–59
331. Lin RZ, Moreno-Luna R, Zhou B, Pu WT, Melero-Martin JM (2012) Equal modulation of endothelial cell function by four distinct tissue-specific mesenchymal stem cells. Angiogenesis 15:443–455
332. Amos PJ et al (2008) IFATS collection: the role of human adipose-derived stromal cells in inflammatory microvascular remodeling and evidence of a perivascular phenotype. Stem Cells 26:2682–2690
333. Crisan M et al (2008) A perivascular origin for mesenchymal stem cells in multiple human organs. Cell Stem Cell 3:301–313
334. Sacchetti B et al (2007) Self-renewing osteoprogenitors in bone marrow sinusoids can organize a hematopoietic microenvironment. Cell 131:324–336
335. Traktuev DO et al (2008) A population of multipotent CD34-positive adipose stromal cells share pericyte and mesenchymal surface markers, reside in a periendothelial location, and stabilize endothelial networks. Circ Res 102:77–85
336. Mulder GD, Lee DK, Jeppesen NS (2012) Comprehensive review of the clinical application of autologous mesenchymal stem cells in the treatment of chronic wounds and diabetic bone healing. Int Wound J 9:595–600
337. Brower J et al (2011) Mesenchymal stem cell therapy and delivery systems in nonhealing wounds. Adv Skin Wound Care 24:524–532; quiz 533–524
338. Falanga V et al (2007) Autologous bone marrow-derived cultured mesenchymal stem cells delivered in a fibrin spray accelerate healing in murine and human cutaneous wounds. Tissue Eng 13:1299–1312
339. Kim EK, Li G, Lee TJ, Hong JP (2011) The effect of human adipose-derived stem cells on healing of ischemic wounds in a diabetic nude mouse model. Plast Reconstr Surg 128:387–394
340. Au P, Tam J, Fukumura D, Jain RK (2008) Bone marrow-derived mesenchymal stem cells facilitate engineering of long-lasting functional vasculature. Blood 111:4551–4558
341. Garcia-Olmo D et al (2005) A phase I clinical trial of the treatment of Crohn's fistula by adipose mesenchymal stem cell transplantation. Dis Colon Rectum 48:1416–1423
342. Garcia-Olmo D et al (2009) Expanded adipose-derived stem cells for the treatment of complex perianal fistula: a phase II clinical trial. Dis Colon Rectum 52:79–86

343. Melero-Martin JM et al (2008) Engineering robust and functional vascular networks in vivo with human adult and cord blood-derived progenitor cells. Circ Res 103:194–202
344. D'Ippolito G et al (2004) Marrow-isolated adult multilineage inducible (MIAMI) cells, a unique population of postnatal young and old human cells with extensive expansion and differentiation potential. J Cell Sci 117:2971–2981
345. Jiang Y et al (2002) Pluripotency of mesenchymal stem cells derived from adult marrow. Nature 418:41–49
346. Yoon YS et al (2005) Clonally expanded novel multipotent stem cells from human bone marrow regenerate myocardium after myocardial infarction. J Clin Invest 115:326–338
347. Sohni A, Verfaillie CM (2011) Multipotent adult progenitor cells. Best Pract Res Clin Haematol 24:3–11
348. Kogler G et al (2004) A new human somatic stem cell from placental cord blood with intrinsic pluripotent differentiation potential. J Exp Med 200:123–135
349. Aranguren XL et al (2007) In vitro and in vivo arterial differentiation of human multipotent adult progenitor cells. Blood 109:2634–2642
350. Aranguren XL et al (2008) Multipotent adult progenitor cells sustain function of ischemic limbs in mice. J Clin Invest 118:505–514
351. Reyes M et al (2002) Origin of endothelial progenitors in human postnatal bone marrow. J Clin Invest 109:337–346
352. Ulloa-Montoya F et al (2007) Comparative transcriptome analysis of embryonic and adult stem cells with extended and limited differentiation capacity. Genome Biol 8:R163
353. Mahpatra S, Firpo MT, Bacanamwo M (2010) Inhibition of DNA methyltransferases and histone deacetylases induces bone marrow-derived multipotent adult progenitor cells to differentiate into endothelial cells. Ethn Dis 20:S1-60–S1-64
354. Highfill SL et al (2009) Multipotent adult progenitor cells can suppress graft-versus-host disease via prostaglandin E2 synthesis and only if localized to sites of allopriming. Blood 114:693–701
355. Luyckx A et al (2010) Mouse MAPC-mediated immunomodulation: cell-line dependent variation. Exp Hematol 38:1–2
356. Evans MJ, Kaufman MH (1981) Establishment in culture of pluripotential cells from mouse embryos. Nature 292:154–156
357. Thomson JA et al (1998) Embryonic stem cell lines derived from human blastocysts. Science 282:1145–1147
358. Odorico JS, Kaufman DS, Thomson JA (2001) Multilineage differentiation from human embryonic stem cell lines. Stem Cells 19:193–204
359. Wang X et al (2012) MIF produced by bone marrow-derived macrophages contributes to teratoma progression after embryonic stem cell transplantation. Cancer Res 72: 2867–2878
360. Moon SH et al (2011) A system for treating ischemic disease using human embryonic stem cell-derived endothelial cells without direct incorporation. Biomaterials 32:6445–6455
361. Jaenisch R (2004) Human cloning – the science and ethics of nuclear transplantation. N Engl J Med 351:2787–2791
362. Puymirat E et al (2009) Can mesenchymal stem cells induce tolerance to cotransplanted human embryonic stem cells? Mol Ther 17:176–182
363. Yamashita J et al (2000) Flk1-positive cells derived from embryonic stem cells serve as vascular progenitors. Nature 408:92–96
364. Sone M et al (2007) Pathway for differentiation of human embryonic stem cells to vascular cell components and their potential for vascular regeneration. Arterioscler Thromb Vasc Biol 27:2127–2134
365. Yamahara K et al (2008) Augmentation of neovascularization [corrected] in hindlimb ischemia by combined transplantation of human embryonic stem cells-derived endothelial and mural cells. PLoS One 3:e1666
366. Descamps B, Emanueli C (2012) Vascular differentiation from embryonic stem cells: novel technologies and therapeutic promises. Vascul Pharmacol 56:267–279

367. Li Z, Han Z, Wu JC (2009) Transplantation of human embryonic stem cell-derived endothelial cells for vascular diseases. J Cell Biochem 106:194–199
368. Volz KS, Miljan E, Khoo A, Cooke JP (2012) Development of pluripotent stem cells for vascular therapy. Vascul Pharmacol 56:288–296
369. Hsiai TK, Wu JC (2008) Hemodynamic forces regulate embryonic stem cell commitment to vascular progenitors. Curr Cardiol Rev 4:269–274
370. Kane NM et al (2010) Derivation of endothelial cells from human embryonic stem cells by directed differentiation: analysis of microRNA and angiogenesis in vitro and in vivo. Arterioscler Thromb Vasc Biol 30:1389–1397
371. Lee AS et al (2009) Effects of cell number on teratoma formation by human embryonic stem cells. Cell Cycle 8:2608–2612
372. Tang C et al (2011) An antibody against SSEA-5 glycan on human pluripotent stem cells enables removal of teratoma-forming cells. Nat Biotechnol 29:829–834
373. Kalka C et al (2000) Transplantation of ex vivo expanded endothelial progenitor cells for therapeutic neovascularization. Proc Natl Acad Sci USA 97:3422–3427
374. Yang C et al (2004) Enhancement of neovascularization with cord blood CD133+ cell-derived endothelial progenitor cell transplantation. Thromb Haemost 91:1202–1212
375. Vatansever HS, Uluer ET, Aydede H, Ozbilgin MK (2013) Analysis of transferred keratinocyte-like cells derived from mouse embryonic stem cells on experimental surgical skin wounds of mouse. Acta Histochem 115:32–41
376. Lee MJ et al (2011) Enhancement of wound healing by secretory factors of endothelial precursor cells derived from human embryonic stem cells. Cytotherapy 13:165–178
377. Takahashi K, Yamanaka S (2006) Induction of pluripotent stem cells from mouse embryonic and adult fibroblast cultures by defined factors. Cell 126:663–676
378. Yu J et al (2007) Induced pluripotent stem cell lines derived from human somatic cells. Science 318:1917–1920
379. Rolletschek A, Wobus AM (2009) Induced human pluripotent stem cells: promises and open questions. Biol Chem 390:845–849
380. Jia F et al (2010) A nonviral minicircle vector for deriving human iPS cells. Nat Methods 7:197–199
381. Li M, Chen M, Han W, Fu X (2010) How far are induced pluripotent stem cells from the clinic? Ageing Res Rev 9:257–264
382. Choi KD et al (2009) Hematopoietic and endothelial differentiation of human induced pluripotent stem cells. Stem Cells 27:559–567
383. Narazaki G et al (2008) Directed and systematic differentiation of cardiovascular cells from mouse induced pluripotent stem cells. Circulation 118:498–506
384. Taura D et al (2009) Induction and isolation of vascular cells from human induced pluripotent stem cells – brief report. Arterioscler Thromb Vasc Biol 29:1100–1103
385. Rufaihah AJ et al (2011) Endothelial cells derived from human iPSCS increase capillary density and improve perfusion in a mouse model of peripheral arterial disease. Arterioscler Thromb Vasc Biol 31:e72–e79
386. Suzuki H et al (2010) Therapeutic angiogenesis by transplantation of induced pluripotent stem cell-derived Flk-1 positive cells. BMC Cell Biol 11:72
387. Iwaguro H et al (2002) Endothelial progenitor cell vascular endothelial growth factor gene transfer for vascular regeneration. Circulation 105:732–738
388. Aranguren XL, Verfaillie CM, Luttun A (2009) Emerging hurdles in stem cell therapy for peripheral vascular disease. J Mol Med (Berl) 87(3–16)
389. Choudhery MS et al (2012) Bone marrow derived mesenchymal stem cells from aged mice have reduced wound healing, angiogenesis, proliferation and anti-apoptosis capabilities. Cell Biol Int 36:747–753
390. Di Rocco G et al (2010) Enhanced healing of diabetic wounds by topical administration of adipose tissue-derived stromal cells overexpressing stromal-derived factor-1: biodistribution and engraftment analysis by bioluminescent imaging. Stem Cells Int 2011:304562
391. Pasha Z et al (2008) Preconditioning enhances cell survival and differentiation of stem cells during transplantation in infarcted myocardium. Cardiovasc Res 77:134–142

392. Marrotte EJ, Chen DD, Hakim JS, Chen AF (2010) Manganese superoxide dismutase expression in endothelial progenitor cells accelerates wound healing in diabetic mice. J Clin Invest 120:4207–4219
393. Song SH et al (2012) Genetic modification of human adipose-derived stem cells for promoting wound healing. J Dermatol Sci 66:98–107
394. Lu D et al (2012) Peroxisome proliferator-activated receptor-gamma coactivator-1alpha (PGC-1alpha) enhances engraftment and angiogenesis of mesenchymal stem cells in diabetic hindlimb ischemia. Diabetes 61:1153–1159
395. Barzilay R, Melamed E, Offen D (2009) Introducing transcription factors to multipotent mesenchymal stem cells: making transdifferentiation possible. Stem Cells 27:2509–2515
396. Duffy GP et al (2010) Mesenchymal stem cells overexpressing ephrin-b2 rapidly adopt an early endothelial phenotype with simultaneous reduction of osteogenic potential. Tissue Eng Part A 16:2755–2768
397. Dufourcq P et al (2008) Secreted frizzled-related protein-1 enhances mesenchymal stem cell function in angiogenesis and contributes to neovessel maturation. Stem Cells 26:2991–3001
398. Moues CM, Heule F, Hovius SE (2011) A review of topical negative pressure therapy in wound healing: sufficient evidence? Am J Surg 201:544–556
399. Labler L et al (2009) Vacuum-assisted closure therapy increases local interleukin-8 and vascular endothelial growth factor levels in traumatic wounds. J Trauma 66:749–757
400. Labanaris AP, Polykandriotis E, Horch RE (2009) The effect of vacuum-assisted closure on lymph vessels in chronic wounds. J Plast Reconstr Aesthet Surg 62:1068–1075
401. Costin GE, Birlea SA, Norris DA (2012) Trends in wound repair: cellular and molecular basis of regenerative therapy using electromagnetic fields. Curr Mol Med 12:14–26
402. Callaghan MJ et al (2008) Pulsed electromagnetic fields accelerate normal and diabetic wound healing by increasing endogenous FGF-2 release. Plast Reconstr Surg 121:130–141
403. Sebastian A et al (2011) Acceleration of cutaneous healing by electrical stimulation: degenerate electrical waveform down-regulates inflammation, up-regulates angiogenesis and advances remodeling in temporal punch biopsies in a human volunteer study. Wound Repair Regen 19:693–708
404. MacNeil S (2007) Progress and opportunities for tissue-engineered skin. Nature 445:874–880
405. Groeber F, Holeiter M, Hampel M, Hinderer S, Schenke-Layland K (2011) Skin tissue engineering – in vivo and in vitro applications. Adv Drug Deliv Rev 63:352–366
406. Griffith CK et al (2005) Diffusion limits of an in vitro thick prevascularized tissue. Tissue Eng 11:257–266
407. Papavasiliou G, Cheng MH, Brey EM (2010) Strategies for vascularization of polymer scaffolds. J Investig Med 58:838–844
408. Glotzbach JP, Wong VW, Gurtner GC, Longaker MT (2011) Regenerative medicine. Curr Probl Surg 48:148–212
409. Phelps EA, Garcia AJ (2010) Engineering more than a cell: vascularization strategies in tissue engineering. Curr Opin Biotechnol 21:704–709
410. Sarkar A, Tatlidede S, Scherer SS, Orgill DP, Berthiaume F (2011) Combination of stromal cell-derived factor-1 and collagen-glycosaminoglycan scaffold delays contraction and accelerates reepithelialization of dermal wounds in wild-type mice. Wound Repair Regen 19:71–79
411. Lugo LM, Lei P, Andreadis ST (2011) Vascularization of the dermal support enhances wound re-epithelialization by in situ delivery of epidermal keratinocytes. Tissue Eng Part A 17:665–675
412. Supp DM, Boyce ST (2002) Overexpression of vascular endothelial growth factor accelerates early vascularization and improves healing of genetically modified cultured skin substitutes. J Burn Care Rehabil 23:10–20
413. Erdag G, Medalie DA, Rakhorst H, Krueger GG, Morgan JR (2004) FGF-7 expression enhances the performance of bioengineered skin. Mol Ther 10:76–85
414. Tremblay PL, Hudon V, Berthod F, Germain L, Auger FA (2005) Inosculation of tissue-engineered capillaries with the host's vasculature in a reconstructed skin transplanted on mice. Am J Transplant 5:1002–1010

415. Halbleib M, Skurk T, de Luca C, von Heimburg D, Hauner H (2003) Tissue engineering of white adipose tissue using hyaluronic acid-based scaffolds. I: in vitro differentiation of human adipocyte precursor cells on scaffolds. Biomaterials 24:3125–3132
416. Burg KJ et al (2000) Comparative study of seeding methods for three-dimensional polymeric scaffolds. J Biomed Mater Res 52:576
417. Lawrence BJ, Madihally SV (2008) Cell colonization in degradable 3D porous matrices. Cell Adh Migr 2:9–16
418. Kannan RY, Salacinski HJ, Sales K, Butler P, Seifalian AM (2005) The roles of tissue engineering and vascularisation in the development of micro-vascular networks: a review. Biomaterials 26:1857–1875
419. Calcagni M et al (2011) In vivo visualization of the origination of skin graft vasculature in a wild-type/GFP crossover model. Microvasc Res 82:237–245
420. Tonello C et al (2003) In vitro reconstruction of human dermal equivalent enriched with endothelial cells. Biomaterials 24:1205–1211
421. Black AF, Berthod F, L'Heureux N, Germain L, Auger FA (1998) In vitro reconstruction of a human capillary-like network in a tissue-engineered skin equivalent. FASEB J 12:1331–1340
422. Hudon V et al (2003) A tissue-engineered endothelialized dermis to study the modulation of angiogenic and angiostatic molecules on capillary-like tube formation in vitro. Br J Dermatol 148:1094–1104
423. Schechner JS et al (2003) Engraftment of a vascularized human skin equivalent. FASEB J 17:2250–2256
424. Sarkanen JR et al (2012) Adipose stromal cell tubule network model provides a versatile tool for vascular research and tissue engineering. Cells Tissues Organs 196:385–397
425. Verseijden F et al (2012) Vascularization of prevascularized and non-prevascularized fibrin-based human adipose tissue constructs after implantation in nude mice. J Tissue Eng Regen Med 6:169–178
426. Frerich B, Winter K, Scheller K, Braumann UD (2012) Comparison of different fabrication techniques for human adipose tissue engineering in severe combined immunodeficient mice. Artif Organs 36:227–237
427. Wu X et al (2004) Tissue-engineered microvessels on three-dimensional biodegradable scaffolds using human endothelial progenitor cells. Am J Physiol Heart Circ Physiol 287:H480–H487
428. Sieminski AL, Hebbel RP, Gooch KJ (2005) Improved microvascular network in vitro by human blood outgrowth endothelial cells relative to vessel-derived endothelial cells. Tissue Eng 11:1332–1345
429. Melero-Martin JM et al (2007) In vivo vasculogenic potential of human blood-derived endothelial progenitor cells. Blood 109:4761–4768
430. Thebaud NB et al (2010) Human progenitor-derived endothelial cells vs. venous endothelial cells for vascular tissue engineering: an in vitro study. J Tissue Eng Regen Med 4:473–484
431. Kung EF, Wang F, Schechner JS (2008) In vivo perfusion of human skin substitutes with microvessels formed by adult circulating endothelial progenitor cells. Dermatol Surg 34:137–146
432. Mujaj S, Manton K, Upton Z, Richards S (2010) Serum-free primary human fibroblast and keratinocyte coculture. Tissue Eng Part A 16:1407–1420
433. Markowicz M et al (2006) Human bone marrow mesenchymal stem cells seeded on modified collagen improved dermal regeneration in vivo. Cell Transplant 15:723–732
434. Liu P et al (2008) Tissue-engineered skin containing mesenchymal stem cells improves burn wounds. Artif Organs 32:925–931
435. Liu S et al (2011) Synergistic angiogenesis promoting effects of extracellular matrix scaffolds and adipose-derived stem cells during wound repair. Tissue Eng Part A 17:725–739
436. Zografou A et al (2011) Improvement of skin-graft survival after autologous transplantation of adipose-derived stem cells in rats. J Plast Reconstr Aesthet Surg 64:1647–1656

437. Altman AM et al (2008) Dermal matrix as a carrier for in vivo delivery of human adipose-derived stem cells. Biomaterials 29:1431–1442
438. Altman AM et al (2009) IFATS collection: human adipose-derived stem cells seeded on a silk fibroin-chitosan scaffold enhance wound repair in a murine soft tissue injury model. Stem Cells 27:250–258
439. Rustad KC et al (2012) Enhancement of mesenchymal stem cell angiogenic capacity and stemness by a biomimetic hydrogel scaffold. Biomaterials 33:80–90
440. Inoue H et al (2008) Bioimaging assessment and effect of skin wound healing using bone-marrow-derived mesenchymal stromal cells with the artificial dermis in diabetic rats. J Biomed Opt 13:064036
441. Vojtassak J et al (2006) Autologous biograft and mesenchymal stem cells in treatment of the diabetic foot. Neuro Endocrinol Lett 27(Suppl 2):134–137
442. Levenberg S, Golub JS, Amit M, Itskovitz-Eldor J, Langer R (2002) Endothelial cells derived from human embryonic stem cells. Proc Natl Acad Sci USA 99:4391–4396
443. Levenberg S et al (2005) Engineering vascularized skeletal muscle tissue. Nat Biotechnol 23:879–884
444. Shen G et al (2003) Tissue engineering of blood vessels with endothelial cells differentiated from mouse embryonic stem cells. Cell Res 13:335–341
445. Huang H et al (2005) Differentiation from embryonic stem cells to vascular wall cells under in vitro pulsatile flow loading. J Artif Organs 8:110–118
446. Nourse MB et al (2010) VEGF induces differentiation of functional endothelium from human embryonic stem cells: implications for tissue engineering. Arterioscler Thromb Vasc Biol 30:80–89
447. Kraehenbuehl TP et al (2011) Human embryonic stem cell-derived microvascular grafts for cardiac tissue preservation after myocardial infarction. Biomaterials 32:1102–1109
448. Lammers G et al (2011) An overview of methods for the in vivo evaluation of tissue-engineered skin constructs. Tissue Eng Part B Rev 17:33–55

Index

M.K. Danquah, R.I. Mahato (eds.), *Emerging Trends in Cell and Gene Therapy*,
DOI 10.1007/978-1-62703-417-3,

N

O

X

MIX
Papier aus verantwortungsvollen Quellen
Paper from responsible sources
FSC® C105338

If you have any concerns about our products, you can contact us on
ProductSafety@springernature.com

In case Publisher is established outside the EU, the EU authorized representative is:
Springer Nature Customer Service Center GmbH
Europaplatz 3, 69115 Heidelberg, Germany

Printed by Libri Plureos GmbH
in Hamburg, Germany